FINITE PRECISION NUMBER SYSTEMS AND ARITHMETIC

Fundamental arithmetic operations support virtually all of the engineering, scientific, and financial computations required for practical applications from cryptography, to financial planning, to rocket science. This comprehensive reference provides researchers with the thorough understanding of number representations that is a necessary foundation for designing efficient arithmetic algorithms.

Using the elementary foundations of radix number systems as a basis for arithmetic, the authors develop and compare alternative algorithms for the fundamental operations of addition, multiplication, division, and square root with precisely defined roundings. Various finite precision number systems are investigated, with the focus on comparative analysis of practically efficient algorithms for closed arithmetic operations over these systems.

Each chapter begins with an introduction to its contents and ends with bibliographic notes and an extensive bibliography. The book may also be used for graduate teaching: problems and exercises are scattered throughout the text and a solutions manual is available for instructors.

All the titles listed below can be obtained from good booksellers or from Cambridge University Press. For a complete series listing visit

http://www.cambridge.org/uk/series/sSeries.asp?code=EOM

80 O. Stormark *Lie's Structural Approach to PDE Systems*
81 C. F. Dunkl and Y. Xu *Orthogonal Polynomials of Several Variables*
82 J. P. Mayberry *The Foundations of Mathematics in the Theory of Sets*
83 C. Foias *et al. Navier–Stokes Equations and Turbulence*
84 B. Polster and G. F. Steinke *Geometries on Surfaces*
85 R. B. Paris and D. Kaminski *Asymptotics and Mellin–Barnes Integrals*
86 R. McEliece *The Theory of Information and Coding, 2nd edn*
87 B. A. Magurn *An Algebraic Introduction to K-Theory*
88 T. Mora *Solving Polynomial Equation Systems I*
89 K. Bichteler *Stochastic Integration with Jumps*
90 M. Lothaire *Algebraic Combinatorics on Words*
91 A. A. Ivanov and S. V. Shpectorov *Geometry of Sporadic Groups II*
92 P. McMullen and E. Schulte *Abstract Regular Polytopes*
93 G. Gierz *et al. Continuous Lattices and Domains*
94 S. R. Finch *Mathematical Constants*
95 Y. Jabri *The Mountain Pass Theorem*
96 G. Gasper and M. Rahman *Basic Hypergeometric Series, 2nd edn*
97 M. C. Pedicchio and W. Tholen (eds.) *Categorical Foundations*
98 M. E. H. Ismail *Classical and Quantum Orthogonal Polynomials in One Variable*
99 T. Mora *Solving Polynomial Equation Systems II*
100 E. Olivieri and M. Eulália Vares *Large Deviations and Metastability*
101 A. Kushner, V. Lychagin and V. Rubtsov *Contact Geometry and Nonlinear Differential Equations*
102 L. W. Beineke and R. J. Wilson (eds.) with P. J. Cameron *Topics in Algebraic Graph Theory*
103 O. J. Staffans *Well-Posed Linear Systems*
104 J. M. Lewis, S. Lakshmivarahan and S. K. Dhall *Dynamic Data Assimilation*
105 M. Lothaire *Applied Combinatorics on Words*
106 A. Markoe *Analytic Tomography*
107 P. A. Martin *Multiple Scattering*
108 R. A. Brualdi *Combinatorial Matrix Classes*
109 J. M. Borwein and J. D. Vanderwerff *Convex Functions*
110 M.-J. Lai and L. L. Schumaker *Spline Functions on Triangulations*
111 R. T. Curtis *Symmetric Generation of Groups*
112 H. Salzmann *et al. The Classical Fields*
113 S. Peszat and J. Zabczyk *Stochastic Partial Differential Equations with Lévy Noise*
114 J. Beck *Combinatorial Games*
115 L. Barreira and Y. Pesin *Nonuniform Hyperbolicity*
116 D. Z. Arov and H. Dym *J-Contractive Matrix Valued Functions and Related Topics*
117 R. Glowinski, J.-L. Lions and J. He *Exact and Approximate Controllability for Distributed Parameter Systems*
118 A. A. Borovkov and K. A. Borovkov *Asymptotic Analysis of Random Walks*
119 M. Deza and M. Dutour Sikirić *Geometry of Chemical Graphs*
120 T. Nishiura *Absolute Measurable Spaces*
121 M. Prest *Purity, Spectra and Localisation*
122 S. Khrushchev *Orthogonal Polynomials and Continued Fractions*
123 H. Nagamochi and T. Ibaraki *Algorithmic Aspects of Graph Connectivity*
124 F. W. King *Hilbert Transforms I*
125 F. W. King *Hilbert Transforms II*
126 O. Calin and D.-C. Chang *Sub-Riemannian Geometry*
127 M. Grabisch *et al. Aggregation Functions*
128 L. W. Beineke and R. J. Wilson (eds.) with J. L. Gross and T. W. Tucker *Topics in Topological Graph Theory*
129 J. Berstel, D. Perrin and C. Reutenauer *Codes and Automata*
130 T. G. Faticoni *Modules over Endomorphism Rings*
131 H. Morimoto *Stochastic Control and Mathematical Modeling*
132 G. Schmidt *Relational Mathematics*
133 P. Kornerup and D. W. Matula *Finite Precision Number Systems and Arithmetic*
134 Y. Crama and P. L. Hammer (eds.) *Boolean Functions*
135 V. Berthé and M. Rigo (eds.) *Combinatorics, Automata and Number Theory*
136 A. Kristály, V. D. Rădulescu and C. Varga *Variational Principles in Mathematical Physics, Geometry, and Economics*

ENCYCLOPEDIA OF MATHEMATICS AND ITS APPLICATIONS

Finite Precision Number Systems and Arithmetic

PETER KORNERUP
University of Southern Denmark, Odense

DAVID W. MATULA
Southern Methodist University, Dallas

CAMBRIDGE
UNIVERSITY PRESS

CAMBRIDGE UNIVERSITY PRESS
Cambridge, New York, Melbourne, Madrid, Cape Town, Singapore,
São Paulo, Delhi, Dubai, Tokyo, Mexico City

Cambridge University Press
The Edinburgh Building, Cambridge CB2 8RU, UK

Published in the United States of America by Cambridge University Press, New York

www.cambridge.org
Information on this title: www.cambridge.org/9780521761352

First published 2010

Printed in the United Kingdom at the University Press, Cambridge

A catalog record for this publication is available from the British Library

Library of Congress Cataloging in Publication data
Kornerup, Peter.
Finite precision number systems and arithmetic / Peter Kornerup, David W. Matula.
p. cm. – (Encyclopedia of mathematics and its applications ; 133)
Includes bibliographical references and indexes.
ISBN 978-0-521-76135-2
1. Arithmetic – Foundations. I. Matula, David W. II. Title.
QA248.K627 2010
513 – dc22 2010030521

ISBN 978-0-521-76135-2 Hardback

Additional resources for this publication at www.cambridge.org/9780521761352

CONTENTS

PREFACE

This book builds a solid foundation for finite precision number systems and arithmetic, as used in present day general purpose computers and special purpose processors for applications such as signal processing, cryptology, and graphics. It is based on the thesis that a thorough understanding of number representations is a necessary foundation for designing efficient arithmetic algorithms.

Although computational performance is enhanced by the ever increasing clock frequencies of VLSI technology, selection of the appropriate fundamental arithmetic algorithms remains a significant factor in the realization of fast arithmetic processors. This is true whether for general purpose CPUs or specialized processors for complex and time-critical calculations. With faster computers the solution of ever larger problems becomes feasible, implying need for greater precision in numerical problem solving, as well as for larger domains for the representation of numerical data. Where 32-bit floating-point representations used to be the standard precision employed for routine scientific calculations, with 64-bit double-precision only occasionally required, the standard today is 64-bit precision, supported by 128-bit accuracy in special situations. This is becoming mainstream for commodity microprocessors as the revised IEEE standard of 2008 extends the scalable hierarchy for floating-point values to include 128-bit formats. Regarding addressing large memories, the trend is that the standard width of registers and buses in modern CPUs is to be 64 bits, since 32 bits will not support the larger address spaces needed for growing memory sizes. It follows that basic address calculations may require larger precision operands.

The rising demand for mobile processors has made realizing "low-power" (less energy consumption) without sacrificing speed the preeminent focus of the next generation of many arithmetic unit designs. Multicore processors allow opportunities in heterogeneous arithmetic unit designs to be realized alongside legacy systems. All of these opportunities require a new level of understanding of number representation and arithmetic algorithm design as the core of new arithmetic architectures.

This book emphasizes achieving fluency in redundant representations to avoid the self imposed design straightjacket of prematurely forcing intermediate values into more familiar non-redundant forms. Exploiting parallelism is crucial for multicore processors and for realizing fast arithmetic on large word-size operands, and employing redundant radix representation of numbers allows addition to be performed in constant time, independent of the word size of the operands. Allowing intermediate results to remain in a redundant representation for use in subsequent calculations has to be exploited, avoiding slow (at best logarithmic time) conversion into non-redundant representations where possible. Fortunately, conversion between redundant representations in most cases (for "compatible" radix values) can be performed in constant time.

Radix representation remains the single most important and fundamental way of representing numbers, even serving as a foundation for most other number representations, some of which are presented in the later chapters of this book. We have chosen the foundations of radix arithmetic as the definitive topic for initiating the study of finite precision number systems and arithmetic. We provide a very thorough treatment of radix representations, looking into the implications of the choice of digit set for a given radix as well as the choice of radix. Properties of the resulting set of representable values in the system (its "completeness") and uniqueness of representations ("redundancy") are investigated in a detail not found elsewhere. Conversions between radix representations are analyzed as a separate topic of significant use and importance in the implementations of arithmetic algorithms. Our objective is to provide a substantive mathematical foundation for radix number systems and their properties rather than ad hoc developments tied to specific limited applications.

It is our belief that a detailed understanding of radix representations and conversions between these is of great importance when developing and/or implementing arithmetic algorithms. The results on these topics presented here form a "toolbox" no arithmetic "algorithm engineer" should be without. We have found these tools extremely useful over the more than 30 years of our own joint research on alternative number representation systems and their arithmetic, on algorithm engineering in general, and on developments for actual processor implementations. Being intimately involved with the organization of the bi-annual IEEE Symposia on Computer Arithmetic for a similar time frame, we have been able to follow the challenges and research in this area, often allowing us to improve on existing algorithms, or to explain fundamental issues. For example, writing this book has spawned ideas for several research papers, some of which appeared first in drafts of the book, but the book also includes results that we first presented at meetings, in journal papers, and in actual processor implementations. Participation in the arithmetic unit design and testing of several generations of commercially successful processors such as the Cyrix $\times 87$ coprocessor and the National Semiconductor/AMD Geode "one Watt" IEEE floating-point compliant processor chosen for the One Laptop per Child (OLPC) project has provided valuable feedback on the real world practicality of new arithmetic algorithms.

It has not been possible to include here all the developments presented on computer arithmetic and number systems over the past years; a selection has had to be made. But we claim in most cases to present both classical and up-to-date algorithms for the problems covered. Over the years of writing the book, we have constantly been monitoring the literature, modifying and updating the text with new results and algorithms as they became known.

The approach used in this book is quite mathematical when presenting and analyzing ideas and algorithms, but we go into very little detail on the logic design, and do not look at all at hardware implementations. Complexities of algorithms and designs are generally only specified in the mathematical O-notation, but occasionally we do count gates, just as hardware designs may be sketched as logic diagrams. However, it is definitely our intent that the presentations here should also be of great value for engineers selecting and designing actual VLSI or FPGA implementations of arithmetic algorithms.

The reader is not expected to have more depth of knowledge of logic design and electronics than would be gained from an undergraduate class on computer architecture. Nor is the reader expected to have a deep mathematical background, no more than is usually acquired from an undergraduate computer science curriculum. We do occasionally use terminology from abstract algebra, e.g., (denoting a mapping a homomorphism), or some number system to be a commutative ring, as a benefit to readers familiar with these concepts. We will not, however, use advanced properties beyond the few defined and described in the text. We extensively use the concept of sets and the standard notation for such, and, of course, assume knowledge of simple Boolean algebra. Our derivations and proofs will most often be based on elementary algebra and elementary number theory without recourse to calculus or analysis. The arguments should be quite accessible to those with a natural affinity for games and mathematical puzzles and the book should be invaluable to the serious student who may want to analyze or design such games and puzzles.

The contents of the book can be seen to consist of three parts. The first part comprising two chapters, covers the fundamentals of radix representation and conversion between these representations. For this part we introduce a formal notation for expressing radix polynomials, allowing us to distinguish between different radix representations of the same value, and of the value itself. This notation is used heavily in the first three chapters, but later our notation is more relaxed, when the interpretation should be implicitly obvious from the context. The second part covers in four chapters the basic arithmetic operations: addition, multiplication, division, and square root. These are the fundamental arithmetic operations that are standardized in the IEEE floating-point standard and are the subject of very competitive hardware implementations and much academic research regarding the practical performance of alternative algorithms. The third part presents examples of some special number systems, usually built on the fundamental radix systems. These are the floating-point systems, residue number systems, and finally rational number systems and arithmetic that were largely developed by the authors. The finite precision rational number systems are built on the number theoretic foundations of fractions and continued fractions. The chapter on residue number representations and modular arithmetic includes an extensive presentation of the basic modular operations, some of which are applicable to and important in cryptographic algorithms. The chapter on floating-point systems seeks to provide a foundation for these systems which have largely been ignored in the mathematical literature on number system foundations. We carefully distinguish floating-point number systems at three levels. First, we distinguish those subsets of real numbers which form a floating-point number system, then we specify individual floating-point numbers as real numbers characterized by their factorizations into component terms constituting a sign factor, a scale (or radix shift) factor, and a significand factor, and lastly we investigate the encodings of the various factors into component bit strings of a floating-point word in compliance with the IEEE floating-point standard. The first two levels treating floating-point numbers as reals characterized by a factorization allow the development of a number theoretic foundation for floating-point arithmetic similar to the foundation for rational arithmetic derived from reals characterized by being representable as fractions. The concept of precise roundings allows for a development of the best radix (and floating-point) approximation similar to the best rational approximation concept in the established number theoretic literature on continued fractions. Noteworthily absent among the special number systems are systems employing logarithmic representations, as well as error tolerant systems.

Each chapter begins with an introduction to its contents, and ends with bibliographic notes and a bibliography of publications and selected patents, pointing to the sources used for the ideas and presentations and to further reading. Most sections end with some problems and exercises, illustrating the material presented or further developing the topics. A solutions manual is available for instructors, describing possible solutions to (most of) the presented problems.

We are well aware that the contents of the book are beyond what can be covered in a normal (graduate) semester course. But it is feasible in that time to cover most of Chapters 1–4 and the early parts of Chapters 5 and 6. The remaining parts and chapters can be used for a follow-up course or for individual studies, possibly serving as an introduction to a master's student project, or as background for research towards a Ph D.

The content of this book is based on the last four decades of research in many of these topics, pursued in response to the explosive growth and omnipresence of digital computers. The content includes some of our own results over this period, although most material is based on what is found in the open literature and learned from active communication with colleagues in the international community of "arithmeticians," in particular through our active participation at the bi-annual IEEE Symposia on Computer Arithmetic attended by at least one of us since its initiation in 1969. We "stand on the shoulders" of many, including the very early pioneers of the field, but also many past and present colleagues and students have inspired our work in general, and in particular influenced the presentations here. We have tried to be very comprehensive in our coverage of the results on the topics presented, but it is, of course, not possible to cover everything.

It is our hope that the book may serve as a valuable resource for further academic research on these topics, and also as a useful bookshelf tool for practitioners in the industry who are building the processors of the future.

Despite our efforts, without doubt there are typos and possibly also more serious errors in this text. We apologize, and encourage the reader to contact us if such are found. We will establish a web page listing corrections to the book.

We would especially like to recognize the students who collaborated and contributed to the development of this book over the last two decades. They include D. DasSarma, M. Daumas, A. Fit-Florea, C. S. Iordache, C. N. Lyu, L. D. McFearin and S. N. Parikh in Dallas; and T. A. Jensen, S. Johansen, A. M. Nielsen, H. Orup in Aarhus and Odense. We also thank other graduate and postdoctoral students who have worked with us on this manuscript including S. Datla, G. Even, L. Li, J. Moore, A. Panhaleux, G. Wei and J. Zhang, as well as faculty collaborators W. E. Ferguson, M. A. Thornton, and P. -M. Seidel in Dallas, and R. T. Gregory, U. Kulisch, and J. -M. Muller at their institutions.

Finally we want to express our gratitude to our wives, Margot and Patricia, for their patience with our absence during the numerous hours spent on writing and discussions over many years, and during our mutual visits where they have generally followed us.

August MMX

Peter Kornerup
Dept. of Math. and Computer Science
University of Southern Denmark
Odense, Denmark
kornerup@imada.sdu.dk

David W. Matula
Dept. of Computer Science and Engineering
Southern Methodist University
Dallas, TX
matula@lyle.smu.edu

1

Radix polynomial representation

1.1 Introduction

From the earliest cultures humans have used methods of recording numbers (integers), by notches in wooden sticks or collecting pebbles in piles or rows. Conventions for replacing a larger group or pile, e.g., five, ten, or twelve objects, by another object or marking, are also found in some early cultures. Number representations like these are examples of *positional number systems*, in which objects have different weight according to their relative positions in the number. The weights associated with different positions need not be related by a constant ratio between the weights of neighboring positions. In time, distance, old currency, and other measuring systems we find varying ratios between the different units of the same system, e.g., for time 60 minutes to the hour, 24 hours to the day, and 7 days to the week, etc.

Systems with a constant ratio between the position weights are called radix systems; each position has a weight which is a power of the radix. Such systems can be traced back to the Babylonians who used radix 60 for astronomical calculations, however without a specific notation for positioning of the unit, so it can be considered a kind of floating-point notation. Manipulating numbers in such a notation is fairly convenient for multiplication and division, as is known for anyone who has used a slide rule. Our decimal notation with its fixed radix point seems to have been developed in India about 600 CE, but without decimal fractions. Decimal notation with fractions appeared later in the Middle East, and in the fifteenth century a Persian mathematician computed the value of π correctly to 16 decimal digits. In the seventeenth century the binary system was developed, and it was realized that any integer greater than 1 could be used as a radix. Knuth in [Knu98, Chapter 4] gives an account of the development of positional systems, and provides further references on their history.

Usually digits used for denoting numbers in a positional system are restricted by the ratios between weights of neighboring positions, in radix systems by the value of the radix. If the factor or radix is β, then the digits normally used are from the set $\{0, 1, \dots, \beta - 1\}$, but other digit sets are possible. If the cardinality of the set of permissible digits is larger than the radix, then some numbers can be represented in more than one way and the number system is said to be redundant. Such redundancy permits some arithmetic operations to be performed faster than with a "normal" (non-redundant) digit set. For example, as we know, when adding a number of distances measured in feet and inches, we can add up the feet and the inches independently, and only at the end convert excess inches into feet. In this way we avoid converting all the intermediate results, and hence the carry transfers, during accumulation of the individual measures.

In this chapter we will discuss positional number systems mainly through standard radix representations, with only a few deviations into other weighted systems. But we will thoroughly investigate systems where the digits may be drawn from fairly general sets of integers, and in particular also redundant systems. Although it is possible to define systems with a non-integral value of the radix, even a complex value, we shall restrict our treatment to integral values β, $|\beta| \geq 2$.

1.2 Radix polynomials

As a foundation for our development of a theory for the representation of numbers in positional notation, we will use the algebraic structure of sets of polynomials. Arithmetic on numbers in positional notation is closely related to arithmetic on polynomials, so a firm foundation for the former can be based on the theory for the latter. We will here be concerned with the characterization of systems employing an integral-valued radix and digits, but our analysis will go beyond the usual radix 2, 8, 10, and 16 systems and the related standard digit sets.

Let $\mathbb{Z}^*[x]$ be the set of *extended polynomials*[1]

$$P(x) = a_m x^m + a_{m-1} x^{m-1} + \cdots + a_\ell x^\ell \tag{1.2.1}$$

with $a_i \in \mathbb{Z}$ (the set of integers), $-\infty < \ell \leq i \leq m < \infty$, considered *formal expressions in the indeterminate variable* x. The coefficients a_m and a_ℓ may be zero, but the zero polynomial is also denoted $P(x) = 0$. However, in general we will only display non-zero coefficients a_i. If $m = \ell$ we call $P(x)$ an *extended monomial*.

It is then possible to define addition and multiplication on polynomials from $\mathbb{Z}^*[x]$, i.e., with $P(x), Q(x) \in \mathbb{Z}^*[x]$,

$$P(x) = a_m x^m + a_{m-1} x^{m-1} + \cdots + a_\ell x^\ell,$$

$$Q(x) = b_n x^n + b_{n-1} x^{n-1} + \cdots + b_k x^k,$$

[1] The extension here is that, in general, we allow negative powers of x.

and $p = \max(m, n)$, $q = \min(\ell, k)$, we may define:

$$S(x) = P(x)+Q(x) = (a_p + b_p)x^p + (a_{p-1} + b_{p-1})x^{p-1} + \cdots + (a_q + b_q)x^q, \tag{1.2.2}$$

$$R(x) = P(x) \times Q(x) = c_{m+n}x^{m+n} + c_{m+n-1}x^{m+n-1} + \cdots + c_{\ell+k}x^{\ell+k}, \tag{1.2.3}$$

where

$$c_j = a_\ell b_{j-\ell} + a_{\ell+1}b_{j-\ell-1} + \cdots + a_{j-k}b_k. \tag{1.2.4}$$

Based on the fact that $(\mathbb{Z}, +, \times)$ is a commutative ring with identity employing integer addition and multiplication as operators, it may now be seen that $(\mathbb{Z}^*[x], +, \times)$ is also a commutative ring with identity satisfying the cancellation law (i.e., an integral domain), when $+$ and $\times$ here is taken as addition and multiplication defined by (1.2.2) respectively (1.2.3) and (1.2.4).

When in (1.2.1) the indeterminate variable x is taken as a real variable, $P(x)$ is a function whose value can be found at any real value b. We will denote this evaluation as

$$P(x)\,\big|_{x=b} \in \mathbb{R}$$

and define the *evaluation mapping* E_b as

$$E_b : P(x) \to P(x)|_{x=b}. \tag{1.2.5}$$

Observation 1.2.1 *For $b \neq 0$, E_b is a homomorphism of $\mathbb{Z}^*[x]$ to the reals:*

$$S(x) = P(x) + Q(x) \Rightarrow S(x)|_{x=b} = P(x)|_{x=b} + Q(x)|_{x=b}, \tag{1.2.6}$$

$$R(x) = P(x) \times Q(x) \Rightarrow R(x)|_{x=b} = P(x)|_{x=b} \times Q(x)|_{x=b}. \tag{1.2.7}$$

Thus for fixed b the extended polynomials $P(x)$ and $Q(x)$ may be used as *representations* of the real numbers $P(x)|_{x=b}$ and $Q(x)|_{x=b}$ respectively. Note that addition of extended polynomials is independent in each position or "carry-free" since the coefficients are unrestricted integers. Also note that E_b is not one-to-one, e.g.,

$$(9x + 3 + 4x^{-1})|_{x=8} = (x^2 + x + 3 + 4x^{-1})|_{x=8} \;(= 75.5).$$

For fixed a and b let us define

$$\mathcal{V}_b(a) = \{P(x) \in \mathbb{Z}^*[x] \mid P(x)|_{x=b} = a\},$$

i.e., the set of polynomials whose value at b is a. Or using different terminology, for fixed b, $\mathcal{V}_b(a)$ is the set of *redundant representations* of a. Also $\mathcal{V}_b(a)$ can be characterized as a *residue class* in the set of extended polynomials.

Theorem 1.2.2 *Given any two extended polynomials $P(x)$, $Q(x)$, then $P(x)|_{x=b} = Q(x)|_{x=b}$ if and only if $P(x) \equiv Q(x)$* (mod $(x - b)$).

Proof For $P(x) \equiv Q(x) \pmod{(x-b)}$ we have $P(x) = Q(x) + R(x) \times (x - b)$. So then $P(x)|_{x=b} = Q(x)|_{x=b} + R(x)|_{x=b} \times (x-b)|_{x=b} = Q(x)|_{x=b}$ since $(x-b)|_{x=b} = 0$.

Alternatively, assume $P(x)|_{x=b} = Q(x)|_{x=b}$. Then the polynomial $S(x) = P(x) - Q(x)$ satisfies $S(x)|_{x=b} = 0$, so b is a root of $S(x)$. This means $(x-b)$ must divide $S(x)$, so we obtain $S(x) = R(x) \times (x-b)$ for some $R(x)$. Hence $P(x) = Q(x) + R(x) \times (x-b)$. □

Example 1.2.1 Let $b = 8$ and consider the extended polynomials

$$P(x) = x^2 - 5x - 6 + 13x^{-1},$$
$$Q(x) = 2x + 4 - 3x^{-1}.$$

Now $P(x)$ has the same value as $Q(x)$ for $x = b = 8$, $P(x)|_{x=8} = Q(x)|_{x=8} = 19\frac{5}{8}$, so $Q(x)$ and $P(x)$ both belong to $\mathcal{V}_8(19\frac{5}{8})$. We further note that

$$P(x) - Q(x) = x^2 - 7x - 10 + 16x^{-1} = (x + 1 - 2x^{-1})(x - 8),$$

so $P(x) \equiv Q(x) \pmod{(x-8)}$ as required by Theorem 1.2.2. □

From the proof of Theorem 1.2.2 we note the following.

Observation 1.2.3 *If $P(x)|_{x=b} = Q(x)|_{x=b}$, then there exists a transfer polynomial*

$$R(x) = \sum_{i=\ell}^{m} c_i x^i$$

satisfying

$$\begin{aligned} P(x) &= Q(x) + R(x)(x-b) \\ &= Q(x) + \sum_{i=\ell}^{m} c_i (x-b) x^i, \end{aligned}$$

where each term $c_i x^{i+1} - c_i b x^i$ performs a transfer (a "carry") of information from position i to position $i+1$ of the polynomial.

In the rest of this book we will consider various systems characterized by the value of b chosen. We will use the symbol β for such a value, termed[2] the *radix* or the *base* of the system. The radix can be positive or negative, even non-integral or complex. In the following we will only consider integral values of the radix β. Since we intend to evaluate extended polynomials at β, but still retain a distinction between the formal expression $P(x)$ and its value obtained

[2] We will, in general, use the term radix rather than base, except where traditionally the latter is used, such as in base conversion.

by the evaluation mapping E_β, we will for $P(x) \in \mathbb{Z}^*[x]$ distinguish between the extended polynomial

$$P([\beta]) = d_m[\beta]^m + d_{m-1}[\beta]^{m-1} + \cdots + d_\ell[\beta]^\ell$$

i.e., an unevaluated expression, and the real value obtained by evaluating

$$P(\beta) = d_m\beta^m + d_{m-1}\beta^{m-1} + \cdots + d_\ell\beta^\ell.$$

Definition 1.2.4 *For β such that $|\beta| \geq 2$, the set of* radix-β polynomials $\mathcal{P}[\beta]$ *is the set composed of the zero-polynomial and all extended polynomials of the form*

$$P([\beta]) = d_m[\beta]^m + d_{m-1}[\beta]^{m-1} + \cdots + d_\ell[\beta]^\ell,$$

where $d_i \in \mathbb{Z}$ for $-\infty < \ell \leq i \leq m < \infty$.

Notation For $P([\beta]) = \sum_{i=\ell}^{m} d_i[\beta]^i \in \mathcal{P}[\beta]$ we introduce the following notation:

β: the radix or base;
d_i: the digit in position i; or
$d_i(P)$: the digit in position i of polynomial $P([\beta])$;
m: the most-significant position: $\mathrm{msp}(P([\beta]))$;
ℓ: the least-significant position: $\mathrm{lsp}(P([\beta]))$;
d_m: the most-significant digit: $d_m \neq 0$;
d_ℓ: the least-significant digit: $d_\ell \neq 0$.

We assume that $d_m \neq 0$ and $d_\ell \neq 0$, except when $P = 0$, where $d_m = d_\ell = m = \ell = 0$. In the following $\mathrm{lsp}(P[\beta])$ will be called be the *last place*.

$\mathcal{P}[\beta]$ thus forms a ring with the same additive and multiplicative structure as the ring $\mathbb{Z}^*[x]$. Particular examples are:

$\mathcal{P}[16]$: the *hexadecimal* radix polynomials;
$\mathcal{P}[10]$: the *decimal* radix polynomials;
$\mathcal{P}[8]$: the *octal* radix polynomials;
$\mathcal{P}[3]$: the *ternary* radix polynomials;
$\mathcal{P}[2]$: the *binary* radix polynomials;
$\mathcal{P}[-2]$: the *nega-binary* radix polynomials.

Our general definition of radix polynomials allows positive and/or negative digits as well as digit values exceeding the magnitude of the radix.

In arithmetic algorithms there is a need to deal with individual terms of a radix-β polynomial, corresponding to individual digits, and, in general, in order to have a kind of "pointer" to a specific position of a radix polynomial. Hence we introduce the following definition.

Definition 1.2.5 *For β such that $|\beta| \geq 2$, the set of* radix-β monomials of order j, $\mathcal{M}^j[\beta]$, *is the set of all extended polynomials of the form*

$$M([\beta]) = i[\beta]^j,$$

where $i, j \in \mathbb{Z}$. The exponent j is called the order *of the monomial.*

Note that i may be any integer, including zero, and may contain β as a factor. The order j serves to position the digit value i, e.g., for adding into a particular position using (1.2.2). Multiplying with a *unit monomial* $[\beta]^j$ using (1.2.3) corresponds to a "shifting" operation on a number in positional notation. Adding a monomial $[\beta]^\ell$ to a radix polynomial $P[\beta]$ with $\text{lsp}(P[\beta]) = \ell$ corresponds to adding a *unit in the last place* (ulp).

By Theorem 1.2.2, the relation for equality-of-value of extended polynomials under the evaluation mapping $E_\beta : P(x) \to P(x)|_{x=\beta}$ yields the residue classes of $\mathbb{Z}^*[x]$ modulo $(x - \beta)$ as equivalence classes. The extended monomials provide for characterizing a useful *complete residue system* for these equivalence classes.

Theorem 1.2.6 *For β with $|\beta| \geq 2$, let*[3] $\mathcal{M}' = \{i\,[\beta]^j \mid i \neq 0, \beta \not| \, i \in \mathbb{Z}, j \in \mathbb{Z}\} \cup \{0[\beta]^0\}$, *i.e., $\mathcal{M}'$ is the set of all radix-β monomials with coefficients not divisible by β, along with the zero polynomial. Then $\mathcal{M}'$ is a complete residue system for $\mathbb{Z}^*[x]$ modulo $(x - \beta)$.*

Proof To show that members of $\mathcal{M}'$ are in distinct residue classes it is sufficient by Theorem 1.2.2 to show the evaluation mapping maps distinct members of $\mathcal{M}'$ into distinct real values. Suppose $i[\beta]^j, k[\beta]^\ell \in \mathcal{M}'$ are distinct non-zero members of $\mathcal{M}'$ of the same value. Now $i[\beta]^j$ evaluates to $i\beta^j$ and $k[\beta]^\ell$ has the value $k\beta^\ell$, so $i\beta^j = k\beta^\ell$. We may assume $\ell \geq j$, so $i = k\beta^{\ell-j}$. Then $\ell = j$, since β does not divide i, and $i = k$, a contradiction, and it follows that members of $\mathcal{M}'$ are congruent modulo $(x - \beta)$. Moreover $P(x) \in \mathbb{Z}^*[x]$ is either the zero polynomial or has a least-significant digit $d_\ell \neq 0$. Then $P(x)|_{x=\beta} = i\beta^l$ for integers i, ℓ. If $i = 0$, then $P(x)$ is congruent to the zero polynomial modulo $(x - \beta)$. If $i = k\beta^n$ for $n \geq 1$ and $\beta \not| \, k$, then $P(x)|_{x=\beta} = k\beta^{\ell+n}$ and $P(x) \equiv k[\beta]^{l+n}$ modulo $(x - \beta)$ with $k[\beta]^{\ell+n} \in \mathcal{M}'$. Thus $\mathcal{M}'$ is a complete residue system for $\mathbb{Z}^*[x]$ modulo $(x - \beta)$. □

The members of $\mathcal{M}'$ provide convenient unique expressions for the real values a such that the redundancy classes $\mathcal{V}_\beta(a)$ are non-vacuous.

Problems and exercises

1.2.1 For $P \in \mathcal{P}[\beta]$, $P \neq 0$ show that $\lfloor P \times [\beta]^{-\text{msp}(P)} \rfloor$ is a radix polynomial whose value is the most-significant digit of P (here $\lfloor\cdot\rfloor$ means truncate

[3] The symbol $\not|$ means "does not divide."

the polynomial to its integer part). Derive a similar formula for the least-significant digit.

1.2.2 For P and Q given as

$$P = 5 \times [\beta]^6 + 3 \times [\beta]^4 + 2 \times [\beta]^3 - 2 \times [\beta]^2,$$
$$Q = 4 \times [\beta]^3 + 2 \times [\beta]^2 - 6 \times [\beta] - 7 \times [\beta]^{-3} - 3 \times [\beta]^{-4} + 5 \times [\beta]^{-7},$$

find msp$(P + Q)$, msp$(P \times P)$, lsp$(P \times Q)$, lsp$(\lfloor P + Q \rfloor)$, and $d_{\text{lsp}(\lfloor P \times Q \rfloor)}(\lfloor P \times Q \rfloor)$.

1.3 Radix-β numbers

The radix polynomials introduced in the previous section provide a representation of numbers in which the evaluation mapping E_β provides a mapping from extended polynomials into the reals. Since β is fixed for a radix polynomial $P([\beta]) \in \mathcal{P}[\beta]$ we will introduce the operator $\|\cdot\|$ defined on $\mathcal{P}[\beta]$ as

$$\|P([\beta])\| = P(x)|_{x=\beta} = P(\beta) = \sum_{i=\ell}^{m} d_i \beta^i$$

for the evaluation mapping.

Restricting β and the digits d_i to integral values for given β with $|\beta| \geq 2$, the real value $v = \sum_{i=l}^{m} d_i \beta^i$ determined by the evaluation operation is a rational number belonging to a subset of rationals characterized by the radix β. The set

$$\mathbb{Q}_\beta = \{k\beta^\ell \mid k, \ell \in \mathbb{Z}\} \tag{1.3.1}$$

is called the *radix-β numbers*, but could equivalenly be termed the *radix-β rationals*. Note that either or both of k and ℓ may be negative in (1.3.1). Observing that $\mathbb{Q}_\beta = \mathbb{Q}_{-\beta}$, it is evident that the set of values of the radix polynomials of $\mathcal{P}[\beta]$ is precisely $\mathbb{Q}_{|\beta|}$. Also $\mathbb{Q}_{|\beta|}$ inherits the algebraic structure of addition and multiplication as defined on reals. Thus polynomial arithmetic in $\mathcal{P}[\beta]$ corresponds to the arithmetic of the real numbers in $\mathbb{Q}_{|\beta|}$.

Observation 1.3.1 *For any integer radix β with $|\beta| \geq 2$ the evaluation mapping $\|P\|$ for $P \in \mathcal{P}[\beta]$ is a homomorphism of $\mathcal{P}[\beta]$ onto $\mathbb{Q}_{|\beta|}$; $(\mathbb{Q}_{|\beta|}, +, \times)$ is a commutative ring with identity.*

Thus reference to $\mathbb{Q}_\beta$ as the *radix-β number system* denotes the set $\mathbb{Q}_\beta$ along with the arithmetic structure provided by the commutative ring $(\mathbb{Q}_\beta, +, \times)$. For the most often used radix values it is customary to employ the following terminology:

$\mathbb{Q}_2$: binary number system;
$\mathbb{Q}_3$: ternary number system;
$\mathbb{Q}_8$: octal number system;
$\mathbb{Q}_{10}$: decimal number system;
$\mathbb{Q}_{16}$: hexadecimal number system;

but note $v \in \mathbb{Q}_{|\beta|}$ just implies that v is a number which *can* be represented as a (finite length) radix-β polynomial.

The factors of a particular *radix factorization* $v = k\beta^\ell$ for $v \in \mathbb{Q}_\beta$ identify two components of the representation of v that are typically handled separately in implementing radix arithmetic. Specifically

- k is the integer *significand*, and
- β^ℓ is the *scale factor* with *exponent* ℓ.

For $\ell > 0$, v is an integer tuple factorization. For $\ell \leq 0$, the factorization is equivalent to the fraction $k/\beta^{-\ell}$ with numerator k and denominator $\beta^{|\ell|}$.

Observation 1.3.2 *The radix-β numbers $\mathbb{Q}_\beta$ form the subset of rationals given by fractions with denominators restricted to powers of the radix β.*

A radix factorization $k\beta^\ell$ is termed reducible when $\beta | k$, and is a unique *irreducible radix factorization* when $\beta \nmid k$, yielding a unique minimum magnitude significand.

Observation 1.3.3 $v = k\beta^\ell$ *is an irreducible radix factorization if and only if* $k \bmod \beta \neq 0$.

Observation 1.3.4 *Let $\mathcal{M}_\beta$ be a complete residue system for $\mathbb{Z}^*[x]$ modulo $(x - \beta)$. Then the evaluation mapping $\| \cdot \|$ is an isomorphism between $\mathcal{M}_\beta$ and the irreducible radix factorizations.*

The relationship between the "scaled significand" radix factorizations $\{k \times \beta^\ell\}$ as tuples, and the set of radix-β numbers $\mathbb{Q}_\beta$ is analogous to the relation between fractions $\{i/j\}$ as tuples, and the set of rational numbers $\mathbb{Q}$, as summarized in Table 1.3.1.

Note that $\mathbb{Q}_{|\beta|}$ is not a field as $1/(|\beta| + 1) \notin \mathbb{Q}_{|\beta|}$, but $\bigcup_p \mathbb{Q}_p$ is a field, the rationals $\mathbb{Q}$. It is well known that $\mathbb{Q}_2 = \mathbb{Q}_8 = \mathbb{Q}_{16}$, and probably also that

Table 1.3.1. *Analogies between the radix-β numbers and the rational numbers*

Number characterization	Tuple terminology	Irreducible form	Number system
Radix factorizations	Significand × scale factor	$k \times \beta^\ell$ $\beta \nmid k$	Radix-β number system $\mathbb{Q}_\beta$
Rational fractions	Numerator / denominator	i/j $\gcd(i, j) = 1$	Rational number system $\mathbb{Q}$

$\mathbb{Q}_2 \subset \mathbb{Q}_{10}$, but what is the relation between $\mathbb{Q}_{12}$ and $\mathbb{Q}_{18}$? The following theorem provides a tool for determining such relations.

Theorem 1.3.5 *If the integers $p \geq 2$ and $q \geq 2$ have prime decompositions $p = \Pi p_i^{n_i}$ and $q = \Pi p_i^{m_i}$, where $\{p_i\}$ is the set of all primes, then:*

(1) If $\forall i : n_i \neq 0 \Rightarrow m_i \neq 0$ (the prime factors of p are factors of q), then $\mathbb{Q}_p \subseteq \mathbb{Q}_q$.
(2) If $\forall i : n_i \neq 0 \Rightarrow m_i \neq 0$ and $\exists j : 0 = n_j < m_j$, then $\mathbb{Q}_p \subset \mathbb{Q}_q$.
(3) If $\forall i : n_i \neq 0 \Leftrightarrow m_i \neq 0$ (p and q contain the same prime factors), then $\mathbb{Q}_p = \mathbb{Q}_q$.

Proof To show (1) first note that for $v = kp^j \in \mathbb{Q}_p$, then also $v \in \mathbb{Q}_q$ if $j \geq 0$, so assume $j < 0$. If t divides u, then for $v \in \mathbb{Q}_t$, $v = kt^j = k(u/t)^{-j}u^j = k'u^j$ with $k' \in \mathbb{Z}$, so $v \in \mathbb{Q}_u$ and hence $\mathbb{Q}_t \subseteq \mathbb{Q}_u$. Specifically with $r = \prod_{p_i \text{ divides } p} p_i$ we have $\mathbb{Q}_r \subseteq \mathbb{Q}_p$. Now let $n = \max\{n_i | p_i{}^{n_i} \; divides \; p\}$ and consider $v \in \mathbb{Q}_p$, so

$$v = kp^j = k\left(\prod p_i{}^{n_i}\right)^j = k\left(\prod p_i{}^{j(n_i - n)}\right)\left(\prod p_i\right)^{jn} = k'r^{jn},$$

where the products are over the p_i dividing p. Since $j < 0$, we have $j(n_i - n) \geq 0$, so $k' = k(\prod p_i{}^{j(n_i-n)}) \in \mathbb{Z}$ hence $v \in \mathbb{Q}_r$ and $\mathbb{Q}_p = \mathbb{Q}_r$. But r also divides q by assumption in (1), so $\mathbb{Q}_p = \mathbb{Q}_r \subseteq \mathbb{Q}_q$, which proves (1).

To prove (2) assume there exists a j such that $0 = n_j < m_j$, then $p_j^{-1} \in \mathbb{Q}_q$ but $p_j^{-1} \notin \mathbb{Q}_p$. Finally (3) follows from (1) by symmetry. □

Example 1.3.1 From Theorem 1.3.5 it follows that $\mathbb{Q}_{12} = \mathbb{Q}_{18}$ with $\mathbb{Q}_2 \subset \mathbb{Q}_{12}$ and $\mathbb{Q}_3 \subset \mathbb{Q}_{12}$. However, $\mathbb{Q}_2 \not\subseteq \mathbb{Q}_3$ and $\mathbb{Q}_3 \not\subseteq \mathbb{Q}_2$ since $\frac{1}{2} \notin \mathbb{Q}_3$ and $\frac{1}{3} \notin \mathbb{Q}_2$. □

The radix numbers are effectively represented with two types of redundancy, both of which are important in efficient implementation of radix arithmetic as operations on digit strings. The choice of exponent, ℓ, in the factorization implicitly recognizes any number of low-order zero digits, and the choice of radix polynomial for the irreducible significand brings flexibility to the choice of coefficients (digits) of the polynomial.

In general, any number $v \in \mathbb{Q}_{|\beta|}$ has an infinity of different representations as a radix-β or radix-$(-\beta)$ polynomial. The evaluation mapping $\|\cdot\| : \mathcal{P}[\beta] \to \mathbb{Q}_{|\beta|}$ partitions the members of $\mathcal{P}[\beta]$ into equivalence classes such that all radix polynomials of a given class have the same real value.

Definition 1.3.6 *For any $v \in \mathbb{Q}_{|\beta|}$ let the* redundancy class $V_\beta(v)$ *be the set*

$$V_\beta(v) = \{P \in \mathcal{P}[\beta] \mid \|P\| = v\}.$$

Example 1.3.2 Let

$$\begin{aligned} P([5]) &= \phantom{1[5]^3 -{}} 2[5]^2 + 3[5] + 1 \in V_5(66), \\ Q([5]) &= 1[5]^3 - 2[5]^2 - 2[5] + 1 \in V_5(66), \\ R([5]) &= \phantom{1[5]^3 -{}} 3[5]^2 - 1[5] - 4 \in V_5(66), \end{aligned}$$

so $P([5])$, $Q([5])$, and $R([5])$ are all in the same redundancy class $V_5(66)$, and form alternative radix-5 representations of the value 66 (which here has been written in ordinary decimal notation).

If the coefficients (the digits) of the polynomials above had to be chosen from the set $\{0, 1, 2, 3, 4\}$, then it is well known that this particular P is the only radix-5 polynomial evaluating to 66. If the digits are drawn from the set $\{-2, -1, 0, 1, 2\}$, then we shall see later that Q is similarly a unique representation. However, if the digit set is $\{-4, -3, -2, -1, 0, 1, 2, 3, 4\}$, then R as well as P and Q are possible representations. □

Problems and exercises

1.3.1 Show that $1/(n+1) \notin \mathbb{Q}_n$ for $n \geq 2$.

1.3.2 List all the members of the redundancy class $V_2(5)$ that can be written with four digits or fewer, using $\{-1, 0, 1\}$ as the permissible set of digit values (for convenience write them in string notation).

1.3.3 For p, q distinct primes, show that $\mathbb{Q}_p \cap \mathbb{Q}_q = \mathbb{Z}$.

1.4 Digit symbols and digit strings

When using radix-β polynomials for the representation of numbers from $\mathbb{Q}_{|\beta|}$:

$$P([\beta]) = \sum_{i=\ell}^{m} d_i [\beta]^i,$$

the representation might also be denoted as a list:

$$((d_m, d_{m-1}, \ldots, d_\ell), \ell)$$

explicitly including all zero-valued digits d_i for $\ell < i < m$, or alternatively only non-zero digits could be listed, e.g., (digit,index)-pairs:

$$((d_{i_1}, i_1), (d_{i_2}, i_2), \ldots, (d_{i_k}, i_k)).$$

The convention is to denote a radix polynomial as a string of symbols drawn from some alphabet, implicitly associated with the chosen radix. We will use the alphabet in Table 1.4.1 in our examples where we only employ "small" radices.

Table 1.4.1. *Digit alphabet for digit values between −15 and 15*

Symbol	Value	Symbol	Value
0	0		
1	1	$\bar{1}$	−1
2	2	$\bar{2}$	−2
3	3	$\bar{3}$	−3
4	4	$\bar{4}$	−4
5	5	$\bar{5}$	−5
6	6	$\bar{6}$	−6
7	7	$\bar{7}$	−7
8	8	$\bar{8}$	−8
9	9	$\bar{9}$	−9
A	10	$\bar{A}$	−10
B	11	$\bar{B}$	−11
C	12	$\bar{C}$	−12
D	13	$\bar{D}$	−13
E	14	$\bar{E}$	−14
F	15	$\bar{F}$	−15

Based on some digit alphabet N we will then for the radix polynomial

$$P([\beta]) = \sum_{i=\ell}^{m} d_i [\beta]^i, \tag{1.4.1}$$

satisfying the restriction that each d_i, $\ell \le i \le m$, has a value representable by a member $\delta_i \in N$, define the *radix digit string*[4] $\xi(P) \in \{N^*.N^*\} \cup \{0\}$ as

$$\xi(P) = \begin{cases} \delta_m \delta_{m-1} \cdots \delta_\ell \overbrace{0 \cdots 0}^{\ell \text{ zeroes}}{}_{[\beta]} & \text{for } \ell \ge 0, \\ .\overbrace{0 \cdots 0}^{-m-1 \text{ zeroes}} \delta_m \delta_{m-1} \cdots \delta_{\ell[\beta]} & \text{for } m < 0, \\ \delta_m \delta_{m-1} \cdots \delta_0 . \delta_{-1} \cdots \delta_{\ell[\beta]} & \text{for } \ell < 0 \le m, \\ 0_{[\beta]} & \text{for } P([\beta]) = 0, \end{cases}$$

where δ_i is the digit symbol for the value d_i of (1.4.1), and 0 is the symbol for the digit value zero. Note the presence of the period (dot) used as the *radix point*, and the string of "significant low-order zeroes" implicitly identifying the exponent ℓ in this *place-value* representation.

Observe the distinction we are making here between a *symbol* like $\delta_i \in N$ and a digit *value* $d_i \in \mathbb{Z}$, where δ_i is used to *denote* some integer value d_i. As it will be too cumbersome to continue this distinction, we shall in the following interpret

4 Recall that N^* denotes zero or more occurrences of a symbol from the alphabet N.

d_i appearing in a radix digit string as a symbol denoting a member of a digit set, whereas when it appears in an expression like $d_i \geq 0$ it denotes a value. If a digit appears repeatedly in a radix digit string, we allow the notation d^n to denote a sequence of n occurrences of the digit d.

To avoid ambiguity in discussions of radix strings and their values we shall abide by the convention that if $\sigma \in \{N^*.N^*\} \cup \{0\}$, then $\sigma_{[\beta]}$ is a symbol string denoting a radix polynomial, whereas $\sigma_\beta = \|\sigma_{[\beta]}\|$ is the value represented, generalizing the notation $\|\cdot\|$ to the domain of radix digit strings. Hence

$$113_{[8]} \neq 93_{[8]} \qquad \text{(strings)},$$

whereas

$$113_8 = \|113_{[8]}\| = \|93_{[8]}\| = 93_8 \qquad \text{(values)}.$$

In the obvious way we also introduce the concepts of *radix integer part* and *radix fraction part* as respectively the strings to the left and the right of the *radix point* in (1.4.2).

For any radix polynomial having only an integer part ($\ell \geq 0$), the value of the polynomial is related to certain sums of the digits, modulo $|\beta - 1|$, $|\beta|$, and $|\beta + 1|$.

Observation 1.4.1 *Let* $P = \sum_{i=0}^{m} d_i [\beta]^i$. *Then*

- $\|P\| \equiv d_0 \pmod{|\beta|}$,
- $\|P\| \equiv \sum_{i=0}^{m} d_i \pmod{|\beta - 1|}$,
- $\|P\| \equiv \sum_{i=0}^{m} (-1)^i d_i \pmod{|\beta + 1|}$,

so then

$$\begin{aligned} &113_8 \equiv 3 \pmod 8, && 93_8 \equiv 3 \pmod 8, \\ &113_8 \equiv (1+1+3) \equiv 5 \pmod 7, && 93_8 \equiv (9+3) \equiv 5 \pmod 7, \\ &113_8 \equiv (1-1+3) \equiv 3 \pmod 9, && 93_8 \equiv (-9+3) \equiv 3 \pmod 9. \end{aligned}$$

A binary-to-decimal integer conversion can then be checked by interpreting the binary number as octal digits, and comparing digit sums modulo 9.

Example 1.4.1 Consider $n = 101011111101_2 = 2813_{10}$. Now $n = 101011111101_2 = 5375_8$ so from the octal digit string $n \equiv -5+3-7+5 \equiv 5 \pmod 9$, and from the decimal string $n \equiv 2+8+1+3 \equiv 5 \pmod 9$, satisfying the check. □

Note that $n\beta^2 \equiv n \pmod{|\beta - 1|}$ and $n\beta^2 \equiv n \pmod{|\beta + 1|}$. This allows Observation 1.4.1 to be generalized to extended polynomials, and thereby to radix digit strings containing a radix point.

Observation 1.4.2 *For any irreducible $k\beta^\ell \in \mathbb{Q}_\beta$, where β does not divide k, and any $P = \sum_{i=\ell}^{m} d_i[\beta]^i$ with $\|P\| = k\beta^\ell$,*

$$\sum_{i=\ell}^{m} d_i \equiv k \pmod{|\beta - 1|},$$

$$\sum_{i=\ell}^{m} (-1)^i d_i \equiv k \pmod{|\beta + 1|}.$$

These results provide an efficient digit sum check on arithmetic operations, e.g.,

$$37.\bar{2}1_9 + 4.26_9 \neq 43.07_9,$$

since $[(3+7-2+1)+(4+2+6)] \equiv 5 \pmod 8$ and $[4+3+0+7] \equiv 6 \pmod 8$.

Problems and exercises

1.4.1 Determine which of the following radix polynomials have the same value:

(a) $\bar{1}9\bar{C}9\bar{2}_{[7]}$; (d) $1\bar{1}1113_{[2]}$; (g) $\bar{1}4_{[16]}$;
(b) $1600.160_{[-6]}$; (e) $\bar{1}A\bar{F}.\bar{9}8_{[8]}$; (h) $1210_{[-3]}$;
(c) $\bar{3}3_{[-10]}$; (f) $\bar{2}0.0_{[6]}$; (i) $1.2_{[-2]}$.

1.4.2 For (a)–(i) of the previous problem, determine the value modulo $|\beta - 1|$ and $|\beta + 1|$.

1.4.3 Which of the following expressions can be shown to be faulty by a digit sum check modulo $|\beta - 1|$ or $|\beta + 1|$:

(a) $142\bar{3}_5 + 2\bar{2}14_5 = 33\bar{2}1_5$;
(b) $23_{10} \times 4\bar{3}_{10} = 10\bar{4}1_{10}$;
(c) $2.436_{-7} + 6.543_{-7} = 6.362_{-7}$;
(d) $8156_{10} \times 3741_{10} = 26433595_{10}$;
(e) $312_{10} \times 8110_{10} = 2620320_{10}$;
(f) $452_{10} - 326_{10} = 125_{10}$;
(g) $259_{10} + 136_{10} = 395_{10}$;
(h) 1386_{10} div $125_{10} = 10_{10}$, remainder 11_{10}?

Which faults were not detected? Why?

1.4.4 Check the following octal–decimal and binary–decimal identities by appropriate digit sums modulo 9:

(a) $21\bar{3}4_8 = 1068_{10}$;
(b) $101001101010101_2 = 21461_{10}$.

1.5 Digit sets for radix representation

As discussed in the previous section, the redundancy classes $V_\beta(a)$ can be restricted if the digits used as coefficients of the radix polynomials in $\mathcal{P}[\beta]$ are drawn from some subset of the integers. So let D be such a *digit set*

$$D \subset \mathbb{Z} \quad \text{with } 0 \in D,$$

where $d \in D$ is called a digit. We will say that D is a *signed digit set* if $\exists d_1, d_2 \in D$ such that $d_1 < 0 < d_2$.

Given a radix β and digit set D we may then define

$$\mathcal{P}[\beta, D] = \{P \in \mathcal{P}[\beta] \mid \forall i \; d_i(P) \in D\},$$

where $d_i(P)$ is the coefficient of $[\beta]^i$ in P.

Applying Definition 1.3.6 of the redundancy class $V_\beta(a)$ and using the notation $|\cdot|$ for the cardinality of a set, we will in the following investigate two fundamental questions concerning radix β and digit set D pairs. The first is the *completeness question*:

$$\forall v \in \mathbb{Q}_{|\beta|} : |V_\beta(v) \cap \mathcal{P}[\beta, D]| \geq 1, \tag{1.5.1}$$

i.e., can every radix-β number be represented by some radix polynomial in $\mathcal{P}[\beta, D]$?

The other question is the *non-redundancy question*:

$$\forall v \in \mathbb{Q}_{|\beta|} : |V_\beta(v) \cap \mathcal{P}[\beta, D]| \leq 1, \tag{1.5.2}$$

i.e., do distinct radix polynomials of $\mathcal{P}[\beta, D]$ have distinct values?

Together the completeness and non-redundancy questions effectively ask whether or not a digit set D provides complete and unique representations of the radix-β numbers. Specifically regarding:

- Theoretical foundations: what are the necessary and sufficient conditions for a digit set D to provide complete and unique representations of the radix-β numbers?
- Computational foundations: can a particular digit set be efficiently verified to provide a complete and unique representation for the radix-β numbers?

In this section we obtain the necessary and sufficient conditions answering the theoretical foundations question and in the next section we resolve the computational foundations question. The existence of a particular complete and unique digit set D for a particular radix β is provided in the following example.

Example 1.5.1 (Radix 2) For the radix $\beta = 2$ and digit set $\{0, 1\}$ it is easy to see that for any radix polynomial $P[2] \in \mathcal{P}[2, \{0, 1\}]$

$$\|P\| = \sum_{i=\ell}^{m} d_i 2^i \geq 0 \quad \text{with } d_i \in \{0, 1\},$$

hence negative numbers in $\mathbb{Q}_2$ cannot be represented. Thus the condition (1.5.1) fails, and the digit set {0, 1} is not "complete" for radix 2. On the other hand, any non-negative number in $\mathbb{Q}_2$ can be represented uniquely by a member of $\mathcal{P}[2, \{0, 1\}]$, so the condition (1.5.2) is satisfied and the digit set {0, 1} is then "non-redundant" for radix $\beta = 2$.

Obviously, if negative numbers have to be representable, either the radix β has to be negative, or the digit set has to include at least one negative digit value. Choosing $\beta = -2$ and $D = \{0, 1\}$ it is easy to see that now both (1.5.1) and (1.5.2) can be answered affirmatively: i.e., $|V_{-2}(v) \cap \mathcal{P}[-2, \{0, 1\}]| = 1$ for all $v \in \mathbb{Q}_2$.

Choosing the digit set $D = \{-1, 0, 1\}$ for radix $\beta = 2$ the completeness question in (1.5.1) can be answered affirmatively: any binary number (in $\mathbb{Q}_2$) can be represented. However, any non-zero number now has an infinity of representations, e.g., in radix string notation:

$$1 = 1\bar{1}_2 = 1\bar{1}\bar{1}_2 = \cdots .$$

Despite the redundancy in representations from $\mathcal{P}[2, \{-1, 0, 1\}]$ this turns out to be a very useful system as we shall see in the following chapters.

A standard method of obtaining non-redundancy and completeness in radix 2 is to employ the digit set $\{-1, 0, 1\}$, with the further restriction that the digit value -1 may only occur in the most-significant position. Furthermore, if $d_m = -1$, then $d_{m-1} = 1$ to insure uniqueness. For fixed-length registers or storage words in computers this particular representation is often used under the name *2's complement*, however modified such that the digit value -1 is positioned in an imaginary position immediately to the left of the "word," exploiting the fact that the polynomial

$$\bar{1}1 \cdots 10b_k \cdots b_1 b_{0[2]}$$

represents the same value as

$$\bar{1}0b_k \cdots b_1 b_{0[2]}$$

so the $\bar{1}$ digit can be "pushed" into the desired position outside the "word." Since the digit value -1, if present, always occurs in a fixed position, uniqueness is maintained. We shall return to such radix-complement representations in Sections 1.7 and 1.8. □

A number of issues have been raised in the previous example, which we will now investigate in a more rigorous way. We will start with the completeness question, introducing the notation

$$\mathcal{P}_I[\beta, D] = \left\{ P \in \mathcal{P}[\beta, D] \;\middle|\; P([\beta]) = \sum_{i=0}^{m} d_i [\beta]^i \right\},$$

i.e., $\mathcal{P}_I$ is the subset of $\mathcal{P}$ containing radix polynomials with least-significant position $\ell \geq 0$, thus representing integer values.

Definition 1.5.1 *A digit set D is* complete *for radix β if and only if*

$$\forall i \in \mathbb{Z} : \mathcal{P}_I[\beta, D] \cap V_\beta(i) \neq \emptyset.$$

Lemma 1.5.2 *If D is a digit set which is complete for radix β, then*

$$\forall v \in \mathbb{Q}_{|\beta|} : \mathcal{P}[\beta, D] \cap V_\beta(v) \neq \emptyset. \tag{1.5.3}$$

Proof By definition any $v \in \mathbb{Q}_{|\beta|}$ can be written as $v = k\beta^\ell$, where $k, \ell \in \mathbb{Z}$. Since D is complete for β there exists a radix polynomial $P \in \mathcal{P}_I[\beta, D] \cap V_\beta(k)$, i.e., $\|P\| = k$. Let $Q = P \times [\beta]^\ell$, then the digits of Q are the same as the digits of P, thus $Q \in \mathcal{P}[\beta, D]$ and $\|Q\| = k\beta^\ell = v$. □

The converse of Lemma 1.5.2 does not hold, i.e., (1.5.3) cannot be used as the definition for completeness. Although with $\beta = 2$ and $D = \{-2, 0, 2\}$ (1.5.3) holds, no odd integer can be represented in $\mathcal{P}_I[2, \{-2, 0, 2\}]$. Definition 1.5.1 was deliberately chosen to exclude such digit sets where all members contain a common factor, which forces "unnatural" representations of integers.

The digit set we are most accustomed to (for $\beta > 1$) is

$$D_\beta = \{0, 1, 2, \ldots, \beta - 1\}, \tag{1.5.4}$$

which is the *standard radix-β digit set*. The most common alternative unique representation system employs the following digit set for a positive, odd β, $\beta = 2n + 1 \geq 3$:

$$B_\beta = \{-n, -n + 1, \ldots, -1, 0, 1, \ldots, n\}, \tag{1.5.5}$$

and is called the *symmetric radix-β digit set*. Such digit sets are characterized by the fact that they contain precisely β digit values, which form a *complete residue system* modulo β. Formally the set of integers ($\mathbb{Z}$) may be partitioned for $n \geq 2$ into n *residue classes*

$$C_i = \{j \mid j \in \mathbb{Z}, j \equiv i \pmod{n}\} = \{i + kn \mid k \in \mathbb{Z}\} \quad \text{for } 0 \leq i \leq n - 1,$$

and a set $S \subset \mathbb{Z}$ of n integers containing exactly one member of each C_i is termed a *complete residue system* modulo n. Thus $\{-1, 0, 1\}$, $\{2, 3, 4\}$, and $\{6, 13, 23\}$ are all complete residue systems modulo 3.

Theorem 1.5.3 *If D is a digit set which is complete for radix β for $|\beta| \geq 2$, then D contains a complete residue system modulo $|\beta|$, and hence contains at least $|\beta|$ digit values.*

Proof Let $1 \leq i \leq |\beta| - 1$, since D is complete there exists a $P \in \mathcal{P}_I[\beta, D] \cap V_\beta(i)$ with

$$P = d_m[\beta]^m + \cdots + d_1[\beta] + d_0.$$

Then $d_0 \neq 0$ since $\|P\| = i$ is not divisible by β, also $\|P\| \equiv d_0 \pmod{|\beta|}$ so $d_0 \equiv i \pmod{|\beta|}$ for the given i. Since $0 \in D$, D contains a complete residue system modulo $|\beta|$, and its cardinality is at least $|\beta|$. □

Again the converse does not hold, e.g., for radix $\beta = 3$ the digit set $D = \{-2, 0, 2\}$ is a complete residue system modulo 3, however, no odd integer can be represented in the system $\mathcal{P}_I[3, \{-2, 0, 2\}]$.

Let us now turn to the other fundamental question for digit sets, the non-redundancy question.

Definition 1.5.4 *A digit set D is* redundant *radix β if and only if*

$$\exists v \in \mathbb{Q}_{|\beta|} : |\mathcal{P}[\beta, D] \cap V_\beta(v)| \geq 2,$$

and is non-redundant *radix β if and only if*

$$\forall v \in \mathbb{Q}_{|\beta|} : |\mathcal{P}[\beta, D] \cap V_\beta(v)| \leq 1.$$

Informally a redundant digit set yields at least two distinct digit strings, $d_m d_{m-1} \cdots d_\ell$ and $d'_{m'} d'_{m'-1} \cdots d'_{\ell'}$, with $d_\ell, d'_{\ell'} \neq 0$ for some $v \in \mathbb{Q}_\beta$, whereas a non-redundant digit set provides at most one digit string for every $v \in \mathbb{Q}_\beta$.

We now immediately find a sufficient condition for redundancy from a "crowding" argument.

Theorem 1.5.5 *Let β be a radix, $|\beta| \geq 2$, and D be a digit set such that $|D| \geq |\beta| + 1$. Then D is redundant for radix β.*

Proof Let $\mathcal{L}_n = \{P \in \mathcal{P}_I[\beta, D] \mid \operatorname{degr}(P) \leq n\}$ for $n \geq 0$ and $\delta = \max\{|d| \mid d \in D\}$, then for $P \in \mathcal{L}_n$

$$\begin{aligned} |\|P\|| &\leq \delta \cdot (|\beta|^n + |\beta|^{n-1} + \cdots + 1) \\ &= \delta \cdot \frac{|\beta|^{n+1} - 1}{|\beta| - 1}, \end{aligned}$$

thus the polynomials of $\mathcal{L}_n$ can represent at most

$$2\delta \frac{|\beta|^{n+1} - 1}{|\beta| - 1}$$

different integer values. But $|\mathcal{L}_n| = |D|^{n+1} \geq \big||\beta| + 1\big|^{n+1}$ and

$$\lim_{n \to \infty} \frac{2\delta \frac{|\beta|^{n+1}-1}{|\beta|-1}}{\big||\beta| + 1\big|^{n+1}} = 0.$$

Hence for sufficiently large n there are more polynomials than values to represent, thus D is redundant radix β. □

Theorem 1.5.5 does not imply that a digit set with fewer than $|\beta|$ digits cannot be redundant, e.g., the digit set $\{0, 1, \beta + 1\}$ is redundant for radix $\beta \geq 2$ since

$1[\beta]^2 + 1[\beta]$ and $(\beta + 1)[\beta]$ are distinct polynomials of the same value. Obviously the problem in this example is that 1 and $\beta + 1$ belong to the same residue class modulo β, and the next theorem shows that if the digit set is a complete residue system, then D is non-redundant.

Theorem 1.5.6 *Let β be a radix with $|\beta| \geq 2$ and D be a complete residue system modulo $|\beta|$ with $0 \in D$. Then D is a non-redundant digit set for radix β.*

Proof Let $P, P' \in \mathcal{P}[\beta, D]$ with $P \neq P'$ and $\|P\| = \|P'\|$. Then with

$$P = \sum_{i=\ell}^{m} d_i[\beta]^i \quad \text{and} \quad P' = \sum_{i=\ell'}^{m'} d_i'[\beta]^i$$

choose k as the minimal index i for which $d_i \neq d_i'$, so

$$\sum_{i=k}^{m} d_i \beta^i = \sum_{i=k}^{m'} d_i' \beta^i$$

and then

$$\sum_{i=k}^{m} d_i \beta^{i-k} = \sum_{i=k}^{m'} d_i' \beta^{i-k} \tag{1.5.6}$$

is an integer. Taking residues modulo $|\beta|$ on both sides of (1.5.6)

$$d_k \equiv d_k' \pmod{|\beta|},$$

but since D is a complete residue system modulo $|\beta|$ this is a contradiction. □

Theorem 1.5.7 *Let D be a digit set which is complete for radix β for $|\beta| \geq 2$. Then any one of the following statements implies the other two.*

- *$|D| = |\beta|$;*
- *D is a complete residue system modulo $|\beta|$;*
- *D is a non-redundant digit set.*

The converse question arises as to what are the additional conditions on a digit set D, beyond simply being a complete residue system, that will allow us to say D is also complete for radix β. The following provides another essential condition for completeness of a digit set.

Lemma 1.5.8 *Let D be a non-redundant digit set which is complete for radix β. Then $j(\beta - 1) \notin D$ for any non-zero integer j.*

Proof Assume $j(\beta - 1) \in D$ for some $j \neq 0$. By assumption on D there exists a $P \in \mathcal{P}[\beta, D]$ such that

$$\|P\| = d_m\beta^m + \cdots + d_1\beta + d_0 = -j$$

with $d_m \neq 0$. Since $d_0 \equiv -j \pmod{|\beta|}$, then $d_0 = j(\beta - 1)$ as D is non-redundant, hence

$$Q = d_m[\beta]^{m-1} + \cdots + d_1$$

is a polynomial in $\mathcal{P}[\beta, D]$ of value

$$\|Q\| = \frac{\|P\| - d_0}{\beta} = \frac{-j - j(\beta - 1)}{\beta} = -j.$$

So $\|P\| = \|Q\| = -j$ but $P \neq Q$ contradicting the non-redundancy of the digit set for radix β. □

Lemma 1.5.8 leads to a family of related necessary conditions in view of the following elementary number theoretic observation.

Observation 1.5.9 *For $|\beta| \geq 2$ and D a complete residue system modulo $|\beta|$ with $0 \in D$, the set of values*

$$D^{(n)} = \left\{ \textstyle\sum_{i=0}^{n-1} d_i \beta^i \;\middle|\; d_i \in D \right\} \tag{1.5.7}$$

is a complete residue system modulo $|\beta|^n$.

Proof Note that $D^{(n)}$ is formed from the values of $|\beta|^n$ distinct $(n-1)$th-order polynomials with $0 \in D^{(n)}$.

$D^{(n)}$ will then be a complete residue system modulo $|\beta|^n$ unless there exist two distinct polynomials $P = \sum_{i=0}^{n-1} d_i[\beta]^i$ and $P' = \sum_{i=0}^{n-1} d_i'[\beta]^i$ with the same value modulo $|\beta|^n$, i.e., $\|P\| \equiv \|P'\| \pmod{|\beta|^n}$.

Assume such P and P' exist, then $\sum_{i=0}^{n-1}(d_i - d_i')\beta^i \equiv 0 \pmod{|\beta|^n}$. Let j be the smallest index such that $d_j \neq d_j'$, then $\sum_{i=j}^{n-1}(d_i - d_i')\beta^i \equiv 0 \pmod{|\beta|^n}$ implies $(d_j - d_j')\beta^j \equiv 0 \pmod{|\beta|^{j+1}}$, and hence $d_j \equiv d_j' \pmod{|\beta|}$, a contradiction. □

We then obtain the following corollary of the lemma.

Corollary 1.5.10 *Let D be a non-redundant digit set which is complete for radix β, $|\beta| \geq 2$, and let $D^{(n)}$ be given by (1.5.7) for every $n \geq 2$. Then $j(\beta^n - 1) \notin D^{(n)}$ for any non-zero j.*

Proof $D^{(n)}$ is complete for radix β^n and also a non-redundant digit set. Since D is complete for radix β, the result then follows from Lemma 1.5.8. □

Example 1.5.2 Note that the digit set $D = \{-1, 0, 19\}$ is a complete residue system modulo $\beta = 3$, with no digit a multiple of $\beta - 1 = 2$. However, $19 \times 3 - 1 = 56 \in D^{(2)}$ so $7 \times (3^2 - 1) = 56 \in D^{(2)}$. Thus $D^{(2)}$ is not complete for β^2, so D is not complete for radix 3. □

We note here that Corollary 1.5.10 provides another necessary condition for the converse question of when a digit set that is a complete residue system is also complete. But we also note the following "sufficiency" result.

Theorem 1.5.11 *Let D be a digit set for radix β with $|\beta| \geq 2$, where*

- *D is a signed digit set for $\beta \geq 2$,*
- *D is a complete residue system modulo $|\beta|$,*
- *$d \in D^{(n)}$, $d \neq 0$, implies $d \not\equiv 0 \pmod{|\beta^n - 1|}$ for all $n \geq 1$.*

Then D is a complete and non-redundant digit set for radix β.

The proof will be given in the next section, employing an algorithm introduced there for determining radix polynomial representations.

Observe that Theorem 1.5.11 is not intended as an algorithm for testing whether a digit set is complete. Rather Theorem 1.5.11 provides a characterization highlighting three straightforward necessary conditions for completeness and/or uniqueness (non-redundancy) that in total provide sufficiency for both conditions. The result provides a foundation for understanding complete and non-redundant digit sets.

Problems and exercises

1.5.1 Show that in Theorem 1.5.6 it is sufficient to require that D is a subset of a complete residue system modulo $|\beta|$ with $0 \in D$.

1.5.2 Determine which of the following digit sets are redundant and which are non-redundant for the specified radix:

(a) $\{-3, -1, 0, 1, 3, 5\}$ for radix 7;
(b) $\{0, 3, 6, 9, 12\}$ for radix -5;
(c) $\{-3, -2, -1, 0, 1\}$ for radix 4;
(d) $\{-4, -1, 0, 1, 3, 7\}$ for radix 8.

Also consider whether any of these digit sets might be complete for the radix.

1.5.3 Determine why the following digit sets cannot be complete for the associated radix:

(a) $\{-1, 0, 1, 2\}$ for radix 5;
(b) $\{-1, 0, 1, 2, 4\}$ for radix -5;
(c) $\{-8, 0, 1, 2, 3, 4, 5\}$ for radix -7;
(d) $\{-7, 0, 13\}$ for radix 3;
(e) $\{-8, -4, -2, -1, 0, 1, 2, 4, 8\}$ for radix 9.

1.6 Determining a radix representation

Having investigated the completeness and non-redundancy question we now turn to the question of efficiently determining a radix polynomial $P \in \mathcal{P}[\beta, D]$ such that $\|P\| = v \in \mathbb{Q}_{|\beta|}$ (if it exists) when given the radix β and digit set D. Theorem 1.5.6

insures that there is at most one such radix polynomial when D is a complete residue system modulo $|\beta|$, but on the other hand D need not be complete for radix β. We shall provide an algorithm which will determine such a radix polynomial if it exists, and, as we shall see, the algorithm may also be used to check whether a digit set D is complete for a radix β.

First we will prove a lemma which is fundamental to the algorithm.

Lemma 1.6.1 *For the radix β, $|\beta| \geq 2$, let D be a digit set which is a complete residue system modulo $|\beta|$. Let $k \in \mathbb{Z}$ and $d(k) \in D$ be the unique element of D such that $k \equiv d(k) \pmod{|\beta|}$. Then there exists $P \in \mathcal{P}_I[\beta, D]$ such that $\|P\| = k$ if and only if there exists $P' \in \mathcal{P}_I[\beta, D]$ with $\|P'\| = (k - d(k))/\beta$.*

Proof Let $P \in \mathcal{P}_I[\beta, D]$ with $\|P\| = k$. Then $P = \sum_{i=0}^{m} d_i[\beta]^i$ and $\|P\| = k \equiv d_0 \pmod{|\beta|}$ so $d(k) = d_0$. Hence

$$\frac{k - d_0}{\beta} = \sum_{i=1}^{m} d_i \beta^{i-1},$$

so

$$P' = \sum_{i=1}^{m} d_i[\beta]^{i-1} \in \mathcal{P}_I[\beta, D]$$

with $\|P'\| = (k - d(k))/\beta$.

Now assume there exists a $P' \in \mathcal{P}_{\mathcal{I}}[\beta, D]$ with $\|P'\| = (k - d(k))/\beta$ and define

$$Q = P' \times [\beta] + d(k).$$

Then $Q \in \mathcal{P}_I[\beta, D]$ and $\|Q\| = \|P'\| \cdot \beta + d(k) = (k - d(k))/\beta \cdot \beta + d(k) = k$ which completes the proof. □

We may now formulate a digit serial, right-to-left (least significant first) algorithm for determining the existence and coefficients of a radix polynomial.

Algorithm 1.6.2 (DGT Algorithm)

Stimulus: *A radix β, $|\beta| \geq 2$.*
A digit set D which is a complete residue system modulo $|\beta|$.
A radix-β number $v \in \mathbb{Q}_{|\beta|}$.

Response: *Either the radix polynomial $P \in \mathcal{P}[\beta, D]$, $P = \sum_{i=\ell}^{m} d_i[\beta]^i$, $\|P\| = v$, or a signal that no such polynomial exists.*

Method: $\ell := 0;\ r := v;$ *OK:=true;*
if $r = 0$ **then** $d_0 := 0;\ m := 0$
else
L1: **while** $r \notin \mathbb{Z}$ **do** $r := r * \beta;\ \ell := \ell - 1$ **end**;
L2: **while** $r \pmod{\beta} = 0$ **do** $r := r/\beta;\ \ell := \ell + 1$ **end**;
$m := \ell;\ r_m := r;$
L3: **while** $r \neq 0$ *and OK* **do**

L4: $\langle$ *find* $d \in D$ *such that* $d \equiv r \pmod{\beta}\rangle$
$d_m := d;\ m := m + 1;$
$r := (r - d)/\beta;$
$r_m := r;$
L5: $OK := \langle \forall j,\ \ell \le j < m :: r \ne r_j \rangle$
end;
end;

Where: {*The boolean variable OK, if true, signals that a radix polynomial* $P = \sum_{i=\ell}^{m} d_i[\beta]^i$ *with* $\|P\| = v$ *has been found, or if false, that no such polynomial exists.*}

The loops L1 and L2 serve to determine ℓ, the lower index of the radix polynomial, and insure that the value of r is integral when the while-loop is entered at L3. At L4 a digit is uniquely determined since D is a complete residue system modulo $|\beta|$. The array r is used to save the value of the "remainder" so that termination can be insured, in case a previous remainder reoccurs.

Before proving the correctness of Algorithm 1.6.2 let us consider an example. Note that the (real or integer) values in the example are denoted in decimal radix representation, but any representation could be used. The only requirements are that the input value b must be available in the chosen representation, and, of course, that the necessary arithmetic operations are available in the chosen host system.

Example 1.6.1 Let $v = 195.64_{10}$ be the value whose representation in $\mathcal{P}[5, \{-2, -1, 0, 1, 2\}]$ is wanted. Since $\beta = 5$ we observe that $v \in \mathbb{Q}_5$, and also that $D = \{-2, -1, 0, 1, 2\}$ is a complete residue system modulo 5. Loop L1 exits with

$$r = 4891_{10} \quad \text{and} \quad \ell = -2$$

and the steps of the loop L3 can then be displayed in a table:

m	$r_m = \dfrac{r_{m-1} - d_{m-1}}{5}$	d_m
−2	4891	1
−1	978	−2
0	196	1
1	39	−1
2	8	−2
3	2	2

so $P = 2 \times [5]^3 - 2 \times [5]^2 - 1 \times [5] + 1 - 2 \times [5]^{-1} + 1 \times [5]^{-2}$ is the desired polynomial in $\mathcal{P}[5, \{-2, -1, 0, 1, 2\}]$ such that $\|P\| = 195.64_{10}$. In string notation

$$195.64_{10} = 2\bar{2}\bar{1}1.\bar{2}1_5.$$

For another example, let $v = 17_{10}$, $\beta = -3$ and $D = \{0, 2, 4\}$ so the input conditions of Algorithm 1.6.2 are satisfied. When entering the loop L3, $r = v = 17_{10}$ and $\ell = 0$ so the following table is obtained:

m	$r_m = \dfrac{r_{m-1} - d_{m-1}}{-3}$	d_m
0	17	2
1	−5	4
2	3	0
3	−1	2
4	1	4
5	1	

so at step L5 it is detected that the value $r_m = 1$ has occurred previously and loop L3 exits with OK = false. Hence there is no polynomial in $\mathcal{P}[-3, \{0, 2, 4\}]$ of value 17_{10}, which is evident since $\|P\|$ is even for all $P \in \mathcal{P}_I[-3, \{0, 2, 4\}]$. □

Theorem 1.6.3 *For a given radix β, digit set D which is a complete residue system modulo $|\beta|$, and any radix-$|\beta|$ number $v \in \mathbb{Q}_{|\beta|}$, the DGT Algorithm will terminate after a finite number of steps. If, furthermore, there exists a $P \in \mathcal{P}[\beta, D]$ with $\|P\| = v$, then the algorithm will terminate with OK = true, and P is the polynomial determined by the coefficients d_i, $i = \ell, \ldots, m$.*

Proof Since $v = i\beta^j$ for some $i, j \in \mathbb{Z}$ the loops L1 and L2 will terminate. In loop L3 with $\delta = \max_{d \in D} |d|$

$$\left|r_{m+1}\right| \le \frac{|r_m| + \delta}{|\beta|} \begin{cases} < \left|r_m\right| & \text{for} \quad |r_m| > \delta, \\ \le \delta & \text{for} \quad |r_m| \le \delta, \end{cases}$$

thus $r_\ell, r_{\ell+1}, \ldots$ decrease in absolute magnitude until an index k is reached where $\left|r_k\right| \le \delta$. Then for all $m \ge k$, $|r_m| \le \delta$, so since δ is finite, either for some m, r_m will equal a value in D and the loop will terminate with $b = 0$, or r_m will equal some value r_i for $i < m$ and the loop terminates with OK = false.

For the second part, assume $P \neq 0$, $P = \sum_{k=s}^{t} b_k[\beta]^k$, $b_k \in D$, $b_s \neq 0$, $b_t \neq 0$. Loops L1 and L2 then determine $\ell = s$, and loop L3 is entered with $b = r_\ell = \sum_{k=\ell}^{t} b_k \beta^{k-s}$. Since D is a complete residue system modulo $|\beta|$, step L4 determines $d_m = b_m$ for $\ell \le m \le t$. Furthermore, the expression for OK at L5 always returns the value true since

$$\sum_{k=m}^{t} b_k[\beta]^{k-m} \neq \sum_{k=j}^{t} b_k[\beta]^{k-j} \quad \text{for} \quad \ell \le j < m \quad \text{implies} \quad r_m \neq r_j,$$

due to D being non-redundant for radix β (by Theorem 1.5.6). Since the algorithm terminates, it must stop with OK = true having determined a polynomial

$$P = \sum_{k=\ell}^{m} d_k[\beta]^k = \sum_{k=s}^{t} b_k[\beta]^k.$$

□

Note that the DGT Algorithm and Theorem 1.6.3 require D to be a complete residue system modulo $|\beta|$, so D is non-redundant for radix $|\beta|$.

We can now return to the proof of Theorem 1.5.11.

Proof of Theorem 1.5.11 Let k be any integer and apply the DGT Algorithm to k for digit set D and radix β, and let $d_i(k)$, for $i = 0, 1, \ldots$ be the sequence of digits obtained. For any $m \geq 1$ we then have

$$k = r_m \beta^m + \sum_{i=0}^{m-1} d_i(k)\beta^i,$$

where r_m is the remainder after m cycles. If $r_m = 0$ the algorithm terminates and a radix representation of k has been found. If $|k|$ is greater than the maximal absolute digit value, say $\delta = \max_{d \in D} |d|$, then by the construction of the remainders these will monotonically decrease in absolute magnitude, hence there can only be a finite number of them greater than δ (for a strict argument see the proof of the following theorem). Now assume that the algorithm cycles, then there is a first remainder r_m such that $r_m = r_j$ for some $j < m$ satisfying

$$k = r_m \beta^m + \sum_{i=0}^{m-1} d_i(k)\beta^i = r_j \beta^j + \sum_{i=0}^{j-1} d_i(k)\beta^i,$$

in which case since $r_m = r_j \neq 0$

$$-r_m(\beta^{m-j} - 1) = \sum_{i=j}^{m-1} d_i(k)\beta^{i-j},$$

hence $-r_m(\beta^n - 1) \in D^{(n)}$ for $n = m - j$, a contradiction. Thus the DGT Algorithm will terminate after a finite number of steps, and D is complete for radix β. □

As mentioned after the statement of the theorem, it does not provide a useful algorithm for determining whether a given digit set is complete for a particular radix, but the following theorem provides a very efficient algorithm for that purpose.

Theorem 1.6.4 *For the radix β with $|\beta| \geq 2$ let D be a digit set which contains a complete residue system modulo β. Then D is complete for radix β if and only if*

$$\forall i : |i| \leq \left\lfloor \frac{\delta}{|\beta| - 1} \right\rfloor :: \exists P \in \mathcal{P}_I[\beta, D] : \|P\| = i, \tag{1.6.1}$$

where $\delta = \max_{d \in D} |d|$.

Proof If D is complete for radix β, then by definition (1.6.1) holds, so assume that (1.6.1) holds and pick any integer v. In analogy with the DGT Algorithm we now choose a sequence of digit values $d_0, d_1, \ldots$ from the values of the remainders $r_0(= v), r_1, r_2, \ldots$ such that $d_j \equiv r_j \pmod{\beta}$ with $d_j \in D$, which is possible since D contains a complete residue system modulo $|\beta|$. Then for $j \geq 0$

$$r_{j+1} = \frac{r_j - d_j}{\beta} \quad \Longrightarrow \quad |r_{j+1}| \leq \frac{|r_j| + \delta}{|\beta|},$$

which implies

$$|r_{j+1}| \begin{cases} < \dfrac{|r_j| + (|\beta| - 1)|r_j|}{|\beta|} = |r_j| & \text{for} \quad |r_j| > \dfrac{\delta}{|\beta| - 1}, \\[2ex] \leq \dfrac{\frac{\delta}{|\beta|-1} + \delta}{|\beta|} = \dfrac{\delta}{|\beta| - 1} & \text{for} \quad |r_j| \leq \dfrac{\delta}{|\beta| - 1}, \end{cases}$$

hence the sequence $r_0, r_1, \ldots$ will monotonically decrease in absolute value, until for some value of k

$$|r_j| \leq \frac{\delta}{|\beta| - 1} \quad \text{for all} \quad j \geq k.$$

By the assumption (1.6.1) there exists a polynomial $P \in \mathcal{P}_I[\beta, D]$ such that $r_k = \|P\|$, and following the construction of the sequences $d_0, d_1, \ldots$ and $r_0, r_1, \ldots$ we have

$$\begin{aligned} n &= \beta \cdot r_1 + d_0 \\ &= \beta(\beta r_2 + d_1) + d_0 \\ &\vdots \\ &= r_k \beta^k + d_{k-1}\beta^{k-1} + \cdots + d_1\beta + d_0. \end{aligned}$$

But $\|P\| = r_k$ with $P \in \mathcal{P}_I[\beta, D]$ implies

$$Q = P \cdot [\beta]^k + \sum_{i=0}^{k-1} d_i [\beta]^i \in \mathcal{P}_I[\beta, D]$$

and $\|Q\| = n$, hence D is complete for radix β. □

Regarding computational complexity, from Theorems 1.5.11 and 1.6.4, and the structure of the DGT Algorithm, we obtain with $\delta = \max_{d \in D} |d|$ the following:

- Whether or not D is a complete and non-redundant digit set for radix β can be resolved in time $O(\delta)$, and if δ grows only polynomially in the radix β, then the result is obtained in a time that is a polynomial in $|\beta|$.
- For D a complete and non-redundant digit set for radix β and integer k, then the unique digit string with $k = \|d_m d_{m-1} \cdots d_0\|$ can be determined by the DGT Algorithm in time $O(\log k)$.

Theorem 1.6.4 together with the DGT Algorithm provides a simple test for the completeness of a digit set, since for "practical" digit sets $\delta/(|\beta| - 1)$ will be quite small. Actually for $\beta \geq 2$ only a smaller set needs to be tested since the bound in (1.6.1) in this case can be tightened to the interval:

$$\left\lceil \frac{-d_{\max}}{\beta - 1} \right\rceil \leq i \leq \left\lfloor \frac{-d_{\min}}{\beta - 1} \right\rfloor \quad \text{for } \beta \geq 2, \tag{1.6.2}$$

where $d_{max} = \max_{d \in D} d$ and $d_{min} = \min_{d \in D} d$. For $\beta \leq -2$ and a non-negative digit set it is similarly possible to show that it is sufficient to test the set

$$0 \leq i \leq \left\lfloor \frac{d_{max}}{-\beta - 1} \right\rfloor \quad \text{for } \beta \leq -2. \tag{1.6.3}$$

The proofs of these tighter bounds are left as exercises.

Example 1.6.2 For $\beta = -2$ the digit set $\{0, 2, 3\}$ contains a complete residue system modulo 2 (the set $\{0, 3\}$), so by Theorem 1.6.4 and (1.6.3) it is sufficient to test whether the integers $0, 1, 2, 3$ can be represented in $\mathcal{P}_I[-2, \{0, 2, 3\}]$. Applying the DGT algorithm for $a = 1$ it is easily seen that the remainder 1 will reoccur after determining the digit 3, hence $\{0, 2, 3\}$ is not complete for $\beta = -2$. □

Whenever D is complete for radix β, by definition any $v \in \mathbb{Q}_{|\beta|}$ has a radix polynomial representation, but Algorithm 1.6.2 can only be applied when D is non-redundant (D is a complete residue system modulo $|\beta|$). Theorem 1.6.4 provides a clue to a modified algorithm which can also be applied for redundant digit sets. The problem here is that there might be infinitely many radix polynomial representations of a given $v \in \mathbb{Q}_{|\beta|}$, so to insure termination a finite representation must be enforced. But for any i satisfying Theorem 1.6.4, or the tighter bounds (1.6.2) or (1.6.3), we can choose a finite, shortest, *canonical* representation $P_i \in \mathcal{P}[\beta, D]$, $\|P_i\| = i$, and these representations can then be tabulated. We may thus formulate a modified DGT Algorithm as follows:

Algorithm 1.6.5 (DGT Algorithm for complete digit sets)

Stimulus: *A radix* β, $|\beta| \geq 2$.
A digit set D which is complete for radix β.
A radix-β *number* $v \in \mathbb{Q}_{|\beta|}$.

Response: *A radix polynomial* $P \in \mathcal{P}[\beta, D]$, $P = \sum_{i=\ell}^{m} d_i [\beta]^i$, $\|P\| = v$.

Method: $\ell := 0$; $b = v$; $\delta := \max_{d \in D} |d|$;
if $b = 0$ **then** $d_0 := 0$; $m := 0$
else
L1: **while** $b \notin \mathbb{Z}$ **do** $b := b * \beta$; $\ell := \ell - 1$ **end**;
L2: **while** $b \bmod \beta = 0$ **do** $b := b/\beta$; $\ell := \ell + 1$ **end**;
$m := \ell$;

L3: **while** $b \neq 0$ **do**
L4: **if** $|b| \leq \lfloor \delta/(|\beta|-1) \rfloor$ **then**
$\langle$*choose d as the least significant digit of the canonical representation of b*$\rangle$
else
$\langle$*choose some* $d \in D$ *such that* $d \equiv b \pmod{\beta}$$\rangle$
$d_m := d;\ m := m + 1;$
$b := (b - d)/\beta;$
end;
end;

The algorithm thus needs a table containing for each i, $|i| \leq \lfloor \delta/(|\beta|-1) \rfloor$, the value of the least-significant digit of the canonical polynomial of value i. If there is more than one polynomial of the lowest possible degree representing i, any one can be chosen as the canonical representation, except when the degree is zero when the canonical representation has to be chosen as a digit in the digit set.

Theorem 1.6.6 *For a given radix β, digit set D which is complete for radix β, and any* $v \in \mathbb{Q}_{|\beta|}$, *Algorithm 1.6.5 will terminate having determined a radix polynomial* $P \in \mathcal{P}[\beta, D]$, $P = \sum_{i=\ell}^{m} d_i [\beta]^i$, *with* $\|P\| = v$.

Proof The only difference from Algorithm 1.6.2 is the question of termination. Following the proof of Theorem 1.6.4, the sequence of remainders (the values of b) will monotonically decrease strictly until for some value of k:

$$|r_j| \leq \frac{\delta}{|\beta| - 1} \quad \text{for all} \quad j \geq k.$$

From this point the digits will be chosen deterministically from the canonical polynomials P_j representing r_j, $j = k, k+1, \ldots$ Let the canonical polynomial for some specific $j \geq k$ be

$$P_j = Q_j * [\beta] + d_j \quad \text{with} \quad \|P_j\| = r_j,$$

so $r_{j+1} = (r_j - d_j)/\beta$ implies $\|Q_j\| = r_{j+1}$. But also $\|P_{j+1}\| = r_{j+1}$, thus $\text{degree}(P_{j+1}) \leq \text{degree}(Q_j) < \text{degree}(P_j)$ since P_{j+1} is canonical for r_{j+1}. Thus the degrees of the chosen sequence of canonical polynomials form a strictly decreasing sequence, and the algorithm will terminate. □

Example 1.6.3 Let $\beta = 3$ and $D = \{-1, 0, 5, 7\}$, which contains a complete residue system modulo 3. To check whether D is complete for radix 3, and to apply Algorithm 1.6.5 we have to find canonical representations of the values $-3, -2, -1, 0, 1, 2, 3$ (since $\lfloor \delta/(|\beta|-1) \rfloor = 3$). Two of these are trivially given by the corresponding digit values, and in attempting to find representations of 2 we find alternative representations

r_m	d_m
2	−1
1	7
−2	7
−3	0
−1	−1

r_m	d_m
2	5
−1	−1

where we note that we simultaneously also derive representations of the remainders, so that we can list the following canonical representations

$$\begin{aligned} -3:&\quad \bar{1}0,\\ -2:&\quad \bar{1}07,\\ -1:&\quad \bar{1},\\ 0:&\quad 0,\\ 1:&\quad \bar{1}077,\\ 2:&\quad \bar{1}5,\\ 3:&\quad \bar{1}0770, \end{aligned}$$

where the representation of 3 is derived from the expansion of 1. Note that any number only has finitely many representations in this system, so termination here would be assured even if the other possible representation of 2 is chosen.

But in the system $\beta = 2$, $D = \{-1, 0, 1\}$, any non-zero number has an infinite number of representations, e.g., $1 = 1\bar{1} = 1\bar{1}\bar{1}\cdots$, hence the algorithm would loop indefinitely if it consistently chose the digit -1 when the remainder is 1. □

For redundant systems it will often be convenient if the representation of zero is unique, i.e., $P = 0$ is the only polynomial satisfying $\|P\| = 0$. As an aid in determining uniqueness of the representation of zero the following lemma may be used.

Lemma 1.6.7 *Let D be a digit set which is complete for the radix β, then $P = 0$ is the only polynomial in $\mathcal{P}[\beta, D]$ with $\|P\| = 0$ if and only if D contains no non-zero multiple of β.*

Proof Assume $\|P\| = \|\sum_{i=\ell}^{m} d_i[\beta]^i\| = 0$ with $d_\ell \neq 0$. Without loss of generality we may assume that $\ell = 0$, so

$$\|d_k[\beta]^k + \cdots + d_1[\beta] + d_0\| = 0$$

implies $d_0 \equiv 0 \bmod \beta$. If $d = 0$ is the only digit in D divisible by β we have a contradiction, so $P = 0$.

Now assume there is a digit $d \in D$, $d = k \cdot \beta$ with $k \neq 0$, $k \in \mathbb{Z}$. Then there exists a polynomial $Q \in \mathcal{P}[\beta, D]$ with $\|Q\| = -k$ such that

$$d = k \cdot \beta = -\|Q\| \cdot \beta = -\left(\sum_{i=0}^{m} d_i \beta^i\right) \cdot \beta,$$

since D is complete for β. But then

$$P = Q \cdot [\beta] + d$$

is a non-zero polynomial of value zero. Thus if $P = 0$ is the only polynomial with $\|P\| = 0$, then no non-zero digit in D can be divisible by β. □

We will conclude this section by observing that although $\mathcal{P}[\beta]$ is closed under formal addition and multiplication of polynomials, $\mathcal{P}[\beta, D]$ is not. If P and Q are polynomials from $\mathcal{P}[\beta, D]$ their sum $S([\beta]) = P([\beta]) + Q([\beta])$ and product $R([\beta]) = P([\beta]) \times Q([\beta])$ (as defined for extended polynomials by (1.2.2) and (1.2.3)) do not necessarily have coefficients in D. But $\|S([\beta])\| \in \mathbb{Q}_{|\beta|}$ and $\|R([\beta])\| \in \mathbb{Q}_{|\beta|}$, hence the DGT Algorithms (Algorithms 1.6.2 or 1.6.5) may be applied if D is complete for β to map the result back into $\mathcal{P}[\beta, D]$. It is the purpose in later chapters to develop more efficient algorithms where the mapping into the appropriate digit set is implicit.

Problems and exercises

1.6.1 Apply the DGT Algorithm to the following radix, digit set and radix 10 number tuples:

	Base	*Digit set*	*Number*
(a)	5	$\{-2, -1, 0, 1, 2\}$	6725
(b)	-8	$\{0, 1, 2, 3, 4, 5, 6, 7\}$	1.109375
(c)	7	$\{-5, -3, -1, 0, 1, 3, 5\}$	3721
(d)	4	$\{-2, 0, 1, 3\}$	-335
(e)	3	$\{-1, 0, 4\}$	-17

1.6.2 Prove that for $\beta \geq 2$ in Theorem 1.6.4 it is sufficient to test for values of i in the interval specified in (1.6.2).

1.6.3 Similarly prove that for $\beta \leq -2$ testing members of the interval in (1.6.3) is sufficient.

1.6.4 Prove that $D_{M,h} = \{km \mid m \in M, 0 \leq k \leq h\}$ for $M = \{\pm 1, \pm 2\}$, $h = \lfloor \beta/3 \rfloor$ is a complete digit set for any odd $\beta \geq 3$.

1.6.5 By utilizing either the theoretical conditions developed, or the digit set test (Theorem 1.6.4), determine which of the following digit sets are complete for the radix indicated:

	Base	*Digit set*
(a)	10	$\{-7, -3, -1, 0, 1, 2, 4, 5, 6, 8\}$
(b)	5	$\{-1, 0, 1, 2\}$
(c)	4	$\{-2, -1, 0, 5, 6\}$
(d)	3	$\{-19, -13, 0, 1\}$
(e)	-5	$\{0, 1, 2, 4, 8\}$
(f)	7	$\{-4, -2, -1, 0, 1, 2, 4\}$
(g)	7	$\{-9, -3, -1, 0, 1, 3, 9\}$
(h)	13	$\{-32, -16, -8, -4, -2, -1, 0, 1, 2, 4, 8, 16, 32\}$
(i)	7	$\{-4, -2, -1, 0, 1, 2, 4\}$
(j)	11	$\{-81, -27, -9, -3, -1, 0, 1, 3, 9, 27, 81\}$

1.6.6 Show that in Exercise 1.6.5(d) no proper subset of D is complete for the radix $\beta = 3$.

1.6.7 For which radices are the following digit sets complete? And if so, redundant?

(a) $\{0, 1, 2\}$
(b) $\{-1, 0, 1\}$
(c) $\{-2, -1, 0, 12, 34\}$
(d) $\{-19, -13, 0, 1\}$.

1.6.8 For the radix and digit set pairs in Exercise 1.6.5(c) and (d) list a minimal set of canonical polynomials for use in Algorithm 1.6.5, and apply the algorithm to find the representation of 137_{10} in these systems.

1.6.9 For which radices β with $3 \leq \beta = 2j + 3 \leq 31$ are the corresponding β-membered binary-power sets

$$\{-2^j, -2^{j-1}, \ldots, -2, -1, 0, 1, 2, \cdots, 2^j\}$$

both complete and non-redundant for radix β. (Hint: β must be a prime.)

1.6.10 For the DGT Algorithm with a digit set which is complete for radix β, give a bound on the number of digits determined, $m - \ell$, in terms of v, β and δ.

1.6.11 For the radix $\beta \geq 2$, let D be a signed digit set which is a complete residue system modulo β. The DGT Algorithm applied to any integer remainder r_m yields r_{m+1}, which can be represented as an edge from vertex r_m to r_{m+1} in a directed graph with vertex set the integers satisfying $|i| \leq \lfloor \delta/(|\beta| - 1) \rfloor$. Discuss how the cyclic and/or acyclic nature of this graph, and any cycle size, relates to verification of D as complete, or to $D^{(n)}$ containing a multiple of $\beta^n - 1$. From the graph can you determine asymptotically in n what proportion of the integers $|i| \leq n$ have non-redundant representations $i = \|d_m d_{m-1} \cdots d_0\|_\beta$?

1.7 Classifying base-digit set combinations

Our definition of digit sets allows very general radix polynomials as representations of numbers. For most practical purposes such generality is not needed, and certain restrictions could be applied.

In practice digit sets are most often composed as *contiguous* sets of integers, i.e., sets of the form

$$D = \{r, r+1, \ldots, s-1, s\}, \tag{1.7.1}$$

where $r \leq 0$ and $s \geq 0$, and at least one of r and s is non-zero. The cardinality of such a set is

$$|D| = s - r + 1 \geq 2,$$

so for $|\beta| = s - r + 1$, D is a complete residue set modulo $|\beta|$ and by Theorem 1.5.6 D is then a non-redundant digit set for radix β. If $|\beta| < s - r + 1$, D is then a redundant digit set for radix β, but note that D need not be complete in either of these cases, e.g., neither $\{0, 1\}$ nor $\{0, 1, 2\}$ is a complete digit set for radix $\beta = 2$.

To identify interesting classes of digit sets we introduce the following *notation* for digit sets of the form of the equation (1.7.1) satisfying further constraints:

Digit set class	r, s restrictions	
Basic	$-\lvert\beta\rvert \leq r \leq 0 \leq s \leq \lvert\beta\rvert$,	$s - r + 1 \geq \lvert\beta\rvert$
Standard	$r = 0, s = \lvert\beta\rvert - 1$,	$s - r + 1 = \lvert\beta\rvert$
Extended	$r = 0, s = \beta > 0$,	$s - r + 1 = \beta + 1$
Symmetric	$r = -s$,	$2s + 1 \geq \lvert\beta\rvert \geq s$
Minimally redundant	$-\lvert\beta\rvert \leq r \leq 0 \leq s \leq \lvert\beta\rvert$,	$s - r + 1 = \lvert\beta\rvert + 1$
Maximally redundant	$r = -\lvert\beta\rvert + 1, s = \lvert\beta\rvert - 1$,	$s - r + 1 = 2\lvert\beta\rvert - 1$

Note that standard, extended, symmetric, minimally or maximally redundant digit sets are all basic digit sets. Also a maximally redundant digit set is necessarily symmetric. Utilizing Theorems 1.5.5, 1.5.6, and 1.6.4, the redundancy and completeness questions for these digit sets may be summarized in the following:

Observation 1.7.1 *All basic digit sets contain complete residue systems modulo $|\beta|$. The conditions under which the various basic digit sets are respectively redundant, non-redundant and complete are:*

Digit set	Redundant	Non-redundant	Complete
Basic	$s - r + 1 > \lvert\beta\rvert$	$s - r + 1 = \lvert\beta\rvert$	$(rs < 0 \wedge \beta > 0) \vee (\beta < 0)$
Standard	False	True	$\beta < 0$
Extended	True	False	False
Symmetric	$2s + 1 > \lvert\beta\rvert$	$2s + 1 = \lvert\beta\rvert$	True
Min. redundant	True	False	$(rs < 0 \wedge \beta > 0) \vee (\beta < 0)$
Max. redundant	True	False	True

Example 1.7.1 For the interesting case $\beta = 2$ we can summarize some particular digit sets:

Name	Digit set	Properties
Standard binary	$\{0, 1\}$	Non-redundant, not complete
"Carry-save"	$\{0, 1, 2\}$	Redundant, not complete
"Borrow-save"	$\{-1, 0, 1\}$	Redundant, complete

all of which have found extensive use. The names "carry-save" and "borrow-save" are actually not the names of just the digit sets themselves, but rather names also denoting specific binary encodings of the digit values. These very important digit sets will be discussed in more detail in Section 2.5 and again in Chapter 3. Note that the "borrow-save" digit set is a digit set which is symmetric as well as minimally and maximally redundant. It is also sometimes called *signed digit binary* when the encoding is not of concern. □

Example 1.7.2 To illustrate the variety of possible basic digit sets that exist for higher values of β, let us list some examples for $\beta = 5$:

Standard	$\{0, 1, 2, 3, 4\}$	Non-redundant, not complete
Symmetric	$\{-2, -1, 0, 1, 2\}$	Non-redundant, complete
Min. redundant	$\{-1, 0, 1, 2, 3, 4\}$	Redundant, complete
Symmetric	$\{-3, -2, -1, 0, 1, 2, 3\}$	Redundant, complete

Note that for an odd radix the smallest symmetric digit set is non-redundant, whereas it is redundant for an even radix. □

However, some non-contiguous digit sets are useful, for these we note the following.

Observation 1.7.2 *Let D' be a subset of the maximally redundant digit set D for $\beta \geq 3$. If $\{-1, 0, 1\} \subseteq D'$, and D' contains a complete residue system modulo β, then D' is a complete digit set for radix β.*

It has been mentioned a few times that for $\beta \geq 2$ the standard digit set $\{0, 1, \ldots, \beta - 1\}$ is not complete, since no negative number can be represented in this system. One way of dealing with this problem is to "factor out" the sign into what is called a *sign-magnitude representation*, as is customary in the every-day (decimal) string notation, e.g., as in -172 for negative values.

Definition 1.7.3 *For $\beta \geq 2$ we define the* sign-magnitude radix polynomials *as*

$$\mathcal{P}^{\pm}[\beta, D] = \{s \cdot P \mid s \in \{-1, 1\},\ P \in \mathcal{P}[\beta, D]\}$$

and define D to be semicomplete *for radix β if and only if*

$$\forall d \in D : d \geq 0 \quad \textit{and} \quad \forall i \in \mathbb{Z} : \mathcal{P}_I^{\pm}[\beta, D] \cap V_\beta(i) \neq \emptyset,$$

where $\mathcal{P}_I^{\pm}$ is defined in analogy with $\mathcal{P}_I$.

We then obtain the following.

Observation 1.7.4 *If D is a digit set which is semicomplete for radix $\beta \geq 2$, then*

$$\forall a \in \mathbb{Q}_\beta : \mathcal{P}^{\pm}[\beta, D] \cap V_\beta(a) \neq \emptyset.$$

Observation 1.7.5 *If D is a digit set which is semicomplete for radix $\beta \geq 2$, then D contains a complete residue system modulo β.*

Since by Observation 1.7.4 for all $a \in \mathbb{Q}_\beta$ either a or $-a$ has a radix polynomial representation when D is semicomplete, by Theorem 1.6.3 and Observation 1.7.5 the DGT Algorithm may be applied to a and $-a$.

Observation 1.7.6 *If D is a digit set which is semicomplete for radix $\beta \geq 2$, and a is any radix-β number, $a \in \mathbb{Q}_\beta$, then there exists a digit set $D' \subseteq D$ such that either $DGT(\beta, D', a)$ or $DGT(\beta, D', -a)$ will determine a radix polynomial $P \in \mathcal{P}^{\pm}[\beta, D']$ with $\|P\| = a$.*

As previously mentioned, *radix-complement* representations (e.g., 2's complement) provide an alternative way of representing negative numbers for standard and extended digit sets when $\beta > 0$. The idea employed is to augment the digit set with negative digit values, but to restrict their use to the leading (most-significant) position. In finite precision representations (limited to a fixed number of digit positions) the possible leading negative digit may then be pushed to a position outside the "word."

Define the *standard complement digit set* C_β for $\beta \geq 2$ to be

$$C_\beta = \left\{ -\left\lfloor \frac{\beta}{2} \right\rfloor, -\left\lfloor \frac{\beta}{2} \right\rfloor + 1, \ldots, 0, \ldots, \beta - 1 \right\}, \tag{1.7.2}$$

e.g., $C_2 = \{-1, 0, 1\}$ and $C_{10} = \{-5, -4, \ldots, 0, \ldots, 9\}$. Also define the *complement polynomials* as

$$\mathcal{P}^c[\beta, C_\beta] = \left\{ P \;\middle|\; P = \sum_{i=\ell}^{m} d_i [\beta]^i, d_i \in C_\beta, d_i \geq 0 \text{ for } \ell \leq i < m \right\},$$

where it is easily seen that this class is redundant, e.g.,

$$-4 = \bar{4}_{10} = \bar{1}6_{10} = \bar{1}96_{10} = \bar{1}996_{10} = \cdots,$$
$$-6 = \bar{1}4_{10} = \bar{1}94_{10} = \bar{1}994_{10} = \cdots,$$

and also note that the sign of $\|P\|$ can be determined from the leading (non-zero) digit of the representation.

Uniqueness can be obtained by restricting the values of the leading digit, e.g., let

$$\mathcal{P}^{cc}[\beta, C_\beta] = \left\{ P \in \mathcal{P}^{c}[\beta, C_\beta] \,\middle|\, (d_m \neq -1) \vee \left((d_m = -1) \wedge 0 < d_{m-1} \leq \left\lfloor \frac{\beta - 1}{2} \right\rfloor \right) \right\} \qquad (1.7.3)$$

be the set of *canonical complement polynomials*. That uniqueness is obtained can be seen from Algorithm 1.6.5 when this is modified such that the digit selection at L4 reads:

L4: **if** $b \in C_\beta$ **then** $d := b$
else
$\langle$choose $d \in C_\beta$, $d \geq 0$ such that $d \equiv b \pmod{\beta}\rangle$

Example 1.7.3

$$\bar{3}241_{[10]} \in \mathcal{P}^{cc}[10, C_{10}] \cap V_{10}(-2759),$$
$$\bar{1}7241_{[10]} \in \mathcal{P}^{c}[10, C_{10}] \cap V_{10}(-2759),$$

so $\bar{3}241_{[10]} = \bar{1}7241_{10} = \bar{1}97241_{10} = \bar{1}997241_{10}$. But note that $\bar{6}131_{[10]}$ is not in $\mathcal{P}^{cc}[10, C_{10}]$, nor is it in $\mathcal{P}^{c}[10, C_{10}]$. □

Observation 1.7.7 *For any $\beta \geq 2$ and any $P \in \mathcal{P}^{cc}[\beta, C_\beta]$ such that $\|P\| < 0$, there is for each $k >$ msp(P) a radix polynomial $P^* \in \mathcal{P}^{c}[\beta, C_\beta]$ with $\|P^*\| = \|P\|$, where* msp$(P^*) = k$ *and $d_k(P^*) = -1$.*

For an odd radix β, the minimally redundant digit set $\{-\lfloor \beta/2 \rfloor, \ldots, 0, \ldots, \lfloor \beta/2 \rfloor\}$ yields non-redundant representations. While for even values of $\beta \geq 4$ the representations are redundant, it is possible to obtain non-redundancy by restricting the use of digits.

Let the set of *canonical symmetric radix polynomials* be defined for even $\beta \geq 4$ and digit set $C_\beta^{can} = \{-\beta/2, \ldots, \beta/2\}$ be defined by

$$\mathcal{P}_\beta^{can} = \left\{ P \,\middle|\, \textstyle\sum_{i=\ell}^{m} d_i [\beta]^i,\ d_i \in C_\beta^{can}, \quad d_i = \pm\frac{\beta}{2} \Rightarrow \operatorname{sgn}(d_i) = -\operatorname{sgn}(\|d_{i-1} \cdots d_\ell\|),\ \text{or } d_\ell = -\frac{\beta}{2} \right\},$$

or, in words, a digit $\pm\beta/2$ must be followed by a non-zero tail of opposite sign, or choose $-\beta/2$ if all zeroes follow. A polynomial of this type appears as the result of *Modified Booth Recoding* of a 2's complement number, used in multipliers, as discussed in Sections 2.4 and 4.3. These polynomials have the interesting property that truncation of the corresponding radix digit strings yields proper rounding of the values represented, as discussed in Section 1.9.

Observation 1.7.8 *Prefixes of a radix digit string with non-zero tail for $P \in \mathcal{P}_\beta^{can}$ are half-ulp approximations.*

Thus this recoding (digit set conversion from 2's complement into $P \in \mathcal{P}_\beta^{can}$) is "shifting" the tails from a one-sided, ulp-size error to a symmetric error of the same size.

A related result for radix digit strings over the maximally redundant digit set $\{-\beta + 1, \ldots, 0, \ldots, \beta - 1\}$ is that a prefix with non-zero tail t is a directed one-ulp prefix, with direction given by sgn(t).

For $\beta = 2$ and digit set $\{-1, 0, 1\}$, non-redundancy can be obtained by requiring that no two consecutive digits are non-zero. This form is called *canonical signed-digit form* or *non-adjacent form* (NAF). This will be discussed in more detail in Section 4.3, as it played an important role in some early multipliers and more recently in modular exponentiation used in encryption, as discussed in Chapter 8.

Problems and exercises

1.7.1 Prove the properties of the various basic digit sets stated in Observation 1.7.1.

1.7.2 Check whether the digit set $D = \{0, 4, 5\}$ is semicomplete and non-redundant for radix $\beta = 3$.

1.7.3 Prove or disprove the following: if the premise $\{-1, 0, 1\} \subseteq D'$ in Observation 1.7.2 is replaced by $\{-1, 1\} \cap D' \neq \emptyset$, then we may conclude that D' is a complete digit set for radix $\beta < 0$.

1.7.4 Apply the DGT algorithm, as modified for canonical complement polynomials, to find $P \in \mathcal{P}^{cc}[10, C_{10}]$ such that $\|P\| = -1234_{10}$.

1.7.5 Prove that the polynomial P^*, whose existence is asserted in Observation 1.7.7 (given k), is uniquely determined.

1.7.6 Show the correctness of Observation 1.7.8.

1.8 Finite-precision and complement representations

Due to the physical limitations and derived architectural constraints, number representations in computers are usually restricted such that only a fixed number of digits is available in the standard number representations. The "word size" of the architecture limits the number of digits which can be transferred as one unit between the memory and arithmetic/logic unit of the CPU, as well as what can be accessed as operands and produced as results of the arithmetic operations.

The interpretation of the contents of such a "word," say as an encoding of a string of digits, is implicitly given by the instruction manipulating the word as an operand. The positioning of the radix point is of no significance to an ADD operation, as long as both operands and the result have the same assumed radix point position. For multiplication and division, however, the assumed radix point position is significant.

Systems with such a limited number of digits and assumed radix point position are called *fixed-point systems*, and may be formally defined as

$$\mathcal{F}_{\ell m}[\beta, D] = \left\{ P \in \mathcal{P}[\beta, D] \,\middle|\, P = \sum_{i=\ell}^{m} d_i [\beta]^i, \forall i : d_i \in D \right\} \tag{1.8.1}$$

for given ℓ and m where $\ell \le m$.

The set of numbers representable in this system is a proper (finite) subset of $\mathbb{Q}_{|\beta|}$, a set of numbers of the form $i\beta^\ell$, where i is in some finite set of integers, which, in general, need not be a contiguous set (i.e., an interval) even if D is complete for β.

Example 1.8.1

(a) $\mathcal{F}_{0,9}[10, \{0, 1, \ldots, 9\}] \sim$ 10 digit unsigned decimal integers.
(b) $\mathcal{F}_{-32,-1}[2, \{0, 1\}] \sim$ 32-bit unsigned, proper binary fractions.
(c) $\mathcal{F}_{0,1}[-3, \{0, 1, 5\}] \sim \{-15, -14, -10, -3, -2, 0, 1, 2, 5\}$.
(d) $\mathcal{F}_{0,1}[2, \{-1, 0, 2, 3\}] \sim \{-3, -2, \ldots, 8, 9\}$.

Example (c) shows that the set of representable numbers does not, in general, form a contiguous set when the digit set is not contiguous, even though the digit set $\{0, 1, 5\}$ is complete for radix -3. On the other hand, example (d) demonstrates that the set of representable numbers may be contiguous even if the digit set is not. □

Our particular interest in basic digit sets arises from the fact that in the interval $[-\beta, \beta]$ a contiguous digit set is necessary and sufficient to obtain a contiguous set of representable fixed-point numbers, in the sense that the set

$$\{\|P\| \cdot \beta^\ell \mid P \in \mathcal{F}_{\ell,m}[\beta, D]\}$$

is a contiguous set of integers. This is formalized in the following theorem.

Theorem 1.8.1 *Let D be a digit set which contains a complete residue system modulo $|\beta|$, and such that for all $d \in D : -|\beta| \le d \le |\beta|$. Then $S_n = \{\|P\| \mid P \in \mathcal{F}_{0n}[\beta, D]\}$, $n \ge 0$, is a contiguous set of integers iff D is a contiguous set of integers.*

Proof If $D = \{r, r+1, \ldots, s-1, s\}$ with $s - r + 1 \ge \beta$, then it is easy to prove by induction in n that S_n is a contiguous set of integers. So assume that S_n is a contiguous set of integers, which for $\beta > 0$ must be of the form:

$$S_n = \left\{ i \,\middle|\, d_{min} \cdot \frac{\beta^{n+1} - 1}{\beta - 1} \le i \le d_{max} \cdot \frac{\beta^{n+1} - 1}{\beta - 1} \right\} \tag{1.8.2}$$

with $d_{min} = \min_{d \in D} d$ and $d_{max} = \max_{d \in D} d$. For $\beta < 0$ the set (1.8.2) just takes a different form when n is odd. Now for a contradiction assume there exists a $d \neq 0$: $d_{min} < d < d_{max}$ such that $d \notin D$. Since D contains a complete residue system there is a unique d', $d' \in D$, such that either $d' = d + \beta$ or $d' = d - \beta$. Now, consider the number

$$k = d\beta^n + \sum_{i=0}^{n-1} d'\beta^i,$$

where it is easily seen from (1.8.2) that $k \in S_n$. Since $d' \equiv k \pmod{\beta}$, any $P \in \mathcal{F}_{0n}[\beta, D]$ with $\| P \| = k$ must be of the form:

$$P = d'' \cdot [\beta]^n + \sum_{i=0}^{n-1} d'[\beta]^i$$

with $d'' \in D$, which is a contradiction since d'' must equal d. □

As a small aside, let us consider the set of representable values obtained with a digit set which need not be contiguous, but for the special case where the digit set is symmetric and non-redundant.

Theorem 1.8.2 *For an odd radix $\beta \geq 3$, let D be a symmetric digit set which is a complete residue system modulo β. With $\delta = \max_{d \in D}\{|d|\}$ and $[-M_n; M_n]$ being the maximal contiguous subsequence of $S_n = \{\| P \| \mid P \in \mathcal{F}_{0n-1}[\beta, D]\}$, $n \geq 1\}$ (the n-digit values representable), such that for some $k \geq 0$, $\delta \leq (M_k + 1)(\beta - 1) - 1$, then*

$$M_{n+1} = (M_n + 1)\beta - (\delta + 1)$$

for $n \geq k$, and $M_{n+1} > M_n$.

Proof For given $k \geq 0$, assume that $\delta \leq (M_n + 1)(\beta - 1) - 1$. Let $j = (M_n + 1)\beta - \delta$, then a radix polynomial of value j needs one more digit than a polynomial representing $M_n + 1$. Since $M_n + 1$ is not an n-digit value, j is not an $(n + 1)$-digit value, so $M_{k+1} \leq j - 1 = (M_n + 1)\beta - (\delta + 1)$. As $\delta \leq (M_n + 1)(\beta - 1) - 1$, we also have $(M_n + 1)\beta - (\delta + 1) \geq (M_n + 1)$.

Let m satisfy $|m| \leq (M_n + 1)\beta - (\delta + 1)$. For $m \equiv d \pmod{\beta}$ with $d \in D$, let $m = i\beta + d$, then $|i\beta + d| \leq (M_n + 1)\beta - (\delta + 1)$ implies $-(M_n + 1)(\beta - 1) + 1 \leq i\beta \leq (M_n + 1)(\beta - 1) - 1$, so then i is an n-digit value and $m = i\beta + d$ is an $(n + 1)$-digit value, hence $M_{n+1} = (M_n + 1)\beta - (\delta + 1)$. □

Example 1.8.2 With $\beta = 7$, the digit set $D = \{-4, -2, -1, 0, 1, 2, 4\}$ is easily seen to be a complete residue system. With $M_1 = 2$ and $\delta = 4$, then $4 \leq (2 + 1)(7 - 1) - 1 = 17$, hence $k = 1$ and by Theorem 1.8.2, $M_2 = (2 + 1)7 - (4 + 1) = 16$. □

In practice one is often interested in choosing the radix and digit set such that the set of representable numbers is symmetric around zero. One way to do this is to use a symmetric digit set, which will then be redundant if the radix is even, independent of the sign of the radix which may be chosen as positive. Only for an odd radix is it possible to choose a symmetric, non-redundant digit set, and thus obtain a symmetric set of representable values. With the standard digit set $\{0, 1, \ldots, |\beta| - 1\}$ it is necessary to have $\beta < 0$ to represent negative numbers, but the set of representable numbers will be unsymmetric: there will be more positive than negative numbers when n is even, and vice-versa if n is odd.

Alternatively, to represent a symmetric set of numbers a sign-magnitude representation may be used, e.g., as is often used in connection with the standard digit set $\{0, 1, \ldots, \beta - 1\}$ for $\beta \geq 2$. We may thus define

$$\mathcal{F}^{\pm}_{\ell m}[\beta, D] = \{s \cdot P \mid s \in \{-1, 1\}, P \in \mathcal{F}_{\ell m}[\beta, D]\} \tag{1.8.3}$$

as the set of *sign-magnitude, fixed-point systems*.

1.8.1 Finite-precision radix-complement representations

In many hardware realizations a "complement" notation is used to represent negative numbers, so by adapting the definition of *radix-complement representations* from Section 1.7 we define

$$\begin{aligned}\mathcal{F}^{rc}_{\ell m}[\beta, C^c_\beta] = \Bigg\{ P \in \mathcal{P}[\beta, C^c_\beta] \;\Bigg|\; & P = \sum_{i=\ell}^{m+1} d_i[\beta]^i,\, d_i \in C^c_\beta \backslash \{-1\} \text{ for } \ell \leq i \leq m \\ & \text{and} \left(d_{m+1} = 0 \wedge \left(0 \leq d_m < \left\lfloor \frac{\beta+1}{2} \right\rfloor \right)\right) \\ & \vee \left(d_{m+1} = -1 \wedge d_m \geq \left\lfloor \frac{\beta+1}{2} \right\rfloor \right) \Bigg\}\end{aligned} \tag{1.8.4}$$

where $C^c_\beta = \{-1, 0, 1, \ldots, \beta - 1\}$ is the standard digit set extended with the digit value -1, which is only allowed in the most-significant position $(m + 1)$. Note that the digit 0 or -1 in this position need not be part of a string representation of a number, the digit in position m uniquely determines the value of the digit in position $m + 1$. Also note that the representation is non-redundant, although C^c_β is redundant for radix β.

Observation 1.8.3 (Radix-complement representation) *For $P \in \mathcal{F}^{rc}_{\ell m}[\beta, C^c_\beta]$ the value of the polynomial may be determined as*

$$\|P\| = \begin{cases} \sum_{i=\ell}^{m} d_i \beta^i & \text{if } d_m < \left\lfloor \frac{\beta+1}{2} \right\rfloor, \\ \sum_{i=\ell}^{m} d_i \beta^i - \beta^{m+1} & \text{if } d_m \geq \left\lfloor \frac{\beta+1}{2} \right\rfloor, \end{cases} \tag{1.8.5}$$

i.e., negative numbers are represented with a bias *of* β^{m+1}, *and*

$$-\left(\beta - \left\lfloor \frac{\beta+1}{2} \right\rfloor\right)\beta^m \le \|P\| \le \left(\left\lfloor \frac{\beta+1}{2} \right\rfloor \beta^m - \beta^\ell\right)$$

with both bounds attainable.

Example 1.8.3

$$\beta = 2, \left\lfloor \frac{\beta+1}{2} \right\rfloor = 1 : \quad -2^m \le \|P\| \le 2^m - 2^\ell.$$

$$\beta = 10, \left\lfloor \frac{\beta+1}{2} \right\rfloor = 5 : -5 \cdot 10^m \le \|P\| \le 5 \cdot 10^m - 10^\ell.$$

$$\beta = 3, \left\lfloor \frac{\beta+1}{2} \right\rfloor = 2 : \quad -3^m \le \|P\| \le 2 \cdot 3^m - 3^\ell.$$

□

From (1.8.5) we note the following concerning the "every-day" complement representation.

Observation 1.8.4 *The standard radix-complement representation of* $||P||$ *(without the leading digit* d_{m+1}*) is obtained by representing the non-negative value* $||P|| \bmod \beta^{m+1}$ *in the standard system* $\mathcal{F}_{\ell m}[\beta, \{0, 1, \cdots, \beta - 1\}]$.

Observe that the binary modulus operator here may be applied to a non-integral value, to bias negative values by adding β^{m+1}, and thus obtain a value in the half-open interval $[0; \beta^{m+1})$. We can generalize the modulus operator to be applicable on expressions $a \bmod b$, where a and/or b may be non-integral, by defining

$$a \bmod b \doteq \left((a\,\beta^{-k}) \bmod (b\,\beta^{-k})\right)\beta^k,$$

where the modulus operator on the right-hand side is the ordinary integer operator, choosing $k = \min(0, \ell)$ where $b = \sum_{i=\ell}^{m} b_i \beta^i$ with $b_\ell \neq 0$. Note that this is based on possibly "left-shifting" b until it is integral, and afterward shifting it back again. Thus the operator depends on the radix β, which we will assume implicitly known from the context.

For the most commonly used case of $\beta = 2$ we obtain the classical *2's complement representation.*

Observation 1.8.5 (2's complement representation) *With* $C_2^c = \{-1, 0, 1\}$ *and*

$$\mathcal{F}^{2c}_{\ell m}[2, C_2^c] = \left\{ P \in \mathcal{P}[2, C_2^c] \,\middle|\, P = \textstyle\sum_{i=\ell}^{m+1} d_i [2]^i,\, d_i \in \{0, 1\} \right.$$

$$\left. \text{for } \ell \le i \le m \text{ and } d_{m+1} = -d_m \right\},$$

then for $P \in \mathcal{F}^{2c}_{\ell m}[2, C_2^c]$ *the value of the polynomial may be obtained as*

$$\|P\| = \begin{cases} \sum_{i=\ell}^{m-1} d_i 2^i & \text{if } d_m = 0 \\ \sum_{i=\ell}^{m-1} d_i 2^i - 2^m & \text{if } d_m = 1 \end{cases}$$

or alternatively

$$\|P\| = \sum_{i=\ell}^{m-1} d_i 2^i - d_m 2^m.$$

Hence the value may be obtained by using the weight -2^m *in the most-significant position.*

The name of the representation, 2's complement, derives from the fact that the additive inverse (the polynomial $-P$) can be obtained from the representation of P by forming the so-called 2's complement of the string representation. If

$$P = \sum_{i=\ell}^{m+1} d_i [2]^i \Rightarrow -P = \sum_{i=\ell}^{m+1} (-d_i)[2]^i$$

and

$$\begin{aligned} -P &= -d_{m+1}[2]^{m+1} + \sum_{i=\ell}^{m} (-d_i)[2]^i \\ &= -d_{m+1}[2]^{m+1} + \sum_{i=\ell}^{m} (1-d_i)[2]^i - \sum_{i=\ell}^{m} [2]^i \\ &= -(d_{m+1}+1)[2]^{m+1} + \left(\sum_{i=\ell}^{m} (1-d_i)[2]^i + [2]^\ell \right), \end{aligned} \tag{1.8.6}$$

then $-P \in \mathcal{F}^{2c}_{\ell m}[2, C^c_2]$ if and only if the right-hand term of (1.8.6) belongs to $\mathcal{F}_{\ell,m}[2, \{0, 1\}]$, i.e., if and only if $(d_m, d_{m-1}, \ldots, d_\ell) \neq (1, 0, \ldots, 0)$. The right-hand term is the polynomial corresponding to the 2's complement of the bit-pattern $d_m, d_{m-1}, \ldots, d_\ell$, obtained by complementing each bit, followed by the addition of a unit in the least significant position.

As demonstrated, the standard digit set $\{0, 1, \ldots, \beta - 1\}$ can be used for a non-redundant radix-complement representation by adding the digit -1, but may only be used in the most-significant position. It is similarly possible to obtain redundant radix-complement representations based on the extended digit set $\{0, 1, \ldots, \beta\}$ by including negative digits, but restricting their use to the most-significant position. For example, for $\beta = 2$ we may define the very important set of *carry-save 2's complement polynomials.*

Definition 1.8.6 (2's complement carry-save polynomials) *The finite precision 2's complement carry-save polynomials are the set*

$$\mathcal{F}^{cs}_{\ell m}[2, \{-2, -1, 0, 1, 2\}] = \left\{ P \in \mathcal{P}[2, \{-2, -1, 0, 1, 2\}] \,\middle|\, P = \textstyle\sum_{i=\ell}^{m+1} d_i [2]^i, \right.$$
$$\left. d_i \geq 0 \textit{ for } \ell \leq i \leq m \textit{ and } d_{m+1} = -d_m \right\}.$$

for which the the value of $P \in \mathcal{F}^{cs}_{\ell m}[2, \{-2, -1, 0, 1, 2\}]$ *may be determined by*

$$\|P\| = \sum_{i=\ell}^{m-1} d_i 2^i - d_m 2^m,$$

with range $-2^{m+1} \leq \|P\| \leq 2(2^m - 2^\ell)$.

In a string representation d_{m+1} need not be present, and only the digits $\{0, 1, 2\}$ are needed. Note, however, that the sign of the number cannot be determined from the leading digit d_m whenever this digit has the value 1. In general, it requires $O(\log(m - \ell))$ time to determine the sign of $||P||$, as will be discussed in Section 3.10. But a limited number of the following digits $d_{m-1}, \ldots$ are sufficient to determine the sign if it is known that $||P||$ is bounded away from zero. Also observe that zero is not unique in this representation; however, -2^ℓ has a unique representation (all ones), thus zero-testing can be realized by subtracting 2^l (or adding all ones) and testing for all ones.

Example 1.8.4 The set $\mathcal{F}^{cs}_{02}[2, \{-2, -1, 0, 1, 2\}]$ of three-digit, carry-save, 2's complement integers, without the leading non-positive digit $d_3 = -d_2$, is

6	022		
5	021		
4	020	012	
3		011	
2	002	010	122
1	001		121
0	000	112	120
−1		111	
−2	222	110	102
−3	221		101
−4	220	212	100
−5		211	
−6	202	210	
−7	201		
−8	200		

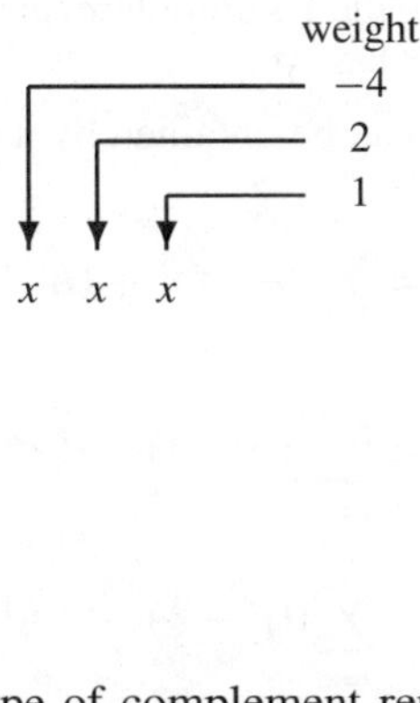

□

To complete the picture, there is another type of complement representation that has been used, the so-called *diminished radix-complement* representation. In particular, for radix 2 the *1's complement* and for radix 10 the *9's complement* representations are examples of this type. However, these cannot be considered radix polynomials as discussed here; rather they are examples of more general weighted number systems, which will be discussed in Section 1.10.

As with radix-complement systems, the digit set

$$C^c_\beta = \{-1, 0, 1, \ldots, \beta - 1\}$$

is used with a digit vector:

$$v = \{d_{m+1}, d_m, \ldots, d_\ell\},$$

where only the digit d_{m+1} may take the value -1. Imposing the restriction

$$\left(d_{m+1} = 0 \wedge \left(0 \le d_m < \left\lfloor \frac{\beta}{2} \right\rfloor\right)\right) \vee \left(d_{m+1} = -1 \wedge d_m \ge \left\lfloor \frac{\beta}{2} \right\rfloor\right)$$

and

$$d_i \ge 0 \text{ for } \ell \le i \le m,$$

then v by definition represents the value:

$$\begin{aligned} \|v\| &= \sum_{i=\ell}^{m} d_i \beta^i + (\beta^{m+1} - \beta^\ell) d_{m+1} \qquad (1.8.7)\\ &= \begin{cases} \sum_{i=\ell}^{m} d_i \beta^i & \text{if } \; 0 \le d_m < \left\lfloor \frac{\beta}{2} \right\rfloor, \\ \sum_{i=\ell}^{m} d_i \beta^i - (\beta^{m+1} - \beta^\ell) & \text{if } \; d_m \ge \left\lfloor \frac{\beta}{2} \right\rfloor. \end{cases} \end{aligned}$$

Hence d_{m+1} need not be represented in the corresponding string notation. Observe that the bias for negative numbers here is $(\beta^{m+1} - \beta^\ell)$ since $d_{m+1} = -1$, thus conversion to the 1's complement representation is performed by taking the residue modulo $(\beta^{m+1} - \beta^\ell)$.

Negation can be obtained by noting that

$$\begin{aligned} -\|v\| &= \sum_{i=\ell}^{m} (-d_i)\beta^i + (\beta^{m+1} - \beta^\ell)(-d_{m+1}) \\ &= \sum_{i=\ell}^{m} ((\beta - 1) - d_i)\beta^i - (\beta - 1)\sum_{i=\ell}^{m} \beta^i + (\beta^{m+1} - \beta^\ell)(-d_{m+1}) \\ &= \sum_{i=\ell}^{m} ((\beta - 1) - d_i)\beta^i + (\beta^{m+1} - \beta^\ell)(-(1 + d_{m+1})). \end{aligned}$$

Hence $-v$ can be represented by the digit vector $\{d'_{m+1}, d'_m, \cdots, d'_\ell)$, where $d'_i = (\beta - 1) - d_i$ for $\ell \le i \le m$ and $d'_{m+1} = -(1 + d_{m+1})$. The name diminished radix-complement derives from these relations.

In particular, for $\beta = 2$ note that $d_{m+1} = -d_m$, $d_m \in \{0, 1\}$ and that the additive inverse can be obtained by boolean complementation (inversion) of the binary digits

$$d'_i = 1 - d_i \text{ for } \ell \le i \le m,\; d'_{m+1} = -(1 + d_{m+1})$$

in the binary string notation of these bits. This representation is, however, redundant since zero has two representations, all zeroes and all ones.

Problems and exercises

1.8.1 Find the range of values obtained by four-digit, carry-save, 2's complement integers.

1.8.2 List all members of $\mathcal{F}^{rc}_{01}[5, C^c_5]$ using radix string notation.

1.8.3 Assume that a 2's complement, carry-save represented number $P = \sum_{i=\ell}^{m} d_i[2]^i$, $d_i \in \{0, 1, 2\}$, is left-normalized, say with $m = 0$ and it is known that $||P|| \geq \frac{1}{4}$ or $||P|| \leq -\frac{1}{4}$. How many digits $d_0, d_{-1}, \cdots$ are needed to determine the sign of $||P||$?

1.8.4 With $\beta = 23$, $D = \{0, \pm1, \pm2, \ldots, \pm32, \pm3, \pm6, \pm12, \pm5, \pm10\}$ can be shown to be a complete residue system. It is easily seen that $M_1 = 6$, so determine M_2 by Theorem 1.8.2.

1.8.5 Show the correctness of the following alternative rule for negating a 2's complement number in string notation:

Invert all bits down to, but not including, the rightmost non-zero bit.

1.8.6 Determine rules for negating a number in the 2's complement, carry-save representation.

1.8.7 Prove that in a fixed-point system with an even radix satisfying Theorem 1.8.1, the midpoint of two representable numbers is always representable using one more least-significant digit.

1.9 Radix-β approximation and roundings

For any $\beta \geq 2$, the radix-β numbers $\mathbb{Q}_\beta$ are dense in the reals. Thus every real number can be approximated arbitrarily well by members of $\mathbb{Q}_\beta$. For any real x and radix β, there is a family of radix-β approximations of x that are of particular interest in that the accuracy of the approximations is related to the size of the significands of the radix-β approximations. This allows us to characterize a series of "best radix-β approximations" approaching any limiting real value x in a manner similar to the characterization of "best rational approximations," which have a well established theory (see Chapter 9).

1.9.1 Best radix-β approximations

For every least-significant position ℓ, there is a unique integer significand k such that $k\beta^\ell \leq x < (k+1)\beta^\ell$. Thus $x - k\beta^\ell < \beta^\ell$, and if $x \neq k\beta^\ell$, then also $|x - (k+1)\beta^\ell| < \beta^\ell$, and these radix-$\beta$ numbers are seen to differ by less than one *unit-in-the-last-place (ulp)* from x.

Definition 1.9.1 *For any real $x \neq 0$, and any radix $\beta \geq 2$, the radix-β number $v = k\beta^\ell$ is a* one-ulp radix approximation *of x if $|x - v| < \beta^\ell$, and v is a* half-ulp radix approximation *of x if $|x - v| \leq \frac{1}{2}\beta^\ell$.*

Observation 1.9.2 *For any even radix $\beta \geq 2$, any real $x \notin \mathbb{Q}_\beta$, and any last position ℓ, there exist two distinct one-ulp approximations, $v = k\beta^\ell$ and $v' = (k+1)\beta^\ell$, of x, with exactly one of these being a half-ulp radix approximation of x. The radix-β number $x = (k + \frac{1}{2})\beta^j$ has a unique half-ulp radix approximation for every last position ℓ other than $\ell = j$.*

For any real x, $\lfloor x \rfloor$ denotes the greatest integer less than or equal to x, and $\lceil x \rceil$ denotes the least integer greater than or equal to x, and $\lfloor x + \frac{1}{2} \rfloor$ denotes the unique nearest integer to x whenever $x + \frac{1}{2}$ is not an integer.

Lemma 1.9.3 *For any real x, radix $\beta \geq 2$, and any last position ℓ, the unique one-ulp radix approximation less than or equal to x is given by $v = \lfloor x\beta^{-\ell} \rfloor \beta^\ell$, and the unique one-ulp radix approximation greater than or equal to x is given by $v = \lceil x\beta^{-\ell} \rceil \beta^\ell$, and the unique half-ulp radix approximation is given by $v = \lfloor (x + \frac{1}{2})\beta^{-\ell} \rfloor \beta^\ell$ iff $|x - v| \neq \frac{1}{2}\beta^\ell$.*

For any real x we can define a canonical sequence of half-ulp, *best (nearest), radix-β approximations* converging with non-increasing error towards x by

$$v = \lfloor (x + \tfrac{1}{2})\beta^{-\ell} \rfloor \beta^\ell \begin{cases} \text{for } \ell = -1, -2, \ldots \text{ when } x \neq (i + \frac{1}{2})\beta^{-j} \text{ for any } i, j, \\ \text{for } \ell = -1, -2, \ldots - (j-1), -(j+1), \ldots \\ \quad \text{when } x = (i + \frac{1}{2})\beta^{-j} \text{ and } \beta \text{ is even.} \end{cases}$$

The nearest sequences for $x = \pi \notin \mathbb{Q}_\beta$ and $x = \frac{25}{16} \in \mathbb{Q}_\beta$ for various β are illustrated in Table 1.9.1. The radix-β numbers are given in fixed-point with the appended sign denoting the sign of $x - v$ for $x \neq v$. Note that for β even, $(i + \frac{1}{2})\beta^{-j}$ has no strict half-ulp approximation only for $l = j$, and the approximation sequence converges to x at $\ell = -(j+1)$. When β is odd, there are two half-ulp radix approximations for $(i + \frac{1}{2})\beta^{-j}$ for all $\ell \leq -j$, and no approximation sequence is defined.

More generally, for any real $x \geq 0$, we can define a canonical sequence of one-ulp, *best (directed-towards-zero), radix-β* approximations

$$v = \lfloor x\beta^{-\ell} \rfloor \beta^\ell, \text{ for } \ell = -1, -2, \ldots$$

converging from below up to x. Generating this sequence is popularly termed developing the "correct" digits of x. Table 1.9.1 also illustrates the directed sequences for $x = \pi$, $\frac{25}{16}$ in decimal and octal (respectively binary) fixed point.

A one-ulp radix-β approximation with last place ℓ partitions any non-zero real according to $x = (k + f)\beta^\ell = k\beta^\ell + f\beta^\ell$, where $v = k\beta^\ell$ is a one-ulp approximation with integer significand k and $t = f\beta^\ell$ is a fractional *tail* with $|f| < 1$.

If x is of the numeric type transcendental, irrational, or a rational not in $\mathbb{Q}_\beta$, then the fractional tail $x - v = f\beta^\ell$ will be of the same type as x. It is sometimes possible to work recursively with expressions for the successive tails and

Table 1.9.1. *Sequence of best radix approximations of nearest and directed-towards-zero type for* $\pi \notin \mathbb{Q}$ *and* $\frac{25}{16} \in \mathbb{Q}$

Value x	Radix β	Approx. type	Least-significant position ℓ: -1	-2	-3	-4	-5
π	10	nearest	3.1^+	3.14^+	3.142^-	3.1416^-	3.14159^+
		directed	3.1	3.14	3.141	3.1415	3.14159
	8	nearest	3.1^+	3.11^+	3.110^+	3.1104^-	3.11040^-
		directed	3.1	3.11	3.110	3.1103	3.11037
$\frac{25}{16}$	10	nearest	1.6^-	1.56^+		1.5625	1.56250
		directed	1.5	1.56	1.562	1.5625	1.56250
	2	nearest	10.0^-	1.10^+		1.1001	1.10010
		directed	1.1	1.10	1.100	1.1001	1.10010

develop an algorithm for generating a sequence of best radix-β approximations. The following examples illustrate these uses of the tails.

Example 1.9.1

$$\tfrac{1}{7} = .142857 + \tfrac{1}{7} \times 10^{-6} \qquad (\ell = -6)$$

$$\tfrac{36}{11} = 3 + \tfrac{3}{11} = 3.27 + \tfrac{3}{11} \times 10^{-2} \qquad (\ell = 0, -2)$$

$$\sqrt{31} = 5 + \tfrac{6}{\sqrt{31}+5} = 5.5 + \tfrac{7.5}{\sqrt{31}+5.5} \times 10^{-1} \qquad (\ell = 0, -1,$$
$$= 5.57 + \tfrac{2.49}{\sqrt{31}+5.57} \times 10^{-2} \qquad -2)$$

$$\log_2 3 = 1.1_2 + [\log_2 (\tfrac{3^2}{2^3})] \times 2^{-1} \qquad (\ell = -1,$$
$$= 1.1001_2 + [\log_2 (\tfrac{3^{16}}{2^{25}})] \times 2^{-4} \qquad -4) \qquad \square$$

For a non-radix-β rational x, extracting an appropriate-size radix-β significand can be shown to identify a fractional part f of the tail $t = f\beta^\ell$ that repeats a previous tail's fractional part f. Such a repetition serves to characterize the significand for a radix-β approximation for any last place ℓ. For the algebraic number $x = \sqrt{31}$, the algebraically valued tail $6/(\sqrt{31}+5)$ for $\ell = 0$ can be employed to extend the radix-β approximation to include a "next term" 5×10^{-1}, yielding a fractional tail $t_{-1} = 7.5/(\sqrt{31}+5.5)$, and similarly then yielding $t_{-2} = 2.49/(\sqrt{31}+5.57)$, and so on. For the transcendental radix-2 logarithm of the integer 3, note that the fractional tails of the binary one-ulp partitions such as shown for $\ell = -1$ and $\ell = -4$ can be determined by the ratio of powers of the integer 3 and the radix 2, where these powers may even be computed in a different host arithmetic radix, e.g., by decimal arithmetic.

In summary, radix-β partitioning of a real number is very useful for expressing input and/or output of digit serial algorithms common for division and square root extraction, and for expressing values computed in an *on-line* fashion, meaning most-significant digit first. Generally any computable real-valued function $y(x)$ may be "evaluated" digit serially by determining a sequence of radix-β partitions $x = (k\beta^\ell, f_\ell(y)\beta^\ell)$ with $\lim_{\ell\to-\infty} f_\ell(y)\beta^\ell = 0$, where each tail is specified indirectly in functional form $f_\ell(y)\beta^\ell$. In formal semantics (e.g., in the lambda calculus) such an expression for evaluating the tail is called a *continuation*.

Example 1.9.1 illustrates three continuation functions for the tails of division, square root, and integer logarithms. For division the tail is given by the continuation function *remainder*/*divisor*, where the remainder will be a radix-β number whenever both the dividend and divisor are radix-β numbers. For square root of a radix-β radicand having the partition *root* = *leading part* + *tail*, with *remainder* = *radicand* − (*leading part*)^2, a continuation function for the tail is *remainder*/(*root* + *leading part*). The continuation function for integer logarithms is discussed further in the problems. Digit serial division and square root algorithms employing these remainder continuation functions are covered in Chapters 5 and 6.

When a function is evaluated most-significant digit first (on-line), typically by a one-ulp radix approximation, the resulting significand may have to be represented in a redundant manner, as it may not be possible to determine whether a non-redundant significand of the final result will be just above or just below a certain radix-β value. For example, consider evaluating the function $y = x^2$, most-significant digit first, with argument $x = 1.414214$ having the partition $1.414 + .214 \times 10^{-3}$. Using the significand 1.414 of the input partition gives $(1.414)^2 = 1.999\,396$, whereas $(1.414214)^2 = 2.000001\,237796$. Knowledge of the sign of the tail can be more useful than determining another digit, which may be a significant zero.

A radix-β, *directed, one-ulp (half-ulp) approximation* of x is given by the pair $(v, \mathrm{sgn}(t))$ where v is a one-ulp (respectively half-ulp) approximation, and

$$\mathrm{sgn}(t) = \begin{cases} -1 & \textit{for} \quad t < 0, \\ 0 & \textit{for} \quad t = 0, \\ +1 & \textit{for} \quad t > 0. \end{cases}$$

Notationally, a radix-β, directed, one-ulp approximation $(v, \mathrm{sgn}(t))$ with $\mathrm{sgn}(t) \neq 0$ is conveniently given by a fixed-point string with a superscript + or − sign of t, as employed in Table 1.9.1. Use of the superscript sign clearly identifies a trailing zero as significant for determining a tighter bound on $x - v$.

Lemma 1.9.4 *For even radix $\beta \geq 2$, a directed one-ulp radix-β approximation $(k\beta^\ell, \mathrm{sgn}(t))$ of x with $\mathrm{sgn}(t) \neq 0$ is sufficient to determine the unique, directed, half-ulp, radix-β approximation for any higher last place $\ell' \geq \ell + 1$.*

With $\ell = -20$ the half-ulp, decimal, directed approximation of π is $3.14159\,26535\,89763\,23846^+$, which provides enough information to determine the half-ulp, decimal, directed approximation of π for any last place

$\ell' \geq -20$. Similarly the half-ulp, octal, directed approximation of π for $\ell = -25$ is 3.11037 55242 10264 30215 14231^{-}, which provides enough information to determine a half-ulp traditional binary directed approximation of π for any last place ℓ in the range $3 \geq \ell \geq -75$. Note that the (undirected) half-ulp octal approximation 3.1104, of π for $\ell = -4$ is not sufficient to determine the half-ulp octal approximation for $l = -3$, so directed approximations are essential for convenient "down sizing" of half-ulp significands.

Computation of a one-ulp undirected approximation of an algebraic or transcendental function value such as $y = \sqrt{x}$ or $y = \ln(x)$ is possible by analytic approximation algorithms as a consequence of the following *Ulp Approximation Lemma*.

Lemma 1.9.5 *For the function value $y(x)$, let $y' = y + \varepsilon$ be an approximation of y with $|\varepsilon| < \frac{1}{2}\beta^{\ell}$, then $v = \lfloor (y' + \frac{1}{2})\beta^{-\ell} \rfloor \beta^{\ell} = k\beta^{\ell}$ is a one-ulp approximation of y.*

Proof $\left| \lfloor (y' + \frac{1}{2})\beta^{-\ell} \rfloor \beta^{\ell} - y' \right| \leq \frac{1}{2}\beta^{\ell}$ and $|y - y'| < \frac{1}{2}\beta^{\ell}$, so $|y - v| < \beta^{\ell}$. □

Analytic algorithms can be refined at little extra implementation cost to obtain an approximation with somewhat smaller error in the target "last place" ℓ (e.g., $|\varepsilon| < (.01)\frac{1}{2}\beta^{\ell}$). This will then allow sgn(t) for last place ℓ to be obtained for the output for "most input values" (e.g., over 95% of output one-ulp approximations will be directed) for $y(x)$ as x ranges over a large discrete set of inputs. However, determining sgn(t) generically requires function-specific algorithms of a more number theoretic flavor, particularly when the discrete set of inputs for the function range is too large for exhaustive evaluation. A function-specific exception is the supplemental computation of the remainder for division or square root, noting that sgn(*tail*) = sgn(*remainder*). The "*last bit problem*" of determining the sign bit when sgn(t) $\neq 0$ can be reduced to executing a single multiply–subtract operation for division and square root, which can be further simplified from some elementary number theory observations as discussed in Chapters 5 and 6.

The last bit problem of determining a table of unique half-ulp transcendental values for input and output of specified last place is sufficiently intractable that it has been called the *table maker's dilemma*.

1.9.2 Rounding into finite representations

Since a finite-size system like the fixed-point system $\mathcal{F}_{\ell m}[\beta, D]$ or $\mathcal{F}^{rc}_{\ell m}[\beta, C^{c}_{\beta}]$ is not closed under arithmetic operations, e.g., for addition and multiplication, there is always the possibility that the result of a computation or an input value is not representable in the system. Thus representing the result may require higher-order most-significant digits (terms of the form $d_i\beta^j$ for $j > m$) than are available (an "overflow" situation). Similarly the result may need more least-significant digits than allocated. However, in the latter case it may be sufficient for the particular application to substitute a value for the result that is close to the exact value, but needing fewer significant digits, e.g., by choosing one of the two representable one-ulp approximations using the minimum allowed last position.

The process of rounding the result of an arithmetic operation is thus an intrinsic necessity in finite precision computations, as required when a result is to be stored in memory according to some bounded-size number representation, e.g., as here in a fixed-point radix representation, or as discussed later a floating point radix or a rational fraction number representation. Various rounding strategies may be used, where ideally a rounding is a mapping of an exactly computed value into a set of representable numbers. It can be into the "nearest representable" number (with various rules in case of a tie), or it can be "directed" towards plus or minus infinity, or towards/away from zero. The following notation for the various roundings will be used here and later:

RN	round to nearest (ties separately specified);
RU	round up (i.e., towards $+\infty$);
RD	round down (i.e., towards $-\infty$);
RZ	round towards zero;
RA	round away from zero (i,e., towards $\pm\infty$).

Formally we can define a *rounding* as a mapping from the set of reals into some (usually finite) set of representable numbers, possibly extended by some representations/encodings of $+\infty$ and $-\infty$.

Definition 1.9.6 *With $\mathcal{F} \subset \mathbb{R}$ being a finite set of reals, a* rounding *is a mapping $\Phi_{\mathcal{F}} : \mathbb{R} \to \mathcal{F}$ satisfying the following three properties:*[5]

monotonic	$x < y \Rightarrow \Phi_{\mathcal{F}}(x) \leq \Phi_{\mathcal{F}}(y)$;
fixed points	$x \in \mathcal{F} \Rightarrow \Phi_{\mathcal{F}}(x) = x$;
antisymmetric	$\Phi_{\mathcal{F}}(-x) = -\Phi_{\mathcal{F}}(x)$.

To simplify the discussion, we assume that we are dealing with a fixed-point system where the digit set is contiguous and forms a complete residue system modulo the radix β, so that Theorem 1.8.1 applies. Hence the set of representable values is contiguous such that neighboring values differ by β^{ℓ}, ℓ being the index of the least significant digit.

The result of adding, subtracting, or multiplying finite precision radix-represented operands is always expressible exactly as a finite precision number, hence an approximation of the result with fewer digits can be obtained by truncating the polynomial, possibly modifying the truncated result by adding or subtracting a unit in the least-significant position (an "ulp"), based on the value of the *finite* sequence of discarded digits. For directed roundings there is no problem

[5] The property of antisymmetry has a different form for directed roundings towards $+\infty$ and $-\infty$, where the rule $\text{RU}(-x) = -\text{RD}(x)$ applies.

in deciding which value to choose. However, for round-to-nearest roundings there is a problem if the exact value happens to be precisely the midpoint between the two representable "neighbors," where some given rule is to be applied concerning which way to round.

Note that this problem of a "tie" situation cannot occur if the radix is odd, but for even radices (like the most frequently used 2 or 10), it is necessary to identify the midpoint. If a finite sequence of digits is to be discarded there is no problem (exercise), but the situation is quite different for the result of a division or square root extraction, whose result may not be representable with a finite number of digits in an even or odd radix. Thus alternative methods need to be developed for these situations, and for calculating transcendental functions (the previously mentioned table maker's dilemma). Some of these issues will be discussed in more detail in Chapter 7 in the context of the floating-point representations most often used for approximating arithmetic over the reals. Hence we will not go into more detail here.

Problems and exercises

1.9.1 Determine the directed half-ulp traditional decimal and octal approximations for last place $\ell = -6$, for:

(a) the rational constants: $x = \frac{1}{3}, \frac{1}{7}, \frac{13}{21}$,
(b) the real constants: $x = \sqrt{2}, \sqrt{5}, e, 1/e, 1/\pi$.

1.9.2 Discuss the cyclic behavior of sgn(t) in the even, radix-β-directed-half-ulp approximations of the rational constant $x = i/j \notin \mathbb{Q}_\beta$, as a function of the last place ℓ. How long can the cycle be in terms of j and β ?

1.9.3 Give a continuation function for computing the tail in the half-ulp traditional decimal partitions of $x = \frac{1}{7}$ as a function of the last place ℓ.

1.9.4 Verify for appropriate n, i, β, and j that the identity $\log_\beta n = i\beta^{-j} + [\log_\beta (n^{\beta^j}/\beta^i)] \times \beta^{-j}$ with $\beta^i \leq n^{\beta^j} < \beta^{i+1}$ determines a one-ulp partition of $\log_\beta n$ at last place ℓ. Use this identity to develop an algorithm for exact digit serial generation of $\log_\beta n$, and discuss the complexity of the algorithm.

1.9.5 Develop a result analogous to Lemma 1.9.4 (or indicate why no similar result is possible) for an odd radix $\beta \geq 3$ given

(a) a directed one-ulp leading part of x,
(b) a directed half-ulp leading part of x.

1.9.6 For a directed half-ulp traditional decimal prefix $d_m \cdots d_l^+$ or $d_m \cdots d_\ell^-$ of x, give the "truncation rule" for obtaining the directed half-ulp prefix for last place ℓ' for any $\ell + 1 \leq l' \leq m - 1$. Apply this rule to the decimal prefix for π for $\ell = -15, -10, -5$.

1.9.7 Would you expect a bias in the proportion of values of ℓ for which sgn(t) of π is $+1$, respectively -1, in directed half-ulp decimal leading parts as $\ell \to -\infty$? How could you determine this? Are the answers any different for the constant e ? Are the answers any different in another radix?

1.10 Other weighted systems

From ancient time, various measuring systems based on a positional notation have been used, where the weights associated with consecutive positions are not always related by a constant ratio. Our measurement of time in seconds, minutes, days, weeks, etc., and the weight and length measures still in use in the USA and elsewhere are examples.

Such systems can be considered generalizations of the radix systems, but cannot be modeled by radix polynomials where the weight associated with each position is a power of the radix. In its most general form (a *weighted number system*) any set of weights can be used, e.g., given a set of weights

$$(w_1, w_2, \ldots, w_n),$$

it is possible to represent a number x by a corresponding digit vector

$$(d_1, d_2, \ldots, d_n),$$

such that

$$x = \sum_{i=1}^{n} d_i w_i,$$

where the digits d_i are chosen from some digit sets D_i, $i = 1, 2, \ldots, n$.

Restricting the digit sets and weights such that

$$w_{i+1} > \sum_{j=1}^{i} |w_j||d_j| \qquad i = 1, 2, \ldots, n-1 \tag{1.10.1}$$

for all $d_j \in D_j$, $j = 1, 2, \ldots, n-1$ makes it possible to determine unique digit vectors (non-redundant representations), and to perform comparisons digitwise. Obviously non-redundant radix systems can be considered such systems when choosing $w_i = \beta^{i-1}$.

1.10.1 Mixed-radix systems

Usually systems satisfying (1.10.1) are constructed as integer mixed-radix systems, as in:

Definition 1.10.1 *A mixed-radix system consists of a* base vector *(or* radix vector*):*

$$\rho = (r_1, r_2, \ldots, r_n), \quad r_i \geq 2, \quad 1 \leq i \leq n,$$

which is an ordered set of integers, together with a digit set vector

$$\mathbb{D}_\rho = (D_0, D_1, \ldots, D_{n-1}),$$

where each set D_{i-1} is a digit set for the radix r_i, $1 \le i \le n$.

The *standard mixed-radix system* is obtained when D_{i-1} is chosen as the standard digit set for radix r_i, $r_i > 0$, i.e.,

$$D_{i-1} = \{0, 1, \ldots, r_i - 1\},$$

in which case the system is non-redundant and of cardinality $R = \prod_1^n r_i$, as we shall see below.

Although it is possible to use negative radices in the radix vector, and allow D_{i-1} to be a redundant digit set for radix r_i, we will restrict our considerations to standard mixed-radix systems. Generalizations obtained by relaxing these restrictions are fairly obvious, using the principles developed for radix polynomials.

Theorem 1.10.2 *Given a standard mixed-radix system with base vector $\rho = b(r_1, r_2, \ldots, r_n)$ and digit set vector $\mathbb{D}_\rho = (D_0, D_1, \ldots, D_{n-1})$, for any $k : 0 \le k < R$ there exists a unique digit vector $(d_0, d_1, \ldots, d_{n-1})$ such that*

$$k = d_0 + d_1(r_1) + d_2(r_1 r_2) + \cdots + d_{n-1}(r_1 r_2 \ldots r_{n-1}),$$

where $d_i \in D_i$, $i = 0, 1, \ldots, n-1$.

The proof of this theorem follows from the following algorithm, whose correctness we leave to the reader to prove.

Algorithm 1.10.3 (MRDGT)
Stimulus: *An integer $k : 0 \le k < R = \prod_{i=1}^n r_i$*
A base vector $\rho = (r_1, r_2, \ldots, r_n)$ where $r_i \ge 2$
A digit set vector $\mathbb{D}_\rho = (D_0, D_1, \ldots, D_{n-1})$
where $D_i = \{0, 1, \ldots, r_{i+1} - 1\}$
Response: *A digit vector $(d_0, d_1, \ldots, d_{n-1})$ such that $k = \sum_{i=0}^{n-1} d_i \prod_{j=1}^{i} r_j$*
Method: $a := k;$
for $i := 0$ **to** $n - 1$ **do**
$d_i := a \bmod r_{i+1};$
$a := (a - d_i) \operatorname{div} r_{i+1}$
end;

Notice that r_n is used solely to determine the digit set D_{n-1}, and is not used in the representation itself.

1.10.2 Two-level radix systems

Another weighted number system with interesting properties can be formed from a radix system $\mathcal{P}[\beta, D]$ by representing the digits of D in a second radix system $\mathcal{P}[\gamma, D_\gamma]$, where $|\gamma| < |\beta|$. We shall refer to such weighted systems as *two-level radix systems* $\mathcal{P}[\beta, (\gamma, D_\gamma)]$ with primary digit set D.

Our particular interest is in those radix systems where the primary radix β and all the non-zero digits in D_γ from the secondary radix system are powers of 2.

Example 1.10.1 Consider the system $\mathcal{P}[32, \{-16, -15, \ldots, 16\}]$ with primary radix 32 and minimally redundant digit set $D = \{-16, -15, \ldots, 16\}$. Note that the members of D can each be represented by two-digit numbers in a second radix system $\mathcal{P}[7, \{-4, -2, -1, 0, 1, 2, 4\}$. The members of the secondary radix system $\mathcal{P}[32, (7, \{-4, -2, -1, 0, 1, 2, 4\})]$ with primary digit set D are formed as shown by the following example. Let

$$P = 16\,[32]^2 + 5\,[32] + (-11) \in P[32, \{-16\,..\,16\}],$$

so then

$$P' = (2\,[7] + 2)\,[32]^2 + (1\,[7] + (-2))\,[32] + ((-1)\,[7] + (-4)),$$

and identifying weights of the forms $[32^i]$ and $[7 \cdot 32^i]$ we obtain

$$P'' = 2\,[7 \cdot 32^2] + 2\,[32^2] + 1\,[7 \cdot 32] + (-2)[32] + (-1)[7 \cdot 32^0] + (-4)[32^0],$$

where all digits in the weighted system are from $\{-4, -2, -1, 0, 1, 2, 4\}$. With digits of D_γ and primary radix $\beta = 32$ all powers of 2, we rewrite $\|P\| = \|P''\|$ combining powers of 2 and distinguishing the factor 7,

$$\|P\| = 7\,(2^{11} + 2^5 - 2^0) + (2^{11} - 2^6 - 2^2).$$ □

There is a potential application of the secondary radix system from the example in binary multiplication. Consider that the multiplicand M may be distributed over the final expression of the example, yielding

$$\|P\| \cdot M = 7\,(2^{11}M + 2^5 M - 2^0 M) + (2^{11}M - 2^6 M - 2^2 M).$$

Such expressions allow alternative multiplier encodings for a binary multiplier.

1.10.3 Double-radix systems

It has been suggested to use a system employing two (or more) mutually prime radices, in particular using the radices 2 and 3, where a polynomial P is represented as

$$P = \sum_{i,j} d_{i,j}[2]^i[3]^j, \quad \text{where} \quad d_{i,j} \in \{0, 1\}.$$

Note that the set of such polynomials, $\mathcal{P}[\{2, 3\}, \{0, 1\}]$, includes as a subset the non-redundant binary polynomials, but that the set itself allows multiple representations of the same value (i.e., it is a redundant representation, and is easily seen to be semicomplete). This is due to the existence of the transformation rules:

$$2^i 3^j + 2^{i+1} 3^j = 2^i 3^{j+1} \text{ (column reduction)},$$
$$2^i 3^j + 2^i 3^{j+1} = 2^{i+2} 3^j \quad \text{(row reduction)},$$

and other more complicated rules of the form $2^i \pm 2^j = 3^n \pm 3^n$, which may allow reductions in the number of non-zero digits needed to represent a given value (the *weight* of the representation). In the case that the weight is minimal, the representation is said to be *canonical*. Note that in a canonical representation, the matrix of digits will never contain two adjacent non-zero digits. Obviously, minimizing the number of non-zero terms will simplify arithmetic operations on such representations (e.g., addition is accomplished by overlaying the bit matrices), but the procedure to find such a canonical representation seems to be very complicated.

A greedy algorithm that for an integer $x > 0$ recursively chooses the largest value of $w = 2^i 3^j$ such that $w \leq x$, and then continues with $x \leftarrow x - w$, can be shown to terminate after $O(\log x / \log \log x)$ steps. Although this algorithm does not, in general, provide a canonical representation, it turns out to deliver sufficiently sparse representations to be useful for addition and multiplication.

The double-base representation has been suggested for use in digital signal processing, where a significant number of computations are calculating inner products, i.e., sums of products used in filtering signals. However, we will not further pursue this representation.

Problems and exercises

1.10.1 Prove the correctness of Algorithm 1.10.3 and thus also Theorem 1.10.2.

1.10.2 Write the number 3761_{10} in the mixed-radix system specified by the base vector $\rho = (2, 3, 5, 7, 11)$.

1.10.3 Add the two mixed-radix numbers $a = (0, 2, 1, 3, 0)$ and $b = (1, 2, 3, 2, 5)$ in the system with the base vector $\rho = (2, 3, 5, 7, 11)$.

1.10.4 For primary radix system $\mathcal{P}^* = \mathcal{P}[256, \{-128 .. 128\}]$, with $\mathcal{P}[128, (11, \{-32, -16, \ldots, 32\})]$, as the secondary radix system show how any $P \in \mathcal{P}^*$ may be reformulated to obtain

$$\|P\| = 11\,(\cdot) + (\cdot),$$

where $(\cdot)$ denotes sums of positive or negative powers of 2.

1.10.5 Write 1245629_{10} in a minimally redundant, symmetric number system, radix 32. Convert the answer to a secondary radix of 7 with a binary powered digit set.

1.10.6 Discuss the potential for using the system defined in Problem 1.8.4 in a two-level radix system.

1.10.7 Discuss in more detail than above the addition of two numbers given in canonical double-base {2, 3} representation.

1.11 Notes on the literature

While many introductory books on computing spend a few pages on introducing the ordinary binary and 2's complement representations of integers, and a few even mention floating-point numbers, very few books on computer architecture go much beyond this level, except possibly spending some effort on converting between binary, octal, hexadecimal, and decimal. And when it comes to more advanced, classical textbooks on computer arithmetic, the situation is not much better. Although redundant representations play a very important role in the implementation of addition, and thus in almost all other arithmetic operations, such representations are often just mentioned "in passing," say when discussing accumulating the partial products of a multiplication, or generation of quotient digits during division. Generally speaking, the knowledge, techniques and algorithms dealing with redundant radix representations were considered "folklore," just something "handed down" as "tricks of the trade" of an algorithm. Most of the notation and results on radix representations presented here originate from a chapter written by Matula in a book published in the 1970s [Mat76], with further results from [Mat82].

Concerning non-redundant representations, as usual Knuth [Knu98] is a good source of information. Also, the remarkable early description of the "von Neumann Architecture" [BGvN46] contains a detailed description of binary, 2's complement representation and the basic arithmetic operations thereupon. Information on radix representations in standard mathematics textbooks is very limited, and only deals with unique representations. Cauchy [Cau40] found that he could do certain calculations faster allowing negative digits. Of course some other various mentions of radix representations can be found in algebra and number theory textbooks, and in publications, e.g., there has been some interest in representing other number systems, in particular complex numbers [Knu60, Pen65, KS75, Gil81, KK81, Gil84, Kat94], employing complex-valued digits and/or a complex radix.

There are a few areas in computer arithmetic where there has been some activity related to radix (and in particular redundant) representations, but in general the discussion of these has usually been intertwined with some particular encoding, and narrowly associated with the particular application, e.g., the "recoding" (conversion) of multipliers, starting with the original paper by Booth [Boo51] for converting a number from 2's complement to a signed digit representation with fewer non-zero digits. Actually, the output of the algorithm is usually stated as a specification of "actions" like: `subtract`, `shift` or `add`, rather than $\{-1, 0, 1\}$. The same also applies to the later papers on "recoding" into radix 4 and 8 [Mac61, SG90]. We shall return to these and other conversions with a minimal number of non-zero digits in Chapter 2.

Another area of activity concerns constant-time addition, in particular that based on carry-save and signed-digit representations, first described in an organized manner by Metze and Robertson [MR59] and Avizienis [Avi61]. As an interesting case in point, although the signed-digit radix-2 representation was described as early as in 1959, it was not until 1985 with [TYY85] that a multiplier based on the accumulation of partial products in this representation was first described. Carry-save had been the preferred representation used for this purpose in multipliers, and in many other places where redundant representations were used for addition.

Radix-2 signed-digit representation was first really promoted in connection with on-line arithmetic [TE77], which was soon followed by a flurry of further publications on the implementation of the arithmetic operations and elementary functions. Only more recently has radix-2 signed-digit representation become more standard material in textbooks on computer arithmetic, e.g., [Kor93, Par00, EL03]. Signed-digit number systems have also been discussed by other authors, e.g., [Che85], in particular Parhami in [Par90, Par93] discusses higher radix signed-digit number systems. Redundant quotient representations have been discussed in connection with many digit-serial division algorithms, as discussed later.

The last-bit problem in the form of determining a unique half-ulp transcendental value of specified accuracy is so intractable that it has been called the *table maker's dilemma*. In published tables of the precomputer era there were generally a handful of entries that were known to have tails so close to a half-ulp in value that the table values were not guaranteed to be the desired half-ulp result in all instances. Indeed, in at least one published table a "difficult to compute" entry was knowingly chosen pragmatically to be the complementary one-ulp value (i.e., having a tail slightly greater than a half-ulp) to guard against copyright violation for that particular table. The use of a mathematical "hard problem" to provide copyright protection is an interesting early example of the approach in current document protection methods like digital watermarking. As undesirable as this sounds, i.e., to have "greater-than-half-ulf-errors" in the tables, current floating-point transcendental function hardware and software function libraries typically have a larger number of "greater-than-half-ulp-errors" than the earlier hand-computed tables. Effectively, the implementation cost of providing the *last-bit guarantee* for most functions is considered too high to provide for all inputs, except for the standard arithmetic operations (addition, multiplication, division, square root), where the IEEE standard has mandated their implementation.

It should be mentioned that beyond the mixed-radix representation, various proposals have been made where groups of bits in a bit-string may be treated differently. Some of these may have weights that are negative, and/or some positions may have digits from a redundant digit set, whereas other are drawn from a non-redundant set, e.g., as in the hybrid signed-digit systems suggested in [PK94, PGK01] where a radix-2^k system has digits encoded with $k - 1$ positions in normal binary $\{0, 1\}$, and one position (say the most-significant) has digits in $\{-1, 0, 1\}$ or $\{-2, -1, 0\}$. The two-level radix systems for multiplier designs were introduced in [SMM05], and the double-base system in [DGM99]. Another interesting proposal is a radix number system using continuously valued digits, based on analog circuitry [SAJ99].

References

[Avi61] A. Avizienis. Signed-digit number representations for fast parallel arithmetic. *IRE Trans. Electronic Computers*, EC-10:389–400, September 1961. Reprinted in [Swa90].

[BGvN46] A. Burks, H. H. Goldstine, and J Von Neumann. *Preliminary Discussion of the Logic Design of an Electronic Computing Instrument*. Technical report, Institute for Advanced Study, Princeton, 1946. Reprinted in C. G. Bell, *Computer Structures, Readings and Examples*, Mc Graw-Hill, 1971.

[Boo51] A. D. Booth. A signed binary multiplication technique. *Q. J. Mech. Appl. Math.*, 4:236–240, 1951. Reprinted in [Swa80].

[Cau40] A. Cauchy. Sur les moyens d'éviter les erreurs dans les calculs numériques. *Comptes Rendus de l'Académie des Sciences, Paris*, 11:789–798, 1840. Republished in: *Augustin Cauchy, oeuvres complètes*, 1ère série, Tome V, pp. 431–442, available at http://gallica.bnf.fr/arh:/12148/bpt6k901859.

[Che85] T. C. Chen. Maximal redundancy signed-digit systems. In *Proc. 7th IEEE Symposium on Computer Arithmetic*, pages 296–300. IEEE Computer Society, 1985.

[DGM99] V. S. Dimitrov, G. A. Jullien, and W. C. Miller. Theory and applications for a double-base number system. *IEEE Trans. Computers*, 48(10):1098–1106, 1999.

[EL03] M. D. Ercegovac and T. Lang. *Digital Arithmetic*. Morgan Kaufmann, 2003.

[Gil81] W. J. Gilbert. Radix representations of quadratic fields, *J. Math. Anal. Appl.*, 83:264–274, 1981.

[Gil84] W. J. Gilbert. Arithmetic in complex bases. *Math. Mag.*, 57(2):77–81, March 1984.

[Kat94] I. Katái. Number systems in imaginary quadratic fields. *Ann. Univ. Budapest, Sect. Comp.*, 14:91–103, 1994.

[KK81] J. Katái and B. Kovács. Canonical number systems in imaginary quadratic fields. *Acta Math. Acad. Sci. Hungaricae*, 37:1–3, 1981.

[Knu60] D. E. Knuth. An imaginary number system. *CACM*, 3(4):245–247, April 1960.

[Knu98] D. E. Knuth. *Seminumerical Algorithms*, volume 2 of *The Art of Computer Programming*. Addison Wesley, first edition 1969, second edition 1981, third edition, 1998.

[Kor93] I. Koren. *Computer Arithmetic Algorithms*. Prentice Hall, first edition, 1993, second edition, 2001.

[KS75] I. Katái and J. Szabo. Canonical number systems for complex integers. *Acta Sci. Math. (Szeged)*, 37:255–260, 1975.

[Mac61] O. L. MacSorley. High-speed arithmetic in binary computers. *Proc. IRE*, 49:67–91, January 1961. Reprinted in [Swa80].

[Mat76] D. W. Matula. Radix arithmetic: digital algorithms for computer architecture. In R. T. Yeh, editor, *Applied Computation Theory: Analysis, Design, Modeling*, chapter 9, pages 374–448. Prentice-Hall, Inc., 1976.

[Mat82] D. W. Matula. Basic digit sets for radix representation. *J ACM*, 29(4):1131–1143, October 1982.

[MR59] G. Metze and J. E. Robertson. Elimination of carry propagation in digital computers. *International Conference on Information Processing, Paris*, pages 389–396, 1959.

[Par90] B. Parhami. Generalized signed-digit number systems: a unifying framework for redundant number representations. *IEEE Trans. Computers*, C-39(1):89–98, January 1990.

[Par93] B. Parhami. On the implementation of arithmetic support functions for generalized signed digit number systems. *IEEE Trans. Computers*, C-42(3):379–384, March 1993.

[Par00] B. Parhami. *Computer Arithmetic, Algorithms and Hardware Designs*. Oxford University Press, 2000.

[Pen65] W. Penney. A 'binary' system for complex numbers. *JACM*, 12(2):247–248, April 1965.

[PGK01] D. S. Phatak, T. Goff, and I. Koren. Constant-time addition and simultaneous format conversion based on redundant binary representations. *IEEE Trans. Computers*, 50(11):1267–1278, 2001.

[PK94] D. S. Phatak and I. Koren. Hybrid signed-digit number systems: a unified framework for redundant number representations with bounded carry propagation chains. *IEEE Trans. Computers*, 43(8):880–891, August 1994.

[SAJ99] A. Saed, M. Ahmadi, and G. A. Jullien. Arithmetic with signed analog digits. In *Proc. 14th IEEE Symposium on Computer Arithmetic*, pages 134–141, IEEE Computer Society, 1999.

[SG90] H. Sam and A. Gupta. A generalized multibit recoding of two's complement binary numbers and its proof with application in multiplier implementations. *IEEE Trans. Computers*, C-39(8):1006–1015, August 1990.

[SMM05] P. -M. Seidel, L. McFearin, and D. W. Matula. Secondary radix recodings for higher radix multipliers. *IEEE Trans. Computers*, 54(2):111–123, February 2005.

[Swa80] E. E. Swartzlander, editor. *Computer Arithmetic,* Volume I. Dowden, Hutchinson and Ross, Inc., 1980. Reprinted by IEEE Computer Society Press, 1990.

[Swa90] E. E. Swartzlander, editor. *Computer Arithmetic,* Volume II. IEEE Computer Society Press, 1990.

[TE77] K. S. Trivedi and M. D. Ercegovac. On-line algorithms for division and multiplication. *IEEE Trans. Computers*, C-26(7):681–687, July 1977. Reprinted in [Swa90].

[TYY85] N. Takagi, H. Yasuura, and S. Yajima. High-speed VLSI multiplication algorithm with a redundant binary addition tree. *IEEE Trans. Computers*, C-34(9):789–796, September 1985.

2

Base and digit set conversion

2.1 Introduction

Conversion of a number from one radix representation into another plays an important role in computing, the most obvious example being the conversion between the binary representation and the everyday decimal representation that occurs in most I/O operations. But, as we shall see later, many arithmetic algorithms depend heavily on the ability to convert between radix systems, because some algorithms may be faster if performed in higher radices, and in particular if redundant digit sets are exploited.

There is amazingly little published in the open literature on these subjects in their own right; most of what can be found is usually either very trivial, or embedded in the description of some particular application.

There are, however, many such results, e.g., in multioperand addition or in multiplication the accumulation of many summands is often performed in a redundant representation, and only at the end converted into a non-redundant representation. This final digit set conversion is traditionally described as an addition process, but one may also consider ordinary addition a special case of digit set conversion.

The characteristic idea in many algorithms for multiplication often lies in converting one of the factors into a redundant representation in a higher radix, but may not really be described as such, rather it is seen as some kind of trick. Similarly, in digit serial division algorithms the quotient may be delivered in a redundant representation, but has to be converted before it is delivered as the result.

Here we shall fairly briefly discuss (base[1] or) radix conversions between non-redundant representations, in particular the "not-quite-trivial" radix conversions

[1] As mentioned before, we prefer to use the word "radix" rather than "base," except for the traditional use in "base conversion."

for the case where the radices are "incompatible," as between binary and decimal, where at best such conversions require logarithmic time. The easy cases are the constant-time conversions between "compatible" radices, like hexadecimal, octal, and binary.

However, most of this chapter is about digit set conversion, either from a redundant into a non-redundant representation, requiring at best logarithmic time, or from a redundant into a redundant representation, which can be realized in constant time.

We have decided to deal with these topics independently of the particular application, and investigate conversions by treating them as independent operations, on a par with the basic arithmetic operations addition, multiplication, division, and square root.

2.2 Base/radix conversion

It is not always possible to perform radix conversions, e.g., not all decimal numbers can be represented in binary, since $\mathbb{Q}_2$ is a proper subset of $\mathbb{Q}_{10}$ ($0.2_{10} \notin \mathbb{Q}_2$), but the following theorem states when it is possible.

Theorem 2.2.1 *Let $\mathcal{P}[\beta, D]$ be given such that $|\beta| \geq 2$ and D is complete for radix β. Given any $Q \in \mathcal{P}[\gamma, E]$ for $|\gamma| \geq 2$ where the prime factors of γ are a subset of the prime factors of β, and any digit set E, then there exists a $P \in \mathcal{P}[\beta, D]$ such that $\|P\| = \|Q\|$.*

Proof By Theorem 1.3.5 $Q \in \mathcal{P}[\gamma, E] \Rightarrow \|Q\| \in \mathbb{Q}_{|\gamma|} \subseteq \mathbb{Q}_{|\beta|}$, hence by Theorem 1.6.6 (DGT Algorithm for complete digit sets) there exists $P \in \mathcal{P}[\beta, D]$ such that $\|P\| = \|Q\|$. □

Note that it is essentially only the "target system" $\mathcal{P}[\beta, D]$ which is restricted, the "source-system" $\mathcal{P}[\gamma, E]$ only has to satisfy the requirement on the prime factors of γ and β, a condition which is not necessary when $Q \in \mathcal{P}_I[\gamma, E]$, i.e., $\|Q\|$ is an integer.

Observation 2.2.2 *For any $Q \in \mathcal{P}_I(\gamma, E)$ there exists a $P \in \mathcal{P}_I(\beta, D)$, such that $\|Q\| = \|P\|$ for any $|\gamma| \geq 2$ and $|\beta| \geq 2$ and any digit set D complete for radix β.*

The problem, in general, is that the "fractional part" cannot always be mapped into the target system. But an "arbitrarily good" approximation can always be found.

Theorem 2.2.3 *Let $\mathcal{P}[\beta, D]$ be given such that $|\beta| \geq 2$ and D is complete for radix β. Given any $Q \in \mathcal{P}[\gamma, E]$ for $|\gamma| \geq 2$ and any digit set E, then for arbitrary*

$\varepsilon > 0$ *there exists a* $P \in \mathcal{P}[\beta, D]$ *such that*

$$|\|P\| - \|Q\|| < \varepsilon.$$

Proof Choose a k such that $|\beta|^{-k} \leq \varepsilon$, define $q = \lfloor \|Q\| \cdot |\beta|^k \rfloor$, and select (possibly by the DGT Algorithm) $P' \in \mathcal{P}_I[\beta, D]$ such that $\|P'\| = q$. Then with $P = P' \cdot [\beta]^{-k}$ we have

$$|\|P\| - \|Q\|| = |q \cdot |\beta|^{-k} - \|Q\|| < |\beta|^{-k} \leq \varepsilon.$$ □

The conversion from $Q \in \mathcal{P}[\gamma, E]$ into $P \in \mathcal{P}[\beta, D]$ may be performed in two steps: first convert from $\mathcal{P}[\gamma, E]$ into $\mathbb{Q}$ (the set of rationals) and then convert from $\mathbb{Q}$ into $\mathcal{P}[\beta, D]$, i.e., first evaluate the polynomial and then perform the DGT Algorithm (e.g., Algorithm 1.6.5). However, if we perform the arithmetic in the target system the second step disappears, and conversion reduces to evaluating the polynomial. On the other hand, if the conversion is taking place in the source system, then the first step disappears since the polynomial (number) to be converted must be assumed to be known in that system, hence what remains is to apply the DGT Algorithm.

Example 2.2.1 Convert $109_{[10]} \in \mathcal{P}[10, \{0, 1, \ldots, 9\}]$ into $\mathcal{P}[2, \{0, 1\}]$ using binary (target system) arithmetic. First convert the radix 10 and the digits $\{0, 1, \ldots, 9\}$ into binary ($\{0, 1\}^*$), and then evaluate by the Horner scheme for computing the value of a polynomial

$$(1_{[2]} \times 1010_{[2]} + 0_{[2]}) \times 1010_{[2]} + 1001_{[2]} = 1101101_{[2]}$$

in binary arithmetic. □

Example 2.2.2 Convert $109_{[10]} \in \mathcal{P}[10, \{0, 1, \ldots, 9\}]$ into $\mathcal{P}[2, \{0, 1\}]$ using decimal (source system) arithmetic. Here the value to be converted is already known in the system to be used, so what remains is to apply the DGT Algorithm, which is most readily illustrated in the following table:

r_m	d_m
109	1
54	0
27	1
13	1
6	0
3	1
1	1

where $d_m = r_m \bmod 2$ and $r_{m+1} = r_m \operatorname{div} 2, m = 0, 1, \ldots$ is computed in decimal arithmetic. □

Example 2.2.3 Find $x \in \mathcal{P}[2, \{0, 1\}]$ such that $|0.2_{[10]} - x| < \varepsilon < 1$ using decimal arithmetic:

> $a := 0.2_{[10]}$; $i := 0$;
> **while** $2^i \geq \varepsilon$ **do**
> $b := 2_{[10]} \times a$; $i := i - 1$;
> $d_i := \lfloor b \rfloor$; $a := b - d_i$
> **end**;

so $a = 0.00110011\cdots$ to the precision specified by ε. □

As is implicit in the algorithms for base-digit set conversion presented so far, the process looks inherently sequential, whether performed in source or target system arithmetic, i.e., the process takes $O(n)$ time, where n is the number of digits converted. As may also be noted when converting an integer from say binary to decimal, the least-significant decimal digit depends on *all* digits of the binary number converted. As we shall see in Chapter 3, under a fairly general computational model, there is then no hope that arbitrary conversions can be performed faster than $O(\log n)$ time. Similarly, when converting an integer from decimal to binary, the most-significant bit of the result may depend on *all* the digits of the decimal number, including the least significant.

On the other hand, it is well known that certain conversions are very trivial, e.g., between binary and octal, and can be performed in parallel "digit-by-digit," hence in $O(1)$ time. Two radices β and γ such that $|\beta| \geq 2$ and $|\gamma| \geq 2$ are said to be *compatible* if and only if there exist integers $\sigma, p, q, |\sigma| \geq 2, p \geq 1, q \geq 1$ such that $|\beta| = |\sigma|^p$ and $|\gamma| = |\sigma|^q$. We will first consider radix conversion, i.e., the digit set of the target system will be derived from that of the source system together with the radices.

First note that $\mathbb{Q}_\beta = \mathbb{Q}_\gamma$ for compatible radices β and γ, so conversion is always possible if the digit set of the target system is suitably chosen. Now consider a conversion from radix $\beta = \sigma$ into radix $\gamma = \sigma^p$, where for any $P \in \mathcal{P}[\sigma, D]$, $P = \sum_{i=\ell}^{m} d_i[\sigma]^i$, terms (or radix-$\sigma$ monomials) from P may be grouped by rewriting P as

$$P = \sum_{i=\left\lfloor \frac{\ell}{p} \right\rfloor}^{\left\lceil \frac{m}{p} \right\rceil} \left(\sum_{j=0}^{p-1} d_{pi+j}[\sigma]^j \right) [\sigma^p]^i. \tag{2.2.1}$$

Therefore conversion into radix $\gamma = \sigma^p$ is realized by evaluating the inner sums of (2.2.1), whose values then are the digits of a polynomial $Q = \sum_{i=\ell'}^{m'} e_i[\sigma^p]^i \in \mathcal{P}[\sigma^p, E]$ where E is the digit set

$$E = D^{(p)} = \left\{ e \;\middle|\; e = \sum_{i=0}^{p-1} d_i \sigma^i \text{ with } d_i \in D \text{ for } i = 0, 1, \ldots, p-1 \right\}. \tag{2.2.2}$$

Note that D need not be complete for radix σ, and the derived digit set E may not be complete for radix σ^p, but any polynomial from $\mathcal{P}[\sigma, D]$ can be converted into $\mathcal{P}[\sigma^p, E]$ if E is defined by (2.2.2).

For the most useful cases where D is basic and complete (or semicomplete) it is easy to see that D non-redundant implies that the derived E is non-redundant, and D redundant implies E redundant. We leave the proof of this as an exercise.

Also notice that this conversion can be performed truly in parallel, a p-digit group can be evaluated to form a new digit completely independent of other groups, hence in a time that is independent of the number of digits in the polynomial to be converted, i.e., $O(1)$ time (actually $O(p)$ where $\beta = \sigma^p$).

Example 2.2.4 Let $\beta = \sigma = 2$ and $\gamma = \sigma^3 = 8$, and let the digit set be $D = \{0, 1\}$, then from (2.2.2) we derive $E = \{0, 1, \ldots, 7\}$, which corresponds to the standard way of "reading" a bit pattern three bits at a time as an octal digit. With $D = \{-1, 0, 1\}$ we can derive $E = \{-7, -6, \ldots, 6, 7\}$ for the same radices.

If we choose $\beta = 3$, $\gamma = \sigma^2 = 9$, and $D = \{-1, 0, 1\}$, then we get $E = \{-4, -3, \ldots, 4\}$, whereas with $D = \{-1, 0, 1, 2\}$ we obtain $E = \{-4, -3, \ldots, 8\}$. □

As may be noted from the examples above, among basic digit sets there are some where the derived digit set is of the same "type," e.g., if D is the standard digit set $\{0, 1, \ldots, \sigma - 1\}$ for radix σ, then the derived digit set will be the standard digit set $\{0, 1, \ldots, \sigma^n - 1\}$ for radix σ^n. In general, if $D = \{r, r+1, \ldots, s\}$ is a basic digit set for radix σ, then, using Theorem 1.8.1, from (2.2.2) we obtain for $\gamma = \sigma^n$

$$E = \left\{ e \,\middle|\, r \cdot \frac{\sigma^n - 1}{\sigma - 1} \le e \le s \cdot \frac{\sigma^n - 1}{\sigma - 1} \right\},$$

so for $\sigma > 0$,

$$D = [0; \sigma - 1] \Rightarrow E = [0; \sigma^n - 1] \qquad \text{(standard)},$$

$$D = [-(\sigma - 1); \sigma - 1] \Rightarrow E = [-(\sigma^n - 1); \sigma^n - 1] \qquad \text{(maximally redundant)},$$

and if furthermore σ is odd:

$$D = \left[-\left\lfloor \frac{\sigma}{2} \right\rfloor; \left\lfloor \frac{\sigma}{2} \right\rfloor \right] \Rightarrow E = \left[-\left\lfloor \frac{\sigma^n}{2} \right\rfloor; \left\lfloor \frac{\sigma^n}{2} \right\rfloor \right],$$

which is the case where D and E are symmetric and non-redundant.

For conversion from radix $\beta = \sigma^p$ to radix $\gamma = \sigma$ the choice of a suitable digit set is not as straightforward as in the previous case. Here a digit in $P = \sum_{i=\ell}^{m} d_i [\sigma^p]^i$ has to be "split," i.e., a term or radix-σ^p digit $d \in D$ has to be written as a finite precision radix-σ polynomial, i.e., in $\mathcal{F}_{0p}[\sigma, E]$ for some digit set E. Hence E must be chosen such that

$$D \subseteq \left\{ \sum_{i=0}^{p-1} e_i \sigma^i \;\middle|\; e_i \in E \right\}. \tag{2.2.3}$$

There are many digit sets E satisfying (2.2.3), but there is no simple way of generating a minimal E given D. However when D is the standard, the maximally redundant, or the symmetric, minimally redundant (β odd) digit set for radix $\beta = \sigma^p$, then E can be chosen as the equivalent digit set for radix σ.

For the actual conversion, given $P = \sum_{i=\ell}^{m} d_i [\beta]^i$ with $\beta = \sigma^p$ find radix-σ polynomials P_i, $i = \ell, \ell+1, \ldots, m$ such that $\|P_i\| = d_i$, $P_i = \sum_{j=0}^{p-1} e_{ij}[\sigma]^j \in \mathcal{P}[\sigma, E]$. Then with

$$Q = \sum_{i=\ell}^{m} \left(\sum_{j=0}^{p-1} e_{ij}[\sigma]^j \right) [\sigma]^{pi}, \tag{2.2.4}$$

$Q \in \mathcal{P}[\sigma, E]$ and $\|P\| = \|Q\|$.

Example 2.2.5 Let $\beta = 2^4 = 16$ and $\gamma = 2$ (i.e., $\sigma = 2$), then with $D = \{0, 1, \ldots, 15\}$ we may choose $E = \{0, 1\}$, which obviously satisfies (2.2.3), but so also does $\{0, 1, 2\}$ and D itself. If $D = \{0, 1, \ldots, 15, 16\}$ (the extended digit set for $\beta = 16$), then $E = \{0, 1, 2\}$ could be chosen, so as to satisfy (2.2.3). □

In the general case of converting between compatible radices $\beta = \sigma^p$ and $\gamma = \sigma^q$, there are two possible approaches, both consisting of two steps:

(a) First convert from radix σ^p to σ, and then from σ to σ^q.
(b) With $r = \mathrm{lcm}(p, q)$ first convert from radix σ^p to σ^r, and then from σ^r to σ^q.

Example 2.2.6 Consider converting from $\mathcal{P}[16, \{0, 1, \ldots, 15\}]$ (i.e., ordinary hexadecimal) to octal representation $\mathcal{P}[8, E]$ for some E. In approach (a) the first conversion is into binary, where the digit set $\{0, 1\}$ may be chosen. In the second step binary digits are just grouped to form octal digits over the digit set $\{0, 1, \ldots, 7\}$.

In approach (b) triples of hexadecimal digits are first grouped to form digits in radix $2^{12} = 4096$ over the digit set $\{0, 1, \ldots, 4095\}$, and next each of these digits is "split" into four octal digits, say over the digit set $\{0, 1, \ldots, 7\}$. □

Notice in both approaches, a group of r/p radix-p digits is converted into r/q radix-q digits, where $r = \mathrm{lcm}(p, q)$. Also conversion of such groups can be performed truly in parallel, hence in time independent of the number of digits of the radix polynomial converted.

We can summarize the results on radix conversion for compatible radices in the following observation.

Observation 2.2.4 *Let β and γ be compatible radices such that $\beta = \sigma^p$ and $\gamma = \sigma^q$, and let D be a digit set for radix β. Then there exists a digit set E for radix γ such that for any $P \in \mathcal{P}[\beta, D]$, $P = \sum_{i=\ell}^{m} d_i[\beta]^i$ there exists a $Q \in \mathcal{P}[\gamma, E]$ such that $\|Q\| = \|P\|$. Furthermore, with $r = \text{lcm}(p, q)$, $p' = r/p$ and $q' = r/q$ the digit set E must satisfy*

$$\sum_{i=0}^{p'-1} d_i \beta^i \in \left\{ k \;\middle|\; k = \sum_{i=0}^{q'-1} e_j \gamma^j, e_j \in E, j = 0, 1, \ldots, q'-1 \right\}$$

for all $(d_0, d_1, \ldots, d_{p'-1}) \in D^{p'}$. If D is non-redundant, then E can be chosen as non-redundant, and if D is redundant, then E will be redundant. Conversion can be performed in a time $O(r)$, hence in a time independent of the number of digits of P and Q.

Equivalent number of digits. When converting an integer from one radix to another, the number of digits needed in the target system is obviously approximately proportional to the number of digits used in representing the value to be converted. For systems with compatible radices the same applies to fractional values, and thus in general when converting fixed-point numbers.

However, the situation is quite different when the two systems have incompatible radices. Since, in general, only approximate results can be obtained when converting fractional parts, the question of how many digits should be allocated in the target system arises.

If the source value to be converted is considered an approximate value, say representing some physical quantity, it makes sense to think of it representing some "interval of uncertainty," and thus to choose the number of fractional digits of the target system to be such that the "density" of representable values matches that uncertainty. For given values of the "last position" $\ell < 0$ and $\ell' < 0$ for say radices β and γ, the densities can be expressed as the width of the intervals between representable values. In two such systems these widths can be expressed as β^ℓ, respectively $\gamma^{\ell'}$, hence one should probably then choose $\ell' = \lfloor \ell \log_\gamma \beta \rfloor$.

Then the conversion effectively becomes a rounding into the target system with the chosen number of fractional digits. For the often cited case of fixed-point decimal-to-binary conversion we thus obtain the *equivalent digit formula*:

$$\#\text{bits} = 3.32 \times \#\text{decimal digits}. \tag{2.2.5}$$

Problems and exercises

2.2.1 Convert $109_{[10]} \in \mathcal{P}[10, \{0, 1, \ldots, 9\}]$ to $\mathcal{P}[3, \{-1, 0, 1\}]$ using target as well as source system arithmetic.

2.2.2 Convert $2E6_{[16]} \in \mathcal{P}[16, \{0, 1, \ldots, 15\}]$ to $\mathcal{P}[8, \{0, 1, \ldots, 7\}]$ using the two methods on page 64.

2.2.3 Prove the claim (page 63 and Observation 2.2.4) that when D is basic and complete, then D non-redundant implies E non-redundant, and similarly that D redundant implies E redundant.

2.3 Conversion into non-redundant digit sets

Having investigated radix conversion we shall now consider the problem of digit set conversion for a fixed radix. Again we shall look in particular for algorithms that can be performed in time $O(1)$, i.e., independently of the number of digits of the polynomial converted. First we shall identify the cases where this is not possible.

Our problem is, given $P \in \mathcal{P}[\beta, D]$, $P = \sum_{i=\ell}^{m} d_i[\beta]^i$, $d_i \in D$, to find a $Q \in \mathcal{P}[\beta, E]$ where the digit sets D and E are different. Let us first assume that E is complete and non-redundant for radix β. Then by Theorem 2.2.1 such a unique Q exists and can be found by evaluating $\|P\|$ and applying the DGT Algorithm in a time proportional to the number of digits in P and Q.

The following theorem shows that when converting into a non-redundant digit set, the most-significant digit of the result may depend on the least-significant digit of the radix polynomial being converted, but we will first illustrate this through an example before stating the theorem.

Example 2.3.1 Let $\beta = 5$ and consider the digit sets $D = \{-1, 0, 1, 2, 3\}$ and $E = \{-2, 0, 1, 2, 9\}$, which are both complete and non-redundant for radix 5. Now consider the number 30_{10} which in $\mathcal{P}[5, D]$ as well as in $\mathcal{P}[5, E]$ has the string representation 110_5. Changing the last digit to -1, i.e., considering $29_{10} = 11\bar{1}_5$ we have $1 \cdot [5]^2 + 1[5] - 1 \in \mathcal{P}[5, D]$. The number 29_{10} has the representation $-2[5]^3 + 9[5]^2 + 9[5] + 9$ in $\mathcal{P}[5, E]$, so $29_{10} = \bar{2}999_5$, hence a change in the least-significant digit changes all the digits of the converted result. Obviously the leading string of 1s could be of arbitrary length, hence a change in one digit can ripple an arbitrary distance to the left. Phrased differently, the value of a digit of the result may depend on digit values arbitrarily far to the right. □

Theorem 2.3.1 *Let E be a digit set which is non-redundant and complete for radix β, and let D be a digit set which is complete for radix β, where $D \neq E$. Then there exists an $n_0 \geq 0$ such that for any $n > n_0$ there is a radix polynomial $P \in \mathcal{P}_I[\beta, D]$, digits $d_1, d_2 \in D$, and radix polynomials $\widehat{Q}, \widetilde{Q} \in \mathcal{P}_I[\beta, E]$ such that $\|\widehat{Q}\| = \|P \cdot [\beta] + d_1\|$ and $\|\widetilde{Q}\| = \|P \cdot [\beta] + d_2\|$, where $d_n(\widehat{Q}) = 0$ and $d_n(\widetilde{Q}) \neq 0$, $d_n(Q)$ being the coefficient of $[\beta]^n$ in Q.*

Proof Let $d_1 \in D$ be chosen such that $d_1 \notin E$, which is possible since $|D| \geq |E|$ and $D \neq E$. Let $d_2 \in D \cap E$; such a d_2 exists since $D \cap E$ is non-empty. Then let e_1 be the unique member of E such that $e_1 = k\beta + d_1$, $k \neq 0$, where we first assume that $k > 0$.

Consider the family of polynomials $\{Q_j\}_{j\in\mathbb{Z}}$, $Q_j \in \mathcal{P}_I[\beta, E]$ chosen uniquely such that $\|Q_j\| = j\beta + e_1$, then $d_n(Q_0) = 0$ for all $n > 0$. Now there must exist an n_0 such that $d_n(Q_j) = 0$ for all $n > n_0$ and $0 \le j \le k$. For any fixed $n > n_0$ consider the sequence $\{Q_j\}_{j=k+1,k+2,\ldots}$. Since $\mathcal{F}_{0n-1}[\beta, E]$ is finite there must be a first $j = j_0$ for which $d_n(Q_{j_0}) \neq 0$, whereas $d_n(Q_j) = 0$ for $0 \le j < j_0$, where now $j_0 > k$. For the j_0 chosen let

$$\widehat{Q} = Q_{j_0-k} \Rightarrow d_n(\widehat{Q}) = 0,$$
$$\widetilde{Q} = Q_{j_0} - e_1 + d_2 \Rightarrow d_n(\widetilde{Q}) \neq 0.$$

Now let $P \in \mathcal{P}_I[\beta, D]$ be such that $\|P\| = j_0$. Then with

$$P_1 = P \cdot [\beta] + d_1 \quad \text{and } P_2 = P \cdot [\beta] + d_2$$

we have $\|\widehat{Q}\| = \|P_1\|$ and $\|\widetilde{Q}\| = \|P_2\|$, with $\widehat{Q}, \widetilde{Q} \in \mathcal{P}_I[\beta, E]$ since $d_2 \in E \cap D$.

For the case $k < 0$ just consider the set of polynomials $\{Q_j\}$ for $j \le 0$, but otherwise proceed as above. □

For a conversion from $\mathcal{P}[\beta, D]$ into $\mathcal{P}[\beta, E]$, where E is assumed complete and non-redundant, any digit $d \in D$ can be rewritten uniquely as $d = c\beta + e$ with $e \in E$ and c belonging to some *carry set*[2] C. In general, an incoming carry has to be added before conversion, so a *conversion mapping* α is a mapping

$$\alpha : C \times D \to C \times E$$

defined along with C such that for all $(c, d) \in C \times D$ there exists $(c', e) \in C \times E$ such that

$$c + d = c'\beta + e. \tag{2.3.1}$$

A conversion mapping is most conveniently described by a table which can be constructed from D and E in an iterative process, also building up the carry set C. Initially C only contains 0, and new members of C are added when a new value of c in (2.3.1) is encountered when rewriting $c + d$ for some known value of c.

Example 2.3.2 With $\beta = 3$, $D = \{0, 1, 2\}$, and $E = \{-1, 0, 1\}$ the following table for the conversion mapping α is constructed row by row, starting with a row for $c = 0$ and later adding a row for $c = 1$, since a carry of 1 is generated in the row for $c = 0$:

		D		
	α	0	1	2
C	0	00	01	$1\bar{1}$
	1	01	$1\bar{1}$	10

(2.3.2)

[2] We use the shorter notation “carry” for the more general term “transfer digit,” and also occasionally “borrow” for a possibly negative-valued carry.

where each entry is of the form $c'e$ representing the pair $(c', e) \in C \times E$ in (2.3.1). □

Clearly, if the final value of the incoming carry is known, the final converted digit in any position can be determined by α. But Theorem 2.3.1 implies that this carry value, in general, depends on all positions to the right when E is non-redundant.

The most obvious way to perform such a conversion into a non-redundant digit set is through a sequential process, starting with the least-significant position and progressing in the direction of the carry propagation. This allows the conversion to be realized by a very simple *transducer*, i.e., a finite state machine that, given a state c (a carry from the previous step) and an incoming digit d, produces a new state c' and an output digit e based on the function α as shown in the following figure:

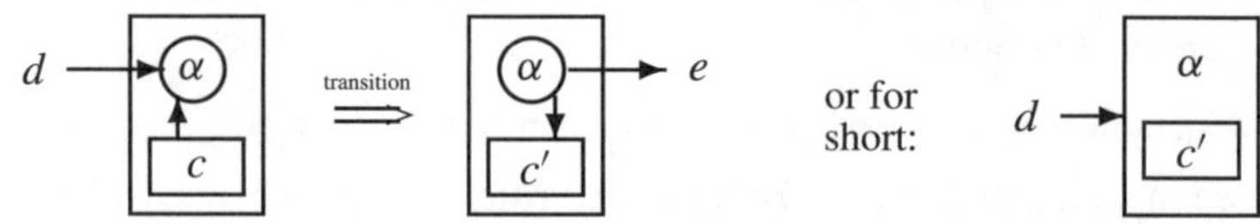

The transducer can then be applied digit by digit, to the conversion of a polynomial $P = \sum_{i=0}^{n-1} d_i \in \mathcal{P}[\beta, D]$ into $Q = \sum_{i=0}^{n} e_i \in \mathcal{P}[\beta, E]$, where E is complete for β:

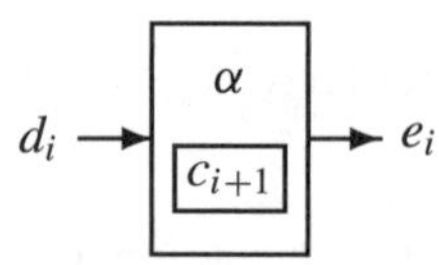

It is assumed that the transducer processes digits in the order $i = 0, 1, \ldots, n$, i.e., least-significant digit first, with boundary values $c_0 = d_n = 0$.

Later we shall illustrate how the set of all final carry values can be computed in parallel. However, it turns out that (even at best) computing the value of the carry into position i takes a time proportional to $\log i$.

Many digit set conversions map into a radix system $\mathcal{P}[\beta, E]$, where E is a complete digit set that is a subset of the maximally redundant digit set radix $\beta \geq 2$. The conversion from standard representation is particularly simple in these cases.

Observation 2.3.2 *For a conversion from the standard system* $\mathcal{P}[\beta, \{0, 1, \ldots, \beta - 1\}]$ *with* $\beta \geq 3$ *into* $\mathcal{P}[\beta, E]$, *where* E *is a complete digit set with bounded digit values,* $d \in E$, *satisfying* $-(\beta - 1) \leq d \leq \beta - 1$, *the carries* c *are limited to the set* $\{0, 1\}$.

For the digit set conversions indicated by Observation 2.3.2, the value $0 \leq d + c \leq \beta$ is either the result with 0 carry-out, or $d + c - \beta$ is the result with a carry-out of 1. The conversions can simply be performed as in traditional right-to-left carry-ripple additions.

Example 2.3.3 The conversion from the standard digit set $\{0\,..\,8\}$ into $\{-7, -5, -3, -1, 0, 1, 3, 5, 7\}$, radix 9 is given by

$$\begin{array}{ccccccc} {\scriptstyle 1} & & & {\scriptstyle 1} & {\scriptstyle 1} & {\scriptstyle 1} & \\ 4 & 6 & 5 & 2 & 1 & 3 & 6_9 \\ 5 & \bar{3} & 5 & 3 & \bar{7} & \bar{5} & \bar{3}_9 \end{array}$$

And similarly from the set $\{0\,..\,4\}$ into $\{-3, -1, 0, 1, 3\}$, radix 5:

$$\begin{array}{ccccccc} {\scriptstyle 1} & & & & {\scriptstyle 1} & {\scriptstyle 1} & \\ & 2 & 0 & 1 & 2 & 3 & 4_5 \\ 1 & \bar{3} & 0 & 1 & 3 & \bar{1} & \bar{1}_5 \end{array}$$

□

In some arithmetic algorithms (e.g., some division algorithms) the result is generated digitwise, most-significant digit first, and may have to be converted into a non-redundant digit set. It is then advantageous if the conversion can take place in parallel with the process generating the digits. Such a conversion is called *on-the-fly conversion* and consists of updating several radix polynomials in $\mathcal{P}[\beta, E]$ (the target system) whenever a new digit becomes known. Each polynomial represents a correct prefix of the result, corresponding to some possible value of the carry-in, i.e., there is one polynomial for each member of the carry set C.

Theorem 2.3.3 (On-the-fly conversion) *Let $P = \sum_{\ell}^{m} d_i[\beta]^i \in \mathcal{P}[\beta, D]$ and C be the carry set for the conversion mapping $\alpha : C \times D \to C \times E$, where E is complete for radix β. For all $c \in C$ define $Q_{m+1}^c \in \mathcal{P}[\beta, E]$ such that $\|Q_{m+1}^c\| = c\beta^{m+1}$, and for $k = m, m-1, \ldots, \ell$ and for all $c \in C$, let Q_k^c be defined by $Q_k^c = Q_{k+1}^{c'} + e_k[\beta]^k$, where $\alpha(c, d_k) = (c', e_k)$. Then $Q = Q_\ell^0 \in \mathcal{P}[\beta, E]$ satisfies $\|P\| = \|Q\|$.*

Proof Let $P_k = \sum_{k}^{m} d_i[\beta]^i \in \mathcal{P}[\beta, D]$ for $\ell \le k \le m$ with $P_{m+1} = 0$, then we want to prove by induction that

$$\forall c \in C : \|Q_k^c\| = \|P_k + c[\beta]^k\|, l \le k \le m+1, \tag{2.3.3}$$

which obviously holds for $k = m+1$. So assume (2.3.3) holds for $j = k+1$, then by the definition of Q_k^c with $\alpha(c, d_k) = (c', e_k)$ we obtain for any $c \in C$

$$\begin{aligned} \|Q_k^c\| &= \|Q_{k+1}^{c'} + e_k[\beta]^k\| \\ &= \|P_{k+1} + c'[\beta]^{k+1} + e_k[\beta]^k\| \\ &= \|P_{k+1} + d_k[\beta]^k + c[\beta]^k\| \\ &= \|P_k + c[\beta]^k\|. \end{aligned}$$

Hence with $Q = Q_\ell^0$ we have $\|Q\| = \|P_\ell\| = \|P\|$. □

Example 2.3.4 With $\beta = 3$, $D = \{0, 1, 2\}$, and $E = \{-1, 0, 1\}$ we found in Example 2.3.2 the carry-set to be $C = \{0, 1\}$, and also derived a table for the conversion mapping α. For convenience we show the table again:

α	0	1	2
0	00	01	$1\bar{1}$
1	01	$1\bar{1}$	10

From $Q_k^c = Q_{k+1}^{c'} + e_k[\beta]^k$ and α we can now derive rules for updating the polynomials $\{Q_k^c\}_{c\in C}$ when a digit d_k becomes known:

$$Q_k^0 = \begin{cases} Q_{k+1}^0 + 0[3]^k & \text{if} \quad d_k = 0, \\ Q_{k+1}^0 + 1[3]^k & \text{if} \quad d_k = 1, \\ Q_{k+1}^1 - 1[3]^k & \text{if} \quad d_k = 2, \end{cases}$$

$$Q_k^1 = \begin{cases} Q_{k+1}^0 + 1[3]^k & \text{if} \quad d_k = 0, \\ Q_{k+1}^1 - 1[3]^k & \text{if} \quad d_k = 1, \\ Q_{k+1}^1 + 0[3]^k & \text{if} \quad d_k = 2. \end{cases}$$

The pattern of which digits are appended to which polynomials to form which new polynomials should now be fairly obvious when comparing with the table for α.

When converting the number 12020_3, with the digits supplied most-significant digit first, we can display the steps of the conversion as follows in string notation:

k	d_k	Q_k^0	Q_k^1
5	–	0	1
4	1	01	$1\bar{1}$
3	2	$1\bar{1}\bar{1}$	$1\bar{1}0$
2	0	$1\bar{1}\bar{1}0$	$1\bar{1}\bar{1}1$
1	2	$1\bar{1}\bar{1}1\bar{1}$	$1\bar{1}\bar{1}10$
0	0	$1\bar{1}\bar{1}1\bar{1}0$	$1\bar{1}\bar{1}1\bar{1}1$

Hence the converted number is $1\bar{1}\bar{1}1\bar{1}0_3$. □

The algorithm for "on-the-fly" conversion as specified in Theorem 2.3.3 is a purely sequential algorithm. In each step a digit (actually a digit plus an incoming carry) is converted, and the outgoing carry value is used to select to which prefix the converted digit is appended. But why just append a single converted digit? Why not similarly append another converted polynomial to the appropriate prefix, chosen by the outgoing carry value?

To develop an algorithm for conversion based on this idea we will initially restrict the considerations to the computation of the carry values. When these have been obtained, they can then be added in parallel to the digit values, and converted in parallel.

Given the conversion mapping $\alpha : C \times D \to C \times E$, we now define a set of functions $\{\gamma_d\}_{d\in D}$, $\gamma_d : C \to C$, called *carry-transfer functions*:

$$\forall c \in C : \gamma_d(c) = c', \quad \text{where} \quad \alpha(c, d) = (c', e),$$

where we note that γ_d is a function describing the mapping of an incoming carry value (c) into its outgoing carry value (c'), when "passing through" the particular digit value (d) being converted. Note also that the values of γ_d as a function of c can be described simply by the column of outgoing carry values in the table describing α, e.g., as in (2.3.2) where for $d = 1$

$$\gamma_1\left(\begin{Bmatrix} 0 \\ 1 \end{Bmatrix}\right) = \begin{Bmatrix} 0 \\ 1 \end{Bmatrix}.$$

Similarly, from the conversion mapping α we may also define another set of functions $\{\xi_d\}_{d\in D}$, $\xi_d : C \to E$, the *digit-mapping functions*:

$$\forall c \in C \; : \; \xi_d(c) = e, \quad \text{where } \alpha(c, d) = (c', e),$$

which describe the conversion of a digit value d when the incoming carry value is c. Similarly here ξ_d can be described as a column of values from the table describing α, e.g., as in (2.3.2) where for $d = 1$

$$\xi_1\left(\begin{Bmatrix} 0 \\ 1 \end{Bmatrix}\right) = \begin{Bmatrix} 1 \\ \bar{1} \end{Bmatrix}.$$

We can now immediately generalize carry-transfer functions to strings of digits:

$$\gamma_{d_k d_{k-1} \cdots d_j}(c) = \gamma_{d_k}(\gamma_{d_{k-1}}(\cdots \gamma_{d_j}(c) \cdots))$$

or

$$\gamma_{d_k d_{k-1} \cdots d_j} = \gamma_{d_k} \circ \gamma_{d_{k-1}} \circ \cdots \circ \gamma_{d_j},$$

where $\circ$ denotes functional composition.

The function $\gamma_{d_k d_{k-1} \cdots d_j} : C \to C$ hence describes the carry transfer through the digit string $d_k d_{k-1} \cdots d_j$, when this is being converted.

Lemma 2.3.4 *Let $P = \sum_{i=\ell}^{m} d_i[\beta]^i \in \mathcal{P}[\beta, D]$ and C be the carry set for the conversion mapping $\alpha : C \times D \to C \times E$, where E is complete for radix β. Let $\{\gamma_d\}_{d\in D}$ be the carry-transfer functions derived from α, and c_i, $i = \ell, \ell + 1, \ldots, m$ be defined by*

$$c_i = \gamma_{d_i d_{i-1} \cdots d_\ell}(0).$$

Then the set $\{c_i\}_{i=\ell,\ell+1,\ldots,m}$ satisfies $c_i = \gamma_{d_i}(c_{i-1})$ with $c_{\ell-1} = 0$, and can be computed in time

$$O(\log(m - \ell) \log |C|),$$

where $|C|$ is the cardinality of C.

Proof Since functional composition is associative, the subexpressions of

$$\gamma_{d_m d_{m-1} \cdots d_\ell} = \gamma_{d_m} \circ \gamma_{d_{m-1}} \circ \cdots \circ \gamma_{d_\ell}$$

may be computed in any order; in particular disjoint subexpressions can be computed in parallel and later combined. Any value of c_i can be computed in a binary tree of height $O(\log(m-\ell))$, where each node performs a functional composition $\circ$ of two carry-transfer functions, hence each node completes in time $O(\log|C|)$. □

However, note that c_{i-1} is a prefix of c_i, so these trees overlap and we may apply the principles of *parallel prefix computation* to compute all the prefixes in parallel by combining the computation in these binary trees. We shall further illustrate this important principle of implementation in Chapter 3.

Before we combine these results into a conversion algorithm, we have to consider the carry coming out of the most-significant position, since in general we cannot assume that $C \subseteq E$. However, the polynomial to be converted can be extended at the most-significant end with some additional monomial terms $0[\beta]^i$, $i > m$, which can be converted along with the incoming carry. And the number of such extra terms needed can be at most $|C|$, since otherwise there would be a cycle, which is impossible when E is complete and hence the converted polynomial must be finite.

Theorem 2.3.5 *Let $P = \sum_{i=\ell}^{m} d_i[\beta]^i \in \mathcal{P}[\beta, D]$ and C be the carry-set for the conversion mapping $\alpha : C \times D \to C \times E$, where E is complete for radix β. Also let $\{\gamma_d\}_{d\in D}$ and $\{\xi_d\}_{d\in D}$ be the carry-transfer functions, respectively the digit-mapping functions, derived from α.*

Let the polynomial $Q = \sum_{i=\ell}^{m+k} e_i[\beta]^i \in \mathcal{P}[\beta, E]$ be defined with digits

$$e_i = \xi_{d_i}(\gamma_{d_{i-1}d_{i-2}\cdots d_\ell}(0)) \text{ for } i = \ell, \ell+1, \ldots, m+k, \tag{2.3.4}$$

where $k = |C|$ is the cardinality of C, and $d_i = 0$ for $m < i \le m+k$. Then $\|Q\| = \|P\|$ and Q can be computed in time

$$O(\log(|C| + m - \ell)\log|C|).$$

Proof By the preceding remarks, extending P with k $(= |C|)$ zero-valued terms is sufficient to absorb any carry coming into position $m+1$. Also, $\gamma_{d_{i-1}d_{i-2}\cdots d_\ell}(0)$ is the value c_{i-1} of the carry coming into position i, and e_i is the value of the converted digit d_i. Formally, by Lemma 2.3.4 we have

$$(c_i, e_i) = \alpha(c_{i-1}, d_i) \text{ for } i = \ell, \ell+1, \ldots, m, \ldots, m+k$$

with $(c_\ell, e_\ell) = \alpha(0, d_\ell)$ using the definitions of γ_d and ξ_d. The computing time follows from the computation of the c_i in Lemma 2.3.4. □

Observation 2.3.6 *The set of functions $\{\gamma_{d_{i-1}d_{i-2}\cdots d_\ell}\}_{i=\ell}^{m+k}$ computed in Theorem 2.3.5 also provides the possibility of obtaining $\gamma_{d_{i-1}d_{i-2}\cdots d_\ell}(c)$ for all i and all $c \in C$, hence a Q^c can simultaneously be found such that $\|Q^c\| = \|P + c[\beta]^\ell\|$.*

Example 2.3.5 For a numeric example consider three-operand addition over the digit set $E = \{-1, 0, 1\}$ for radix 3. By formal addition of radix polynomials, the sum of three operands from $\mathcal{P}[\beta, E]$ is a polynomial in $\mathcal{P}[\beta, D]$ with

$D = \{-3, -2, \ldots, 3\}$, so we need the conversion mapping $\alpha : C \times D \rightarrow C \times E$, where C is seen to be $C = \{-1, 0, 1\}$:

	α	D: -3	-2	-1	0	1	2	3
C	-1	$\bar{1}\bar{1}$	$\bar{1}0$	$\bar{1}1$	$0\bar{1}$	00	01	$1\bar{1}$
	0	$\bar{1}0$	$\bar{1}1$	$0\bar{1}$	00	01	$1\bar{1}$	10
	1	$\bar{1}1$	$0\bar{1}$	00	01	$1\bar{1}$	10	11

The carry-transfer functions γ_d and digit-mapping functions ξ_d can be found columnwise in the table for α. Now consider the sum of three numbers x, y, and z in radix 3, digit set $\{-1, 0, 1\}$, where three digits in one position x_i, y_i, z_i are considered a coding of a digit d_i in $\{-3, -2, \ldots, 3\}$. Then build up a *parallel prefix structure* of $\circ$-compositions, here a minimal height tree for eight-digit operands:

x_i		1	$\bar{1}$	1	$\bar{1}$	0	$\bar{1}$	1	1
y_i		1	0	0	1	1	$\bar{1}$	0	1
z_i		1	$\bar{1}$	0	1	1	0	0	$\bar{1}$
d_i	0	3	$\bar{2}$	1	1	2	$\bar{2}$	1	1

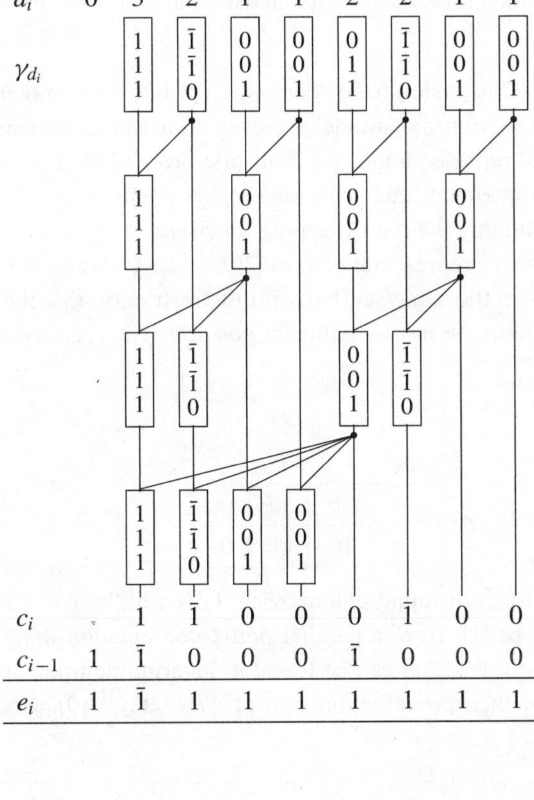

c_i		1	$\bar{1}$	0	0	0	$\bar{1}$	0	0
c_{i-1}	1	$\bar{1}$	0	0	0	$\bar{1}$	0	0	0
e_i	1	$\bar{1}$	1	1	1	1	1	1	1

where e_i is found from (2.3.4) as $e_i = \xi_{d_i}(c_{i-1})$, assuming the carry-in is 0. Each box in the diagram represents the combined carry-transfer function at that node (the result of the functional composition of the nodes above it in the tree). The functions are represented by tables, with entries corresponding to the function value for each of the three input values in $\{-1, 0, 1\}$. □

Observe from the previous example that the operation of digit addition has been hidden in the encoding of digit sums. For example, a triple of digits is just considered a representation of their sum as a digit value. The properties of radix representations and the algorithms for digit set conversion have been formulated exclusively in terms of the radix and the digit sets involved. The actual encoding of the digits, i.e., the way particular digit values may be "represented" in the binary or possibly multi-valued logic being used in physical implementations is immaterial to the algorithms and their complexity. The particular encoding chosen will only influence the specifics of the logic design, and can only change the speed of the circuitry by constant factors.

Addition of radix polynomials can thus be seen as a special case of digit set conversion. In the next chapter we shall investigate addition as a fundamental operation, and develop algorithms and structures for the specific purpose of addition. However, as demonstrated above, addition and digit set conversion are very closely related processes.

We shall conclude this section with an example of a more standard conversion task.

Example 2.3.6 (Conversion from "borrow-save" into 2's complement and "sign-magnitude") Since "sign-magnitude" representation cannot be considered a proper radix polynomial representation, we shall first proceed as if the target representation is 2's complement, and then rewrite the result as the "sign-magnitude" representation utilizing the digit-mapping functions.

The "borrow-save" representation is in $\mathcal{P}[2, \{-1, 0, 1\}]$ and the 2's complement representation is in the same set, but with the restriction that the digit value -1 may only be used in the most-significant position. The conversion mapping α is thus:

		D		
	α	-1	0	1
C	-1	$\bar{1}0$	$\bar{1}1$	00
	0	$\bar{1}1$	00	01

in which γ_d and ξ_d are found columnwise. Given a "borrow-save" polynomial $P \in \mathcal{F}_{0n}[2, \{-1, 0, 1\}]$, from a parallel prefix computation the functions $\gamma^{(i)} = \gamma_{d_i d_{i-1} \cdots d_0}$ for $i = 0, 1, \ldots, n$ can be found in logarithmic time, and the carry-out $\gamma^{(n)}(0) \in \{-1, 0\}$ then provides the sign of $\|P\|$. If $\gamma^{(n)}(0) = 0$, then $\|P\| \geq 0$

and

$$Q = \sum_{i=0}^{n} e_i[2]^i \quad \text{with} \quad e_i = \xi_{d_i}(\gamma^{(i-1)}(0))$$

is the desired result such that $\|Q\| = \|P\|$. When $\gamma^{(n)}(0) = -1$, then $\|P\|$ is negative and

$$Q' = \sum_{i=0}^{n} (1 - e'_i)[2]^i \quad \text{with} \quad e'_i = \xi_{d_i}(\gamma^{(i-1)}(-1))$$

is such that $\|Q'\| = -\|P\|$. Note that Q and $\gamma^{(n)}(0)$ are sufficient to construct the 2's complement representation, but the sign-magnitude representation requires an extra level of selection to determine whether the resulting digits are e_i or $1 - e'_i$. □

Problems and exercises

2.3.1 For conversion from a standard into a complete digit set E radix β, let $\delta(E) = \max_{d \in E} |d|$. What is then the smallest digit value bound δ^* such that the carries in the conversion mapping are limited to $\{-1, 0, 1\}$ whenever $\delta(E) \le \delta^*$? Similarly, what bound $\delta(E) \le \delta^*$ limits the carries to $\{-2, -1, 0, 1, 2\}$?

2.3.2 Perform on-the-fly conversion of $\bar{1}1\bar{1}\bar{1}1_{[2]} \in \mathcal{P}[2, \{-1, 0, 1\}]$ into $\mathcal{F}^{2c}_{04}[2, C^c_2]$, i.e., conversion from "borrow-save" into 2's complement. (Note: In some division algorithms the quotient is delivered digitwise in the digit set $\{-1, 1\}$ so this conversion applies.)

2.3.3 Develop the conversion mapping α (in the form of a table) for the conversion from 2's complement "carry-save" to ordinary 2's complement.

2.3.4 Draw a parallel prefix computation tree which computes all the prefixes of a 12-term expression given in some associative operator. Either apply the restriction that the maximal allowed fan-out of any node is 2, or consider the modifications needed due to this restriction.

2.4 Digit set conversion for redundant systems

The situation is quite different when the target system of a conversion has a redundant digit set. Here we shall see that under certain conditions conversion can take place in constant time, i.e., the "ripple effect" is bounded. Such conversions can take place truly in parallel, but in general through more than one level because conversion in one position influences neighboring positions. However, due to the redundancy, this influence can be absorbed.

First consider conversion of the radix-3 polynomial given by the digit string $12\bar{1}20133_3$ into the system with digits from the set $E = \{-1\,..\,2\}$, i.e., a contiguous and redundant digit set. Rewriting in parallel each digit into a digit in the set $\{-1, 0, 1\}$ of the same weight (a "place digit"), and a carry in $\{0, 1\}$ moved one position to the left, and then adding these digit strings, we find that no further carries are generated:

$$
\begin{array}{ccccccccc l}
 & 1 & 2 & \bar{1} & 2 & 0 & 1 & 1 & 3 & \in \{-1\,..\,3\} \\
\hline
 & 1 & \bar{1} & \bar{1} & \bar{1} & 0 & 1 & 1 & 0 & \in \{-1\,..\,1\} \\
0 & 1 & 0 & 1 & 0 & 0 & 0 & 1 & & \in \{0, 1\} \\
\hline
0 & 2 & \bar{1} & 0 & \bar{1} & 0 & 1 & 2 & 0 & \in \{-1\,..\,2\}
\end{array}
$$

Now consider conversions of $11\cdots12_3$, respectively $11\cdots11_3$, into the digit set $\{-1, 0, 1, 4\}$ for radix 3:

$$
\begin{array}{r}
1\,1\cdots\,1\,2 \\
\hline
1\,\bar{1}\,\bar{1}\cdots\,\bar{1}\,\bar{1} \\
\text{or } \bar{1}\,4\,\bar{1}\,\bar{1}\cdots\,\bar{1}\,\bar{1}
\end{array}
\qquad\qquad
\begin{array}{r}
1\,1\cdots\,1\,1 \\
\hline
1\,1\cdots\,1\,1 \\
\text{or } 1\,1\cdots\,0\,4
\end{array}
$$

where in the latter case any digit string 11 can be substituted by 04. Thus the effect of changing the least-significant digit can ripple arbitrarily far to the left.

However, adding an extra digit 5 makes constant-time conversion possible from any string over the digit set $\{-3, -2, 0, 1, 2, 5\}$, so here $11\cdots12_3$ converts into $11\cdots05_3$ with bounded carry propagation. This is due to[3] $\{0, 1, 5\} + 3\{-1, 0\} = \{-3, -2, 0, 1, 2, 5\}$ and that $\{0, 1, 5\} + \{-1, 0\} = \{-1, 0, 1, 4, 5\}$, so that any carry can be absorbed. There is also an alternative decomposition of the digit set $\{-1, 0, 1, 4, 5\} = \{-1, 0, 4\} + \{0, 1\}$, allowing parallel conversion from $\{-1, 0, 2, 3, 4, 7\} = \{-1, 0, 4\} + 3\{0, 1\}$. Observe that here conversion from $D = \Sigma + \beta C$ into $E = \Sigma + C$, for suitable sets Σ and C, looks as if it could be used as the basis of a general principle.

Surprisingly, changing the digit 4 in $\{-1, 0, 1, 4\}$ into 5 also allows parallel conversion into the digit set $\{-1, 0, 1, 5\}$ from any string over the set $\{-1\,..\,3\}$, although there is no apparent structure (decomposition) like the above. Due to the redundancy there are many possible results, depending on the use of intermediate digit sets, and choices in rewriting the digits of the source operand, e.g.,

$$
\begin{array}{r l}
1\,2\,\bar{1}\,2\,0\,1\,1\,3 & \in \{-1\,..\,3\} \\
\hline
1\,\bar{1}\,\bar{1}\,0\,\bar{1}\,\bar{1}\,5\,\bar{1}\,0 & \in \{-1, 0, 1, 5\} \\
\text{or } 1\,\bar{1}\,\bar{1}\,0\,\bar{1}\,1\,\bar{1}\,\bar{1}\,0 & \in \{-1, 0, 1, 5\}.
\end{array}
$$

Normally in such a conversion the outgoing carry is chosen such that the remaining *place digit* (belonging to that place/position in a digit string) can absorb

[3] Recall the following notation for set addition: $A + B = \{a + b \mid a \in A,\ b \in B\}$.

whatever incoming carry is being determined by the right context. Alternatively, the carry could be determined by the right as well as the left context, or possibly exclusively by the left context. Note that the carry into position i uniquely determines the residue class of the new digit at position i. Since the target digit set may contain more than one member of a given residue class, there may be a choice of outgoing carry. The essential problem is to limit the carry propagation, i.e., determine a minimal constant k such that the carry into position $i+k$ is independent of the carry into position i. But there does not seem to be any simple way to determine such a k by inspection of the digit sets involved.

In the following we shall explore the possibility of introducing context by grouping radix-β digits into digits of a higher radix β^k, and performing the digit set conversion in radix β^k. Thus we will need a notation for digits and digit sets obtained by base conversion into radix β^k for some $k > 1$. With radix β and digit set D define as previously in Section 1.5

$$D^{(k)} = \left\{ \textstyle\sum_{i=0}^{k-1} d_i \beta^i \,\middle|\, d_i \in D \right\},$$

i.e., the set of radix-β^k digits formed by grouping k digits from D. Digits from $D^{(k)}$ will be denoted similarly with an upper index, $d^{(k)}$.

We shall show that it is possible to perform conversion based on the right context with bounded carry propagation if and only if for some value of k the radix converted digit sets of radix β^k allow a suitable decomposition as above. With the left context it can be shown similarly that provided a (more complicated) decomposition of the target digit set exists, then such a conversion is also possible.

Note that in such parallel conversions, a source digit set is determined by the target digit set, although there might be several source digit sets that can be converted into a given target set. If conversion from another, possibly larger, digit set is needed, then several "layers" of conversions will be necessary.

2.4.1 Limited carry propagation

For a discussion of parallel, constant-time conversion, it is necessary to define what is meant by this. We shall here assume that it is possible to implement some kind of logic which will calculate the values of a set of consecutive output digits in terms of a bounded set of consecutive input digits. An array of such identical logic units will then in parallel be able to perform the conversion in constant time. This is formalized as follows.

Definition 2.4.1 *Parallel, constant-time conversion of $P \in \mathcal{P}[\beta, D]$ into $Q \in \mathcal{P}[\beta, E]$ is possible if and only if there exists a bounded "input window" defined by constants ℓ, r, $\ell \geq 0, r \geq 1$, an "output window" defined by a constant $k \geq 1$, together with a vector function $f : D^{\ell+k+r} \to E^k$ such that for $P = \sum_{i=0}^{n-1} d_j \beta^j$*

and $Q = \sum_{i=0}^{mk-1} e_j \beta^j$

$$[e_{ki+k-1}, \ldots, e_{ki}] = f(d_{ki+k+\ell-1}, \ldots, d_{ki-r}), \tag{2.4.1}$$

for $i \in [0\,..\,m-1]$, *with* $e_j \in E$ *and* $d_j \in D$, *where* P *is possibly extended with some zero-valued digits.*

The mapping from an input window to an output window may be pictured as

$$\cdots \underbrace{\overbrace{d_{ki+k+\ell-1} \cdots d_{ki+k}}^{\text{left context}}\ d_{ki+k-1} \cdots d_{ki}\ \overbrace{d_{ki-1} \cdots d_{ki-r}}^{\text{right context}}} \cdots$$
$$\overbrace{e_{ki+k-1} \cdots e_{ki}}$$

In the definition we may assume that $k \geq \ell + r$, since otherwise we may combine several "units" into one, such that this condition is satisfied. We may also assume that $k = \ell + r$, since we can just increase the value of r. Since E may be redundant there may be more than one valid conversion from P to Q, but f selects a particular resulting vector $[e_{ki+k-1}, \ldots, e_{ki}]$.

Now base convert $P \in \mathcal{P}[\beta, D]$ into $P^{(k)} \in \mathcal{P}[\beta^k, D^{(k)}]$

$$P^{(k)} = \sum_{i=0}^{\lceil \frac{n}{k} \rceil - 1} d_i^{(k)} [\beta^k]^i \quad \text{where} \quad d_i^{(k)} = \sum_{j=0}^{k-1} d_{ki+j} \beta^j \in D^{(k)},$$

and $Q = \sum_{i=0}^{mk-1} e_i [\beta]^i$ can similarly be base converted into

$$Q^{(k)} = \sum_{i=0}^{m-1} e_i^{(k)} [\beta^k]^i \quad \text{where} \quad e_i^{(k)} = \sum_{j=0}^{k-1} e_{ki+j} \beta^j \in E^{(k)},$$

and by (2.4.1), $e_i^{(k)} = \widehat{f}(d_{ki+k+\ell-1}, \ldots, d_{ki}, \ldots, d_{ki-r})$, where $\widehat{f}(\cdot)$ is the equivalent of $f(\cdot)$, mapping into $E^{(k)}$ by evaluation of the polynomial specified by the digit string $e_{ki+k-1} \cdots e_{ki}$ resulting from $f(\cdot)$.

The digits and carries $c_i^{(k)}$ satisfy the *carry-relation*

$$e_i^{(k)} + \beta^k c_{i+1}^{(k)} = d_i^{(k)} + c_i^{(k)}$$

for $i = 1, \ldots, m-1$.

Obviously, $c_i^{(k)}$ can only depend on a bounded window of digits d_j, say

$$c_i^{(k)} = \gamma(d_{ki+u}, \ldots, d_{ki}, \ldots, d_{ki-v})$$

but by (2.4.1) we have

$$\begin{aligned} \beta^k c_{i+1}^{(k)} - c_i^{(k)} &= d_i^{(k)} - e_i^{(k)} \\ &= g(d_{ki+k+\ell-1}, \ldots, d_{ki}, \ldots, d_{ki-r}) \end{aligned}$$

for some function g. Thus we find that $u = \ell - 1$ and $v = r$, hence

$$c_i^{(k)} = \gamma(d_{ki+\ell-1}, \ldots, d_{ki-r}), \tag{2.4.2}$$

and since $k \geq \ell + r$, then $c_{i+1}^{(k)}$ is independent of $c_i^{(k)}$. The following diagram shows the conversion with carries:

$$\begin{array}{ccc}
\cdots\ d_{ki+k+\ell-1} \cdots d_{ki+k} & d_{ki+k-1} \cdots d_{ki} & d_{ki-1} \cdots d_{ki-r}\ \cdots \\
c_{i+1}^{(k)} \swarrow & & \swarrow c_i^{(k)} \\
& e_{ki+k-1} \cdots e_{ki} &
\end{array}$$

Observe that (2.4.2) only defines carries occurring at positions with an index of the form ki. Since the digit set E may be redundant, carries occurring at other positions may even belong to a larger set as we will see in Section 2.4.2.

Example 2.4.1 For an example where the left context is needed consider the following:

$$\beta = 7 \quad D = \{-3, -2, -1, 0, 1, 2, 3\} \quad E = \{-6, -5, -3, -2, -1, 0, 2, 3, 8, 9\}$$

Producing possible mappings for digits from D into E, we find:

D	$C\beta + E$		
$\bar{3}$	$0\bar{3}$		
$\bar{2}$	$0\bar{2}$		
$\bar{1}$	$0\bar{1}$		
0	00		
1	$1\bar{6}$	$\bar{1}8$	
2	02	$1\bar{5}$	$\bar{1}9$
3	03		

The only possible carries here are $C = \{-1, 0, 1\}$. We then have $D + C = \{-4, \ldots, 4\}$, which means we have two more digits to map into E:

$D + C$	$C\beta + E$
$\bar{4}$	$\bar{1}3$
4	$1\bar{3}$

With these digit sets, it is not possible to perform a constant-time conversion using only the right context, even by grouping k digits.

On the contrary, assume constant-time conversion is possible with only the right context. If we fix a value of k we have:

$$\underbrace{33\cdots 33}_{k} = d_+^{(k)} \in D^{(k)} \quad \text{and} \quad \underbrace{\bar{3}\bar{3}\cdots\bar{3}\bar{3}}_{k} = d_-^{(k)} \in D^{(k)},$$

and since 3 and -3 can only be represented by themselves in E, then $d_+^{(k)}$ and $d_-^{(k)}$ also map into themselves in $E^{(k)}$.

It is then easy to see that the digit $\overbrace{13\cdots 3}^{k} = d_p^{(k)} \in D^{(k)}$ will produce a ripple-carry problem, because the digit 1 has two possible representations in E, producing a carry of either 1 or -1. A carry of 1 introduced in a string of 3s will change it into a string of $\bar{3}$s, so if we choose an outgoing carry of 1 for $d_p^{(k)}$, then conversion of the number $d_+^{(k)} \cdots d_+^{(k)} d_p^{(k)}$ will cause a carry to ripple all the way through. Similarly a carry of -1 will ripple through $d_-^{(k)} \cdots d_-^{(k)}$.

Hence it is not possible to perform a constant-time conversion using only the right context. But it is possible to perform constant-time conversion if we look at one digit to the left.

For any digit $d \in D$ by looking at a digit l to the left, we want to produce an outgoing carry c_{out} and a digit e', such that the left digit can absorb the carry whatever it is, and the converted digit e' will absorb any carry-in c_{in} coming from the right digit r:

$$\cdots ldr \cdots \longrightarrow (c_{out}, e') \qquad e' + c_{in} \in E.$$

We can thus use the following mapping to obtain a constant-time conversion:

l	d	(c_{out}, e')
$\forall l$	$d \neq 1$	$(0, d)$
$l \in \{-2, -1, 0, 3\}$	$d = 1$	$(\bar{1}, 8)$
$l \in \{-3, 1, 2\}$	$d = 1$	$(1, \bar{6})$

since for $e' \in \{\bar{6}, 8\}$ the only possible incoming carry from the right (its left context is 1) is in $\{0, 1\}$. So for example:

	$\bar{1}$	3	1	1	$\bar{3}$	$\bar{3}$	1	1	3	3	3	1	0	2	$D = \{-3, \cdots, 3\}$
	$\bar{1}$	3	8	$\bar{6}$	$\bar{3}$	$\bar{3}$	$\bar{6}$	$\bar{6}$	3	3	3	8	0	2	e'
0	0	$\bar{1}$	1	0	0	1	1	0	0	0	$\bar{1}$	0	0	0	$C = \{-1, 0, 1\}$
	$\bar{1}$	2	9	$\bar{3}$	$\bar{3}$	$\bar{2}$	$\bar{5}$	$\bar{6}$	3	3	2	8	0	2	$E = \{-6, -5, -3, -2, -1, 0, 2, 3, 8, 9\}$

□

2.4.2 Carry determined by the right context

We will here only consider the case where $\ell = 0$, i.e., we assume there is no dependence on the left context.

Theorem 2.4.2 *Given digit sets D and E for radix-β with $|\beta| \geq 2$, then parallel and constant-time conversion from $\mathcal{P}[\beta, D]$ into $\mathcal{P}[\beta, E]$ is possible with carries determined by a bounded right context if and only if there exists an integer $k \geq 1$ and digits sets Σ and C such that*

$$D^{(k)} \subseteq \Sigma + \beta^k C \text{ and } \Sigma + C \subseteq E^{(k)}. \tag{2.4.3}$$

Furthermore, if D contains a complete residue set modulo β, then Σ contains a complete residue set modulo β^k. If there exist $d \in D,\ d \notin E$, then $C \neq \{0\}$, and thus if E is a complete digit set for radix-β, then it is also redundant.

Proof For the if part, let P be any polynomial in $\mathcal{P}[\beta, D]$ and base convert it into a polynomial $P^{(k)} \in \mathcal{P}[\beta^k, D^{(k)}]$ by grouping digits of P and evaluating these into digits from $D^{(k)}$. By (2.4.3) there exist functions σ and γ such that for any $d_i^{(k)} \in D^{(k)}$ from $P^{(k)} = \sum_{i=0}^{n-1} d_i^{(k)}[\beta^k]^i$, digits $s_i^{(k)}$ and $c_{i+1}^{(k)}$ can be chosen by $s_i^{(k)} = \sigma(d_i^{(k)}) \in \Sigma$ and $c_{i+1}^{(k)} = \gamma(d_i^{(k)}) \in C$ satisfying

$$d_i^{(k)} = \sum_{j=0}^{k-1} d_{ki+j}\beta^j = s_i^{(k)} + \beta^k c_{i+1}^{(k)},$$

where it may be observed that $s_i^{(k)}$ and $c_{i+1}^{(k)}$ are chosen independent of any left context.

By (2.4.3), $s_i^{(k)} + c \in E^{(k)}$ for any $c \in C$, and then in particular also for $c = c_i^{(k)} = \gamma(d_{i-1}^{(k)})$. Hence $e_i^{(k)} = s_i^{(k)} + c_i^{(k)} \in E^{(k)}$ and it is thus possible to convert $P^{(k)}$ into a polynomial $Q^{(k)} \in \mathcal{P}[\beta^k, \Sigma + C] \subseteq \mathcal{P}[\beta^k, E^{(k)}]$ by a parallel, constant-time conversion of digits, where carries are absorbed in a neighboring position. Each digit $e_i^{(k)} \in \Sigma + C \subseteq E^{(k)}$ of $Q^{(k)}$ can now be represented (possibly not uniquely) by a polynomial in $\mathcal{P}_{0,k-1}[\beta, E]$, thus concluding the if part.

Now assume that conversion of $P \in \mathcal{P}[\beta, D]$ into $Q \in \mathcal{P}[\beta, E]$ can be performed by a parallel, constant-time algorithm, using functions f, γ with $\ell = 0$ and $k = r \geq 1$, where the converted digits e_i of Q can be expressed by (2.4.1) and the carry $c_i^{(k)}$ by (2.4.2) so that

$$c_i^{(k)} = \gamma(d_{ki-1}, \ldots, d_{ki-k}).$$

From the carry-relation of the base converted polynomials $P^{(k)} \in \mathcal{P}[\beta^k, D^{(k)}]$ and $Q^{(k)} \in \mathcal{P}[\beta^k, E^{(k)}]$

$$d_i^{(k)} - \beta^k\gamma(d_{k(i+1)-1}, \ldots, d_{ki}) = e_i^{(k)} - \gamma(d_{ki-1}, \ldots, d_{k(i-1)}).$$

We note that the left-hand side is exclusively dependent on $d_{k(i+1)-1}, \ldots, d_{ki}$, hence when defining the function σ by

$$\sigma(a_{k-1}, \ldots, a_0) = \sum_{i=0}^{k-1} a_i\beta^i - \beta^k\gamma(a_{k-1}, \ldots, a_0),\ a_i \in D, \tag{2.4.4}$$

$e_i^{(k)}$ can be written as

$$e_i^{(k)} = \sigma(d_{k(i+1)-1}, \ldots, d_{ki}) + \gamma(d_{ki-1}, \ldots, d_{k(i-1)}),$$

and there exist sets Σ and C,

$$\Sigma = \{s \mid s = \sigma(a_{k-1}, \cdots, a_0), a_i \in D\},$$
$$C = \{c \mid c = \gamma(a_{k-1}, \cdots, a_0), a_i \in D\},$$

satisfying $\Sigma + C \subseteq E^{(k)}$, since the arguments of σ and γ are disjoint. Note that $0 \in C$ since $\gamma(0, \ldots, 0) = 0$.

By definition (2.4.4) of σ, for any sequence $a_{k-1}, \ldots, a_0$ of digits from D

$$\sum_{i=0}^{k-1} a_i \beta^i = \sigma(a_{k-1}, \ldots, a_0) + \beta^k \gamma(a_{k-1}, \ldots, a_0),$$

so $D^{(k)} \subseteq \Sigma + \beta^k C$, which concludes the only-if part.

For the final part of the theorem, first note that there are trivial cases of constant-time conversion, e.g., where D is a subset of E. If D contains a complete residue system modulo β, then $D^{(k)}$ contains a complete residue system modulo β^k. Since $D^{(k)} \subseteq \Sigma + \beta^k C$, then Σ must contain a complete residue system modulo β^k. Let $d \in D$ where $d \notin E$, and consider conversion of the polynomial $P = d\,[\beta]^{k-1}$. Thus after base conversion $P^{(k)} = d\,\beta^{k-1}[\beta]^0$, and since $d \notin E$ implies $\gamma(d, 0, \ldots, 0) \neq 0$, then $C \neq \{0\}$. If E is a complete digit set for radix β, then by Theorem 1.5.5, E must be redundant. □

Note that the function γ introduced in the proof is actually the carry-transfer function introduced previously, but here the carry-transfer function is independent of the incoming carry. The digit-mapping function ξ in radix β^k is here a linear function of the carry, $\xi_{d^{(k)}}(c) = \sigma(d^{(k)}) + c$, defined in terms of the function σ of the proof.

Returning to the example on page 76 of conversion into the digit set $E = \{-1, 0, 1, 5\}$ for radix $\beta = 3$, we notice that with $k = 2$, $E^{(2)} = \{-4..5, 8, 14..16, 20\}$, so $\Sigma + C \subset E^{(2)}$ when choosing $\Sigma = \{-4..4\}$ and $C = \{0, 1\}$. Thus constant-time conversion is possible in radix 9 from any $D^{(2)}$, $D^{(2)} \subseteq \Sigma + \beta^2 C = \{-4..4\} + 9\{0, 1\} = \{-4..13\}$ into $E^{(2)}$.

In particular it is then possible to convert from $D = \{-1..3\}$ into $E = \{-1, 0, 1, 5\}$ since then $D^{(2)} = \{-4..12\}$, so returning to the previous example in radix 3, we may perform it through base conversion into radix 9:

	1	2	$\bar{1}$	2	0	1	1	3		$\in \{-1..3\}$	$\beta = 3$
		5		$\bar{1}$		1		6		$\in \{-4..12\}$	$\beta = 9$
		$\bar{4}$		$\bar{1}$		1		$\bar{3}$		$\in \{-4..4\}$	$\beta = 9$
1		0		0		1				$\in \{0, 1\}$	$\beta = 9$
1		$\bar{4}$		$\bar{1}$		2		$\bar{3}$		$\in \{-4..5\}$	$\beta = 9$
0 1	$\bar{1}$	$\bar{1}$	0	$\bar{1}$	$\bar{1}$	5	$\bar{1}$	0		$\in \{-1, 0, 1, 5\}$	$\beta = 3$

where we note that the chosen radix-9 conversion is unique, since $\{-4\,..\,4\}$ is non-redundant, but the conversion back to radix 3 is non-unique. The equivalent conversion in radix 3 is shown below, together with an another possible conversion:

$$
\begin{array}{ccccccccc l}
 & 1 & 2 & \bar{1} & 2 & 0 & 1 & 1 & 3 & \in \{-1\,..\,3\} \\
\hline
 & \bar{2} & \bar{1} & \bar{1} & \bar{1} & 0 & 4 & \bar{2} & 0 & \\
1 & 1 & 0 & 1 & 0 & \bar{1} & 1 & 1 & & \\
\hline
1 & \bar{1} & \bar{1} & 0 & \bar{1} & \bar{1} & 5 & \bar{1} & 0 & \in \{-1, 0, 1, 5\}
\end{array}
\qquad
\begin{array}{ccccccccc l}
 & 1 & 2 & \bar{1} & 2 & 0 & 1 & 1 & 3 & \in \{-1\,..\,3\} \\
\hline
 & \bar{2} & \bar{1} & \bar{1} & \bar{1} & 0 & \bar{2} & \bar{2} & 0 & \\
1 & 1 & 0 & 1 & 0 & 1 & 1 & 1 & & \\
\hline
1 & \bar{1} & \bar{1} & 0 & \bar{1} & 1 & \bar{1} & \bar{1} & 0 & \in \{-1, 0, 1, 5\}
\end{array}
$$

where it may be noted that a carry of -1 occurs internally in one of the groups in the left example.

For practical applications of Theorem 2.4.2, it is in general fairly easy to perform a search for a suitable value of k, generating $E^{(k)}$ for increasing values of k, looking for possible decompositions of the form $\Sigma + C \subseteq E^{(k)}$. Also, it is easy to spot situations where no such decomposition will be possible for any value of k, e.g., this is easily seen to be the case for the digit set $\{-1, 0, 1, 4\}$ discussed previously; this is here left as an exercise.

2.4.3 Conversion into a contiguous digit set

We will now restrict our considerations to conversions in the usual situation where the target system has a contiguous and redundant digit set. In this situation it turns out that the conditions of Theorem 2.4.2 can always be satisfied for some $k \geq 1$.

Lemma 2.4.3 *For $\beta \geq 2$ and $r \leq 0 \leq s$ such that $s - r + 1 > \beta$, let E be the redundant digit set $\{r, r+1, \ldots, s\}$. Let u and v be such that*

$$r \leq u \leq 0 \leq v \leq s \quad \text{and } v - u \leq (s - r + 1) - \beta.$$

Then for any d such that

$$u(\beta - 1) + r \leq d \leq v(\beta - 1) + s \tag{2.4.5}$$

there exists c, f such that $d = c \cdot \beta + f$, where

$$u \leq c \leq v \quad \text{and} \quad r - u \leq f \leq s - v. \tag{2.4.6}$$

Also for any c, f satisfying (2.4.6), $c + f \in E$.

Proof Let us denote the sets determined by (2.4.5) and (2.4.6) in the lemma as follows:

$$
\begin{aligned}
D &= \{d \mid u(\beta - 1) + r \leq d \leq v(\beta - 1) + s\}, \\
C &= \{c \mid u \leq c \leq v\}, \\
F &= \{f \mid r - u \leq f \leq s - v\}, \\
E &= \{e \mid r \leq e \leq s\},
\end{aligned}
$$

where we note that $D = F + \beta C$ and $F + C = E$, thus the conditions of Theorem 2.4.2 are satisfied with $k = 1$ and $\Sigma = F$. □

To explain the relation between the digit sets involved, we can illustrate a conversion (application of the lemma) graphically as a two-step process in the following *conversion diagram*, using the notation for the sets introduced above

$$D = \{u(\beta - 1) + r, \ldots, v(\beta - 1) + s\}$$

$$C = \{u, \ldots, v\} \qquad\qquad C = \{u, \ldots, v\}$$

$$F = \{r - u, \ldots, s - v\}$$

$$E = \{r, \ldots, s\}$$

where the left arrow describes the set of carries emitted in the first step, and the right arrow the received carries being absorbed in the second step.

Example 2.4.2 Let $\beta = 5$ and $E = \{-1, 0, 1, 2, 3, 4\}$, then with $u = 0, v = 1$ for all $d \in D = \{-1, 0, 1, \ldots, 8\}$ there exist c, f such that $d = c \cdot 5 + f$. Obviously $c \in C = \{0, 1\}$ and $f \in F = \{-1, 0, 1, 2, 3\}$ and for all $(c, f) \in C \times F$ we have $c + f \in E$. Now consider the conversion of a polynomial from $\mathcal{P}[5, D]$ into $\mathcal{P}[5, E]$, e.g., in string form the conversion of $73\bar{1}28_5$. By (2.4.6) we can rewrite the individual digits in the form $c \cdot 5 + d$ and align the value of c with the position to the left for a final digitwise parallel addition:

$$\begin{array}{cccccc l}
 & 7 & 3 & \bar{1} & 2 & 8 & \in \{-1..8\} \\
\hline
 & 2 & 3 & \bar{1} & 2 & 3 & \in \{-1..3\} \\
1 & 0 & 0 & 0 & 1 & & \in \{0, 1\} \\
\hline
1 & 2 & 3 & \bar{1} & 3 & 3 & \in \{-1..4\}
\end{array}$$

utilizing the conversion diagram:

$$D = \{-1, 0, 1, 2, 3, 4, 5, 6, 7, 8\}$$

$$C = \{0, 1\} \qquad\qquad C = \{0, 1\}$$

$$F = \{-1, 0, 1, 2, 3\}$$

$$E = \{-1, 0, 1, 2, 3, 4\}$$

We chose $u = 0, v = 1$ and since $E = \{-1, 0, 1, 2, 3, 4\}$ is minimally redundant for radix 5 the only other non-trivial choice would be $u = -1$ and $v = 0$, which would allow conversion from the digit set $\{-5, -4, \ldots, 0, \ldots, 4\}$ into E in a two-level process as in the example. Hence if the target system digit set is minimally redundant it is only possible to obtain a "one-sided" reduction or "shift" of the digit values employing a two-level conversion. □

Observation 2.4.4 *If in Lemma 2.4.3, u and v are chosen such that $v - u = (s - r + 1) - \beta$, then the cardinality of D is maximal and equals $\beta + (v - u)\beta$. In this case F is non-redundant and the conversion is unique.*

Note that F may be redundant for radix β, but along the proof of Theorem 2.4.2 functions σ and γ may be defined (possibly not uniquely) such that for all $d \in D$, $\sigma(d) \in \Sigma$ and $\gamma(d) \in C$, where

$$d = \sigma(d) + \beta\gamma(d).$$

The conversion mapping $\alpha : C \times D \to C \times E$ from Section 2.2 is then defined by

$$\alpha(c, d) = (c', e) = (\gamma(d), \sigma(d) + c), \tag{2.4.7}$$

noting that the outgoing carry $c' = \gamma(d)$ is independent of the incoming carry c. Defining the carry-transfer functions $\{\gamma_d\}_{d\in D}$ and digit-mapping functions $\{\xi_d\}_{d\in D}$ from α as in Section 2.2 we have the following observation.

Observation 2.4.5 *If the conditions of Lemma 2.4.3 are satisfied and the conversion mapping α is defined by (2.4.7), then the carry-transfer functions $\{\gamma_d\}_{d\in D}$ are constant functions:*

$$\gamma_d(c) = \gamma_d(0) = \gamma(d) \text{ for all } c \in C \text{ and } d \in D$$

and the digit-mapping functions $\{\xi_d\}_{d\in D}$ are linear functions:

$$\xi_d(c) = \xi_d(0) + c = \sigma(d) + c \text{ for all } c \in C \text{ and } d \in D,$$

where γ and σ are the functions defined in the proof of Theorem 2.4.2.

Lemma 2.4.6 *Let D_S be a source digit set for radix $\beta \geq 2$ with $d_{max} = \max_{d\in D_S} d$ and $d_{min} = \min_{d\in D_S} d$ such that $d_{min} \leq 0 \leq d_{max}$. Then for any u, v, r and s, with $r \leq u \leq 0 \leq v \leq s$ also satisfying*

$$u(\beta - 1) + r \leq d_{min} \leq 0 \leq d_{max} \leq v(\beta - 1) + s \tag{2.4.8}$$

and

$$s - r + 1 = v - u + \beta > \beta \tag{2.4.9}$$

there exists a target digit set $D_T = E = \{d \mid r \leq d \leq s\}$, redundant for radix β, such that for any $P \in \mathcal{P}[\beta, D_S]$ a polynomial $Q \in \mathcal{P}[\beta, D_T]$ with $\|Q\| = \|P\|$ can be found in time $\mathcal{O}(1)$ by a digit-parallel, two-level process.

Proof With $k = 1$ and

$$\begin{aligned} D_S \subseteq D &= \{u(\beta - 1) + r, \ldots, v(\beta - 1) + s\}, \\ F = \Sigma &= \{r - u, \ldots, s - v\}, \\ C &= \{u, \ldots, v\}, \\ D_T = E &= \{r, \ldots, s\}, \end{aligned}$$

the result follows from Theorem 2.4.2. □

Observation 2.4.7 *If $u = \lfloor d_{min}/\beta \rfloor$ and $v = \lceil d_{max}/\beta \rceil$ and r, s satisfy (2.4.9) with $r \leq u \leq 0 \leq v \leq s$, then condition (2.4.8) is also satisfied.*

Usually the target digit set E is given, and possibly several different digit sets D can be determined, such that conversion from subsets of D into E is possible. But the cardinality of such a source digit set is bounded, hence the amount of "reduction" is limited. However, the lemma may be applied recursively, thus conversion into a redundant system from any source system with finite digit set can be performed truly in parallel in a constant number of conversion steps ($O(1)$ time). The number of applications required depends on the *redundancy index* $\rho = |D| - \beta$ of the source digit system D.

In the proof above we have embedded the digit set D_S in a digit set D of cardinality $|D| = (s - r + 1) + (v - u)(\beta - 1) = (v - u)\beta + \beta$, i.e., D's index of redundancy is $\rho_S = (v - u)\beta$. The target system has index of redundancy $\rho_T = v - u$, as can be seen from (2.4.9), hence the ratio of the indices is $\rho_S/\rho_T = \beta$, and from D to D_T we get reduction of the redundancy by a factor β in a single conversion. Note, however, that D_S need not be contiguous, and that D is not necessarily the smallest contiguous digit set containing D_S, there may also be other choices of (u, v) and (r, s) that satisfy (2.4.8).

In the case that the target digit system $E = D_T = \{d \mid r \leq d \leq s\}$ is given, it may be necessary to go through several conversions, which by Observation 2.4.4 can be chosen such that each reduces the cardinality by a factor of β.

Theorem 2.4.8 *Let D_S be a digit set for radix $\beta \geq 2$ with $\delta = \max_{d \in D_S} |d|$. Then for any contiguous digit set E which is redundant for radix β, and for all $P \in \mathcal{P}[\beta, D_S]$, P can be converted to a polynomial $Q \in \mathcal{P}[\beta, E]$ with $\|Q\| = \|P\|$ in time $O(\log_\beta \delta)$.*

Proof With $E = \{r, \ldots, s\}$, $d_{max} = \max_{d \in D_S} d$, and $d_{min} = \min_{d \in D_S} d$, choose constants k, u, v, where $k \geq 1$, $u < v$, and $r \leq u \leq 0 \leq v \leq s$ such that

$$u(\beta - 1) + r\beta^k \leq d_{min} \leq 0 \leq d_{max} \leq v(\beta - 1) + r\beta^k.$$

It is easy to prove (exercise) that for $i > 0$, sets F_i and C_i can be found such that the contiguous digit set

$$D_i = \{u(\beta - 1) + r\beta^i, \ldots, v(\beta - 1) + s\beta^i\}$$

is converted into

$$E_i = \{u(\beta - 1) + r\beta^{i-1}, \ldots, v(\beta - 1) + s\beta^{i-1}\}.$$

Starting with $D_S \subseteq D_k$ and by repeated applications with $D_{i-1} = E_i$ for $i = k, k-1, \ldots 0$, we end up with $D_0 = E_1 = D$, where D is the digit set of Lemma 2.4.3. Finally D can then be converted into the digit set $E = \{r, \ldots, s\}$. By the choice of k, $k = \lceil \log_\beta \delta \rceil$ is sufficient for $D_S \subseteq D_k$. □

Since this conversion can be performed in constant time, it is obvious that we may directly apply Theorem 2.4.2.

Corollary 2.4.9 *Under the conditions of Theorem 2.4.8 there exists a constant $k \geq 1$ such that conversion from D_S into E can be performed in a single step by base converting into radix β^k, i.e., there exists digit sets Σ and C such that*

$$D_S^{(k)} \subseteq \Sigma + \beta^k C \qquad \textit{and} \qquad \Sigma + \beta C \subseteq E^{(k)}.$$

Proof It is easy to see that $D_S^{(k)} \subseteq D_k^{(k)}$, with $k \geq 1$ and D_k as defined in the previous proof, and

$$D_k^{(k)} = \left\{[u(\beta - 1) + r\beta^k]\frac{\beta^k - 1}{\beta - 1}, \ldots, [v(\beta - 1) + s\beta^k]\frac{\beta^k - 1}{\beta - 1}\right\}$$

and

$$E^{(k)} = \left\{r\frac{\beta^k - 1}{\beta - 1}, \ldots, s\frac{\beta^k - 1}{\beta - 1}\right\}.$$

Thus the conditions of Theorem 2.4.2 are satisfied with

$$\Sigma = \{r\beta^{k-1} - u, \ldots, s\beta^{k-1} - v\} \text{ and } C = \left\{r\frac{\beta^{k-1} - 1}{\beta - 1} + u, \ldots, s\frac{\beta^{k-1} - 1}{\beta - 1} + v\right\}$$

for $k \geq 1$ (exercise). □

We will conclude this section with a few examples of digit set conversions for some practical applications.

Example 2.4.3 (Conversion from 2's complement into redundant, borrow-save binary) This is a conversion used in the implementation of multiplication to reduce the number of additions needed.

The source system is $\mathcal{F}^{2c}_{\ell m}[2, C^c_2]$, where $C^c_2 = \{-1, 0, 1\}$, however it is restricted such that the digit value -1 is only used in position $m + 1$ (which is not represented), where $d_m = -d_{m+1}$ then carries the sign information. The target system is $\mathcal{F}_{\ell m}[2, \{-1, 0, 1\}]$ so if position $m + 1$ is included there is nothing to convert.

To avoid too many additions in the process of multiplication it is advantageous to avoid long strings of 1s, e.g., to rewrite $\cdots 011 \cdots 10 \cdots$ into $\cdots 10 \cdots 0\bar{1}0 \cdots$, which is possible due to the redundancy of the target system.

Therefore, assume we have a 2's complement number in a register where the leftmost position may be interpreted as having the weight -1. However, we may assume that there is a "hidden position" further to the left, and thus consider the following conversion by rewriting a digit 1 into a carry of 1 and a place digit of -1, but leaving zeroes unchanged:

$\bar{1}$	1	0	0	1	1	1	1	1	1	0	1
$\bar{1}$	$\bar{1}$	0	0	$\bar{1}$	$\bar{1}$	$\bar{1}$	$\bar{1}$	$\bar{1}$	$\bar{1}$	0	$\bar{1}$
1	0	0	1	1	1	1	1	1	0	1	
0	$\bar{1}$	0	1	0	0	0	0	0	$\bar{1}$	1	$\bar{1}$

where the leftmost column does not participate in the conversion, but anyway provides the correct result. Note that the result does not have a minimal number of non-zero digits, but as we shall see shortly there are optimal algorithms. □

A frequently applied technique in such digit set conversions is to avoid the two-level process of Lemma 2.4.3 by creating conversion rules for the individual positions which do not wait until the second level to see which carry to add in when it arrives. Instead the position investigates (by "carry anticipation") whether its rightmost neighbor will generate a carry, and then takes this into account, immediately generating the final digit value.

Example 2.4.4 ("Booth Recoding") Continuing the previous example, the rules of the combined process may be described by a table, defining the new digit e_i in terms of bits b_i and b_{i-1}:

	e_i	b_{i-1} = 0	b_{i-1} = 1
b_i	0	0	1
	1	$\bar{1}$	0

where it may be noted that the leftmost bit is treated just like a bit in any other position. When converting b_0 an imaginary bit, $b_{-1} = 0$, must be considered. Note also that algebraically $e_i = b_{i-1} - b_i$. □

A Booth-recoded number has some interesting properties, some of which are highlighted in the following.

Theorem 2.4.10 *Let $P \sim b_m b_{m-1} \cdots b_\ell$, be a 2's complement polynomial in string notation, with b_m having negative weight. Then the borrow-save polynomial $Q = e_m e_{m-1} \cdots e_\ell$, where $e_i = b_{i-1} - b_i$, with $b_{\ell-1} = 0$, satisfies $\|P\| = \|Q\|$. The non-zero digits of Q have alternating signs, i.e., if $e_k \neq 0$ and $e_j \neq 0$, where these digits for $m \geq k > j \geq \ell$, are separated by zeroes (i.e., $e_i = 0$ for $k > i > j$), then $e_k e_j = -1$. Furthermore, if the polynomial Q is truncated at position j, $m > j > \ell$, giving $Q' = e_m e_{m-1} \cdots e_j$, then the value $\|Q'\|$ is the correctly rounded (round-to-nearest) value of $\|Q\|$ and $\|P\|$.*

Proof To see that $\|Q\| = \|P\|$ just notice that $\|Q\| = 2\|P\| - \|P\|$. Let us show that $e_k \neq 0$ and $e_j \neq 0$ have opposite signs when $e_i = 0$ for $j < i < k$.

Assume $e_k = 1$ (the case $e_k = -1$ is similar). From $e_k = b_{k-1} - b_k$ we deduce that $b_{k-1} = 1$ and $b_k = 0$. If $j < k - 1$, then since $e_{k-1} = 0$, $b_{k-2} = b_{k-1} = 1$ and by induction we deduce that $b_i = 1$ for any i between $k - 1$ and j. This is also obviously true if $j = k - 1$. From $e_j = b_{j-1} - b_j \neq 0$, we deduce that $b_{j-1} = 0$, hence $e_j = -1$.

Owing to the alternating signs it follows that $\left| \|Q\| - \|Q'\| \right| = \left| \sum_{i=\ell}^{j-1} e_i 2^i \right| \leq \frac{1}{2} 2^j$, but we also note that

$$
\begin{aligned}
\|Q'\| &= \sum_{i=j}^{m} e_i 2^i = \sum_{i=j}^{m} (b_{i-1} - b_i) 2^i \\
&= 2 \sum_{i=j}^{m-1} b_i 2^i + b_{j-1} 2^j - \sum_{i=j}^{m-1} b_i 2^i - b_m 2^m \\
&= -b_m 2^m + \sum_{i=j}^{m-1} b_i 2^i + b_{j-1} 2^j,
\end{aligned}
$$

i.e., the value of the 2's complement polynomial P, rounded to nearest at position j. □

Observe that this rounding is unbiased, in the sense that the exact midpoint rounds up or down, depending on the sign of the part truncated away, or equivalently on the sign of the last non-zero digit retained. The IEEE 754 standard for floating-point is similarly unbiased, but in a different way since here the midpoint is rounded to *nearest even*, i.e., ending with a zero-bit.

It is interesting that the truncated (Booth-recoded) borrow-save polynomial $Q' \sim e_m e_{m-1} \cdots e_j$ is thus equivalent to a representation consisting of a truncated 2's complement polynomial $P' \sim b_m b_{m-1} \cdots b_j$, supplemented with the "round-bit" b_{j-1}, both representing the same value non-redundantly. Obviously, conversion from the extended 2's complement representation to the Booth-recoded representation can be performed in constant time, whereas conversion the other way cannot, as seen from the example $10\cdots 0\bar{1}$, with 2's complement representation $01\cdots 11$ with round-bit 0, representing the same value.

Observation 2.4.11 *Let* $\mathrm{RN}_j(\cdot)$ *represent the operation of rounding the operand in the Booth-recoded representation by truncation at position* j*. Then for* $j_1 \geq j_2$

$$rn_{j_1}(x) = rn_{j_1}(rn_{j_2}(x)),$$

hence no errors are introduced by double-rounding.

In the implementation of multiplication it is often advantageous to convert the multiplier from radix 2 to a higher radix, say $\beta = 4$ or 8, using the appropriate symmetric, minimally redundant digit set, as this will reduce the number of "partial products" to be added.

Example 2.4.5 ("Modified Booth Recoding" of 2's complement into a minimally redundant, symmetric radix-4 digit set). Here we combine a base conversion with a digit set conversion by first pairing bits of the digit set $\{0, 1\}$ into radix-4 digits from the set $\{0, 1, 2, 3\}$, and then converting into the digit set $\{-2, -1, 0, 1, 2\}$. The leading digit of negative weight is handled separately. We assume that the operand to be converted contains an even number of bits.

Again we will combine the two levels by carry anticipation, basing the conversion rules on bits b_{2i+1}, b_{2i} forming a radix-4 digit of value $d_i = 2b_{2i+1} + b_{2i}$, combined with information from the right-hand neighboring radix-4 digit. Applying Observation 2.4.7 we may use $u = \lfloor d_{min}/\beta \rfloor = 0$ and $v = \lceil d_{max}/\beta \rceil = 1$ and choose

$$r = -2, \quad s = 2 \quad \text{with } s - r + 1 = 5,$$

so the first rewriting emitting a carry converts into the digit set $\{-2, -1, 0, 1\}$ possibly receiving a carry to be added next. Realizing that a carry is only generated if the digit is 2 or 3, we find that we can determine the value of the outgoing carry as b_{2i+1} and the incoming carry by the value of b_{2i-1} (with $b_{-1} = 0$), hence we obtain the converted digit $e_i = d_i - 4b_{2i+1} + b_{2i-1} = -2b_{2i+1} + b_{2i} + b_{2i-1}$, as shown in the following rules for the combined process:

	e_i	b_{2i-1}: 0	b_{2i-1}: 1
b_{2i+1}, b_{2i}	00	0	1
	01	1	2
	10	$\bar{2}$	$\bar{1}$
	11	$\bar{1}$	0

where this table is also valid in the most significant position e_m, since b_{2m+1} has negative weight. □

In the next example we will consider a similar conversion where the source digit is 2's complement, carry-save, e.g., for use in multiplications where the multiplier is provided in redundant carry-save form.

Example 2.4.6 (Conversion from redundant carry-save (binary) into minimally redundant, symmetric radix-4.) Grouping two radix-2 digits from the digit set $\{0, 1, 2\}$ we get the radix-4 digit set $D = \{0, 1, 2, 3, 4, 5, 6\}$, which we want to convert into the digit set $E = \{-2, -1, 0, 1, 2\}$. Note that we cannot perform this conversion in one step. Since the target digit set is minimally redundant, the only possibilities for (u, v) are $(0, 1)$ and $(-1, 0)$. The first choice $(u, v) = (0, 1)$ allows conversion from $\{-2\,..\,5\}$ through the set $\{-2\,..\,1\}$, and the other choice from $\{-5\,..\,2\}$ through the set $\{-1\,..\,2\}$. But neither of these choices will allow conversion of all members from the set D.

Let us first convert through two steps. With $u = 0$, $v = 1$ we may choose $r = -1$ and $s = 3$ in Theorem 2.4.6, so we first map into the digit set $\{-2\,..\,3\}$ using $\{-2\,..\,2\}$ as the intermediate digit set with carries in $\{0, 1\}$. Through another application we map from $\{-2\,..\,3\}$ into $\{-2\,..\,2\}$ using $\{-2\,..\,1\}$ as the intermediate, and $\{0, 1\}$ as the carry digit set, resulting in the following:

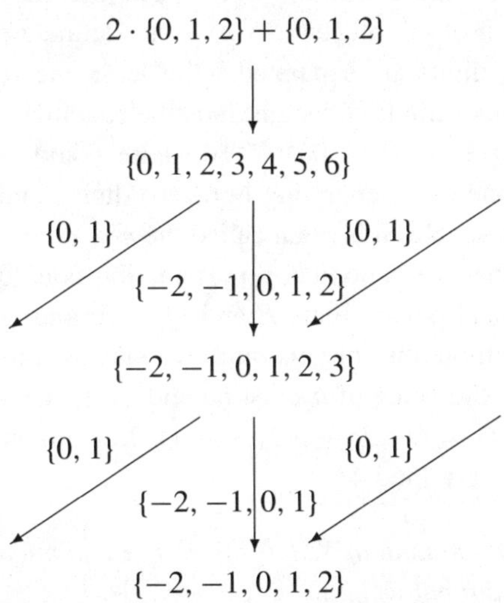

To perform the conversion in one step we base convert into radix 16, i.e., we want to convert from $D^{(2)} = \{0\,..\,30\}$ into $E^{(2)} = \{-10\,..\,10\}$. It turns out that there are now plenty of choices for Σ and C, e.g., we may choose $\Sigma = \{-8\,..\,8\}$ and $C = \{-2\,..\,2\}$, which will allow conversion from any subset of $\{-40\,..\,40\}$ into $E^{(2)} = \{-10\,..\,10\}$. □

In Section 2.5 we shall show that there are some very simple ways of implementing these conversions, in particular that the Booth and Modified Booth Recodings can be performed without any logic transformation at all, utilizing particular encodings of the digits. Also the conversion in the last example above can be implemented in a very simple way. But before that we will look into conversion of a binary number into a form with a minimal number of non-zero digits.

2.4.4 Conversion into canonical, non-adjacent form

It was noted in Example 2.4.4 that conversion into the digit set $\{-1, 0, 1\}$ transforms a string of repeated ones $011 \cdots 10$ into a string $100 \cdots \bar{1}0$, hence potentially eliminating some add operations in a sequential multiplication. But the converted number representation need not be the optimal one, in the sense of having the minimal number of non-zero digits. In fact the number of non-zero digits may even increase, since an isolated 1 is converted into two non-zero digits. For example, $\cdots 010 \cdots$ becomes $\cdots 1\bar{1}0 \cdots$. Minimizing the number of non-zero digits also plays a significant role in exponentiation, where it similarly reduces the number of multiplications. We will thus investigate this problem in more detail.

Let us look for a representation of integers in $\mathcal{P}_I[2, \{-1, 0, 1\}]$ with a minimal number of non-zero digits, and define the *weight*, $\omega(i)$, as the number of such digits in the representation of an integer i. Also let us define a representation to be *canonical* if non-zero digits are separated by at least one zero-valued digit. Obviously representations of minimal weight must be canonical, since if we had sequences of the form $011 \cdots 10$ or $0\bar{1}\bar{1} \cdots \bar{1}0$, these could be rewritten into sequences with at least one zero separating non-zero digits, and of the same or smaller weight. This representation is often called the *non-adjacent form* (*NAF*).

It is also easy to see that the canonical representation is unique. Assume we have two different canonical polynomials P and Q representing the same value $\|P\| = \|Q\| = a$, then without loss of generality we may assume that they differ in the unit position. Then the value of a must be odd, so let us assume that $P = P'[2]^2 + 0 \cdot [2] + 1$ and $Q = Q'[2]^2 + 0[2] + (-1)$. But this is a contradiction since then $\|P\| \bmod 4 \neq \|Q\| \bmod 4$.

Observation 2.4.12 *A polynomial in NAF is the unique canonical representation of the number, and is of minimal weight.*

Since non-zero digits in the canonical (NAF) representation are separated by at least one zero-valued digit, the number of non-zero digits is at most $n/2$ in an n-digit word. It can be shown that the average number of non-zero digits in the canonical representation is $n/3$. Hence a quite substantial decrease in the number of add/subtract operations can be obtained by converting the multiplier, or the number of multiplications in an exponentiation by converting the exponent.

The original algorithm developed by Reitwiesner operates in a sequential, right-to-left mode, starting at the least-significant end.

Algorithm 2.4.13 (Reitwiesner's canonical recoding)
Stimulus: *A 2's complement n-bit integer polynomial* $P = [b_{n-1} \cdots b_1 b_0]_{2c}$.
Response: *The canonical* $(n+1)$*-digit, polynomial* $Q = \sum_{i=0}^{n} e_i 2^i \in \mathcal{P}_I[2, \{-1, 0, 1\}]$ *with* $\|P\| = \|Q\|$, *such that* $e_i e_{i-1} = 0$ *for* $i = 0, 1, \cdots, n-1$ *(*e_{-1} *assumed* 0*).*
Method: $g_{-1} := b_{-1} := 0;\ b_n := b_{n-1};$
for $i := 0$ **to** n **do**
$g_i := \overline{g_{i-1}}(b_i \oplus b_{i-1});$
$e_i := (1 - 2b_{i+1})g_i$
end

We shall not formally prove the validity of this algorithm, just provide a sketch. Whenever $b_i \oplus b_{i-1} = 0$ it follows that $g_i = 0$, and thus $e_i = 0$ during strings of identical bits. Initially, and after a string of zeroes, a non-zero bit generates a digit -1, and after a string of 1s a zero bit generates a digit 1, thus utilizing the rule that $011 \cdots 10$ represents the same value as $010 \cdots \bar{1}0$. Since the invariant $g_i g_{i-1} = 0$ obviously holds for $i = 0, \ldots, n$, it follows that $e_i e_{i-1} = 0$, hence the result is a canonical string (NAF) and thus optimal.

The right-to-left conversion algorithm is usually given by a decision table, but to be able to prove its correctness, it has here been converted to an equivalent algorithmic form, describing it with the help of carries.

Algorithm 2.4.14 (RL canonical recoding)
Stimulus: *A 2's complement n-bit integer polynomial* $P = [b_{n-1} \cdots b_1 b_0]_{2c}$.
Response: *The canonical* $(n+1)$*-digit polynomial* $Q = \sum_{i=0}^{n} e_i 2^i \in \mathcal{P}_I[2, \{-1, 0, 1\}]$ *with* $\|P\| = \|Q\|$, *such that* $e_i e_{i-1} = 0$ *for* $i = 0, 1, \cdots, n-1$ *(*e_{-1} *assumed* 0*).*
Method: $c_0 := 0;\ b_{n+1} := b_n := b_{n-1};$
for $i := 0$ **to** n **do**
$c_{i+1} := \lfloor (c_i + b_i + b_{i+1})/2 \rfloor;$
$e_i := c_i + b_i - 2c_{i+1}$
end;

Theorem 2.4.15 *Given a 2's complement radix polynomial* $P = \sum_{i=0}^{n-1} b_i[2]^i \in \mathcal{F}_{0n-1}^{rc}[2, C_2^{rc}]$, *Algorithm 2.4.14 correctly computes the canonical polynomial* $Q = \sum_{i=0}^{n} e_i[2]^i \in \mathcal{P}_I[2, \{-1, 0, 1\}]$ *with* $\|P\| = \|Q\|$, *satisfying* $e_i e_{i-1} = 0$ *for* $i = 0, 1, \ldots, n-1$ *(*e_{-1} *assumed* 0*).*

Proof From the expression for c_{i+1} we find by induction that $c_i \in \{0, 1\}$ since $c_0 = 0$ and $b_i \in \{0, 1\}$. It also follows that

$$b_i = b_{i+1} \Rightarrow c_{i+1} = \lfloor c_i/2 \rfloor + b_i = b_i \Rightarrow e_i = c_i - b_i,$$

and

$$b_i \neq b_{i+1} \Rightarrow b_i + b_{i+1} = 1 \Rightarrow c_{i+1} = \lfloor (c_i + 1)/2 \rfloor = c_i \Rightarrow e_i = b_i - c_i,$$

such that $e_i \in \{-1, 0, 1\}$. Then from the definitions of e_i and c_{i+1},

$$\begin{aligned} e_i \neq 0 \Rightarrow c_i + b_i &= 1 \\ \Rightarrow c_{i+1} &= \lfloor (1 + b_{i+1})/2 \rfloor = b_{i+1} \\ \Rightarrow e_{i+1} &= c_{i+1} + b_{i+1} - 2c_{i+2} \\ &= 2(c_{i+1} - c_{i+2}) \\ &= 0, \end{aligned}$$

since $e_{i+1} \in \{-1, 0, 1\}$. Hence $e_i e_{i+1} = 0$ and the polynomial is in NAF, thus canonical. To prove that $\|Q\|$ has the correct value consider

$$\begin{aligned} \|Q\| &= \sum_{i=0}^{n} e_i 2^i = \sum_{i=0}^{n} (c_i + b_i - 2c_{i+1}) 2^i \\ &= \sum_{i=0}^{n} b_i 2^i + \sum_{i=0}^{n} (c_i 2^i - c_{i+1} 2^{i+1}) \\ &= \sum_{i=0}^{n-1} b_i 2^i + b_n 2^n + c_0 - c_{n+1} 2^{n+1} \\ &= \sum_{i=0}^{n-1} b_i 2^i - b_n 2^n = \|P\|, \end{aligned}$$

since $c_0 = 0$ and $c_{n+1} = \lfloor (c_n + b_n + b_{n+1})/2 \rfloor = b_n$ by the sign extension $b_{n+1} = b_n$. □

Since the NAF is unique, it may be expected that it is not possible to perform the conversion in constant time. In fact, comparing the binary value $85 = 01010101_2$, which is, in NAF form, with $86 = 01010110_2 = 10\bar{1}0\bar{1}01\bar{1}0$, by generalization it is seen that a change of the least-significant bits can have an effect arbitrarily far to the left. Thus conversion into NAF must take logarithmic time.

Actually an $O(\log n)$-time parallel, as well as an $O(n \log n)$-time, left-to-right version of the conversion to canonical form have been discovered, both of which are fairly complicated. However, just requiring the output to be of minimal weight, and not necessarily canonical, the following linear time, left-to-right algorithm is surprisingly similar to Algorithm 2.4.14.

Algorithm 2.4.16 (LR optimal recoding)
Stimulus: *A 2's complement n-bit integer polynomial* $P = [b_{n-1} \cdots b_1 b_0]_{2c}$.
Response: *A minimal weight* $(n+1)$*-digit polynomial* $Q = \sum_{i=0}^{n} e_i 2^i \in \mathcal{P}_I[2, \{-1, 0, 1\}]$ *with* $\|P\| = \|Q\|$.
Method: $c_n := b_n := b_{n-1};\ b_{-1} := b_{-2} := 0;$
for $i := n$ **downto** 0 **do**

$c_{i-1} := \lfloor (c_i + b_{i-1} + b_{i-2})/2 \rfloor;$
$e_i := b_i + c_{i-1} - 2c_i$
end;

The proof of the correctness of this algorithm is fairly lengthy, so we will omit it. The algorithm may deliver sequences $\cdots 0110 \cdots$ or $\cdots 0\bar{1}\bar{1}0 \cdots$ which are of optimal (minimal) weight, but not canonical since they are not in NAF. But for all practical purposes this algorithm may be employed.

Problems and exercises

2.4.1 Try to form $E^{(k)}$ for $k = 1, 2 \ldots$ with the digit set $E = \{-1, 0, 1, 4\}$ for $\beta = 3$, and thus determine that for no value of k is it possible to obtain a decomposition of the form $\Sigma + C \subseteq E^{(k)}$, where Σ is complete and $C \neq \{0\}$.

2.4.2 Prove Observation 2.4.7.

2.4.3 Describe by means of a conversion diagram a possible conversion from 2's complement, "carry-save" into $\mathcal{P}[8, \{-4, -3, \ldots, 4\}]$ such that conversion can take place in parallel.

2.4.4 Perform the conversion of 101110, considered as a 2's complement (negative) number in string notation, into $\mathcal{P}[8, \{-4, -3, \ldots, 4\}]$ according to the scheme developed in the previous problem.

2.4.5 Determine sets F_k and C_k such that the contiguous digit set

$$D_k = \{u(\beta - 1) + r\beta^k, \ldots, v(\beta - 1) + s\beta^k\}$$

is converted into

$$E_k = \{u(\beta - 1) + r\beta^{k-1}, \ldots, v(\beta - 1) + s\beta^{k-1}\},$$

i.e., $D_k \subseteq F_k + \beta C_k$ and $F_k + C_k \subseteq E_k$.

2.4.6 Prove that that the sets $D_k^{(k)}$, $E^{(k)}$, Σ, and C in the proof of Corollary 2.4.9 satisfy the conditions of Theorem 2.4.2.

2.4.7 To prove that $e_i \in \{-1, 0, 1\}$ also holds for Algorithm 2.4.16, the ideas used in the proof of Theorem 2.4.15 do not work. Find an alternative proof for this case.

2.5 Implementing base and digit set conversions

In this section we shall show that there exist some remarkably simple base and digit set conversion algorithms, for the particular case where digits of a radix-2 system are grouped, forming digits in a maximally redundant digit set, and then converted to form digits of a minimally redundant radix-2^k system. As we will see in Chapter 4 such conversions are very useful in the implementation of multiplication. We shall first analyze these conversions at the digit level using three very simple radix-2 digit set conversions, and then later show how these can be implemented in digital logic, employing suitable binary encodings of the digit values.

Definition 2.5.1 *The* P-mapping *is the digit set conversion for* $\beta = 2$, *mapping digits from* $D = \{-1, 0, 1\}$ *onto* D *itself, employing carries from* $C = \{0, 1\}$, *defined by*

$$\alpha_P : C \times D \to C \times D,$$

where

	α_P	D: -1	0	1
C	0	$0\bar{1}$	00	$1\bar{1}$
	1	00	01	10

(2.5.1)

with the outgoing carry being independent of the incoming carry.

Similarly we make the following definition.

Definition 2.5.2 *The* N-mapping *is the digit set conversion for* $\beta = 2$, *mapping digits from* $D = \{-1, 0, 1\}$ *onto* D *itself, employing carries from* $C = \{-1, 0\}$, *defined by*

$$\alpha_N : C \times D \to C \times D,$$

where

	α_N	D: -1	0	1
C	-1	$\bar{1}0$	$0\bar{1}$	00
	0	$\bar{1}1$	00	01

(2.5.2)

with the outgoing carry being independent of the incoming carry.

Furthermore, let $\hat{\alpha}_P$ and $\hat{\alpha}_N$ be generalized mappings where α_P, respectively α_N, is applied in parallel to all positions of a radix polynomial.

Intuitively the P-mapping moves a non-negative carry forward and leaves behind -1 or 0 (first row of the table in (2.5.2)), where of course -1 may be incremented into 0, and 0 incremented into 1 by an incoming carry (second row). Similarly the N-mapping moves a non-positive carry forward. As in the previous section, the mappings may be pictured as two-level processes performed in parallel on all digits of a radix polynomial. For example, the P-mapping may be pictured by the following conversion diagram:

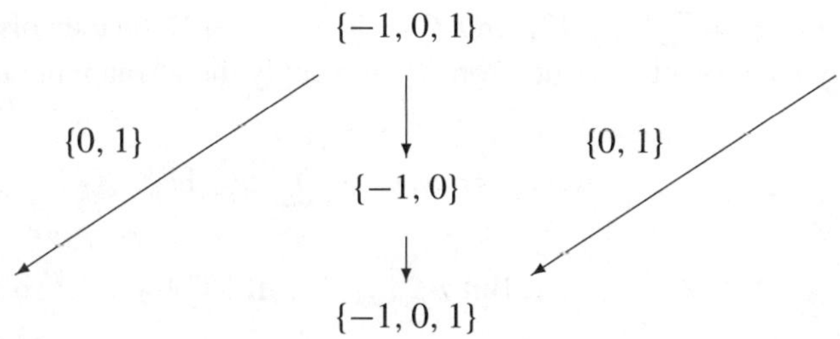

Example 2.5.1 Let us apply the P-mapping to the digit string $\bar{1}100111_{[2]}$ and write it as a two-level conversion, i.e., the ce-pairs from the first row of the table ($c_{in} = 0$) are split and displayed in the proper position, and then the rows are added without further carries:

		$\bar{1}$	1	0	0	1	1	1
		$\bar{1}$	$\bar{1}$	0	0	$\bar{1}$	$\bar{1}$	$\bar{1}$
	0	1	0	0	1	1	1	
$\hat{\alpha}_P$	0	0	$\bar{1}$	0	1	0	0	$\bar{1}$

The value of the polynomial represented by the string is -25, which is also the value of the string to the right of the dotted line, when interpreted as a 2's complement number. Thus it appears that we can apply the P-mapping to a 2's complement number and obtain the same value represented in the digit set $\{-1, 0, 1\}$ by just discarding the outgoing carry.

Now consider any string of $k \geq 2$ consecutive digits from the original 2's complement string. These can be interpreted as a digit in a radix-2^k system with a value in the digit set $\{0, 1, \ldots, 2^k - 1\}$. Then consider the same k positions in the converted string, and it appears that these can be interpreted as a digit in the minimally redundant set $\{-2^{k-1}, \ldots, 0, \ldots, 2^{k-1}\}$. □

The observation of the example is no coincidence as we shall show below, thus we have effectively converted a 2's complement number into a minimally redundant radix-2^k system, for any suitable $k \geq 2$ (note that no conversion is necessary for $k = 1$).

Lemma 2.5.3 *Let* $P \in \mathcal{P}_{On-1}[2, \{0, 1\}]$ *and* $Q = \hat{\alpha}_P(P) \in \mathcal{P}_{On}[2, \{-1, 0, 1\}]$. *With* $Q = \sum_{i=0}^{kn-1} q_i [2]^i$ *for* $k \geq 2$ *let*

$$U_j = \sum_{i=0}^{k-1} q_{j+i} [2]^i$$

be formed by grouping any k consecutive digits from Q. Then the value $\|U_j\|$ satisfies

$$-2^{k-1} \leq \|U_j\| \leq 2^{k-1}.$$

Proof Let $P = \sum_{i=0}^{n-1} p_i[2]^i$ and $T_j = \sum_{i=0}^{k-1} p_{j+i}[2]^i$ be the polynomial formed from any k consecutive digits from P, and apply the P-mapping to T_j, obtaining

$$\hat{\alpha}_P(T_j) = c_k[2]^k + \sum_{i=0}^{k-1} p'_{j+i}[2]^i,$$

where $c_k = 1$ iff $p_{j+k-1} = 1$. But $p_{j+k-1} = 1$ iff $\|T_j\| \geq 2^{k-1}$ so from

$$0 \leq \|\hat{\alpha}_P(T_j)\| = \|T_j\| \leq 2^k - 1$$

we obtain

$$-2^{k-1} \leq \|\hat{\alpha}_P(T_j) - c_k[2]^k\| = \left\|\sum_{i=0}^{k-1} p'_{j+i}[2]^i\right\| \leq 2^{k-1} - 1$$

and since any carry coming into position j is in $\{0, 1\}$, the lemma has been proved. □

Theorem 2.5.4 *Let $P \in \mathcal{F}^{rc}_{0,n-1}[2, C^c_2]$ be a 2's complement polynomial with $n = mk$ for some $m \geq 1, k \geq 2$, and let*

$$Q = \sum_{i=0}^{n-1} q_i[2]^i = \hat{\alpha}_P(P \bmod [2]^n) \bmod [2]^n$$

be the result of the P-mapping applied to the non-negative part of P, discarding any outgoing carry of the conversion. Then by grouping the digits q_i of Q, $q_i \in \{-1, 0, 1\}$ in groups of k, a radix-2^k polynomial $Q' \in \mathcal{P}_{0m-1}[2^k, \{-2^{k-1}, \ldots, 0, \ldots, 2^{k-1}\}]$ is obtained, with $\|P\| = \|Q'\|$ and digits in the minimally redundant digit set.

Proof This follows immediately from the previous lemma. □

Obviously, the above result also applies to standard, non-negative binary numbers. Also, sign-magnitude binary numbers can easily be converted into a minimally redundant radix-2^k digit set. In the case that the sign is positive, just apply the P-conversion to the magnitude, and by symmetry if it is negative apply the N-conversion to the string of negated digits.

Observation 2.5.5 *A binary sign-magnitude number can be converted into the minimally redundant radix-2^k representation by applying conditionally on the sign either the P-mapping on the digits of the magnitude part or the N-mapping on the negated digits, followed by combining k-digit groups into radix-2^k digits.*

We have now established very simple conversions from non-redundant binary representations into radix-2^k, minimally redundant representations, noting that these are $O(1)$ truly parallel conversions, i.e., independent of k.

Our next goal is to find simple conversions from redundant binary representations into radix-2^k, minimally redundant representations. It can be shown that this can be done using a $O(\log k)$ time parallel process, but as k is usually fairly small we will here be satisfied with $O(k)$ time conversions.

So let us start by finding an algorithm for converting a number in "borrow-save" binary representation, i.e., from $\mathcal{P}[2, \{-1, 0, 1\}]$ into $\mathcal{P}[2^k, \{-2^{k-1}, \ldots, 0, \ldots, 2^{k-1}\}]$. For this purpose we will now look at the N-mapping, and the effect of repeated applications of it.

Example 2.5.2 We want in particular to see what happens to digits of value -1:

$$
\begin{array}{c|cc:ccccccccc}
 & & & \bar{1} & 0 & 0 & \bar{1} & 0 & 1 & 0 & \bar{1} & \bar{1} \\
\cline{2-12}
 & & & 1 & 0 & 0 & 1 & 0 & 1 & 0 & 1 & 1 \\
\hat{\alpha}_N & & \bar{1} & 0 & 0 & \bar{1} & 0 & 0 & 0 & \bar{1} & \bar{1} & \\
\cline{2-12}
 & & \bar{1} & 1 & 0 & \bar{1} & 1 & 0 & 1 & \bar{1} & 0 & 1 \\
\cline{2-12}
 & & 1 & 1 & 0 & 1 & 1 & 0 & 1 & 1 & 0 & 1 \\
\hat{\alpha}_N & \bar{1} & 0 & 0 & \bar{1} & 0 & 0 & 0 & \bar{1} & 0 & 0 & \\
\cline{2-12}
 & \bar{1} & 1 & 1 & \bar{1} & 1 & 1 & 0 & 0 & 1 & 0 & 1
\end{array}
$$

Note how the value -1 moves left over strings of zeroes until it is possibly wiped out by hitting a digit of value 1. A -1 leaves behind it a 0 or 1 in the first mapping, and then only isolated digits of value -1 exist, so that further mappings of -1 leave digits of value 1 behind. In the worst case it might require n steps to eliminate all internal digits of value -1 from an n-digit number (the number containing all zeroes except for a -1 in the rightmost position). However, this would convert the number into a non-redundant representation, and that is not our purpose here. Also note that the value of the string to the left of the vertical dashed line remains -1, even after further applications of the N-mapping. □

We can summarize these observations without proof, as they follow trivially from the definition of α_N.

Observation 2.5.6 *By $k \geq 2$ repeated applications of the N-mapping $\hat{\alpha}_N$ to the polynomial $P = \sum_{i=0}^{n-1} d_i [2]^i \in \mathcal{P}[2, \{-1, 0, 1\}]$, a polynomial*

$$
\hat{\alpha}_N^k(P) = \sum_{i=0}^{n+k-1} d_i^{(k)} [2]^i \tag{2.5.3}
$$

is obtained, wherein any digit -1 is followed by at least $k-1$ ones, i.e.,

$$
d_i^{(k)} = -1 \Rightarrow d_{i-j}^{(k)} = 1 \quad \textit{for} \quad j = 1, 2, \ldots, k-1.
$$

In particular, if $P = -1[2]^n + \sum_{i=0}^{n-1} d_i [2]^i$, then

$$
\|\hat{\alpha}_N^k(P)\| = -2^n + \left\| \sum_{i=0}^{n-1} d_i^{(k)} [2]^i \right\|.
$$

Now look at any *k-segment* polynomial formed by extracting a segment of k consecutive digits from $\hat{\alpha}^k_N(P)$. Applying the observation we immediately obtain the following lemma.

Lemma 2.5.7 *The value of any k-segment polynomial* $P_j^{(k)} = \sum_{i=0}^{k-1} d_{i+j}^{(k)}[2]^i$ *extracted from* $\hat{\alpha}^k_N(P) = \sum_{i=0}^{n+k-1} d_i^{(k)}[2]^i$*, obtained by* $k \geq 2$ *repeated applications of the N-mapping to a polynomial* $P \in \mathcal{P}[2, \{-1, 0, 1\}]$*, satisfies*

$$-1 \leq \|P_j^{(k)}\| \leq 2^k - 1.$$

We have now reduced the range of values of such k-segments, but we want to shift the range to obtain minimally redundant digits of a radix-2^k system. As in Lemma 2.5.3 we can achieve such a shift by applying the P-mapping. Let

$$\widetilde{P}_j^{(k)} = \sum_{i=0}^{k-1} \tilde{d}_{j+i}^{(k)}[2]^i$$

be a k-segment of the polynomial $\hat{\alpha}_P(\hat{\alpha}^k_N(P))$, corresponding to $P_j^{(k)}$ of Lemma 2.5.7. Then we may express a relation between the values of these k-segments by

$$\|\widetilde{P}_j^{(k)}\| = -c_{out}2^k + \|P_j^{(k)}\| + c_{in}, \tag{2.5.4}$$

where $c_{in} \in \{0, 1\}$ and $c_{out} \in \{0, 1\}$ with $c_{out} = 1$ iff $d_{i+k-1}^{(k)} = 1$. To find bounds on $\|\widetilde{P}_j^{(k)}\|$ we split into three cases, depending on the value of $d_{j+k-1}^{(k)}$:

$d_{j+k-1}^{(k)} = -1$ By Observation 2.5.6 it follows that $d_{j+i}^{(k)} = 1$ for $i = 0, 1, \ldots, k-2$, hence $\|\widetilde{P}_j^{(k)}\| = \|P_j^{(k)}\| + c_{in} = -1 + c_{in} \in \{-1, 0\}$.

$d_{j+k-1}^{(k)} = 0$ Here $-1 \leq \|P_j^{(k)}\| \leq 2^{k-1} - 1$, hence $-1 \leq \|\widetilde{P}_j^{(k)}\| - c_{in} \leq 2^{k-1} - 1$. If $-1 = \|P_j^{(k)}\|$, then necessarily $P_j^{(k)}$ must have the form $0^i\bar{1}1^{k-i-1}$ in string notation for some $i > 0$. Hence $c_{in} = 1$ since the digit to the right of this k-segment, $d_{j-1}^{(k)} = 1$. Thus here $0 \leq \|\widetilde{P}_j^{(k)}\| \leq 2^{k-1}$.

$d_{j+k-1}^{(k)} = 1$ Here we have $2^{k-1} - 1 \leq \|P_j^{(k)}\| \leq 2^k - 1$ and hence $-2^{k-1} - 1 \leq \|\widetilde{P}_j^{(k)}\| - c_{in} \leq -1$. As in the previous case for the lower bound, we find that $c_{in} = 1$, thus for this case we have $-2^{k-1} \leq \|\widetilde{P}_j^{(k)}\| \leq 0$.

Observe the following to be used later:

$$d_{j+k-1}^{(k)} = 0 \text{ implies } \|\widetilde{P}_j^{(k)}\| \geq 0 \text{ and } d_{j+k-1}^{(k)} \neq 0 \text{ implies } \|\widetilde{P}_j^{(k)}\| \leq 0. \tag{2.5.5}$$

Combining the cases we have now proven that any k-segment from $\hat{\alpha}_P(\hat{\alpha}^k_N(P))$ has a value satisfying the bounds for a minimally redundant digit set, radix-2^k.

Theorem 2.5.8 *Let* $P \in \mathcal{P}_{0,n-1}[2, \{-1, 0, 1\}]$ *be a borrow-save polynomial with* $n = mk$ *for some* $m \geq 1, k \geq 2$*, and let*

$$Q = \sum_{i=0}^{n+k-1} \tilde{d}_i [2]^i = \hat{\alpha}_P(\hat{\alpha}_N^k(P))$$

be the result of k applications of the N-mapping followed by a single P-mapping. Then by grouping the digits $\tilde{d}_i$ *of* Q *in groups of k, an* $(m+1)$*-digit radix-*2^k *polynomial*

$$Q' = \sum_{i=0}^{m} q_i [2^k]^i \in \mathcal{P}_{0m}[2^k, \{-2^{k-1}, \ldots, 0, \ldots, 2^{k-1}\}]$$

is obtained, with $\|P\| = \|Q'\|$ *and digits in the minimally redundant digit set. Furthermore, the value of the most-significant digit* q_m *is restricted such that* $q_m \in \{-1, 0, 1\}$.

Observe by symmetry that we may also first apply the P-mapping k times followed by a single N-mapping, to achieve the same effect (but not the same resulting polynomials).

A short remark on the complexity is appropriate here. Since the process above is an $O(k)$ time parallel conversion, it is only useful for moderate values of k. By just grouping k radix-2 digits from $\{-1, 0, 1\}$, we obtain radix-2^k digits from the maximally redundant set $\{-2^k + 1, \ldots, 0, \ldots, 2^k - 1\}$, and the process above reduces the redundancy index from $2^k - 1$ to 1. However, the value of such a radix-2^k digit can be obtained in non-redundant, 2's complement form just by subtracting the k-digit polynomial composed of the negative digits from the corresponding k-digit polynomial consisting of the positive digits. As we shall see in the next chapter this subtraction can be performed in time $O(\log k)$. With the result of the subtraction as a $(k+1)$-digit, 2's complement number, the leading digit being in $\{-1, 0\}$ can be added into the converted k-digit group to the left by a trivial modification of the least-significant digit of that group, such that the result is in $\mathcal{P}[2, \{-1, 0, 1\}]$. It is then easy to see that a single P-mapping will be sufficient to bring the k segments into the minimally redundant digit set (exercise).

Let us now look at the conversion from 2's complement, carry-save into radix-2^k, minimally redundant. Recall that the source digit set here is $\{-2, -1, 0, 1, 2\}$ with negative digits only permitted in the most-significant position. Here we need one further trivial conversion.

Definition 2.5.9 *The* Q-mapping *is the digit set conversion for* $\beta = 2$*, mapping digits from* $D = \{-2, -1, 0, 1, 2\}$ *into* $D' = \{-1, 0, 1\}$*, employing carries from* $C = \{0, 1\}$*, defined by*

$$\alpha_Q : C \times D \to C \times D',$$

where

		D				
	α_Q	-2	-1	0	1	2
C	0	na	na	00	$1\bar{1}$	10
	1	$0\bar{1}$	00	01	10	11

(2.5.6)

with the outgoing carry being independent of the incoming carry.

The restrictions on the use of negative digits in a 2's complement polynomial are such that in $P = \sum_{i=0}^{n} d_i[2]^i$, the leading digit d_n satisfies $d_n = -d_{n-1}$, whereas $d_i \geq 0$ for $i = 0, 1, \ldots, n-1$. Thus $d_n < 0$ implies that the carry into this position will always be 1, and the mapping for $c = 0$ need not be defined. Also the transformed polynomial $\hat{\alpha}_Q(P) \in \mathcal{P}_{0n}[2, \{-1, 0, 1\}]$, since the carry-out of position n is always zero.

The next observation we can make from the definition of the Q-mapping is that in $\hat{\alpha}_Q(P) = \sum_{i=0}^{n} d_i'[2]^i$, the digit value -1 always appears isolated, $d_i' = -1 \Rightarrow d_{i-1}' \in \{0, 1\}$. This is the same property as found for the N-mapping, which implies that if we subsequently apply the N-mapping $k-1$ times, the resulting polynomial

$$\hat{\alpha}_N^{k-1}(\hat{\alpha}_Q(P)) = \sum_{i=0}^{n+k-1} d_i^{(k)}[2]^i$$

satisfies the same condition:

$$d_i^{(k)} = -1 \Rightarrow d_{i-j}^{(k)} = 1 \quad \text{for} \quad j = 1, 2, \cdots, k-1$$

as if we had applied the N-mapping k times to a polynomial from $\mathcal{P}[2, \{-1, 0, 1\}]$. Thus we immediately obtain the following theorem in analogy with Theorem 2.5.8.

Theorem 2.5.10 *Let $P \in \mathcal{F}_{0,n}^{cs}[2, \{-2, -1, 0, 1, 2\}]$ be a 2's complement carry-save polynomial with $n = mk$ for some $m \geq 1, k \geq 2$, and let*

$$Q = \sum_{i=0}^{n+k-1} \tilde{d}_i[2]^i = \hat{\alpha}_P(\hat{\alpha}_N^{k-1}(\hat{\alpha}_Q(P)))$$

be the result of a Q-mapping, followed by $k-1$ N-mappings, followed by a P-mapping. Then by grouping the digits $\tilde{d}_i$ of Q in groups of k, an $(m+1)$-digit radix-2^k polynomial

$$Q' = \sum_{i=0}^{m} q_i[2^k]^i \in \mathcal{P}_{0m}[2^k, \{-2^{k-1}, \ldots, 0, \ldots, 2^{k-1}\}]$$

is obtained, with $\|P\| = \|Q'\|$ and digits in the minimally redundant digit set. Furthermore, the value of the most-significant digit q_m is restricted such that $q_m \in \{-1, 0, 1\}$.

Theorems 2.5.8 and 2.5.10 describe how k-segments of borrow-save or carry-save binary numbers can be reduced from the range of $[-2^k+1\,..\,2^k-1]$, respectively $[0\,..\,2^{k+1}-2]$, to the minimally redundant range $[-2^{k-1}\,..\,2^{k-1}]$ by appropriate PN^k or $PN^{k-1}Q$-mappings for $k \geq 2$. Actually, as is easily seen, they also hold for $k = 1$.

Consider now the range of the "tail" of a binary borrow-save or carry-save represented number, i.e., the fractional parts of a polynomial of the form $P = \sum_{i=-jk}^{mk-1} p_i[2]^i$ for $m \geq 0$ and $k, j \geq 1$. Applying these mappings moves carries into the integer part, thus reducing the range of the fractional part and hence the error when the number is truncated to its integer part. Note that truncation of a number over a digit set of the form $[-2^{k-1}\,..\,2^{k-1}]$ corresponds to round-to-nearest.

Observation 2.5.11 *By applying the PN^k or $PN^{k-1}Q$-mappings to a borrow-save, respectively a carry-save, represented polynomial $P = \sum_{i=-jk}^{mk-1} p_i[2]^i$ for $m \geq 0$ and $k, j \geq 1$, the transformed polynomial $Q = \sum_{i=-jk}^{mk+k-1} q_i[2]^i$ with $q_i \in \{-1, 0, 1\}$ has a fractional part with a range satisfying*

$$\Big|\|Q - \lfloor Q \rfloor\|\Big| \leq \frac{1}{2}(1 + 2^{-k} + \cdots + 2^{-(j-1)k}) < \frac{1}{2} \cdot \frac{1}{1-2^{-k}} = \frac{2^{k-1}}{2^k - 1}.$$

2.5.1 Implementation in logic

In Section 1.7 we introduced the names "carry-save" and "borrow-save" for the digit sets $\{0, 1, 2\}$ and $\{-1, 0, 1\}$ respectively, noting that these names are actually derived from particular binary encodings of the digit values, employing the standard two-state (bit) values usually denoted by the symbols 0 and 1.

Since both digit sets contain three values, two bits are needed to encode the values distinctively, leaving one pair of bit values unnecessary (or redundant). Although it is possible to rule out one such pair as illegal, it is often advantageous to allow one of the digit values to have two binary encodings, as this may simplify the logic needed for the implementation of operations on digit strings.

The name *"carry-save"* derives from the *encoding*:

$$\begin{array}{l} 00 \quad \sim 0, \\ \left.\begin{array}{l} 01 \\ 10 \end{array}\right\} \sim 1, \\ 11 \quad \sim 2, \end{array} \tag{2.5.7}$$

where one of the bits may be thought of as a "place digit" ($p \in \{0, 1\}$), whereas the other is a "carry" ($c \in \{0, 1\}$) brought in from the right-hand neighboring position

$$\begin{array}{ccc} \cdots & p & \cdots \\ \cdots & c \swarrow & \cdots \end{array},$$

which just has not been added in (hence "carry-save"). The digit value is thus the sum of the two bit values, when these are interpreted as integers. Note that given two bit-strings placed next to one another

$$\begin{array}{cccccc} a_{n-1} & a_{n-2} & \cdots & a_1 & a_0 & a_i \in \{0,1\} \\ b_{n-1} & b_{n-2} & \cdots & b_1 & b_0 & b_i \in \{0,1\} \end{array}$$

these provide a carry-save encoding of the sum of two numbers A and B, by just pairing the bits a_i, b_i, where A and B are given by the bit strings as ordinary binary (or 2's complement binary) numbers. Observe that this also applies to position $n-1$ in the case that the two bit-strings are to be interpreted in 2's complement, thus providing the 2's complement carry-save representation of $A + B$.

Similarly *"borrow-save"* derives its name from the following *encoding* of the digit set $\{-1, 0, 1\}$:

$$\begin{array}{ll} 10 & \sim -1, \\ \left.\begin{array}{l} 11 \\ 00 \end{array}\right\} & \sim 0, \\ 01 & \sim 1, \end{array} \tag{2.5.8}$$

where the rightmost bit of the pair may be thought of as a "place" value ($p \in \{0, 1\}$) and the leftmost bit of negative weight is a "borrow" value ($b \in \{-1, 0\}$) brought in from the right-hand neighboring position:

$$\begin{array}{ccccc} \cdots & p & \swarrow & \cdots & (+) \\ \cdots & b & & \cdots & (-) \end{array}.$$

Again given two bit-strings placed next to one another, this provides an encoding in "borrow-save" of the difference between the two numbers, given as ordinary binary numbers. However, if the two strings represent 2's complement values, then the leading bits of the strings have to be swapped, to compensate for this position having negative weight.

Besides ruling out, say, the encoding 11 of the digit value 0, there are other useful encodings of the digit set $\{-1, 0, 1\}$, e.g.,

$$\begin{array}{ll} 11 & \sim -1, \\ \left.\begin{array}{l} 10 \\ 00 \end{array}\right\} & \sim 0, \\ 01 & \sim 1, \end{array}$$

which has been denoted the *sign-magnitude* or *signed-digit* encoding. Again here the encoding 10 could be ruled out to provide a *non-redundant encoding* of the redundant digit set $\{-1, 0, 1\}$.

Having introduced these encodings of the digit sets, let us turn our attention to possible logic implementations of the P-, N-, and Q-mappings when applied to strings of encoded digit values. So let

$$\begin{array}{cccccl} p_{n-1} & p_{n-2} & \cdots & p_1 & p_0 & (+) \\ b_{n-1} & b_{n-2} & \cdots & b_1 & b_0 & (-) \end{array} \tag{2.5.9}$$

be the borrow-save encoding of a string of digits $d_{n-1}d_{n-2}\cdots d_1d_0$, each pair (p_i, b_i) being the encoding of a digit $d_i \in \{-1, 0, 1\}$, p_i having positive weight and b_i negative weight. Notice that the value of the digit d_i can be obtained as $d_i = p_i - b_i$.

Then the P-mapping applied to the digit string encoded as (2.5.9) yields a digit string with borrow-save encoding

$$\begin{array}{ccccccl} p'_n & p'_{n-1} & p'_{n-2} & \cdots & p'_1 & 0 & (+) \\ 0 & b'_{n-1} & b'_{n-2} & \cdots & b'_1 & b'_0 & (-) \end{array}, \tag{2.5.10}$$

where it is seen from the conversion table (2.5.1) and (2.5.8) that

$$\textit{P-mapping} \qquad p'_{i+1} = p_i\bar{b}_i \quad \text{and} \quad b'_i = p_i \oplus b_i, \tag{2.5.11}$$

where $\oplus$ is the exclusive OR-function, XOR. The mapping (2.5.1) first rewrites a digit into a carry and a place digit, and then adds these. Here the carry-out ($p'_{i+1} = 1$ iff $d_i = 1$) is represented as a bit of positive weight moved over one position, leaving behind a place digit as a bit of negative weight ($b'_i = 1$ iff $d_i = \pm 1$). No addition has to be performed, as this is implicit in the interpretation of the bit vectors including their weights. Here the fact that zero has two encodings in exploited. Observe the correlation that $p'_{i+1} = 1$ implies that $b'_i = 1$.

Similarly we obtain by symmetry that the N-mapping of (2.5.9) yields a digit string with borrow-save encoding:

$$\begin{array}{ccccccl} 0 & p'_{n-1} & p'_{n-2} & \cdots & p'_1 & p'_0 & (+) \\ b'_n & b'_{n-1} & b'_{n-2} & \cdots & b'_1 & 0 & (-) \end{array},$$

where

$$\textit{N-mapping} \qquad p'_i = p_i \oplus b_i \quad \text{and} \quad b'_{i+1} = \bar{p}_i b_i. \tag{2.5.12}$$

These mappings are remarkably simple, requiring only a single level of logic. It turns out that when using the sign-magnitude (signed-digit) encoding, two levels of logic are necessary. For example, for the P-mapping with $d_i \sim (s_i, m_i)$ and $d'_i = \alpha_P(d_i) \sim (s'_i, m'_i)$,

$$m'_i = m_i \oplus (m_{i-1}\bar{s}_{i-1}) \quad \text{and} \quad s'_i = m_i,$$

where $m_{i-1}\bar{s}_{i-1}$ is the carry-in.

Finally for the Q-mapping, with input

$$\begin{array}{cccccl} p_{n-1} & p_{n-2} & \cdots & p_1 & p_0 & (+) \\ c_{n-1} & c_{n-2} & \cdots & c_1 & c_0 & (+) \end{array}$$

in carry-save encoding (but only the positions with non-negative digits), and output in borrow-save encoding:

$$\begin{array}{ccccccc} p'_n & p'_{n-1} & p'_{n-2} & \cdots & p'_1 & 0 & (+) \\ b'_n & b'_{n-1} & b'_{n-2} & \cdots & b'_1 & b'_0 & (-) \end{array},$$

we find that

$$\textit{Q-mapping} \qquad \begin{aligned} p'_{i+1} &= \begin{cases} p_i + c_i & \text{for } 0 \le i \le n-2 \\ p_i \oplus c_i & \text{for } i = n-1 \end{cases}, \\ b'_i &= \begin{cases} p_i \oplus c_i & \text{for } 0 \le i \le n-1 \\ p_{i-1} + c_{i-1} & \text{for } i = n \end{cases}. \end{aligned} \tag{2.5.13}$$

Thus here also we have a simple one-level logic implementation of the mapping. Note that the roles of the bits have just been interchanged in position n, corresponding to a negation of the digit value constructed from the digit in position $n - 1$.

There is an interesting trick obtainable by combining the result of a P-mapping as encoded by (2.5.11) in borrow-save, and the results of converting borrow-save or carry-save polynomials as expressed in Theorems 2.5.8 and 2.5.10. Both of these use a P-mapping as the last step in the conversion, and by (2.5.5) in combination with $b'_i = p_i \oplus b_i$ of (2.5.11) we obtain a simple way of determining the signs of the converted radix-2^k digits.

Observation 2.5.12 *If a borrow-save or a 2's complement carry-save polynomial by Theorem 2.5.8, respectively Theorem 2.5.10, is converted into the radix-2^k, minimally redundant representation, where each digit q_j is obtained by grouping k converted radix-2 digits:*

$$q_j = \sum_{i=0}^{k-1} \tilde{d}_{jk+i} 2^i,$$

then sign information on q_j can be obtained from the borrow-save encoding of $\tilde{d}_{jk+k-1}$ by

$$\begin{aligned} b'_{jk+k-1} &= 0 \textit{ implies } q_j \ge 0, \\ b'_{jk+k-1} &= 1 \textit{ implies } q_j \le 0, \end{aligned} \tag{2.5.14}$$

provided that the last P-mapping employs (2.5.11) to yield the encoding (p'_{jk+i}, b'_{jk+i}) of the digits $\tilde{d}_{jk+i}$.

Let us finally return to the particularly simple case of applying the P-mapping to a 2's complement number, as used in Theorem 2.5.4. With the trivial encoding of the digits (bits) b_i in $\{0, 1\}$, we observe that a 2's complement number $[b_{n-1}b_{n-2}\cdots b_1b_0]_{[2c]}$ can be represented in borrow-save encoding, with b_{n-1}

interpreted with negative weight as

$$\begin{array}{cccccc} 0 & b_{n-2} & \cdots & b_1 & b_0 & (+) \\ b_{n-1} & 0 & \cdots & 0 & 0 & (-) \end{array}. \tag{2.5.15}$$

Hence according to (2.5.10) the P-mapping of the number is in borrow-save encoding:

$$\begin{array}{cccccc} b_{n-1} & b_{n-2} & b_{n-3} & \cdots & b_0 & 0 & (+) \\ b_n & b_{n-1} & b_{n-2} & \cdots & b_1 & b_0 & (-) \end{array} \tag{2.5.16}$$

with $b_n = b_{n-1}$ by sign extension. We then conclude with the following results on the classical Booth Recodings.

Observation 2.5.13 (Booth Recoding) *A 2's complement number* $[b_{n-1}b_{n-2}\cdots b_1b_0]_{[2c]}$ *can be converted by the P-mapping without any logic transformation to the borrow-save representation* $[e_{n-1}e_{n-2}\cdots e_1e_0]_{[2]}$, *where the digit e_i of the converted number is encoded in borrow-save as the pair b_i, b_{i-1}, where $e_i = b_{i-1} - b_i$ with $b_{-1} = 0$.*

Note that the digit e_n is not needed to represent the range of a converted n-digit 2's complement number (its value is always zero).

By Theorem 2.5.4, strings obtained by grouping k of these converted radix-2 digits provide a representation of the form $[e_{jk+k-1}\cdots e_{jk}]_2$ of a radix-2^k digit in the minimally redundant digit set $\{-2^{k-1}, \ldots, 0, \ldots, 2^{k-1}\}$. Considering the encoding of the digits e_i and rewriting the value of a k-digit group

$$\begin{aligned} \sum_{i=0}^{k-1} e_{jk+i}\, 2^i &= \sum_{i=0}^{k-1} (b_{jk+i-1} - b_{jk+i})\, 2^i \\ &= -b_{jk+k-1}\, 2^{k-1} + \sum_{i=0}^{k-2} b_{jk+i}\, 2^i + b_{jk-1}, \end{aligned}$$

we obtain the following result.

Observation 2.5.14 (Modified Booth Recoding) *Using P-mapping, an $n = mk$ bit 2's complement number* $[b_{n-1}b_{n-2}\cdots b_1b_0]_{[2c]}$ *can be converted without any logic transformation to the minimally redundant radix-2^k digit set* $\{-2^{k-1}, \ldots, 0, \ldots, 2^{k-1}\}$, *where the jth digit is encoded by the following string of $k+1$ bits: $b_{jk+k-1}\cdots b_{jk}b_{jk-1}$, for $j = 0, 1, \ldots, m-1$ with $b_{-1} = 0$. The value of the jth digit is then* $-b_{jk+k-1}2^{k-1} + \sum_{i=0}^{k-2} b_{jk+i}\, 2^i + b_{jk-1}$.

Although an encoding of the digits of the converted number is obtained without any logic needed (just "wiring"), that particular encoding may not be the most suitable for a specific purpose, e.g., for use in an actual implementation of a multiplier where the encoding of a recoded multiplier digit is to be used to generate multiples of the multiplicand. We shall discuss other suitable encodings in Chapter 4.

2.5.2 On-line digit set conversion

We have seen how it is possible to convert a radix polynomial represented in one digit set into a polynomial over any other complete digit set by processing the digits one by one from the least-significant end to the most-significant end. Under suitable conditions for redundant target digit sets (e.g., a contiguous set), it is possible to perform the conversion in digit parallel. But when a conversion can be performed in digit parallel, it can also be performed sequentially, starting from the most-significant end, and progressing towards the least-significant end.

There are situations where it is natural to process values represented by radix polynomials in order starting from the most-significant end. Notice that these algorithms must necessarily make decisions based on incomplete knowledge, not knowing which digits and carries may come later. But as we know, this is precisely why certain conditions on the digit sets are necessary, to allow the conversion to depend only on a bounded context. Algorithms which have to make the best decisions on incomplete knowledge about what may happen in the future are generally called *on-line algorithms*; here we will use that term specifically for algorithms that are on-line with respect to generating digits from the most-significant end.

A closer look at the examples in Section 2.3 reveals that a "look-ahead" of k digits will allow the algorithm to determine the incoming carry, provided that k and the digit sets involved satisfy the conditions of Theorem 2.4.2. Thus the carry function γ can compute the incoming carry based on the look-ahead (a k-digit right context), and the digit-mapping function[4] σ can determine the digit value generated at the position, after the carry was emitted.

To simplify the description of the on-line conversion algorithm we will, however, assume that $k = 1$. Note that this does not limit the applicability of the algorithm, as we can just assume that the digit sets D and E have been formed by grouping k digits of the original digit sets, and that β is the kth power of the original radix that is, as it may have been necessary to satisfy the conditions, of Theorem 2.4.2.

Algorithm 2.5.15 (On-line digit set conversion)

Stimulus: *A radix β, $|\beta| \geq 2$, and digit sets D, E, Σ and $C \neq \{0\}$, satisfying $D \subseteq \Sigma + \beta C$ and $\Sigma + C \subseteq E$, where Σ contains a complete residue system modulo β.*

A conversion mapping α composed of a digit mapping function $\sigma : D \to \Sigma$ and carry function $\gamma : D \to C$, such that $d = \sigma(d) + \beta\gamma(d), \forall d \in D$.

A Polynomial $P = \sum_{i=0}^{n-1} d_i[\beta]^i \in \mathcal{P}[\beta, D]$.

[4] Actually the digit-mapping function is ξ, but in this case ξ is defined in terms of σ owing to (2.4.7).

Response: *A Polynomial* $Q = \sum_{i=0}^{n} e_i[\beta]^i \in \mathcal{P}[\beta, E]$ *satisfying* $\|P\| = \|Q\|$.
Method: $s_n := d_{-1} := 0;$
for $i := n$ **downto** 0 **do**
$e_i := s_i + \gamma(d_{i-1});$
$s_{i-1} := \sigma(d_{i-1});$
end;

Note that, as we have seen in Section 2.3, the algorithm actually bases it decision on two digits, the digit at the current position and here a one-digit right context, corresponding to the value $k = 1$. This look-ahead is called the *on-line delay*, which can be as little as $k = 0$, such as in the case where there would be no conversion at all, e.g., when $D \subseteq E$.

The algorithm above may be thought of as a transducer, which, based on an internal state s_i, and given an input d_{i-1}, goes to a new state $s_{i-1} = \sigma(d_{i-1})$ producing output $e_i = s_i + \gamma(d_{i-1})$, for $i = n, n-1, \ldots, 0$, as pictured below:

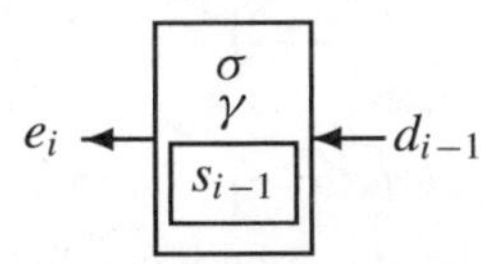

where we note that the state here is represented by a digit value, wheras in the transducers used in Section 2.3 the state represented a saved carry value. The difference is obviously caused by reversing the direction of the processing of the digits.

Instead of converting from radix β to radix β^k when grouping the right context together, it is of course possible to consider this context as k digits in radix β. Just perform the conversion as the result of a sequential composition of k transducers as in Figure 2.5.1, each of them of delay 1 and defined appropriately, so that they together perform the correct conversion.

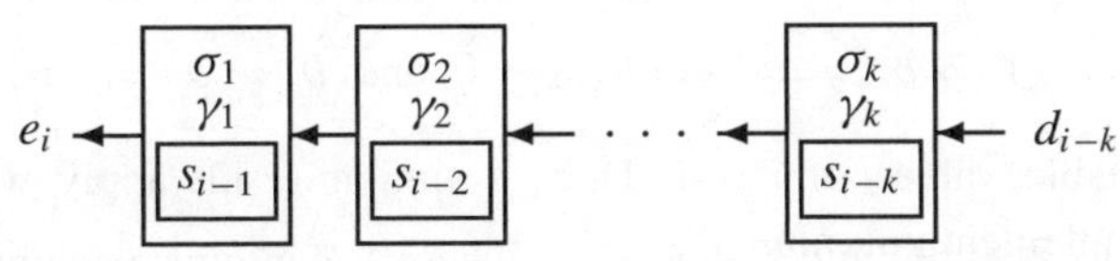

Figure 2.5.1. A composition of k transducers as a delay-k on-line converter.

Example 2.5.3 Let us consider the situation in which two radix polynomials in $\mathcal{P}[2, \{-1, 0, 1\}]$ are being added, digit serially, most-significant digit first, resulting in a polynomial in $\mathcal{P}[2, \{-2, -1, 0, 1, 2\}]$. The result is then to be converted back into the same digit set as the operands were received in, and in an on-line fashion. That is, we need a conversion as in one of the following two diagrams, since two steps are needed, and the on-line delay is thus 2:

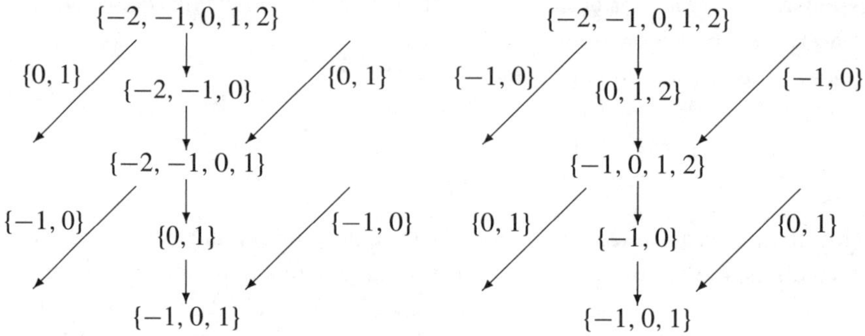

Note that the order of the two conversions here is immaterial. The corresponding diagram for a single conversion in radix 4 is:

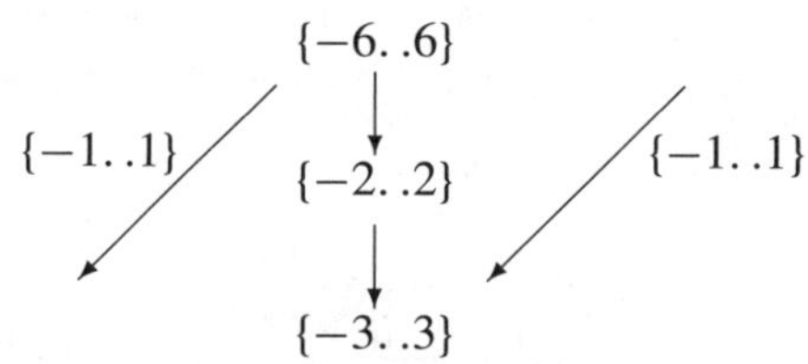

Note that the corresponding transducer here would have a delay of 1, but of course now the delay is measured in terms of digits from a larger digit set. □

In the next chapter we shall see how these conversions can be implemented by some fairly simple logic "building blocks."

Problems and exercises

2.5.1 Show that if $P \in \mathcal{P}_{0,n-1}[2, \{0, 1\}]$ and $Q = \hat{\alpha}_P(P) \in \mathcal{P}_{0,n}[2, \{-1, 0, 1\}]$, where $Q = \sum_{i=0}^{n} b_i [2]^i$, then the non-zero digits of Q alternate in sign:

$$b_i = 1 \Rightarrow b_{i-1} = \cdots = b_{j+1} = 0 \text{ and } b_j \neq 0 \;\Rightarrow\; b_j = -1,$$
$$b_i = -1 \Rightarrow b_{i-1} = \cdots = b_{j+1} = 0 \text{ and } b_j \neq 0 \;\Rightarrow\; b_j = 1$$

for suitable values of $i > j$. Does a similar result apply when P is a 2's complement polynomial?

2.5.2 As discussed on page 101 it is also possible to convert into minimally redundant radix 2^k by ordinary subtraction of k-bit segments with the result in 2's complement, possibly with a negative carry-out. Show how this carry can be absorbed in the neighboring segment.

2.5.3 Given $A = 100111_{[2c]}$ and $B = 000101_{[2c]}$ as 2's complement numbers, interpret their sum $A + B$ as a single number in the 2's complement carry-save representation and convert it into minimally redundant, radix-4 representation using Theorem 2.5.10.

2.5.4 With the same values as in the previous problem, interpret the difference $A - B$ as a borrow-save number, this time applying Theorem 2.5.8 for conversion into minimally redundant radix 4.

2.5.5 Observation 2.5.11 assumes that $k \geq 2$, but what is the effect of applying the PN-mapping to a borrow-save polynomial (i.e., for $k = 1$)?

2.5.6 A correlation was noticed in the binary realization of the P-mapping (2.5.11). Are there similar correlations for the N-mapping and Q-mapping?

2.5.7 Prove Observation 2.5.12.

2.6 The additive inverse

There have been occasional comments in the text on the problem of negating, or finding a representation of the additive inverse. We shall conclude this chapter with a more thorough treatment, exploiting the results of Sections 2.3 and 2.4. The problem can be formally stated as follows: given a radix polynomial $P \in \mathcal{P}[\beta, D]$, find (if possible) another radix polynomial $Q \in \mathcal{P}[\beta, D]$ such that $\|Q\| = -\|P\|$.

The problem obviously is trivial if the digit set D is symmetric, $d \in D \Longrightarrow -d \in D$, since then with $P = \sum_{i=\ell}^{m} d_i[\beta]^i$ the polynomial $Q = \sum_{i=\ell}^{m}(-d_i)[\beta]^i$ is the result wanted and can be obtained in constant time. Since such a symmetric digit set necessarily is of odd cardinality, only for an odd radix can such a system be non-redundant and complete.

In general, for the additive inverse to be representable we must assume that the digit set D is complete for the radix of the system in question. If by $\overline{D}$ we denote the *negated digit set*:

$$\overline{D} = \{-d \mid d \in D\},$$

we may initially note the following symmetries.

Lemma 2.6.1 *D is complete for radix β if and only if $\overline{D}$ is complete for radix β.*

Proof This follows trivially from:

$$\forall a \in \mathbb{Q}_{|\beta|}\ \exists P \in \mathcal{P}[\beta, D] : \|P\| = -a \Longrightarrow -P \in \mathcal{P}[\beta, \overline{D}]. \qquad \square$$

Lemma 2.6.2 *D is redundant (non-redundant) for radix β if and only if $\overline{D}$ is redundant (non-redundant) for radix β.*

Proof Trivial. $\square$

Since the problem of finding the additive inverse basically is a problem of digit set conversion (from $\overline{D}$ to D) we find that the effect of a change of a single digit can ripple an arbitrary distance to the left if the digit set is non-redundant. By applying Theorem 2.3.1 we immediately have the following corollary.

Corollary 2.6.3 *Let D be a digit set which is non-redundant and complete for radix β such that $D \neq \overline{D}$. Then for any $n > 0$ there exists a radix polynomial $P \in \mathcal{P}_I[\beta, D]$, digits $d_1, d_2 \in D$, and radix polynomials $Q_1, Q_2 \in \mathcal{P}_I[\beta, D]$ such that $-\|Q_1\| = \|P[\beta] + d_1\|$ and $-\|Q_2\| = \|P[\beta] + d_2\|$, where $d_n(Q_1) \neq d_n(Q_2)$.*

Similarly other results of Sections 2.3 and 2.4 apply directly to the problem of negating a radix polynomial, in particular note that for redundant and complete digit sets the additive inverse can be found in constant time, independent of operand length. Also observe that negation can take place most-significant digit first by the on-the-fly conversion of Theorem 2.3.3, and in logarithmic time when a suitable tree structure is applied to exploit parallelism in the non-redundant case.

Let us then turn to the peculiarities of the finite precision fixed-point systems. First note that negation in two of these systems is actually very simple.

Observation 2.6.4 *Sign-magnitude and the diminished radix-complement systems both allow trivial negation, in the latter case by substituting each digit by its complement with respect to the "diminished radix," $\beta - 1$. Both systems are completely symmetric and allow constant time negation, but both suffer from a non-unique representation of zero.*

The situation is more complicated in the fixed-point, radix-complement systems. By Observation 1.8.3 the range of values representable by polynomials in $\mathcal{F}^{rc}_{\ell m}[\beta, C^c_\beta]$ is

$$-\left(\beta - \left\lfloor \frac{\beta+1}{2} \right\rfloor\right) \beta^m \leq \|P\| \leq \left(\left\lfloor \frac{\beta+1}{2} \right\rfloor \beta^m - \beta^\ell\right),$$

so for β even there is one more negative value than there are positive values, whereas for odd β there are $\beta^m - 1$ extra positive values; hence some numbers do not have representable additive inverses.

From the definition (1.8.4) of $\mathcal{F}^{rc}_{\ell m}[\beta, C^c_\beta]$ we can then derive the conversion mapping α which will map a polynomial $-P$, $P \in \mathcal{F}^{rc}_{\ell m}[\beta, C^c_\beta]$ into another polynomial $Q \in \mathcal{F}^{rc}_{\ell m}[\beta, C^c_\beta]$. With $D = \{0, 1, \ldots, \beta - 1\}$ and $C^c_\beta = D \cup \{-1\}$, where the digit -1 is only allowed in the most significant position $(m + 1)$, the conversion mapping α is a mapping which can be defined in two parts α' and α'':

$$\alpha' : \ \{-1, 0\} \times \overline{D} \to \{-1, 0\} \times D,$$

where for $\ell \leq i \leq m$, $d_i \geq 0$

$$\begin{aligned} \alpha'(0, 0) &= (0, 0), \\ \alpha'(0, -d_i) &= (-1, \beta - d_i) \qquad \text{for } d_i > 0, \\ \alpha'(-1, -d_i) &= (-1, (\beta - 1) - d_i), \end{aligned}$$

and for the most-significant position $i = m + 1$, $d_{m+1} \in \{-1, 0\}$;

$$\alpha'' : \{-1, 0\} \times \{-1, 0\} \to \{-1, 0\} \times \{-1, 0\}$$

is defined by

$$\alpha''(0, 0) = (0, 0),$$
$$\alpha''(-1, -d_{m+1}) = (0, -1 - d_{m+1}).$$

Note that $\alpha''(0, 1)$ need not be defined since there must always be a carry into position $m + 1$ if $d_{m+1} \neq 0$, since then $d_m \geq \lfloor(\beta + 1)/2\rfloor$.

Now, the conversion mapping defined above could also be obtained by rewriting the value of $\| - P\|$. Let $P = \sum_{i=\ell}^{m+1} d_i[\beta]^i$, $P \in \mathcal{F}_{\ell m}^{rc}$ as defined by (1.8.4), then

$$\begin{aligned}
\| - P\| &= \sum_{i=\ell}^{m+1}(-d_i)\beta^i \\
&= -d_{m+1}\beta^{m+1} + \sum_{i=\ell}^{m}(-d_i)\beta^i \\
&= -d_{m+1}\beta^{m+1} + \sum_{i=\ell}^{m}(\beta - 1 - d_i)\beta^i - (\beta - 1)\frac{\beta^{m+1} - \beta^\ell}{\beta - 1} \\
&= -(d_{m+1} + 1)\beta^{m+1} + \sum_{i=\ell}^{m}(\beta - 1 - d_i)\beta^i + \beta^\ell.
\end{aligned}$$

Hence the digits of the negated polynomial in general are found by complementing all digits $d_i' = \beta - 1 - d_i$ with a unit in the least-significant position. If $d_\ell = 0 - 1$, the term β^ℓ will cause a carry ripple, changing all least-significant digits of value $d_i' = \beta - 1$ (say into position $k - 1$) into zeroes, terminating with $d_k' = \beta - d_k$.

Observation 2.6.5 *For $d_\ell \neq 0$ the additive inverse of a polynomial in radix-complement representation, $P = \sum_{i=\ell}^{m+1} d_i[\beta]^i \in \mathcal{F}_{\ell m}^{rc}[\beta, C_\beta^c]$, is obtained as the polynomial*

$$-P = \sum_{i=\ell}^{m+1} d_i'[\beta]^i,$$

where $d_{m+1}' = -(d_{m+1} + 1)$, $d_i' = \beta - 1 - d_i$ for $i = \ell + 1, \ldots, m$ and $d_\ell' = \beta - d_\ell$.

For β even, the smallest possible representable number has the form $P = -1[\beta]^{m+1} + \lfloor(\beta + 1)/2\rfloor\,[\beta]^m$, which negated is the radix polynomial $-P = 1[\beta]^{m+1} - \lfloor(\beta + 1)/2\rfloor\,[\beta]^m$. Applying the conversion mapping to $-P$, we obtain $Q = \lfloor(\beta + 1)/2\rfloor\,[\beta]^m$, which is correct since $-\|P\| = \|Q\| = \beta/2\beta^m$. But if the digit d_{m+1} is not represented, and not included in the conversion, then that number is mapped into itself by the algorithm, which then has failed for this case. However, the situation can be identified since the digit d_m has not changed its value, which it otherwise must do when the value of d_{m+1} changes.

Similarly for odd β, any number of the form $P = (\lfloor(\beta+1)/2\rfloor - 1)[\beta]^m + P'$ with $P' \neq 0$ and $\mathrm{msp}(P') < m$ has additive inverse on the form

$$Q = -1[\beta]^{m+1} + (\lfloor(\beta+1)/2\rfloor - 1)[\beta]^m + Q',$$

since

$$\alpha'(-1, -(\lfloor(\beta+1)/2\rfloor - 1)) = (-1, \lfloor(\beta+1)/2\rfloor - 1)$$

as the carry coming into this position m must be -1 when $P' \neq 0$. Thus these cases can also be identified by the fact that the digit value d_m does not change.

Observation 2.6.6 *The situation where $P = \sum_{i=\ell}^{m+1} d_i[\beta]^i \in \mathcal{F}^{rc}_{\ell m}[\beta, C^c_\beta]$ has no additive inverse in the radix-complement system can be identified when $-P$ is converted into a polynomial $Q = \sum_{i=\ell}^{m+1} d'_i[\beta]^i$, where $0 \le d'_i < \beta$ for $\ell \le i \le m$. Then*

$$Q \notin \mathcal{F}^{rc}_{\ell m}[\beta, C^c_\beta] \quad \text{iff} \quad d'_m = d_m \neq 0.$$

Let us conclude by illustrating the situation for $\beta = 3$.

Example 2.6.1 With $\beta = 3$, $m = 1$, and $\ell = 0$ the following table shows exhaustively the negation of all representable numbers in $\mathcal{F}^{rc}_{01}[3, \{-1, 0, 1, 2\}]$:

$\|P\|$	P	$-P$	Q	$\|Q\|$	v
-3	$\bar{1}20$	$1\bar{2}0$	010	3	3
-2	$\bar{1}21$	$1\bar{2}\bar{1}$	002	2	2
-1	$\bar{1}22$	$1\bar{2}\bar{2}$	001	1	1
0	000	000	000	0	0
1	001	$00\bar{1}$	$\bar{1}22$	-1	-1
2	002	$00\bar{2}$	$\bar{1}21$	-2	-2
3	010	$0\bar{1}0$	$\bar{1}20$	-3	-3
4	011	$0\bar{1}\bar{1}$	$\bar{1}12$	-4	$\boxed{5}$
5	012	$0\bar{1}\bar{2}$	$\bar{1}11$	-5	$\boxed{4}$

where the value v in the last column is the value of Q formed by considering only d_1 and d_0, interpreted as a 3's complement number. Note that $d'_1 = d_1 = 1$ in the last two rows, thus identifying the "overflow" in the conversion. □

Problems and exercises

2.6.1 Find the additive inverse of $\bar{1}2\bar{1}021_{[4]} \in \mathcal{P}[4, \{-1, 0, 1, 2\}]$ by on-the-fly conversion.

2.6.2 Negate the following radix-complement numbers shown in string notation:

(a) 26475_{10}; (b) 73520_8; (c) 2314_5.

2.6.3 Show that an odd integer in 2's complement representation can be negated in constant time.

2.7 Notes on the literature

Many of the results presented in this chapter were developed along with the writing of this book, or have been further developed by Munk Nielsen and Kornerup in [Kor94, Nie97, NK97a, NK97b, NK99, Kor99]. There is amazingly little published in the open literature on these subjects in their own right; most of what can be found is embedded in the description of some application, e.g., as in Booth Recodings [Boo51, Mac61] used in multiplication. These will be discussed and referenced in later chapters. In most of the original descriptions of these algorithms, the description is based on particular encodings of digit values. A fundamental thesis for this description has been the understanding that base and digit set conversion can be described independently of the chosen encoding of digits. It is only in the implementation phase, and thus when designing the (binary) logic that the actual encoding plays a role. Other publications more directly covering these and related topics are [EL96, Fro96, Fro97, ACG+97, PGK01]. The properties of Theorem 2.4.10 are from [KM05]. Some early work on base and digit set conversions was presented by Matula in [Mat67, Mat68, Mat70, MM73, Mat76].

The "on-the-fly" conversion algorithm, presented here in Theorem 2.3.3, is a generalization of the original algorithm by Ercegovac and Lang [EL87].

The parallel prefix structure used in Section 2.4 is well known from the theory of parallel computing: the classical reference there is by Ladner and Fisher [LF80]. The example built using this structure is also known as the *conditional-sum adder* [Skl60], which will be further discussed in Chapter 3. But as this example points out, logarithmic time carry computations were known very much earlier, e.g., [WS58] contains structures very similar to the parallel prefix computations, and MacSorley [Mac61] introduced the carry look-ahead structure and its name. Kogge and Stone [KS73] also discussed the general parallel prefix problem.

The very general discussion on limited carry propagation and carries determined by the right context is based on [Kor99], which also explores the possibility of dependence on the left context. Only much later did Panhaleux provide Example 2.4.1 showing such dependence.

The round-to-nearest property of Booth-recoded numbers of Theorem 2.4.10 was discussed by Kornerup, Muller, and Panhaleux in [KM05, KM06, KMP10] as a property of a more general class under the name *RN codings*.

Reitwiesner [Rei60] gave the first algorithm for rewriting a binary number into the unique canonical form with a minimal number of non-zero digits (NAF).

This minimal form has been the subject of renewed interest in connection with minimizing the number of multiplications needed for exponentiation, in particular for use in RSA encryption and decryption and other cryptographic algorithms needing modular exponentiation of very large numbers. Koç [Koç96] developed an $\mathcal{O}(\log n)$-time algorithm for parallel conversion into NAF using parallel prefix computation of carries, and [JY00] provides algorithms working from left to right. Some further references to the recoding of multipliers are [Rob70, SG90, AW93, EL96]. A quite extensive discussion on minimal weight digit set conversions was provided by Phillips and Burgess in [PB04].

The P-, N-, and Q-mappings of Section 2.5 were originally introduced by Matula and Daumas in [DM97, DM03], however, there they were more directly based on the implementations than is described here in Section 2.6.

References

[ACG+97] J.-P. Allouche, E. Cateland, W. J. Gilbert, H.-O. Peitgen, J. O. Shallit, and G. Skordev. Automatic maps in exotic number systems. *Theory Comp. Syst.*, 30:285–331, 1997.

[AW93] S. Arno and F. S. Wheeler. Signed digit representations of minimal Hamming weight. *IEEE Trans. Computers*, C-42(8):1007–1010, August 1993.

[Boo51] A. D. Booth. A signed binary multiplication technique. *Q. J. Mech. Appl. Math.*, 4:236–240, 1951. Reprinted in [Swa80].

[DM97] M. Daumas and D. W. Matula. *Recoders for Partial Compression and Rounding*. Technical Report RR97-01, LIP, Ecole Normale Superieure de Lyon, 1997. Available at: http://www.ens-lyon.fr/LIP.

[DM03] M. Daumas and D. W. Matula. Further reducing the redundancy of notation over a minimally redundant digit set. *VLSI Signal Proc.*, 33(1/2):7–18, 2003.

[EL87] M. D. Ercegovac and T. Lang. On-the-fly conversion of redundant into conventional representations. *IEEE Trans. Computers*, C-36(7):895–897, July 1987. Reprinted in [Swa90].

[EL96] M. D. Ercegovac and T. Lang. On recoding in arithmetic algorithms. *J. VLSI Signal Proc.*, 14:283–294, 1996.

[Fro96] C. Frougny. Parallel and on-line addition in negative base and some complex number systems. In *Proc. Euro-Par 96, Lyon*, LNCS-1124, pages 175–182. Springer Verlag, 1996.

[Fro97] C. Frougny. On-the-fly algorithms and sequential machines. In *Proc. 13th IEEE Symposium on Computer Arithmetic*, pages 260–265. IEEE Computer Society, 1997.

[JY00] M. Joye and S.-M. Yen. Optimal left-to right binary signed-digit recoding. *IEEE Trans. Computers*, 49(7):740–748, 2000.

[KM05] P. Kornerup and J.-M. Muller. RN-coding of numbers: definition and some properties. In *Proc. IMACS 2005*, July 2005.

[KM06] P. Kornerup and J.-M. Muller. RN-codings: new insights and some applications. In *Proceedings of RNC7, Nancy*, July 2006.

[KMP10] P. Kornerup, J.-M. Muller, and A. Panhaleux. Performing arithmetic operations on round-to-nearest representations. *IEEE Trans. Computers*, to appear.

[Koç96] Ç. K. Koç. Parallel canonical recoding. *Electron. Lett.*, 32(22):2063–2065, October 1996.

[Kor94] P. Kornerup. Digit-set conversions: generalizations and applications. *IEEE Trans. Computers*, C-43(6):622–629, May 1994.

[Kor99] P. Kornerup. Necessary and sufficient conditions for parallel and constant time conversion and addition. In *Proc. 14th IEEE Symposium on Computer Arithmetic*. IEEE Computer Society, April 1999.

[KS73] P. M. Kogge and H. S. Stone. A parallel algorithm for the efficient solution of a general class of recurrence equations. *IEEE Trans. Computers*, 22(8):783–791, August 1973.

[LF80] R. E. Ladner and M. J. Fischer. Parallel prefix computation. *J. ACM*, 27(4):831–838, Oct. 1980.

[Mac61] O. L. MacSorley. High-speed arithmetic in binary computers. *Proc. IRE*, 49:67–91, January 1961. Reprinted in [Swa80].

[Mat67] D. W. Matula. Base conversion mappings. In *Spring Joint Computer Conference*, pages 311–318, 1967.

[Mat68] D. W. Matula. In-and-out conversions. *Commun. ACM*, 11(1):47–50, January 1968.

[Mat70] D. W. Matula. A formalization of floating-point numeric base conversion. *IEEE Trans. Computers*, C-19(8):681–692, August 1970. Reprinted in [Swa90].

[Mat76] D. W. Matula. Radix arithmetic: digital algorithms for computer architecture. In R. T. Yeh, editor, *Applied Computation Theory: Analysis, Design, Modeling*, chapter 9, pages 374–448. Prentice-Hall, Inc., 1976.

[MM73] J. D. Marasa and D. W. Matula. A simulative study of correlated error propagation in various finite-precision arithmetics. *IEEE Trans. Computers*, C-22:587–597, 1973.

[Nie97] A. Munk Nielsen. Number Systems and Digit Serial Arithmetic. PhD thesis, Odense University, Denmark, October 1997. Available at: www.imada.sdu.dk/~kornerup/research/asger.ps.gz.

[NK97a] A. Munk Nielsen and P. Kornerup. Generalized base and digit set conversion. In *Proc. SCAN 97, Lyon*, pages XII-8–11. GAMM/IMACS, September 1997. Extended abstract.

[NK97b] A. Munk Nielsen and P. Kornerup. On radix representation of rings. In *Proc. 13th IEEE Symposium on Computer Arithmetic*, pages 34–43. IEEE Computer Society, July 1997.

[NK99] A. Munk Nielsen and P. Kornerup. Redundant radix representation of rings. *IEEE Trans. Computers*, 48(11):1153–1165, November 1999.

[PB04] B. Phillips and N. Burgess. Minimal weight digit set conversions. *IEEE Trans. Computers*, 53(6):666–677, June 2004.

[PGK01] D. S. Phatak, T. Goff, and I. Koren. Constant-time addition and simultaneous format conversion based on redundant binary representations. *IEEE Trans. Computers*, 50(11):1267–1278, 2001.

[Rei60] G. W. Reitwiesner. Binary arithmetic. In A. D. Booth and R. E. Meager, editors, *Advances in Computers*, volume 1, pages 231–308. Academic Press, 1960.

[Rob70] J. E. Robertson. The correspondance between methods of digital division and multiplier recoding procedures. *IEEE Trans. Computers*, C-19:692–701, 1970.

[SG90] H. Sam and A. Gupta. A generalized multibit recoding of two's complement binary numbers and its proof with application in multiplier implementations. *IEEE Trans. Computers*, C-39(8):1006–1015, August 1990.

[Skl60] J. Sklansky. An evaluation of several two-summand binary adders. *IRE Trans. Electronic Computers*, EC-9:213–226, 1960. Reprinted in [Swa80].

[Swa80] E. E. Swartzlander, editor. *Computer Arithmetic,* volume I. Dowden, Hutchinson and Ross, Inc., 1980. Reprinted by IEEE Computer Society Press, 1990.

[Swa90] E. E. Swartzlander, editor. *Computer Arithmetic,* volume II. IEEE Computer Society Press, 1990.

[WS58] A. Weinberger and J. L. Smith. A logic for high-speed addition. *Nat. Bur. Stand. Circ.*, 591:3–12, 1958. Reprinted in [Swa80].

3

Addition

3.1 Introduction

Just as the radix representation is the ubiquitous representation of numbers, so radix integer addition is the most fundamental operation for all computer arithmetic; it is used in the implementation of all the other standard arithmetic operations: subtraction, multiplication, division, and square root. For a discussion of the implementation of the standard arithmetic operators on simple radix-represented operands, it is sufficient to consider integer operands. The modifications necessary to extend the algorithms and hardware implementations to other types of fixed-point operands are very trivial. Also the problem of subtraction is trivially reduced to addition, so we shall only briefly cover subtraction in the following.

It is, however, crucial to the discussions in this chapter that we restrict ourselves to a finite set of integer values, as typically defined by the limitations of computer words or registers used in hardware implementations. Radix integer addition on such restricted operand domains then forms the most basic "building block" upon which other radix integer arithmetic operations can be built. Eventually we will also build other finite precision arithmetic systems on these radix integer operations, e.g., floating-point arithmetic and more unusual or exotic types of arithmetic based on other number representations, like rational representations. Even the types of "exact" or unlimited precision arithmetic found in symbolic and algebraic systems are based on these basic "building block" integer arithmetic operations.

Limiting the discussion to operands from such restricted operand domains does not imply that we will not be concerned with the asymptotic behavior of the algorithms. When using small radices, the number of digits needed may be quite large, and the following treatment of addition may often be applicable to the types of unlimited precision arithmetic mentioned above.

The fundamental problem in radix addition is the carry problem, where it is well known that the standard method of adding two n-digit non-redundant operands is an $O(n)$ process. A formal model by Winograd [Win65] tells us that $O(\log n)$ running times are possible and we shall show how. However, introducing redundancy will permit $O(1)$ running times, albeit with larger constant factors and increased storage and transmission requirements due to the redundancy. This confirms our results from Chapter 2, since we have seen that the problem of addition is fundamentally equivalent to the problem of digit set conversion.

We shall here and in the following chapters be more relaxed about using the formal notation for radix polynomials, e.g., we shall not always be strict about distinguishing between a polynomial $P = \sum_\ell^m d_i[\beta]^i$ and its value $\|P\| = \sum_\ell^m d_i\beta^i$, and we shall often use the string notation for specifying a polynomial. We will, in general, do so when the interpretation is fairly obvious from the context.

3.2 How fast can we compute?

Ultimately any arithmetic function in present technology will have to be built out of *combinational circuits* whose basic primitives will correspond to boolean operators like NOT, AND, OR, or NAND and NOR, where operands coded in binary are transmitted to and from *gates* capable of performing such primitive operations. Technology permits gates realizing boolean operators with a limited number of arguments (*fan-in restriction*), and the multiple but limited use of an output value (*fan-out restriction*). When storage elements are included a *sequential circuit* is obtained, and feed-back becomes possible. For the present analysis we do not need such a capability, and will restrict our considerations to pure combinational circuits taking n inputs and producing m outputs. For simplicity we will furthermore assume that the circuit can be separated into m disjoint circuits where each node of the circuit has a fan-out of 1, i.e., the circuit consists of m trees. Each tree then computes one bit of the result, based on input supplied at the leaves of the tree. In general, however, it is necessary that several trees "read" the same input values which requires an unrestricted fan-out of input values. The restriction to a fan-out of 1 at internal nodes implies that there might be an unnecessary replication of internal nodes, but note that this replication does not change the length of paths from the input (leaves) to the output (root) of any tree, and hence does not change the timing bounds which are the purpose of these considerations.

Following Winograd and Spira we will model circuits built of r-input, one-output devices where each input or output line carries values from the set $\{0, 1, \ldots, \beta - 1\}$ for some $\beta \geq 2$, the so-called (r, β)-*devices*:

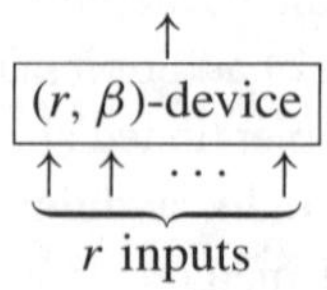

Such a device is assumed to have unit delay, i.e., if input is available on all input lines at time t, then output will be available at time $t + 1$. A circuit in the form of a disjoint set of trees built out of such devices (as in Figure 3.2.1) will then have a delay equal to the longest path from a leaf to the root, i.e., the height of the tree containing the path.

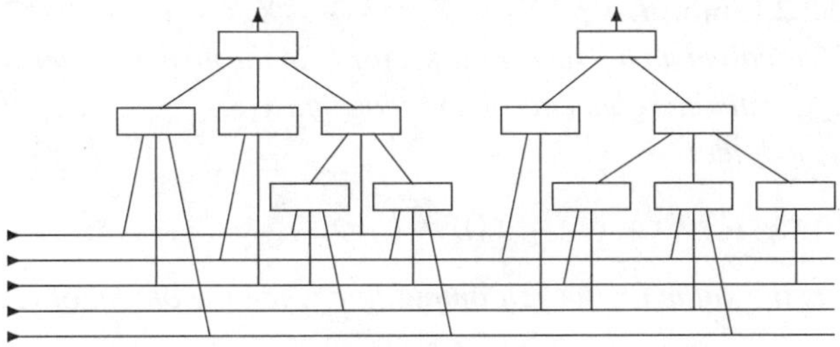

Figure 3.2.1. An example of a circuit of (r, β)-devices.

For some circuit C let O_C be the set of possible configurations (states) of the output lines, and assume that the input lines are partitioned into n sets with $I_{C,j}$ being the set of possible configurations of the jth set, $j = 1, 2, \ldots, n$. To relate the computation in the circuit C to the computation of some function ϕ we need the following definition.

Definition 3.2.1 (Spira) *Let $\phi\colon X_1 \times X_2 \times \cdots \times X_n \to Y$ be a function on finite sets. A circuit C is said to compute ϕ in time τ if there are maps $g_j\colon X_j \to I_{C,j}$ $(j = 1, 2, \ldots, n)$ and a one-to-one function $h\colon Y \to O_c$ such that if C receives constant input $[g_1(x_1), g_2(x_2), \ldots, g_n(x_n)]$ from time 0 through time $\tau - 1$, then the output at time τ will be $h(\phi(x_1, x_2, \ldots, x_n))$.*

The mapping g_j is used as a coding of an operand $x_j \in X_j$ into a configuration in $I_{C,j}$, i.e., a particular set of values of the input lines of the jth partition of C's input lines. Similarly h is an encoding of results $y \in Y$, or equivalently h^{-1} is a decoding of output configurations of C. This has been illustrated in the following diagram:

$$\begin{array}{ccc}
X_1 \times X_2 \times \cdots \times X_n & \xrightarrow{\ \phi\ } & Y \\
\Big\downarrow (g_1, g_2, \cdots, g_n) & & h \Big\downarrow \Big\uparrow h^{-1} \\
I_{C,1} \times I_{C,2} \times \cdots \times I_{C,n} & \xrightarrow{\ C\ } & O_C
\end{array}$$

For the timing considerations, note that the mappings g_j, $j = 1, 2, \ldots, n$ and the function h are assumed not to be part of the computation, and hence do not consume time. This is also the reason that h has to be a one-to-one mapping, otherwise we could choose C to be a circuit performing the identity transformation in zero time, and assume the actual computation of ϕ takes place in the function h^{-1}. Thus by this assumption redundant representations are excluded from the model.

To express succinctly the dependence between input and output of a circuit which computes a function ϕ we introduce the following definition.

Definition 3.2.2 (Spira) *Let $\phi\colon X_1 \times X_2 \times \cdots \times X_n \to Y$ and let C compute ϕ. Then $S \subseteq X_m$ is called an h_j-separable set for C in the mth argument of ϕ if whenever s_1 and s_2 are distinct elements of S we can find $x_1, x_2, \ldots, x_{m-1}, x_{m+1}, \ldots, x_n$ with $x_i \in X_i$ such that*

$$h_j(\phi(x_1, \ldots, x_{m-1}, s_1, x_{m+1}, \ldots, x_n)) \neq h_j(\phi(x_1, \ldots, x_{m-1}, s_2, x_{m+1}, \ldots, x_n)),$$

where $h_j(y)$ is the value on the jth output line when the output of C is $h(y)$.

Loosely speaking, an h_j-separable set S is a subset of values of X_m which influences the jth output line of C. The cardinality of such a maximal set $S = S_m(j)$ thus tells us how many members of X_m actually influence the jth output of C. Furthermore, in any (r, β)-circuit the output of an element at time τ can depend on at most r^τ input lines, since a tree of height τ and fan-in r cannot have more than r^τ leaves. Based on this we may now prove the following lemma.

Lemma 3.2.3 (Spira) *Let C be a (r, β)-circuit which computes ϕ in time τ. Then*

$$\tau \geq \max_j\{\lceil \log_r(\sum_{i=1}^{n} \lceil \log_\beta |S_i(j)| \rceil) \rceil\},$$

where $S_i(j)$ is an h_j-separable set for C in the ith argument of ϕ.

Proof $\lceil \log_\beta |S_i(j)| \rceil$ is the minimal number of input lines from $I_{C,i}$ needed to express the dependence of $h_j(y)$ on elements from X_i. The sum then expresses the minimal number of input lines whose values affect the value of $h_j(y)$, considering all arguments of $y = \phi(x_1, \ldots, x_n)$. But for all j the sum is at most r^τ, hence the lemma. □

First we consider the problem of digit set conversion where the target system digit set is non-redundant, so let ϕ be the function which maps polynomials from $\mathcal{F}_{0n}[\beta, D]$ into polynomials of $\mathcal{P}_I[\beta, E]$, where E is non-redundant and complete. So ϕ is a mapping:

$$\phi\colon\ D^{n+1} \to Y,$$

and let us for simplicity assume we use a β-valued logic, hence the functions g_i are simple mappings and so is $h = \{h_0, h_1, \ldots, h_m\}$, where m is chosen appropriately.

Theorem 3.2.4 *Any (r, β)-circuit performing the digit set conversion of mapping polynomials from $\mathcal{F}_{0n}[\beta, D]$ into polynomials in $\mathcal{P}_I[\beta, E]$, where $E \neq D$ and E is non-redundant and complete for radix β, requires a computing time of at least*

$$\lceil \log_r(n+1) \rceil.$$

Proof It follows immediately from Theorem 2.3.1 that there exists a subset $S \subseteq D$ containing at least two elements such that S is an h_j-separable subset of $D = X_i$, where $0 \le i \le n$ and $i \le j \le m$ for m chosen such that

$$m = \max\{\text{degr}(Q) \mid Q \in \mathcal{P}_I[\beta, E] : \exists P \in \mathcal{F}_{0n}[\beta, D] : \|Q\| = \|P\|\}.$$

Applying Lemma 3.2.3 we find that any (r, β)-circuit C computing ϕ requires a computing time of at least $\lceil \log_r(n+1) \rceil$. □

Now let us consider addition of non-negative numbers, e.g., let ϕ_1 and ϕ_2 be defined by

$$\phi_1 \colon \mathbb{Z}_{2^n} \times \mathbb{Z}_{2^n} \to \mathbb{Z}_{2^{n+1}-1} \qquad \phi_1(a, b) = a + b,$$
$$\phi_2 \colon \mathbb{Z}_{2^n} \times \mathbb{Z}_{2^n} \to \mathbb{Z}_{2^n} \qquad \phi_2(a, b) = (a + b) \bmod 2^n.$$

The input codings g_1 and g_2 need not be specified, but let the output coding be the ordinary radix-2 polynomials. So let

$$\phi_1(a, b) = y \quad \Rightarrow \quad h(y) = (h_n(y), h_{n-1}(y), \ldots, h_0(y)),$$
$$\phi_2(a, b) = y \quad \Rightarrow \quad h(y) = (h_{n-1}(y), \ldots, h_0(y)),$$

where $h_j(y) \in \{0, 1\}$ and $y = \sum h_j(y) 2^j$. We can now show that for any $(r, 2)$-circuit C computing ϕ_1 or ϕ_2 the domain $\mathbb{Z}_{2^n}$ is h_{n-1}-separable in both arguments of ϕ_1 or ϕ_2.

Theorem 3.2.5 (Winograd) *Any $(r, 2)$-circuit performing addition $\phi_1 \colon \mathbb{Z}_{2^n} \times \mathbb{Z}_{2^n} \to \mathbb{Z}_{2^{n+1}-1}$ or modular addition $\phi_2 \colon \mathbb{Z}_{2^n} \times \mathbb{Z}_{2^n} \to \mathbb{Z}_{2^n}$, $\phi_2(a, b) = (a + b) \bmod 2^n$, with the result represented in $\mathcal{P}_I[2, \{0, 1\}]$, requires a computing time of at least*

$$\lceil \log_r(2n) \rceil.$$

Proof Since addition is commutative we need only consider separability in one argument, say the leftmost. As we shall prove h_{n-1} separability we may consider ϕ_1 and ϕ_2 together, so let ϕ be either ϕ_1 or ϕ_2.

Let $a_1 < a_2$ be arbitrary members of $\mathbb{Z}_{2^n}$ as distinct left arguments of ϕ, and distinguish two cases:

Case 1 $a_2 - a_1 \geq 2^{n-1}$: Choose a right operand $b = 0$, then

$$h_{n-1}(\phi(a_1, b)) = h_{n-1}(a_1) = 0,$$
$$h_{n-1}(\phi(a_2, b)) = h_{n-1}(a_2) = 1,$$

since necessarily we have $a_1 < 2^{n-1} \leq a_2$.
Case 2 $a_2 - a_1 < 2^{n-1}$: With $b = 2^n - a_2$ certainly $0 < b \leq 2^n - 1$ so $b \in \mathbb{Z}_{2^n}$ and

$$\begin{aligned}\phi(a_1, b) = 2^n - (a_2 - a_1) &\Rightarrow 2^{n-1} < \phi(a_1, b) \leq 2^n - 1\\ &\Rightarrow h_{n-1}(\phi(a_1, b)) = 1.\end{aligned}$$

Also

$$\phi(a_2, b) = \begin{cases} 2^n & \text{if } \phi = \phi_1 \\ 0 & \text{if } \phi = \phi_2 \end{cases} \Rightarrow h_{n-1}(\phi(a, b)) = 0.$$

So $\mathbb{Z}_{2^n}$ is an h_{n-1} separable set in both arguments for any circuit C computing ϕ_1 or ϕ_2 and by Lemma 3.2.3 we obtain the theorem. □

The previous theorem can easily be generalized to addition in radix β using (r, β)-devices, or combinations like radix-β_1 addition using (r, β_2)-devices. Another generalization is to consider k-addend addition, where it is straightforward to obtain the lower bound $\lceil \log_2(kn) \rceil$ using binary radix coding of the result. Also it is easily seen that "overflow testing" of modular addition has the same lower bound as addition (consider h_n-separability, for $a_1 < a_2$ choose $b = 2^n - a_2$).

For another example consider multiplication with the result represented as binary radix polynomials

$$\phi_1 : \mathbb{Z}_{2^n} \times \mathbb{Z}_{2^n} \to \mathbb{Z}_{2^{2n} - 2^{n+1} + 1} \qquad \phi_1(a, b) = a \cdot b$$

or

$$\phi_2 : \mathbb{Z}_{2^n} \times \mathbb{Z}_{2^n} \to \mathbb{Z}_{2^n} \qquad \phi_2(a, b) = (a \cdot b) \bmod 2^n,$$

where it is also possible to show h_{n-1}-separability. Just choose $a_1 \neq a_2$ from $\mathbb{Z}_{2^n}$, then there is at least one position in their binary representation where they differ, say the kth position. Then choose $b = 2^{n-k-1}$ so $h_{n-1}(\phi(a_1, b)) \neq h_{n-1}(\phi(a_2, b))$, and there is an h_{n-1}-separable set ($\mathbb{Z}_{2^n}$) of size 2^n in both arguments. In analogy with the previous theorem we then have the following.

Theorem 3.2.6 *Any $(r, 2)$-circuit performing multiplication: $\phi_1 : \mathbb{Z}_{2^n} \times \mathbb{Z}_{2^n} \to \mathbb{Z}_{2^{2n} - 2^{n+1} + 1}$, $\phi_1(a, b) = a \cdot b$, or modular multiplication $\phi_2 : \mathbb{Z}_{2^n} \times \mathbb{Z}_{2^n} \to \mathbb{Z}_{2^n}$, $\phi_2(a, b) = (a \cdot b) \bmod 2^n$, with the result represented in $\mathcal{P}_I[2, \{0, 1\}]$, requires a computing time of at least*

$$\lceil \log_r(2n) \rceil.$$

Note that the bounds of Theorems 3.2.5 and 3.2.6 are theoretical lower bounds for the computing time of addition and multiplication, when the result is to be expressed as a *non-redundant* binary radix polynomial. As we shall see in Section 3.6 the lower bound for addition can actually be realized, using ordinary binary logic gates. In Chapter 4, we shall also show that the bound for binary multiplication is within a constant factor, i.e., we shall show implementations with an execution time of $O(\log n)$.

To obtain faster execution times it is necessary to look for alternative representations of the output. Lemma 3.2.3 provides the clue towards this goal: we must look for representations where the cardinality of the h_j-separable sets is minimal. If we still require non-redundant output coding, one (optimal) way for addition modulo N is to employ *multiple modulus residue representation*, which will be discussed in Chapter 8. If N can be factored $N = m_1 \cdot m_2 \cdots m_n$, where m_i is relatively prime to m_j for all $i \neq j$, then addition can be realized in time $\lceil \log_r(2 \max_i \lceil \log_\beta m_i \rceil) \rceil$ using (r, β)-circuits. The representation employed for input as well as output takes an integer p into a vector $(p_1, p_2, \ldots, p_n)$, where $p_i = p \bmod m_i$. Then addition and multiplication can be performed componentwise in parallel, e.g., if $r = p + q$ then the ith component r_i can be computed as $r_i = (p_i + q_i) \bmod m_i$, where p_i and q_i are the ith components of p and q respectively. We shall not pursue this topic further, as it rather belongs to the theory of complexity.

3.3 Digit addition

The basic primitive to realize radix addition is an algorithm for a digit-by-digit addition of radix polynomials over some digit set D. By formal addition of two polynomials P_1 and P_2 from $\mathcal{P}[\beta, D]$ we obtain a polynomial $R = P_1 + P_2$ whose digits do not in general belong to the digit set D, but are rather in the digit set $E = \{d_1 + d_2 \mid d_1, d_2 \in D\}$, so $R = P_1 + P_2 \in \mathcal{P}[\beta, E]$. For example, if we add two polynomials $P_1, P_2 \in \mathcal{P}[2, \{0, 1\}]$ we obtain a polynomial $R = P_1 + P_2 \in \mathcal{P}[2, \{0, 1, 2\}]$, i.e., the digits of the result are in the carry-save digit set, which may be fine if that is what we want, but often we want the result to be a polynomial over the same digit set as the operands. Also there might be cases where the two operands are represented with different digit sets, e.g., $P_1 \in \mathcal{P}[2, \{0, 1\}]$ and $P_2 \in \mathcal{P}[2, \{0, 1, 2\}]$, where say a result in $\mathcal{P}[2, \{0, 1, 2\}]$ is wanted. This particular combination is very frequently used for the accumulation of many summands, which we shall investigate in Chapter 4.

Since the problem of addition (and equivalently subtraction) then reduces to the problem of digit set conversion we may employ the results and techniques of Chapter 2 to handle the most general cases. But let us initially just assume that the two operands are represented in the same digit set D, with the result also to be represented in D, and let us assume that D is complete. Adding a pair of digits

$d_1, d_2 \in D$ yields an integer value which may be written as

$$d_1 + d_2 = c\beta + d,$$

where the digit $d \in D$ and the carry c are not necessarily unique. Choosing particular pairs (c, d) for each (d_1, d_2) yields an *addition mapping*

$$\alpha_0 \colon D \times D \to C_1 \times D, \tag{3.3.1}$$

where C_1 is a set of possible carries. To assimilate carries it will be necessary to add digits and carries, hence another mapping is needed:

$$\alpha_1 \colon C_1 \times D \to C_2 \times D, \tag{3.3.2}$$

which can be chosen such that it coincides with α_0 whenever $c \in C_1 \cap D$. Continuing, we may recursively define

$$\alpha_i \colon C_i \times D \to C_{i+1} \times D$$

with the restriction that $\alpha_i(c, d) = \alpha_{i-1}(c, d)$ for $c \in C_i \cap D$. Since D is finite, there exists an n such that $C_{n+1} = C_n$, and extending α_0 to the domain $C = D \cup C_n$ it is then possible to define a mapping

$$\alpha \colon C \times D \to C \times D. \tag{3.3.3}$$

We term C the *carry set* for α, but note that we have chosen C here such that $D \subseteq C$.

Since C and D are finite, α may be described by an *addition table* whose entries are pairs (c, d), which for convenience will be written as strings cd using standard symbols for digits, e.g., as in Table 3.3.1.

Table 3.3.1. *α: standard binary; γ: balanced ternary*

α	0	1
0	00	01
1	01	10

γ	-1	0	1
-1	$\bar{1}1$	$0\bar{1}$	00
0	$0\bar{1}$	00	01
1	00	01	$1\bar{1}$

Note that $C = D$ in both examples chosen in Table 3.3.1; such (quadratic) addition tables are called *regular*.

Example 3.3.1 With $D = \{-1, 0, 7\}$, which is complete for $\beta = 3$, we find for $\alpha_0 : D \times D \to C_1 \times D$ that it is necessary to represent the digit sums $-2, -1, 0, 6, 7, 14$ for which carries in the set $C_1 = \{-3, 0, 2, 5\}$ are needed. Trying to represent the sum of a carry from C_1 and a digit from D we find that $-2, 1, 3$, and 4 are also needed. The process is most conveniently expressed by filling out the addition table row by row starting with the table for $D \times D$, adding new carry-values as needed:

α	-1	0	7
-1	$\bar{3}7$	$0\bar{1}$	20
0	$0\bar{1}$	00	07
7	20	07	$5\bar{1}$
-3	$\bar{1}\bar{1}$	$\bar{1}0$	$\bar{1}7$
2	$\bar{2}7$	$1\bar{1}$	30
5	$\bar{1}7$	$2\bar{1}$	40
-2	$\bar{1}0$	$\bar{3}7$	$2\bar{1}$
1	00	$\bar{2}7$	$3\bar{1}$
3	$1\bar{1}$	10	17
4	10	$\bar{1}7$	$4\bar{1}$

Hence $C = \{-3, -2, -1, 0, 1, 2, 3, 4, 5, 7\}$ is the carry-set and α the (non-regular) addition table for addition in $\mathcal{P}[3, \{-1, 0, 7\}]$. □

If a negative radix $\beta \leq -2$ is chosen with the standard digit set $D_{|\beta|} = \{0, 1, \ldots, |\beta| - 1\}$ a carry digit of -1 is needed so $C = D_{|\beta|} \cup \{-1\}$, as in Table 3.3.2.

Table 3.3.2. *Radix −4 addition table*

α	0	1	2	3
-1	13	00	01	02
0	00	01	02	03
1	01	02	03	$\bar{1}0$
2	02	03	$\bar{1}0$	$\bar{1}1$
3	03	$\bar{1}0$	$\bar{1}1$	$\bar{1}2$

The following observations are now immediately seen from the DGT Algorithm (Algorithm 1.6.2).

Observation 3.3.1 *If D is complete and non-redundant, then the carry set C and addition mapping α are uniquely determined. Also $\alpha(0, d) = (0, d)$ for $d \in D$.*

Observation 3.3.2 *If D is basic and non-redundant, then $C = D \cup \{1\} \cup \{-1\}$ and the addition mapping α is unique.*

A radix-β addition mapping α may now be extended to radix polynomials by applying α digit-wise. To simplify the discussion we will only consider radix integer polynomials; the generalization to the general case is trivial. Let the *radix polynomial addition mapping*

$$\hat{\alpha}: \mathcal{P}_I[\beta, C] \times \mathcal{P}_I[\beta, D] \to \mathcal{P}_I[\beta, C] \times \mathcal{P}_I[\beta, D]$$

be defined so that for $P_0 \in \mathcal{P}_I[\beta, C]$ and $Q_0 \in \mathcal{P}_I[\beta, D]$ the digits of $\hat{\alpha}(P_0, Q_0) = (P_1, Q_1)$ are defined by

$$\alpha(d_i(P_0), d_i(Q_0)) = (d_{i+1}(P_1), d_i(Q_1)). \tag{3.3.4}$$

Obviously, in general it is necessary to apply $\hat{\alpha}$ repeatedly to absorb carries, e.g., recursively to compute

$$\hat{\alpha}(P_j, Q_j) = (P_{j+1}, Q_{j+1}) \qquad j = 0, 1, 2, \ldots$$

If P_k (the carry polynomial) becomes the zero polynomial,

$$\hat{\alpha}^k(P_0, Q_0) = (0, Q_k),$$

then

$$\|P_0 + Q_0\| = \|Q_k\|,$$

i.e., Q_k is a radix-β representation of the sum of P_0 and Q_0.

Example 3.3.2 (Balanced ternary addition (radix $\beta = 3$, Table 3.3.1γ))

$$\begin{array}{rl}
 & \begin{array}{lcccccc} Q_0 & & 1 & 1 & 0 & \bar{1} & \\ P_0 & \bar{1} & 1 & 1 & \bar{1} & 0 & \end{array} \\
\hat{\alpha}(P_0, Q_0) = & \left\{ \begin{array}{lcccccl} Q_1 & \bar{1} & \bar{1} & \bar{1} & \bar{1} & \bar{1} & \leftarrow d \\ P_1 \; 0 & 1 & 1 & 0 & 0 & & \leftarrow c \end{array} \right. \\
\hat{\alpha}(P_1, Q_1) = & \left\{ \begin{array}{lcccccl} Q_2 & 0 & 0 & \bar{1} & \bar{1} & \bar{1} & \leftarrow d \\ P_2 \; 0 & 0 & 0 & 0 & 0 & & \leftarrow c \end{array} \right.
\end{array}$$

□

If we relax the condition that D is complete (or semicomplete) the process of carry absorption may not stop, as shown in the following example:

Example 3.3.3 With $\beta = 3$ and $D = \{-1, 0, 4\}$ we find the addition table:

α	-1	0	4
-2	$\bar{1}0$	$\bar{2}4$	$1\bar{1}$
-1	$\bar{2}4$	$0\bar{1}$	10
0	$0\bar{1}$	00	04
1	00	$\bar{1}4$	$2\bar{1}$
2	$\bar{1}4$	$1\bar{1}$	20
3	$1\bar{1}$	10	14
4	10	04	$3\bar{1}$

but we may also notice that -2 is not representable in $\mathcal{P}_I[3, \{-1, 0, 4\}]$, so with $Q_0 = P_0 = -1$ we obtain

$$Q_1 = 4,$$
$$P_1 = -2[3]$$

and

$$Q_2 = 4[3] + 4,$$
$$P_2 = -2[3]^2,$$

hence the process never terminates. □

The number of repeated applications of $\hat{\alpha}$ obviously depends on the actual polynomials to be added and on the digit set.

Problems and exercises

3.3.1 Construct addition tables for the following systems:

(a) $\beta = -2$ $D = \{0, 1\}$;
(b) $\beta = 7$ $D = \{-4, -3, -1, 0, 1, 2, 5\}$;
(c) $\beta = 4$ $D = \{-1, 0, 1, 6\}$.

3.3.2 Carefully describe an algorithm that takes as input a radix β and a digit set D, which is a complete residue set modulo $|\beta|$, and gives as output the minimal carry set C with $C \subset D$ and the addition table for the uniquely determined addition mapping $\alpha : C \times D \longrightarrow C \times D$.

3.3.3 Add 7_3 and $\bar{1}_3$ in the system $\beta = 3$, $D = \{-1, 0, 7\}$ using the table on page 127.

3.4 Addition with redundant digit sets

Whereas α and the carry set C are uniquely determined by the digit set D if the latter is non-redundant, it is possible to find several addition mappings for redundant digit sets. Table 3.4.1 displays two regular radix-3 addition mappings for the redundant digit set $D = \{-2, -1, 0, 1, 2\}$.

Table 3.4.1. *Regular radix-3 addition mappings*

α	-2	-1	0	1	2
-2	$\bar{1}\bar{1}$	$\bar{1}0$	$\bar{1}1$	$0\bar{1}$	00
-1	$\bar{1}0$	$\bar{1}1$	$0\bar{1}$	00	01
0	$\bar{1}1$	$0\bar{1}$	00	01	$1\bar{1}$
1	$0\bar{1}$	00	01	$1\bar{1}$	10
2	00	01	$1\bar{1}$	10	11

γ	-2	-1	0	1	2
-2	$\bar{1}\bar{1}$	$\bar{1}0$	$0\bar{2}$	$0\bar{1}$	00
-1	$\bar{1}0$	$0\bar{2}$	$0\bar{1}$	00	01
0	$0\bar{2}$	$0\bar{1}$	00	01	02
1	$0\bar{1}$	00	01	02	10
2	00	01	02	10	11

The existence of several addition mappings permits us to combine such mappings when performing addition, as in the following example.

Example 3.4.1 (Radix $\beta = 3$ redundant addition)

$$
\begin{array}{r|l|ccccccc|l}
 & P_0 & & 2 & \bar{1} & 0 & 2 & 0 & 1 & \\
 & Q_0 & & & \bar{1} & \bar{2} & 2 & 2 & 0 & \\
\hline
\hat{\alpha}(P_0, Q_0) = & P_1 & & \bar{1} & 1 & 1 & 1 & \bar{1} & 1 & \leftarrow d \\
 & Q_1 & 1 & \bar{1} & \bar{1} & 1 & 1 & 0 & & \leftarrow c \\
\hline
\hat{\gamma}(\hat{\alpha}(P_0, Q_0)) = & P_2 & 1 & \bar{2} & 0 & 2 & 2 & \bar{1} & 1 & \leftarrow d \\
 & Q_2 & 0 & 0 & 0 & 0 & 0 & 0 & & \leftarrow c
\end{array}
$$

Note from the tables for α and γ that two "levels" will always be sufficient here. This follows from the fact that the digits -2 and 2 are never generated by the α-mapping, and a non-zero carry is never generated by the γ-mapping when the operands are in the set $\{-1, 0, 1\}$. □

As illustrated in the example, addition in a system with a redundant digit set can be performed in constant time, $O(1)$, in a digit-by-digit parallel process. This is an immediate consequence of Theorem 1.11.8 since the target digit system is basic (contiguous) and redundant. In the following we will restrict our considerations to basic digit sets for $\beta \geq 2$ which are complete or semi-complete. If the digit set has the form $D = \{r, r+1, \ldots, s\}$, then for an addition mapping $\alpha : D \times D \to C \times D$ we need a conversion from $D^+ = \{2r, 2r+1, \cdots, 2s\}$ back into D, assuming D is redundant, i.e., $s - r + 1 > \beta$.

To find conditions under which this is possible let

$$\Sigma = \{r+1, \ldots, r+\beta\} \quad \text{and} \quad C = \{-1, 0, 1\}$$

so that our target digit set is $D = E = \Sigma + C = \{r, \ldots, r+\beta+1\}$ and our source digit set D^+ thus should satisfy

$$
\begin{aligned}
D^+ \subseteq \Sigma + \beta C &= \{r+1, \ldots, r+\beta\} + \beta\{-1, 0, 1\} \\
&= \{-\beta+r+1, \ldots, r, r+1, \ldots, r+\beta, r+\beta+1, \ldots, r+2\beta\}.
\end{aligned}
$$

But $D^+ = \{2r, \ldots, 2(r+\beta+1)\}$, hence the conditions for conversion as a two-level process ($k = 1$) are

$$
\begin{array}{rcr}
-\beta + r + 1 \leq 2r & \text{and} & 2(r+\beta+1) \leq r + 2\beta, \\
\beta \geq 1 - r & \text{and} & r \leq -2,
\end{array}
$$

i.e., β must be at least 3 and $r \leq -2$ (actually we must require $-\beta + 1 \leq r \leq -2$).

Theorem 3.4.1 (Avizienis) *For $\beta \geq 3$ and $-\beta + 1 \leq r \leq -2$ let D be the basic and redundant digit set $\{r, r+1, \ldots, r+\beta+1\}$. Then there exist digit sets*

$D' \subset D$ and $C \subset D$, with radix-β addition mappings $\alpha\colon D \times D \to C \times D'$ and $\gamma\colon C \times D' \to \{0\} \times D$ such that for any $P, Q \in \mathcal{P}[\beta, D]$

$$\hat{\gamma}(\hat{\alpha}(P, Q)) = (0, R),$$

where $R \in \mathcal{P}[\beta, D]$.

Proof (constructive) $D' = \{r+1, \ldots, r+\beta\}$ is basic and non-redundant for radix β. Thus there exists $\alpha\colon D \times D \to C \times D'$ with carry set $C = \{-1, 0, 1\}$ uniquely determined such that $\alpha\colon D \times D \to \{-1, 0, 1\} \times D'$. The addition mapping γ can then be defined such that $\gamma(c, d) = (0, c + d)$ for $c \in \{-1, 0, 1\}$ and $d \in D'$. Then for $P, Q \in \mathcal{P}[\beta, D]$:

$$\hat{\alpha}(P, Q) = (P_1, Q_1) \text{ with } P_1 \in \mathcal{P}[\beta, \{-1, 0, 1\}] \qquad Q_1 \in \mathcal{P}[\beta, D'],$$
$$\hat{\gamma}(P_1, Q_1) = (0, R) \text{ with } R \in \mathcal{P}[\beta, D]. \qquad \square$$

Note from the proof that γ need only be defined on $\{-1, 0, 1\} \times D'$, the remaining entries will never be consulted. Here follow α and γ addition tables for redundant radix-10 addition with $D = \{-2, -1, \ldots, 8, 9\}$.

α	−2	−1	0	1	2	3	4	5	6	7	8	9
−2	$\bar{1}6$	$\bar{1}7$	$\bar{1}8$	$0\bar{1}$	00	01	02	03	04	05	06	07
−1	$\bar{1}7$	$\bar{1}8$	$0\bar{1}$	00	01	02	03	04	05	06	07	08
0	$\bar{1}8$	$0\bar{1}$	00	01	02	03	04	05	06	07	08	$1\bar{1}$
1	$0\bar{1}$	00	01	02	03	04	05	06	07	08	$1\bar{1}$	10
2	00	01	02	03	04	05	06	07	08	$1\bar{1}$	10	11
3	01	02	03	04	05	06	07	08	$1\bar{1}$	10	11	12
4	02	03	04	05	06	07	08	$1\bar{1}$	10	11	12	13
5	03	04	05	06	07	08	$1\bar{1}$	10	11	12	13	14
6	04	05	06	07	08	$1\bar{1}$	10	11	12	13	14	15
7	05	06	07	08	$1\bar{1}$	10	11	12	13	14	15	16
8	06	07	08	$1\bar{1}$	10	11	12	13	14	15	16	17
9	07	08	$1\bar{1}$	10	11	12	13	14	15	16	17	18

γ	−1	0	1
−1	$0\bar{2}$	$0\bar{1}$	00
0	$0\bar{1}$	00	01
1	00	01	02
2	01	02	03
3	02	03	04
4	03	04	05
5	04	05	06
6	05	06	07
7	06	07	08
8	07	08	09

To perform addition in two levels it is necessary to have two extra redundant digits in the digit set D. Limiting redundancy to one extra digit (e.g., for minimally redundant systems) it is also possible to limit carry propagation, but then three levels are needed.

Theorem 3.4.2 (Avizienis) *For $\beta \geq 2$ and $-\beta \leq r \leq 0$ let D be the basic and minimally redundant set $\{r, r+1, \ldots, r+\beta\}$. Then there exists radix β addition* *mappings* $\alpha, \gamma, \xi : D \times D \to D \times D$ *such that for any* $P, Q \in \mathcal{P}[\beta, D]$*:*

$$\hat{\xi}(\hat{\gamma}(\hat{\alpha}(P, Q))) = (0, R),$$

where $R \in \mathcal{P}[\beta, D]$.

Proof The mappings γ and ξ need only be defined on certain subsets of $D \times D$, as found below. First assume $-\beta < r < 0$ and let $C' = \{-1, 0, 1\}$ and $D' = \{r+1, \ldots, r+\beta\}$, which is non-redundant and complete. Any a, $2r \leq a \leq 2r + 2\beta$ can be written uniquely as $a = c'\beta + d'$ with $(c', d') \in C' \times D'$, since $2r + \beta \geq r + 1$ and $2r + 2\beta - \beta \leq r + \beta$. This determines α. Now let $C'' = \{0, 1\}$ and $D'' = \{r, r+1, \ldots, r+\beta-1\}$ and consider $E = \{c' + d' | c' \in C, d' \in D\} = \{r, r+1, \ldots, r+\beta\}$. Any $e \in E$ can be written uniquely as $e = c''\beta + d''$ with $c'' \in C''$ and $d'' \in D''$, which determines γ and ξ. Hence for all $P, Q \in \mathcal{P}[\beta, D]$ we have

$$\hat{\alpha}(P, Q) = (P_1, Q_1) \text{ with } P_1 \in \mathcal{P}[\beta, C'], Q_1 \in \mathcal{P}[\beta, D'],$$
$$\hat{\gamma}(P_1, Q_1) = (P_2, Q_2) \text{ with } P_2 \in \mathcal{P}[\beta, C''], Q_2 \in \mathcal{P}[\beta, D''],$$
$$\hat{\xi}(P_2, Q_2) = (0, R) \text{ with } R \in \mathcal{P}[\beta, D].$$

For $r = 0$ choose $C' = \{0, 1, 2\}$ and $D' = \{0, 1, \ldots, \beta - 1\}$, and for $r = -\beta$, choose $C' = \{-2, -1, 0\}$, $D' = \{-\beta + 1, \ldots, 0\}$, and $C'' = \{-1, 0\}$, but otherwise proceed as above, noting that D is semicomplete in these cases. □

Example 3.4.2 Let us construct regular addition mappings α, γ, and ξ, each mapping $D \times D$ into $D \times D$ for $D = \{-1, 0, 1, 2\}$ with $\beta = 3$, such that

$$\hat{\xi}(\hat{\gamma}(\hat{\alpha}(P, Q))) = (0, R)$$

for all $P, Q \in \mathcal{P}[3, D]$, with result $R \in \mathcal{P}[3, D]$. Following the proof above we have $r = -1$ and define $C' = \{-1, 0, 1\}$ and $D' = \{0, 1, 2\}$ implying that we will avoid using the digit value -1 in the "place" position, which defines the α table. The γ table must then be able to add carries in $C' = \{-1, 0, 1\}$ to digits in $D' = \{0, 1, 2\}$, where we avoid generating the carry value -1, so $C'' = \{0, 1\}$ with $D'' = \{-1, 0, 1\}$. We can finally define the ξ table absorbing carries.

α	-1	0	1	2
-1	$\bar{1}1$	$\bar{1}2$	00	01
0	$\bar{1}2$	00	01	02
1	00	01	02	10
2	01	02	10	11

γ	-1	0	1	2
-1		$0\bar{1}$	00	01
0		00	01	$1\bar{1}$
1		01	$1\bar{1}$	10
2				

ξ	-1	0	1	2
-1				
0	$0\bar{1}$	00	01	
1	00	01	02	
2				

But instead of eliminating the digit value -1 in D' we may choose to avoid the value 2 and then obtain the following mappings:

α	-1	0	1	2
-1	$\bar{1}1$	$0\bar{1}$	00	01
0	$0\bar{1}$	00	01	$1\bar{1}$
1	00	01	$1\bar{1}$	10
2	01	$1\bar{1}$	10	11

γ	-1	0	1	2
-1	$\bar{1}1$	$\bar{1}2$	00	
0	$\bar{1}2$	00	01	
1	00	01	02	
2				

ξ	-1	0	1	2
-1		$0\bar{1}$	00	01
0		00	01	02
1				
2				

□

For radix $\beta = 2$ the minimally redundant (symmetric) digit set $D = \{-1, 0, 1\}$ ("borrow-save" or "signed-digit") satisfies Theorem 3.4.2. Table 3.4.2 shows three such addition mapping, α, γ, ξ, for a three-level addition over the digit set D. Since the last table (ξ) of Tables 3.4.2 always acts as a *carry-absorption table* there is no need to display the zero-valued carry, hence in the following these zeroes will not be shown.

Table 3.4.2. *Redundant symmetric radix-2 addition tables*

α	-1	0	1
-1	$\bar{1}0$	$\bar{1}1$	00
0	$\bar{1}1$	00	01
1	00	01	10

γ	-1	0	1
-1		$0\bar{1}$	00
0		00	$1\bar{1}$
1		$1\bar{1}$	10

ξ	-1	0	1
-1			
0	$0\bar{1}$	00	
1	00	01	

The addition table for α shows that the carry digit set is $\{-1, 0, 1\}$, whereas the sum digits belong to $\{0, 1\}$. The table for γ has carry digits in $\{0, 1\}$ and sum digits in $\{-1, 0\}$, thus the carry digit in the table for ξ will always be 0.

Note that in the table for γ only one entry cannot be written with a zero carry, the $(1, 1)$ entry has to be 10. The γ table is used to add a sum digit in this position with a carry digit from the rightmost neighbor. The α table determines which sum digit is to be generated in this position, and a closer look reveals that the α mapping need not generate a 1 digit; it could alternatively choose to generate a -1 digit, if it were known that it was later to be added to a carry digit of value 1. Thus the $(1, 1)$ entry would never be consulted, and the γ table could be redefined only to generate 0 as the carry, hence the ξ table would not be needed.

Assuming that some partial information on the rightmost neighboring position is known, allowing the sign of the incoming carry to be determined, then a choice between two different α tables as in Table 3.4.3 can be made by *carry anticipation*:

Here α^- is used when the incoming carry is known to be non-positive ($c_{in} \leq 0$), and α^+ is used when it is non-negative ($c_{in} \geq 0$), hence the entries $\gamma(-1, -1)$ and $\gamma(1, 1)$ will never be consulted and γ is carry-free. It is assumed that the choice

Table 3.4.3. *Carry anticipating redundant symmetric radix 2 addition tables*

α^-	-1	0	1
-1	$\bar{1}0$	$\bar{1}1$	00
0	$\bar{1}1$	00	01
1	00	01	10

α^+	-1	0	1
-1	$\bar{1}0$	$0\bar{1}$	00
0	$0\bar{1}$	00	$1\bar{1}$
1	00	$1\bar{1}$	10

γ	-1	0	1
-1		$\bar{1}$	0
0	$\bar{1}$	0	1
1	0	1	

of table can be based on some partial information on the digits to be added in the rightmost neighboring position. We shall return to an implementation using this idea in Section 3.7.

Comparing the tables for α^- and α^+ it is seen that the result of adding a zero digit to a non-zero digit is different in the two tables. Hence the choice of which table to use is *non-deterministic* when the incoming carry is zero, since then either one of the tables may be chosen. In an implementation the actual choice may depend on how information on a zero-valued incoming carry c_{in} has been determined and is represented (encoded). If some kind of sign-magnitude encoding is used, the choice of table will depend on the actual sign, and if zero carries no sign, a fixed choice will have to be made. Of course, both results are correct, they are just different redundant representations of the same value.

The other "limited carry" radix-2 addition is the well known *carry-save addition*, often used in multipliers for accumulating several addends. One "double" register (the accumulator) contains the result of previous additions encoded in "carry-save" form with two bits per position:

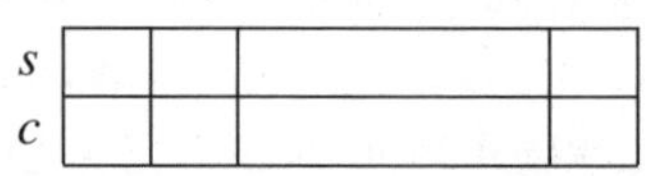

where the bits in the c register represent carries which have not yet been added in ("saved"). Each position of the combined registers of the accumulator is then an encoding of the semicomplete digit set $\{0, 1, 2\}$ with two possible encodings of the digit 1. Writing a (c, s)-pair as a string cs, the encodings are 00 for 0, 01, or 10 for 1 and 11 for 2.

The carry-save adder takes one such operand and another operand in standard binary (e.g., 2's complement) and returns the result in the accumulator, effectively "counting" the number of ones in the same position of the three registers:

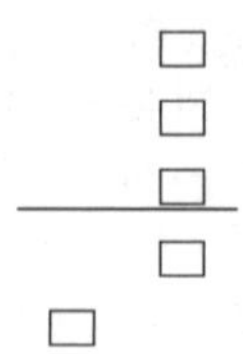

placing the resulting (c, s) pair in accumulator registers, the c-value shifted one position to the left.

In our terminology carry-save addition uses two addition mappings $\gamma\colon \{0, 1, 2\} \times \{0, 1\} \to \{0, 1\} \times \{0, 1\}$ and $\xi\colon \{0, 1\} \times \{0, 1\} \to \{0\} \times \{0, 1, 2\}$, as described in Tables 3.4.4 γ and Table 3.4.4 ξ.

Table 3.4.4. *"Carry save" addition tables*

α	0	1	2
0	00	01	10
1	01	10	11
2	10	11	20

γ	0	1	2
0	00	01	
1	01	10	
2	10	11	

ξ	0	1	2
0	0	1	
1	1	2	
2			

The addition mapping α of Table 3.4.4 has been included to allow the symmetric case where two "carry-save" operands are to be added in a three-level process, in analogy with the redundant radix-2 addition as in Table 3.4.2. With the chosen encoding of digit values the mapping ξ is just the identity mapping (pairing of bits), and as we shall see in Section 3.5, the mapping γ may be realized by a circuit called the *full-adder*. In Section 3.7 we shall implement the combined mappings in two levels, using carry anticipation in analogy with Table 3.4.3.

Carry-save addition is an example of a non-symmetric addition, where operands need not be from the same set of radix polynomials. In general, it is possible to define radix-β addition where operands are over digit sets D_1 and D_2, and the result is represented in a digit set D_3, with D_1, D_2, and D_3 possibly being different. Addition tables can easily be defined for such *non-regular addition*, and again the process can be performed in constant time if D_3 is redundant, employing the techniques of digit set conversions.

Carry-save addition is often termed 3-to-2 addition, as its effect is to rewrite the sum of three addends into two addends with the same sum, all addends in binary. The principle may be generalized, e.g., seven addends may be rewritten as three addends having the same sum, in effect counting the number of ones in a particular position of seven registers (a column), writing the result as a three-bit number. A circuit performing this type of addition is henceforth often called a *7-to-3 counter*, just as the full-adder may be considered a *3-to-2 counter*,

In general, we may consider a generalized addition mapping α in radix β as a mapping

$$\alpha \;:\; D_\beta^n \to D_\beta^m,$$

where for $s_i, c_i \in D_\beta$

$$\alpha(s_1, s_2, \ldots, s_n) = (c_1, c_2, \ldots, c_m),$$

such that

$$s_1 + s_2 + \cdots + s_n = c_m\beta^{m-1} + \cdots + c_2\beta + c_1.$$

The relation between n and m depends on the digit set D, but with $\delta = \max_{d \in D} |d|$ obviously

$$n \times \delta \le \delta(\beta^{m-1} + \cdots + \beta + 1)$$

or

$$n \le \frac{\beta^m - 1}{\beta - 1} \tag{3.4.1}$$

must hold.

For the fastest possible addition of multiple addends it is advantageous to reduce the problem to a number of smaller addition problems performed in parallel, whose results are then again added in parallel, etc., so the process takes the form of a tree structure. For example, reducing to ordinary dyadic additions a binary tree is obtained:

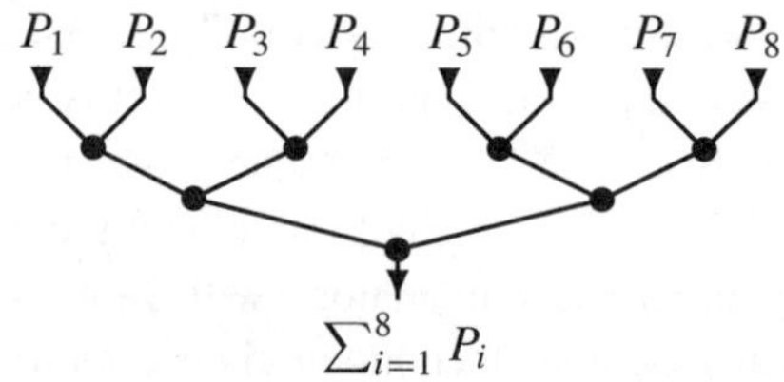

Obviously, it is here advantageous if the additions at the nodes can be performed in digit parallel, i.e., the results of the operations at the nodes are allowed to be in a redundant representation and are then permitted as operands further down the tree. Thus the time for adding p operands will be proportional to $\log(p)$ and independent of the number of digits of the operands, not including a possible final conversion to a non-redundant representation, which can also be performed in $O(\log n)$ time for n-digit operands.

Problems and exercises

3.4.1 With $\beta = 3$ and $D = \{-2, -1, 0, 1, 2\}$ how many regular radix-3 addition mappings $\alpha : \; D \times D \longrightarrow D \times D$ are there?

3.4.2 Apply both of the mappings $\hat{\xi}\hat{\gamma}\hat{\alpha}$ found in Example 3.4.2 to $P = 1221.201_{[3]}$ and $Q = 201\bar{1}.112_{[3]}$.

3.5 Basic linear-time adders

Addition of radix-β integers represented over some digit set D (redundant or non-redundant) can be realized by hardware implementations of the addition mapping as a table look-up. This may be feasible for moderate values of $|\beta|$; however, for small values of $|\beta|$, or for radices of the form β^k for small $|\beta|$ combinational circuitry is generally used.

As computers presently are almost exclusively built with two-state (binary) logic, we will assume that the digits of the digit set D are coded as *bit-patterns*, i.e., we assume there is a $k \geq 1$ such that $2^k \geq |D|$, a set $R \subseteq \{0, 1\}^k$, $|R| \geq |D|$ and a mapping $\chi : R \rightarrow D$ which is onto D, but need not be one-to-one. χ acts as a *decoding function* mapping a k-bit pattern $(b_{k-1}, \ldots, b_1, b_0) \in R, b_i \in \{0, 1\}$, into a digit $d \in D$. Since χ is onto D, for all $d \in D, \exists r \in R$ such that $\chi(r) = d$, and R may be partitioned into equivalence classes

$$R_d = \{r \in R \mid \chi(r) = d, d \in D\} \text{ with } R = \bigcup_{d \in D} R_d$$

so that

$$R_{d_1} \cap R_{d_2} = \emptyset \text{ for } d_1 \neq d_2 \in D.$$

If for any equivalence class R_d a particular member $r_d \in R_d$ is chosen, an *encoding function* $\chi' : D \rightarrow \{r_d \mid d \in D\}$ may be defined, mapping digits into bit-patterns. For convenience we will use the term *coding* collectively for a pair of decoding and encoding functions. Also we will often write a bit-pattern $(b_{k-1}, \ldots, b_1, b_0) \in R, b_i \in \{0, 1\}$ as a string $b_{k-1} \cdots b_1 b_0$.

The simplest case occurs when $\beta = 2^k$, $k \geq 0$, and the non-redundant digit set $D = \{0, 1, \ldots, \beta - 1\}$ is chosen. Here digits can be coded as bit-strings when represented as radix-2 integers, i.e., with $r = b_{k-1} \cdots b_1 b_0 \in \{0, 1\}^k = R$, then $\chi(r) = \chi(b_{k-1} \cdots b_1 b_0) = \sum_{i=0}^{k-1} b_i 2^i \in D$, and χ' becomes the inverse mapping from integers into their ordinary binary radix representation, which can be realized by the DGT Algorithm. In this case the coding is *non-redundant* and *complete*, in analogy with the terminology used for digit sets. Note, however, that a redundant digit set can have a non-redundant encoding and vice versa. We shall return later to the implications of redundancy and non-redundancy in digit codings.

For signed digit sets, in particular symmetric digit sets, one possible encoding is a *sign-magnitude* encoding, e.g., for the symmetric, redundant radix-2 digit set $\{-1, 0, 1\}$ the following encoding is often used:

$$\begin{aligned} -1 &\sim 11, \\ 0 &\sim 10 \quad \text{or} \quad 00, \\ 1 &\sim 01, \end{aligned}$$

which is a redundant encoding of a redundant digit set. By ruling out say the encoding 10, a non-redundant coding is obtained, but another way to achieve this is to use the following:

$$\begin{aligned} -1 &\sim 100, \\ 0 &\sim 010, \\ 1 &\sim 001, \end{aligned}$$

or equivalent codings, sometimes termed a *direct coding*.

As another case consider the BCD code for representing the decimal digits in four-bit binary, shown in Table 3.5.1 together with two other codes, the *excess-3 code* and the BCD4221 code. Note that in the two first codings some bit-patterns do not have an interpretation as decimal digits, i.e., they are illegal. In the last (redundant) coding the four bits have weights as indicated in the name of the code.

Table 3.5.1. *BCD and excess-3 coded decimals*

d	BCD(d)	Exs3(d)	BCD4221(d)
0	0000	0011	0000
1	0001	0100	0001
2	0010	0101	0010 or 0100
3	0011	0110	0011 or 0101
4	0100	0111	0110 or 1000
5	0101	1000	0111 or 1001
6	0110	1001	1010 or 1100
7	0111	1010	1011 or 1101
8	1000	1011	1110
9	1001	1100	1111

These four-bit codes are clearly not very compact, but other encodings are possible where a larger number of decimal digits are encoded in binary. For example, encoding three decimal digits in ten bits is very efficient, and has found use in decimal floating-point representations in the encoding of "declets" under the name *densely packed decimal* (DPD). There are then 24 non-canonical bit-patterns which map into the range 0–999; however, these should not be generated. The encoding and decoding can be implemented fairly efficiently, requiring only a few gate levels of logic.

The *carry-save encoding* introduced in the previous section is another example with a redundant encoding, where 01 and 10 both decode into the digit value 1. It is customary to permit both of these two encodings as they allow a simpler circuitry.

It is essential to notice the distinction between redundancy in a digit set (here $\{0, 1, 2\}$ in radix 2), and redundancy in the coding of the digits (the codings 01 and 10 for the digit value 1). Redundancy in the digit set is fundamental to the algorithm, e.g., to allow constant-time addition irrespective of the coding of the digits. Redundancy in the digit coding does influence the logic design of the adder, but can only change its speed by a constant factor.

The simplest adder structure is the *half-adder* which realizes the addition table shown in Table 3.3.1α, i.e., it takes two operands x and y in the standard radix-2 digit set $\{0, 1\}$ with the trivial encoding into $\{0, 1\}$. From the addition table it is easy to derive the following logical expressions for c and s:

$$\begin{aligned} c &= x\,y, \\ s &= x\,\bar{y} + \bar{x}\,y \\ &= x \oplus y \\ &= (x+y)\,\bar{c}, \end{aligned}$$

where three alternative expressions for s have been listed. A possible circuit is shown in Figure 3.5.1.

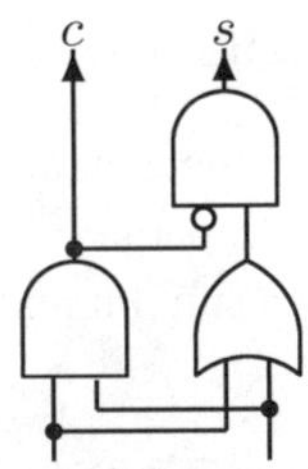

Figure 3.5.1. Half-adder circuit.

Combining n half-adders, each realizing an addition mapping α, we obtain a realization of a radix polynomial addition mapping $\hat{\alpha}$, capable of adding two radix-2 polynomials (integers), producing the output in the form of a pair of radix-2 polynomials, as defined by (3.3.4). Note that the carry polynomial is "shifted" one position to the left. It is now possible to apply $\hat{\alpha}$ repeatedly by employing another set of half-adders for each application of $\hat{\alpha}$, but due to the "shifting" of the carry polynomial, each level of adders need only contain one half-adder less than the previous level, since there is no carry coming into the least-significant position. Thus a triangularly shaped array of half-adders suffices, but a little observation will show that even two levels are sufficient. From Table 3.3.1α it is immediately seen that $(c, s) \neq (1, 1)$ always holds for the half-adder. Thus in two levels of half-adders:

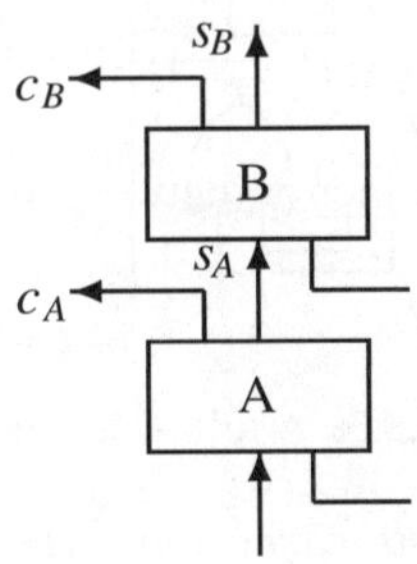

we obtain $(c_A, c_B) \neq (1, 1)$, since

$$c_B = 1 \Rightarrow s_A = 1 \Rightarrow c_A = 0 \ \text{ and } \ c_A = 1 \Rightarrow s_A = 0 \Rightarrow c_B = 0.$$

Hence c_A and c_B may be combined (added) in an OR gate Figure 3.5.2, and used as a common carry out of the A, B pair of half-adders:

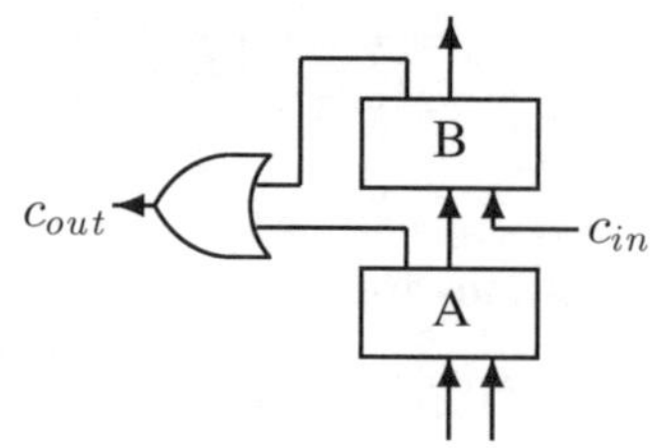

Figure 3.5.2. A full-adder composed of two half-adders.

A *full-adder* is thus a circuit realizing a mapping $\alpha : \{0,1\} \times \{0,1\} \times \{0,1\} \to \{0,1\} \times \{0,1\}$, i.e., a *3-to-2 adder* capable of adding two binary digits together with an incoming carry, producing a sum bit and a carry bit which may be defined as

$$\begin{aligned} s &= a \oplus b \oplus c_{in}, \\ c_{out} &= ab + (a+b)c_{in} \\ &= ab + ac_{in} + bc_{in} \\ &= ab + (a \oplus b)c_{in}. \end{aligned}$$

thus realizing the addition Table 3.4.4γ.

One of several possible ways of implementing the full-adder is shown in Figure 3.5.3.

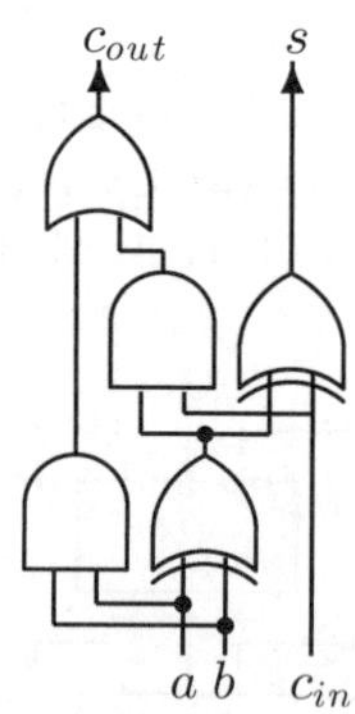

Figure 3.5.3. A full-adder circuit.

Cascading n full-adders a *ripple-carry adder* (Figure 3.5.4) thus realizes the repeated applications of the $\hat{\alpha}$-mapping by feed-back of the carries. As seen from the figure a carry from position 0 may "ripple" through all positions, and thus influence all positions to the left. The timing of this type of adder is

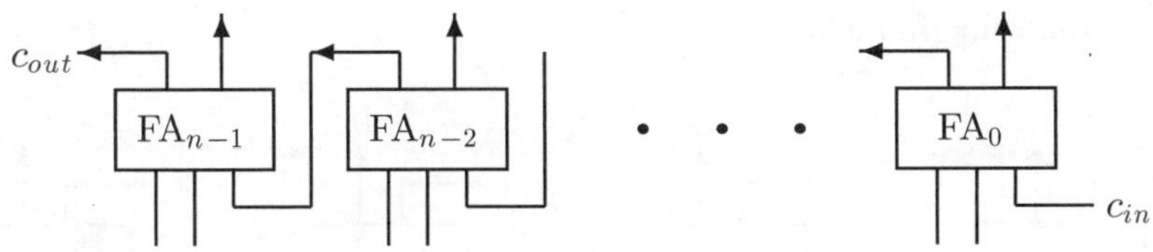

Figure 3.5.4. Ripple-carry adder.

then $O(n)$. The carry into the least-significant position (c_{in} in Figure 3.5.4) will usually just be a constant zero, in which case a half-adder is sufficient in that position.

For $\beta = 2^k$ with non-redundant digit set $D = \{0, 1, \ldots, \beta - 1\}$ and binary encoding of digit values, a radix-2^k-digit adder can then be constructed as a k-bit ripple-carry adder. Then m such radix-2^k-digit adders can be cascaded to form an m-digit adder, which is of course the hardware equivalent of an $m \times k$ binary ripple-carry adder.

For β not a power of 2, radix-β digit adders can be constructed from full-adders when a binary encoding of digit values is employed. For example, for $\beta = 10$ and BCD encoding a decimal digit adder can be constructed from four full-adders, when supplied with additional circuitry for reducing the sum modulo 10 and generating the appropriate carry. It is, however, simpler to generate the carry (and know when to reduce modulo 10) when the excess-3 encoding is used (see Table 3.5.1). Here each digit d is encoded as the binary representation of $d + 3$. Hence the carry-out is identical to the carry-out of the four-bit adder, and reduction modulo 10 is performed by adding 3, whereas 3 has to be subtracted whenever no carry is generated.

For technical reasons adders may be constructed with some of the inputs or outputs inverted, which turns out to be useful in certain situations. We will describe such adders by denoting their "terminals" according to their relative "weights," respectively with a "1" or a "2," and a "sign" "p" for "positive" (i.e., "direct"), respectively "n" for "negative" (or "inverted"). Including the standard half- and full-adders (pp and ppp), we have the following half-adders:

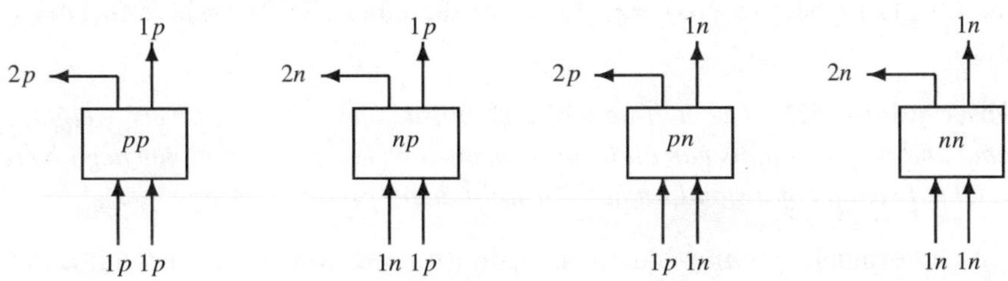

and the following full-adders:

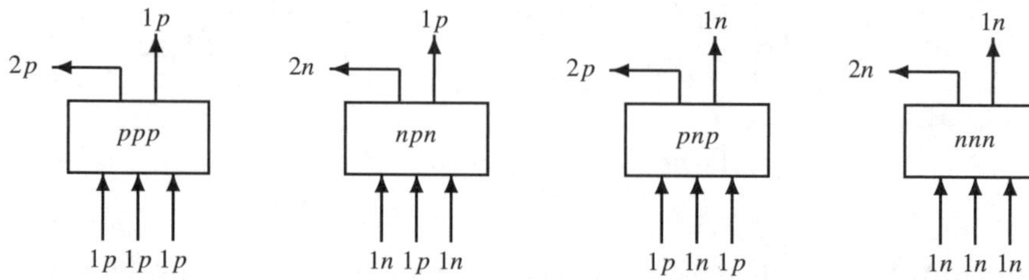

where an n-output is to be connected to an n-input, etc. To prove the claimed relationship consider the following.

Lemma 3.5.1 *Let a binary signal $b \in \{0, 1\}$ have associated weight w, so that the value of the signal is $v = w\,b$. Inverting the signal into $1 - b$ while at the same time negating the sign of the weight, changes the value into $v' = v - w$, i.e., the value is being biased by the amount $-w$.*

Proof This is trivial, since $v' = (-w)(1 - b) = b\,w - w = v - w$. □

Since the domain of combined input values must equal the range of the output values, it is possible to state certain equalities which must be satisfied. First consider the full-adders, denoting the input signals s_1, s_2, s_3 and output signals o_1, o_2. Using the extremal values $s_i \in \{-1, 1\}$ and $o_i \in \{-1, 1\}$, we seek solutions to the equation

$$s_1 + s_2 + s_3 = 2\,o_1 + o_2,$$

from which the four combinations above (not including permutations of input signals) are easily found, considering various biases caused by changes on the left-hand side, and solving for the right-hand side with the same bias. For example, changing the input from *ppp* to *pnp* changes the input domain from $\{0, 1, 2, 3\}$ to $\{-1, 0, 1, 2\}$ (bias -1), hence the output o_2 must change from p to n, to be biased by the same amount.

Half-adders are slightly more complicated, since certain combinations of output signals cannot occur. Obviously the *pn* and *np* adders have the same input domain $\{-1, 0, 1\}$, but allow two different solutions to the equation $s_1 + s_2 = 2\,o_1 + o_2$, one $(2n, 1p)$ where $(o_1, o_2) \neq (-1, 0)$ and the other $(2p, 1n)$ where $(o_1, o_2) \neq (1, 0)$.

Observation 3.5.2 *The half-adders and full-adders with negatively weighted input and output signals can all be implemented by complementing the negatively weighted signals of a standard half- or full-adder (pp or ppp).*

As an example, an *npn*-adder can be constructed from a standard full-adder, i.e., a *ppp*-adder:

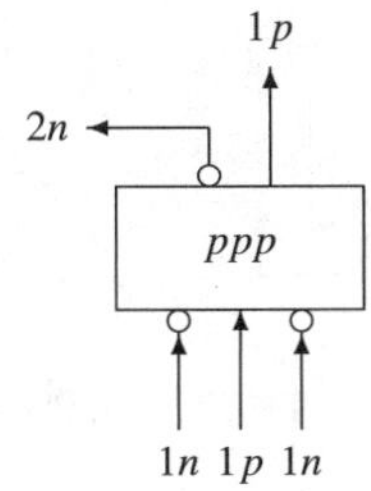

These circuits turn out to be very useful when implementing logic based on the radix-2, "borrow-save" or "signed-digit" symmetric digit set $\{-1, 0, 1\}$ using the following particular encoding of the digits:

$$\begin{aligned} -1 &\sim 10, \\ 0 &\sim 00 \quad \text{or} \quad 11, \\ 1 &\sim 01, \end{aligned} \tag{3.5.1}$$

where the leftmost bit carries a negative weight and the rightmost one a positive weight. Note that this is a redundant encoding, since zero has two legal encodings.

However, let us start with the standard full-adder, the *ppp*-adder, denoting its three inputs from left to right by x, y, and c, and the two outputs by c' and s, satisfying the adder relation

$$x + y + c = 2c' + s.$$

As we have seen, if we take an array of such *ppp*-adders and connect their outgoing carries c' to the incoming carries c of their leftmost neighboring cells, etc., we obtain the ripple-carry adder. Or as we may also interpret it, a converter which takes a carry-save encoded operand in $\mathcal{P}[2, \{0, 1, 2\}]$ and converts it into the set $\mathcal{P}[2, \{0, 1\}]$.

With the same labeling of the terminals we find the following relation for the *npn*-adder:

$$-x + y - c = -2c' + s, \tag{3.5.2}$$

where x, y, c, c', and s all belong to $\{0, 1\}$. If in (3.5.2) for an array of similarly carry-connected *npn*-adders we consider the pair (x, y), the encoding of a radix-2, signed-digit operand in the digit set $\{-1, 0, 1\}$ using the encoding (3.5.1), the operands c-bits act as "borrows" due to the negative weight, and the circuit itself generates "borrow" bits and "sum" bits. In effect we have a converter from signed digit $\mathcal{P}[2, \{-1, 0, 1\}]$ into 2's complement, since the digits generated are in $\{0, 1\}$, with carries in $\{-1, 0\}$, so possibly a carry of -1 is pushed outside the finite representation. But on the other hand it can also be considered a subtractor, computing $y - x$ in 2's complement.

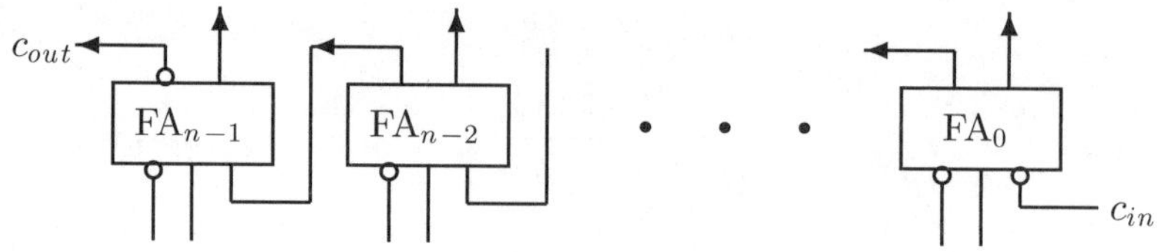

Figure 3.5.5. 2's complement ripple-carry subtractor.

Instead of using *npn*-adders, by complementing one of the input bits to each full-adder in Figure 3.5.4, together with c_{in} and c_{out}, we obtain a subtractor such as that shown in Figure 3.5.5.

Note that it is not necessary internally in the array to invert the carry-out of the full-adders, as the carry-in to the next adder also has to be inverted.

3.5.1 Digit serial and on-line addition

An alternative implementation of an addition circuit can be obtained by observing in Figure 3.5.4, that the computations in the adders may be performed in sequential order, starting with the least-significant digit. Thus a single full-(*ppp*-)adder is sufficient, if only the outgoing carry value is fed back and used as incoming carry, simultaneously with a new set of operand bits being supplied. A buffer is needed to retain the value of the carry from one cycle to the next.

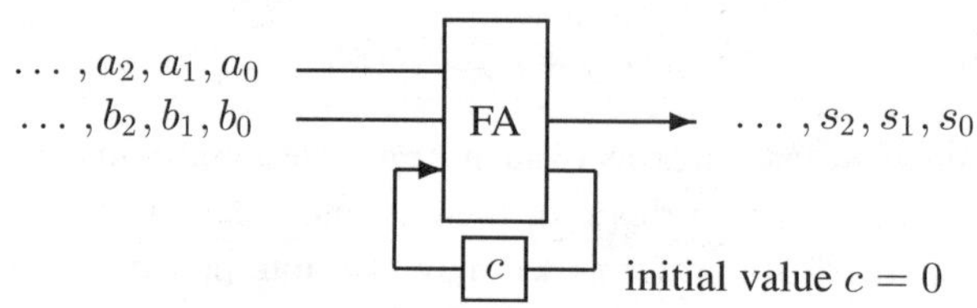

Figure 3.5.6. Digit-serial adder.

Such a *digit-serial adder* is shown in Figure 3.5.6, where operands A and B are supplied digit by digit (least-significant digit first) and the result S is produced digit by digit. Note that this principle can be applied to addition in any radix. As in Section 2.3 we may picture this as a transducer, now taking two digit streams and producing their sum as one digit stream, where α is the addition mapping, and the state information is the carry value c_{i+1}.

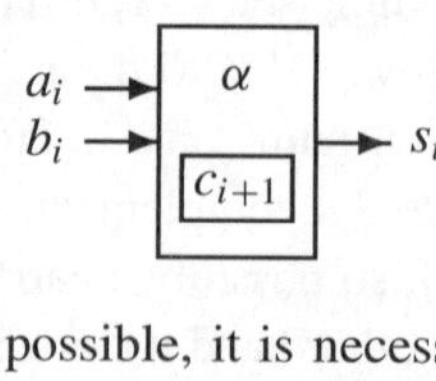

For on-line addition to be possible, it is necessary that the output digit set is redundant, and we will furthermore assume that it is basic and symmetric, hence also contiguous. Let us then investigate under which conditions it is possible to perform a digit set conversion from $D^+ = \{-2r, \dots, 2r\}$ back into $D = E = \{-r, \dots, r\} = \Sigma + C$ in a two-level conversion (i.e., with an on-line delay of 1), for suitably chosen β, r, Σ, assuming that $C = \{-1, 0, 1\}$. Since $\Sigma = \{-r + 1, \dots, r - 1\}$ with $|\Sigma| = 2r - 1$ must be a complete digit set, we must require $2r > \beta$. However, from this condition it follows that

$$\begin{aligned}\Sigma + \beta C &= \{-r+1, \dots, r-1\} + \beta\{-1, 0, 1\} \\ &= \{-r - \beta + 1, \dots, r - 1 - \beta, -r + 1, \dots, r - 1, -r + 1 \\ &\quad + \beta, \dots, r - 1 + \beta\}\end{aligned}$$

is a contiguous digit set. But we must also require that $D^+ \subseteq \Sigma + \beta C$, or equivalently that $2r \le r - 1 + \beta$, hence we have shown the following.

Theorem 3.5.3 *On-line addition with a delay of* 1 *is possible for radix* β *if the digit set* D *has the form* $D = \{-r, \dots, r\}$, *if* β *and* r *satisfy*

$$r < \beta < 2r.$$

Note that these conditions cannot be satisfied for $\beta = 2$, but as we saw in Section 2.4 it is possible to perform on-line conversion in this case with an on-line delay of 2. Since radix 2 is the most important case, we will not pursue the higher-radix cases further, but now look into the realization of the radix-2 case for the digit set $\{-1, 0, 1\}$.

Let us first look at a *pnp*-adder with inputs coming in from the right:

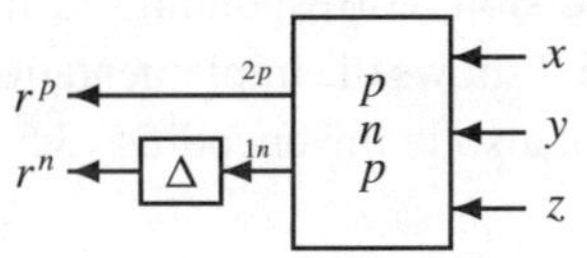

where the result bit r_n of negative weight $1n$ is being delayed in a buffer (delay element Δ), so that it is paired with the next bit of twice the numeric weight $2p$. This now represents a transducer, converting from the digit set $D = \{0, 1\} + \{-1, 0\} + \{0, 1\} = \{-1, 0, 1, 2\}$ into the set $E = \{-1, 0, 1\}$ using the borrow-save encoding (r^p, r^n) of (3.5.1), as may easily be seen by considering the sets to which x, y, z, r^p and r^n, respectively belong. The following figure shows the *pnp*-transducer together with its conversion diagram:

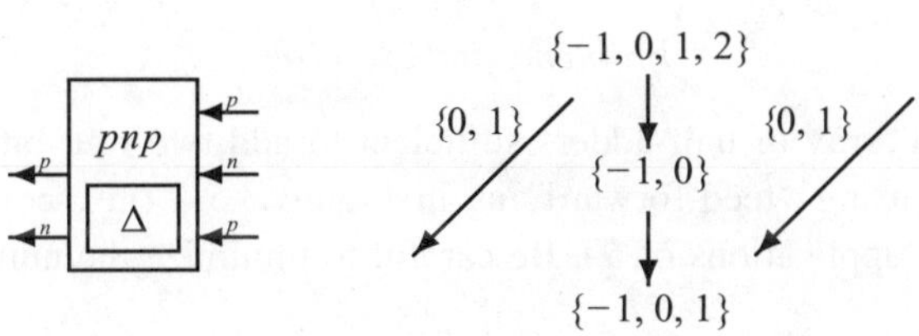

Similarly we can derive the *npn*-transducer with its conversion diagram:

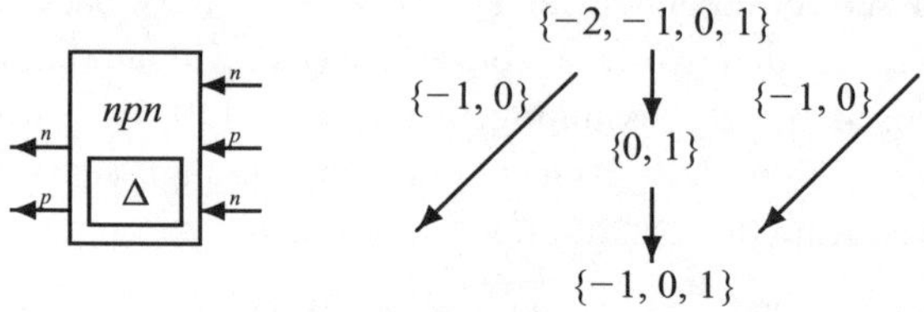

where we may recall from Example 2.5.3 that either one of these may be used as the last transducer for an on-line adder of delay 2.

However, neither of them will suffice as the initial transducer; both will need an extra input. Say we want to extend an *pnp* transducer, then an extra *n* input is needed, corresponding to the four inputs from two operands $A = (a^n, a^p)$ and $B = (b^n, b^p)$. This input then has to be delayed to be synchronized with the other signals. Incorporating the digit values $\{-1, 0\}$ of the extra signal into the digit sets we find the digit set conversion diagram shown:

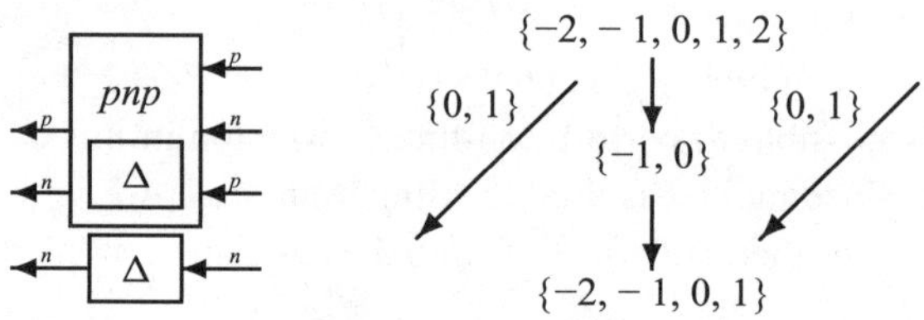

whose output digit set fits precisely with the *npn* transducer as the next transducer to complete the digit set conversion, corresponding to the on-line addition. To complete the picture, Figure 3.5.7 shows the implementation in the form of adders, producing the sum $S = (s^n, s^p)$, also in the encoding (3.5.1).

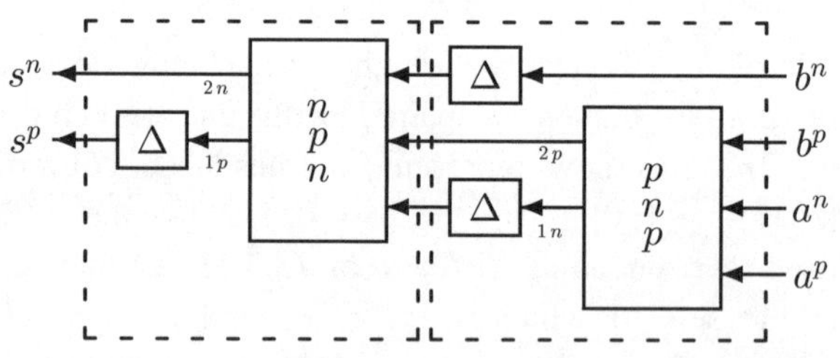

Figure 3.5.7. Radix 2 on-line adder of delay 2

Problems and exercises

3.5.1 Draw an array of half-adders sufficient to add two four-bit numbers, but without using "feed-forward" as in Figure 3.5.4 (i.e., corresponding to repeated applications of $\hat{\alpha}$). Be careful to minimize the number of adders used.

3.5.2 Show that the full-adder of Figure 3.5.3 actually implements addition table γ of Table 3.4.4.

3.5.3 Construct a 7-to-3 counter as discussed in Section 3.4, using 3-to-2 counters (i.e., full-adders) as building blocks.

3.5.4 Show that the N-, P-, and Q-mappings of Section 2.5 can be realized by half-adders.

3.6 Sub-linear time adders

In the ripple-carry adder c_{out} depends on c_{in} so one might think that the process of adding two numbers x and y is an inherently sequential process. However, if in the ith adder the input satisfies $x_i = y_i$, then c_{out} does not depend on c_{in}, since $x_i = y_i = 0$ implies $c_{out} = 0$ and $x_i = y_i = 1$ implies $c_{out} = 1$ irrespective of the value of c_{in}. On the other hand, if $x_i \neq y_i$, then c_{out} will equal c_{in}, and the carry is propagated through the adder. So carries *propagate* through sequences of positions where $x_i \neq y_i$, the incoming carry is *absorbed* or *killed* in an adder where $x_i = y_i = 0$, and a new carry is *generated* when $x_i = y_i = 1$.

Observation 3.6.1 *The input–output properties of the carries entering and leaving a single adder cell in a ripple-carry adder can be described by the following table:*

x_i	y_i	Carry relation	Cell status
0	0	$c_{out} = 0$	Kill
0	1	$c_{out} = c_{in}$	Propagate
1	0	$c_{out} = c_{in}$	Propagate
1	1	$c_{out} = 1$	Generate

(3.6.1)

It is possible to show that the average length of the longest sequence of positions, where $x_i \neq y_i$, is $\log_2 n$, when adding two n-bit numbers chosen randomly (uniformly). Hence the *average* running time of the ripple-carry adder is actually only $O(\log n)$, but for most applications it is inconvenient to design a system based on a primitive which has a variable execution time. Also some additional circuitry is needed to detect and signal the completion of the ripple-carry, and hence the availability of the result. Circuitry with such a capability is called *self-timed*, and has been exploited and is finding use in some high-speed designs.

However, in most applications it is convenient if the timing of the circuitry is clocked, which requires that results are guaranteed to be available at specific points in time, hence designs are based on the worst-case running time of aggregated components. Hence let us proceed to find adder constructions where we exploit the above observations on carry behavior.

3.6.1 Carry-skip adders

This class of adders is based on the observation that the ability of a single cell to propagate, or not c_{in} may be generalized to a block of cells. In the full-adder of Figure 3.5.3, the expression

$$p_i = x_i \oplus y_i, \tag{3.6.2}$$

which is computed anyway, is the signal determining if a cell will propagate c_{in}. Then

$$P_{\ell m} = \prod_{\ell}^{m} p_i \tag{3.6.3}$$

will similarly determine whether the carry-in to position ℓ will propagate through as the carry-out of position m. Thus we can construct an n-bit adder by concatenating n/w blocks, each of width w, using additional circuitry to allow the carry to *skip* around each block if $P_{i,i+w}$ for that block is true. Such a block with additional circuitry is shown, in Figure 3.6.1.

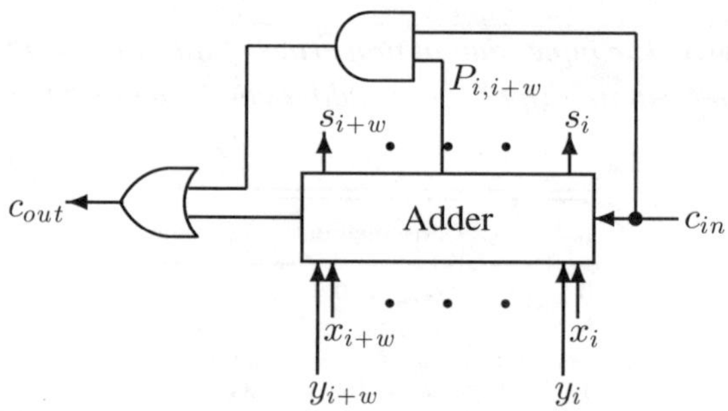

Figure 3.6.1. A carry-skip block.

Note that within the block, full-adders are combined as an ordinary ripple-carry adder, operating in time $O(w)$. However, the additional circuitry computes $P_{i,i+w}$ directly from the values of (x_j, y_j), $j = i, \ldots, i + w$, hence independently of c_{in} and intermediate carry values. So for moderate values of w the carry can skip around the block in constant time to the OR gate defining c_{out}. If $P_{i,i+w} = 0$, then the block itself defines the value of c_{out} in time $O(w)$.

For moderate values of n we may choose n/w blocks of width w, where $w \approx \sqrt{n}$. If all carry-skip signals are true, then the longest carry-skip chain is of length w, and if all carry-skip signals are false, then all blocks operate in full parallel with internal carry ripples of at most length w. The worst-case situations occur if a carry has to start in the rightmost position of the rightmost block, and then has to skip around all $w - 1$ blocks to the left, or has to start in the rightmost position of some block and has to ripple $w - 1$ positions into the next block.

Hence for moderate values of n a running time of $O(\sqrt{n})$ may be achieved, using space $\mathcal{O}(n)$.

One problem with the carry-skip adder as described above is that it does not "scale-up," because of the fan-in restrictions in the computation of $P_{i,i+w}$ for a block. It is possible to increase the size of the blocks by allowing more levels of AND gates for computing $P_{\ell m}$ for the block, at the expense of increasing the skip time. Another possibility is to group a number of blocks into "superblocks," thus performing the carry-skip operation in two (or more) levels. The size of the individual blocks should not be the same everywhere; a careful analysis is needed involving the skip time versus the ripple time to define the block (and possibly the superblock) sizes of an optimal design, in particular in VLSI where the length of wires may have to be accounted for.

Note that the amount of additional hardware needed is very minimal: since the signals $p_i = x_i \oplus y_i$ may be computed anyway in each full-adder, only one extra AND gate (of high fan-in) plus one OR-gate is needed per block. Much research has gone into optimizing the design of this type of adder for particular technologies, and usually these adders compete well with other types that theoretically (asymptotically) should be superior. In particular, a special implementation technology ought to be mentioned here, the so-called *Manchester Carry Chain*, which is a special transistorized version obtained by chaining a sequence of pass-transistors (up to about eight), controlled by the values of p_is such that if they are all open, the carry will pass through very fast. Combined with forcing the value of the g_is onto the chain, the final carry values can be obtained faster than if generated by the adders themselves, but of course still in linear time.

Before investigating adders of higher speed we will, for completeness, describe another $O(\sqrt{n})$-time adder, however of higher gate count.

3.6.2 Carry-select adders

The idea in this type of adder is to perform two ripple-carry additions in parallel in a block, one in which the incoming carry has value 0, and the other in which it has value 1. When the incoming carry is known it is used to select the output of the appropriate adder as the final output. Such a block is shown in Figure 3.6.2, where the small boxes marked "s" are binary selectors.

As in the previous adder, an n-bit adder can now be constructed from n/w such blocks of width w, and by choosing $w = \sqrt{n}$ we get an approximately running time $O(\sqrt{n})$, but the gate count is more than twice that of the ripple-carry adder. A little consideration also shows that the blocks should not be of identical width: the most-significant block will have more computing time available before the incoming carry arrives, hence it can be wider than its neighbor. Continuing this argument we find that the blocks should be designed with decreasing width towards the least-significant end, where the difference in neighboring block sizes should

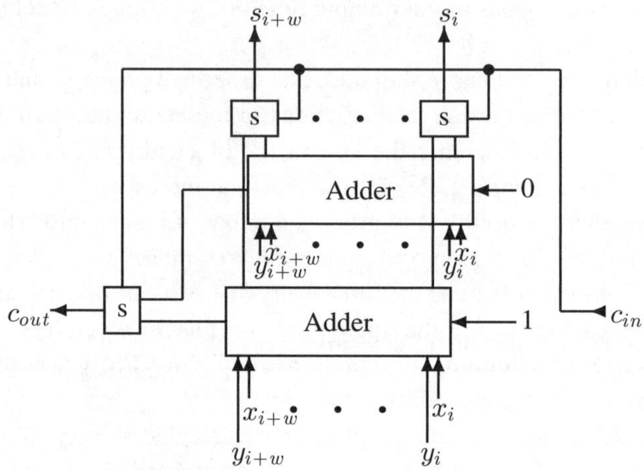

Figure 3.6.2. A carry-select block.

be symmetric with the delay in the selector determining c_{out} of the block. The technological problem in this type of adder is the high fan-out required for c_{out} to drive many selectors.

A variation of this type of adder is obtained when multilevel selection is introduced. Starting with a single bit adder as the cell at the lowest level, and then grouping in pairs to form blocks of width 2, this process may be repeated until at height $\lceil \log_2 n \rceil$ the final selection is made. This so-called *conditional-sum adder* then achieves a running time proportional to $\log_2 n$. We can derive this structure as an example of using the techniques of digit set conversion from Section 2.4.

Example 3.6.1 (Binary conditional-sum addition) The addition of two binary integers may be considered a conversion from $\mathcal{P}[2, \{0, 1, 2\}]$ into $\mathcal{P}[2, \{0, 1\}]$ by interpreting the two addends as a coding of a single polynomial in $\mathcal{P}[2, \{0, 1, 2\}]$ using the carry-save encoding:

$$\begin{aligned} 0 &\sim 00, \\ 1 &\sim 01 \quad \text{or} \quad 10, \\ 2 &\sim 11. \end{aligned}$$

The conversion mapping α can be described by the table:

		D		
	α	0	1	2
C	0	00	01	10
	1	01	10	11

from which we derive the carry-transfer functions $\{\gamma_d\}$ and digit-mapping functions $\{\xi_d\}$:

		γ_0	γ_1	γ_2
C	0	0	0	1
	1	0	1	1

		ξ_0	ξ_1	ξ_2
C	0	0	1	0
	1	1	0	1

where each column describes a function from C into C.

We note that γ_0 is the carry-transfer function that absorbs (kills) any incoming carry, γ_1 propagates the incoming carry, and finally γ_2 generates a carry-out irrespective of the value of the carry-in. Assuming that we have two carry-transfer functions:

$$\gamma' = \begin{Bmatrix} a' \\ b' \end{Bmatrix} \text{ and } \gamma'' = \begin{Bmatrix} a'' \\ b'' \end{Bmatrix},$$

we can combine these by functional composition:

$$\gamma = \begin{Bmatrix} a \\ b \end{Bmatrix} = \gamma' \circ \gamma'' = \begin{Bmatrix} a'\overline{a''} + b'a'' \\ a'\overline{b''} + b'b'' \end{Bmatrix}. \tag{3.6.4}$$

Observe how a'' is used to select the value of a as either a' or b', and similarly b'' is used to select the value of b. For an implementation it may be noted that the function $\begin{Bmatrix} 1 \\ 0 \end{Bmatrix}$ never occurs initially, and it can easily be seen never to occur at all. This implies that (3.6.4) can be simplified to

$$\gamma = \gamma' \circ \gamma'' = \begin{Bmatrix} a' + b'a'' \\ a' + b'b'' \end{Bmatrix}. \tag{3.6.5}$$

For a construction of a tree structure of $\circ$-composition nodes, the technique of *parallel prefix computation* may be employed to compute $c_i = \gamma_{d_i d_{i-1} \cdots d_\ell}(0)$ (the a-components above), which can then be used to compute the final sum digits by $\xi_{d_i}(c_i)$ as $s_i = c_i \oplus x_i \oplus y_i$. The tree structure will also compute $c_i' = \gamma_{d_i d_{i-1} \cdots d_\ell}(1)$, the value of the carry when $c_{\ell-1} = 1$. Such a structure was developed in Section 2.4. □

3.6.3 Carry-look-ahead adders

In the carry-skip adder only the propagate status p_i of the full-adder is exploited since it is easy to compute the propagate status $P_{\ell m}$ for a block of full-adders. The *carry-look-ahead adders* furthermore generalize the generate status of a full-adder to the generate status $G_{\ell m}$ of a block. So with

$$p_i = x_i \oplus y_i \qquad \text{the propagate status of cell } i$$

and

$$g_i = x_i y_i \qquad \text{the generate status of cell } i$$

we introduce a binary operator $\circ$, defined as

$$(g, p) \circ (g', p') = (g + pg', pp'),$$

which expresses the (generate, propagate) status of a block composed of two full-adders in terms of the statuses of the two adders. We then obtain the following lemma.

Lemma 3.6.2 *Let*

$$(G_i, P_i) = \begin{cases} (g_0, p_0) & \text{if } i = 0, \\ (g_i, p_i) \circ (G_{i-1}, P_{i-1}) & \text{if } 1 \leq i \leq n, \end{cases} \tag{3.6.6}$$

then for $i = 1, 2, \ldots, n$, *the value of the carry,* c_i*, into the* i*th adder is*

$$c_i = G_{i-1} + c_0 P_{i-1},$$

or

$$c_i = \begin{cases} G_{i-1} & \text{for } c_0 = 0, \\ G_{i-1} + P_{i-1} & \text{for } c_0 = 1, \end{cases} \tag{3.6.7}$$

where c_0 *is the carry entering at the least-significant position.*

Proof We prove the lemma by induction in i. In general we have

$$c_{i+1} = g_i + p_i c_i, \tag{3.6.8}$$

so in particular for $i = 0$ and the definition of (G_0, P_0)

$$c_1 = g_0 + p_0 c_0 = G_0 + c_0 P_0.$$

If $i > 0$ and assuming that $c_i = G_{i-1} + c_0 P_{i-1}$ holds, then by (3.6.8) and (3.6.6)

$$\begin{aligned} c_{i+1} &= g_i + p_i c_i \\ &= g_i + p_i (G_{i-1} + c_0 P_{i-1}) \\ &= (g_i + p_i G_{i-1}) + c_0 (p_i P_{i-1}) \\ &= G_i + c_0 P_i, \end{aligned}$$

and the result then follows. □

We are thus able to express the value of the carry c_i entering the ith full-adder, which is to be added to x_i and y_i. Our problem now concerns the efficient computation of the pairs (G_i, P_i) of (3.6.6). As defined these pairs are to be computed sequentially, corresponding to the natural carry ripple, i.e., in linear time; but by introducing the $\circ$ operator we will be able to exploit parallelism, and thus reduce the time to logarithmic. By showing that $\circ$ is an associative operator it is possible to apply the principle of "divide and conquer" to the problem.

Lemma 3.6.3 *The operator $\circ$ is associative.*

Proof For any (g_3, p_3), (g_2, p_2), and (g_1, p_1) we have

$$\begin{aligned}&[(g_3, p_3) \circ (g_2, p_2)] \circ (g_1, p_1)\\ &\quad = (g_3 + p_3 g_2, p_3 p_2) \circ (g_1, p_1)\\ &\quad = (g_3 + p_3 g_2 + p_3 p_2 g_1, p_3 p_2 p_1),\end{aligned}$$

but also

$$\begin{aligned}&(g_3, p_3) \circ [(g_2, p_2) \circ (g_1, p_1)]\\ &\quad = (g_3, p_3) \circ (g_2 + p_2 g_1, p_2 p_1)\\ &\quad = (g_3 + p_3(g_2 + p_2 g_1), p_3 p_2 p_1)\\ &\quad = (g_3 + p_3 g_2 + p_3 p_2 g_1, p_3 p_2 p_1),\end{aligned}$$

□

Due to the associativity of the operator $\circ$ it is then possible to compute

$$(G_i, P_i) = (g_i, p_i) \circ (g_{i-1}, p_{i-1}) \circ \cdots \circ (g_0, p_0) \tag{3.6.9}$$

in any order from the (g_j, p_j) of the individual full-adders; in particular we want to exploit parallelism to compute subexpressions of (3.6.9) corresponding to blocks of full-adders, and then later combine these. So let us define

$$(G_{\ell m}, P_{\ell m}) = (g_m, p_m) \circ (g_{m-1}, p_{m-1}) \circ \cdots \circ (g_\ell, p_\ell) \tag{3.6.10}$$

and use these to compute the prefixes (G_i, P_i) of (G_n, P_n) in order to find $c_1, c_2, \ldots, c_n, c_{n+1}$, which can then be combined with the operand bits $x_1, \ldots, x_n$ and $y_1, \ldots, y_n$ in a totally parallel addition process. We will thus concentrate on the computation of (G_i, P_i) for all i, utilizing *parallel prefix computation* and noting that

$$(G_{\ell m}, P_{\ell m}) = (G_{\ell,k-1}, P_{\ell,k-1}) \circ (G_{km}, P_{km}) \quad \text{for } 0 \le \ell < k \le m \le n. \tag{3.6.11}$$

Assuming that n has the form $n = 2^t - 1$, it is possible to compute expressions of the form (3.6.10) in the order of a binary tree spanning the input nodes, and then later combine these by (3.6.11) to compute all (G_i, P_i) according to (3.6.9). Following Brent and Kung we will show how these computations can be organized and how the nodes can be laid out suitably for a VLSI design.

The computation takes place in a regular structure consisting of *white nodes*:

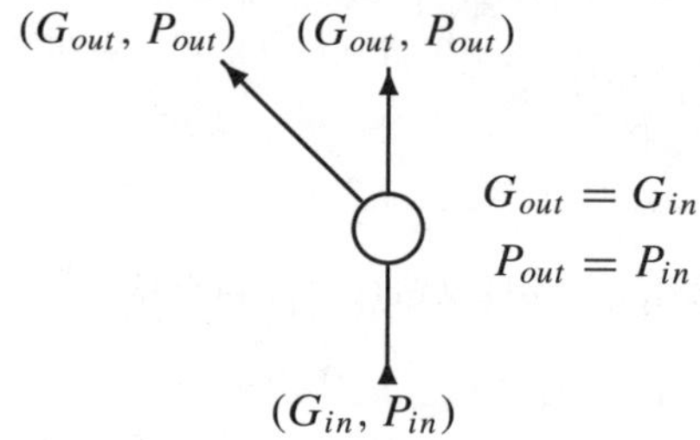

which just copy input to output, together with *black nodes* which compute the ∘-operator on their input:

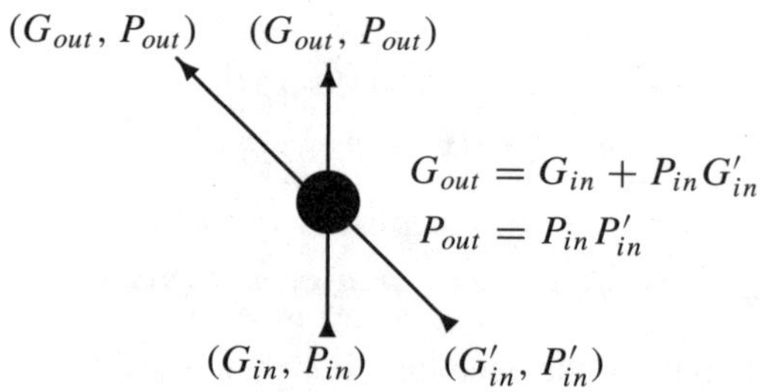

These operators can then be combined in various binary tree structures, two of which are shown in Figure 3.6.3 for $n = 7$, both using only a fan-out of 2. Note that the height of the tree in Figure 3.6.3(a) is smaller than that in Figure 3.6.3(b), but has more processing nodes.

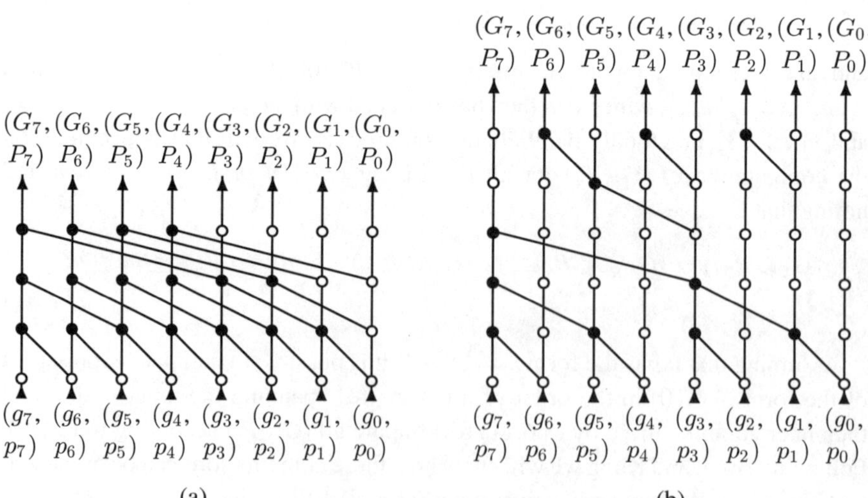

Figure 3.6.3. Carry-look-ahead trees: (a) Kogge–Stone, (b) Brent–Kung.

Theorem 3.6.4 *The carries of an n-bit binary addition can be computed in a time proportional to* $\log_2 n$ *and an area proportional to* $n \cdot \log_2 n$*, and so can the complete addition.*

After the carries c_i are found from the values of G_i (or as $G_i + P_i$, see below), the sum bits are found as

$$s_i = c_i \oplus p_i = c_i \oplus x_i \oplus y_i .$$

Observe that the computations at the nodes here are different from those of the nodes in the conditional-sum adder tree, which composes the selector tuples (a, b) according to the rules:

$$a = a' + b'a'',$$
$$b = a' + b'b'',$$

whose logic is slightly more complex, but of the same time complexity.

Note that in the conditional-sum adder tree the carries with assumed carry-in of both 0 and 1 are available, hence $x + y$ as well as $x + y + 1$ are easily computed from the carries. However, in the carry-look-ahead tree only the carries with assumed carry-in of 0 are directly available (the values of G_i), but the values for a carry-in of 1 by Lemma 3.6.2 can be obtained as $G_i + P_i$. To emphasize this we state it formally in the following observation.

Observation 3.6.5 *In logarithmic time adders using conditional-sum addition and carry-look-ahead addition it is possible at the end of the carry computation to select either one of, or provide both of, the results* $x + y$ *and* $x + y + 1$*, at no or only marginal extra cost.*

Alternatively, for the carry-look-ahead adder it is also possible to feed the carry-in directly into the tree by noting the following variant of Lemma 3.6.2, the proof of which follows by a simple modification of the proof of the lemma.

Observation 3.6.6 *Let*

$$(G_i^*, P_i^*) = \begin{cases} (g_0 + p_0 c_0, p_0) = (g_0, p_0) \circ (c_0, 1) & \text{if } i = 0, \\ (g_i, p_i) \circ (G_{i-1}^*, P_{i-1}^*) & \text{if } 1 \le i \le n, \end{cases}$$

then for $i = 1, 2, \ldots, n$

$$c_i = G_{i-1}^*,$$

where c_i *is the value of the carry-in into the* i*th adder, as determined by* c_0*, the carry entering at the least-significant position.*

By changing a white node into a black, Figure 3.6.4 shows how a carry-in can be fed into a carry-look-ahead tree, e.g., in a Kogge–Stone tree.

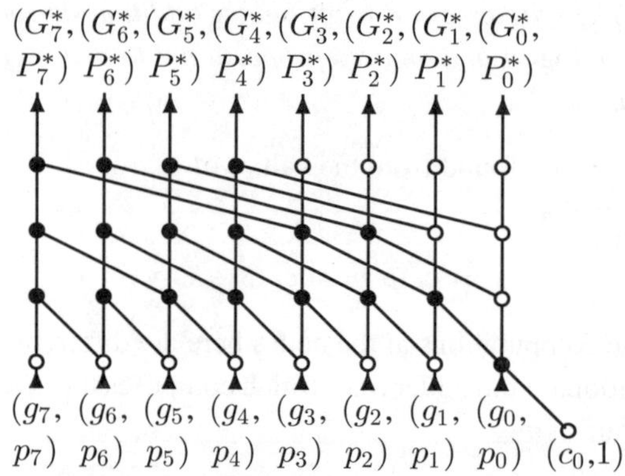

Figure 3.6.4. Kogge–Stone tree with carry-in.

Recall from (3.6.7) that it is also possible to add in a "late" carry-in, c_0, by forming the final carry as $c_i = G_{i-1} + c_0 P_{i-1}$ instead of $c_i = G_{i-1}$. Hence it is possible to combine these possiblities, thus to add in an extended carry in the set $\{0, 1, 2\}$.

Consider now the possibility of using such a carry-look-ahead adder as a subtractor. As usual the subtrahend can be obtained by complementing its bits, and supplying a carry-in $c_0 = 1$. Thus it can also be used to convert a borrow-save represented number to non-redundant 2's complement form, by supplying the negatively weighted bits in complemented form, with carry-in $c_0 = 1$. Again combining the two possibilities it is possible to add in an extended carry in the set $\{-1, 0, 1\}$ as follows. Let the extended carry be encoded as a borrow-save digit employing two bits b^p, b^n of respectively positive and negative weight. Then supplying $\overline{b}^n$ as the carry-in at the bottom will (if non-zero) effectively supply a carry of -1, while using b^p on the output to yield $c_i = G_{i-1} + b^p P_{i-1}$ will add in the positively weighted bit. Note that for the zero encoding $b^p b^n = 11$ this will yield the same effect as using the other zero encoding $b^p b^n = 00$.

Observation 3.6.7 *By combining the possibilities of supplying a carry-in at the bottom of the tree with selection at the top, the carry-look-ahead adder can be used to convert redundant carry-save and borrow-save represented numbers into non-redundant 2's complement representation, employing extended carry values in the sets* $\{0, 1, 2\}$, *respectively* $\{-1, 0, 1\}$.

In certain situations a conversion from 2's complement carry-save or borrow-save representations into sign-magnitude will be needed. The trivial but slow way would be first to convert into 2's complement, possibly followed by a negation. However, the possibility of supplying a carry-in after the sign has been determined can be utilized as follows. Consider adding the two components of a 2's complement carry-save represented number, employing either a conditional-sum or a carry-look-ahead adder. Let us for simplicity assume that the two operands A and B are given as 2's complement integers, where we want the sign of $A + B$ as well as the value of $|A + B|$.

Lemma 3.6.8 (Conversion from borrow-save to sign-magnitude) *Let A and B be the values of the two components of the encoding of an n-digit, borrow-save represented integer. Provided that there is no overflow, the sign of $A - B$ can be determined as $c_n = G_{n-1} + P_{n-1}$, based on the output of a carry-look-ahead tree for calculating $A + \overline{B}$.*

From the same tree, if $A - B < 0$, then the magnitude $|A - B|$ can be determined as

$$|A - B| = -(A - B) = \overline{(A + \overline{B})},$$

by complementing the output of the adder, using carries $c_i = G_{i-1}$. Then for $A - B \geq 0$ the absolute value is found as

$$|A - B| = A - B = (A + \overline{B}) + 1,$$

using carries $c_i = G_{i-1} + P_{i-1}$ for $i = 1, 2, \ldots, n-1$.

Proof Since $-X = \overline{X} + 1$, then $A - B = A + (\overline{B} + 1) = (A + \overline{B}) + 1$ and hence by Lemma 3.6.2 the sign of $A + B$ can be determined as claimed. Also $\overline{A + \overline{B}} = \overline{A - B - 1} = (-A + B + 1) - 1 = -(A - B)$, thus for $A - B < 0$ the carries are to be calculated for $c_0 = 0$ and by the same lemma are found as $c_i = G_{i-i}$ for $i = 1, 2, \ldots, n-1$. When $A - B \geq 0$ a unit has to be added, which can be realized by choosing $c_0 = 1$ and hence using the carries $c_i = G_{i-1} + P_{i-1}$. □

Observation 3.6.9 *Conversion from carry-save to sign-magnitude can be performed by first converting the operand into borrow-save, which can be achieved using the Q-mapping of Definition 2.5.9, which is realizable in constant time by (2.5.13). The subsequent conversion into sign-magnitude can then be performed using Lemma 3.6.8.*

Very fast and wide adders are often built as hybrid adders, combining various techniques, e.g., where the lowest level may be implemented as say eight-bit Manchester carry-chain adders, with overlapping carry-look-ahead trees of low height, providing fast communication of the carries over the longer distances. Note that the distances the carries have to move up the tree then become extremely short in adders of common size.

Traditionally carry-look-ahead adders have been built using *carry-look-ahead generators*, typically realized in early integration technology to accommodate four positions, which can then be applied in several levels, e.g., two levels for a 16-bit adder and three levels for a 64-bit adder. Each carry-look-ahead generator receives four (G, P)-pairs and a carry-in, and produces four carries and a (G, P)-pair.

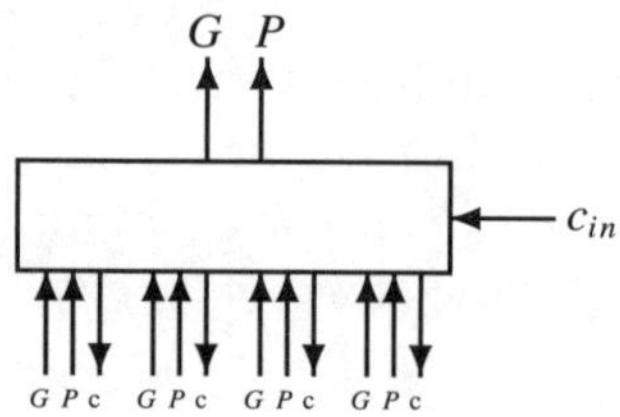

For a 16-bit adder we show in Figure 3.6.5 a partially connected structure, which also contains the (slightly modified) full-adders which are supplied with the final carry values coming down from the look-ahead generators.

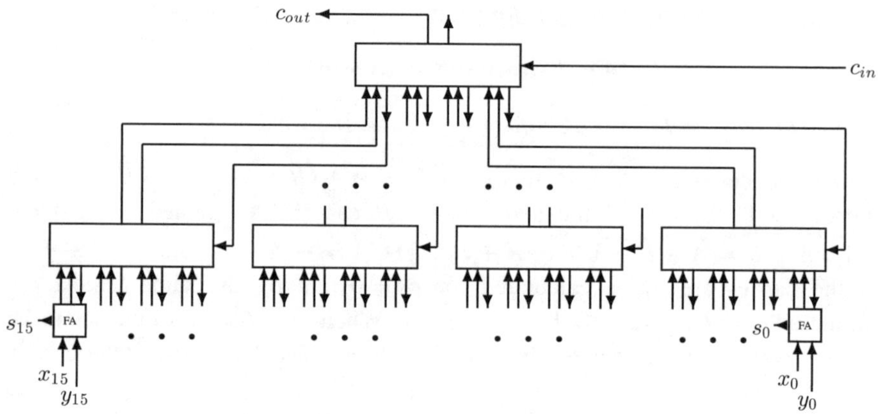

Figure 3.6.5. Carry look-ahead generator tree.

Problems and exercises

3.6.1 Show that the definition (3.6.2) of p_i could alternatively be written as $p_i = x_i + y_i$, and give an argument why $p_i = x_i \oplus y_i$ may be preferred.

3.6.2 At the lowest level of a conditional-sum adder the constant carry-values 0 and 1 enter a single bit adder cell. Show that the two adders of the cell can be combined and realized by a minor extension of the half-adder of Figure 3.5.1 (a single NOT gate is added), as a two-input, four-output cell.

3.6.3 Prove that (3.6.4) reduces to (3.6.5) by showing that $\gamma \neq \left\{ \begin{matrix} 1 \\ 0 \end{matrix} \right\}$.

3.6.4 Show that the operator $\circ$ of the carry-look-ahead adder is idempotent, i.e., $(g, p) = (g, p) \circ (g, p)$.

3.6.5 For the conditional-sum addition, traditionally a tree structure as shown in the example on page 73 has been used. Compare the complexities of this structure and the two structures of Figure 3.6.3.

3.6.6 It was mentioned in Observation 3.6.5 that conditional-sum adders and carry-look-ahead adders by proper selection can compute $A + B$ as well as $A + B + 1$. If, furthermore, the possibility of inverting the input B is provided, what are the four possible combinations of results obtainable when operands and results are interpreted in 2's complement? What results would be obtainable if the output could also be conditionally inverted?

3.6.7 Develop a log-time circuit that can increment a binary number.

3.6.8 Find general expressions for the number of (black) processing nodes in the carry-look-head trees of Figure 3.6.3(a) and (b) as functions of n, where the width of the adders is 2^n. Also find expressions for the longest paths in the two types of adders.

3.6.9 Develop the internal logic of the four-bit carry-look-ahead generators, and in the slightly modified full-adders, denoted FA in Figure 3.6.5.

3.7 Constant-time adders

For multiple addend summation it is advantageous to avoid carry propagation when the result of the addition of a pair of addends is going to be used as the input to another addition process. As we have seen in Section 3.4, addition can be performed in constant time if redundancy is introduced, and if further processing of the result is needed the result may often be used directly in such redundant form. In this section we shall investigate only the repeated additions of multiple addend summation, but as we shall see later it is sometimes possible to use redundant representations also as input to multiplication and division algorithms, just as these may produce their result in redundant representations.

3.7.1 Carry-save addition

In Section 3.4 we introduced the carry-save representation, e.g., radix 2 over the digit set {0, 1, 2} using the encoding employing two bits of equal weight:

$$\begin{aligned} 0 &\sim 00, \\ 1 &\sim 01 \quad \text{or} \quad 10, \\ 2 &\sim 11, \end{aligned} \tag{3.7.1}$$

Carry-save addition is also often termed *3-to-2 addition* or *deferred carry-assimilation*. The *carry-save adder* is just an n-array of full-adders where the set of $3n$ input lines is partitioned such that $2n$ input lines (two for each adder) are used for the input of an n-digit carry-save represented number in the above encoding, and the remaining n input lines form the input of an ordinary binary addend. The $2n$ output lines then contain the $n + 1$ digit result in carry-save representation, obtained by pairing bits of equal weight, i.e., by shifting carries one position to the left, as shown in Figure 3.7.1.

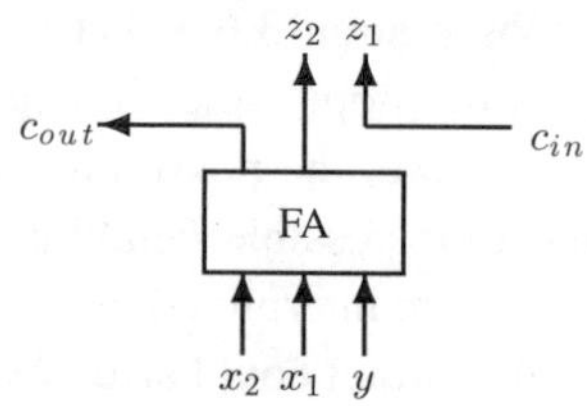

Figure 3.7.1. 3-to-2 carry-save adder

Note that the encoding employed also allows the $3n$ input lines to be partitioned such that input to the carry-save adder could be three ordinary binary numbers. Similarly the $2n$ output lines can be considered two ordinary binary numbers, whose sum can be computed by (say) a carry-look-ahead adder, if the ordinary (non-redundant) binary representation is wanted. For an iterative type of accumulation N addends in ordinary binary can thus be summed by accumulation (feed-back) in $N - 2$ steps of constant time. In a tree structure approximately $\log_{3/2} N$ levels are required, each level being also of constant time equivalent to the time of the full-adder circuit. In both cases conversion into non-redundant binary will add at least $\log_2 n$ time if the width of the registers is n bits.

To reduce the height of the tree structure, and possibly to allow operands which are both in carry-save representation, an extended form of adder is needed which can take two carry-save represented operands as input. A straightforward way of implementing such a 4-to-2 adder would be to employ two full-adders, as shown in Figure 3.7.2.

Observe from Figure 3.5.3 of the full-adder that the path from c_{in} to s is only through a single XOR gate, and that the path from a and b to c_{out} is through one XOR, one AND, and one OR gate, whose delay is approximately the same as two

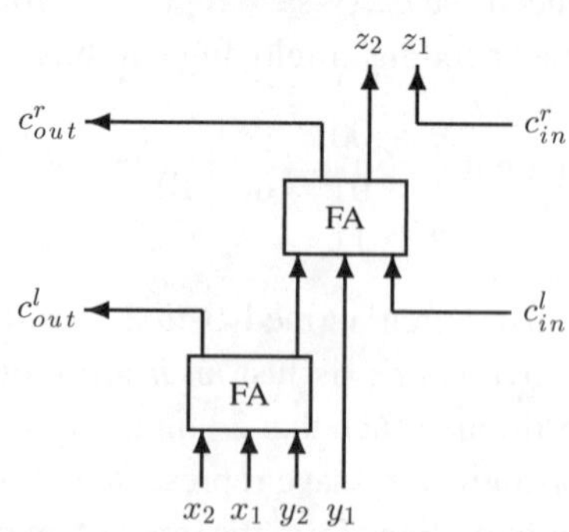

Figure 3.7.2. 4-to-2 carry-save adder composed from full-adders.

XOR gates. Thus in Figure 3.7.2 the longest path only has a delay of three XOR gates, when appropriately connected.

But as indicated in Table 3.4.4 it is possible to define addition tables for the direct addition of carry-save represented operands. Following the idea outlined in Table 3.4.3 for radix-2 signed-digit, the α addition table of Table 3.4.4 may be split in two tables α^- and α^+, to be applied when it is known that the incoming carry is "low" (≤ 1) or "high" (≥ 1) respectively, exploiting the fact that the digit set employed for the output of the α table need not be the same as the set used on input (since the output is to be used as input for another addition table for carry absorption)

In Table 3.7.1α^- the sum digits belong to the set $\{0,1\}$, i.e., they are "prepared" for an incoming carry in the set $\{0,1\}$, whereas in Table 3.7.1α^+ the sum digits in $\{-1, 0\}$ are "prepared" for incoming carries in $\{1, 2\}$, so in the succeeding addition table carries can be absorbed without generating a new carry (i.e., sum digits belong to $\{0,1,2\}$).

Table 3.7.1. *Two addition tables for carry-save input*

α^-	0	1	2
0	00	01	10
1	01	10	11
2	10	11	20

α^+	0	1	2
0	00	$1\bar{1}$	10
1	$1\bar{1}$	10	$2\bar{1}$
2	10	$2\bar{1}$	20

Whereas the encoding of input and output digits in carry-save is given by (3.7.1) we have the freedom to choose encodings for the sum and carry values to be output from a circuit realizing the α^- and α^+ addition tables. This freedom may be exploited to find alternative circuit implementations beyond those that can be found by rewriting logic expressions.

One possible choice for the encodings of the carry output of the α tables is to use the standard carry-save encoding (3.7.1), since the carry belongs to the set $\{0,1,2\}$. If the bit pair in (3.7.1) is considered an ordered pair (c^l, c^r), the leftmost bit c^l may be considered an encoding with 0 for $c \leq 1$ and 1 for $c \geq 1$. The rightmost bit c^r then further separates the cases with $0 \sim$ "low" and $1 \sim$ "high." The leftmost bit c^l may then be directly used to choose between the α^- and α^+ tables in the position to the left, to prepare it for an incoming "low" or "high" carry value.

The sum or place digit of the α^- and α^+ tables is a signed digit (i.e., in $\{-1, 0, 1\}$) which can be encoded in "sign magnitude":

$$\begin{aligned} -1 &\sim 11, \\ 0 &\sim 10 \quad \text{or} \quad 00, \\ 1 &\sim 01. \end{aligned} \tag{3.7.2}$$

Denoting the sum digit s with encoding (s^s, s^m) it may be noted from Table 3.7.1 that the incoming c^l_{in} bit choosing between the α^- and α^+ tables can be directly used as the sign bit s^s. The magnitude bit s^m can be found as the XOR (parity) of the four input bits x_1, x_2, y_1, y_2 encoding the operands x and y:

$$\begin{aligned} s^s &= c^l_{in}, \\ s^m &= (x_1 \oplus x_2) \oplus (y_1 \oplus y_2). \end{aligned} \tag{3.7.3}$$

For the carry $c = (c^l, c^r)$ we chose the carry-save encoding (3.7.1), which permitts a simple selection upon the "high" or "low" value of the incoming carry. In the tables for α^- and α^+ it may be noted that the carry to be generated is always "high" (1 or 2) in the diagonal and below. Using small 3×3 *dot-matrices* to indicate where a 1 is to be generated, we would like to find a logic expression for c^l generating the pattern 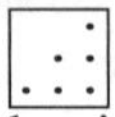; one such expression is $x_1x_2 + y_1y_2 + (x_1 + x_2)(y_1 + y_2)$. Note that c^l is the signal to be propagated to the left, but the value of c^r then has to be chosen in conformance with the chosen value for c^l to form a correct encoding of c. The value of c to be generated depends on which table α^- or α^+ is to be used, i.e., on the value of an *incoming* carry signal c^l_{in}. However, the coding redundancy permits us to chose a simpler equation for c^l for propagating to the left, and then compensate for it in the choice of an expression for c^r:

$$c^l = x_1x_2 + y_1y_2 =$$ 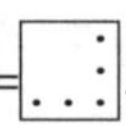. (3.7.4)

If for c^r we can generate 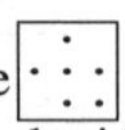 in α^+ and 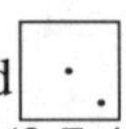 in α^-, then a correct encoding is obtained when this is paired with c^l from (3.7.4). The incoming carry c^l_{in} is used to "select" from these patterns:

$c^r = c^l_{in}$  $+ \overline{c^l_{in}}$ 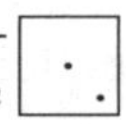

$= c^l_{in} \Big($ 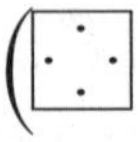 $+$ 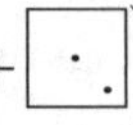 $\Big) + \overline{c^l_{in}}$ 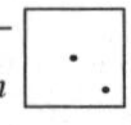

$= c^l_{in}$ 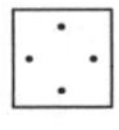 $+$ 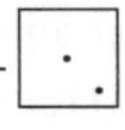.

Recognizing the dot-matrix 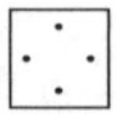 for s^m from (3.7.3) we then obtain

$$c^r = c^l_{in} s^m + (x_1 \oplus x_2)(y_1 \oplus y_2) + x_1x_2y_1y_2. \tag{3.7.5}$$

Finally the sum or place digit $s = (s^s, s^m)$ has to be added to the incoming carry $c = (c^l, c^r)$ without generating a new carry. This carry absorption is realized through Table 3.7.2, where the output $z = (z_1, z_2)$ is to be encoded as standard carry-save, e.g., following (3.7.1).

Note that when $c^l_{in} = 1$ then $c \in \{1, 2\}$ and $s \in \{-1, 0\}$, and only the upper right-hand 2×2 part of Table 3.7.2 is used, and when $c^l_{in} = 0$ then $c \in \{0, 1\}$ and $s \in \{0, 1\}$ and only the lower left-hand 2×2 part of the table is used. A possible

Table 3.7.2. *Carry-absorption addition table for carry-save*

γ	0	1	2
$\bar{1}$		0	1
0	0	1	2
1	1	2	

solution is then:

$$\begin{aligned} z_1 &= c_{in}^r, \\ z_2 &= c_{in}^l \overline{s^m} + \overline{c_{in}^l} s^m \\ &= c_{in}^l \oplus s^m. \end{aligned} \tag{3.7.6}$$

Finally, combining (3.7.3), (3.7.4), (3.7.5), and (3.7.6), we obtain Figure 3.7.3.

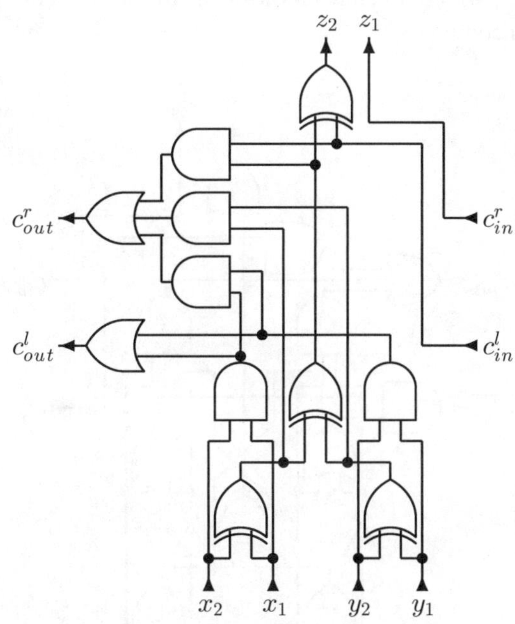

Figure 3.7.3. 4-to-2 adder for carry-save operands.

Observe that this circuit is completely symmetric in the four input bits x_1, x_2, y_1, y_2, and that both output bits z_1 and z_2 depend on precisely eight input bits. Hence in an array of such adders the output at any position depends only on the input into the same adder and its rightmost two neighbors.

3.7.2 Borrow-save addition

For the redundant radix-2 representation over the symmetric digit set $\{-1, 0, 1\}$ Tables 3.4.2 and 3.4.3 described addition tables for constant-time addition. If we choose the following encoding of operands and result:

$$\begin{aligned} -1 &\sim 10, \\ 0 &\sim 00 \quad \text{or} \quad 11, \\ 1 &\sim 01, \end{aligned} \tag{3.7.7}$$

it may be noted that this encoding corresponds to the leftmost bit having weight -1 and the rightmost $+1$. Based on Table 3.4.3 it is possible to derive logic expressions for an adder circuit based on this representation, using the same method as used above for the carry-save adder.

Let the operands be encoded as (x^n, x^p) and (y^n, y^p), with result (z^n, z^p) and carry (c^n, c^p), where the leftmost signals have negative weight and the rightmost positive weight. Then we may obtain the circuit in Figure 3.7.4. However, as we saw in Section 3.5 it is also possible to derive a circuit for the borrow-save adder from Figure 3.7.3, by complementation of all the negatively weighted signals when using the encoding (3.7.7).

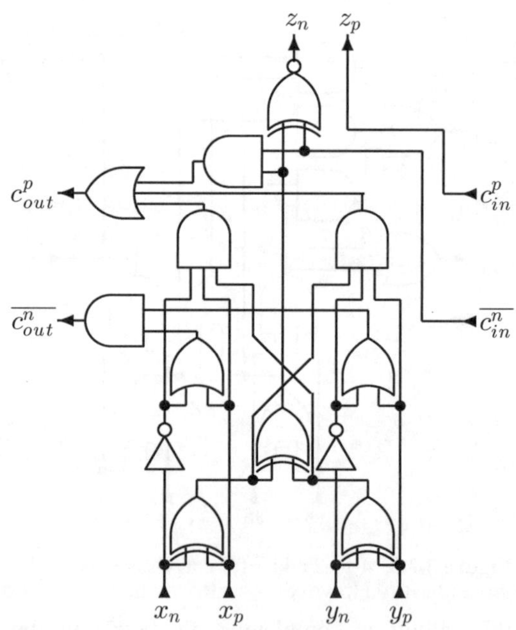

Figure 3.7.4. 4-to-2 adder for borrow-save operands.

From the adder in Figure 3.7.4 it is possible to derive a simpler adder accepting one operand in borrow-save and another in ordinary binary. Assuming that, say, y_n is always zero the circuit can be simplified, accepting y_p as an operand in ordinary binary. However, a simpler circuitry may be found directly by observing that this is a *pnp*-adder, which can be derived from the full-adder (*ppp*).

Problems and exercises

3.7.1 In the implementations in Figures 3.7.2 and 3.7.3 the choice between Table 3.7.1α^- and Table 3.7.1α^+ depends on the value of c^l. If the incoming carry has the value 1, then the encoding as a pair (c^l, c^r) can be either (0, 1) or (1, 0), and thus the result of adding two digits in this position depends on the actual coding of the incoming carry in this case. Show some examples where the choice between the two addition tables differs when using the implementation of Figure 3.7.2, even when the value of the carry-out is the same: $c_{out} = 1$.

3.7.2 Repeat the previous problem but this time using the implementation of Figure 3.7.3.

3.7.3 Prove (3.7.6). (Consider only those combinations of $(s^s, s^m, c^l_{in}, c^r_{in})$ where $s^s = c^l_{in}$.)

3.7.4 Expanding the two full-adders, compare the logic of Figure 3.7.2 with the logic of Figure 3.7.3. Be careful to combine the two adders in such a way that the worst-case path is minimal.

3.7.5 Develop another 4-to-2 borrow-save adder by changing signal weights in Figure 3.7.3, using Lemma 3.5.1 and Observation 3.5.2. (Just show Figure 3.7.3 as a box.)

3.8 Addition and overflow in finite precision systems

Addition in $\mathcal{F}_{\ell m}[\beta, D]$ is not a closed operation in such a finite system, so precautions have to be taken if the result is not representable. Notice that very often the basic integer addition operation of a computer does not notify the user of such an overflow situation, the system just discards digits (carry-outs) that cannot be represented and thus the computer implements a modular addition $(a + b) \bmod \beta^m$.

It is obvious that if the digit set D is non-redundant for the radix β, then for addition in $\mathcal{F}_{\ell,m}[\beta, D]$, any non-zero digit generated in position k, with $k > m$, is a signal that the result is not representable. Thus the carry generated in position m directly acts as an overflow signal. For addition in the non-redundant 2's complement system, the situation is different, since position $m + 1$ in practice is not included in the representation. However, overflow can be detected when the carry-in to position m is different from the carry-out of that position, see Corollary 3.8.7 below.

3.8.1 Addition in redundant digit sets

If D is redundant for radix β the situation is more complicated. First, note that if the value of the carry generated in the most-significant position m is zero, then the result is certainly representable in $\mathcal{F}_{\ell m}[\beta, D]$, but a non-zero carry does not necessarily imply that an overflow situation has arisen.

Example 3.8.1 Consider the following addition in $\mathcal{F}_{04}[2, \{-1, 0, 1\}]$:

$$\begin{array}{cccccc} & 0 & 1 & \bar{1} & \bar{1} & \bar{1} \\ & 1 & 1 & 1 & 1 & 0 \\ \hline 1 & 0 & 0 & 0 & 0 & \bar{1} \end{array} \sim \quad 1\ 1\ 1\ 1\ 1$$

and note that the value of the result is actually representable in the system. □

As the example indicates, a proper signalling of overflow must take at least $O(\log(m - \ell))$ time, hence the advantage of constant-time addition is lost if it is necessary to test for overflow following each addition. Fortunately this may not be necessary in a composite computation, as long as a sufficient number of *leading guard digits* are carried along in the computation to avoid loss of any significant information. Hence, with a little planning, overflow testing and possibly scaling of the result can be postponed to a point in the computation where a conversion into a non-redundant representation may be needed anyway.

Since a number representation in a redundant digit set may have a prefix string of non-zero digits which can be reduced (e.g., $1\bar{1} \sim 01$ in signed-digit radix-2), it may be necessary to be aware of the presence of such strings. From Theorems 3.4.1 and 3.4.2 it may be noted that up to two additional leading digits may be generated in a redundant addition. Fortunately the value represented by such a prefix string is always bounded.

Lemma 3.8.1 *Let $P \in \mathcal{P}[\beta, D]$ with D a basic digit set of the form $D = \{d \mid -\beta \le r \le d \le s \le \beta\}$, $s - r + 1 = \beta + \rho$, where the redundancy index $\rho \ge 1$, and such that the value of P is representable in $\mathcal{F}_{\ell m}[\beta, D]$:*

$$r\frac{\beta^{m+1} - \beta^{\ell}}{\beta - 1} \le \|P\| \le s\frac{\beta^{m+1} - \beta^{\ell}}{\beta - 1}.$$

Then with Q defined by

$$P = Q[\beta]^{m+1} + R,$$

the value of Q satisfies

$$|\|Q\|| < 1 + \frac{\rho}{\beta - 1},$$

and thus if $\rho \le \beta - 1$ it follows that $|\|Q\|| \le 1$.

Proof From the definition of Q we have $\|P\| = \|Q\|\beta^{m+1} + \|R\|$, hence

$$r\frac{\beta^{m+1} - \beta^{\ell}}{\beta - 1} - \|R\| \le \|Q\|\beta^{m+1} \le s\frac{\beta^{m+1} - \beta^{\ell}}{\beta - 1} - \|R\|.$$

Since $R \in \mathcal{F}_{\ell m}[\beta, D]$ implies

$$r\frac{\beta^{m+1} - \beta^{\ell}}{\beta - 1} - \|R\| \le 0 \text{ and } s\frac{\beta^{m+1} - \beta^{\ell}}{\beta - 1} - \|R\| \ge 0,$$

then

$$\begin{aligned}\left|\|Q\|\beta^{m+1}\right| &\le (s-r)\frac{\beta^{m+1}-\beta^{\ell}}{\beta-1}\\ &= (\beta-1+\rho)\frac{\beta^{m+1}-\beta^{\ell}}{\beta-1}\\ &< \left(1+\frac{\rho}{\beta-1}\right)\beta^{m+1},\end{aligned}$$

from which the results follow, noting that $\|Q\|$ is integral. □

Observe that only in the very extreme cases of $\rho = \beta$ or $\rho = \beta + 1$ is it possible that $|\|Q\|| = 2$, as confirmed in the case $\beta = 2$, $D = \{-1, 0, 1, 2\}$ with $2\bar{1}\bar{1}_2 = 5 = 21_2$. In the next subsection we shall discuss the 2's complement carry-save representation. Note also that the situation where $\rho \le \beta - 1$ includes cases with non-unique zero representation, e.g., $\beta = 3$, $D = \{-1, 0, 1, 2, 3\}$ with $\rho = \beta - 1 = 2$.

Theorem 3.8.2 (Guard digits for $\beta \ge 3$) *Let $P \in \mathcal{P}[\beta, D]$ be a polynomial with $\|P\|$ representable in $\mathcal{F}_{\ell m}[\beta, D]$, where D is a basic digit set for $\beta \ge 3$ with redundancy index $\rho \le \beta - 1$. With Q being the prefix of P defined by $P = Q[\beta]^{m+1} + R$, the value of $\|Q\|$ is bounded by $|\|Q\|| \le 1$ and can be uniquely determined by the relation $\|Q\| \equiv d_{m+1} \pmod{\beta}$, where d_{m+1} is the $(m+1)th$ digit of P.*

Proof This follows trivially from Lemma 3.8.1 and the DGT Algorithm since $\|Q\| \in \{-1, 0, 1\}$ implies that the only permissible values of d_{m+1} are $-\beta, -\beta + 1, -1, 0, 1, \beta - 1, \beta$. □

The theorem shows that for $\beta \ge 3$ and $\rho \le \beta - 1$ one leading *guard digit* is sufficient to insure that the value of Q and hence $\|P\|$ can be identified. From the proof also notice that for larger values of β it is possible to uniquely identify the value of $\|Q\|$ from the value of d_{m+1}, in the case where $\rho > \beta - 1$ and $\|Q\|$ may take the values ± 2.

There are two particularly interesting cases for $\beta = 2$, the carry-save and the borrow-save representations, of which we will first consider the former where the following result is obtained trivially from Lemma 3.8.1 since $\rho = \beta - 1$.

Theorem 3.8.3 (Guard digits for unsigned carry-save) *Let $P = \sum_{i=\ell}^{m+k} d_i[2]^i \in \mathcal{P}[2, \{0, 1, 2\}]$ for $k \ge 2$ be a polynomial whose value $\|P\|$ is representable in $\mathcal{F}_{\ell,m}[2, \{0, 1, 2\}]$. With Q defined by $P = Q[2]^{m+1} + R$, then Q is either the zero polynomial or the polynomial $Q = 1 \cdot [2]^0$. In the latter case d_m must be zero since the value of $\|P\|$ is representable, hence the polynomial can be rewritten by changing d_m into $d'_m = 2$.*

Note that this result does not cover the 2's complement carry-save representation. We shall see later (Theorem 3.8.9) that two guard digits are necessary and sufficient to determine the value of the prefix Q in this case.

However, for the signed-digit, radix-2 (borrow-save) representation it is easy to see that a single guard digit is necessary and sufficient:

Theorem 3.8.4 (Guard digits for borrow-save) *Let* $P = \sum_{i=\ell}^{m+k} d_i[2]^i \in \mathcal{P}[2, \{-1, 0, 1\}]$ *for some* $k \geq 2$ *be a borrow-save polynomial whose value* $\|P\|$ *is representable in* $\mathcal{F}_{\ell,m}[2, \{-1, 0, 1\}]$. *With prefix* Q *defined by* $P = Q[2]^{m+1} + R$, *the value of* Q *satisfies* $\|Q\| \in \{-1, 0, 1\}$ *and can be determined as*

$$\begin{aligned}
\|Q\| &= -1 \Leftrightarrow (d_{m+2} \neq 0 \wedge d_{m+1} = 1) \vee (d_{m+2} = 0 \wedge d_{m+1} = -1),\\
\|Q\| &= \ \ \, 0 \Leftrightarrow d_{m+1} = 0,\\
\|Q\| &= \ \ \, 1 \Leftrightarrow (d_{m+2} \neq 0 \wedge d_{m+1} = -1) \vee (d_{m+2} = 0 \wedge d_{m+1} = 1).
\end{aligned}$$

Thus two guard digits are necessary and sufficient to determine $\|Q\|$.

Proof By Lemma 3.8.1, $|\|Q\|| \leq 1$. Let S be defined by $Q = S[2]^1 + d_{m+1}$, then again by the lemma $\|S\| \in \{-1, 0, 1\}$ and $\|Q\| = \|S\| \cdot 2 + d_{m+1}$, hence $\|S\| = (\|Q\| - d_{m+1})/2$ from which the result follows by checking the possible solutions to this equation. □

The result of this theorem may be summarized into a simple rule for the conversion of an arbitrary length prefix into a single guard digit. If d_{m+2} is non-zero just negate the value of the digit d_{m+1}, and then discard all digits of weight higher than 2^{m+1}, leaving only one guard digit. We would like to eliminate that guard digit also, i.e., convert it into $\mathcal{F}_{m\ell}[2, \{-1, 0, 1\}]$, but in general this is not possible in constant time, as is evident from Example 3.8.1 in which logarithmic time is needed for such a rewriting, when $d_m = 0$.

Corollary 3.8.5 *Under the conditions of Theorem 3.8.4, and using the value of* Q *obtained, the polynomial* $P = \sum_{i=\ell}^{m+k} d_i[2]^i$ *can be converted into* $P' = \sum_{i=\ell}^{m+1} d_i'[2]^i \in \mathcal{F}_{\ell,m+1}[2, \{-1, 0, 1\}]$ *in constant time, where*

$$d_i' = \begin{cases} d_i & \text{if } \ell \leq i \leq m,\\ \|Q\| & \text{if } i = m+1. \end{cases}$$

Elimination of the digit $d'_{m+1} = \|Q\| \neq 0$ in constant time is only a problem if $d_m = 0$. For $d_m \neq 0$ just consider the case $\|Q\| = 1$. The value $\|P\|$ then satisfies

$$\begin{aligned}
\|P\| &= 2^{m+1} + \textstyle\sum_{i=\ell}^{m} d_i 2^i\\
&\geq 2^{m+1} + d_m 2^m - (2^m - 2^\ell)\\
&= (1 + d_m)2^m + 2^\ell,
\end{aligned}$$

but P is representable in $\mathcal{F}_{\ell,m}[2, \{-1, 0, 1\}]$, thus $-2^{m+1} + 2^{\ell} \leq \|P\| \leq 2^{m+1} - 2^{\ell}$, from which it follows that

$$(1 + d_m)2^m + 2^{\ell} \leq 2^{m+1} - 2^{\ell},$$

so $d_m < 1$. If $d_m = -1 = -\|Q\|$, then to eliminate Q of weight 2^{m+1}, it is sufficient to set $d'_m = 1 = \|Q\|$ or equivalently to change the sign of d_m. The case $\|Q\| = -1$ is equivalent.

There is an interesting "automatic" conversion taking place when adding a number which has zeroes in guard positions, e.g.,

$$\begin{array}{cccccccc}
 & & 1 & \bar{1} & \bar{1} & \bar{1} & \cdots \\
 & & 0 & 0 & 0 & 0 & \cdots \\
\cline{3-6}
 & & 1 & 1 & 1 & 1 & \cdots \\
\hat{\alpha} & 0 & \bar{1} & \bar{1} & \bar{1} & \mathrm{c} & \cdots \\
\cline{2-6}
 & 0 & 0 & 0 & 0 & \mathrm{d} & \cdots
\end{array}$$

when applying Table 3.4.2α for the first level addition. Hence, in the case of multioperand addition, such reducible leading strings of non-zero digits tend to disappear, while of course new such strings may reoccur further to the right. Observe that adding a string of zeroes by Table 3.4.2α is equivalent to applying the N-mapping of Section 2.6.

The previous results insure that in binary multioperand addition and in many other computations it is sufficient to use one or two *leading guard digits* beyond the digits needed to represent the result, *even in the case that the result is known to have fewer digits than the operands*. And in the case of the summation of k operands, when nothing in particular is known about these, it is sufficient with one guard digit beyond the at most $\lceil \log_\beta k \rceil$ extra digits needed to accommodate the worst-case growth of the sum compared to the operands.

3.8.2 Addition in radix-complement systems

Let us now turn to the finite precision radix-complement representations, i.e., the operands are from $\mathcal{F}^{rc}_{\ell m}[\beta, C^c_\beta]$ and the result is to be delivered in the same system. Recall that these systems utilize an extra position $m + 1$, where the digit $d_{m+1} \in \{-1, 0\}$ is not represented, but its value can be identified from the value of the digit d_m:

$$d_{m+1} = 0 \iff 0 \leq d_m < \left\lfloor \frac{\beta + 1}{2} \right\rfloor, \tag{3.8.1}$$

$$d_{m+1} = -1 \iff d_m \geq \left\lfloor \frac{\beta + 1}{2} \right\rfloor.$$

If digit positions $m + 1$ for both operands are included in the addition, then the remaining problem is only whether the result is representable in $\mathcal{F}^{rc}_{\ell m}[\beta, C^c_\beta]$, i.e., whether $d_{m+1} \in \{-1, 0\}$ and the pair $d_{m+1}d_m$ satisfies (3.8.1). This then becomes precisely the overflow condition, but since the digits in position $m + 1$

are actually not represented, the condition must be expressed in terms of digits and carries of the represented parts.

Recall that the radix-complement representation of some value v, without the leading digit d_{m+1}, is the standard radix representation of $P \in \mathcal{F}_{\ell m}[\beta, \{0, 1, \ldots, \beta - 1\}]$, where $\|P\| = v \bmod \beta^{m+1}$. Addition is thus here to be performed as addition modulo β^{m+1}, and the carry-out must be interpreted appropriately.

First notice that the carry coming out of position m is in the set $\{0, 1\}$, so potentially the resulting digit in position $m + 1$ is in $\{-2, -1, 0, 1\}$. However it is easily seen from (3.8.1) that the values -2 and 1 cannot occur when adding two numbers. Since the values of the digits in position $m + 1$ can be deduced from those in position m of the operands, it is possible to deduce the value of the digit in position $m + 1$ of the result, and a condition for overflow can be found as follows.

Theorem 3.8.6 *When two radix-complement operands $P = \sum_{i=\ell}^{m+1} a_i[\beta]^i$ and $Q = \sum_{i=\ell}^{m+1} b_i[\beta]^i$ from $\mathcal{F}_{\ell m}^{rc}[\beta, C_\beta^c]$ are added producing a result $R = \sum_{i=\ell}^{m+1} d_i[\beta]^i$, then addition can be accomplished and overflow tested solely based on the digits from positions ℓ through m as*

$$\|R\| = \left(\sum_{i=\ell}^{m} a_i\beta^i + \sum_{i=\ell}^{m} b_i\beta^i\right) \bmod \beta^{m+1},$$

with overflow condition

$$OF \equiv \bar{a}\bar{b}d + ab\bar{d}$$

where

$$a = \left(a_m \geq \left\lfloor\frac{\beta+1}{2}\right\rfloor\right), b = \left(b_m \geq \left\lfloor\frac{\beta+1}{2}\right\rfloor\right) \text{ and } d = \left(d_m \geq \left\lfloor\frac{\beta+1}{2}\right\rfloor\right).$$

Proof The digit values a_{m+1} and b_{m+1} can be derived from a and b, and d_{m+1} from a, b together with the carry-out from position m, c_{out}. There are 16 combinations of a, b, d, and c_{out} to consider, however, taking into account the symmetry in a, b and the fact that four cases cannot occur (e.g., adding two positive numbers cannot produce a carry) eight combinations are left:

Operands a	*Operands* b	c_{out} c	Result d	Status
p	p	0	p	OK
p	p	0	n	Overflow
(p,n) or (n,p)		0	p	NA
(p,n) or (n,p)		0	n	OK
(p,n) or (n,p)		1	p	OK
(p,n) or (n,p)		1	n	NA
n	n	1	p	Overflow
n	n	1	n	OK

where we have used n to indicate negative (i.e., $\geq \lfloor(\beta+1)/2\rfloor \sim$ true) and p for positive ($< \lfloor(\beta+1)/2\rfloor \sim$ false). The status is found as follows.

Consider the first entry $(p, p, 0, p)$ where both operands are positive, e.g., both operands have zeroes in position $m+1$. The carry-out is also zero, so the result will have a zero in position $m+1$ which agrees with d being positive. The second entry is obvious, and the last two entries are found similarly. Obviously, adding two numbers of opposite sign cannot produce overflow. Finally there are two entries marked NA which cannot occur, the first because an n- and a p-digit producing a p-digit must also produce a carry, and the second because an n-digit and a carry cannot be produced simultaneously in this situation.

We are left with the fairly obvious result that an overflow only occurs when adding two operands of the same sign, and the result apparently has a different sign (as seen from the value of d_m), which is the expression for OF. □

For the most used case of 2's complement it turns out there is an even simpler expression for the overflow condition.

Corollary 3.8.7 *For addition in 2's complement $\mathcal{F}^{2c}_{\ell m}[2, C^c_2]$ overflow can be tested as*

$$OF \equiv c_{in} \oplus c_{out},$$

where c_{in} is the carry coming into position m, and c_{out} the carry going out of position m.

Proof This follows from the table in the preceding proof by noting that the value of c_{in} can be derived from the values of a, b, and d when the digit set is $\{0, 1\}$. Only in the cases of overflow do the values of c_{in} and c_{out} differ. □

In practice overflow testing in 2's complement is sometimes performed by sign-extending both operands into one extra position and then checking the two leading bits of the extended result, which then mirrors the c_{in}, c_{out} of the corollary. Obviously this is more costly in time and circuitry than directly using the corollary.

The non-equivalence (XOR) of the two leading bits of a register containing a 2's complement operand plays a role in many algorithms where a *normalization* of an operand is wanted (e.g., in division and in floating-point representations). Normalization is a scaling of an operand such that it is positioned as far to the left as possible in a register (in the most-significant end). For a normal radix polynomial it is just a question of scaling the operand by a power of the radix corresponding to the number of leading zeroes in the operand (assumed non-zero). For a radix-complement operand it is a question of scaling by left-shifting until its two leading digits are different.

3.8.3 1's complement addition

For the diminished radix-complement systems we will just discuss the radix 2 case of 1's complement addition. By the general definition (1.8.7) of the diminished radix-complement, for $\beta = 2$ the digit string $\{a_m, a_{m-1}, \ldots, a_\ell\}$, $a_i \in \{0, 1\}$ for $\ell \leq i \leq m$, in 1's complement represents a value

$$A = \sum_{i=\ell}^{m} a_i 2^i + (2^{m+1} - 2^\ell) a_{m+1},$$

where $a_{m+1} = -a_m$, since negative numbers have a bias of $(2^{m+1} - 2^\ell)$. In other words, the 1's complement representation of a value V is the standard radix representation of $V \bmod (2^{m+1} - 2^\ell) = \|P\|$, where $P \in \mathcal{F}_{\ell m}[2, \{0, 1\}]$, and 1's complement addition is to be performed modulo[1] $2^{m+1} - 2^\ell$.

Without loss of generality we may assume that $\ell = 0$ so addition is to be performed modulo $2^{m+1} - 1$. For this note the equation

$$a \bmod (k-1) = \big((a \bmod k) + (a \operatorname{div} k)\big) \bmod (k-1), \tag{3.8.2}$$

whose proof we leave as an exercise. The carry-out after addition of $A = \sum_{i=0}^{m} a_i 2^i$ and $B = \sum_{i=0}^{m} b_i 2^i$ in an $(m+1)$-bit adder is $c_{out} = (A + B) \operatorname{div} 2^{m+1}$, and what is left in the adder is $(A + B) \bmod 2^{m+1}$. By (3.8.2) we then have

$$(A + B) \bmod (2^{m+1} - 1) = \big((A + B) \bmod 2^{m+1} + c_{out}\big) \bmod (2^{m+1} - 1),$$

i.e., c_{out} is to be added in the least-significant position (called an *end-around carry*). Now the right-hand side may still have to be modulo reduced, but observe that this can only happen when $c_{out} = 1$ and $(A + B) \bmod 2^{m+1} + c_{out} = 2^{m+1}$. But then $(A + B) \bmod 2^{m+1} = 2^{m+1} - 1$, in which case a pattern of all ones is changed into all zeroes with another carry-out. Since all zeroes and all ones are both legal representations of zero, we can just avoid the outer modulo reduction.

What remains is to consider the possibility of overflow, but it turns out that the analysis is exactly as for 2's complement, resulting in the very same condition for overflow. Hence we can summarize the results for 1's complement in the following theorem.

Theorem 3.8.8 *When two 1's complement operands given as digit strings over* $\{0, 1\}$, $P = \{a_m, a_{m-1}, \ldots, a_\ell\}$ *and* $Q = \{b_m, b_{m-1}, \ldots, b_\ell\}$, *are to be added producing a result* $R = \{r_m, r_{m-1}, \ldots, r_\ell\}$, *then the addition can be accomplished using end-around carry as*

$$\|R\| = \left(\sum_{i=\ell}^{m} a_i \beta^i + \sum_{i=\ell}^{m} b_i \beta^i\right) \bmod 2^{m+1} + c_{out} 2^\ell$$

with the overflow condition

$$OF \equiv c_{in} \oplus c_{out},$$

where c_{in} *is the carry coming into position* m, *and* c_{out} *the carry going out of position* m.

[1] The modulo operator is assumed generalized to non-integral values as defined in Chapter 1.

Note that an end-around carry can be fed into an adder as c_{in} in the least-significant position, but may cause a carry ripple there. However, if the adder is built using a look-ahead tree, then it can be fed into the tree as discussed in Observation 3.6.5 and in the next section.

3.8.4 2's complement carry-save addition

The combination of a redundant digit set with a radix-complement representation has turned out to be very useful for the case of radix 2, in the form of the 2's complement carry-save system introduced in Chapter 1. Formally the system is $\mathcal{F}^{2c}_{\ell m}[2, \{-2, -1, 0, 1, 2\}]$ with the restriction that negative digits are only allowed in position $m + 1$, and only such that $d_{m+1} = -d_m$, i.e., $d_{m+1} \in \{-2, -1, 0\}$.

But this system is a subset of $\mathcal{F}_{\ell,m+1}[2, \{-2, -1, 0, 1, 2\}]$, hence we can apply Lemma 3.8.1 since restricting the use of some of the digits cannot increase the bound on the prefix $\|Q\|$, which turns out to be $|\|Q\|| \leq 2$. Also recall that zero is not uniquely represented in this system when $m \geq \ell + 1$, and the range of numbers represented is $[-2^{m+1}; 2(2^m - 2^\ell)]$.

Example 3.8.2 Adding 6 and 0 in some of the non-zero representations of zero in respectively a three-, a four-, and a five-digit system using the carry-save addition tables of Table 3.4.4 yields

$\hat{\alpha}$	0	0	2	2
	$\bar{1}$	1	1	2
$\hat{\gamma}$	$\bar{1}$	1	1	0
	0	1	2	
$\hat{\xi}$	$\bar{1}$	0	1	0
	1	1	0	
	0	1	1	0

0	0	0	2	2
$\bar{1}$	1	1	1	2
$\bar{1}$	1	1	1	0
0	0	1	2	
$\bar{1}$	1	0	1	0
0	1	1	0	
$\bar{1}$	2	1	1	0

0	0	0	0	2	2
$\bar{1}$	1	1	1	1	2
$\bar{1}$	1	1	1	1	0
0	0	0	1	2	
$\bar{1}$	1	1	0	1	0
0	0	1	1	0	
$\bar{1}$	1	2	1	1	0

where the leftmost digits are not part of the actual computation in the bounded length registers, but have shown to illustrate the correct computation, and hence where the bounded length computations fail.

Note that the value 6 is representable in the three-digit system, but only in the form 022, not in the form 110. If position $m + 1$ is included, it is easy to see that all results correctly represent the sum 6; however only in the five-digit system is the result in $\mathcal{F}^{2c}_{\ell m}[2, \{-2, -1, 0, 1, 2\}]$, and only in this system is the string without the leading non-positive digit a correct representation of the value 6. Note that it is not possible from the leading digit values to determine the sign of the number in this representation. The prefix string $\bar{1}12$ is just another representation of zero, and the example illustrates that such leading guard digits must be included if the computation is performed without the digit in position $m + 1$. Also note that the guard digits must be beyond *all possible representations* of the value, excluding prefixes of value zero, which is equivalent to saying that the guard digits must be beyond those needed to represent the value in *non-redundant 2's complement.* □

The observations in the example tell us that two guard digits may be necessary, but they also turn out to be sufficient as shown below.

Theorem 3.8.9 (Guard digits for 2's complement carry-save) *Let* $P = \sum_{i=\ell}^{m+g+1} d_i[2]^i \in \mathcal{F}^{2c}_{\ell,m+g}[2, \{-2, -1, 0, 1, 2\}]$, $g \geq 2$, *be a 2's complement, carry-save polynomial whose value is representable non-redundantly in* $\mathcal{F}^{2c}_{\ell,m}[2, \{-1, 0, 1\}]$*. Let* $Q \in \mathcal{F}^{2c}_{0,g-1}[2, \{-2, -1, 0, 1, 2\}]$ *be the prefix of P defined by* $P = Q\,[2]^{m+1} + \sum_{i=\ell}^{m} d_i[2]^i$*. Then the value of Q can be determined by solving:*

$$\|Q\| \equiv 2d_{m+2} + d_{m+1} \pmod 4 \quad \textit{and} \quad \|Q\| \in \{-2, -1, 0\} \tag{3.8.3}$$

and the polynomial $P' = Q\,[2]^{m+2} - Q\,[2]^{m+1} + \sum_{i=\ell}^{m} d_i[2]^i$ *has the same value as* P*.*

Hence two leading guard digits are sufficient to convert the representation of P into $P' \in \mathcal{F}^{2c}_{\ell,m+1}[2, \{-2, -1, 0, 1, 2\}]$*, where the conversion can be performed in constant time by a simple rewriting of the guard digits.*

Proof By Lemma 3.8.1, $|\|Q\|| \leq 2$, but necessarily $\|Q\| \in \{-2, -1, 0\}$. Let S be defined as the prefix of P by $P = S\,[2]^{m+3} + \sum_{i=\ell}^{m+2} d_i[2]^i$, then $\|Q\| = 4\|S\| + 2d_{m+2} + d_{m+1}$, which by the lemma implies that $\|S\| \in \{-2, -1, 0\}$, where it is easily seen that $\|S\| = -2$ is impossible. Writing the polynomial P in digit-string representation for the various possibilities of the prefix Q we find the following table:

$\|Q\|$	$\|S\|$	d_{m+2}	d_{m+1}	tail
0	0	0	0	$d_m d_{m-1} \cdots d_\ell$
	$\bar{1}$	2	0	$d_m d_{m-1} \cdots d_\ell$
	$\bar{1}$	1	2	$d_m d_{m-1} \cdots d_\ell$
-1	$\bar{1}$	1	1	$d_m d_{m-1} \cdots d_\ell$
-2	$\bar{1}$	1	0	$d_m d_{m-1} \cdots d_\ell$
	$\bar{1}$	0	2	$d_m d_{m-1} \cdots d_\ell$

As is easily seen from the table, (3.8.3) for $\|Q\|$ holds, and P can be represented in $\mathcal{F}^{2c}_{\ell,m+1}[2, \{-2, -1, 0, 1, 2\}]$ by a simple rewriting of the digits d_{m+2} and d_{m+1}, such that $d'_{m+1} = -\|Q\| = d'_{m+2}$. □

Note that it is not possible just to prepend the value of $\|Q\|$ as a digit in the set $\{-2, -1, 0\}$ to the string $d_m d_{m-1} \cdots d_\ell$, to obtain a polynomial in $\mathcal{F}^{2c}_{\ell,m}[2, \{-2, -1, 0, 1, 2\}]$, as this may violate the rule that $d_{m+1} = -d_m$. And as we shall see below, a rewriting may require a change arbitrarily far to the right. Recall that it is required that the value $\|P\|$ is representable in non-redundant

2's complement, hence $-2^m \leq \|P\| \leq 2^m - 2^\ell$ or

$$-2^m \leq \|P\| = \|Q\| \cdot 2^{m+1} + d_m 2^m + \sum_{i=\ell}^{m-1} d_i 2^i < 2^m. \qquad (3.8.4)$$

If $\|Q\| = 0$, then $\|P\|$ must be non-negative, hence, changing the left bound to zero then,

$$0 \leq d_m 2^m + \sum_{i=\ell}^{m-1} d_i 2^i < 2^m,$$

and thus d_m must be zero. For $\|Q\| = -2$, where $\|P\| < 0$, it follows similarly from (3.8.4) by changing the upper bound that

$$3 \cdot 2^m \leq d_m 2^m + \sum_{i=\ell}^{m-1} d_i 2^i < 2^{m+2},$$

which implies that $2 \leq d_m < 4$, hence $d_m = 2$. Thus for $\|Q\| = 0, -2$ we have the correct value of d_m, but this is not always the case for $\|Q\| = -1$, where the sign of $\|P\|$ cannot be determined from that of $\|Q\|$. Here $-2^m \leq \|P\| \leq 2(2^{m-1} - 2^\ell)$ so

$$-2^m \leq -2^{m+1} + (2d_m + d_{m-1})2^{m-1} + \sum_{i=\ell}^{m-2} d_i 2^i < 2^m,$$

where $0 \leq \sum_{i=\ell}^{m-2} d_i 2^i < 2^m$, hence $\|Q\| = -1$ implies $1 \leq 2d_m + d_{m-1} \leq 5$. Now there are three possible values of d_m, where $d_m = 1$ satisfies $d_m = -\|Q\|$, $d_m = 2$ is possible since this combination is equivalent to $d_{m+1} = d_m = 0$, and finally $d_m = 0$ is possible, together with $d_{m-1} = 1$ or 2. The combination $d_m d_{m-1} = 02$ can easily be changed into $d_m d_{m-1} = 10$, and $d_m d_{m-1} = 01$ is possible provided that $d_{m-1} = \cdots = d_{j+1} = 1$ with $d_j = 2$ for some $j \geq \ell$, because the string representation can then be changed into $\bar{1}100 \cdots 0 d_{j-1} \cdots d_\ell$ representing the same value. However, this would require an arbitrary "look-ahead" into the trailing set of digits, which cannot be done in constant time.

To complete the picture, we will now show that, for $\|Q\| = -1$, if $d_m d_{m-1} = 01$ there is an integer j, $m - 1 \geq j \geq \ell$ where $d_{m-1} = \cdots = d_{j+1} = 1$ with $d_j = 2$. From the bounds above we have

$$2^m \leq \sum_{i=j+1}^{m-1} 2^i + d_j 2^j + \sum_{i=\ell}^{j-1} d_i 2^i.$$

Now assume there is no such j, i.e., $d_i = 1$ for $i = m - 1, \ldots, \ell$. Then there is a contradiction, since the right-hand side of the inequality has the value $2^m - 2^\ell$. Hence there must be such a $d_j = 2$ following the sequence of ones, as $d_j = 0$ would make the right-hand side even smaller.

We must then conclude that there is no constant-time algorithm mapping such a result with guard digits into the set $\mathcal{F}^{2c}_{\ell,m}[2, \{-2, -1, 0, 1, 2\}]$. For practical purposes, the two guard digits should be retained during further computations, until the result is to be mapped into a non-redundant representation.

Corollary 3.8.10 *Provided that the result of an addition performed in 2's complement, carry-save arithmetic is representable in ordinary non-redundant 2's complement representation in* $\mathcal{F}^{2c}_{\ell,m}[2, \{-1, 0, 1\}]$, *then it is sufficient to express the result with two guard digits as a carry-save number in* $\mathcal{F}_{\ell,m+2}[2, \{0, 1, 2\}]$, *i.e., without the digits in position* $m + 3$ *and higher.*

It is then possible in constant time to rewrite the guard digits such that the result is a correct polynomial P *in 2's complement, carry-save representation with*

$$P = \sum_{\ell}^{m+3} d_i [2]^i \in \mathcal{F}^{2c}_{\ell,m+1}[2, \{-2, -1, 0, 1, 2\}],$$

satisfying $d_{m+2} = -d_{m+1}$.

Conversion to remove the guard digits, i.e., mapping into the set $\mathcal{F}^{2c}_{\ell,m}[2, \{-2, -1, 0, 1, 2\}]$, *will in general require at least logarithmic time, i.e., time proportional to that required for conversion into non-redundant representation in* $\mathcal{F}^{2c}_{\ell,m}[2, \{-1, 0, 1\}]$.

The observation also applies to multioperand addition, where it is sufficient to perform addition of the operands in a system with two leading guard digits, beyond what is needed for the result. All "higher-order" digits are not needed, as they are assumed to cancel out. This can be summarized as follows.

To complete the discussion, observe that for a redundant representation there are no simple (constant-time) overflow checks, so a computation must be planned such that there is no overflow in the final result. The previous results on guard digits for redundant representations presented in Theorems 3.8.2, 3.8.3, 3.8.4, 3.8.9, and Observation 3.8.10 can now be summarized.

Observation 3.8.11 *Given bounds on the result of a composite, additive fixed-point computation using a redundant number representation, it is sufficient to perform the operations with one or two leading guard digits, beyond those needed to represent the result. It is always possible in constant time to rewrite the leading digits of the result such that a single guard digit is sufficient. However, it may take at least logarithmic time to rewrite the result such that it contains no guard digits at all. Hence removal of guard digits should be postponed until the result has anyway to be converted to a non-redundant representation.*

The use of a few guard digits insures correct constant-time additions in a redundant representation, but may also be needed in the case of right-shifting, where a sign-extension is necessary.

Corollary 3.8.12 *When a right-shift of a fixed-point number in redundant representation is used to realize division by the radix, then either an additional guard digit must be present, or a recoding of the guard digit(s) must be performed, to insure correct sign-extension.*

We leave it as a problem to show how a proper sign-extension can be realized in the case of the 2's complement, carry-save representation, other redundant representations follow similarly.

Problems and exercises

3.8.1 Prove (3.8.2).

3.8.2 Since zero is not unique in the 2's complement carry-save representation, testing for the value zero directly is complicated. However, it was noted in Section 1.8 that -1 is uniquely represented in the integer domain. It is easy to see that, in general, -2^ℓ is uniquely represented in $\mathcal{F}^{2c}_{\ell,m}[2, \{-2, -1, 0, 1, 2\}]$ by $-[2]^{m+1} + \sum_{i=\ell}^{m}[2]^i$, i.e., in string representation as all ones. Thus zero testing can be performed by first subtracting 2^ℓ, or equivalently adding all ones, followed by testing for all ones.

Show that the logic for adding all ones to a carry-save represented number, considered as a 3-to-2 addition problem, can be significantly simplified. Also show that this addition can be performed in constant time.

3.8.3 Consider the problem of using a right shift for performing integer division by 2 of a 2's complement carry-save polynomial $P \in \mathcal{F}_{0,m+g}[2, \{0, 1, 2\}]$, $g \geq 2$, where only the non-negative digits are available, but at least two leading guard digits d_{m+2}, d_{m+1} are included. Show how the necessary sign-extension can be realised using the results of Theorem 3.8.9.

3.9 Subtraction and sign-magnitude addition

Having defined the additive inverse in Section 2.6, subtraction in $\mathcal{P}[\beta, D]$ can be realized through addition of the minuend and the additive inverse of the subtrahend. But note that this should not be interpreted such that subtraction should be implemented as a two-step process, both steps of which would be at least logarithmic time operations if the result is to be in a non-redundant representation. Most often the negation of the subtrahend can be integrated in the addition process at no time penalty whatsoever.

In analogy with our definition of addition we can define subtraction in $\mathcal{P}[\beta, D]$ by defining $\overline{D} = \{d \mid -d \in D\}$ and a subtraction mapping $\alpha : C \times D \to C \times D$, where $\overline{D} \subseteq C$ and C is constructed in analogy with the method in Section 3.3. If D is basic and symmetric then the subtraction mapping coincides with the addition mapping. Table 3.9.1 shows two subtraction mappings:

Table 3.9.1. *Subtraction tables for α standard binary and γ balanced ternary*

α	0	1
−1	$\bar{1}1$	00
0	00	01

γ	−1	0	1
−1	$\bar{1}1$	$0\bar{1}$	00
0	$0\bar{1}$	00	01
1	00	01	$1\bar{1}$

If D is complete for radix β, then all properties of addition in $\mathcal{P}[\beta, D]$ carry over to subtraction, since $P \in \mathcal{P}[\beta, D] \Rightarrow -P \in \mathcal{P}[\beta, D]$ and addition is closed in $\mathcal{P}[\beta, D]$. A (digit) subtraction mapping α thus generalizes to a *radix polynomial subtraction mapping* $\hat{\alpha}$, which possibly has to be applied repeatedly to absorb carries. In a system with a redundant digit set subtraction can be performed in constant time in a digit-by-digit parallel process.

However, if the digit set is not complete for the radix, the result of a subtraction may not be representable. A simple example is the system $\mathcal{P}[2, \{0, 1\}]$ using Table 3.9.1α.

Example 3.9.1 Subtracting $101_{[2]}$ from $11_{[2]}$ we obtain

$$
\begin{array}{rccccl}
P = & & & 1 & 1_{[2]} & \text{minuend} \\
\overline{Q} = & & \bar{1} & 0 & \bar{1}_{[2]} & \text{inverse of subtrahend} \\
\hline
 & & 1 & 1 & 0_{[2]} & \text{sum} \\
\hat{\alpha}(P, Q) & \bar{1} & 0 & 0_{[2]} & & \text{carries (borrows)} \\
\hline
 & 1 & 1 & 1 & 0_{[2]} & \text{sum} \\
\hat{\alpha}(\hat{\alpha}(P, Q)) \quad \bar{1} & 0 & 0 & 0_{[2]} & & \text{carries (borrows)} \\
\hline
 & & & \vdots & &
\end{array}
$$

thus the process never terminates. Note how this process leads to the 2's complement representation where the negative carry is pushed outside a finite representation. □

Cascading a set of *pnp*-adders in a ripple-carry/borrow fashion we obtain a "ripple-borrow subtractor" (Figure 3.9.1) for standard binary (or 2's complement), when the "borrow-in" to the least-significant position is a constant 0. The "borrow" ($\in \{-1, 0\}$) coming out of the most-significant position corresponds again to the 2's complement carry being pushed outside the finite representation. But note that the borrows coming in and sent out are in complemented form for *npn*-adders.

In many situations it is useful to have a circuit which can realize addition as well as subtraction, where the choice is based on a suitable selection signal. Such an adder/subtractor is shown for simplicity in Figure 3.9.2 in the form of a ripple-carry n-bit structure, but it can of course be constructed with any faster carry completion circuitry, capable of accommodating a carry-in of zero or one.

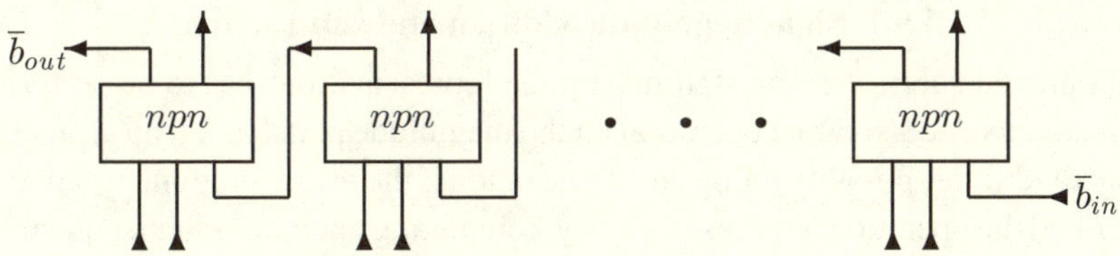

Figure 3.9.1. Ripple-borrow subtractor.

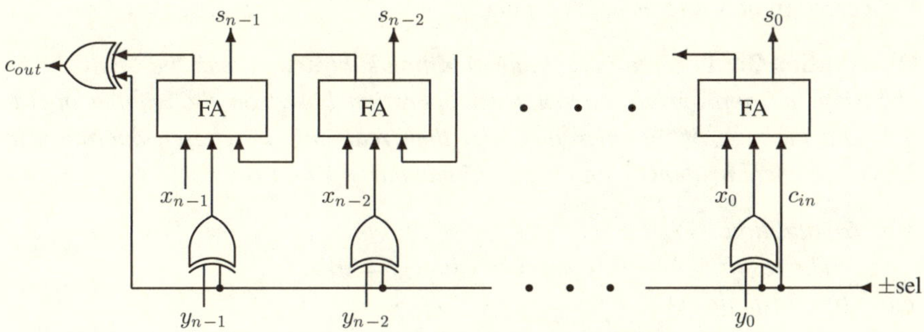

Figure 3.9.2. A ripple-carry add/subtract circuit.

The idea here is conditionally to add the 2's complement, based on the selection signal (1 $\Rightarrow$ subtraction), by inverting all y-bits and simultaneously supplying $c_{in} = 1$ in the case of subtraction. Note that in this case the carry-out is to be complemented, which is clear from examples, but can also be seen by the following argument.

We may also consider the selection of a reconfiguration, where the *ppp*-adder is conditionally changed into a subtractor (an *npn*-adder), by complementing the y-input and changing signs of the weights of the carries. But where carry signals should have been complemented, this is not necessary since it applies to the output of one adder, as well as to the input of the next. The only places where it matters are at the carry-in and -out, but at the carry-in the complement of a zero is precisely what is delivered by the selection signal, and at the left the carry-out is delivered in complemented form when subtracting.

With appropriate modifications these techniques immediately also apply to the redundant carry-save representation. Just note that the additive inverse can be obtained *either* by negating all digits (assuming negative weight of the encoded bits) *or* by complementing the bits and feeding a carry-in as the value 2 (realized as two ones fed into the least significant 4-to-2 adder of Figure 3.7.3).

For the symmetric signed-digit borrow-save representation the additive inverse is trivially obtained in the standard encoding (2.7.7) by interchanging the role of the encoded bits ("twisting wires") when feeding a borrow-save operand.

3.9.1 Sign-magnitude addition and subtraction

Addition and subtraction on sign-magnitude representations has to be performed as an effective add/subtract operation on the magnitude parts, based on an analysis of operand signs, possibly followed by a negation if the result turns out negative. As described the operation requires two carry-completing operations, one sequentially following the other.

However, the negation can be integrated into the adder structure at only a moderate extra cost, if the adder is capable of simultaneously computing the result for a carry-in of a zero as well of a one.

Observation 3.9.1 *Sign-magnitude addition/subtraction can be realized by choosing the appropriate add/subtract operation based on the wanted operation and the sign of the operands, and then realizing the chosen operation in 2's complement arithmetic on the magnitude parts A and B as:*

case (a) addition $A + B$*:*

The result is $A + B$ *(computed with* $c_{in} = 0$*)*

case (b) subtraction $A - B$*:*

case (b_1) *The result is* $\overline{A + \overline{B}}$ *if* $A + \overline{B}$ *(computed with* $c_{in} = 0$*) is negative*

case (b_2) *The result is* $A + \overline{B} + 1$ *(computed with* $c_{in} = 1$*) otherwise*

where $\overline{X}$ *is the operand obtained by complementing each bit. The resulting sign then has to be negated in case* (b_1).

Thus bitwise inversion turns out to be sufficient when the sign is to be reversed. Since negation in 2's complement is achieved by inverting all bits while supplying a carry-in of 1 in the least-significant position, the correctness of the choice in case b_1 follows from

$$A + \overline{B} < 0 \Rightarrow A + \overline{B} + 1 = A - B \leq 0$$

in which case the sign has to be negated and the result is chosen as

$$\overline{A + \overline{B}} = \overline{A + (-B - 1)} = \overline{A - B - 1} = B - A \geq 0.$$

Case (b_2) then is $A + \overline{B} \geq 0 \Rightarrow A - B > 0$ and the result is chosen to be

$$A + \overline{B} + 1 = A - B > 0.$$

Now recall from Section 3.6.3 that in conditional-sum addition, and in the carry-look-ahead adder with a constant-time overhead (cf. equation (3.6.7)), it is possible to compute the sum for a carry-in of 0 and in parallel as well as for a carry-in of 1.

To develop the logic for a sign-magnitude adder/subtractor based on a carry-look-ahead adder, let γ be a binary signal specifying whether an addition or a subtraction is to be performed, say $R = A + (-1)^{\gamma} B$. Let $\sigma(X)$ be the sign of a sign-magnitude represented number X (where 1 represents a negative value), and $m(X)$ be the magnitude as an non-negative integer, so that $X = (-1)^{\sigma(X)}m(X)$. Since

$$R = A + (-1)^{\gamma} B = (-1)^{\sigma(A)} \left(m(A) + (-1)^{\gamma+\sigma(B)-\sigma(A)} m(B)\right), \qquad (3.9.1)$$

then

$$\tau = \gamma \oplus \sigma(A) \oplus \sigma(B) = \begin{cases} 0 \Rightarrow \text{form magnitude} & m(R) = (m(A) + m(B)) \\ 1 \Rightarrow \text{form magnitude} & m(R) = |\, m(A) - m(B)|, \end{cases}$$

is the signal to distinguish between cases (a) and (b) of Observation 3.9.1. Following the observation we want to compute $m(A) + m(B)$ or $m(A) - m(B)$ depending on τ, i.e., to form an intermediate result $D = m(A) + \tau \oplus m(B)$, using the expression $\tau \oplus m(B)$ for the conditional inversion of the bits of $m(B)$.

Since potentially one extra bit may be needed in the case of an effective addition, let us assume that the most-significant positions in both operands (possibly originally containing the sign) have been filled with zero-valued bits. Assume that $m(A)$ has the string representation $a_{n-1} \cdots a_0$ and $m(B)$ similarly has the representation $b_{n-1} \cdots b_0$, where $a_{n-1} = b_{n-1} = 0$. Let D be the intermediate result, similarly with the string representation $d_{n-1} \cdots d_0$, assuming the carry-input $c_0 = 0$.

In case (a) ($\tau = 0$), overflow can then be tested by checking position $n - 1$ of the result. Thus

$$OF = \overline{\tau} d_{n-1} = \overline{\tau} c_{n-1}$$

is the flag denoting overflow, since in this case $d_{n-1} = c_{n-1}$, the carry coming into position $n - 1$.

In case (b) ($\tau = 1$), where $D = m(A) + \tau \oplus m(B) = m(A) + \overline{m(B)}$ is computed, it is necessary to further distinguish the two subcases (b_1) and (b_2), as determined by the sign of the result when interpreted as a 2's complement number. We can express the sign $\sigma = d_{n-1}$ as

$$\sigma = d_{n-1} = a_{n-1} \oplus \overline{b_{n-1}} \oplus c_{n-1},$$

but we assumed initially $a_{n-1} = b_{n-1} = 0$, hence

$$\sigma = \overline{c_{n-1}} \qquad (3.9.2)$$

is the signal that $D < 0$. Hence $\sigma = 1$ implies case (b_1) where the resulting magnitude $m(R)$ is obtained by inverting the bits of D. To complete the discussion of signs, observe from (3.9.1) that

$$\sigma(R) = \sigma(A) \oplus (\tau\sigma)$$

covers both $\tau = 0$ and $\tau = 1$.

For the carry-look-ahead adder and case (a) of the observation, the input $p_i = a_i \oplus b_i$ and $g_i = a_i\, b_i$, $i = 0, 1, \ldots, n - 1$, have to be fed into the look-ahead structure. For case (b) $p_i = a_i \oplus \bar{b}_i$ and $g_i = a_i\,\bar{b}_i$ are to be delivered, thus

$$p_i = a_i \oplus (\tau \oplus b_i) \quad \text{and} \quad g_i = a_i\,(\tau \oplus b_i)$$

can be used as the general input. Recall that p_i also will be needed together with the computed carry c_i to form the final sum bits s_i.

The carry-look-ahead structure can now compute the values of G_i and P_i for $i = 0, 1, \ldots, n-1$. Note that these values depend on τ, and recall from Lemma 3.6.2 that

$$c_i = \begin{cases} G_{i-1} & \text{for } c_0 = 0, \\ G_{i-1} + P_{i-1} & \text{for } c_0 = 1 \end{cases} \tag{3.9.3}$$

is the carry entering the ith position, and thus in particular $\sigma = \overline{c_{n-1}} = \overline{G}_{n-2}$ is the value signalling a negative result in the case of a subtraction, i.e., the sign of $A + \overline{B}$ in case b). In these cases $\tau = 1$ and if $\sigma = 1$ then the bits r_i of $m(R)$ can be computed as specified for case (b_1):

$$r_i = \overline{(a_i \oplus \bar{b}_i) \oplus c_i} = \overline{p_i \oplus c_i} = \sigma \oplus (p_i \oplus c_i),$$

with c_i computed for $c_0 = 0$, so by (3.9.3) we have $c_i = G_{i-1}$ for $i > 0$. For $i = 0$ we may choose $G_{-1} = 0 = \bar{\sigma}$. When $\sigma = 0$ the bit r_i in case (b_2) can be expressed as

$$r_i = (a_i \oplus \bar{b}_i) \oplus c_i = p_i \oplus c_i = \sigma \oplus (p_i \oplus c_i),$$

where c_i now is to be computed for $c_0 = 1$. Hence by (3.9.3) we have $c_i = G_{i-1} + c_0 P_{i-1}$ as a common expression for cases (b_1) and (b_2) and

$$r_i = \sigma \oplus p_i \oplus (G_{i-1} + \bar{\sigma}\,P_{i-1}) = p_i \oplus (\sigma \oplus G_{i-1} + \bar{\sigma}\,P_{i-1}),$$

which is valid for $i = 1, \ldots, n-1$, observing that $c_0 = G_{-1} = \bar{\sigma} = G_{n-2}$. For $i = 0$ we have $r_0 = p_0 \oplus c_0 = p_0 \oplus G_{n-2}$. Note that the "end-around" carry, $G_{-1} = G_{n-2}$, here does not delay the computation since it only affects the value of r_0.

For case (a) when $\tau = 0$, $m(R) = m(A) + m(B)$ we have

$$r_i = p_i \oplus c_i = p_i \oplus G_{i-1},$$

since c_i here has to be computed for $c_0 = 0$.

Finally, we obtain the following expression for the bits r_i of the magnitude part, $m(R)$, covering all the cases for $i > 0$:

$$r_i = p_i \oplus (\bar{\tau} G_{i-1} + \tau(\sigma \oplus G_{i-1} + \bar{\sigma}\,P_{i-1})),$$

since the value of σ in case (a) is a "don't care" because here $\tau = 0$. But for $i = 0$ there is a complication if $G_{-1} = G_{n-2} = \bar{\sigma}$ is used as the end-around carry as in cases (b_1) and (b_2) above, because in case (a) $\bar{\sigma} = G_{n-2}$ is the signal for overflow, which will then influence the value of $r_0 = p_0 \oplus G_{n-2}$. But $i = 0$ in case (a) can then be treated separately using the expression $r_0 = p_0 \oplus (\tau G_{n-2})$, covering case (a) as well as cases (b_1) and (b_2).

Due to the importance of sign-magnitude representations we will summarize the above in the form of a theorem.

Theorem 3.9.2 (Sign-magnitude addition and subtraction) *Given operands* $A = (-1)^{\sigma(A)}m(A)$ *and* $B = (-1)^{\sigma(B)}m(B)$*, where* $m(A) = \sum_{i=0}^{n-1} a_i 2^i$ *with* $a_{n-1} = 0$ *and* $m(B) = \sum_{i=0}^{n-1} b_i 2^i$*, with* $b_{n-1} = 0$*, the bits* r_i *of* $m(R) = \sum_{i=0}^{n-1} r_i 2^i$*, where*

$$R = A + (-1)^{\gamma} B = (-1)^{\sigma(R)} m(R),$$

can be expressed as $r_0 = p_0 \oplus (\tau\bar{\sigma})$ *and*

$$r_i = p_i \oplus (\bar{\tau} G_{i-1} + \tau(\sigma \oplus G_{i-1} + \bar{\sigma} P_{i-1})) \text{ for } 0 < i \le n-1,$$

with $\tau = \gamma \oplus \sigma(A) \oplus \sigma(B)$ *and* $\sigma = \overline{G}_{n-2}$.

The sign of result is $\sigma(R) = \sigma(A) \oplus (\tau\sigma)$*, and the overflow signal is* $OF = \bar{\tau}\bar{\sigma}$.

Here $\{(G_i, P_i)\}_{i=0,1,\cdots,n-1}$ *is the set of (generate, propagate)-pairs produced by a carry-look-ahead tree on the set* $\{(g_i, p_i)\}_{i=0,1,\cdots,n-1}$*, defined by:* $p_i = a_i \oplus (\tau \oplus b_i)$*,* $g_i = a_i\,(\tau \oplus b_i)$ *for* $i = 0, 1, \cdots, n-1$.

From the above analysis leading to the theorem Corollary 3.9.3 follows.

Corollary 3.9.3 *Given two unsigned binary integers* $A = \sum_{i=0}^{n-1} a_i 2^i$ *and* $B = \sum_{i=0}^{n-1} b_i 2^i$*, the absolute value of their difference,* $D = |A - B| = \sum_{i=0}^{n-1} d_i 2^i$*, can be found as the digits* $d_0 = p_0 \oplus G_{n-2}$ *and*

$$d_i = p_i \oplus (\bar{\sigma} \oplus G_{i-1} + \bar{\sigma} P_{i-1}) \text{ for } 0 < i \le n-1,$$

where $p_i = a_i \oplus b_i$*,* $g_i = a_i b_i$*, and* (G_i, P_i)*,* σ *are found as in Theorem 3.9.2.*

3.9.2 Bit-serial subtraction

In Section 3.5 we saw how digit serial addition could be realized by a single full-adder (Figure 3.5.6), where the carry is fed back (but delayed) such that operands could be delivered sequentially, least-significant digit first, and the result delivered similarly least-significant first. It is trivial to modify this serial adder into a *bit-serial subtractor* accepting operands and producing the result in 2's complement, by just inverting the incoming bits of the subtrahend B, if the initial carry is set to one:

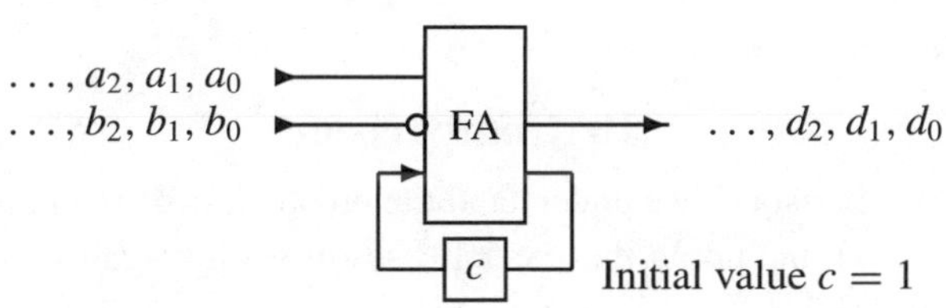

Alternatively, with initial carry set to zero, it is also possible to invert the minuend A if the result is also inverted:

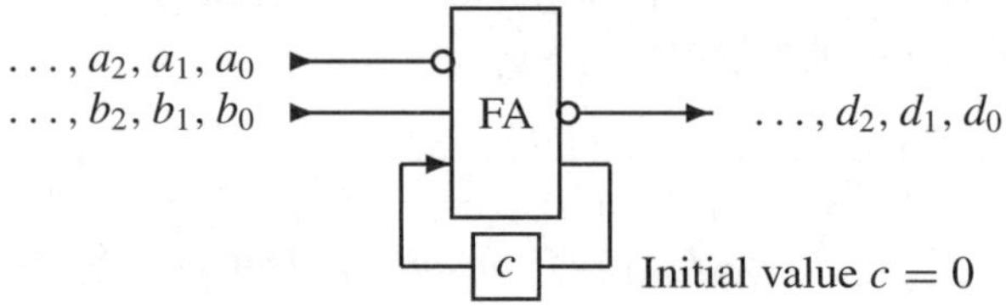

which is, of course, the *pnp*-adder. The correctness of this is seen from the following:

$$\overline{\overline{A} + B} = \overline{-A - 1 + B} = A + 1 - B - 1 = A - B,$$

where again $\overline{X}$ is the complemented bit pattern of X.

Problems and exercises

3.9.1 Find another argument why the carry-out in Figure 3.9.2 has to be complemented when subtraction is realized by adding the 2's complement of the subtrahend.

3.9.2 Given operands $A = +011010$ and $B = -011101$ in binary sign-magnitude representation, compute the sum $A + B$ using the ideas in Observation 3.9.1.

3.9.3 Repeat the computation of Exercise 3.9.2, using Theorem 3.9.2.

3.9.4 Show that the value of P_{-1} in Theorem 3.9.2 is a "don't care."

3.9.5 What is the resulting sign of the value zero obtained by subtracting a sign-magnitude number from itself, using Theorem 3.9.2.

3.9.6 Consider the necessary modifications to Theorem 3.9.2 if a result of zero magnitude should always have a positive sign associated with it.

3.9.7 Would it be possible to evaluate the three condition codes $R < 0$, $R = 0$, and $R > 0$ in parallel with the computation $R := A + (-1)^{\gamma} B$ in Theorem 3.9.2, i.e., without further delaying the execution of the add/subtract instruction?

3.9.8 Hand simulate both of the above digit serial subtraction methods to compute $A - B$ with $A = 011011$ and $B = 011101$. It is not necessary to simulate the internal logic of the adders, just employ the carry and digit mappings of the full-adder.

3.10 Comparisons

By the term "comparisons" we understand the broad class of *relational operators* $\{<, \leq, =, \neq, \geq, >\}$, including the special cases of testing against zero, i.e., sign and zero detection. Actually overflow detection also falls into this category, but has already been discussed in Section 3.8. In general, these latter "monadic" operators, where testing is against a constant, turn out to be somewhat simpler than the more general "dyadic" relational operators. Through a proper subtraction such a dyadic operator can be reduced to one of the simpler monadic operators, but this is not always the best way to implement the operators.

It is essential to notice that comparisons involving one or both operands in a redundant representation cannot, in general, be performed in constant time. This follows trivially from the results of Section 3.2, since the boolean result may depend on all digits of the operands and hence must take logarithmic time. In general a comparison on redundantly represented numbers may implicitly or explicitly involve a conversion to non-redundant representation. Certain comparisons may be performed based on a few or a single leading digit whether from a non-redundant or a redundant representation, but the information thus provided only tells us that the operand belongs to some interval. However, such interval testings play an important role in many algorithms, e.g., in digit-serial division, where quotient digits from a redundant digit set are chosen, based on a few leading digits of the partial remainder.

3.10.1 Equality testing

For any system with unique zero representation the test for $P = 0$ with $P = \sum_{i=\ell}^{m} p_i[\beta]^i$ just becomes

$$T \equiv \prod_{i=\ell}^{m} (p_i = 0),$$

which for ordinary binary or 2's complement reduces to $\prod_{i=\ell}^{m} \overline{p}_i$. Since the AND operator is associative, a tree of AND gates can then compute the result in logarithmic time. Similarly, for a non-redundant system the test for $P = Q$ can be performed in a tree of AND gates in logarithmic time, where the leaves of the tree just test the pair-wise equality of the digits. Note that this is faster than computing the difference, and then testing the result against zero.

For the binary case an even more complicated comparison can be performed with a tree of AND gates, based on some constant-time preprocessing of operands:

Theorem 3.10.1 *The comparison $A + B = D$ with A, B, and D unsigned binary numbers from $\mathcal{F}_{\ell m}[2, \{0, 1\}]$ (or 2's complement numbers from $\mathcal{F}_{\ell m}^{2c}[2, C_2^c]$) can be performed by evaluating*

$$T \equiv \prod_{i=\ell}^{m} \left[(a_i \oplus b_i \oplus \overline{d_i}) \oplus (a_{i-1}b_{i-1} + (a_{i-1} \oplus b_{i-1})\overline{d_{i-1}}) \right],$$

where $A = \sum_{i=\ell}^{m} a_i[2]^i$, $B = \sum_{i=\ell}^{m} b_i[2]^i$ *and* $D = \sum_{i=\ell}^{m} d_i[2]^i$, *together with* $a_{\ell-1} = b_{\ell-1} = d_{\ell-1} = 0$.

Note The expressions in the logical product are the sum, respectively carry, bits from a set of full-adders with input a_i, b_i and $\overline{d_i}$ into the ith full-adder.

Proof Let c_i be the carry coming into position i when adding A and B, then from our analysis of binary adders in Section 3.6 we have

$$c_i = g_{i-1} + p_{i-1}c_{i-1}, \quad i = \ell + 1, \ldots, m,$$

where g_{i-1} and p_{i-1} are the generate and propagate signals at the right-hand neighbor. Equivalently for $c_\ell = 0$, expressing g_{i-1} and p_{i-1} in terms of a_{i-1} and b_{i-1}:

$$\begin{aligned} c_\ell &= 0, \\ c_i &= a_{i-1}b_{i-1} + (a_{i-1} \oplus b_{i-1})c_{i-1}, \quad i = \ell + 1, \ldots, m. \end{aligned} \tag{3.10.1}$$

Comparing $A + B$ with D bitwise, a necessary and sufficient condition for $A + B = D$ is $a_i \oplus b_i \oplus c_i = d_i$ for $i = \ell, \ldots, m$, which is equivalent to $a_i \oplus b_i \oplus d_i = c_i$. Inserting these into (3.10.1) we obtain the following set of conditions:

$$c_\ell = 0 \Longrightarrow a_\ell \oplus b_\ell = d_\ell,$$

and for $i = \ell + 1, \ldots, m$

$$\begin{aligned} a_i \oplus b_i \oplus d_i &= c_i = a_{i-1}b_{i-1} + (a_{i-1} \oplus b_{i-1})c_{i-1} \\ &= a_{i-1}b_{i-1} + (a_{i-1} \oplus b_{i-1})(a_{i-1} \oplus b_{i-1} \oplus d_{i-1}) \\ &= a_{i-1}b_{i-1} + (a_{i-1} \oplus b_{i-1})\overline{d_{i-1}}, \end{aligned}$$

since $x(x \oplus y) = x\overline{y}$, from which the result follows. □

For the special case of $D = 0$ (testing $A = -B$ in 2's complement) the test reduces to

$$T \equiv \prod_{i=\ell}^{m} \overline{(a_i \oplus b_i) \oplus (a_{i-1} + b_{i-1})}, \tag{3.10.2}$$

with $a_{\ell-1} = b_{\ell-1} = 0$. And similarly testing $A = D$ (i.e., $B = 0$) it reduces to

$$T \equiv \prod_{i=\ell}^{m} [(a_i \oplus \overline{d_i}) \oplus (a_{i-1}\overline{d_{i-1}})] = \prod_{i=\ell}^{m} \overline{a_i \oplus d_i}, \tag{3.10.3}$$

where the last reduction is left as an exercise. Note that the right-hand side is the trivially correct value of T obtained by bitwise comparison.

Two particularly useful special cases are formalized in the following two corollaries, the first without proof.

Corollary 3.10.2 *The comparison $A = D$, where A is in redundant carry-save representation, $A \in \mathcal{F}_{\ell m}[2, \{0, 1, 2\}]$, and D is in non-redundant representation, $D \in \mathcal{F}_{\ell m}[2, \{0, 1\}]$ (or both in 2's complement form), can be performed by evaluating*

$$T = \prod_{i=\ell}^{m} \left[(a_i^c \oplus a_i^s \oplus \overline{d_i}) \oplus (a_{i-1}^c a_{i-1}^s + \overline{d_{i-1}}(a_{i-1}^c \oplus a_{i-1}^s)) \right],$$

where $A = \sum_{i=\ell}^{m} a_i [2]^i$, $D = \sum_{i=\ell}^{m} d_i [2]^i$ and the carry-save encoding $a_i = (a_i^c, a_i^s)$ with $0 \sim 00$, $1 \sim 10$ or 01 and $2 \sim 11$ is used for the digits of A. Also it is assumed that $a_{\ell-1}^c = a_{\ell-1}^s = d_{\ell-1} = 0$.

Corollary 3.10.3 *The comparison $A = D$, where A is in redundant signed-digit (borrow-save) representations, $A \in \mathcal{F}_{\ell m}[2, \{-1, 0, 1\}]$, and D is in 2's complement non-redundant representation, $D \in \mathcal{F}_{\ell m}^{2c}[2, C_2^c]$, can be performed by evaluating*

$$T = \prod_{i=\ell}^{m} \left[(a_i^p \oplus \overline{a_i^n} \oplus \overline{d_i}) \oplus (a_{i-1}^p \overline{a_{i-1}^n} + \overline{d_{i-1}}(a_{i-1}^p \oplus \overline{a_{i-1}^n})) \right],$$

where $A = \sum_{i=\ell}^{m} a_i [2]^i$, $D = \sum_{i=\ell}^{m} d_i [2]^i$, and the borrow-save encoding $a_i = (a_i^p, a_i^n)$ with $-1 \sim 01$, $0 \sim 00$ or 11 and $1 \sim 10$ is used for the digits of A. Also $a_{\ell-1}^p = a_{\ell-1}^n = d_{\ell-1} = 0$.

Proof The value $\|A\|$ in 2's complement representation can be obtained as the sum of the two polynomials $A^p = \sum_{i=\ell}^{m} a_i^p [2]^i$ and $A^n = \sum_{i=\ell}^{m} \overline{a_i^n} [2]^i + [2]^\ell$, by disregarding the carry-out of position m (addition modulo 2^m). A carry-in of 1 (the term $[2]^\ell$) is insured by the initialization in position $\ell - 1$, yielding $c_\ell = (a_{\ell-1}^p \overline{a_{\ell-1}^n} + \overline{d_{\ell-1}}(a_{\ell-1}^p \oplus \overline{a_{\ell-1}^n})) = 1$. □

Note that the above test expressions can easily be rewritten for other digit encodings, and can all be evaluated in logarithmic time by a tree structure.

Example 3.10.1 Consider testing $A = D$, where $A = 00\bar{1}110\bar{1}_{[2]}$ is in $\mathcal{F}_{06}[2, \{-1, 0, 1\}]$ in the borrow-save encoding $A = A^p - A^n$ with

$$A^p = 0101110_{[2]} \sim 46,$$
$$A^n = 0110011_{[2]} \sim 51, \qquad \text{and}$$
$$D = 1111011_{[2c]} \sim -5 \quad \text{(2's complement).}$$

In the notation of Corollary 3.10.3 we then have

a_i^p	:	0	1	0	1	1	1	0	(0)
$\overline{a_i^n}$	:	1	0	0	1	1	0	0	(1)
d_i	:	1	1	1	1	0	1	1	(0)
$a_i^p \oplus \overline{a_i^n} \oplus \overline{d_i}$	:	1	1	0	0	1	1	0	
$a_{i-1}^p \overline{a_{i-1}^n}$	:	0	0	1	1	0	0	0	
$a_{i-1}^p \oplus \overline{a_{i-1}^n}$	:	1	0	0	0	1	0	1	
$\overline{d_{i-1}}(a_{i-1}^p \oplus \overline{a_{i-1}^n})$	:	0	0	0	0	0	0	1	
$a_{i-1}^p \overline{a_{i-1}^n} + \overline{d_{i-1}}(a_{i-1}^p \oplus \overline{a_{i-1}^n})$	:	0	0	1	1	0	0	1	
t_i	:	1	1	1	1	1	1	1	

where the initializations in position -1 have been added in the rightmost column. The result $T = \prod_0^6 t_i$ confirms that $A = D$. □

3.10.2 Ordering relations

Recall from Lemma 1.5.7 that if D is a digit set which is complete for the radix β, then $P = 0$ is the only polynomial in $\mathcal{P}[\beta, D]$ with $\|P\| = 0$ if and only if D contains no non-zero multiple of β. This condition then becomes necessary and sufficient to insure that the sign of a number can be determined from a constant number of non-zero leading digits, since then there can be no leading prefixes of zero value. If, furthermore, $D \subseteq \{d \mid -\beta < d < \beta\}$, then the sign is determined by the leading non-zero digit.

However, in a fixed-point system $\mathcal{F}_{\ell m}[\beta, D]$ the leading non-zero digit can be positioned anywhere in the range ℓ through m (anywhere in a register), hence the above rules are of no help since the information on the sign is not located at a fixed position.

Incidentally, this is precisely the reason why radix-complement representations are so useful. Here the inherent redundancy in the digit set C_β^c is exploited by the rules for the digits d_{m+1} and d_m, such that a particular representation with sign information in position m is chosen from the many redundant representations.

Similarly, in a radix-complement system $\mathcal{F}_{\ell m}^{rc}[\beta, C_\beta^c]$ some other particular comparisons based on a few leading digits are very simple, e.g., for bounded (small) k in the following observation these can be performed in constant time.

Observation 3.10.4 *For $P \in \mathcal{F}_{\ell m}^{rc}[\beta, C_\beta^c]$ any test of the form*

$$\|P\| \geq a\beta^{m-k}, \quad k \geq 0, a \in \mathbb{Z}$$

can be performed in time $O(\log k)$.

Relaxing our notation we note that $P \geq 0$ can be tested with $a = k = 0$ in constant time, simply by testing the sign bit, but note the non-symmetry in the test. Although the complementary test $P < a\beta^{m-k}$ is just as simple, the general test for $P \leq Q$ may involve all the digits of P and Q. The simplest way of testing $P \leq 0$ is to combine a zero-detection with the sign information from d_m.

If an n-bit adder is available, and a two-way test of the form $P < Q$ is wanted with P and Q given as n-bit 2's complement operands, then the test may be performed using a subtraction. Note first that if the operands are of different sign, then the result can be determined by the most significant bits alone, and is simply given by the sign bit of P. Otherwise assume $P = p_{n-1}p_{n-2}\cdots p_0$ and $Q = q_{n-1}q_{n-2}\cdots q_0$, then $s = \overline{p_{n-1} \oplus q_{n-1}}$ is the condition for determining whether the operands are of the same sign and a subtraction is to be performed. Provided that there is no overflow, the combined result is then

$$(P < Q) \equiv s r_{n-1} + \bar{s} p_{n-1}, \tag{3.10.4}$$

where $R = P - Q = r_{n-1}r_{n-2}\cdots r_0$ is the result of the subtraction.

Since the result of the adder is only used when P and Q have the same sign, $|P - Q| \leq \max(|P|, |Q|)$, so $\max(|P|, |Q|)$ representable implies that $P - Q$ is representable, hence there is no problem if $P \geq 0$ and $Q \geq 0$, nor if $P < 0$ and $-2^n < Q < 0$. But there is a problem if $Q = -2^n$ since then $-Q$ is not representable in 2's complement, and the subtraction with sign-extension yields

$$\begin{array}{rl} P = & 1\ 1\ p_{n-2}\ \cdots\ p_0 \\ \text{2's complement of } Q = -2^n = & 1\ 1\ \ 0\ \ \ \cdots\ \ 0 \\ \hline R = P - Q = & 1\ 0\ p_{n-2}\ \cdots\ p_0 \end{array}$$

In this case $P \geq Q$ with an overflow indication; however, expression (3.10.4) yields 0 hence it is also correct in this situation.

Observation 3.10.5 *For operands $P = p_{n-1}p_{n-2}\cdots p_0$ and $Q = q_{n-1}q_{n-2}\cdots q_0$ in 2's complement representation, the test $P < Q$ can be expressed by subtracting Q from P as*

$$(P < Q) \equiv s r_{n-1} + \bar{s} p_{n-1},$$

where $R = P - Q = r_{n-1}r_{n-2}\cdots r_0$ is the result of the subtraction and $s = \overline{p_{n-1} \oplus q_{n-1}}$.

The non-symmetry in the radix-complement systems is due to the non-negative digits in the trailing part of the number. For a digit set which is complete for the radix of a system with positive radix, there must necessarily be digits of both signs, hence the trailing part can "pull" the value of a prefix both up and down. It thus makes most sense to investigate comparisons in the form of three-way tests, e.g., with outcome in $\{n, 0, p\}$ for respectively < 0, $= 0$, and > 0 when testing against zero.

For the most commonly used digit sets D satisfying $D \subseteq \{d \mid -\beta < d < \beta\}$, the trailing part of a number cannot change the sign of a non-zero prefix of the number. We can thus define a *sign-determination composition operator* $\circ : S \times S \rightarrow S$, where $S = \{n, 0, p\}$ by the following table:

$\circ$	n	0	p
n	n	n	n
0	n	0	p
p	p	p	p

(3.10.5)

and use it to determine the sign of a digit string based on the following definition of the *signum function* $\sigma : \mathbb{Z} \to S$ defined as

$$\sigma(d) = \begin{cases} n & \text{if} \quad d < 0, \\ 0 & \text{if} \quad d = 0, \\ p & \text{if} \quad d > 0. \end{cases} \tag{3.10.6}$$

From this we can define $\hat{\sigma}(\cdot)$ as the extension from digits to digit strings, the *string signum function* $\hat{\sigma} : D^* \to S$ by:

$$\hat{\sigma}(d_{k-1}d_{k-2}\cdots d_0) = \sigma\left(\sum_{i=0}^{k-1} d_i\beta^i\right), \tag{3.10.7}$$

with $\hat{\sigma}(\varepsilon) = 0$ for ε being the empty string.

Lemma 3.10.6 *If D is a complete digit set satisfying $D \subseteq \{d \mid -\beta < d < \beta\}$, then the string signum function $\hat{\sigma}$ satisfies*

$$\hat{\sigma}(d_{m-1}\cdots d_k) \circ \hat{\sigma}(d_{k-1}\cdots d_0) = \hat{\sigma}(d_{m-1}\cdots d_0),$$

where $\circ$ is defined by (3.10.5). It then follows that the operator $\circ$ is associative.

Proof Since

$$\hat{\sigma}(d_{m-1}\cdots d_0) = \sigma\left(\left(\sum_{i=k}^{m-1} d_i\beta^{i-k}\right)\beta^k + \sum_{i=0}^{k-1} d_i\beta^i\right)$$

and

$$\left|\sum_{i=0}^{k-1} d_i\beta^i\right| \leq (\beta - 1)\frac{\beta^k - 1}{\beta - 1} < \beta^k$$

the results follow from the $\circ$-composition table (3.10.5). □

The sign of a radix polynomial can thus be found by a tree structure.

Theorem 3.10.7 *If P is a polynomial, $P \in \mathcal{F}_{\ell m}[\beta, D]$, where $D \subseteq \{d \mid -\beta < d < \beta\}$, then $\sigma(\|P\|)$ can be found in time $O(\log(m - \ell))$ by a tree structure of $\circ$-composition nodes applied to the string of digits of P.*

There is an interesting alternative derivation of these results obtained by observing that pairing digits from D, $D \subseteq \{d \mid -\beta < d < \beta\}$ yields digits in radix β^2 satisfying $d_1\beta + d_2 \in D_2 \subseteq \{d \mid -\beta^2 < d < \beta^2\}$, where $\sigma(d_1\beta + d_2) = \sigma(d_1)$ if

$d_1 \neq 0$, otherwise $\sigma(d_1\beta + d_2) = \sigma(d_2)$. Thus recursively performing base conversion by grouping pairs of digits yields the same results on the composition of σ values.

For an implementation note that there is a trivial mapping from the digit set D into the set $S = \{n, 0, p\}$, and also a simple way of implementing the $\circ$ composition. By using the following encoding of $\sigma(\cdot)$ and $\hat{\sigma}(\cdot)$:

$$\begin{aligned} n &\sim 11, \\ 0 &\sim 10 \text{ or } 00, \\ p &\sim 01, \end{aligned} \tag{3.10.8}$$

i.e., a "sign-magnitude" encoding, each $\circ$-composition node

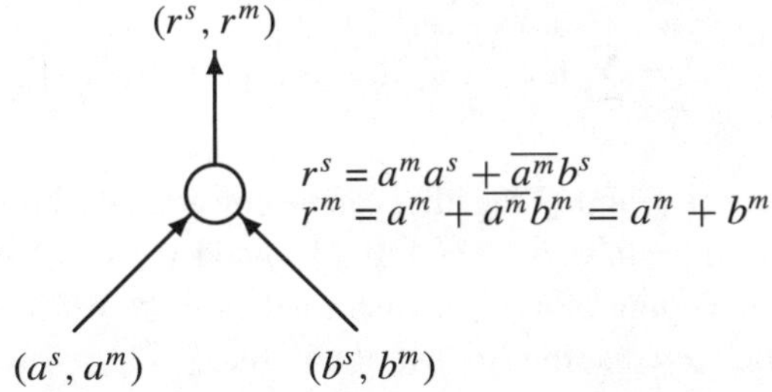

in the tree is simply a selector which, based on the value of a^m, selects whether to forward as the result (r^s, r^m) either the leftmost operand (a^s, a^m), or the rightmost operand (b^s, b^m). The logic for obtaining the nodes is slightly simpler than which is needed for a node in a binary conditional sum adder, and slightly more complex than the nodes needed in a carry-look-ahead tree, but they are all of the same time complexity. However, if the choice of which sign signal to forward is made by selector circuits, then note that the magnitude signals will travel faster and thus be available in the selectors when the sign signals arrive.

By a simple generalization it is possible to generate the signs of all the suffixes ("trailing parts") of a polynomial, i.e., $\sigma_k = \sigma\left(\| \sum_{i=\ell}^{k} d_i \beta^i \|\right)$ for $k = \ell, \ell + 1, \ldots, m$. Simply employ σ-nodes as the black nodes in the kind of look-ahead trees known from adders, shown here as eight-digit Kogge–Stone trees:

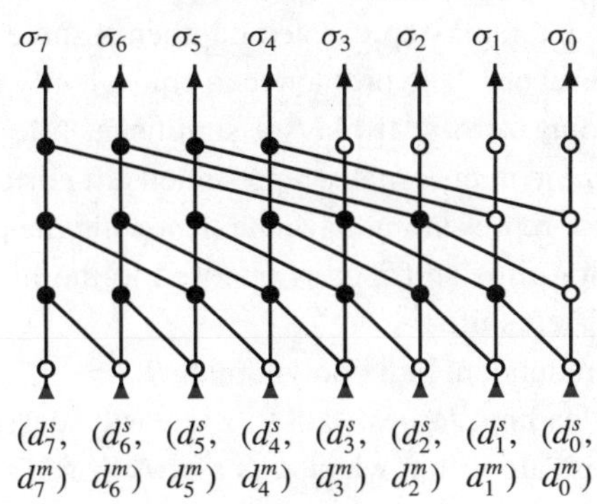

where the digits are input at the bottom and signs delivered at the top in sign-magnitude endoding (3.10.8).

Corollary 3.10.8 *If P is a polynomial $P \in \mathcal{F}_{\ell m}[2, \{-1, 0, 1\}]$, where $P_k = \sum_{i=\ell}^{k} d_i \beta^i$, then the suffixes $\sigma_k = \sigma(\|P_k\|)$ for $k = \ell, \ell+1, \ldots, m$, can all be found in time $O(\log(m-\ell))$ by a look-ahead tree of $\circ$-composition nodes applied to the string of digits of P.*

Observe also that a σ-tree can trivially be used for a three-way comparison of two unsigned binary numbers from $\mathcal{F}_{\ell m}[2, \{0, 1\}]$. Since for P and Q in this set, $P = \sum_{i=\ell}^{m} a_i [2]^i$ and $Q = \sum_{i=\ell}^{m} b_i [2]^i$, then $\|P - Q\| = \|R\|$ with

$$R = \sum_{i=\ell}^{m} (a_i - b_i)[2]^i \in \mathcal{F}_{\ell m}[2, \{-1, 0, 1\}],$$

so the digit strings of P and Q together can be considered a borrow-save encoding of R with digits $d_i = a_i - b_i \in \{-1, 0, 1\}$ encoded as $(d_i^b, d_i^s) = (a_i, b_i)$. The sign-magnitude encoding of $\sigma(d_i)$ is then easily seen to be $(\overline{a}_i, a_i \oplus b_i)$, which is the value to be delivered at the ith leaf of the tree.

Also a three-way comparison of two 2's complement numbers can be obtained by the tree structure. The only difference from the unsigned case is that the complements of the bits a_m and b_m should be used when computing σ in the most significant position.

Observation 3.10.9 *The three-way comparison of two binary numbers, in either unsigned or 2's complement notation, can be obtained in logarithmic time in a single tree structure of $\circ$-composition nodes, i.e., in approximately the same time as an equivalent add or subtract operation, but with significantly smaller area.*

3.10.3 Leading zeroes determination

A similar problem to sign determination is the identification of the position of the first non-zero digit in a radix-represented number in some fixed-point representation, say a computer word. The problem can equivalently be described as finding the number of leading zeroes. It plays a significant role in floating-point arithmetic, since the *significand* is usually represented left normalized in the part of the word(s) allocated for representing its value. Knowing the position of the "leading one" allows for a log-time shifting, as opposed to the linear-time normalization realizable by single left shifts.

For a given non-redundant radix polynomial $P_m = \sum_0^{m-1} d_i [\beta]^i$, determination of the number of leading zeroes, $\mathrm{lzd}(P_m)$ is easily described recursively by a "divide and conquer" algorithm when m is a power of 2, say $m = 2^k$.

Algorithm 3.10.10 (Leading zeroes determination, LZD)

Stimulus: *An integer k and a radix polynomial* $P_{2^k} = \sum_0^{2^k-1} d_i[\beta]^i \in \mathcal{F}[\beta, D]$, *where D is a non-redundant digit set for radix* β.

Response: *The value of* $\mathrm{lzd}(k, P_{2^k})$, *the number of leading zero-valued digits in* P_{2^k}.

Method: **if** $k = 0$
  then if $d_0 = 0$ **then** $t := 1$ **else** $t := 0$ **end**;
    return t;
  else $L := \sum_0^{2^{k-1}-1} d_{i+2^{k-1}}[\beta]^i$; $R := \sum_0^{2^{k-1}-1} d_i[\beta]^i$;
    $l := \mathrm{lzd}(k-1, L)$; $r := \mathrm{lzd}(k-1, R)$; {*Can be done in parallel*}
    if $l = 2^{k-1}$ **then return** $l + r$
            else return l
  end;

The algorithm can be implemented directly as a tree structure where each node receives the number of leading zeroes from two sibling nodes, possibly together with a bit signaling whether the substring is non-zero to simplify the test $l = 2^{k-1}$. When the length is calculated in binary, note that the left signaling bit inverted can then just be concatenated (by $\circ$) as the new most-significant bit to the value of either r or l. The new signalling bit is then the OR of the two incoming bits,

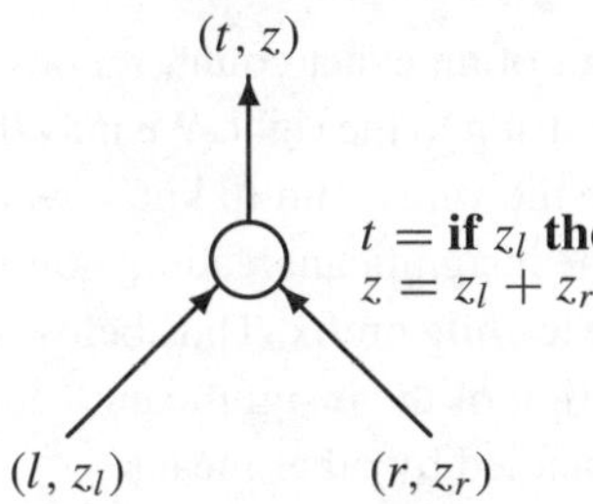

where at the leaf nodes $z \equiv (d \neq 0)$, d being the digit value and $t =$' ' is the empty string.

Quasi-normalization. For a number given in a redundant representation, it is necessary first to eliminate a possible leading string of non-zero digits of no significance. For the two important redundant binary representations, it turns out to be possible by a simple constant time conversion to obtain a bit pattern having either the same or one less leading zero than the non-redundant binary representation of the value. This makes it possible to "quasi-normalize" a redundant binary number, say to be in the interval [1; 4).

For borrow-save representations we noticed in Section 2.5 that the N-mapping is capable of transforming a digit string $01\bar{1}\cdots\bar{1}$ into the string $000\cdots 1$, and the P-mapping will transform a string $0\bar{1}1\cdots 1$ into $000\cdots\bar{1}$. If the number is in carry-save 2's complement, the Q-mapping will convert the representation into borrow-save. Hence if the sign of the number is known we have the following.

Observation 3.10.11 *If the sign of a binary radix polynomial in borrow-save or carry-save representation is known, then it is possible in constant time to transform it into a digit string having the same or one less leading zeroes than the value in non-redundant representation. For borrow-save representations this is obtained by applying the N-mapping to a positive number, and the P-mapping to a negative one. A carry-save represented polynomial is first to be transformed by a Q-mapping, and the result in borrow-save is to be mapped as before.*

To handle the general situation, recall that a constant-time conversion can only be based on a bounded left and right context of each position. Consider the following digit string in borrow-save representation $A = 01\overline{1}^j 0^k \overline{1}$ for $j \geq 1$ and $k \geq 2$. Let the result of applying the N-mapping be B, and the non-redundant result be C, then

$$\begin{aligned} A &= 01\overline{1}\cdots\overline{1}0\cdots 00\overline{1}, \\ B &= 000\cdots 10\cdots 0\overline{1}1, \\ C &= 000\cdots 01\cdots 111. \end{aligned}$$

Since using a bounded context it is not possible to identify the sign of the right context, the presence of the extra leading zero in the non-redundant representation cannot be detected in constant time.

However, relaxing the requirement of an exact count, it does not really matter what happens to any less-significant string to the right. We may thus seek transformations which do not even preserve the value. But to know whether to apply the N- or the P-mapping to eliminate the insignificant leading non-zero digits, it will be necessary to know the sign of the leading prefix. Thus below we shall combine sign determination with the elimination of the insignificant leading digits, and as it turns out the elimination can be handled by other means.

Noting that the N- and P-mappings only use a one-digit right context, we will now include the left context also, as this may be used to identify the sign when dealing with the most-significant end of a string. Considering again a borrow-save representation, assume that an extra zero-valued digit is appended to the left, just for the purpose of having a left context in the most-significant position, but is not to be counted. Similarly append a zero-valued digit on the right as a right context.

For the purpose of the following analysis assume that the thus extended borrow-save string is $d_n d_{n-1} \cdots d_0 d_{-1}$, $d_n = d_{-1} = 0$, with $d_i \in \{-1, 0, 1\}$, but is represented as three bit-strings:

$$e_n e_{n-1} \cdots e_0 e_{-1}, \quad p_n p_{n-1} \cdots p_0 p_{-1}, \quad \text{and} \quad m_n m_{n-1} \cdots m_0 m_{-1},$$

where

$$e_i \equiv (d_i = 0) \quad p_i \equiv (d_i = 1) \text{ and } m_i \equiv (d_i = -1). \tag{3.10.9}$$

This will allow us to use a complement notation where, e.g., $\overline{e}_i = p_i + m_i$ is a shorthand for d_i being either 1 or -1.

Recall that the problem of not obtaining the correct position of the leading non-zero digit is due to a prefix, equivalent to an isolated such non-zero digit, being followed by a string of the opposite sign. This sign cannot be determined in constant time; however, the determination of the approximate number of leading zeroes is also a log-time operation. Hence it may be possible in parallel to check whether a unit prefix is followed by a string of the opposite sign, so that whether an adjustment is needed is available at the same time as the approximate shift amount.

Correcting the normalization shift. We now want to generate bit-strings, identifying "interesting" positions, in particular a leading position containing a unit digit, as this will be needed for the LZD algorithm. But we want to determine if this is an isolated unit, because if it is followed (after a non-empty sequence of zeroes) by a string of the opposite sign, this is precisely the situation in which a correction is needed.

Hence we will try to identify situations with isolated units like $\cdots 010 \cdots$, i.e., in the notation above, such values of i for which $e_{i+1}p_ie_{i-1}$ holds. Or alternatively strings of the form $\cdots 01\bar{1}0 \cdots$ which can be identified by $p_{i+1}m_ie_{i-1}$, or more generally $\cdots 01\bar{1} \cdots \bar{1}0 \cdots$, i.e., other strings which reduce to isolated units. Note that the resulting unit will occur in position i for which $m_{i+1}m_ie_{i-1}$ holds, provided that the leading part is eliminated. However, if the latter pattern is found isolated, i.e., in the context $e_{i+2}m_{i+1}m_ie_{i-1}$ with a leading zero, it will be identified as just some string of negative sign (see t_i below). Thus we can express positions where isolated positive units occur, as positions i where the following expression holds true:

$$\begin{aligned} u_i &= e_{i+1}p_ie_{i-1} + p_{i+1}m_ie_{i-1} + m_{i+1}m_ie_{i-1} \\ &= (e_{i+1}p_i + \overline{e_{i+1}}m_i)e_{i-1}. \end{aligned}$$

Equivalently by symmetry we can express positions where isolated negative units occur by

$$\begin{aligned} v_i &= e_{i+1}m_ie_{i-1} + m_{i+1}p_ie_{i-1} + p_{i+1}p_ie_{i-1} \\ &= (e_{i+1}m_i + \overline{e_{i+1}}p_i)e_{i-1}. \end{aligned}$$

Obviously, there are other positions of positive value; they can be expressed as

$$\begin{aligned} s_i &= e_{i+1}p_ip_{i-1} + p_{i+1}m_ip_{i-1} + m_{i+1}m_ip_{i-1} \\ &= (e_{i+1}p_i + \overline{e_{i+1}}m_i)p_{i-1}, \end{aligned}$$

where the term $m_{i+1}m_ip_{i-1}$ is the termination of a string of the form $\cdots 01\bar{1} \cdots \bar{1}1 \cdots$, provided that the leading part is eliminated; similarly for positions of negative value:

$$\begin{aligned} t_i &= e_{i+1}m_im_{i-1} + m_{i+1}p_im_{i-1} + p_{i+1}p_im_{i-1} \\ &= (e_{i+1}m_i + \overline{e_{i+1}}p_i)m_{i-1}. \end{aligned}$$

As these expressions are mutually exclusive (exercise), we can express positions where zeroes occur by

$$z_i = 1 - (u_i + s_i + v_i + t_i), \tag{3.10.10}$$

whose complement $w_i = u_i + s_i + v_i + t_i$ defines a bit pattern to which the LZD Algorithm can be applied. This is summarized in the following result.

Observation 3.10.12 *Given a number x in borrow-save representation $d_{n-1}\cdots d_0$, then the bit-string $w_{n-1}\cdots w_0$, $w_i = e_{i+1}(p_i\overline{m_{i-1}} + m_i\overline{p_{i-1}}) + \overline{e_{i+1}}(m_i\overline{m_{i-1}} + p_i\overline{p_{i-1}})$, has the same number of or one fewer leading zeroes than the non-redundant representation of x.*

Since the bit patterns defined above are mutually exclusive, for any position i we can thus encode these using three bits in sign-magnitude

$$\begin{aligned} s_i &\sim 001,\\ u_i &\sim 010,\\ x_i &\sim 011,\\ z_i &\sim 100,\\ y_i &\sim 101,\\ v_i &\sim 110,\\ t_i &\sim 111, \end{aligned} \tag{3.10.11}$$

where two new truth values x_i and y_i have been added for later use.

For an implementation note that the three-bit encoding in position i is a function only of the four bits u_i, s_i, t_i, v_i, which are defined in terms of the nine bits e_j, p_j, m_j for $j = i+1, i, i-1$, which in turn are defined by the six bits encoding the three digit values d_{i+1}, d_i, d_{i-1}. Thus a six-bit-in, three-bit-out PLA[2] structure as in Figure 3.10.1 is capable of forming the encoding in a few logic levels. This PLA can simultaneously also define w_i to be used in Observation 3.10.12.

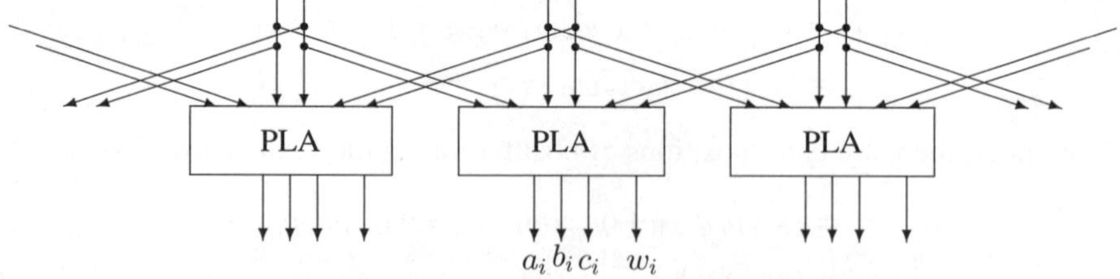

Figure 3.10.1. Encoding of a_i, b_i, c_i, and w_i.

We are now able to define a log-time search to find unit prefixes followed by strings of the opposite sign, i.e., using notation from formal language theory,

[2] PLA: programmed logic array.

strings of the forms

$$X = z^*uz^+v\,\sigma \quad \text{or} \quad X = z^*uz^+t\,\sigma \quad \text{and} \quad Y = z^*vz^+u\,\sigma \quad \text{or} \quad Y = z^*vz^+s\,\sigma,$$

written as regular expressions, where σ denotes an arbitrary string and a^* and a^+ denote strings of consecutive identical symbols a_i. Note that a correction is needed if and only if strings X or Y are found, so these will be the goals for which to search.

To specify the grammar we further define substrings

$$\begin{aligned} U &= z^*uz^* \quad \text{and} \quad S = z^*sz^*\sigma, \\ Z &= z^+, \\ V &= z^*vz^* \quad \text{and} \quad T = z^*tz^*\sigma. \end{aligned}$$

It is now possible to define grammar rules for recursively building these strings in binary tree structures, e.g., $Z \Rightarrow z \mid Z\,Z$, $U \Rightarrow u \mid Z\,U \mid U\,Z$, etc. Using initial values $X = x$, $U = u$, $S = s$, $Z = z$, $T = t$, $V = v$, $Y = y$, it is easier to specify the rules for the nodes as a table:

left \ right	T	V	Y	Z	X	U	S
S	S	S	S	S	S	S	S
U	X	X	X	U	S	S	S
X	X	X	X	X	X	X	X
Z	T	V	Y	Z	X	U	S
Y	Y	Y	Y	Y	Y	Y	Y
V	T	T	T	V	Y	Y	Y
T	T	T	T	T	T	T	T

(3.10.12)

Example 3.10.2 The following example forming the difference $A - B = 448 - 443$ (left normalized) illustrates a situation where a correction is needed:

A		1	1	1	0	0	0	0	0	0	0	0	0	0	0	0	0	
B		1	1	0	1	1	1	0	1	1	0	0	0	0	0	0	0	
$A - B$		0	0	0	0	0	0	1	0	1	0	0	0	0	0	0	0	
P	0	0	0	1	0	0	0	0	0	0	0	0	0	0	0	0	0	0
E	1	1	1	0	0	0	0	1	0	0	1	1	1	1	1	1	1	1
N	0	0	0	0	1	1	1	0	1	1	0	0	0	0	0	0	0	0
U		0	0	0	0	0	1	0	0	1	0	0	0	0	0	0	0	
S		0	0	0	0	0	0	0	0	0	0	0	0	0	0	0	0	
Z		1	1	1	1	1	0	1	0	0	1	1	1	1	1	1	1	
T		0	0	0	0	0	0	0	1	0	0	0	0	0	0	0	0	
V		0	0	0	0	0	0	0	0	0	0	0	0	0	0	0	0	
R		Z	Z	Z	Z	Z	U	Z	T	U	Z	Z	Z	Z	Z	Z	Z	
		Z		Z		U		T		U		Z		Z		Z		
			Z				X				U				Z			
					X								U					
									X									

Recall that the rows R and Z can be obtained directly from A and B by PLA look-up, and that the complement of Z is the bit pattern upon which the LZD algorithm is to work. Also that the three-valued sign of $A - B$ can be derived from the root value: it is positive for X, U, S, zero for Z and negative for T, V, Y, as easily derived from the encoding. □

The rules of (3.10.12) can be expressed as an algorithm for a tree node taking two three-bit encodings (3.10.11) of a left, respectively a right, symbol as input, and producing a three-bit encoding of the result.

Algorithm 3.10.13 (LZD correction node)
Stimulus: *A left symbol encoded as* (a_l, b_l, c_l) *and a right symbol as* (a_r, b_r, c_r).
Response: *An encoding of the result symbol* (a, b, c).
Method: **if** $b_l\bar{c}_l$ **return** $(a_l, a_r, b_r + c_r)$ {*Left symbol is U or V*}
elseif $\bar{c}_l$ **return** (a_r, b_r, c_r) {*Left symbol is Z*}
else return (a_l, b_l, c_l) {*Left symbol is X, S, Y or T, or right is Z*}
end;

For the first case when the left symbol is U or V, according to the rules in (3.10.12) and the encodings (3.10.11) the following is to be returned:

Left is	Right is negative	Right is zero	Right is positive
U	$X \sim 011$	$U \sim 010$	$S \sim 001$
V	$T \sim 111$	$V \sim 110$	$Y \sim 101$

Hence the signs of the symbols can be used as the first two bits, with the last bit on if the right symbol is non-zero ($\neq Z$). The remaining cases are trivial. Figure 3.10.2 shows a possible implementation of the node, whose critical path will be through two selectors.

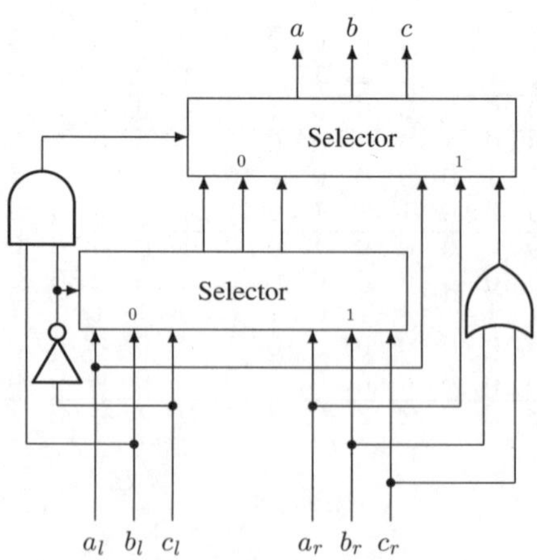

Figure 3.10.2. LZD correction node based on three-bit encoding.

Note that the expression at the root of the tree that decides if a correction is needed is $(a \oplus b)c$ (X or Y has been found), and that the sign of the number is given by a. Finally the expression $\overline{b+c}$ signals that the result is zero.

Observation 3.10.14 (Normalization of redundant binary numbers) *The number of left-shifts necessary to normalize a binary, redundantly represented number can be determined in logarithmic time from a recoded representation of the number. Based on a constant-time transformation, one binary tree structure can determine the number of shifts to obtain a quasi-normalized representation, and in parallel another tree structure can determine whether a correction is needed to obtain the proper normalization. The latter tree simultaneously determines the three-way sign of the number.*

Problems and exercises

3.10.1 Prove Theorem 3.10.1 by considering the equation $A + B = D$ and adding $\overline{D}$ on both sides, thus on the right-hand side obtaining the value $2^n - 1$, where n is the number of bits in the registers containing A, B, and D. You are likely not to obtain the same condition as expressed in Theorem 3.10.1, so explain whether that difference matters.

3.10.2 Prove that $x(x \oplus y) = x\bar{y}$.

3.10.3 Prove the reduction in (3.10.3). Hint: Use induction in i starting with the fact that for $i = \ell$, then $a_{\ell-1} = d_{\ell-1} = 0 \Rightarrow \overline{d}_{\ell-1}a_{\ell-1} = 0$.

3.10.4 As an alternative to the sign-magnitude encoding (3.10.8), consider the usual borrow-save encoding of a digit and use that encoding directly as the encoding of S:

$$\begin{aligned} n &\sim 01, \\ 0 &\sim 00 \quad \text{or} \quad 11, \\ p &\sim 10. \end{aligned}$$

Then determine the necessary logic in the o-nodes. Will the overall gate count be reduced if instead the encoding is chosen to be non-redundant, where 11 does not occur?

3.10.5 Prove that u_i, s_i, v_i, and t_i are mutually exclusive such that (3.10.10) holds.

3.10.6 Derive optimal logic expressions for a, b, and c as specified by Algorithm 3.10.13, utilizing the encodings given by (3.10.11). Hint: Express the entries of Karnaugh maps first in the inputs (a_l, b_l, c_l) and (a_r, b_r, c_r), and then reduce these.

3.11 Notes on the literature

The model of computation we have used in Section 3.2 is the classical one by Winograd [Win65] and employed in particular by Spira [Spi69]. Addition tables and mappings were introduced by Matula in [Mat76]. The constructive proofs of Theorems 3.4.1 and 3.4.2 are based on on the much cited treatment [Avi61] by Avizienis, the first published thorough discussion of redundant number systems. However, the ideas of using saved carries and borrows had been presented earlier by Metze and Robertson in [MR59]. [BGvN46] has a fairly detailed discussion of 2's complement addition.

We have chosen not to go into much detail on the implementation of basic circuits like half- and full-adders, trying instead to emphasize functionality rather than implementation, which anyway depends on the technology available. Hence our diagrams are there just to illustrate some possible implementations, knowing quite well that things look different at the transistor level. There are plenty of references on the engineering of basic adder cells in VLSI, but we shall here concentrate on the composite adders. An important reference for anyone wanting to implement the logic for addition circuits is the very thorough discussion in [CR90], "The set theory of arithmetic decomposition." The senior author of this paper was James (Jim) Robertson, often called the "grandfather" of the many active researchers in the field of computer arithmetic. The paper discusses the design space available for adders, when considering various radices, digit sets, and their encoding. The notation *npn*-adder, *pnp*-adder, etc., were introduced in [GHM89] in connection with some simplified transistorized versions, but the adders were known before, see, e.g., [Hwa79, Table 6.3]. The description of the weighted signals and Lemma 3.5.1 is modeled after [ASH99]. The densely packed decimal coding is based on ideas in [CH75], which were further developed in [Cow02]. It is one of the two possible encodings employed for decimal representation in the revised IEEE floating-point standard [IEE08]. The BCD4221 coding was introduced in [VAM07] by Vàzquez, Antelo, and Montuschi; it illustrates how decimal carries may be developed and used in a multioperand, carry-save addition for use in a decimal multiplier.

On-line arithmetic was introduced by Trivedi and Ercegovac in [TE77], while digit serial (least-significant digit first) had already been used some time within the signal processing community. Addition being fairly simple, most work in on-line arithmetic has been concerned with the more complicated operations, and computation of standard functions. A general methodology for developing on-line algorithms is described in [EL88].

Carry-skip adders [GPW55] have been the topic of much research since [LB61] by Lehman and Burla suggested the simple group sizing of $\sqrt{n}$. In 1967 Majerski [Maj67] developed the first general model, and since then many more and more general models have been developed, e.g., see [GHM87, Tir89, CS90]. The

Manchester Carry Chain was described by Kilburn, Edwards, and Aspinall in [KEA59]. An interesting historical remark may be due here. Metze presented the paper [MR59] in 1959 at the IFIP Congress in Paris which contains the idea of redundant representations (in the form of carry/borrow storage) to be used during accumulation of partial products for multiplication. After the presentation there was a discussion at which Edwards and Lehman were present. Both of them were working on fast carry-propagation techniques and they argued against the "extreme cost" of the extra register storage needed, which could "only produce a 10% speed-up."

The carry-select adder [Bed62] has not been the subject of as much research, due to its inherently higher complexity. MacSorley's conditional-sum adder [Skl60] has continued to be used owing to its speed, and the fact that it simultaneously provides the sum for a carry-in of 0 as well as for 1, which is useful in many situations. This possibility also exists for the carry-look-ahead structures, but this seems to have been overlooked for a long time, despite it being simpler to implement. Only comparatively recently did Tyagi [Tya93] design a hybrid system based on carry-select structures, but in which the dual ripple-carry adders are substituted by single carry-look-ahead blocks, providing dual results, corresponding to a carry-in of 0 as well as 1. This paper appears to be the first to point out this dual output possibility in the open literature. In [LS91] Lynch and Swartzlander describe the design of a carry-look-ahead adder, obtained by combining Manchester Carry Chains with the carry-select concept.

Since the very early examples of [WS58] by Weinburger and Smith and the conditional sum adder [Skl60] by Sklanski, a number of log-time adder structures have been developed, based on the idea of computing the group (generate, propagate)-pairs and parallel prefix computation. The now-classical reference to the general class of parallel prefix computations is [LF80] by Ladner and Fischer. An earlier reference to such adders is [KS73] by Kogge and Stone, but the proposal by Brent and Kung [BK82] is probably the most cited, due to its regular structure and bounded (binary) fan-out. Knowles in [Kno01] describes a spectrum of prefix adders between the Kogge–Stone and Ladner–Fischer adders. In [CS05] Choi and Swartzlander describe a general method of designing such adders. Various other proposals, including hybrid structures have been put forward, e.g., [Lin81, NI85, Dor88, WT90, LS92, Kan93].

Generally parallel multipliers have been built with accumulation of partial products using standard half- and full-adders. Many organizations have been proposed for the quite irregular structure of the reduction trees based on 3-to-2 adders, we shall return to these in the next chapter. Only more recently have more regular structures based on binary trees been exploited, based on 4-to-2 adders like the ones discussed in Section 3.7, the first ones [Wei81, SH88] using carry-save representation. [TYY85, KNE$^+$87] were apparently the first to describe multipliers based on 4-to-2 borrow-save adders. It was proposed to use the digit set {0, 1, 2, 3} in connection with a three-bit encoding of the digits in [EL97], and with a 2's complement encoding in [PGK01]. However, it can be shown [Kor05] that all these 4-to-2 carry-save and signed-digit adders are mutually equivalent: they can all be implemented by the same 4-to-2 carry-save adder, possibly by complementing the negatively weighted signals using Lemma 3.5.1.

Whether the "symmetric" 4-to-2 type (e.g., like the one shown in Figure 3.7.3), is preferable to the type composed of standard full adders (e.g., as in Figure 3.7.2), in which the fact that some paths are shorter than others is used, is discussed in [SMOR98]. [OSY+95] presents a fast carry-save based, 4-to-2 adder using pass transistors for the implementation of XOR gates, obtained by combining and optimizing the combination of two full-adders.

The results on leading guard digits of Lemma 3.8.1, Theorem 3.8.2, and the special cases for radix 2, Theorems 3.8.3, 3.8.4, and 3.8.9, are from [KM06]. They may well exist hidden in implementations, but we have not found them expressed in the open before. The problem of right-shifting a 2's complement, carry-save represented number is discussed in [TPL06]. Sign-magnitude addition and subtraction is described without explanation in [Hwa79, Figure 3.3] using an end-around carry as feed-back in a carry-ripple adder, and discussed in [VLP89] employing a carry-look-ahead adder as here. The comparison $A + B = D$ as in Theorem 3.10.1 is presented in [CJ92]. The algorithm for log-time determination of the number of leading zero digits in a non-redundant number repesentation is from [Okl99] by Oklobdzija. The mapping of a borrow-save or carry-save represented number into a bit-string having the same or one fewer leading zeroes is described in [BL99] by Bruguera and Lang, which is also describes how to obtain the exact count in logarithmic time. For this purpose they proposed using two parallel trees, each operating on four-bit tuples, to determine the need for a correction. The description of the correction of the count, based on a single tree operating on three-bit tuples, is from [Kor09] by Kornerup.

References

[ASH99] T. Aoki, Y. Sawada, and T. Higuchi. Signed-weight arithmetic and its application to a field-programmable digital filter architecture. *IEICE Trans. Electron.*, E82(9):1687–1698, 1999.

[Avi61] A. Avizienis. Signed-digit number representations for fast parallel arithmetic. *IRE Trans. Electronic Computers*, EC-10:389–400, September 1961. Reprinted in [Swa90].

[Bed62] O. J. Bedrij. Carry-select adder. *IEEE Trans. Electronic Computers*, EC-11:340–346, 1962.

[BGvN46] A. Burks, H. H. Goldstine, and J. von Neumann. *Preliminary Discussion of the Logic Design of an Electronic Computing Instrument*. Technical report, Institute for Advanced Study, Princeton, 1946. Reprinted in C. G. Bell, *Computer Structures, Readings and Examples*, Mc Graw-Hill, New York, 1971.

[BK82] R. P. Brent and H. T. Kung. A regular layout for parallel adders. *IEEE Trans. Computers*, C-31(3):260–264, March 1982. Reprinted in [Swa90].

[BL99] J. D. Bruguera and T. Lang. Leading-one prediction with concurrent position correction. *IEEE Trans. Computers*, 48(10):1083–1097, October 1999.

[CH75] T. C. Chen and I. T. Ho. Storage-efficient representation of decimal data. *Commun. ACM*, 18(1):49–52, January 1975.

[CJ92] J. Cortadella and J. M. Llaberia. Evaluation of $A + B = K$ conditions without carry propagation. *IEEE Trans. Computers*, C-41(11):1484–1488, November 1992.

[Cow02] M. Cowlishaw. Densely packed decimal encoding. *IEEE Proc. Computers and Digital Techniques*, 149(3):102–104, May 2002.

[CR90] T. M. Carter and J. E. Robertson. The set theory of arithmetic decomposition. *IEEE Trans. Computers*, C-39(9):993–1005, August 1990.

[CS90] P. K. Chan and M. D. F. Schlag. Analysis and design of CMOS Manchester adders with variable carry-skip. *IEEE Trans. Computers*, C-39(9):983–992, August 1990.

[CS05] Y. Choi and E. S. Swartzlander. Parallel prefix adder design with matrix representation. In *Proc. 17th IEEE Symposium on Computer Arithmetic*, IEEE Computer Society, 2005.

[Dor88] R. W. Doran. Variants of an improved carry look-ahead adder. *IEEE Trans. Computers*, C-37:1110–1113, 1988.

[EL88] M. D. Ercegovac and T. Lang. On-line arithmetic: a design methodology and applications in digital signal processing. *VLSI Signal Processing, IEEE*, III:252–263, 1988. Reprinted in [Swa90].

[EL97] M. D. Ercegovac and T. Lang. Effective coding for fast redundant adders using the radix-2 digit set {0, 1, 2, 3}. In *Proc. 31st Asilomar Conference on Signals Systems and Computers*, pages 1163–1167, IEEE Computer Society, 1997.

[GHM87] A. Guyot, B. Hochet, and J.-M. Muller. A way to build efficient carry-skip adders. *IEEE Trans. Computers*, C-36(10):1144–1152, October 1987.

[GHM89] A. Guyot, B. Hochet, and J.-M. Muller. JANUS, an on-line multiplier/divider for manipulating large numbers. In *Proc. 9th IEEE Symposium on Computer Arithmetic*, pages 106–111. IEEE Computer Society, 1989.

[GPW55] B. Gilchrist, J. H. Pomerene, and S. Y. Wong. Fast carry logic for digital computers. *IRE Trans. Electronic Computers*, EC-4:133–136, 1955.

[Hwa79] K. Hwang. *Computer Arithmetic: Principles, Architecture and Design*. John Wiley and Sons, 1979.

[IEE08] IEEE. *IEEE Std. 754^{TM}-2008 Standard for Floating-Point Arithmtic*. IEEE, 3 Park Avenue, NY 10016-5997, USA, August 2008.

[Kan93] V. Kantabutra. A recursive carry-lookahead/carry-select hybrid adder. *IEEE Trans. Computers*, C-42(12):1495–1499, December 1993.

[KEA59] T. Kilburn, D. B. G. Edwards, and D. Aspinall. Parallel addition in digital computers: a new fast 'carry' circuit. *Proc. IEE*, 106, pt. B:464–466, 1959.

[KM06] P. Kornerup and J.-M. Muller. Leading guard digits in finite precision redundant representations. *IEEE Trans. Computers*, 55(5):541–548, May 2006.

[KNE+87] S. Kuninobu, T. Nishiyama, H. Edamatsu, T. Taniguchi, and N. Takagi. Design of high speed MOS multiplier and divider using redundant binary representation. In *Proc. 8th IEEE Symposium on Computer Arithmetic*, pages 80–86. IEEE Computer Society, 1987.

[Kno01] S. Knowles. A family of adders. In *Proc. 15th IEEE Symposium on Computer Arithmetic*, pages 277–278. IEEE Computer Society, 2001.

[Kor05] P. Kornerup. Reviewing 4-to-2 adders for multi-operand addition. *J. VLSI Signal Processing*, 40(1):143–152, May 2005. Previously presented at ASAP2002.

[Kor09] P. Kornerup. Correcting the normalization shift of redundant binary representations. *IEEE Trans. Computers*, 58(10):1435–1439, October, 2009.

[KS73] P. M. Kogge and H. S. Stone. A parallel algorithm for the efficient solution of a general class of recurrence equations. *IEEE Trans. Computers*, 22(8):783–791, August 1973.

[LB61] M. Lehman and N. Burla. Skip technique for high speed carry propagation in binary arithmetic units. *IRE Trans. Electronic Computers*, EC-10:691–698, 1961.

[LF80] R. E. Ladner and M. J. Fischer. Parallel prefix computation. *J. ACM*, 27(4):831–838, Oct. 1980.

[Lin81] H. Ling. High-speed binary adder. *IBM J. Res. Devel.*, 25(3):156–166, May 1981.

[LS91] T. Lynch and E. Swartzlander. The redundant cell adder. In P. Kornerup and D. W Matula, editors, *Proc. 10th IEEE Symposium on Computer Arithmetic*. IEEE Computer Society, 1991.

[LS92] T. Lynch and E. Swartzlander. A spanning tree carry lookahead adder. *IEEE Trans. Computers*, C-41(8):931–939, August 1992.

[Maj67] S. Majerski. On the determination of optimal distribution of carry skips in adders. *IEEE Trans. Computers*, EC-16:45–58, 1967.

[Mat76] D. W. Matula. Radix arithmetic: digital algorithms for computer architecture. In R. T. Yeh, editor, *Applied Computation Theory: Analysis, Design, Modeling*, chapter 9, pages 374–448. Prentice-Hall, Inc., 1976.

[MR59] G. Metze and J. E. Robertson. Elimination of carry propagation in digital computers. *IFIP International Conference on Information Processing, Paris*, pages 389–396, 1959.

[NI85] T.-F. Ngai and M. J. Irwin. Regular, area-time efficient carry-lookahead adders. In *Proc. 7th IEEE Symposium on Computer Arithmetic*, pages 9–15. IEEE Computer Society, June 1985. Reprinted in [Swa90].

[Okl99] V. G. Oklobdzij. An algorithm and novel design of a leading zero detector circuit: comparison with logic synthesis. *IEEE Trans. VLSI Systems*, 2(1):124–128, March 1999.

[OSY+95] N. Ohkubo, M. Shinbo, T. Yamanaka, A. Shimizu, K. Sasaki, and Y. Nakagome. A 4.4 ns CMOS 54 × 54-b multiplier using pass transistor multiplexer. *IEEE J. Sol. State Circuits*, 30(3):251–257, 1995.

[PGK01] D. S. Phatak, T. Goff, and I. Koren. Constant-time addition and simultaneous format conversion based on redundant binary representations. *IEEE Trans. Computers*, 50(11):1267–1278, 2001.

[SH88] M. R. Santoro and M. R. Horowitz. A pipelined 64×64b iterative array multiplier. *Proc. IEEE International Solid-State Circuit Conference*, pages 36–37. IEEE, 1988.

[Skl60] J. Sklansky. Conditional-sum addition logic. *IRE Trans. Electronic Computers*, EC-9:226–231, 1960. Reprinted in [Swa80].

[SMOR98] P. F. Stelling, C. U. Martel, V.G. Oklobdzija, and R. Ravi. Optimal circuits for parallel multipliers. *IEEE Trans. Computers*, 47(3):273–285, March 1998.

[Spi69] P. M. Spira. On the computation time of certain classes of Boolean functions. In *Annual ACM Symposium on Theory of Computing*, pages 271–272, ACM, 1969.

[Swa80] E. E. Swartzlander, editor. *Computer Arithmetic,* volume I. Dowden, Hutchinson and Ross, Inc., 1980. Reprinted by IEEE Computer Society Press, 1990.

[Swa90] E. E. Swartzlander, editor. *Computer Arithmetic,* volume II. IEEE Computer Society Press, 1990.

[TE77] K. S. Trivedi and M. D. Ercegovac. On-line algorithms for division and multiplication. *IEEE Trans. Computers*, C-26(7):681–687, July 1977. Reprinted in [Swa90].

[Tir89] S. Tirrini. Optimal group distribution in carry-skip adders. In *Proc. 9th IEEE Symposium on Computer Arithmetic*, pages 96–103. IEEE Computer Society, September 1989.

[TPL06] A. F. Tenca, S. Park, and L. A. Tawalbeh. Carry-save representation is shift-unsafe: the problem and its solution. *IEEE Trans. Computers*, 55(5):630–635, May 2006.

[Tya93] A. Tyagi. A reduced-area scheme for carry-select adders. *IEEE Trans. Computers*, C-42(10), October 1993.

[TYY85] N. Takagi, H. Yasuura, and S. Yajima. High-speed VLSI multiplication algorithm with a redundant binary addition tree. *IEEE Trans. Computers*, C-34(9):789–796, September 1985.

[VAM07] A. Vazquez, E. Antelo, and P. Montuschi. A new family of high performance parallel decimal multipliers. In *Proc. 18th IEEE Symposium on Computer Arithmetic*, pages 195–204, June 2007. IEEE, 2007.

[VLP89] S. Vassiliadis, D. S. Lemon, and M. Putrino. S/370 sign-magnitude floating-point adder. *IEEE J. Sol. State Circuits*, 24:1062–1070, 1989.

[Wei81] A. Weinberger. 4–2 carry-save adder Module. *IBM Technical Disclosure Bulletin*, 23, January 1981.

[Win65] S. Winograd. On the time required to perform addition. *JACM*, 12(2):277–285, 1965.

[WS58] A. Weinberger and J. L. Smith. A logic for high-speed addition. *Nat. Bur. Stand. Circ.*, 591:3–12, 1958. Reprinted in [Swa80].

[WT90] B. W. Y. Wei and C. D. Thompson. Area-time optimal adder design. *IEEE Trans. Computers*, C-39(5):666–675, May 1990.

4

Multiplication

4.1 Introduction

The product of two radix polynomials is a radix polynomial whose digits can be expressed as sums of digit products from the operand polynomials. The digits thus expressed are in general from some much larger digit set, and a conversion must be performed in some way to deliver the result in the same digit set as that of the operands. The order in which digit products are generated, and later accumulated, is algorithm specific.

Often intermediate products (called *partial products*) are formed by multiplying a digit from one operand (the multiplier) by the complete other operand (the multiplicand), followed by accumulation of shifted versions of these, to account for the different weights of the multiplier digits. Hence there are three recognizable steps in forming the product:

(i) formation of digit or partial products,
(ii) summation of digit or partial products,
(iii) final digit set conversion,

where these steps individually or in combination can be performed in parallel or sequentially in various ways.

The order in which digit products or partial products are formed and how accumulation is performed thus characterize the algorithms. Some implicitly perform a base and digit set conversion, utilizing a higher radix to reduce the number of terms that have to be added. The different algorithms can also be characterized according to the way operands are delivered/consumed, and how the result is produced, whether word parallel or digit serial, and in the last case in which order.

As multiplication is a very important operation, a wide variety of algorithms have been devised to satisfy various demands, from simplicity of implementation to high speed requirements. We have seen in Chapter 3 that we cannot expect to be able to multiply two radix polynomials in time faster than $O(\log n)$ when the result is to be in non-redundant representation, and this is indeed possible using space $O(n^2)$ (actually $O(n^2/\log n)$ is feasible). On the other hand it is not possible to multiply in space less than $O(n)$, in which case there are several algorithms operating in time $O(n)$.

After classifying the various algorithms, we shall look at ways of reducing the number of terms that have to be added, by converting ("recoding") the multiplier into a higher radix, and how partial products can then be formed. The next general problem is to handle 2's complement operands, and the combination of these with recoded multipliers.

The class of linear time multipliers is then discussed; these are all characterized by sequential formation and summation of the partial products, starting with the classical iterative multipliers, possibly using radix-2^k recoding. Typically such multipliers will employ carry-completing adders, hence the result will be in non-redundant form. The same class also contains array multipliers, and a wide variety of serial multipliers, which will be derived using retiming of simple systolic arrays. Digit-serial multipliers delivering their result least-significant digit/bit first (LSD/LSB-first) produce their result in non-redundant form, whereas on-line (MSD/MSB-first) multipliers necessarily must deliver their result in redundant form.

Finally, logarithmic time multipliers are presented in which partial products are formed in parallel, and adder trees are employed to perform the accumulation in a redundant representation. If necessary, a final conversion into a non-redundant representation can be performed using any log-time, carry-completing adder.

4.2 Classification of multipliers

In general we will here assume that operands are integer radix polynomials from the same finite precision set $\mathcal{F}_{0,n-1}[\beta, D]$, with the result in $\mathcal{F}_{0,2n-1}[\beta, D]$, where the digit set D is basic for the radix β. Although the product is symmetric in the two operands, the *multiplicand* P and the *multiplier* Q play different roles in most algorithms, as also known from the pencil-and-paper methods taught in grade school.

We can use the diagram in Figure 4.2.1 of a five-digit by five-digit multiplication for illustrating those digit products that have to be formed and accumulated to obtain the resulting product.

Notice that the digit products are digits themselves, but possibly from some larger digit set, since a digit set is generally not closed under multiplication. However, some very useful digit sets like $\{0, 1\}$ and $\{-1, 0, 1\}$ are actually closed. Each row in this scheme is a radix polynomial, which then may be converted through a digit set conversion before it is added together with other rows. The value of this polynomial is traditionally called a *partial product*, and can also

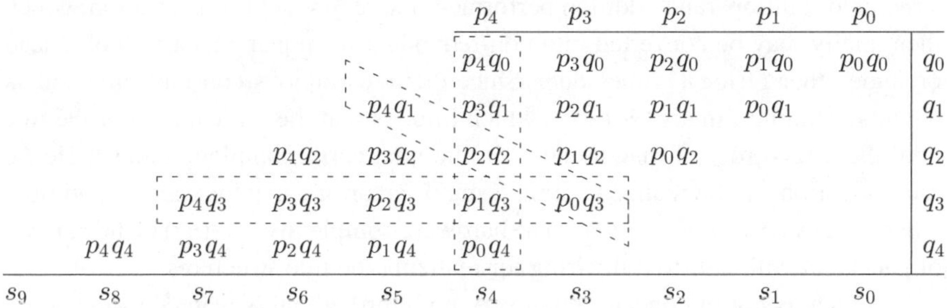

Figure 4.2.1. Array of digit product terms for a 5×5 multiplication.

be obtained as the product of the multiplicand by a single digit of the multiplier (actually a monomial). It is essential for the fastest parallel multipliers that the partial products can be constructed in constant time, e.g., as is the case if the digit set is closed under multiplication, or if the digit set employed for the partial products is redundant.

In the traditional pencil-and-paper method all the partial products are first generated, digit by digit, starting at the least significant end of each row, and forwarding carries in each partial product. Then the result is obtained by columnwise addition of digits from left to right, bringing carries forward from column to column. In the methods we will investigate in the following, different orders of generating and accumulating the digit products and the partial products will be used.

It is not necessary, however, that multiplicand and multiplier are specified in the same digit set, nor is it necessary that they share a common radix, it is sufficient that the two radices are compatible as defined in Section 2.2. This will be exploited when rewriting the multiplier into a higher radix, by grouping digits and simultaneously also performing a digit set conversion. Such a rewriting will reduce the number of partial products that have to be formed and accumulated. Also, the multiplicand and the multiplier need not have the same number of digits, but if they are of different lengths we shall assume that the multiplier is the shorter operand for such a *rectangular multiplier*, since this will yield the fastest implementation.

Multipliers can be classified according to the order in which digits from the operands (multiplicand/multiplier) must be available, and digits of the result are being produced; i.e., whether in parallel or serially, and in the latter case whether digits are needed/produced *least-significant digit first (LSD-first)* or *most-significant digit first (MSD-first)*. The following combinations are recognized:

input:	serial/serial	parallel/serial	parallel/parallel
output:	serial	serial or parallel	parallel

The fastest possible implementation of multiplication is obtained when all partial products are generated in parallel (in constant time), followed by a tree structured multioperand addition performed in a redundant representation, which then finally may be converted into non-redundant form using a carry look-ahead or some other $O(\log n)$ time adder. Since the tree employs constant-time adders the total running time is $\mathcal{O}(\log n)$, which implies that the critical path of the tree is of the same order of magnitude as in the final carry-completion adder. Hence multiplication can be realized within a small factor of carry look-ahead addition, when using such *tree-multipliers*. The hardware complexity is $O(n^2)$ in the number of gates, but with non-trivial wiring for realizing the tree structures.

If latency is not of concern, but a very high throughput is wanted as in certain signal or image processing applications, then an $n \times n$ array of full adders can be employed. Each adder is supplied with digit product circuitry, and if supplied with a suitable set of latches between each row of adders, then such pipelined *array multipliers* can produce a result for every clock tick, where the cycle time is determined by the combined digit product and full-adder delays. Such structures exhibit a very high regularity, but also require $O(n^2)$ space.

The classical way of performing multiplication is successively to produce partial products and to accumulate them into a sum of previous partial products. This *sequential* or *iterative multiplication* is often performed under microprogram control, processing the digits of the multiplier LSD-first, and performing the accumulation in non-redundant form. The hardware needed is a double-length register with a shift capability, together with a single-length adder and a register for the multiplicand, as shown in Figure 4.2.2. The accumulator part is initially empty. During each of n cycles the least-significant digit of the multiplier is used to form a partial product with the multiplicand, which is then added into the accumulator. Then the double-length register AQ with accumulator and multiplier is right shifted one position, and the cycle repeats. After n cycles the result is in AQ. The hardware complexity is $O(n)$ and the timing is $O(n \log n)$, assuming that a fast carry look-ahead adder is used.

For serial output the digits are most often produced LSD-first, in which case the result can be produced directly in non-redundant form. Producing the result

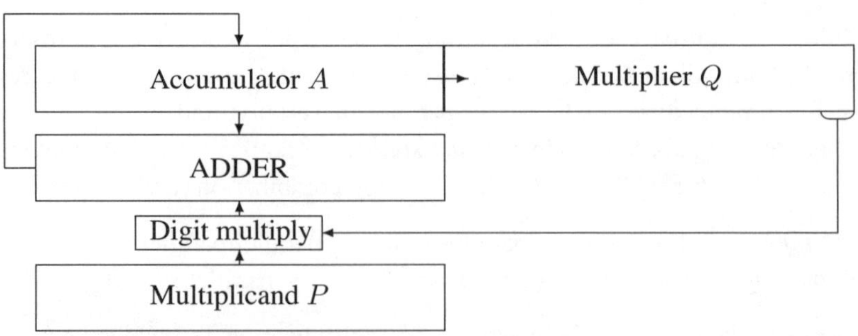

Figure 4.2.2. Hardware structure for iterative multiplication.

MSD-first requires the representation of the result to be in a redundant representation. This is evident when considering the results of Section 3.2 on h_{n-1}-separability.

LSD-first *serial multipliers* generate digit products starting in the upper right corner of Figure 4.2.1, and produce output by columnwise addition in a sweep from right to left in the array. Several different organizations have been tried in the past, same parallel/serial and some serial/serial. *On-line multipliers* take operands and produce the result MSD-first, thus the array is processed starting at the lower left end. Obviously redundant additions are needed here to perform the column summations.

Serial multipliers necessarily take time $O(n)$, actually $2n - 1$ cycles if all digits of the product are wanted. But note that often only n digits are needed, thus depending on the location of the radix point there might be a latency of n cycles before the first useful digit arrives. For example, if digits are delivered and the result produced LSD-first, and only the most-significant digits are used as the result, then such a latency occurs. The space complexity is at least $\Omega(n)$, since all previously read digits of the operands must be memorized, to be able to multiply these with the last incoming digits of the operands.

Multiplication is often used repeatedly for the computation of inner products (or *dot-products*), where the result of the individual multiplication is added into a previously accumulated sum of products. Thus it is often advantageous to combine the multiplier with the ability to add a third operand, i.e., to implement a *multiply-add* (or *fused multiply-add*) instruction, where the extra addend is fed into the process of adding partial products.

Problems and exercises

4.2.1 Show the digit product terms corresponding to Figure 4.2.1 for the multiplication of 317 by 582 in decimal.

4.2.2 Hand simulate the operation of an iterative multiplier like the one in Figure 4.2.2 for the unsigned binary multiplication of 1011 by 0101. Show the contents of the combined double-length register AQ for each step of the algorithm.

4.2.3 Prove formally that any radix multiplication algorithm requires space of at least $\Omega(n)$.

4.3 Recoding and partial product generation

The number of partial products to be added plays a significant role in determining the time of a multiplication, so reducing this number will be beneficial in particular for those algorithms that rely on sequential addition of the partial products. On the other hand, halving the number of terms to be added will only reduce the height of a binary tree by one level. So in this case it is not obvious that spending extra hardware to eliminate half of the potential products pays off in the form of a reduced total critical path.

For binary multiplication it is easy to convert the multiplier to a radix of the form 2^k, while retaining the multiplicand in radix 2. Thus the number of partial products (or digit products) to be added is reduced by a factor k. While digit products and hence a partial product are trivially formed by AND gates for ordinary binary operands, we need to also to consider the construction of *partial product generators* (PPGs) when performing a conversion (recoding) of the multiplier. The complexity of the PPGs then also has to be considered when evaluating a design for a higher-radix multiplier.

The conversion of 2's complement into redundant, borrow-save binary was introduced in Sections 2.4 and 2.5 under the name *Booth Recoding*. Note that in the literature this conversion is traditionally called a *recoding*, which is not quite consistent with the notation used in this book. Here we prefer to use the word "conversion" when mapping between digit sets, since we use the word "coding" in connection with mappings between digit values (integers) and machine states (usually binary states). In the following we shall also consider various suitable digit encodings, and to remain consistent with the usual notation we will retain the word recoding when naming the methods.

4.3.1 Radix-2 multiplication

In early (micro)programmed multiplication algorithms, possibly even employing fairly slow carry completion adders, it was essential to "skip over zeroes" in the multiplier. Any zero-valued digit in the multiplier implies that the corresponding partial product of value zero need not be added in, thus possibly saving some cycles. In randomly chosen binary multipliers on average half of the bits are zero, and by converting the multiplier into the digit set $\{-1, 0, 1\}$, the number of zero-valued digits can be increased. The conversion of a 2's complement multiplier into this digit set then also takes care of the sign, and hence the conversion is useful in a more general context.

Recall from Sections 2.4 and 2.5 that Booth Recoding was formulated as a parallel process, using the value of the right-hand neighboring digit q_{i-1} together with q_i of the multiplier Q to form a borrow-save digit e_i in $\{-1, 0, 1\}$. The conversion process then leads to the following decision table of actions, assuming that $q_{-1} = 0$ and that the operand to be converted is in 2's complement (a digit $q_n = 0$ has to be prepended if the operand is unsigned):

q_i	q_{i-1}	e_i	*Action*
0	0	0	Do nothing
0	1	1	Add multiplicand
1	0	-1	Subtract multiplicand
1	1	0	Do nothing

Note that the tuple (q_i, q_{i-1}) corresponds to a borrow-save encoding of the digit value $e_i = q_{i-1} - q_i$, which needs to be recoded into a sign-magnitude encoding to directly control the adder (add/subtract control) and to form the digit products to be fed into the adder. The sign-magnitude encoding (s_i, m_i) is easily seen to be given by $s_i = q_i$ and $m_i = q_i \oplus q_{i-1}$.

Using $\overline{m_i}$ as the signal to "skip over zero," the sign-bit s_i can, with a broadcast to all positions, be used directly to conditionally invert a 2's complement multiplicand $P = [p_{n-1}p_{n-2}\cdots p_1p_0]_{[2c]}$ and thus effect a subtraction, if it is simultaneously supplied as the carry-in to a 2's complement adder. At each position the digit product is formed as

$$r_{ij} = \bar{s}_i p_j + s_i \bar{p}_j = p_j \oplus s_i.$$

It was noted in Section 2.4 that this conversion transforms a string of repeated ones $011\cdots10$ into a string $100\cdots\bar{1}0$, hence potentially eliminating some add operations. But the converted number representation need not be the optimal one, in the sense of having the minimal number of non-zero digits. In Section 2.4.4 algorithms were given for conversion into canonical or NAF form, and it was shown that this form is unique and optimal.

The following modified DGT Algorithm also determines the canonical representation in a right-to-left conversion process, operating in radix 4.

Algorithm 4.3.1 (Minrep)
Stimulus: *An integer* $a \geq 0$,
Response: *A radix polynomial* $P = \sum_{i=0}^{n-1} e_i 2^i \in \mathcal{P}_I[\beta, \{-1, 0, 1\}]$ *with* $\|P\| = a$, *such that* $e_i e_{i-1} = 0$ *for* $i = 0, 1, \ldots, n-1$ *(*e_{-1} *assumed 0).*
Method: $b := a$; $i := 0$;
repeat
  case b mod 4 **of**
  0: $e_{i+1} := e_i := 0$; $b := b$ div 4; $i := i + 2$;
  1: $e_{i+1} := 0$; $e_i := 1$; $b := (b - 1)$ div 4; $i := i + 2$;
  2: $e_i := 0$; $b := b$ div 2; $i := i + 1$;
  3: $e_{i+1} := 0$; $e_i := -1$; $b := (b + 1)$ div 4; $i = i + 2$;
  end
until $b = 0$;
$n := i - 1$;

It is clear that the generated digit string is canonical, and also that it is a correct representation of the given integer, since it chooses a correct digit or pair of digits for each of the residue classes. Also, the updating of the remainder b is correct; in case 3 a carry is brought forward into the remainder corresponding to the digit -1 generated in radix 4.

This method has traditionally been called the *Modified Booth Recoding*, however, there is some confusion in the literature about this name. In Sections 2.4

and 2.5 we used the same name for the digit set conversion from 2's complement into the minimally redundant, symmetric radix-4 digit set, i.e., into $\mathcal{P}[4, \{-2, -1, 0, 1, 2\}]$. Whereas Algorithm 4.3.1 converts into radix 2 with digit set $\{-1, 0, 1\}$, it does so by considering pairs of digits from the source digit set, and in half the cases also generating a pair of digits radix-2. Pairing the digits two by two, starting from the least-significant end, the output is obviously in $\mathcal{P}[4, \{-2, -1, 0, 1, 2\}]$.

The alternative interpretation of Modified Booth Recoding is the base and digit set conversion from radix 2 into a radix of the form 2^k with associated minimally redundant, symmetric digit set, as discussed in detail in Section 2.5.

4.3.2 Radix-4 multiplication

Modified Booth Recoding into radix 4 is the most frequently employed method of reducing the number of terms to be added during multiplication. In this case the conversion is into $\mathcal{P}[4, \{-2, -1, 0, 1, 2\}]$, hence multiples of the multiplicand P can easily be generated by shifts, so partial products are found simply by selection from $0 \cdot P$, $1 \cdot P$ and $2 \cdot P$, in combination with an add/subtract control. Also recall that the conversion of Q can take place in parallel since the target system is redundant. This is essential for tree-structured multipliers where all partial products have to be generated in parallel. To be explicit, we may reinterpret Observation 2.5.14 on conversion from binary (2's complement) into digits e_i of a minimally redundant radix-4 representation. Triples of bits from a 2's complement multiplier $Q = [q_{2m-1}q_{2m-2} \cdots q_1 q_0]_{[2c]}$ then determine actions as follows:

q_{2i+1}	q_{2i}	q_{2i-1}	e_i	*Action*
0	0	0	0	Do nothing
0	0	1	1	Add multiplicand
0	1	0	1	Add multiplicand
0	1	1	2	Add 2×multiplicand
1	0	0	−2	Subtract 2×multiplicand
1	0	1	−1	Subtract multiplicand
1	1	0	−1	Subtract multiplicand
1	1	1	0	Do nothing

assuming that $q_{-1} = 0$ and that the number of bits is even. For unsigned operands zeroes may have to be prepended to insure that the operand is interpreted as non-negative. The triple $(q_{2i+1}, q_{2i}, q_{2i-1})$ may not be the most convenient encoding of the digit value e_i for specifying the action. The simplest PPG logic is obtained for the case when the adder has an add/subtract capability (as in most iterative multipliers). Here the sign (add/subtract control) is given by $s_i = q_{2i+1}$ and the absolute value of digit product $|r_{2i,j}|$ of weight 2^{2i+j} can be found as

$$|r_{2i,j}| = p_j u_i + p_{j-1} v_i, \qquad (4.3.1)$$

i.e., selecting between the jth bit from P or $2P$, using

$$\begin{aligned} u_i &= \quad q_{2i} \oplus q_{2i-1} \quad \sim |e_i| = 1, \\ v_i &= \bar{s}_i q_{2i} q_{2i-1} + s_i \bar{q}_{2i} \bar{q}_{2i-1} \sim |e_i| = 2. \end{aligned}$$

Note that $u_i = v_i = 0$ implies that $e_i = 0$ and the value of s_i is then immaterial. The logic for constructing u_i, v_i corresponds to the recoding of the multiplier digit e_i into a suitable encoding, and these two signals are then broadcast to all PPGs, each forming a bit $r_{2i,j}$ by (4.3.1) to be sent to the adder. But recall that it is assumed here that each input to the adder contains an add/subtract control, which is essentially an XOR gate to conditionally invert the input signals delivered by the PPGs. This gate could, of course, be incorporated in the PPG logic itself, but an alternative solution is possible.

For array- or tree-structured multipliers $\Theta(n^2)$ PPGs are needed, but only $\Theta(n)$ recoders are necessary, so it is essential to move as much logic as possible to the recoders. Hence instead of conditionally inverting the output of the PPGs, it is feasible to invert the input, i.e., the bits p_j and p_{j-1} of the multiplicand, by selecting from these or their inverted values, which are likely to be available anyway.

By substituting p_j by $p_j \oplus s_i$ respectively p_{j-1} by $p_{j-1} \oplus s_i$ in (4.3.1) and a simple rewriting it is possible to minimize the logic at each PPG. This then requires broadcasting four signals (u_i, u_i', v_i, v_i') to all PPGs, each of these forming a digit product:

$$r_{2i,j} = p_j u_i + \bar{p}_j u_i' + p_{j-1} v_i + \bar{p}_{j-1} v_i' \tag{4.3.2}$$

using the following four selection signals:

$$\begin{aligned} u_i &= \bar{s}_i (q_{2i} \oplus q_{2i-1}) \sim e_i = +1, \\ u_i' &= s_i (q_{2i} \oplus q_{2i-1}) \sim e_i = -1, \\ v_i &= \quad \bar{s}_i q_{2i} q_{2i-1} \quad \sim e_i = +2, \\ v_i' &= \quad s_i \bar{q}_{2i} \bar{q}_{2i-1} \quad \sim e_i = -2, \end{aligned}$$

where again $s_i = q_{2i+1}$ is the sign, which is also to be used as the carry-in to the rightmost adder. Again $u_i = u_i' = v_i = v_i' = 0$ implies $e_i = 0$, but also note that the sign s_i does not have to be broadcast to each PPG, the sign information has been encoded in the four values.

It is obviously sufficient with three bits to encode the five digit values, and among many others an encoding (s_i, u_i, v_i), where $s_i = q_{2i+1}$ is the sign and $u_i = q_{2i} \oplus q_{2i-1}$, $v_i = q_{2i+1} \oplus q_{2i}$, would be a possibility. However, in this case the logic in each PPG will be more complicated and slower than using the four signals above. We leave it as an exercise to develop the logic for this case.

Another essential observation is that the multiplier Q may be used directly in a redundant representation, so in cases where Q is the result of a previous operation it can be used without first converting it into a non-redundant representation. Recall that the recodings above from 2's complement into radix 4 were performed based on a P-mapping realized by a trivial "wiring" into a radix-2, borrow-save encoding (q_i, q_{i-1}) of a digit d_i of value $q_{i-1} - q_i$, and then grouping these in pairs to obtain an encoding of a minimally redundant radix-4 digit e_i, here with the positively weighted bits written above those of negative weight:

$$e_i \sim \begin{pmatrix} q_{2i} \\ q_{2i+1} \end{pmatrix} \begin{pmatrix} q_{2i-1} \\ q_{2i} \end{pmatrix}, \quad e_i \in \{-2, -1, 0, 1, 2\}. \tag{4.3.3}$$

Note that the digit values ± 3 cannot occur here owing to the replication of the bit q_{2i}.

As discussed in Section 2.5 a conversion into the same digit set from an arbitrary borrow-save or carry-save represented number can be obtained in constant time by respectively PN^2 or PNQ-mappings. As an example, if the multiplier Q is the sum or difference of two numbers, provided in standard binary or 2's complement, then $Q = (A \pm B)$ can be directly converted in parallel into the minimally redundant digit set of radix 4. First, group bits (a_i, b_i) from (A, B), considered as carry-save or borrow-save encodings of digits radix-2, then apply the above three-level mappings, and finally group two of the resulting digits to form a radix-4 digit. If the converted radix-2 digits q_i are encoded in borrow-save by (q_i^p, q_i^n) then the radix-4 digits e_i are encoded by grouping pairs of these:

$$e_i \sim \begin{pmatrix} q_{2i+1}^p \\ q_{2i+1}^n \end{pmatrix} \begin{pmatrix} q_{2i}^p \\ q_{2i}^n \end{pmatrix}, \quad e_i \in \{-2, -1, 0, 1, 2\}, \tag{4.3.4}$$

where certain combinations of $q_{2i+1}^p, q_{2i+1}^n, q_{2i}^p, q_{2i}^n$ cannot occur since the digit values ± 3 do not occur. Also by Observation 2.5.14 the sign information of e_i can be obtained from the value of the bit q_{2i+1}^n in the sense that

$$\begin{aligned} q_{2i+1}^n &= 0 \text{ implies } e_i \geq 0, \\ q_{2i+1}^n &= 1 \text{ implies } e_i \leq 0. \end{aligned}$$

Noting the similarity between (4.3.3) and (4.3.4), it is now possible to recode the encoding (4.3.4) into direct control of the PPGs, specifically in the form of the four signals needed for (4.3.2). From (4.3.4) it is easily seen that the expression $u_i = q_{2i}^p \oplus q_{2i}^n$ provides the parity of e_i just as above. Thus with $s_i = q_{2i+1}^n$ as the sign, we obtain the following four selection signals:

$$\begin{aligned} u_i &= \overline{s}_i(q_{2i}^p \oplus q_{2i}^n) \sim e_i = +1, \\ u_i' &= s_i(q_{2i}^p \oplus q_{2i}^n) \sim e_i = -1, \\ v_i &= \overline{s}_i\overline{(q_{2i}^p \oplus q_{2i}^n)} \sim e_i = +2, \\ v_i' &= s_i\overline{(q_{2i}^p \oplus q_{2i}^n)} \sim e_i = -2. \end{aligned}$$

Note that the same expressions (albeit these are slightly more complicated) can also be used in the previous case by a suitable interpretation of the input signals.

Hence the very same radix-4 recoder can be used for input of a non-redundant (2's complement) multiplier Q, as well as for a converted redundant multiplier $\hat{\alpha}_P(\hat{\alpha}_N(\hat{\alpha}_N(Q)))$ for Q in borrow-save, respectively $\hat{\alpha}_P(\hat{\alpha}_N(\hat{\alpha}_Q(Q)))$ for Q in 2's complement carry-save.

4.3.3 High-radix multiplication

For parallel multipliers where all multiples of the multiplicand (partial products) have to be available at the same time to be fed into an adder tree, we will now investigate whether we can further reduce the number of these terms to be added, by converting the multiplier to a higher radix of the form 2^k, e.g., 8, 16, … . In this subsection we will not go into the details of converting the multiplier, nor will we discuss logic for PPGs.

Increasing the value of k from 1 to 2 decreases the height of a binary adding tree by one since it halves the number of inputs to the tree, but increasing k from 2 to 3 only reduces the number of terms to be added from $n/2$ to $n/3$. On the other hand, it is then necessary to be able to generate a larger set of multiplicand multiples, due to the larger digit set of the multiplier. When converting into $\mathcal{P}[8, \{-4, -3, -2, -1, 0, 1, 2, 3, 4\}]$ using the minimally redundant digit set, there is a need for the multiples $\pm 3P$, and for radix 16 there is similarly a need for multiples $\pm 3P$, $\pm 5P$ and $\pm 7P$, where P is the multiplicand. Note, however, that these can all be generated as the sum or difference of two shifted versions of the multiplicand. Thus we can alternatively feed such a pair into the tree, since in general it does not make sense to perform a carry-completion add/subtract requiring $\mathcal{O}(\log n)$ time. Going to even higher radices will require larger multiples to be generated, and thus more shifted versions of the multiplicand will be needed.

Let us define the *weight of a digit set D* as $\tilde{\omega}(D) = \max_{d \in D} \omega(d)$, then $\tilde{\omega}(D)$ is the number of shifted versions of the multiplicand that may have to be added or subtracted for each converted digit of the multiplier. The following result shows that the number of such multiplicand multiples is minimized if we employ the minimally redundant digit set $\{-2^{k-1}, \ldots, 0, 1, \ldots, 2^{k-1}\}$ for $\beta = 2^k$:

Lemma 4.3.2 *For $\beta = 2^k$ the minimally redundant digit set $D = \{-2^{k-1}, \ldots, 0, 1, \ldots, 2^{k-1}\}$ is of minimal weight, i.e., $\tilde{\omega}(D) \le \tilde{\omega}(D')$ for any other digit set D', redundant radix β.*

Proof Obviously increasing the redundancy of D by adding extra digits cannot decrease the weight $\tilde{\omega}(D)$. Also substituting any digit by some other member of the same redundancy class, i.e., substituting d by $d + i2^k$, $i \neq 0$, cannot decrease the weight since $\omega(d + i2^k) \ge \omega(d)$. □

Lemma 4.3.3 *For $D = \{-2^{k-1}, \ldots, 0, 1, \ldots, 2^{k-1}\}$, $\tilde{\omega}(D) = \lceil k/2 \rceil$.*

Proof For k even, $k = 2j$, the digit $d \in D$ of value $0101\cdots01_2$ with j groups of the string 01 is of weight $\omega(d) = j = \lceil k/2 \rceil$. On the other hand this is the maximal weight that can be obtained by any canonical bit-string of length $k = 2j$. For $k = 2j + 1$ similarly consider the digit string consisting of a $\bar{1}$ followed by j groups of 01. □

Assume that a partial product $d \times P$ is to be formed, where d is a digit of the multiplier and P is the multiplicand. Then $\omega(d)$ shifted versions of the multiplicand are needed to form $d \times P$, and these can be fed directly into the adder tree of the multiplier.

We can combine the previous two lemmas into a result on the number of terms that have to be accumulated in such a tree-structured multiplier.

Observation 4.3.4 *Let n be the number of bits in the multiplier. Converting (recoding) the multiplier into radix 2^k, where k divides n, requires at least*

$$\frac{n}{k} \cdot \left\lceil \frac{k}{2} \right\rceil$$

shifted versions of the multiplicand to be formed and accumulated. The minimum is obtained for D minimally redundant and k even, in which case the value is $n/2$, independent of k.

We may thus conclude that for such parallel multipliers there is nothing to be gained by converting the multiplier to a radix higher than 4, since the number of terms to be added is not being further reduced. It is even implementation-dependent whether radix 4 is feasible. The height of the tree is reduced by halving the number of inputs, but additional circuitry is needed for the recoding and selection of the input to the tree. However, if the multiplier is supplied in redundant form there is a definite advantage in using radix 4, since then the multiplier must be converted in some way.

However, if the multiplier is part of a pipeline, possibly itself pipelined, then it may be feasible to precompute a suitable set of multiples of the multiplicand in a previous step of the pipeline, and then select from these for the input of partial products for the tree of adders. If such multiples (say for radix 8, only $3P$) are computed by carry-completion adders in a separate stage, they can be available in non-redundant form for the next pipe stage, and thus reduce the height of the adder tree in the multiplier. A similar approach for radix 8 was used in early Pentium (non-pipelined) processors to save space, where the calculation of $3P$ and the summation in the multiplier tree were excuted within a double-clock cycle.

Also a hybrid scheme is possible where a radix-4 approach is used at the top of the tree, while the special radix-8 multiples are generated in parallel, to be used and fed into the lower part of the tree.

If the multiplier is designed to be able to accept the multiplicand in redundant form (say borrow-save or carry-save), then the adder tree might as well also be able to accept some simple multiples of the multiplicand. Then it is feasible to recode the multiplier (also assumed redundant) into a higher radix, since such multiples can be generated in redundant form in constant time, and then be available for selection for the tree.

The situation is very different for iterative multipliers where the running time is proportional to the number of digits in the multiplier. Here it is advantageous to use the highest possible radix that still allows a partial product to be found and accumulated within each cycle of the iteration. Two approaches can then be identified:

(i) look-up from a precomputed table of multiplicand multiples;
(ii) generation of partial products using a rectangular multiplier.

Approach (i) is most likely to be useful when exploiting a carry-completion adder whose critical path is close to the machine cycle time, and then only for fairly small values of the radix. This is because of the initial overhead of generating the table of multiples, which possibly requires one cycle of the adder per multiple, assuming these multiples are needed in non-redundant form.

Approach (ii) assumes that an $n \times k$ rectangular multiplier is available and capable of completing in one machine cycle. Then a radix-2^k recoding is feasible, where the multiplier is feeding its result every cycle into a carry-completion adder for accumulation of the partial products. Note that the output of the rectangular multiplier need not be in non-redundant form, if the adder is capable of adding the partial product in redundant form into the previously accumulated sum. By latching the output of the rectangular multiplier, and balancing its timing against the timing of the adder a very high radix 2^k can be chosen, allowing the full multiplication to complete in a few cycles.

Example 4.3.1 A 64-bit arithmetic unit is to be designed, to run at a very short cycle time, but using only moderate space or transistor count. A 64-bit carry-completion adder has been designed with a critical path closely matched against the target machine cycle time, corresponding to the approximately 16 logic levels needed for the carry look-ahead circuitry. Within the same number of logic levels it is possible to realize a 64×16 tree structured multiplier with redundant output. Latching the output of this multiplier, its result is available for the adder in the next cycle, thus a full 64×64 bit multiplication can be completed in five cycles. But notice that the adder as well as the rectangular multiplier are only busy during four of these cycles. □

A "small" multiplier can also be used to construct the product of larger operands, e.g., a double-precision product can be constructed using a single-precision multiplier. The trivial way to do this uses four single-precision products, with

$A = A_1 2^n + A_0$ and $B = B_1 2^n + B_0$

$$AB = 2^{2n} A_1 B_1 + 2^n (A_1 B_0 + A_0 B_1) + A_0 B_0,$$

but Karatsuba has devised a way of doing it with three, as shown in the following lemma:

Lemma 4.3.5 (Karatsuba) *If* $A = A_1 2^n + A_0$ *and* $B = B_1 2^n + B_0$, *then*

$$\begin{aligned} AB &= 2^n(2^n - 1)A_1 B_1 + 2^n(A_1 + A_0)(B_1 + B_0) - (2^n - 1)A_0 B_0 \\ &= 2^{2n} A_1 B_1 + 2^n(A_1 + A_0)(B_1 + B_0) + A_0 B_0 - 2^n(A_0 B_0 + A_1 B_1). \end{aligned}$$

Thus one multiplication has been substituted by three extra additions. Note that this approach may be used recursively to speed up multiplication of very large operands.

Problems and exercises

4.3.1 Show that the table for "Booth Recoding" can also be derived from the trivial identity $x = 2x - x$, by insertion of the 2's complement radix polynomial for x.

4.3.2 Apply Algorithm 4.3.1 to the integers 25 and −7.

4.3.3 For the radix-4 case with $Q = [q_{2m-1} q_{2m-2} \cdots q_1 q_0]_{[2c]}$ it was mentioned on page 215 that a broadcast of the three signals (s_i, u_i, v_i), where $s_i = q_{2i+1}$ is the sign, and $u_i = q_{2i} \oplus q_{2i-1}$, $v_i = q_{2i+1} \oplus q_{2i}$ would be sufficient. Develop the logic for the PPGs for this design.

4.3.4 It was noted that the digit values $e_i = \pm 3$ could not occur in (4.3.4). Show by using (2.5.11) that the value 3 cannot occur.

4.3.5 Describe by means of half-adders the conversion in which the difference of two unsigned binary numbers $A - B$ is to be converted directly into minimally redundant radix 4 to be used as a multiplier. What changes are necessary if A and B are both given in 2's complement?

4.3.6 For each value of k, $k = 2, \ldots, 6$ list those digit values from the minimally redundant digit sets radix 2^k that need two or more shifted values of the multiplicand for the formation of partial products (i.e., $\omega(d) \geq 2$).

4.4 Sign-magnitude and radix-complement multiplication

Treating operands specified in sign-magnitude representation is quite straightforward since the sign of the result is given as the exclusive or (XOR) of the signs of the two operands. Thus multiplication is just performed on the extracted unsigned magnitude parts, and the sign is then later attached to the resulting magnitude part.

The handling of radix-complement represented operands is slightly more complicated; however, in principle they can be handled as any other radix polynomial operands by utilizing the results of Section 1.7 on the possible use of the digit value -1 in an extra, not represented position. Since the value of this extra digit (-1 or 0) is uniquely determined by the most-significant digit among those represented, this extra digit can easily be reconstructed and prepended to the operand digit string. However, this is not really feasible for hardware realizations, since the possibility of a negative digit unnecessarily complicates the encoding of digit values, as well as the circuitry for digit product or partial product formation. And as we shall see below for the case of 2's complement there are better ways of dealing with complement representations.

4.4.1 Mapping into unsigned operands

The first and trivial (but "expensive") way is, of course, to negate negative operands, and then perform the multiplication on unsigned (positive) operands, keeping track of operand signs such that the unsigned result can be negated if necessary. But since negation is an "expensive" operation in general, this solution is not to be recommended unless the negations can be carefully integrated in the multiplication algorithm. So let us briefly discuss the possibilities of such integration of the conditional negation for the multiplier, for the multiplicand, and for the result.

Recall from Section 2.6 that negation of a radix-complement number is performed by complementation of digit values (a constant-time operation) followed by the addition of a unit in the least-significant position. If this unit addition (carry-in) can be suitably incorporated in the multiplication algorithm then there is only a constant time overhead.

Conditional negation of the multiplier can be handled easily for the iterative-type multiplications, since here the multiplier is read serially from the least-significant end, thus negation can be performed serially as a simple digit set conversion. Also for any kind of Booth Recoding of the multiplier the conditional negation is trivially realized with constant time overhead. Either because processing is serial, SD-first, or because the result is in non-redundant representation.

If the adder employed for an iterative multiplication is designed for radix-complement operands, and is capable of performing addition as well as subtraction, then a negative multiplicand can be implicitly negated by inverting the add/subtract control of the adder. The situation is more complicated for array- or tree-structured multipliers, since multiplicand multiples not only have to be found from the complemented multiplicand but also multiples of the "carry-in" need to be fed appropriately into the array or tree. Actually the problem is much like the one we shall investigate below when dealing directly with radix-complement operands.

Finally, the conditional negation of the result is usually fairly straightforward. First notice that the information on whether to negate or not is available very early.

Also most adders are capable of producing simultaneously the proper result and its negated value, hence the final result is easy to select.

4.4.2 2's complement operands

The above discussion, although very sketchy, has pointed out that there are a number of complications when trying to integrate the negation of negative operands into algorithms or hardware for unsigned multiplication. It turns out that dealing directly with the (possibly negative) radix-complement represented operands is not that complicated. For the discussion of such situations we will specifically deal with the 2's complement representation, and we will assume that the operands are integers.

First, consider the iterative multiplier of Figure 4.2.2 for the case of 2's complement operands where the multiplier is read directly in radix 2. With a 2's complement adder, the multiplicand can be loaded directly into the P register, so that it can be accumulated in the right shifted contents of the combined AQ register. The only requirement here is that the right shift of AQ is an *arithmetic shift*, i.e., that a sign extension is taking place at the left end of the register. Bits of the multiplier are used to form partial products corresponding to the digit values 0 and 1, except for the most-significant bit which has negative weight. For the last cycle the adder must be changed to perform a subtraction of the last partial product.

4.4.3 The Baugh and Wooley scheme

When generating digit products in parallel for accumulation in an array or tree multiplier we want to avoid dealing with digit products containing sign extensions. In 1973 Baugh and Wooley devised a scheme to minimize the number of bits that have to be generated and accumulated for 2's complement multiplication. To deal with the general case we now consider the situation in which the multiplicand P and the multiplier Q need not be of the same length:

$$P = -[2]^{n-1}p_{n-1} + \sum_{j=0}^{n-2} p_j[2]^j, \qquad Q = -[2]^{m-1}q_{m-1} + \sum_{i=0}^{m-2} q_i[2]^i,$$

hence

$$P \cdot Q = [2]^{n+m-2}p_{n-1}q_{n-1} \tag{4.4.1}$$
$$-[2]^{n-1}p_{n-1}\textstyle\sum_{i=0}^{m-2} q_i[2]^i \tag{4.4.2}$$
$$-[2]^{m-1}q_{m-1}\textstyle\sum_{j=0}^{n-2} p_j[2]^j \tag{4.4.3}$$
$$+\textstyle\sum_{i=0}^{m-2}\sum_{j=0}^{n-2} p_jq_i[2]^{i+j}. \tag{4.4.4}$$

Utilizing the complements $\overline{p}_i = 1 - p_i$ and $\overline{q}_i = 1 - q_i$ we can rewrite the terms (4.4.2) and (4.4.3)

$$-[2]^{n-1} p_{n-1} \sum_{i=0}^{m-2} q_i [2]^i = [2]^{n-1} \Big(- [2]^{m-1} + \overline{p}_{n-1} [2]^{m-1} + p_{n-1} \sum_{i=0}^{m-2} \overline{q}_i [2]^i + p_{n-1} \Big),$$

$$-[2]^{m-1} q_{m-1} \sum_{j=0}^{n-2} p_j [2]^j = [2]^{m-1} \Big(- [2]^{n-1} + \overline{q}_{m-1} [2]^{n-1} + q_{m-1} \sum_{j=0}^{n-2} \overline{p}_j [2]^j + q_{m-1} \Big),$$

and all contributions can be arranged in a diagram as shown in Figure 4.4.1 for $n > m$.

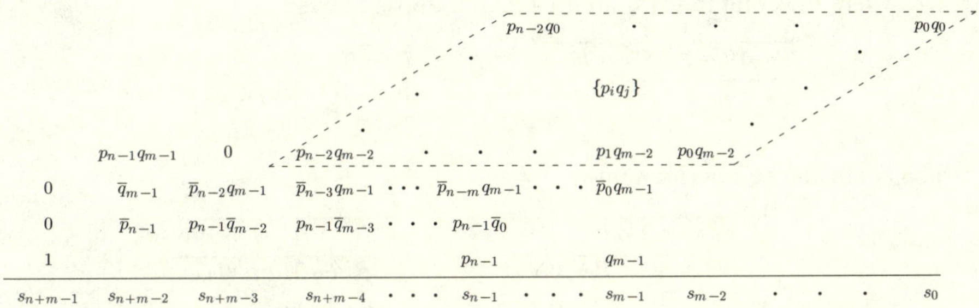

Figure 4.4.1. Baugh and Wooley scheme for 2's complement multiplication, $n > m$.

The terms in the two bottom rows can be rearranged into the rest of the scheme (exercise) by extending the length of the rows as shown in Figure 4.4.2, reducing the number of rows to m, which is usually a power of 2, say 2^k, in which case a binary tree structure of height k can accumulate the rows in time $\Theta(k)$.

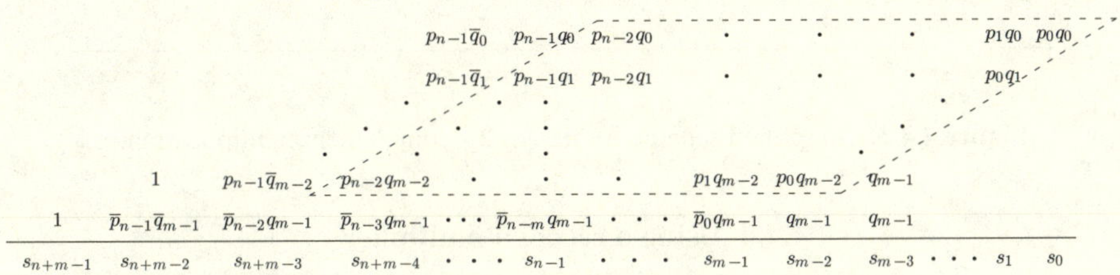

Figure 4.4.2. Compressed Baugh and Wooley scheme.

However, there is an alternative scheme based directly on the 2's complement equations (4.4.1)–(4.4.4) for the case $m = n$ (the case $m \neq n$ does not work out nicely):

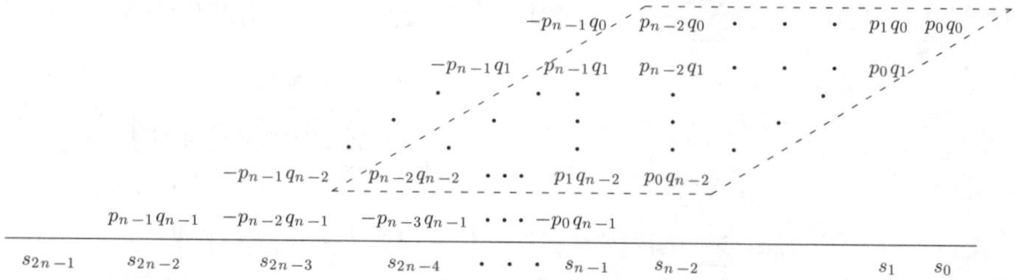

Looking separately at the negative terms, placing terms of equal weight above one-another,

$$\begin{array}{llcll} -p_{n-1}q_{n-2} & -p_{n-1}q_{n-3} & \cdots & -p_{n-1}q_1 & -p_{n-1}q_0 \\ -p_{n-2}q_{n-1} & -p_{n-3}q_{n-1} & \cdots & -p_1q_{n-1} & -p_0q_{n-1} \end{array},$$

and using the identity $-ab = \overline{ab} - 1$, we can rewrite these two rows into three rows, where the column sums are identical to those above:

$$\begin{array}{ccccc} \overline{p_{n-1}q_{n-2}} & \overline{p_{n-1}q_{n-3}} & \cdots & \overline{p_{n-1}q_1} & \overline{p_{n-1}q_0} \\ \overline{p_{n-2}q_{n-1}} & \overline{p_{n-3}q_{n-1}} & \cdots & \overline{p_1q_{n-1}} & \overline{p_0q_{n-1}} \\ -2 & -2 & \cdots & -2 & -2 \end{array},$$

which again can be rewritten into

$$\begin{array}{cccccc} & \cdot & \overline{p_{n-1}q_{n-2}p_{n-1}q_{n-3}} & \cdots & \overline{p_{n-1}q_1} & \overline{p_{n-1}q_0} \\ & \cdot & \overline{p_{n-2}q_{n-1}p_{n-3}q_{n-1}} & \cdots & \overline{p_1q_{n-1}} & \overline{p_0q_{n-1}} \\ -1 & 0 & 0 \quad 0 & \cdots & 1 & \end{array}.$$

Finally combining these rewritings into the total scheme, recalling that the 1 in position $2n - 1$ has negative weight, we find the structure in Figure 4.4.3.

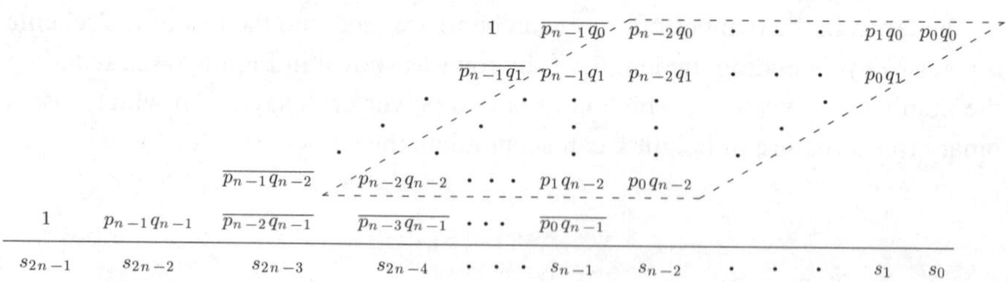

Figure 4.4.3. Simplified scheme for $n \times n$, 2's complement multiplication.

4.4.4 Using a recoded multiplier

When recoding a 2's complement multiplier into minimally redundant radix-2^k representation the sign of the multiplier is handled automatically, since conversion is into signed digits. But now we have to be able to form partial products as products of a signed digit and a 2's complement multiplicand. For accumulation it is most convenient if here we can also avoid explicit sign extensions, but still be able to add bits of positive weight only.

Let us restrict our consideration to the case where the multiplier is converted into radix 4, i.e., partial products can be formed by possibly shifting and/or 2's complementing the multiplicand. Let us assume that the multiplier is represented in $2m$ bits and has been converted (recoded) into

$$Q = \sum_{i=0}^{m-1} d_i[4]^i, \qquad d_i \in \{-2, -1, 0, 1, 2\},$$

and that the multiplicand has n bits in 2's complement:

$$P = -p_{n-1}[2]^{n-1} + \sum_{j=0}^{n-2} p_j[2]^j.$$

The ith partial product $T_i = P \cdot d_i[4]^i$ can then be rewritten in 2's complement:

$$\|T_i\| = \|P\| \cdot d_i 2^{2i} = \left(-t_{i,n}2^n + \sum_{j=0}^{n-1} t_{i,j}2^j + c_i\right) 2^{2i}, \qquad t_{i,j} \in \{0, 1\},$$

where c_i is a carry-in from a possible 2's complementation due to a negative d_i. Note that T_i potentially requires a 2^n term due to a shifting corresponding to $d_j = \pm 2$, hence a sign extension into this position is performed.

To avoid the term of negative weight now consider instead

$$T_i' = P \cdot d_i[2]^{2i} + 3[2]^{n+2i},$$

$$\|T_i'\| = \left(2^{n+1} + (1 - t_{i,n})\, 2^n + \sum_{j=0}^{n-1} t_{i,j}2^j + c_i\right) 2^{2i}, \tag{4.4.5}$$

i.e., complement the "sign bit" and prepend a 1. Then accumulate these modified partial products, together with an extra correction:

$$\begin{aligned} R &= [2]^n + \sum_{i=1}^{m-1} T_i' \\ &= [2]^n + \sum_{i=1}^{m-1} (P \cdot d_i[2]^{2j} + 3 \cdot [2]^{n+2i}), \end{aligned}$$

$$\begin{aligned} \|R\| &= 2^n + \|P \cdot Q\| + 3 \cdot 2^n \cdot \sum_{i=1}^{m-1} 4^i \\ &= \|P \cdot Q\| + 2^{2m+n}; \end{aligned}$$

hence $\|R\| - 2^{2m+n}$ will be the correct result. Figure 4.4.4 is a diagram of the bits to be accumulated where the subtraction of the term 2^{2m+n} has been included in the form of the 1 in the bottom row, position $n + 2m$ having negative weight. This position is not included in the computations anyway, allowing the rest of the result to be interpreted as a 2's complement number.

The scheme of Figure 4.4.4 has $m + 1$ rows to be added, two of which only require single bits to be added. The isolated 1 in position n can easily be combined with $\bar{t}_{0,n}$ and the unit bit to the left, extending the length of the row by one (exercise). However, it turns out that the carry c_{m-1} cannot, in general, be incorporated in the remaining rows (exercise), hence its presence becomes a critical issue in a hardware implementation. But note that c_{m-1} is zero if the multiplier is non-negative, since then the recoded, most-significant digit d_{m-1} is non-negative and thus the last partial product is never negated.

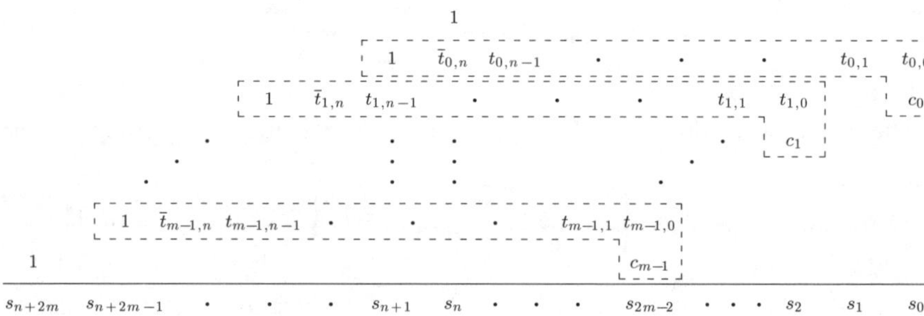

Figure 4.4.4. 2's complement multiplication with radix-4 recoded multiplier.

Hence the scheme of Figure 4.4.4 also applies to the multiplication of unsigned numbers, and to the case of forming the product of two sign-magnitude represented numbers, where the sign is handled separately.

All the rows of Figure 4.4.4 have positive weight, hence accumulation can be performed by a suitable tree structure, employing 3-to-2 or 4-to-2 adders. Often in practice $2m$ will be a power of 2, thus $m + 1$ will be an odd number, so the extra bit to be added will also slightly complicate the design of the tree structure when employing 4-to-2 adders.

Let us conclude this section with two examples.

Example 4.4.1 Let us multiply the six-bit number $-32 \sim 100000_{2c}$ by the four-bit number $-8 \sim 1000_{2c}$, both in 2's complement. Recoding the latter into radix 4 we get $\bar{2}0$ and thus we obtain the following scheme:

where the inverted digits are shown in boldface. Note that the leftmost position of negative weight need not be calculated, and that the result is $+256 = 2^8$ as expected. □

Example 4.4.2 Here we multiply two six-bit numbers, $26 \sim 011010_{2c}$ and $-25 \sim 100111_{2c}$. The latter recodes into $\bar{2}2\bar{1}$:

						1							
					1	**0**	1	0	0	1	0	1	$26 \times (-1) \times 1$
			1	**1**	1	1	0	1	0	0		1	$26 \times 2 \times 4$
	1	**0**	0	0	1	0	1	1		0			$26 \times (-2) \times 16$
1								1					
1	1	1	0	1	0	1	1	1	0	1	1	0	

which correctly represents $26 \times -25 = -650$ in 2's complement. □

Problems and exercises

4.4.1 Prove the correctness of the rearrangement of terms in the compressed Baugh and Wooley scheme of Figure 4.4.2.

4.4.2 Perform the multiplication 26×-25 in binary using the Baugh and Wooley scheme.

4.4.3 Show that the extra 1 in the top row of Figure 4.4.4 can be incorporated in the row below by extending that row to the left.

4.4.4 Show that the carry c_{m-1} in Figure 4.4.4 cannot be substituted by filling in copies of it in the lower right "triangle," so that at most m bits are to be added in each column. (Hint: Show that the weighted sum of ones in the "triangle" is strictly less than 4^{m-1}, the weight of c_{m-1}.)

4.4.5 In Figure 4.4.4 the entry in each position may be generated by some PPG logic using (4.3.2), based on the recoding of multipler digits into four binary signals. However, when using the rewriting of (4.4.5) some modifications are needed in the two leftmost positions of each row and those at the right-hand end (including the carry-in obtained from 2's complementation). Design the PPG logic for these positions using the same four selection signals as used in the other positions.

4.5 Linear-time multipliers

This class encompasses parallel as well as serial multipliers, but all cases are based on consuming at least one of the operands (the multiplier) in a serial fashion. Actually, not all of them can complete with a non-redundant result in time $O(n)$; some of them require time $O(n \log n)$ since they may employ a $O(\log n)$ time adder during each of n steps, as seen below in the first example.

4.5.1 The classical iterative multiplier

The structure of this type of multiplier is shown in Figure 4.5.1, in which we see that the multiplier actually computes $R = A + P \cdot Q$, where A is the original contents of the accumulator register A, P is the multiplicand provided in parallel, and Q is the multiplier, whose digits are used to form partial products while they are gradually shifted out of the Q-register.

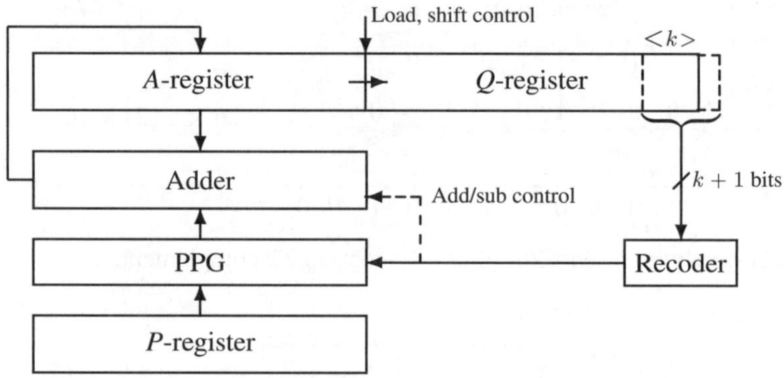

Figure 4.5.1. Iterative multiplier with radix-2^k recoding.

The combined A,Q-register has a shift capability such that in each cycle its content is right shifted k positions (possibly with sign extension), assuming the multiplier is interpreted in radix 2^k. The Q-register contains possibly one extra bit position (initially zero) used by the recoder as described in Section 3.3. The output of the recoder then controls the selection or construction of partial products in the PPG, forming one input to the adder. The other input to the adder is the content of the A-register. The output of the adder is then fed back to the A-register, followed by a right-shift of the combined A,Q-register by k positions. Alternatively, the content of the Q-register can be shifted while the adder is operating. Then the shifted output of the adder can be loaded into the A-register and the k high-order positions of the Q-register, where the shifting of the adder output is realized by a simple wiring.

For higher radices (≥ 8) the PPG logic and P-register could be realized by a small table of precomputed multiples of the multiplicand, in which a look-up is performed employing the recoded multiplier digit as the address. This is most convenient in combination with an add/subtract control to reduce the table size.

In the past multipliers were often implemented such that the adder and suitable registers were connected to some kind of bus structure, and supplied with additional special logic for the recoding and shift capabilities, so that multiplication could be realized by resources also available for other purposes. In particular the adder would then be a "general purpose" ALU, capable of performing a carry-completion add/subtract operation (as well as other parallel logical functions by reconfiguration, e.g., disconnecting carry connections).

4.5.2 Array multipliers

If the loop of the iterative multiplier is unrolled in hardware, each iteration of the loop is realized by its own adder which could be a redundant carry-save adder. In its simplest form the intermediate storing of temporary results is not needed, so one row of adders may directly feed the inputs of the next row, with the carries shifted one position to the left. Digit products may be generated in separate cells, or in cells combined with full-adders, as shown in Figure 4.5.2

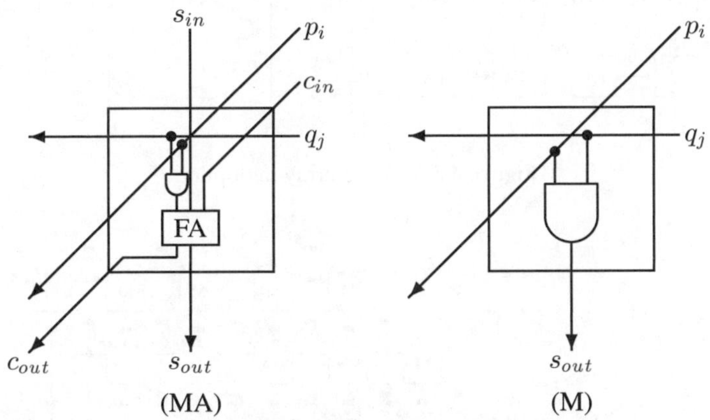

Figure 4.5.2. Multiply-accumulate cell (MA) and digit-multiply cell (M).

These cells can then be combined in an array structure as shown in Figure 4.5.3, also employing cells just containing full-adders (the FA cells). For simplicity the array shown only accepts unsigned binary operands, but it could easily be adapted to the Baugh and Wooley scheme for 2's complement operands, to higher radices, or to a recoded multiplier.

The full-adders of the bottom row have been combined into a ripple-carry structure but obviously other and faster adders could be used. Considering that the area is $\Theta(n^2)$, the timing $\Theta(n)$ is not very optimal and the individual adders will only be active temporarily, while the computation is moving down the array as a "wave".

However, it is very simple to pipeline the structure by inserting latches (buffers) between each row. Then the arrival of bits q_i from the multiplier Q must be delayed such that they arrive at the correct time. Also the final carry assimilation must be expanded (unrolled) into rows of half-adders, as discussed in Chapter 3. Thus with a structure as shown in Figure 4.5.4, it is possible in each clock cycle to feed all the bits of two new operands P and Q in parallel.

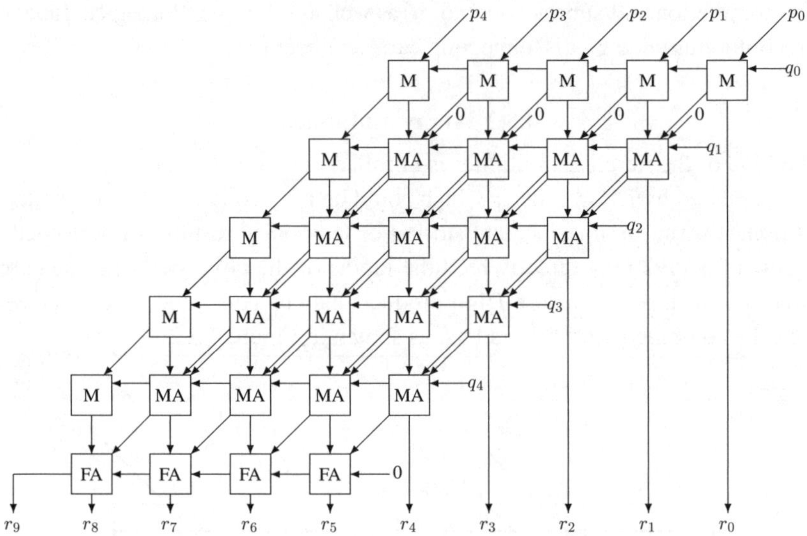

Figure 4.5.3. 5 × 5 array multiplier.

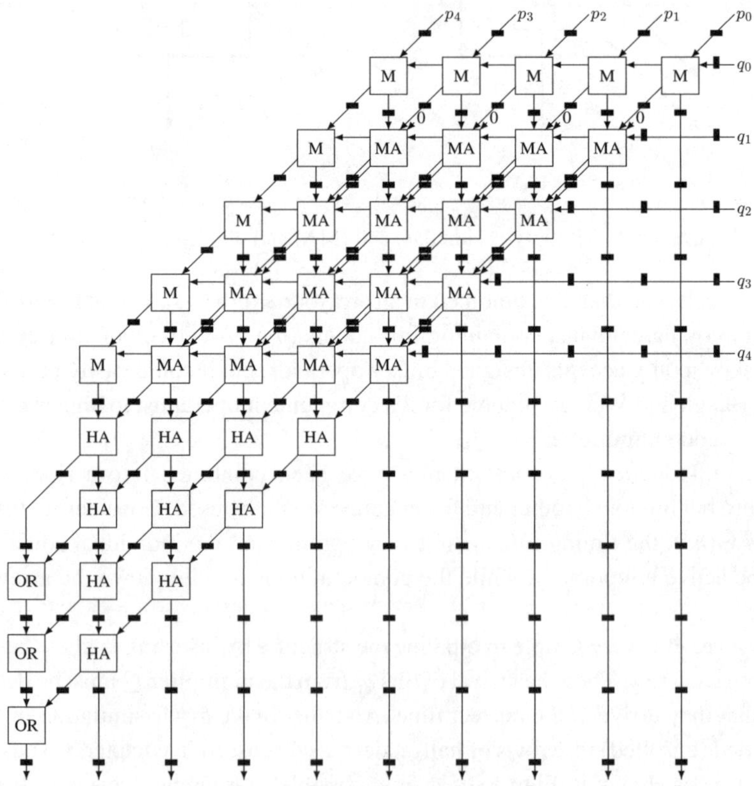

Figure 4.5.4. Pipelined 5 × 5 array multiplier.

In Figure 4.5.4 the buffers are shown as small black rectangles on each line. Note that the triangular array of half-adders (HAs) can produce at most a single carry-out bit, thus simple OR gates are sufficient in the leftmost column.

This array is capable of producing one $(2n - 1)$-bit product for each clock cycle, but at a delay (*latency*) of $2n$ cycles from the time that the two n-bit operands were delivered. For applications, such as signal processing, where often many multiplications have to be performed and latency may not be of concern, such an array can offer a high throughput since the clock frequency can be very high, due to the short critical path in the cells.

Again, the array of Figure 4.5.4 only accepts unsigned binary operands, but generalizations to 2's complement can be realized by appropriate modifications. Also such arrays can be modified to handle operands of differing lengths (i.e., rectangular multiplication), and cells could be designed to accept operands of a higher radix.

One problem with the design in Figure 4.5.4 is the broadcast of the multiplier bits along each row of cells, requiring a high fan-out or repeated amplification of signals. It can be avoided by the insertion of buffers along the broadcast lines, but this requires a *retiming* of the whole array, i.e., other buffers have to be inserted or moved around to insure correct timing of signals. We shall discuss such retiming below for some simpler cases of linear arrays.

Other modifications are also possible, e.g., the arrays above could easily be changed to accept an additive constant (by changing M cells into MA cells), thus computing $R = P \cdot Q + A$, and they could be reduced to deliver only either the least-significant or the most-significant half of the result, in the latter case truncated as correct rounding will require calculation of the full length product. When only the least-significant part is delivered, corresponding to an $n \times n \to n$ integer multiplication, a significant saving can be obtained, but in general it may require the calculation of an overflow signal. We shall return to this possibility in Section 4.6, where we shall develop the logic for log-time overflow detection.

4.5.3 LSB-first serial/parallel multipliers

In this subsection we will discuss situations in which the multiplier is assumed to be delivered and consumed serially, LSB-first, and the result is similarly produced and delivered LSB-first, whereas the multiplicand is assumed to be available in parallel. We assume that the multiplier and multiplicand have the same number of digits. For simplicity we assume that operands are in standard (unsigned) binary, since generalizations to higher radices and other digit sets are fairly straightforward. The handling of 2's complement will also be considered.

Given the multiplicand in parallel, the classical iterative multiplier of Figure 4.5.1 consumes the multiplier by one digit each cycle and produces one result digit per cycle during the first n cycles, shifting the digits of the result into the Q-register, and leaving the high-order part in the A-register. We thus here have the basic unit for a simple serial/parallel multiplier; what remains is just to deliver the high-order digits of the result in serial form also. In this situation we do not need a standard carry-completion adder, so let us consider a simplified version of Figure 4.5.1 where we now substitute the carry-completion adder with a carry-save adder, and remove the Q-register so that the result is delivered directly in serial form, LSD-first. Figure 4.5.5 presents the overall design of such a multiplier, where each cell forms the bit product using an AND gate and accumulates the result in the (c_i, s_i) pair of buffers by means of a full-adder acting as a serial adder.

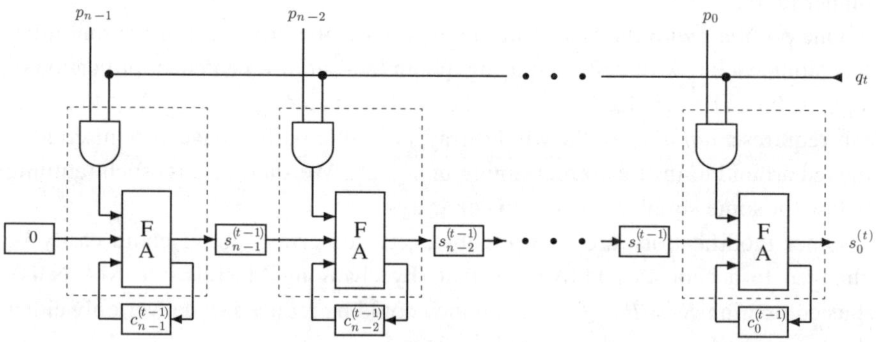

Figure 4.5.5. A simple serial/parallel multiplier

Assuming that all buffers have been initially cleared, at time step t, $t = 0, 1, \ldots, n-1$, the multiplier bit q_t is broadcast to all cells. All cells then compute digit products $p_i q_t$ in parallel and accumulate these into a previously computed, but right-shifted, sum of partial products represented in carry-save form (i.e., digit set $\{0, 1, 2\}$). The interconnection of cells realizes the right-shift corresponding to the increasing weight of the multiplier bit q_t, and the result bit r_t is delivered at the right-hand end of the array. This accounts for the first n bits of the result.

During the following n time steps, $t = n, n+1, \ldots, 2n-1, q_t = 0$ is broadcast to all cells, and result bits $r_n, r_{n+1}, \ldots, r_{2n-1}$ are delivered. This is just a ripple-carry addition of the carry-save represented result residing in the (c_i, s_{i+1}) buffers, except that here it is the sum bits that "ripple" to the right while the carries are put back in the same position.

It is now also obvious how to handle a 2's complement multiplier: just sign-extend it, i.e., for $t = n, n+1, \ldots, 2n-1$ let $q_t = q_{n-1}$. Then 2's complement addition can be handled by sign-extension of the accumulation, which is simply realized by feedback of the output from the leftmost adder into the input buffer of the leftmost cell. This also automatically takes care of a 2's complement multiplicand.

The multiplier of Figure 4.5.5 relies on a broadcast of the signal q_t, and is thus not a *systolic* design using only nearest-neighbor communication, which is advantageous for very high speed – and in particular for very large word-length, e.g., as needed for cryptographic applications.

We can obtain a systolic design if we can insert buffers along the wire by which q_t is distributed. That, however, will change the timing of the arrival of the values of q_t at the different cells, thus a *retiming* of the whole array is needed.

Consider the following simplified version of Figure 4.5.5 (with buffers for p_i, c_i included in cell i):

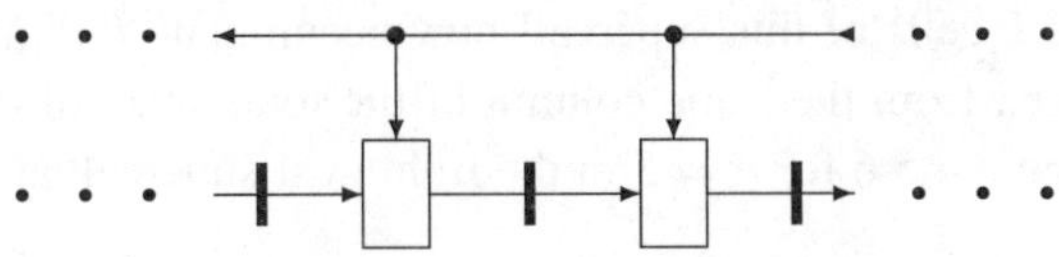

where the vertical bar on a line is a delay (a buffer). Then we can add a delay on a left-going signal between two cells (implicitly assuming that the broadcast is going from right to left), if we simultaneously remove a delay on the right-going signal between the same two cells:

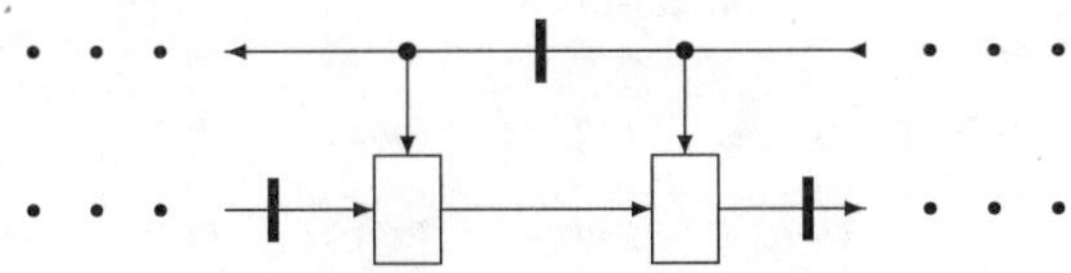

This obviously does not change the functionality of the array as observed at the right-hand end. Hence we may at suitable places move delays (buffers) from right-going signals to left-going signals; however, the obtained array is still not quite systolic in the sense that there should be delays on all lines between cells. Although not needed, it is possible to move the delay on the left-going line to the left if simultaneously a delay is added on the line going down to the cell:

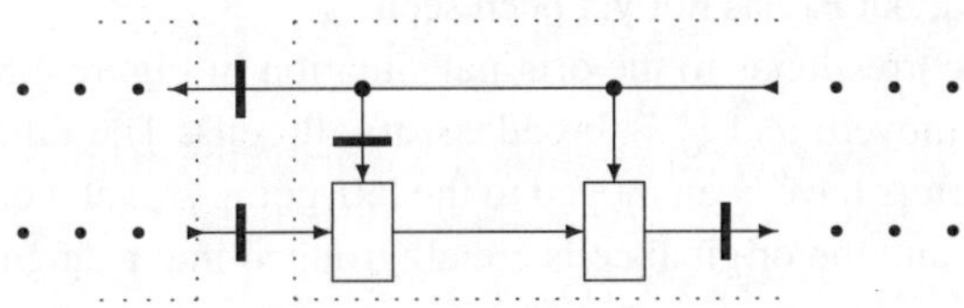

As indicated by the dotted lines we have now created an interface between two "supercells" with delays between these on all signals, i.e., a proper interface between systolic cells, but at a higher cost in the number of buffers.

Returning to the simpler approach, we may now create a variety of serial/parallel multipliers by moving delays at different places from the right-going signals to the equivalent places on the left-going signals. If the $k-1$ buffers between k consecutive cells are moved, a supercell of these is constructed with no intermediate delays between the cells on the right-going signal:

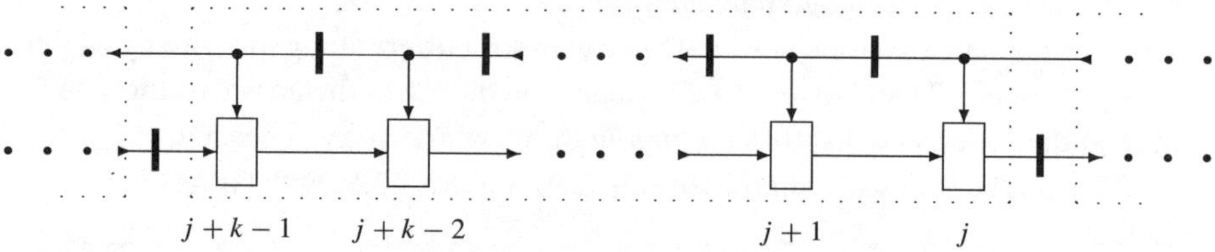

Observe that the k cells of this supercell now accumulate k-digit products of the same weight, i.e., from the same column of the total array of digit products as illustrated in Figure 4.5.6 for $n = 5$ in the rightmost supercell at $t = 3$ for the cases:

$k = 1$ the term p_0q_3 from the horizontal dotted frame;
$k = 2$ the term $p_0q_3 + p_1q_2$ from the slanted dotted frame;
$k = 5$ the term $p_0q_3 + p_1q_2 + p_2q_1 + p_3q_0$ from the vertical dotted frame.

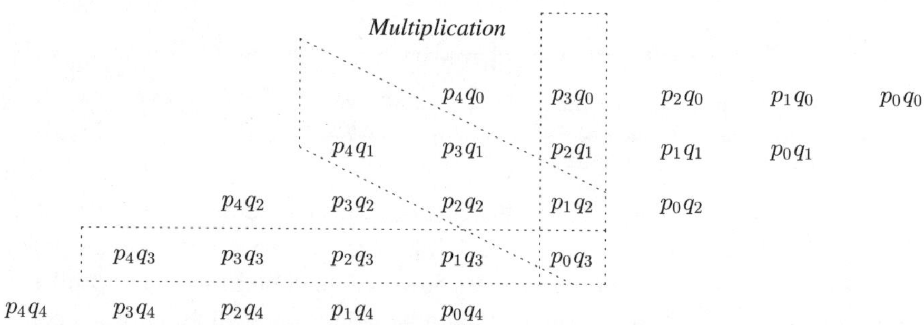

Figure 4.5.6. Array of partial product terms.

Note that the figure illustrates the situation at $t = 3$, when q_0, q_1, q_2, and q_3 have been delivered, but q_4 has not yet been seen.

The case $k = 1$ corresponds to the original situation in Figure 4.5.5, where no buffers have been moved, and q_t is broadcast to all cells. The other extreme is $k = n$ where all buffers have been moved to the left-going signal, i.e., q_t is shifted into a shift register and the original cells are all grouped into a single supercell:

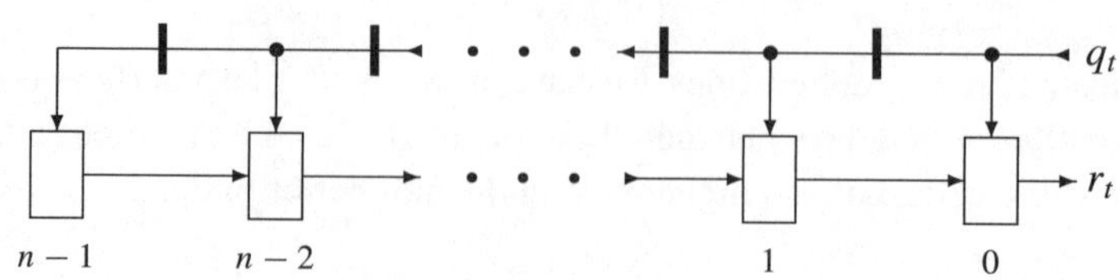

The n cells now form a linear time circuit adding the digit products of a single column, i.e., it is a counter performing a count of the number of ones in a column, where the carries are retained in "unary" form in the buffers of the cells. This is obviously not optimal, neither in the delay of the circuit (linear in n), nor in the amount of storage of the carries (also linear in n).

The same functionality can bc obtained by a tree structure of adders with only $O(\log n)$ space for carry buffering as described in Figure 4.5.7. Note that the number of full-adders needed in the tree is still $O(n)$, but the critical path is reduced to $O(\log n)$. We leave it as an exercise to design such a network using full-adders. The timing of such a multiplier is then $O(n \log n)$.

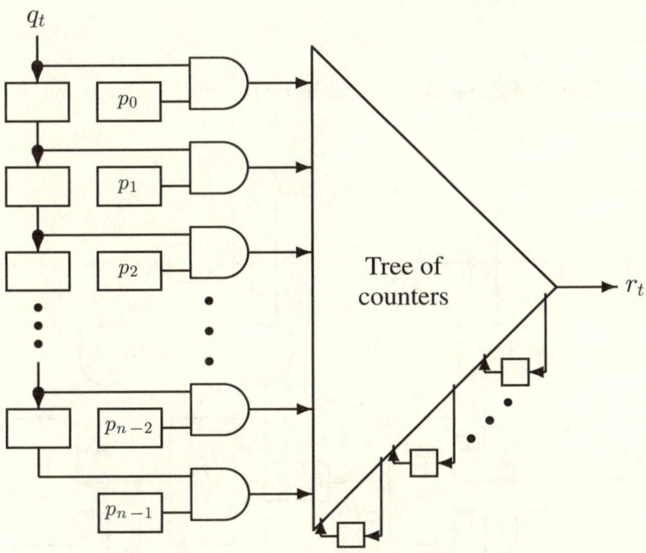

Figure 4.5.7. Column-counter-based serial/parallel multiplier.

Since any multiplier with digit-serial output has to perform $2n$ cycles to produce all $2n$ digits of the result, optimization for speed will have consider the critical path in each cycle. The simple multiplier of Figure 4.5.5 only has an AND gate and a full-adder in its critical path, but requires a broadcast of each multiplier bit q_t to all cells.

Choosing $k = 2$, two product terms from each column have to be added in each cycle, but now there is no broadcast since there is a buffer for every second cell. Figure 4.5.8 shows the product terms formed and accumulated at each step.

Expanding the logic of the cells, Figure 4.5.9 shows the structure of such a multiplier in terms of two of the supercells obtained by the grouping for $k = 2$. Although the buffers are not positioned between the supercells, this still has the functionality of a systolic array, with nearest-neighbor communication. Each

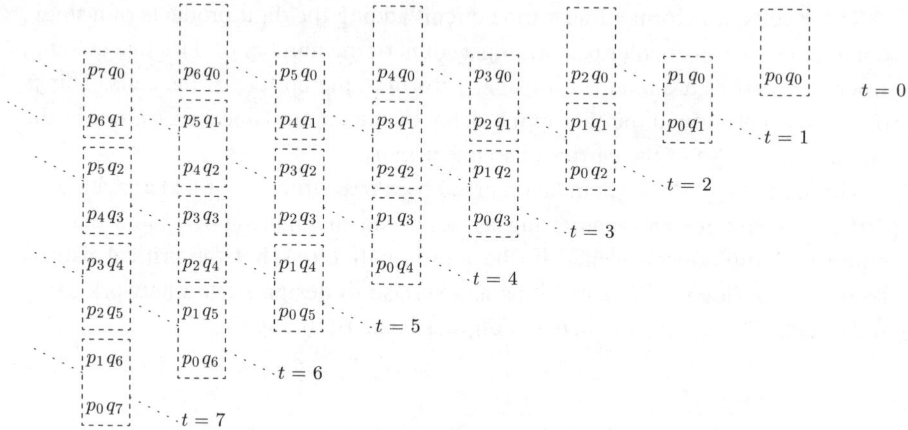

Figure 4.5.8. Grouping of terms by two from each column.

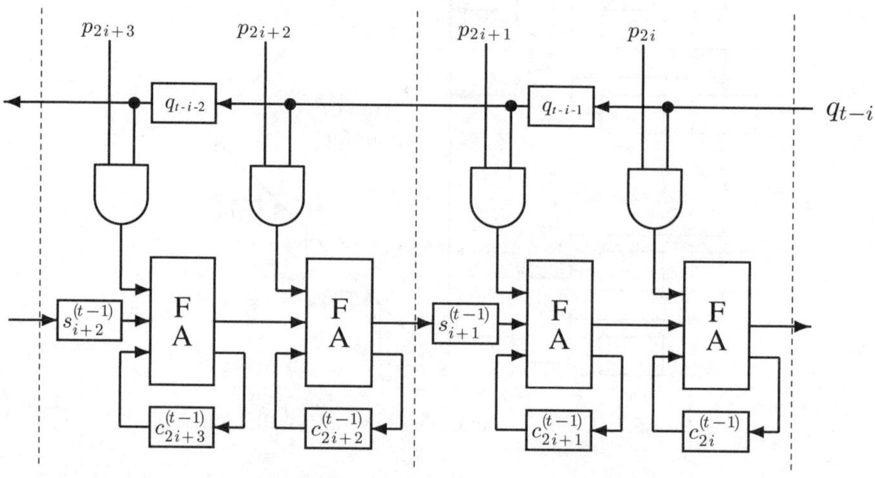

Figure 4.5.9. Systolic multiplier cells.

supercell contains a digit of the product computed so far, represented in the digit set {0, 1, 2, 3}. Two digit products are added to the previous contents (digit) by a 5-to-3 adder.

Each supercell now has twice as long a critical path as the cells of Figure 4.5.5, but only $\lfloor n/2 \rfloor$ cells are needed and the broadcast has been eliminated. Again here q_t, $t = 0, 1, \ldots, n-1$, is delivered during the first n cycles, and $q_t = 0$ (or $q_t = q_{n-1}$ for sign-extension) is delivered during the next n cycles $t = n, n+1, \ldots, 2n-1$. For a 2's complement multiplicand the leftmost cell can be modified to realize the sign-extension of the accumulated partial sums by a feed-back as shown for $n = 2m + 2$ in Figure 4.5.10.

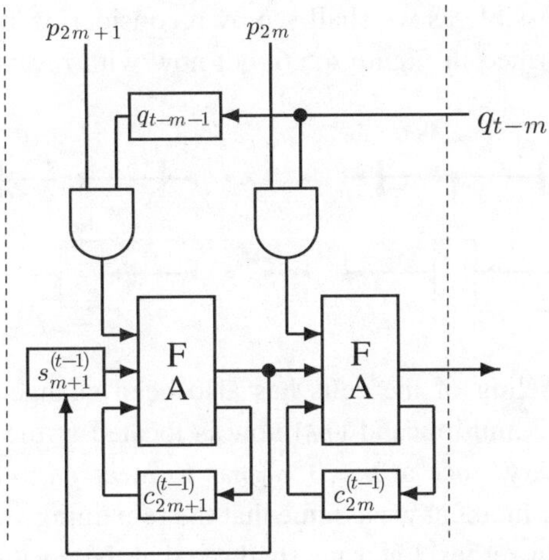

Figure 4.5.10. Sign extension for 2's complement multiplication.

The multiplier design of Figure 4.5.9 is feasible for very large multipliers, e.g., as needed in cryptographic applications. There does not seem to be any point in using values of k beyond 2, the critical path just increases and the number of cycles is still $2n$, since $2n$ bits of result have to be delivered. But it is fairly easy to increase the radix to 4 or 8, and thus reduce the number of cycles needed, while still keeping a fairly short critical path within each (super)cell.

4.5.4 A pipelined serial/parallel multiplier

So far we have implicitly assumed that the broadcast of q_t had a direction (see page 233) opposite to the direction of the results computed in the adders. This allowed us to move buffers at suitable places, and thereby design a range of different multipliers, all producing the full $2n$-bit product. But observe that the adders are not busy all the time, furthermore for many applications (e.g., signal processing) only the most-significant n bits of the result (possibly rounded) are wanted for further processing. This corresponds to interpreting the operands as scaled such that they have the radix point at the leftmost end, e.g., operands are in a range $[-1; 1)$, or $[0; 1)$ for non-negative operands.

If the flow of the multiplier bits q_t is changed to proceed in the same direction as the results of the adders (accumulation of partial products), then it looks like it should be possible to *pipeline* several multiplications since cells become free when the last bit of Q passes down the array, and thus can be utilized for another multiplication. That is, after n cycles delivering one operand, then during the next n cycles while a new operand is delivered, at the other end of the array the n most-significant bits of the product should appear bit sequentially.

This is indeed possible as we shall see by reconsidering first the columnwise accumulation performed in Figure 4.5.6, but now with reversed direction of the flow for q_t:

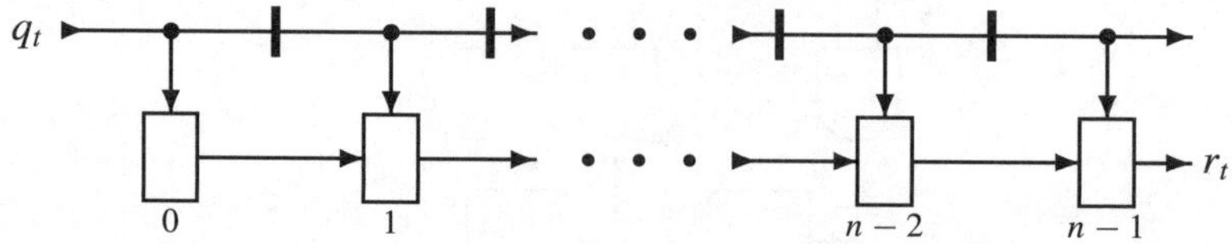

Note that the labeling of the cells has also been changed so that the least-significant bit of the multiplicand (p_0) now is located at the leftmost end. This is obviously necessary since at $t = 0$, q_0 has to meet p_0 to form the first digit product p_0q_0. At the moment we assume that the remaining values along the shift register for q contain zeroes. Later we shall see that the positions to the right can be participating in some other multiplication, but for now the adders in the array are chained together in one long addition, delivering the result at the right-hand end. The following table shows the computations for the first few cycles:

Time	Cell_0	Cell_1	Cell_2	$\cdots$	Output
0	q_0p_0				r_0
1	q_1p_0	q_0p_1			r_1
2	q_2p_0	q_1p_1	q_0p_2		r_2
$\vdots$	$\vdots$	$\vdots$	$\vdots$	$\ddots$	$\vdots$

where each row shows the digit products from a column of Figure 4.5.6 to be accumulated, together with carries retained from the previous cycles.

Now we want to "kill" the n least-significant bits of the result before these reach the right-hand end, so for this purpose we introduce a control signal (a token) which travels along the array together with q_t:

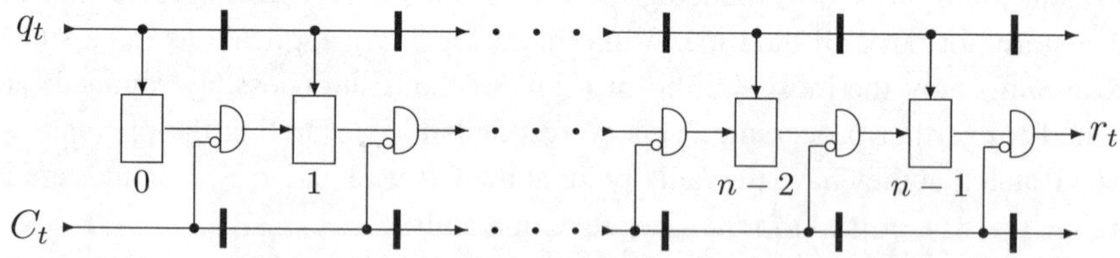

The token C_t, where $C_t = 1$ for $t \bmod n + 1 = 0$ and $C_t = 0$ otherwise, will act as the "killer" and as a separator between distinct multiplications going through the array if we properly dimension the length of the array. To remain consistent with the previous discussions we will assume that the multiplier $Q = \sum_{i=0}^{n} q_i[2]^i$ and the multiplicand $P = \sum_{i=0}^{n-1} p_i[2]^i$, but we want the result $R \text{ div } [2]^n = \sum_{i=n}^{2n} r_i[2]^{i-n}$. Note that we assume Q has an extra digit $q_n = 0$ so that $r_{2n} = 0$, we shall see below how to treat 2's complement numbers in which position n is used for the sign extension.

The bits $p_0, p_1, \ldots, p_{n-1}$ are available in parallel in the respectively labeled cells, and during the first n cycles $q_0, q_1, \ldots, q_{n-1}$ are delivered sequentially. At time step $t = n$, $q_n = 0$ is delivered, and at the other end q_0 appears together with result bit r_n. Also a token value $C_{n+1} = 1$ is delivered at $t = n + 1$ so that whatever is delivered for q_t, $t \geq n + 1$, does not interfere with the computations of r_t during the next n cycles delivering $r_{n+1}, \ldots, r_{2n}$, where finally $r_{2n} = 0$ due to the token arriving at the rightmost AND gate. Notice that the total multiplication takes $2n$ cycles, but any component is only busy with this computation during $n + 1$ of these cycles. The following table shows the computations taking place for $n = 4$:

Time	Input	Cell$_0$	Cell$_1$	Cell$_2$	Cell$_3$	Output	
0	q_0	p_0q_0					
1	q_1	p_0q_1	p_1q_0				
2	q_2	p_0q_2	p_1q_1	p_2q_0			
3	q_3	p_0q_3	p_1q_2	p_2q_1	p_3q_0		
4	0	0	p_1q_3	p_2q_2	p_3q_1	q_0	r_4
5			0	p_2q_3	p_3q_2	q_1	r_5
6				0	p_3q_3	q_2	r_6
7					0	q_3	r_7
8						0	0

It is now easy to see that it is possible to pipeline several multiplications, one after the other, while the individual multiplications are themselves pipelined at the bit level. Note that the cells are then kept very busy.

The next problem to attack is to eliminate the $O(n)$ critical path through all the adders, which is realized simply by inserting extra buffers on all lines between each cell. This will increase the latency of the result, but in addition the rate at which results appear since the critical path is reduced:

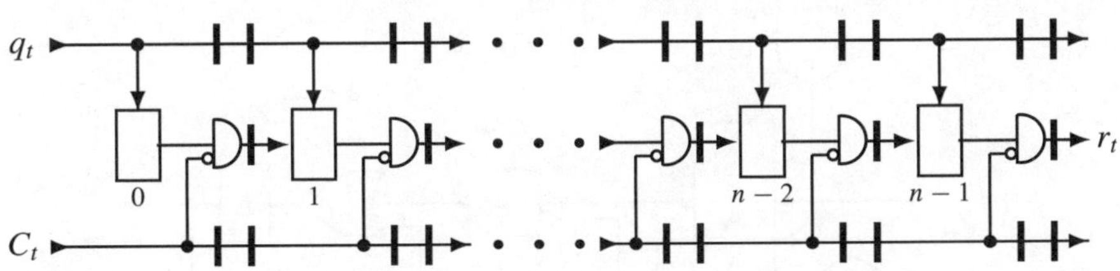

The multiplier presented so far just truncates the result by discarding the low order bits. A rounding can be achieved by inserting a one at the vacant input to the leftmost cell at $t = n - 1$, thus effectively adding a unit to the last discarded bit.

This is the classical Lyon multiplier in its serial/parallel form, but Lyon also extended it for serial/serial processing as we shall see in the next subsection. Furthermore, he also adapted it for 2's complement by noting that instead of forcing a zero as the "killer" and separator, it is possible to realize a sign-extension by substituting the circuit

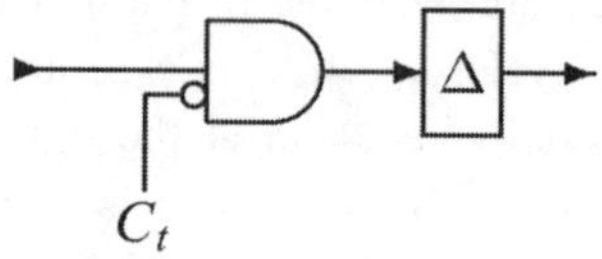

by the following circuit to feed back the (moving) most-significant bit position:

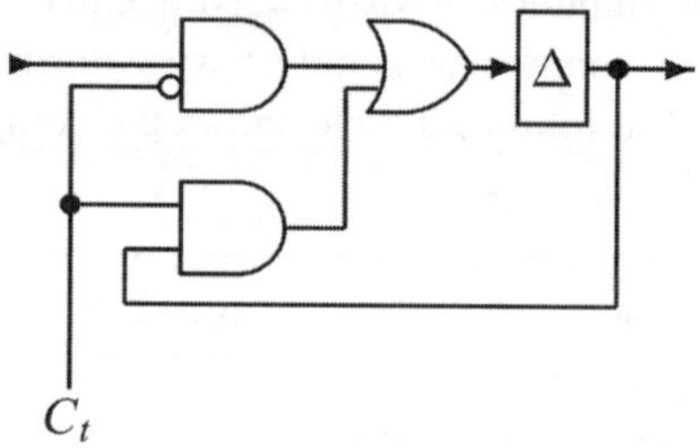

This will sign-extend the accumulated result for one cycle (the separating cycle), but possibly also generate a carry deposited in the cell to the right. This carry has to be cleared before that cell starts to participate in a new multiplication.

Thus we can finally draw the general cell of the Lyon multiplier as shown in Figure 4.5.11.

With these modifications a 2's complement multiplier is handled by accumulation in 2's complement, provided that $q_n = q_{n-1}$ is delivered as a sign-extension

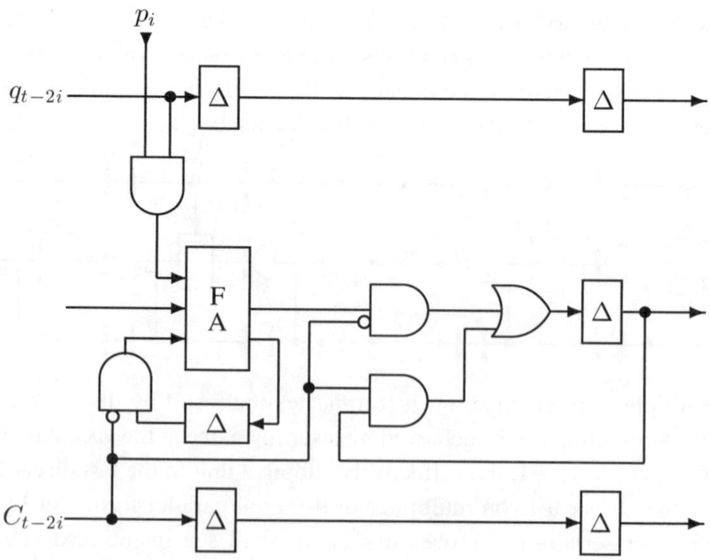

Figure 4.5.11. Cell of serial/parallel Lyon multiplier.

of the multiplier Q. To handle a 2's complement multiplicand P, recall that p_{n-1} resides in the rightmost cell of the array and now $p_{n-1} = 1$ (of negative weight) has to insure an effective subtraction of the digit products formed in that cell, hence a serial subtractor is then needed here. Thus the full-adder (of type ppp) must be substituted by a pnp-type adder, or equivalently inverters must be inserted before and after that cell.

4.5.5 Least-significant bit first (LSB-first) serial/serial multipliers

While a parallel/serial multiplier may be adequate in many situations where the multiplicand is available in parallel, and is possibly a constant for a number of multiplications, there are situations in where there are composite computations where both factors are the result of previous operations delivering their results digit serially.

Thus a serial multiplier may be needed in which both operands are delivered serially. The multiplier of Figure 4.5.9 could be modified such that the multiplicand is also delivered serially, LSB-first, since a little consideration will show that bits p_{2i} and p_{2i-1} are only needed when the first non-zero q_{t-i} arrives. Thus p_{2i} and p_{2i-1} need only be deposited at time step $t = i$, which means that the multiplicand P can also be delivered bit serially; however, its bits will be needed at twice the speed that bits are needed from the multiplier Q.

In 1962 Atrubin devised a cellular automaton, which when it received two bit streams could produce another bit stream representing the LSB-first, bit serial product of the two input streams. For many years this multiplier remained something of a mystery since no explanation or proof was given, just an example was used to show its functionality. Here we shall derive it from considerations similar to those of the previous section, by initially using a broadcast and then by retiming obtaining an equivalent systolic design which is the Atrubin multiplier.

Again let us consider an array of digit products to be formed and accumulated, as shown in Figure 4.5.12 where the terms becoming available at time step $t = 3$ have been framed.

Now assume that at time t columns 0 through $t - 1$ have been accumulated and bits r_0 through r_{t-1} have been emitted. Previously $p_0, p_1, \ldots, p_{t-1}$ and

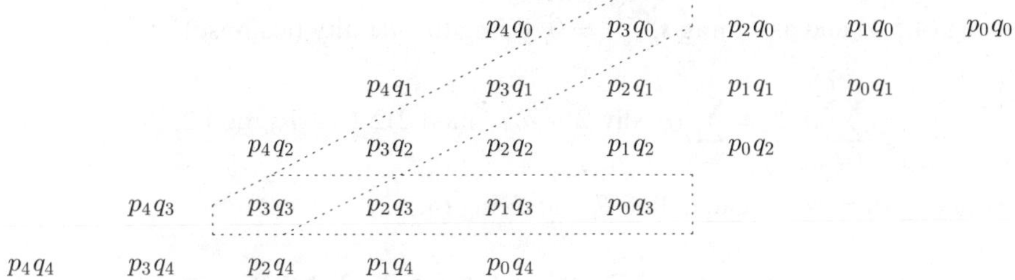

Figure 4.5.12. Digit product terms becoming available at $t = 3$.

$q_0, q_1, \ldots, q_{t-1}$ have been received and recorded, and now p_t, q_t have become available, so that the new terms shown in boldface in following diagram can be formed:

$$
\begin{array}{cccccccc}
 & q_{t-1} & q_{t-2} & \bullet\bullet\bullet & q_2 & q_1 & q_0 & \\
 & \mathbf{p_t q_{t-1}} & \mathbf{p_t q_{t-2}} & \bullet\bullet\bullet & \mathbf{p_t q_2} & \mathbf{p_t q_1} & \mathbf{p_t q_0} & \longleftarrow \mathbf{p_t} \\
\mathbf{p_t q_t} & \boxed{0} & s_{t-2}^{(t-1)} & \bullet\bullet\bullet & s_2^{(t-1)} & s_1^{(t-1)} & s_0^{(t-1)} & \longrightarrow r_t \\
 & \mathbf{p_{t-1} q_t} & \mathbf{p_{t-2} q_t} & \bullet\bullet\bullet & \mathbf{p_2 q_t} & \mathbf{p_1 q_t} & \mathbf{p_0 q_t} & \longleftarrow \mathbf{q_t} \\
 & p_{t-1} & p_{t-2} & \bullet\bullet\bullet & p_2 & p_1 & p_0 &
\end{array}
$$

and these terms can be accumulated into the previous sum of partial products represented in $s_t^{(t-1)}, \ldots, s_{2t-2}^{(t-1)}$, to produce a new digit r_t of the product. To establish recursion formulas for $s_i^{(t)}$ in terms of the digit products $p_t q_i$, $q_t p_i$ and $s_i^{(t-1)}$, we define

$$Q_{t-1} = \sum_{i=0}^{t-1} q_i [2]^i \quad \text{and} \quad P_{t-1} = \sum_{i=0}^{t-1} p_i [2]^i$$

to be the polynomials representing the parts of the factors that have been seen so far for $t = 1, 2, \ldots, n-1$, and we define S_{t-1} as a polynomial of value $\|S_{t-1}\| = \|P_{t-1} Q_{t-1}\|$:

$$S_{t-1} = \sum_{i=0}^{t-2} s_i^{(t-1)} [2]^{t+i} + \sum_{i=0}^{t-1} r_i [2]^i, \tag{4.5.1}$$

where the last sum represents the digits emitted during the previous steps ($r_i \in \{0, 1\}$). As we shall see below, the digits $s_i^{(t-1)}$ belong to the set $\{0, 1, 2, 3\}$ which is redundant for radix 2.

At time step t, S_t is a polynomial of value $\|P_t Q_t\|$:

$$
\begin{aligned}
\|S_t\| &= (p_t 2^t + \|P_{t-1}\|)(q_t 2^t + \|Q_{t-1}\|) \\
&= p_t q_t 2^{2t} + \sum_{i=0}^{t-1} (p_t q_i + p_i q_t + s_i^{(t-1)}) 2^{t+i} + \sum_{i=0}^{t-1} r_i 2^i
\end{aligned} \tag{4.5.2}
$$

using (4.5.1) and assuming $s_{t-1}^{(t-1)} = 0$. Using the identity (exercise)

$$\sum_{i=0}^{t-1} a_i 2^i = \sum_{i=0}^{t-1} (a_i \text{ div } 2 + a_{i+1} \bmod 2) 2^{i+1} + a_0 \bmod 2, \tag{4.5.3}$$

assuming $a_t = 0$, we can define S_t satisfying (4.5.2)

$$S_t = \sum_{i=0}^{t-1} s_i^{(t)} [2]^{t+i+1} + \sum_{i=0}^{t} r_i [2]^i$$

by defining its digits as

$$s_i^{(t)} = (p_t q_i + p_i q_t + s_i^{(t-1)}) \text{ div } 2 + (p_t q_{i+1} + p_{i+1} q_t + s_{i+1}^{(t-1)}) \text{ mod } 2 \quad (4.5.4)$$

for $i = 0, 1, \ldots, t-2$, and

$$s_{t-1}^{(t)} = (p_t q_{t-1} + p_{t-1} q_t) \text{ div } 2 + p_t q_t \quad (4.5.5)$$

with the new digit r_t being

$$r_t = (p_t q_0 + p_0 q_t + s_0^{(t-1)}) \text{ mod } 2. \quad (4.5.6)$$

Now assuming that $s_i^{(t-1)} \in \{0, 1, 2, 3\}$, it is easily seen by induction in t from the updating (4.5.4) of $s_i^{(t)}$ that $s_i^{(t)}$ belongs to the same set for all $t \geq 0$, since $s_i^{(-1)} = 0$.

For a hardware realization, $s_i^{(t)}$ can be represented in cell number i in three buffers, one buffer containing the right-hand term of (4.5.4), $(p_t q_{i+1} + p_{i+1} q_t + s_{i+1}^{(t-1)}) \text{ mod } 2$, received from cell number $i + 1$ to the left. Since $s_i^{(t-1)}$ is represented in three bits, the left-hand term $(p_t q_i + p_i q_t + s_i^{(t-1)}) \text{ div } 2$, represented in two bits, corresponds to the two carries of a 5-to-3 adder structure, whose unit bit is sent forward to the neighbor to the right, as shown in Figure 4.5.13.

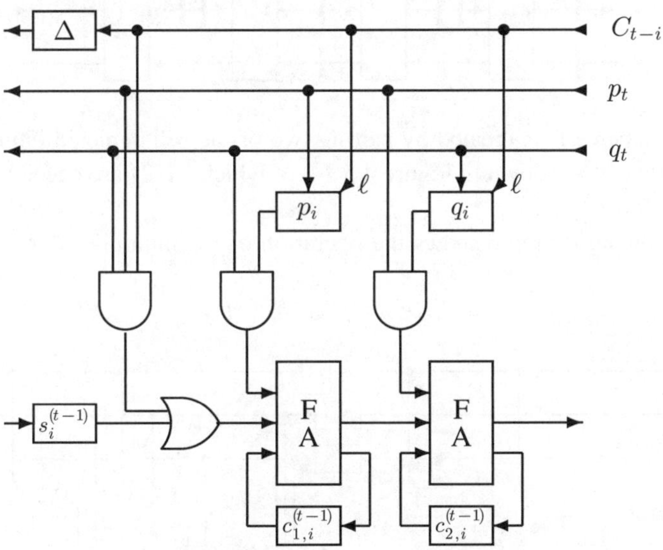

Figure 4.5.13. Cell of a simple serial/serial multiplier using broadcast.

The rightmost cell of an array of such cells then emits r_t at time step t, corresponding to (4.5.6). The array also needs an init signal (a token, $C_0 = 1$ and $C_t = 0$ for $t > 0$) so that p_t and q_t can be deposited in cell number t during step t. It is assumed that the load only takes effect such that p_t and q_t are available during the next cycle. This init signal then also handles the special term $p_t q_t$ of

(4.5.5) to be formed as input to cell number t. Initially all buffers are assumed to be cleared. Observe that n of these cells are needed, as seen from Figure 4.5.12, each containing two full adders, i.e., twice as many as in the serial/parallel adders.

However, an array of the cells depicted in Figure 4.5.13 is still based on a broadcast of the p_t, q_t bits during cycle t, as shown in the following simplified figure:

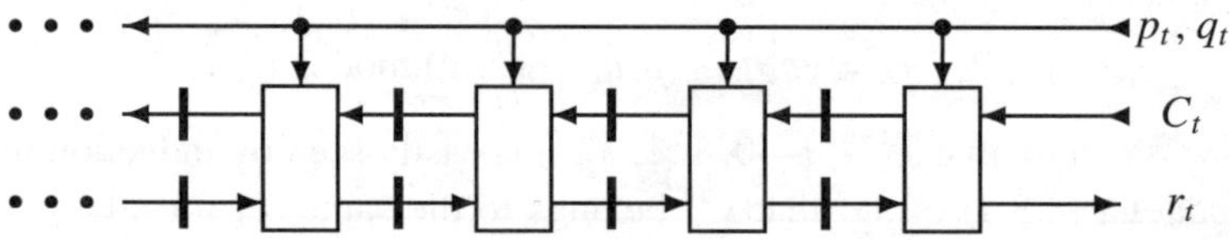

where again we can remove buffers from right-going signals while adding them to left-going signals at suitable places, creating new supercells. Choosing to move buffers for every second cell, we minimize the worst-case critical path, while eliminating the broadcast, as in the following figure:

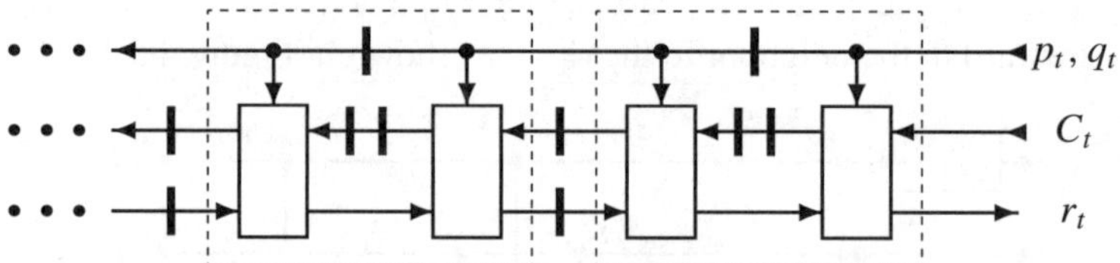

The new "supercell" is created by pairing two of the cell depicted Figure 4.5.13 into a single cell as shown in Figure 4.5.14, of which $\lfloor n/2 \rfloor$ are needed to form a $n \times n$ multiplier.

The following example shows the operation of the multiplier during the first four steps.

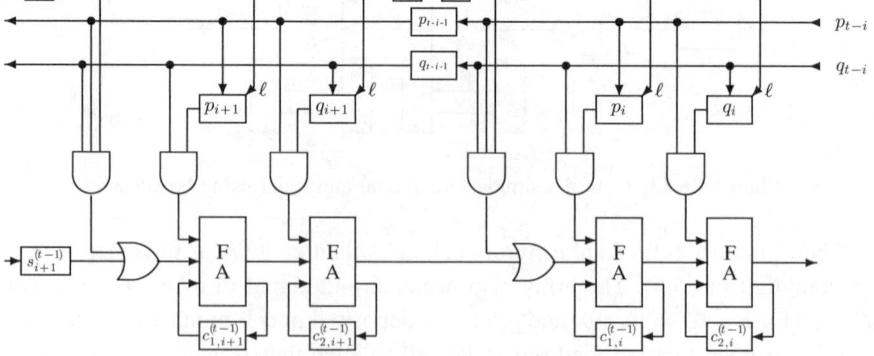

Figure 4.5.14. Cell of an Atrubin multiplier.

Example 4.5.1 To illustrate the propagation of signals in an Atrubin multiplier, below are the first four steps with values shown on each line, except for the control signal where only the value one (the token) is shown, the rest are assumed to be zero.

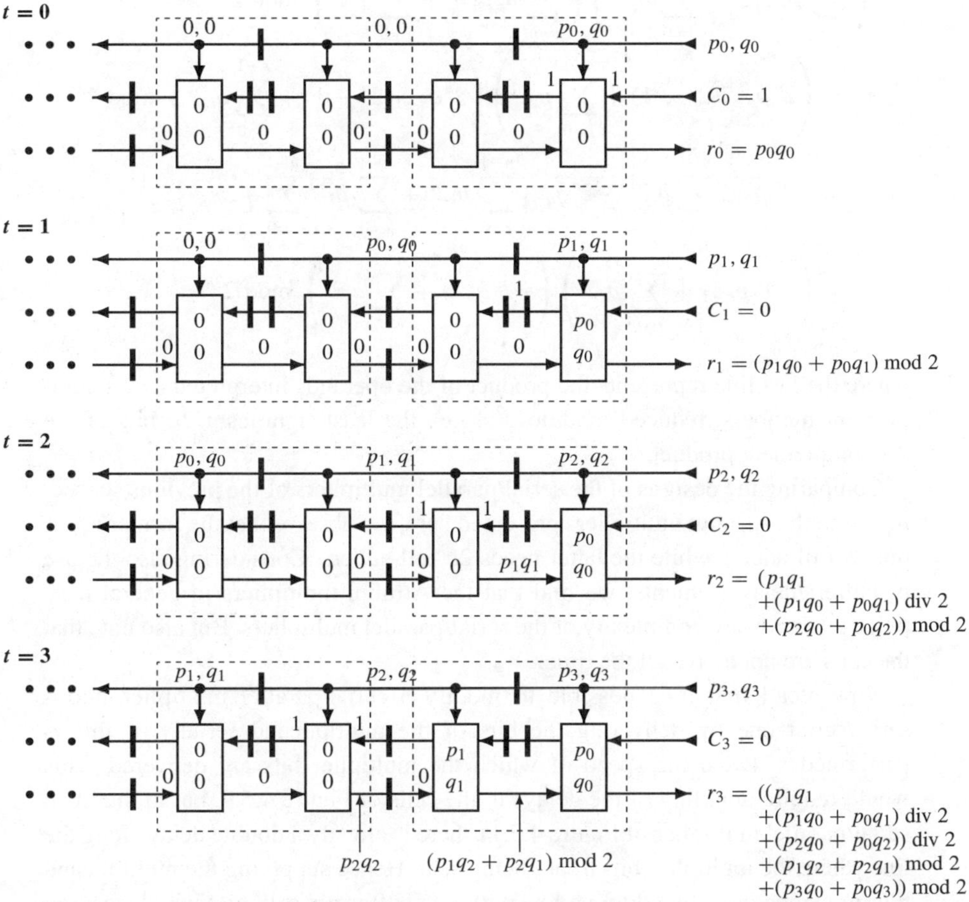

For space reasons, only the values of p_i, q_i are shown loaded into cell i, the carries are not shown. With input $P = 0011_{[2]}$ and $Q = 0010_{[2]}$ (the two leading zeroes being sign fill to empty the array) the output is $R = 0110_{[2]}$. Note that only the rightmost supercell actually contributes to the result, corresponding to $n = 2$ implying only one cell is needed. With $P = 1111_{[2c]}$ and $Q = 1110_{[2c]}$ the output is $R = 0010_{[2c]}$, indicating that 2's complement operands seem to be properly handled when operands are sign-extended. □

Note that the Atrubin multiplier also needs zero-fill for n cycles after the initial n cycles where the operands are delivered. As indicated in the example, in addition, sign-extensions can be delivered in the case of 2's complement operands to produce a 2's complement result. This is easily seen by considering the $2n$ bits produced as the product of the sign-extended operands:

$$\left(\sum_{i=n}^{2n-1} p_{n-1}2^i + \sum_{i=0}^{n-1} p_i 2^i\right)\left(\sum_{i=n}^{2n-1} q_{n-1}2^i + \sum_{i=0}^{n-1} q_i 2^i\right) \bmod 2^{2n}$$

$$= \left(2^n p_{n-1}(2^n - 1) + \sum_{i=0}^{n-1} p_i 2^i\right)\left(2^n q_{n-1}(2^n - 1) + \sum_{i=0}^{n-1} q_i 2^i\right) \bmod 2^{2n}$$

$$= -2^n p_{n-1} \sum_{i=0}^{n-1} q_i 2^i - 2^n q_{n-1} \sum_{i=0}^{n-1} p_i 2^i + \sum_{i=0}^{n-1} p_i 2^i \sum_{i=0}^{n-1} q_i 2^i$$

$$= \left(-2^n p_{n-1} + \sum_{i=0}^{n-1} p_i 2^i\right)\left(-2^n q_{n-1} + \sum_{i=0}^{n-1} q_i 2^i\right) \bmod 2^{2n}$$

where the last line represents the product of the operands interpreted as 2's complement numbers, reduced modulo 2^{2n}, i.e., the least-significant $2n$ bits of the 2's complement product.

Comparing the designs of the serial/parallel multipliers of the previous subsection with the Atrubin multipliers presented here, we observe that the former needs only n full-adders while the latter needs $2n$ full-adders. Considering also the use of buffers/delay elements, we find that the Atrubin multipliers in general have twice the hardware complexity of the serial/parallel multipliers. But also note that the cells are not active all the time.

Now recall that it is possible to modify a serial/parallel multiplier into a serial/serial one by delivering the bits of the multiplicand serially, if this is performed at twice the speed of which the multiplier bits are delivered. This would require doubling of the delays in the cells of Figure 4.5.9, but in the Lyon serial/parallel multiplier of Figure 4.5.11 there is already a double delay along the line where the multiplier bits q_t are being sent. Hence supplying the multiplicand bits p_t can be done by adding a line with one buffer per cell, and just depositing p_t when the token arrives in the cell. This is shown in Figure 4.5.15, where again the loaded value is only available in the next clock cycle, but the original content is used in the present cycle.

The following table shows the timing of this multiplier for the case $n = 4$, where it should be recalled that the Lyon multiplier requires one extra (sign-extension) bit to be delivered for the operands and delivers a sign-extended result (but only the $n + 1$ most-significant bits). Hence the Lyon multiplier allows proper pipelining of multiple sets of operands in this extended format. The value of p_i is deposited in cell number i at time $t = 2i$ (shown framed in the table) when the token is in the front of the cell, otherwise the table displays the formation of the digit products.

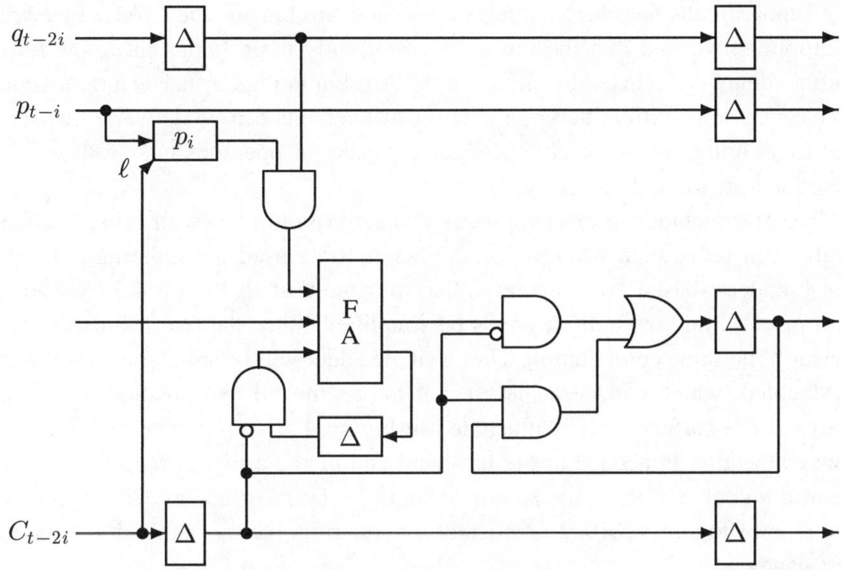

Figure 4.5.15. Cell of serial/serial Lyon multiplier.

The diagram next to the table depicts digit product formation and shows that these are generated along backwards slanted lines.

Time	Cell_0	Cell_1	Cell_2	Cell_3	Result
0	p_0				
1	p_0q_0				
2	p_0q_1	p_1			
3	p_0q_2	p_1q_0			
4	p_0q_3	p_1q_1	p_2		
5	p_0q_3	p_1q_2	p_2q_0		
6		p_1q_3	p_2q_1	p_3	
7		p_1q_3	p_2q_2	p_3q_0	
8			p_2q_3	p_3q_1	r_4
9			p_2q_3	p_3q_2	r_5
10				p_3q_3	r_6
11				p_3q_3	r_7
12					r_8

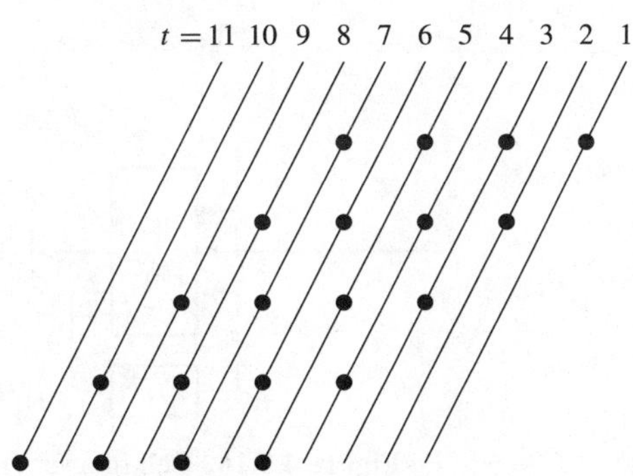

Observe that a one-bit sign-extension is delivered for q_n as well as for p_n, the first of which is visible in the table ($q_4 = q_3$). The extra multiplicand bit p_n is also delivered, but is not deposited since there is no cell number n in which to load it (the token would only catch up with it outside the array). Also recall that cell number $n - 1$ has to perform a subtraction, and that it is possible to effect a proper rounding by supplying a unit bit at the vacant input to the leftmost cell at time $t = n - 1$.

Comparing the hardware complexities of the Atrubin and the Lyon serial/serial multipliers, we find that the former needs slightly more logic, but a few fewer buffers/delay elements. Also note that the Atrubin multiplier needs a reset signal to clear various buffers before a new set of operands can be delivered, whereas the Lyon multiplier is capable of receiving pairs of operands back-to-back in a pipeline flow through the array.

We will conclude this discussion on serial/serial multipliers with a simplification which can be applied when the multiplier is to be used for squaring only. For the broadcast-based Atrubin multiplier we note that in Figure 4.5.13 $p_t = q_t$ and $p_i = q_i$, and the cell may thus be simplified, since the two full-adders will perform the same computation. Thus a single adder will be sufficient if its weight is doubled, which can be achieved if it is just moved one position to the left (beyond the buffer). This is illustrated in Figure 4.5.16, just note that a special interconnection is needed at the left-hand end of the array, to feed the delayed control signal ANDed with $p_t = p_t p_t$ into the leftmost adder. Thus n cells of the type of Figure 4.5.16 are sufficient for squaring, with a small addition at the left-hand end.

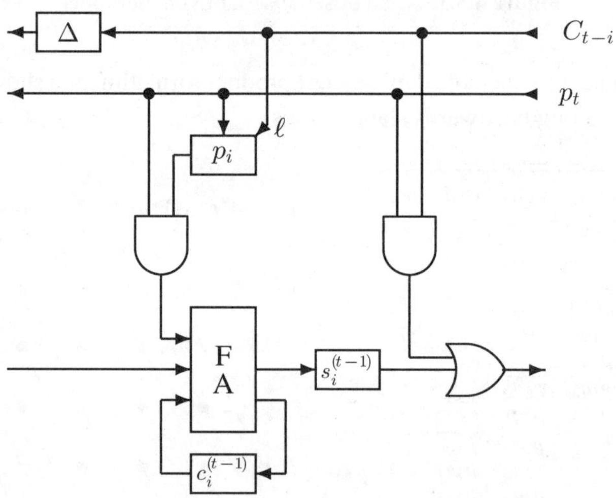

Figure 4.5.16. Cell of a serial squarer using broadcast.

The modifications necessary to eliminate the broadcast should be evident by now, just form a supercell from two of these cells by removing every second buffer from the right-going line, and adding one to each of the two left-going lines. We leave it as an exercise to develop the details of such a design.

4.5.6 On-line or most-significant bit first (MSB-first) multipliers

It turns out to be convenient here to consider polynomials representing proper fractions, i.e., operands $-1 \le P < 1$, and $-1 \le Q < 1$, are given in the form

$$P = \sum_{i=1}^{n} p_i \beta^{-i} \quad \text{and} \quad Q = \sum_{i=1}^{n} q_i \beta^{-i},$$

where we assume that the digit sets are symmetric and of the form

$$D_a = \{-a \,..\, a\} \quad \text{with} \quad \frac{\beta}{2} \le a \le \beta - 1,$$

so that the digit sets are redundant. Define the prefixes of the operands

$$P_j = \sum_{i=1}^{j} p_i \beta^{-i} \quad \text{and} \quad Q_j = \sum_{i=1}^{j} q_i \beta^{-i},$$

then

$$P_j = P_{j-1} + p_j \beta^{-j} \quad \text{and} \quad Q_j = Q_{j-1} + q_j \beta^{-j},$$

so that

$$P_j Q_j = P_{j-1} Q_{j-1} + (P_j q_j + Q_{j-1} p_j) \beta^{-j} \tag{4.5.7}$$

are the intermediate products, $-1 < P_j Q_j < 1$. After some initial delay, to operate on-line, we want to extract and emit a leading digit from $P_j Q_j$, and from then on continue to do so while updating with further digits of the two operands.

Delaying by δ steps we have from (4.5.7) for $j = \delta$

$$P_\delta Q_\delta = P_{\delta-1} Q_{\delta-1} + (P_\delta q_\delta + Q_{\delta-1} p_\delta) \beta^{-\delta},$$

as the polynomial from which to extract the first digit d_1. If $P_\delta Q_\delta$ is shifted one position to the left, we can extract its first digit by rounding its value to an integer. Hence define a sequence of polynomials $\{W_j\}_{j=0,1,\ldots}$, starting with $W_0 = P_{\delta-1} Q_{\delta-1}$, and $d_0 = 0$, and defining W_j and digits d_j by the recursion

$$\begin{aligned} W_j &= (W_{j-1} - d_{j-1})\beta + (P_{j+\delta} q_{j+\delta} + Q_{j-1+\delta} p_{j+\delta}) \beta^{-\delta}, \\ d_j &= \operatorname{round}(W_j) \;=\; \operatorname{sign}(W_j) \left\lfloor |W_j| + \tfrac{1}{2} \right\rfloor, \end{aligned} \tag{4.5.8}$$

i.e., they are shifted (scaled) products, where the extracted digits have been gradually subtracted. Note that instead of adding the term $P_j q_j + Q_{j-1} p_j$ in a position moving to the right as in (4.5.7), it is here added in a fixed position, while the reduced products are now being left-shifted.

To determine the value of δ, we observe that it has to be chosen in such a way that the digits $d_j \in \{-a \,..\, a\}$, i.e.,

$$|W_j| < a + \tfrac{1}{2}$$

must hold. From (4.5.8) and $|\sum_{i=1}^{k} d_i\beta^{-i}| < a\sum_{i=1}^{k}\beta^{-i} < a/(\beta-1)$ we obtain the bound

$$|W_j| < \frac{1}{2}\cdot\beta + 2a\frac{a}{\beta-1}\beta^{-\delta} = \frac{\beta}{2} + \frac{2a^2}{\beta^{\delta}(\beta-1)} \le a + \frac{1}{2},$$

hence for $\{-a\,..\,a\}$ minimally redundant ($a = \lceil\beta/2\rceil$) and β even, δ must satisfy

$$\frac{\beta^2/2}{\beta^{\delta}(\beta-1)} \le \frac{1}{2},$$

implying that $\delta \ge 2$ is a sufficient delay. For $\beta = 3$ and $a = 2$, then $\delta \ge 2$ is also sufficient.

Observe, however, this requires that the determination of the digits d_j (by rounding) is performed in a non-redundant arithmetic. In practice, the computations of (4.5.8) will take place using some redundant representation and thus a larger delay may be necessary.

Let us therefore look at a possible implementation for $\beta = 2$ with $D_a = \{-1, 0, 1\}$. Note that the digit set is closed under multiplication. A sketch of a hardware solution based on broadcast of p_j and q_j at time step $t = j$ is shown in Figure 4.5.17.

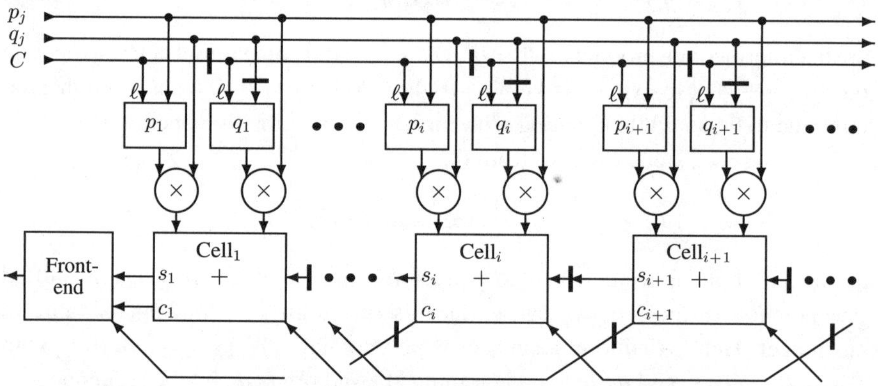

Figure 4.5.17. Broadcast based, radix-2, on-line multiplier.

So far the accumulation and digit extraction in the figure is unspecified, however, the left-shifting corresponds to the requirements of the recursion equation (4.5.8). A token signal is assumed to be sent down the control line C to successively load the buffers for $p_1, q_1, \ldots,$ which are initially assumed to contain zeroes. Note that the loading of the q buffers takes place one cycle after that of the p buffers, corresponding to the index difference of the polynomials in (4.5.7) and (4.5.8). The circles $\otimes$ represent simple digit multipliers $\{-1, 0, 1\} \times \{-1, 0, 1\} \to \{-1, 0, 1\}$.

Let us look at the addition taking place in cell number i at time step $t = j$, where $t > i$. Then the small buffers above have both been loaded with p_i and q_i at some earlier time. Denoting the sum output s_i and the carry c_i, we have the following relation:

$$2c_i + s_i = p_i q_j + p_j q_i + s_{i-1} + c_{i-2}. \tag{4.5.9}$$

If we assume that $c_i \in \{-\alpha \,..\, \alpha\}$ and $s_i \in \{-\sigma \,..\, \sigma\}$, comparing the left- and right-hand the sides of (4.5.9) it is easy to see that $\alpha = 2$ and $\sigma = 1$.

This now allows us to determine the task of the front-end. Its job is to emit digits in the set $\{-1, 0, 1\}$, but what it receives is a value in the digit set

$$\{-1, 0, 1\} + \{-2, -1, 0, 1, 2\} + \{-2, -1, 0, 1, 2\} = \{-5 \,..\, 5\}$$

since it receives s_1, c_1, and c_2. As we saw in Chapter 2, such digit set conversions can be performed on-line by sequential composition of transducers. In this case conversion from the set $\{-2, -1, 0, 1, 2\}$ to $\{-1, 0, 1\}$ is precisely what an on-line adder does, so we can use the adder as a building block. Without going into details we can then realize the front-end as shown in Figure 4.5.18.

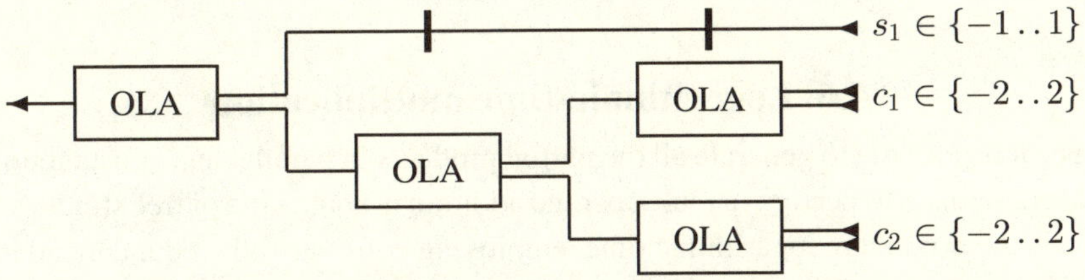

Figure 4.5.18. Front-end as a network of on-line adders.

Note that since an on-line adder has a delay of 1, the sum signal has been delayed correspondingly, and thus the total delay of this radix-2 on-line multiplier is 3. We have, of course, here assumed that no other buffers exist between the front-end and the cell array. Note also that, for consistency, in the figure the signals c_1 and c_2 have been drawn as double lines.

Due to the delay, when the last digits p_n and q_n have been delivered, only $n - 3$ digits of the result have been emitted. To obtain more digits of the result, zeroes may be delivered to empty the array of cells and the front-end. Appending such zeroes to the input, of course, cannot change the value of what remains. Note that an unbiased rounding of the result is obtained by truncation, due to the symmetry of the digit set. Usually in on-line arithmetic, it is expected that the multiplication of two n-digit numbers will return an n-digit result, which can then be obtained after $n + 3$ clock cycles.

Problems and exercises

4.5.1 Design a radix-4 array multiplier by adapting Figure 4.5.3 to operands in minimally redundant radix-4 representation, in terms of appropriate cells, whose logic need not be specified in detail.

4.5.2 The leftmost cell of Figure 4.5.5 can obviously be simplified for unsigned operands. Show how to modify it such that it performs a sign-extension of the result, and such that the multiplicand can be interpreted as a 2's complement number.

4.5.3 Design a tree structure of full-adders to perform the counting of ones in Figure 4.5.7, for the case of $n = 8$.

4.5.4 Prove the correctness of (4.5.3).

4.5.5 Hand simulate the operation of a serial/serial Lyon multiplier with the (unsigned) operands $0.110_{[2]}$ and $0.011_{[2]}$ at the functional level (i.e., in the form of a table showing terms being generated and their accumulation).

4.5.6 Eliminate the broadcast in an array formed of the squarer cells in Figure 4.5.16.

4.5.7 Hand simulate the algorithm for radix-2 on-line multiplication of .11 by .11 with $\delta = 2$ and digit set $\{-1, 0, 1\}$, using (4.5.8).

4.6 Logarithmic-time multiplication

Since it is possible to generate all the partial products in parallel and constant time, what remains is to perform a multioperand addition in some kind of tree structure to obtain the product in logarithmic time, employing constant-time redundant adders in the tree. For a final conversion into non-redundant form, some kind of log-time carry-completion structure (adder) may then be used to complete the result. The issues to be considered in this section are thus primarily of an architectural and technological nature, such as:

- which kind of primitive adders (counters) to use;
- how to organize these into tree structures; and
- the physical layout and wiring of the structures.

It turns out that these issues are very intimately connected with and to some extent dependent on the actual implementation technology.

The task can be reduced to the problem of accumulating a set of numbers represented by bitstrings, where the bits are arranged in the form of parallelograms, e.g., as in the following *partial product array*

(a) (b) (c)

where case (a) represents an ordinary 8×8 multiplication (no recoding), case (b) is the same but with recoding into radix 4, and (c) is an 8×4 rectangular multiplication.

A logarithmic time accumulation can then be realized by a tree structure of constant-time adders, each level compressing a set of rows into a smaller set of rows, which again in another set of adders can be compressed into fewer rows, until eventually only two rows, representing the result in carry-save form, are left. A structure based on 3-to-2 adders was proposed by Wallace in 1964, as illustrated in Figure 4.6.1 for the case of 12 rows.

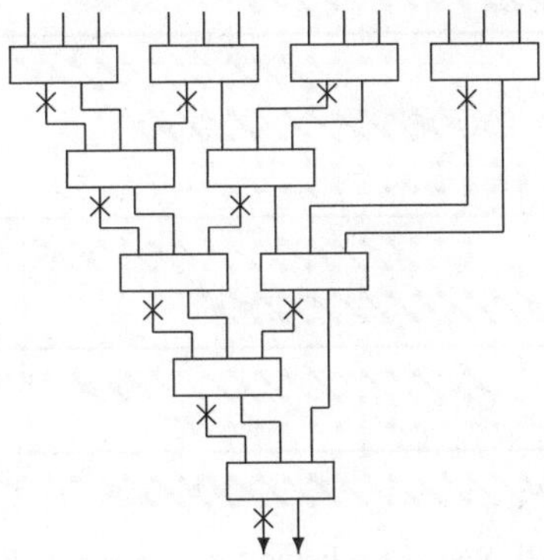

Figure 4.6.1. Wallace-type 12×12 multiplier.

Each "box" here represents a row of mostly full-adders, but with a few half-adders at the ends due to the slanted edges of the array of bits to be accumulated. As denoted by the $\times$s in the figure, the output of weight 2 (the carry) is shifted one position over in the row. The actual allocation of adders is most conveniently described in a diagram as shown in Figure 4.6.2, where the output of each full-adder compressing the sum of three bits into two is shown in the level below by a line connecting the two bits like this ⁄ , and the output of a half-adder similarly by ⌀ .

In 1965 Dadda suggested an improvement, reducing the number of adders needed, observing that it is sufficient at each level only to reduce the number of bits in the tallest columns, in such a way that the new reduced height of the array is the next lower number in the sequence

$$2, 3, 4, 6, 9, 13, 19, 28, 42, 63, \ldots,$$

which is defined starting with $m_0 = 2$ (the goal) and the recursion $m_{i+1} = \lfloor \frac{3}{2} m_i \rfloor$.

The equivalent of Figure 4.6.2 using the Dadda scheme is shown in Figure 4.6.3, where only 111 counters are needed, compared with 136 for the Wallace scheme. For a 24×24 multiplier the equivalent numbers are 505 versus 575.

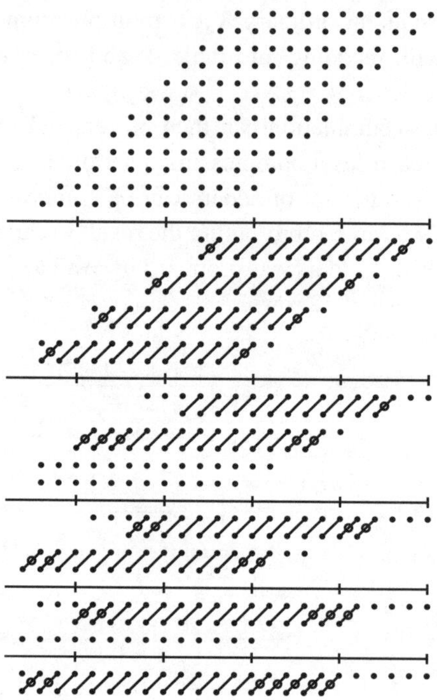

Figure 4.6.2. Wallace scheme for 12 × 12 multiplication.

Observe that in both schemes the least-significant bits pass through fewer levels of logic than the central part, which may have implications for the speed at which the subsequent final carry-completion adder can finish.

Another matter to consider when designing the trees is to take into account that a full-adder is not symmetric in its three inputs, one of which has a shorter path to the output than the other two when the adder is composed of two half-adders. These shorter paths may also be exploited to accommodate the extra row containing a single bit which arises from recoding a 2's complement multiplier as in Figure 4.4.4. Parts of the tree may also be constructed using other types of counters, e.g., 7-to-3 counters, where again the speeds of the different paths through the counters may be exploited to minimize the worst-case critical path.

However, note that the adder structure to be built is actually a set of trees, one tree for each column of the bit-array, where each tree receives carries from and emits carries to a neighboring tree. Mapping such a set of trees (a sort of three-dimensional structure) into the two-dimensional surface of a VLSI chip is no simple task, possibly requiring several levels of crossing wires.

The layout is somewhat more regular and well structured if 4-to-2 adders are employed, yielding binary trees which are easier to map. These adders could be either the 4-to-2 carry-save or the 4-to-2 borrow-save adders of Section 3.7, supplemented with suitable 3-to-2 or 2-to-2 (half-)adders at both ends of the

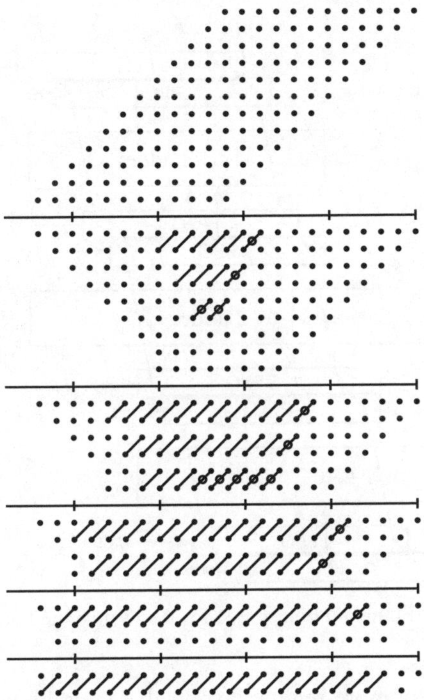

Figure 4.6.3. Dadda scheme for 12 × 12 multiplication.

array. As an example let us consider an 8 × 8 multiplier where eight partial products can be accumulated in two layers of carry-save 4-to-2 adders as shown in Figure 4.6.4.

In the top layer the triangles just pair two partial products into the carry-save encoding of their sum, i.e., the first level of addition is realized "by wiring." Note that this tree can be realized in four levels of ordinary (3-to-2) full-adders since each 4-to-2 adder can be realized with two full-adders, although other designs do exist, as discussed in Chapter 3.

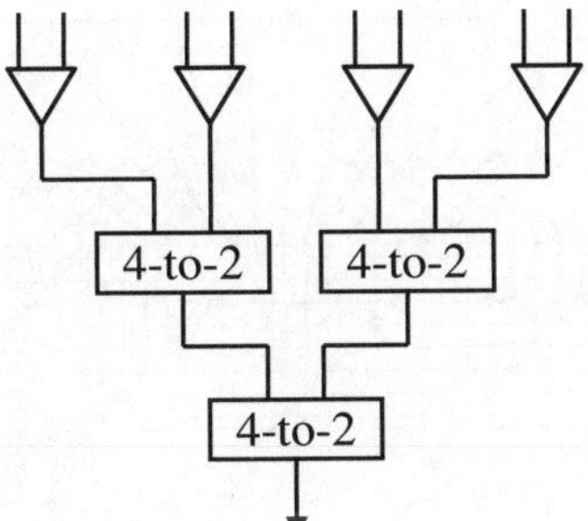

Figure 4.6.4. Binary adder tree for 8 × 8 multiplication.

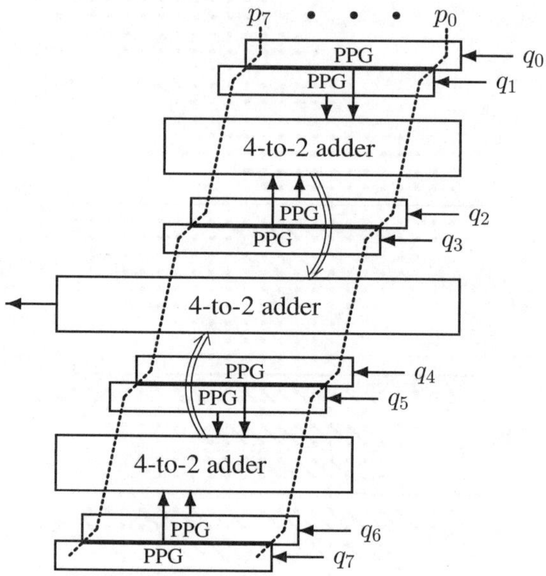

Figure 4.6.5. Flattened binary tree for 8×8 unsigned multiplication.

Binary trees allow for more-regular layouts than trees based on 3-to-2 counters. One possible layout of an 8×8 multiplier for unsigned operands is sketched in Figure 4.6.5.

The interconnect structure of this layout looks fairly complicated, possibly requiring several layers of wiring (metal layers), so we will investigate a different version of the design. Figure 4.6.6 depicts a possible layout of such a multiplier,

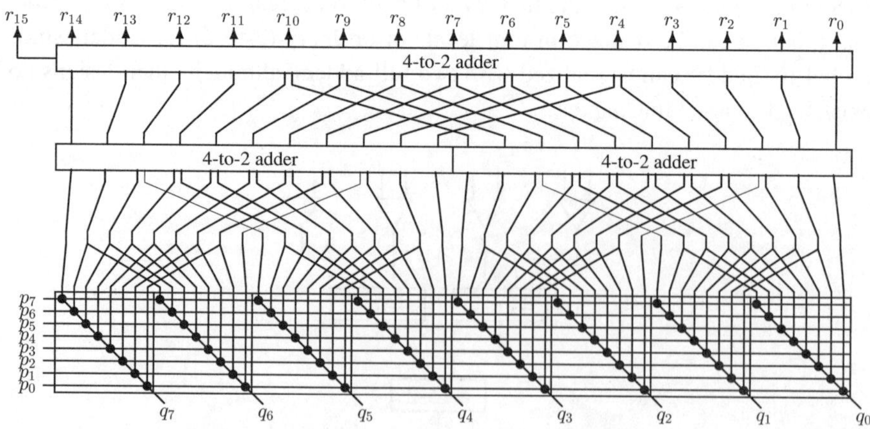

Figure 4.6.6. Multiplier tree for 8×8 unsigned multiplication.

showing that only two layers are needed. Note that the number of layers does not grow with the width of the multiplicand P, since increasing the number of p-bits will just widen the adders.

To keep the figure simple the multiplier only accepts unsigned binary operands, and the PPGs then just contain simple AND gates, which are shown as black circles at the bottom of the figure. The next layer up then pairs partial products as bit vectors into carry-save represented sums of pairs of partial products, this pairing realizing the first level of additions of the tree. Note that some lines carry a single bit, while others carry two bits.

At the third level, two arrays of 4-to-2 carry-save adders each compress the sum of two carry-save operands into one sum in carry-save, the two results of which are finally added together at the fourth level. Notice that the operands going into the adders are shifted relative to one another, corresponding to the different alignments of the operands. This also implies that only the center of the adders has to consist of 4-to-2 adders, while at each end a simpler adder suffices. Since the alignment shifts are respectively 1, 2, and 4, the widths of the adders are 9, 11, and 15.

The output at the top is the product in carry-save representation, which finally may be converted into non-redundant binary by a carry-lookahead adder. Notice that the structure of the layout yields progressively more space towards the top, allowing progressively more powerful (wider) transistors to be employed to drive the signals along the longer wires, so that the speed of the levels can be kept almost constant. Note that a carry-out (r_{15}) is shown at the top level, whereas the handling of possible carries at the level below is not shown. It is left as an exercise to consider how to handle carries at that level.

Now consider increasing the dimension of the multiplier, e.g., doubling the number of bits of Q. Placing two of these structures next to one another, each taking $p_0, p_1 \cdots p_7$ along the vertical dimension, one taking $q_0, q_1 \cdots q_7$ and the other $q_8, q_9 \cdots q_{15}$, then their output can be combined in a new 4-to-2 carry-save adder above, in analogy with the structure shown. As expected, a change in the multiplier dimension changes the height of the structure, whereas a change of the multiplicand width just changes the width of the adders. And in none of the cases does the number of wire levels increase.

Operands in 2's complement representation can be handled by suitable modifications of the PPGs at the bottom level, e.g., corresponding to the Baugh and Wooley scheme. Recoding of the multiplier to radix 4 is similarly possible by a change of the PPGs, and using the scheme of Figure 4.4.4 for generating the rows to be added. However, this introduces some irregularity into the design to accommodate the carry-in bits introduced by 2's complementation, corresponding to a recoded negative multiplier digit. As discussed in Section 4.4, there is a particular problem with the carry-in bit of the bottom row that originates from a negative multiplier; a problem which is absent for unsigned operands, e.g., for sign-magnitude multiplication.

4.6.1 Integer multipliers with overflow detection

As described above we have mostly been interested in the general case of multipliers generating the full double-length product of a "square" multiplier, whereas in practice one is often only interested in either the most-significant (probably rounded) half of the digits, or in the least-significant half of the digits. The former situation occurs when implementing floating-point multipliers, where the operands are the fixed-point mantissas, usually normalized such that they belong to an interval like [0.5; 1) or [1; 2), where for an exact rounding it is necessary to generate the full double length product.

However, the latter may be the case say when implementing an instruction for multiplying two n-bit integer operands to yield an n-bit product with an overflow detection, signaling the user that the resulting integer product cannot be represented in the same format as the operands. If only the least-significant part of the product is needed, about half of the logic (area) can be eliminated in fast array- or tree-structured multipliers, by simply not implementing the parts generating the most-significant digits, which are normally expected to be non-significant (zeroes for unsigned and sign-extensions for radix-complemented operands).

If overflow signaling is requested, a method of detection must be provided, hopefully with minimal area, and in a time that does not exceed that of the product generation, assuming the detection can take place in parallel with product formation.

Let us first consider the product of two unsigned n-bit operands, $A = \sum_{i=0}^{N_A-1} a_i 2^i$ and $B = \sum_{i=0}^{N_B-1} b_i 2^i$ with product $A \times B = P = \sum_{i=0}^{N_P-1} p_i 2^i$. Here N_A and N_B are the number of significant bits in A and B respectively, so assuming that the operands are non-zero, then

$$\begin{aligned} 1 \le N_A \le n \quad &\text{with} \quad a_{N_A-1} \neq 0, \\ 1 \le N_B \le n \quad &\text{with} \quad b_{N_B-1} \neq 0, \end{aligned}$$

hence

$$\begin{aligned} 2^{N_A-1} &\le A < 2^{N_A}, \\ 2^{N_B-1} &\le B < 2^{N_B}, \\ 2^{N_A+N_B-2} &\le P = A \times B < 2^{N_A+N_B}. \end{aligned} \tag{4.6.1}$$

From the left inequality of (4.6.1) it follows that there is overflow if $2^{N_A+N_B-2} \ge 2^n$, or equivalently if

$$N_A + N_B \ge n + 2, \tag{4.6.2}$$

and from the right inequality it follows that the product is guaranteed not to overflow if

$$2^{N_A+N_B} \leq 2^n \quad \text{or equivalently} \quad N_A + N_B \leq n,$$

which is then a sufficient condition for $P = A \times B < 2^n$ to hold.

When $N_A + N_B = n + 1$, from (4.6.1) it follows that $2^{n-1} \leq P < 2^{n+1}$ and hence that only bits $p_0, \ldots, p_n$ need to be generated, and overflow occurs in this case if and only $p_n = 1$. Thus to summarize we have

$$\begin{array}{ll} N_A + N_B \leq n & \text{no overflow;} \\ N_A + N_B = n + 1 & \text{overflow iff } p_n = 1; \\ N_A + N_B \geq n + 2 & \text{overflow.} \end{array}$$

The last condition for overflow is equivalent to checking if there are any non-zero product terms of weight 2^n or greater, which can be expressed as

$$\begin{aligned} F &= \sum_{i=1}^{n-1} \sum_{j=1}^{i} a_{n-j} b_i \\ &= a_{n-1}b_1 + (a_{n-1} + a_{n-2})b_2 + \cdots + (a_{n-1} + a_{n-2} + \cdots + a_1)b_{n-1}. \end{aligned}$$

This can be seen as follows: if the term $a_{n-j}b_i$ (of weight $2^{n-j}2^i = 2^{n+(i-j)} \geq 2^n$) is non-zero, then A has at least $n - j + 1$ and B at least $i + 1$ significant bits, and thus $N_A + N_B \geq n + 2 + (i - j) \geq n + 2$. Note that terms for i or j being zero cannot contribute, nor can A or B being zero cause overflow.

Observation 4.6.1 *The product $P = A \times B$ of two unsigned n-bit operands A and B can be represented with n bits, provided that the overflow condition*

$$OF = p_n + F, \quad \textit{where} \quad F = \sum_{i=1}^{n-1} b_i \sum_{j=1}^{i} a_{n-j},$$

is not raised. Here $p_i = \text{bit}(i, P)$, $a_i = \text{bit}(i, A)$ and $b_i = \text{bit}(i, B)$, hence only the product bits $p_0, p_1, \ldots, p_n$ and F need be calculated, which can all be found in $O(\log n)$ time.

Obviously, the preliminary overflow condition F can be calculated in linear time. But it remains to be shown that F can be found in logarithmic time. However, by rewriting

$$F = A_{n-1}b_1 + A_{n-2}b_2 + \cdots + A_1 b_{n-1},$$

it is easily seen that F can be calculated by a tree structure over the terms $A_k b_{n-k}$, where the coefficients $A_k = \sum_{j=k}^{n-1} a_j$ for $k = 1, \ldots, n-1$, can be calculated by a parallel prefix structure, hence also in logarithmic time. The calculation of F

can thus be performed as shown for eight-bit operands in Figure 4.6.7, when employing a Brent–Kung prefix tree.

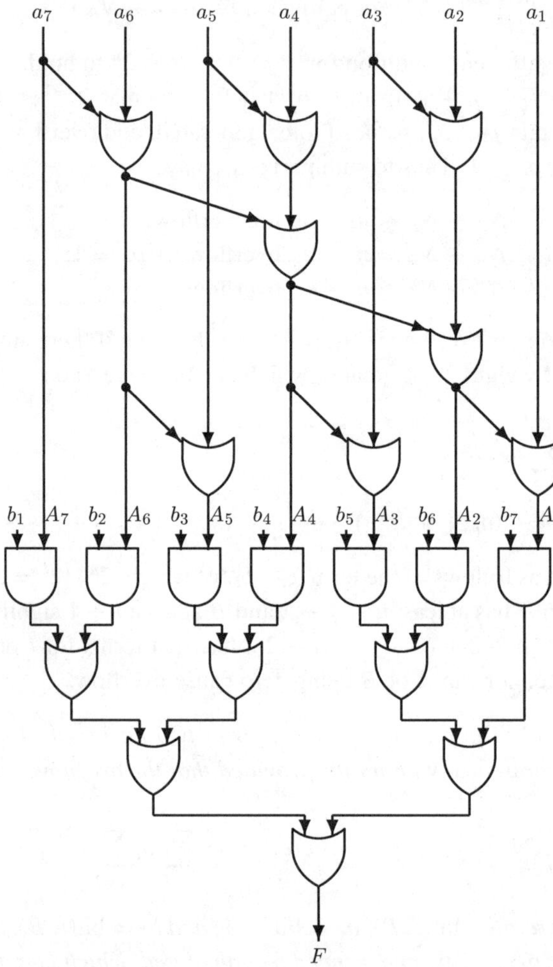

Figure 4.6.7. A log-time calculation of the preliminary overflow condition F for $n = 8$.

Evidently, the calculation of F can take place in parallel with a tree-structured *half-multiplier*, and in less area than would be required for the unimplemented most-significant parts of the partial product array.

Overflow detection for 2's complement integer multiplication. For 2's complement numbers the most-significant bit is determined as the rightmost bit that differs from the sign bit, e.g., $X = 11110101$ has four significant digits. Thus conditionally inverting the bits of the operands forming $\widehat{A} = \hat{a}_{n-1} \cdots \hat{a}_0$, where for

$i = 0, \ldots, n-1$, $\hat{a}_i = a_{n-1} \oplus a_i$, and similarly for B, the number of significant bits N_A and N_B in A and B can be determined from $\widehat{A}$ and $\widehat{B}$ as for unsigned operands above. Furthermore

$$\widehat{A} = \begin{cases} A & \text{for} \quad A \geq 0 \\ -A-1 & \text{for} \quad A < 0 \end{cases}.$$

The number of significant bits determines the following bounds on $\widehat{A}$, $\widehat{B}$, and $\widehat{P} = \widehat{A} \times \widehat{B}$,

$$\begin{aligned} 2^{N_A-1} &\leq \widehat{A} \leq 2^{N_A} - 1, \\ 2^{N_B-1} &\leq \widehat{B} \leq 2^{N_B} - 1, \\ 2^{N_A+N_B-2} &\leq \widehat{P} \leq 2^{N_A+N_B} - 2^{N_A} - 2^{N_B} + 1 \leq 2^{N_A+N_B} - 1, \end{aligned} \tag{4.6.3}$$

since $-2^{N_A} - 2^{N_B} + 1 \leq -1$, and, e.g.,

$$\begin{aligned} A < 0 &\Leftrightarrow -2^{N_A} \leq A \leq -2^{N_A-1} - 1 \\ &\Leftrightarrow 2^{N_A-1} + 1 \leq -A \leq 2^{N_A} \\ &\Leftrightarrow 2^{N_A-1} \leq -A - 1 \leq 2^{N_A} - 1 \\ &\Leftrightarrow 2^{N_A-1} \leq \widehat{A} \leq 2^{N_A} - 1, \end{aligned}$$

in particular note that (4.6.3) is equivalent to

$$\left.\begin{aligned} 2^{N_A+N_B-2} &\leq P \leq 2^{N_A+N_B} - 1 && \text{for } P \geq 0, \\ -2^{N_A+N_B} &\leq P \leq -2^{N_A+N_B-2} - 1 && \text{for } P < 0. \end{aligned}\right\} \tag{4.6.4}$$

In both cases there is overflow if $2^{N_A+N_B-2} \geq 2^{n-1}$, or equivalently

$$N_A + N_B \geq n + 1,$$

which can be tested by the condition

$$\begin{aligned} F &= \sum_{i=1}^{n-2} \sum_{j=1}^{i} \hat{a}_{n-j-1} \hat{b}_i \\ &= \hat{a}_{n-2}\hat{b}_1 + (\hat{a}_{n-2} + \hat{a}_{n-3})\hat{b}_2 + \cdots + (\hat{a}_{n-2} + \cdots + \hat{a}_1)\hat{b}_{n-2}. \end{aligned}$$

On the other hand, it is guaranteed that there is no overflow if $\widehat{P} \leq 2^{N_A+N_B} - 1 \leq 2^{n-1}$, or equivalently

$$N_A + N_B \leq n - 1.$$

What remains is the situation in which $N_A + N_B = n$, where (4.6.4) becomes

$$\begin{aligned} 2^{n-2} &\leq P \leq 2^n - 1 && \text{for } P \geq 0, \\ -2^n &\leq P \leq -2^{n-2} - 1 && \text{for } P < 0, \end{aligned}$$

Table 4.6.1. *2's complement integer multiply overflow status for $N_A + N_B = n$*

	Interval	Status	$p_n p_{n-1}$
$P \geq 0$	$[2^{n-2}; 2^{n-1})$	No overflow	00
	$[2^{n-1}; 2^n)$	Overflow	01
$P < 0$	$[-2^n; -2^{n-1})$	Overflow	10
	$[-2^{n-1}; -2^{n-2})$	No overflow	11

both intervals covering two binades. If $P \geq 0$ there is no overflow if $2^{n-2} \leq P < 2^{n-1}$ and overflow if $2^{n-1} \leq P < 2^n - 1$, and if $P < 0$ there is no overflow if $-2^{n-1} \leq P \leq -2^{n-2} - 1$ and overflow if $-2^n \leq P \leq -2^{n-1} - 1$. Realizing that these four situations can be distinguished by the bits p_n and p_{n-1} we get Table 4.6.1.

From the table it is easily seen that the overflow status for $N_A + N_B = n$ can be determined by $p_n \oplus p_{n-1}$, and thus we have the following observation.

Observation 4.6.2 *The product $P = A \times B$ of two 2's complement n-bit operands A and B can be represented with n bits, provided that the overflow condition*

$$OF = p_n \oplus p_{n-1} + F, \quad \text{where} \quad F = \sum_{i=1}^{n-2} \sum_{j=1}^{i} \hat{a}_{n-j} \hat{b}_i,$$

is not raised. Here $p_i = \text{bit}(i, P)$, $\hat{a}_i = a_{n-1} \oplus a_i$, $a_i = \text{bit}(i, A)$ and $\hat{b}_i = b_{n-1} \oplus b_i$, $b_i = \text{bit}(i, B)$, and hence only the product bits $p_0, p_1, \ldots, p_n$ and F need be calculated, which can all be found in $O(\log n)$ time.

We can finally observe that a multiplier accepting unsigned as well as 2's complement integer operands can easily be implemented as a single unit.

Problems and exercises

4.6.1 In Figure 4.6.6 a carry-out is shown emitted from the top level adder. Consider whether there may be carries from the two adders below.

4.6.2 Now consider how such carries may be accommodated in the tree when increasing the dimension of the multiplier Q.

4.7 Squaring

Squaring is the symmetric case of multiplication where multiplicand and multiplier are the same operand, reducing squaring to a unary operation. This symmetry allows partial product arrays for squaring to be reduced to about half the size and half the depth of comparable multiplication partial product arrays for both radix-2 and recoded radix-4 arrays. There are many applications such as fast exponentiation, Euclidean distance computation, and rounding of floating-point square root, where incorporating a supplemental squaring circuit may be attractive to avoid the time and power requirements of performing all squaring operations in a large multiplier.

4.7.1 Radix-2 squaring

For radix-2 squaring of the n-bit operand $Q = q_{n-1}q_{n-2}\dots q_0$, the n^2-bit product terms of the multiplication array may be reduced to $(n+1)n/2$ terms employing $q_i^2 = q_i$ and $q_jq_i + q_iq_j = 2q_jq_i$ for $i > j$. For example, the 5×5-bit multiplication partial product array of Figure 4.2.1 then reduces for squaring to the skewed triangular array of Figure 4.7.1.

					q_4	q_3	q_2	q_1	q_0	
				q_4q_0	q_3q_0	q_2q_0	q_1q_0	0	q_0	q_0
			q_4q_1	q_3q_1	q_2q_1	0	q_1			q_1
		q_4q_2	q_3q_2	0	q_2					q_2
	q_4q_3	0	q_3							q_3
0	q_4									q_4
s_9	s_8	s_7	s_6	s_5	s_4	s_3	s_2	s_1	s_0	

Figure 4.7.1. Skewed triangular bit product array for five-bit squaring.

Note that each row of the triangular array for n-bit squaring may be realized as an "ith multiplicand prefix" modified partial product conditionally selected by the multiplier bit q_i according to the following.

Observation 4.7.1 *Let* $Q = q_{n-1}q_{n-2}\cdots q_0$, *with* $Q_i = \lfloor Q/2^i \rfloor = q_{n-1}q_{n-2}\cdots q_{i+1}$ *for* $i = 1, 2, \dots, n$. *Then*

$$Q^2 = \sum_{i=0}^{n-1}(4Q_{i+1} + q_i)q_i4^i. \tag{4.7.1}$$

Proof With $Q = 2Q_1 + q_0$, then $Q^2 = 4Qq_1^2 + (4Q_1 + q_0)q_0$. With $Q_i = 2Q_{i+1} + q_i$, then $Q_i^2 = 4Q_{i+1} + (4Q_{i+1} + q_i)q_i)$, and iterative substitution yields (4.7.1). □

The skewed triangular array of modified partial products $4Q_{i+1} + q_i$ illustrated in Figure 4.7.1 may be reorganized to form fewer rows by moving up the diagonals of the left half of the $2n \times n$-bit array. Incorporating the half-adder relation $q_{i+1}q_i + q_{i+1} = 2q_{i+1}q_i + q_{i+1}\overline{q_i}$ for $0 \le i \le n-2$, we can obtain a $2n \times \lceil n/2 \rceil$ nearly symmetric triangular array of $\binom{n+1}{2}$-bit product terms.

						q_4	q_3	q_2	q_1	q_0	
q_4	q_4q_3	$q_4\overline{q_3}$	q_4q_2	q_4q_1	q_4q_0	q_3q_0	q_2q_0	$q_1\overline{q_0}$	0	q_0	q_0
q_3			q_3q_2	$q_3\overline{q_2}$	q_3q_1	$q_2\overline{q_1}$	q_1q_0				q_1
q_2					q_2q_1						q_2

							q_5	q_4	q_3	q_2	q_1	q_0	
q_5	q_5q_4	$q_5\bar{q}_4$	q_5q_3	q_5q_2	q_5q_1	q_5q_0	q_4q_0	q_3q_0	q_2q_0	$q_1\bar{q}_0$	0	q_0	q_0
q_4			q_4q_3	$q_4\bar{q}_3$	q_4q_2	q_4q_1	q_3q_1	$q_2\bar{q}_1$	q_1q_0				q_1
q_3					q_3q_2	$q_3\bar{q}_2$	q_2q_1						q_2

Figure 4.7.2. Partial product consolidation for five- and six-bit squaring.

Figure 4.7.2 illustrates the reorganized triangular arrays for the five- and six-bit squarings. For partial product design of this consolidated $\lceil n/2 \rceil$ row array note that the n-bit right halves of the $2n$-bit rows may be formed by using bits q_i for $i = 0, 1, \ldots, \lceil n/2 \rceil - 1$ as select bits for modified n-bit partial products and the n-bit left halves of the rows may be similarly formed by using $q_{\lfloor n/2 \rfloor}$ through q_{n-1} as select bits for appropriately modified n-bit partial products.

4.7.2 Recoded radix-4 squaring

Let $P = d_{\lfloor n/2 \rfloor} d_{\lfloor n/2 \rfloor - 1} \cdots d_0$ with $d_i = -2q_{2i+1} + q_{2i} + q_{2i-1} \in \{-2, -1, 0, 1, 2\}$ be the Booth recoded radix-4 representation of $Q = q_{n-1}q_{n-2} \cdots q_0$, extended with $q_{-1} = 0$. Let P_i be the integer formed by shifting the radix-4 digit string right i places deleting the low-order i digits to give

$$P_i = d_{\lfloor n/2 \rfloor} d_{\lfloor n/2 \rfloor - 1} \cdots d_i = \sum_{j=i}^{\lfloor n/2 \rfloor} d_j 4^{j-i}$$

for $i = 0, 1, \ldots, \lfloor n/2 \rfloor$, with $P_{\lfloor n/2 \rfloor + 1} = 0$ and $P_0 = P = Q$. Since then $P_i = 4P_{i+1} + d_i$ and $P_i^2 = 16P_{i+1}^2 + (8P_{i+1} + d_i)d_i$ for $i = 0, 1, \ldots, \lfloor n/2 \rfloor$, we obtain

$$Q^2 = \sum_{i=0}^{\lfloor n/2 \rfloor - 1} (8P_{i+1} + d_i)d_i 16^i. \tag{4.7.2}$$

Recall that PPGs for Booth recoded multiplication employ asymmetric representation of operands in that the multiplicand is input to the PPG in binary with the multiplier digits given in a recoded radix-4-select format. Thus it is desirable to reconvert P_{i+1} back to binary to formulate a suitable modified partial product input for a PPG. Note that the Booth radix digit d_i determined by the three bits $q_{2i+1}q_{2i}q_{2i-1}$ can be negative only when $q_{2i+1} = 1$, so we must have $(8P_{i+1} + d_i)d_i = ((-1)^{q_{2i+1}} 8P_{i+1} + |d_i|)|d_i|$, where $|d_i| \in \{0, 1, 2\}$. Furthermore,

$q_{2i+1} = 1$ generates a carry forward, so, in general, with $Q_j = \lfloor Q/2^j \rfloor$,

$$P_{i+1} = Q_{2i+2} + q_{2i+1} \text{ for } i = 0, 1, \ldots, \lfloor n/2 \rfloor - 1.$$

Let Q^*_{2i+2} denote the signed modified partial product given by

$$Q^*_{2i+2} = (-1)^{q_{2i+1}} P_{i+1} = (-1)^{q_{2i+1}} (Q_{2i+2} + q_{2+1}).$$

Note that Q^*_{2i+2} is simply Q_{2i+2} when $q_{2i+1} = 0$. When $q_{2i+1} = 1$, Q^*_{2i+2} is the 2's complement of $(Q_{2i+2} + 1)$ or equivalently the 1's complement of Q_{2i+2}.

Theorem 4.7.2 *Let $P = d_{\lfloor n/2 \rfloor} d_{\lfloor n/2 \rfloor - 1} \cdots d_0$ be the Booth radix-4 recoded representation of $Q = q_{n-1} q_{n-2} \cdots q_0$. Let Q^*_{2i+2} be the conditionally 1's complemented leading bits of Q given for $i = 0, 1, \ldots, \lfloor n/2 \rfloor - 1$ by*

$$Q^*_{2i+2} = \begin{cases} q_{n-1} q_{n-2} \cdots q_{2i+2} & \text{for} \quad q_{2i+1} = 0, \\ (-1) \bar{q}_{n-1} \bar{q}_{n-2} \cdots \bar{q}_{2i+2} & \text{for} \quad q_{2i+1} = 1. \end{cases}$$

Then

$$Q^2 = \sum_{0}^{\lfloor n/2 \rfloor - 1} (8 Q^*_{2i+2} + |d_i|) |d_i| 16^i).$$

From Theorem 4.7.2 we observe that the PPGs for radix-4 squaring are simpler than for a radix-4 multiplier since the complements are simply 1's complements. Figure 4.7.3 illustrates for $n = 12$ the skewed triangular array of roughly $n^2/4$-bit product terms for recoded radix-4 squaring with the $\lfloor n/2 \rfloor$ negative complement bits gathered in the first row.

			$q_{11} q_{10} q_9 q_8$	$q_7 q_6 q_5 q_4$	$q_3 q_2 q_1 q_0$	
$-(q_9\ 0q_7$	$0q_5\ 0q_3$	$0q_1)$ **	** **	** **	** 0*	d_0
		** **	** **	** 0*		d_1
	**	** **	** 0*			d_2
	** **	** 0*				d_3
**	** 0*					d_4
** 0*						d_5

Figure 4.7.3. Partial product array for 12-bit recoded radix-4 squaring.

The $\lceil n/2 \rceil$ radix-4 partial products yielding the square Q^2 according to (4.7.2) may be reorganized as illustrated in Figure 4.7.4. Let us consider the case $n = 4k$, where it is sufficient to consider that the $2k$ partial products selected by digits d_i and d_{k+i} may be consolidated in a single $2n$-bit row, yielding $k + 1$ rows to be accumulated. The negative leading portion of row zero can be handled by a 2's complement, yielding an extra unit to be added in column $n + 2$, as shown in

				$q_{11}q_{10}q_9q_8$	$q_7q_6q_5q_4$	$q_3q_2q_1q_0$	
	$1\bar{q}_9\ 1\bar{q}_7$	$1\bar{q}_5\ 1\bar{q}_3$	$1\bar{q}_1{*}{*}$	** **	** **	** 0*	d_0
d_4	**	** 0*	** **	** **	** 0*		d_1
d_5	** 0*	00 **	** **	** 0*			d_2
		** **	** 0*				d_3
			1				

Figure 4.7.4. Consolidated partial product array for 12-bit recoded radix-4 squaring.

Figure 4.7.4. Note that if $|d_k| \leq 1$, then the unit bit may be inserted in column $n+2$ of the row for d_k. If $|d_k| = 2$, then $d_k^2 = 4$, and the extra unit may be absorbed by inserting 1000 at the four low-order bits of the row for d_k.

Observation 4.7.3 *For $n = 4k$ with $k \leq 1$, the $2n$-bit square of $Q = q_{n-1}q_{n-2}\cdots q_0$ may be obtained as the sum of a $2n \times ((n/4)+1)$-bit array of modified partial products obtained by radix-4 recoded selection.*

Observation 4.7.4 *The n-bit integer square given by $\left|Q^2\right|_{2^n}$ may be obtained as the sum of an $\lceil n/4 \rceil$ row array of modified partial products, where negative products are obtained by 1's complementation without the need for sign-extension.*

Note that the array entries here may be generated in parallel, hence in constant time.

4.7.3 Radix-4 squaring by operand dual recoding

While the recoding of the operand playing the role of the multiplicand (we may here denote it the *squarand*) above took place from right to left, peeling off bits from the least-significant end, we shall now recode it from left to right, which may be advantageous if we are only interested in the most-significant part of the result, e.g., when squaring a floating-point number or forming an approximate square.

The catalyst for the left-to-right recoding is a sequence of "rounded off" *tails* of the operand x.

Definition 4.7.5 *Given an n-bit normalized operand $x = 01.b_1b_2\cdots b_{n-1}$, the radix-4, 2's complement* tails *of x are:*

$$t_0 = x = 1.b_1b_2\cdots b_{n-1},$$
$$t_i = \hat{b}_{2i-1}b_{2i}.b_{2i+1}\cdots b_{n-1} \quad \text{for} \quad 1 \leq i \leq \lfloor (n-1)/2 \rfloor,$$
$$t_{\lfloor \frac{n+1}{2} \rfloor} = 0,$$

where $\hat{b}_{2i-1} = -b_{2i-1}$, assuming that $b_n = b_{n+1} = 0$.

Note that the negative weight of the most-significant bit is here made explicit. We then immediately find (trailing parts cancelling) that for $0 \le i \le \lfloor (n-1)/2 \rfloor$:

$$t_i - t_{i+1}4^{-1} = -2b_{2i-1} + b_{2i} + b_{2i+1} = d_i \in \{-2, -1, 0, 1, 2\},$$

where d_i is the ith radix-4 recoded digit of x. Furthermore t_i has the recoded radix-4 representation $t_i = d_i.d_{i+1}d_{i+2}\cdots d_{\lfloor (n-1)/2 \rfloor}$.

Lemma 4.7.6 *For* $0 \le i \le \lfloor (n-1)/2 \rfloor$ *with* $\hat{b}_{2i-1} = -b_{2i-1}$

$$q_i = t_i + t_{i+1}4^{-1} = \hat{b}_{2i-1}b_{2i}.b_{2i+2}b_{2i+3}\cdots b_{n-1}.$$

Proof When adding the terms of equal weight from the expansions of t_i and $t_{i+1}4^{-1}$ using their definition, it is found that the terms of weight $2^{-(2i+1)}$ (digits b_{2i+1}) cancel, while the terms of lower weight appear twice, and hence move up one position. □

We now combine the definition of $q_i = t_i + t_{i+1}4^{-1}$ with the observation above that $d_i = t_i - t_{i+1}4^{-1}$ and obtain the following theorem.

Theorem 4.7.7 *With* $q_i = t_i + t_{i+1}4^{-1}$ *for* $0 \le i \le \lfloor (n-1)/2 \rfloor$

$$x^2 = \sum_{i=0}^{\lfloor \frac{n-1}{2} \rfloor} q_i d_i 16^{-i} = \sum_{i=0}^{\lfloor \frac{n-1}{2} \rfloor} |q_i|\,|d_i|16^{-i},$$

where the q_i *act as recoded squarands. The modified partial products* $q_i d_i = |q_i||d_i|$ *are all positive and can be formed without sign-extension.*

Proof The result follows in a straightforward way from

$$q_i d_i = (t_i - t_{i+1}4^{-1})(t_i - t_{i+1}4^{-1}) = (t_i^2 - t_{i+1}^2 16^{-1}),$$

hence $x^2 = t_0^2 = \sum_{i=0}^{\lfloor (n-1)/2 \rfloor} (t_i^2 - t_{i+1}^2 16^{-1})16^{-i}$. Also, q_i and d_i have the same sign, $(-1)^{b_{2i-1}}$. □

For an implementation note that while the digits d_i trivially can be converted to $|d_i|$, the recoded squarands q_i may have to be sign inverted by 2's complementation:

$$|q_i| = \begin{cases} b_{2i}.b_{2i+2}b_{2i+2}\cdots b_{n-2}b_{n-1} & \text{for} \quad b_{2i-1} = 0, \\ \overline{b}_{2i}.\overline{b}_{2i+2}\overline{b}_{2i+2}\cdots \overline{b}_{n-2}b''_{n-1} & \text{for} \quad b_{2i-1} = 1, \end{cases}$$

where $b''_{n-1} = 2 - b_{n-1}$ and $\overline{b}_j = 1 - b_j$ for $0 \le j \le n-2$.

The terms $|q_i|\,|d_i|16^{-i}$ to be accumulated may in analogy with similar terms in the general multipliers be termed *partial squarands*, and are generated by a *partial squarand generator*.

When n is even the last tail for $k = (n/2) - 1$ is $t_k = \hat{b}_{n-3}b_{n-2}.b_{n-1}$ and its square term $t_k^2 \in \{0, 1, 4, 9, 16\}$ can be entered directly for the squarand $|q_k||d_k|$, when weighted by 16^{-k}. Similarly for odd n with $k = (n-1)/2$, the term

$t_k^2 = (\hat{b}_{n-2}b_{n-1})^2 \in \{0, 1, 4\}$ may be entered. Note that these terms will be non-negative, hence the 2's complementation issue does not occur for the last term.

Example 4.7.1 Consider the 16-bit normalized operand $x = 01.1000\,1011\,0001\,011$, then since the least-significant bit is 1, 2's complementation is simplified so that the terms to be added are:

		d_i
$x =$	01.100010110001011	
$q_0d_0 =$	10.001011000101 1	2
$q_1d_1 =$	1101001110101	-2
$q_2d_2 =$	00110001011	1
$q_3d_3 =$	101110101	-1
$q_4d_4 =$	0110101	-1
$q_5d_5 =$	00000	0
$q_6d_6 =$	111	1
$q_7d_7 =$	1001	$t_7 = 3$
$x^2 =$	10.0110000110111100111001011111001	

□

Note from the example that less-significant contributions may be "pulled up behind" more-significant terms, thus essentially cutting the number of terms to be added in half. Let us now analyze how such a compression may be realized. There are no sign-extension bits, which facilitates the compression, however, we must consider how the addition of the 2's complement bits may be accommodated, a problem that is not present in the above example, but which needs to be considered for the general case.

Consider first only the term $|q_i||d_i|16^i$, where $q_i = b_{2i}.b_{2i+2}b_{2i+2}\cdots b_{n-2}b_{n-1}$ (or its 2's complement assuming that it takes the same positions), as it appears in Theorem 4.7.7 in the sum forming x^2. Its most-significant bit will appear in the position with weight 2^{-4i}, and its least-significant bit in the position of weight $2^{-(4i+(n-2i-2))} = 2^{-(2i+n-2)}$.

For the term $|q_{i+k}||d_{i+k}|16^{-(i+k)}$ to "fit behind" the term $|q_i||d_i|16^{-i}$ for all $i \geq 0$ and some value of $k > 0$, it is necessary to require that

$$4(i+k)+1 \geq 2i+n-2 \quad \text{or} \quad k \geq \left\lceil \frac{n-2i-3}{4} \right\rceil,$$

allowing for the extreme combination $|d_i| = 1$ and $|d_{i+k}| = 2$. In particular, we want it to hold for $i = 0$, in which case we must require that $k \geq \lceil (n-3)/4 \rceil$.

Now consider the situation where the operand x has trailing zeroes and 2's complementation of q_i forces an increment of the unit bit in the least-significant position. This increment bit could possibly be inserted in the row above. Note that the first row ($|q_o||d_o|$) originates from the operand, which is assumed to be non-negative, hence it will not need such an increment.

However, in general, 2's complementation of $|q_{i+1}|$ will then generate a unit bit to be inserted "behind" the term $|q_i||d_i|16^{-i}$, which has its least-significant bit in the position of weight $2^{-(2i+n-2)}$ if $|d_i| = 1$, or one position to the left if $|d_i| = 2$.

If 2's complementation of $|q_{i+1}|$ generates such a unit bit, it will have to be added in the ith row in the position of weight $2^{-(2(i+1)+n-2)} = 2^{-(2i+n)}$ if $|d_{i+1}| = 1$, or one position to the left if $|d_{i+1}| = 2$. Hence in both cases will it "fit" behind the least-significant bit in the ith row.

For the term $|q_{i+k}||d_{i+k}|16^{-(i+k)}$ again to fit behind such an inserted unit bit in the ith row we require

$$4(i+k)+1 \geq 2i+n \quad \text{or} \quad k \geq \left\lceil \frac{n-2i-1}{4} \right\rceil,$$

where the worst case again is for $i = 0$, so we require $k \geq \lceil (n-1)/4 \rceil \geq \lceil (n-3)/4 \rceil$, where the last bound is the one obtained above.

This implies that there will be only half as many terms to add compared with a standard radix-4 recoded normal multiplier. We can then conclude the following.

Observation 4.7.8 (Radix-4 squaring) *Based on Theorem 4.7.7 for the squaring of a non-negative, n-bit operand, all* $\lfloor (n+1)/2 \rfloor$ *rows to be added may be generated in parallel, but the rows* $\lceil (n-1)/4 \rceil$ *to* $\lfloor (n-1)/2 \rfloor$ *may be consolidated into the first rows, so that there are only* $\lceil (n-1)/4 \rceil$ *rows to be added.*

Problems and exercises

4.7.1 Illustrate the consolidated partial product array employing radix-2 squaring for the 16-bit square of an eight-bit operand.

4.7.2 For the 16-bit normalized binary operand $x = 1.1010\ 1110\ 0101\ 100$, illustrate the left-to-right radix-4 recoded partial square array as in Example 4.7.1, extending the array formation to include the needed 2's complement bits.

4.7.3 Illustrate the consolidated partial product array for 16-bit left-to-right radix-4 recoding (similar to Figure 4.7.4) following the guidelines discussed after Example 4.7.1.

4.7.4 Derive a left-to-right leading-digit-first dual operand recoding in radix 8 utilizing the recurrence $(x')^2 = x^2 - d(d + 2x')$, where d is a minimally redundant leading radix-8 digit and x' is the resulting rounded off tail. Specifically, show that $x^2 = \sum_{i=0}^{\lfloor p/3 \rfloor} d_i q_i 64^{-i}$ with $q_i = \bar{b}_{3i-1} b_{3i} b_{3i+1}.b_{3i+3} b_{3i+4} \ldots b_{p-1}$.

4.8 Notes on the literature

The earliest computers often had a register structure like that illustrated in Figure 4.2.2, containing three registers and an adder. Two of the registers can be connected into one double-length register with a shift capability, and one of these registers is able to act as an accumulator for the output from the adder. This is a convenient configuration for the implementation of multiplication and division, as explained in [BGvN46]. As the adder is a fairly slow circuit, it was a short step to the idea of skipping over zeroes, and Booth's suggestion of "recoding" the multiplier into the digit set $\{-1, 0, 1\}$ to further reduce the number of adder cycles [Boo51] then followed. MacSorley also reported this idea, as well as base converting into radix 4 and radix 8 [Mac61] ("Modified Booth" Recoding). Lemma 4.3.5 has been adapted from [KO62].

The Baugh and Wooley method of dealing with 2's complement operands [BW73] is frequently used for integer arithmetic, whereas floating-point now uses sign-magnitude representation, and thus avoids the complications of handling the sign information. In [Dad85a] Dadda described several ways of handling 2's complement operands, which seem to have been somewhat overlooked.

Array multipliers have been employed in pipelined form [MJ82] in digital signal processing (DSP), where often a major part of the computations is in the form of "foldings" (digital filters: $\sum w_i x_i$) at fairly low precision. An interesting rewriting of such expressions, where multiplications and additions are substituted by table look-ups, shifts, and additions, can be found in [PL74]. Typically in these applications, latency (delay) is not a major issue, but there are often very high throughput requirements. The Lyon multiplier [Lyo76] has also been used in DSP applications, where data sometimes are processed digit serially, LSB-first, when the computational demand is not very high.

The Atrubin multiplier, as described in [Atr65], is a linear array of cellular automata. Each cell is described by its functionality and state transitions, but no logic diagrams are given. Our description above is but one possible realization, employing an explicit control signal traveling along the array, where Atrubin used a state change from a "quiescent state" in the neighboring cell. For this reason his rightmost cell is different from the rest. Also he describes two variants: one with immediate output and one where output is delayed one cycle. For many years this multiplier remained a mystery, as he did not explain its general behavior, but in 1991 Even and Litman [EL91] published an explanation and a proof of its correctness.

In 1979, Chen and Willoner [CW79] described another serial/serial multiplier, claiming superiority over the Atrubin multiplier. Actually their design corresponds closely to an array of cells like that in Figure 4.5.13, relying on a broadcast of p_t, q_t, but with a non-shifted result. The cell interconnection is such that the output of a 5-to-3 adder is deposited in the cell itself, and the carries are propagated to the left-hand neighbor. This implies that $2n$ cells are needed, and the result bit r_i on completion is found in cell number i for $t = i, i+1, \ldots, 2n$. Hence their design requires about twice as much logic, in addition to the logic needed to emit the result in a serial fashion, given that it is deposited distributed over the cells of the array (a fact they do not discuss, although they claim serial output).

Dadda [Dad83] also presented serial/serial multipliers based on broadcast of p_t, q_t; he considered various ways of depositing these (queues and stacks) and associated adder structures for accumulation. The paper also discusses serial/parallel adders based on column summation. In two follow-up papers he discussed 2's complement operands for these adders [Dad85a], and squarers [Dad85b]. The particular serial/parallel multiplier of Figure 4.5.5 by Kornerup is from [Kor94], but many other such multipliers have been proposed, see, e.g., [Swa73, MG75, Gna85, CS85, BO87, IV94]. In [Eve97] Guy Even presented alternative, systolic array serial/parallel designs, including a variation where the $2n$-bit product is produced during the n cycles where the serial operand is delivered, thus making the array immediately available for the next operand.

On-line multipliers were first introduced by Ercegovac and Trivedi in [TE77], using basically the same algorithm as in (4.5.8), but without a delay and thus restricted in operand range. They claimed incorrectly that operands can be in the interval $(-\frac{1}{2}; \frac{1}{2})$ for radix 2, however they must be restricted to $(-\frac{1}{4}; \frac{1}{4})$, as can be shown by a slightly modified example. Ercegovac and Lang in particular have been working extensively with various on-line algorithms, some general methodologies have been reported in [Erc84] and [EL88]. Our radix-2 multiplier in Figure 4.5.17 is partially based on [GHM89].

As early as in 1964 Wallace [Wal64] suggested ways of organizing tree structures of half- and full-adders to accumulate partial products, and Dadda [Dad65] soon after provided the improved organization. But such "expensive" implementations have not often been used, except in the supercomputers of that time. The use of 4-to-2 adders seems to have been introduced by Weinburger in [Wei81] and later described in [SH88], and borrow-save based adders in [TYY85, KNE$^+$87]. See also the discussion on 4-to-2 adders in the Notes on the literature in Chapter 3. In modern microprocessors the tree structures of the multipliers take up significant parts of the "real estate" of the chips. Various publications discussing multiplier organizations are [Vui83, MJ92, OSY$^+$95, OV95, OVL96, EL97, SMOR98, YJ00]. The description of overflow detection in integer $n \times n$ multipliers is from [GSA06] by Gok, Schulte, and Arnold.

Specialized circuits for squaring may be used in DSP, in graphics processors, in transcendental function evaluation, and even for the implementation of multiplication, based on the relation $a \times b = ((a + b)^2 - (a - b)^2)/4$, see, e.g., [Che71]. Squarer designs for radix-2 have been discussed for quite some time, early references are [Dad85b] and [IV94]. Later radix-4 designs and implementations are described in [Erc03, SC03], the description here is based on [MMT08]. In [Mat09] these ideas are used in the design of a radix-8 squarer, also employing dual recoding.

References

[Atr65] A. J. Atrubin. A one-dimensional real-time iterative multiplier. *IEEE Trans. Electronic Computers*, EC-14:394–399, 1965.

[BGvN46] A. Burks, H. H. Goldstine, and J. von Neumann. *Preliminary Discussion of the Logic Design of an Electronic Computing Instrument*. Technical report, Institute for Advanced Study, Princeton, 1946. Reprinted in C. G. Bell, *Computer Structures, Readings and Examples*, Mc Graw-Hill, New York, 1971.

[BO87] P. T. Balsara and R. M. Owens. Systolic and semi-systolic digit serial multipliers. In *Proc. 8th IEEE Symposium on Computer Arithmetic*, pages 169–173. IEEE Computer Society, 1987.

[Boo51] A. D. Booth. A signed binary multiplication technique. *Q. J. Mech. Appl. Math.*, 4:236–240, 1951. Reprinted in [Swa80].

[BW73] C. R. Baugh and B. A. Wooley. A two's complement parallel array multiplication algorithm. *IEEE Trans. Computers*, C-22:1045–1047, 1973. Reprinted in [Swa80].

[Che71] T. C. Chen. A binary multiplication scheme based on squaring. *IEEE Trans. Computers*, C-20:678–680, 1971.

[CS85] L. Ciminiera and A. Serra. Efficient serial-parallel arrays for multiplication and addition. In *Proc. 7th IEEE Symposium on Computer Arithmetic*, pages 28–35. IEEE Computer Society, June 1985.

[CW79] I-Ngo Chen and R. Willoner. An $O(N)$ parallel multiplier with bit-sequential input and output. *IEEE Trans. Computers*, C-28(10):721–727, October 1979.

[Dad65] L. Dadda. Some schemes for parallel multipliers. *Alta Freq.*, 34:349–356, 1965. Reprinted in [Swa80].

[Dad83] L. Dadda. Some schemes for fast serial input multipliers. In *Proc. 6th IEEE Symposium on Computer Arithmetic*, pages 52–59. IEEE Computer Society, 1983.

[Dad85a] L. Dadda. Fast multipliers for two's-complement numbers in serial form. In *Proc. 7th IEEE Symposium on Computer Arithmetic*, pages 57–63. IEEE Computer Society, 1985.

[Dad85b] L. Dadda. Squarers for binary numbers in serial form. In *Proc. 7th IEEE Symposium on Computer Arithmetic*, pages 173–180. IEEE Computer Society, 1985.

[EL88] M. D. Ercegovac and T. Lang. On-line arithmetic: a design methodology and applications in digital signal processing. *VLSI Signal Processing, IEEE*, III:252–263, 1988. Reprinted in [Swa90].

[EL91] S. Even and A. Litman. Systematic design and explanation of the Atrubin multiplier. In *Sequences II, Methods in Communication, Security and Computer Sciences.* Springer, 1991.

[EL97] M. D. Ercegovac and T. Lang. Effective coding for fast redundant adders using the radix-2 digit set {0, 1, 2, 3}. In *Proc. 31st Asilomar Conf. Signals Systems and Computers*, pages 1163–1167. IEEE, 1997.

[Erc84] M. D. Ercegovac. On-line arithmetic: an overview. *Real Time Signal Processing VII*, SPIE-495:86–93, 1984.

[Erc03] M. Ercegovac. Left-to-right squarer with overlapped LS and MS parts. In *Asilomar Conference on Signals, Systems and Computers*, volume 2, pages 1451–1455. IEEE, 2003.

[Eve97] G. Even. A real-time systolic integer multiplier. *Integration*, 22:22–38, 1997.

[GHM89] A. Guyot, B. Hochet, and J.-M. Muller. JANUS, an on-line multiplier/divider for manipulating large numbers. In *Proc. 9th IEEE Symposium on Computer Arithmetic*, pages 106–111. IEEE Computer Society, 1989.

[Gna85] R. Gnanasekaran. A fast serial-parallel binary multiplier. *IEEE Trans. Computers*, C-34:741–744, 1985.

[GSA06] M. Gok, M. J. Schulte, and M. G. Arnold. Integer multipliers with overflow detection. *IEEE Trans. Computers*, 55(8):1062–1066, August 2006.

[IV94] P. Ienne and M. A. Viredaz. Bit-serial multipliers and squarers. *IEEE Trans. Computers*, 43(12):1445–1450, December 1994.

[KNE+87] S. Kuninobu, T. Nishiyama, H. Edamatsu, T. Taniguchi, and N. Takagi. Design of high speed MOS multiplier and divider using redundant binary representation. In *Proc. 8th IEEE Symposium on Computer Arithmetic*, pages 80–86. IEEE Computer Society, 1987.

[KO62] A. Karatsuba and Y. Ofman. Multiplication of many-digital numbers by automatic computers. *Doklady Akad. Nauk SSSR*, 145: 293–294, 1962. (Translation in Physics-Doklady 7: 595–596, 1963.)

[Kor94] P. Kornerup. A systolic, linear-array multiplier for a class of right-shift algorithms. *IEEE Trans. Computers*, C-43(8):892–898, August 1994.

[Lyo76] R. F. Lyon. Two's complement pipeline multipliers. *IEEE Trans. Communication*, 418–425, April 1976.

[Mac61] O. L. MacSorley. High-speed arithmetic in binary computers. *Proc. IRE*, 49:67–91, January 1961. Reprinted in [Swa80].

[Mat09] D. W. Matula. Higher radix squaring operations employing dual operand recoding. In *Proc. 19th IEEE Symposium on Computer Arithmetic*, pages 39–47, IEEE Computer Society, 2009.

[MG75] T. G. Mcdaneld and R. K. Guha. The two's complement quasi-serial multiplier. *IEEE Trans. Computers*, C-24:1233–1235, 1975.

[MJ82] J. V. McCanny and J. G. McWhirter. Completely iterative, pipelined multiplier array suitable for VLSI. *IEE Proc.*, 129(2):40–46, April 1982.

[MJ92] Z. J. Mou and F. Jutand. "Overturned-stairs" adder trees and multiplier design. *IEEE Trans. Computers*, 41(8):940–948, August 1992.

[MMT08] J. Moore, D. W. Matula, and M. L. Thornton. A low power radix-4 dual recoded integer squaring implementation for use in the design of application specific arithmetic circuits. In *Asilomar Conference on Signals, Systems and Computers*. IEEE, 2008.

[OSY+95] N. Ohkubo, M. Shinbo, T. Yamanaka, *et al.* A 4.4 ns CMOS 54 × 54-b multiplier using pass transistor multiplexer. *IEEE J. Sol. State Circuits*, 30(3):251–257, 1995.

[OV95] V. G. Oklobdzija and D. Villeger. Improving multiplier design by using improved column compression tree and optimized final adder in CMOS technology. *IEEE Trans. VLSI*, 3(2):292–301, 1995.

[OVL96] V. G. Oklobdzija, D. Villeger, and S. S Liu. A method for speed optimized partial product reduction and generation of fast parallel multipliers using an algorithmic approach. *IEEE Trans. Computers*, 45(3):294–306, March 1996.

[PL74] A. Peled and B. Liu. A new realization of digital filters. *IEEE Trans. Acoustics, Speech and Signal Processing*, 22(6):456–462, December 1974.

[SC03] A. G. H Strollo and D. De Caro. Booth folding encoding for high performance squarer circuits. *IEEE Trans. Circuits and Systems-II, Analog and Digital Signal Processing*, 50(5):250–254, 2003.

[SH88] M. R. Santoro and M. R. Horowitz. A pipelined 64×64b iterative array multiplier. *Proc. IEEE International Solid-State Circuit Conference*, pages 36–37. IEEE, 2008.

[SMOR98] P. F. Stelling, C. U. Martel, V. G. Oklobdzija, and R. Ravi. Optimal circuits for parallel multipliers. *IEEE Trans. Computers*, 47(3):273–285, March 1998.

[Swa73] E. E. Swartzlander. The quasi-serial multiplier. *IEEE Trans. Computers*, C-22:317–321, 1973. Reprinted in [Swa80].

[Swa80] E. E. Swartzlander, editor. *Computer Arithmetic,* volume I. Dowden, Hutchinson and Ross, Inc., 1980. Reprinted by IEEE Computer Society Press, 1990.

[Swa90] E. E. Swartzlander, editor. *Computer Arithmetic,* volume II. IEEE Computer Society Press, 1990.

[TE77] K. S. Trivedi and M. D. Ercegovac. On-line algorithms for division and multiplication. *IEEE Trans. Computers*, C-26(7):681–687, July 1977. Reprinted in [Swa90].

[TYY85] N. Takagi, H. Yasuura, and S. Yajima. High-speed VLSI multiplication algorithm with a redundant binary addition tree. *IEEE Trans. Computers*, C-34(9):789–796, September 1985.

[Vui83] J. Vuillemin. A very fast multiplication algorithm for VLSI implementation. *INTEGRATION, the VLSI Journal*, 1:39–52, 1983. Reprinted in [Swa90].

[Wal64] C. S. Wallace. A suggestion for a fast multiplier. *IEEE Trans. Electronic Computers*, EC-13:14–17, 1964. Reprinted in [Swa80].

[Wei81] A. Weinberger. 4-2 carry-save adder module. *IBM Technical Disclosure Bulletin*, 23, January 1981.

[YJ00] W.-C. Yeh and C.-W. Jen. High-speed Booth encoded parallel multiplier design. *IEEE Trans. Computers*, 49(7):692–701, 2000.

5

Division

5.1 Introduction

Radix division pertains to the division operation in which the arguments and results are represented by radix polynomials. Thus radix division must be defined over the set $\mathbb{Q}_\beta$ of radix-β numbers. Division is the reverse problem to multiplication. Recall that multiplication is closed for reals, rationals, integers, and radix-β number types in the sense that the product of two numbers of a given type is a unique single number of that same type. Such a $2:1$ closed mapping characterization of division is possible for reals and rationals, but not for integer or radix-β number types.

For radix division of the dividend $x = i\beta^j$ by the divisor $y = k\beta^l$, the value x/y is an exact rational referred to as the *infinitely precise quotient*. The terminology is suggested by the fact that the non-terminating radix-β representation of the rational value x/y has an infinitely repeating trailing digit sequence, as noted in the decimal example $\frac{12.5}{1.1} = 11.363636\ldots$. The closed form finite precision characterization of the result of radix division over $\mathbb{Q}_\beta$ involves specification of a quotient and remainder *pair* both in $\mathbb{Q}_\beta$, analogous to integer division. Radix-β division is characterized by a mapping β-**div**: $\mathbb{Q}_\beta \times \mathbb{Q}_\beta \to \mathbb{Q}_\beta \times \mathbb{Q}_\beta$ where all (*quotient, remainder*) pairs must satisfy the fundamental *division invariant* equation

$$\text{dividend} = \text{quotient} \times \text{divisor} + \text{remainder}. \qquad (5.1.1)$$

For integer division the mapping **div**: $\mathbb{Z} \times \mathbb{Z} \to \mathbb{Z} \times \mathbb{Z}$ is one-to-one, with the quotient and remainder each being uniquely defined integers. For radix division there is a discrete family of (quotient, remainder) pairs with the quotient parameterized by both its last place position and the remainder sign, characterizing a series of *ulp accurate quotients*. For decimal division of 12.5 by 1.1 we have, for example, the decimal ulp accurate quotients $q = 11.36$ and $q = 11.363$ or 11.364.

Table 5.1.1. *Radix division terminology and precision hierarchies*

Result name	Type	Result set
Infinitely precise quotient	Exact rational	Unique value
Quotient, remainder	Exact radix-β pair	Discrete family parameterized by quotient last place
Quotient and remainder sign	Rounding ready	
Ulp accurate quotient	2-ulp interval	
Approximate quotient	Interval	Family parameterized by error bounds

Each ulp accurate quotient $q = n\beta^m$ has a corresponding remainder r satisfying $|r| < \text{ulp}(q) = \beta^m$. Ulp accurate quotients are said to become *more precise* as the last place unit $\text{ulp}(q) = \beta^m$ decreases forcing corresponding remainders to converge toward zero.

The terminology associated with radix division yields the precision hierarchy illustrated in Table 5.1.1. The context of the division problem must include, explicitly or implicitly, the particular type of quotient and precision level required. The hierarchy includes the results of integer, fixed-point, and floating-point division. The hierarchy of Table 5.1.1 also characterizes the intermediate quotients that are obtained by iterative radix division algorithms, including *approximate quotients* which must satisfy implicit or explicit error bounds relative to the infinitely precise quotient. The rounding ready and ulp accurate quotients are fundamental to obtaining precisely rounded floating-point quotients as discussed in Chapter 7.

Known arbitrary precision division methods rely primarily on dependent iterative sequences of additions and/or multiplications. The hardware dedicated to the implementation of a division operation typically comprises microcode to govern the iterative utilization of shared adders and/or multipliers, and look-up tables and normalization procedures crafted to reduce the number of iterative passes through the shared adder/multiplier. Facility for the shared adder/multiplier to accept feedback results in redundant form is another valuable property for expediting division latency.

The unary operation of divisor reciprocal formation is a special case of division with a separate literature for efficient determination of low to moderate precision approximate reciprocals by direct non-iterative table-assisted procedures.

Efficient design of hardware for implementing a radix division operation, and/or a divisor reciprocal operation, requires familiarity with a variety of division and reciprocal algorithms each finely tuned to the nature of the adder or multiplier and look-up table resources available to support the iterative division steps and/or the direct reciprocal evaluation.

In Section 5.2 we survey the popular algorithms for division and reciprocal formation. Most of these algorithms are sufficiently competitive that they have been implemented in various processors. The principal elements of each algorithm are provided so that the casual reader may find Section 5.2 a sufficient introduction to the subject of implementing the division operation in an arithmetic unit of a contemporary processor.

Section 5.3 provides a brief formalization of the radix-β division mapping and the elementary conversion operations between alternative quotient, remainder pairs corresponding to a particular dividend divisor operand pair. The intention is to rely on the known foundation of integer division as the motivation for the extension to radix-β division.

Sections 5.4–5.7 present details of the popular division algorithms for the serious student interested in the challenges of implementing division in an arithmetic unit. Section 5.8 does the same for the popular reciprocal operations that are prevalent for multimedia and graphics data ALUs.

5.2 Survey of division and reciprocal algorithms

Algorithms for division to an arbitrary precision level generally fall into one of two broad classes possessing different methodology and goals. The algorithms described herein provide instances of this classification.

Digit serial division These algorithms provide for each iteration a radix-β quotient, remainder pair (q_i, r_i) exactly satisfying *the division invariant*

$$\text{dividend} = q_i \times \text{divisor} + r_i,$$

by determining a new digit and remainder such that the remainder stays bounded. Convergence is guaranteed by $\text{ulp}(q_{i+1}) < \text{ulp}(q_i)$. Typically $\text{ulp}(q_{i+1}) = \beta^{-1}\text{ulp}(q_i)$, and the methods are said to converge linearly. Each iteration requires the same finite computing resource, and can continue to output successive low-order digits of the quotient indefinitely.

Iterative refinement division These algorithms provide for each iteration an ith value differing at most by a constant from a value q_i, where q_i approximates the ratio dividend/divisor with relative error ε_i, where $\varepsilon_{i+1} \ll \varepsilon_i$. A final postiterative multiplication allows us to obtain the desired quotient approximation.

Typically $\varepsilon_{i+1} = O(\varepsilon_i^2)$, and the methods are said to converge quadratically. When the values q_i, q_{i+1} are given as radix-β numbers, quadratic convergence is said to "double the number of accurate digits" each iteration. Successive iterations providing this performance require increasing computing resources.

The general division methods to be discussed in this chapter are listed and classified in Table 5.2.1.

Table look-up algorithms for determining a divisor reciprocal approximation of moderate precision are valuable to provide both an efficiently computable

Table 5.2.1. *Classification of division algorithms*

Digit-serial algorithms		Iterative refinement algorithms
Deterministic digit selection	*Non-deterministic digit selection*	
Restoring	SRT (higher radix)	Newton–Raphson
Non-restoring	Short reciprocal	Convergence
Binary SRT	Prescaled	Postscaled

SRT: the class of algorithms named after Sweeney, Robertson, and Tucker.

primitive unary arithmetic operation and a seed reciprocal value for the Newton–Raphson and other convergence iterative refinement division algorithms. Table look-up procedures for reciprocals generally fall into two classes that may be characterized as follows.

Direct reciprocal look-up These procedures employ one or more direct look-up tables (LUTs) each indexed by a comparable length index, with the sum providing the approximate reciprocal value of accuracy from one to two times the digit length of the table index.

Multiplicative interpolation These procedures employ table look-up followed by a low-precision multiply to obtain an approximate reciprocal value of accuracy from two to three times the digit length of the table index.

The divisor reciprocal formation table look-up algorithms discussed in this chapter are listed and classified in Table 5.2.2.

Table 5.2.2. *Classification of reciprocal look-up algorithms*

Direct	Multiplicative interpolation
Single index	Linear interpolation
Bipartite	Quadratic interpolation
Multi-partite	

Direct multiplicative binary division. For practical implementations of binary division to a modest range of precision levels, several direct (non-iterative) multiplicative binary division algorithms are available. These direct division methods express the approximate quotient as a product of two-to-four factors, where the factors are formed employing table look-up, addition, and/or a small multiplier. The direct multiplicative division methods to be discussed provide from k to $4k$ bits of accuracy in the approximate quotient employing k-bit table indices, with all methods employing at most two full precision multiply latencies.

5.2.1 Digit-serial algorithms

The *digit-serial paradigm* itemizes the steps implicitly performed sequentially or concurrently in each iteration for all these digit-serial algorithms:

(i) quotient digit selection;
(ii) divisor multiple formation;
(iii) remainder formation.

The steps are conveniently illustrated by a traditional decimal long division example.

Example 5.2.1 Consider the division state for 6.3820 divided by 2.17 after the two-digit quotient 2.9 with remainder 0.089 has been determined:

$$\begin{array}{r l}
 & 2.9[4]^{(i)} \\
2.17 & \overline{)6.3820} \\
 & \underline{434} \\
 & 2042 \\
 & \underline{1953} \\
 & \quad\;\; 890 \\
 & \quad\; \underline{[868]}^{(ii)} \\
 & \qquad\; [22]^{(iii)}
\end{array}$$

The next iteration must then successively (i) determine the next quotient digit 4, (ii) determine the divisor multiple $4 \times 217 = 868$, and (iii) form the next remainder by alignment and subtraction $890 - 868 = 22$. The result is then the three-digit quotient 2.94 with remainder 2.2×10^{-3}, and another iteration may commence. □

The digit-serial algorithms can then be summarized with reference to the steps of the digit-serial paradigm.

Restoring division The aligned divisor is iteratively subtracted from the remainder until the result becomes negative, and is then added back in once to form the next remainder (completing (ii) and (iii)). The net count of the number of subtractions is the next quotient digit (completing (i)). There are, in general, $d + 2$ primitive subtract/add steps employed in concurrently generating the next quotient digit d and new remainder for $0 \leq d \leq \beta - 1$. The applicability of restoring division is limited to relatively small radices.

Non-restoring division The aligned divisor is iteratively subtracted from a positive remainder until the result becomes non-positive, yielding the next remainder (completing (ii) and (iii)). The number of subtractions is the next quotient digit (completing (i)). For a negative remainder, the divisor is iteratively added until the result becomes non-negative, the negative of the count being the next signed quotient digit. There are, in general, $|d|$ primitive add/subtract steps employed in concurrently generating the next remainder and next quotient digit for $-\beta \leq d \leq \beta$, $d \neq 0$. This limits the applicability of non-restoring division to relatively small radices.

SRT division The leading digits of the divisor and remainder are used to determine explicitly the next quotient digit, possibly by table look-up (completing (i)). The next remainder is typically found by subtracting from or adding to the current remainder an appropriately shifted value selected from a precomputed store of selected divisor multiples (completing (ii) and (iii)). SRT division is generally limited to relatively small radices $2 \leq \beta \leq 8$ due to both the excessive size of quotient digit tables for larger radices and the concurrent need for a greater store of selected divisor multiples to realize step (ii).

Short reciprocal division A preprocessing step is used to determine a low-precision "short reciprocal" approximating the divisor reciprocal with relative error less than $1/\beta^k$. Here β^k is a high radix such as $\beta^k = 2^k$ with $6 \leq k \leq 16$, where the "short reciprocal" has k-bit accuracy. Digit selection is performed by multiplying the remainder by the short reciprocal and appropriately rounding the result to a single radix-β^k digit (completing (i)). The divisor multiple is formed by multiplying the selected digit by the divisor (completing (ii)), and the product is subtracted from the remainder to yield the updated remainder (completing (iii)). Note that both digit selection and divisor multiple formation can serially share a single "short-by-long" multiplier, where the shorter dimension effectively limits the choice of high radix size β^k for the quotient.

Prescaled division A first preprocessing step is used to determine a low-precision "short reciprocal" ρ approximating the divisor reciprocal with relative error less than $1/\beta^k$. In a second preprocessing step the dividend and divisor are multiplied by the short reciprocal to obtain a scaled dividend and scaled divisor. In the prescaled division procedure the remainder (which inherits the same scale factor) is appropriately rounded to a single high radix-β^k digit providing the next quotient digit (completing step (i)). Note that the quotient digits of the prescaled division problem identically correspond to quotient digits of the original division problem. The divisor multiple is formed by multiplying the selected digit by the scaled divisor (completing (ii)), and the product is subtracted from the scaled remainder to yield the updated scaled remainder (completing (iii)). Divisor multiple formation and the dividend and divisor prescaling can both share a single "short-by-long" multiplier, where the shorter dimension effectively limits the choice of radix size for the quotient. For moderate precision levels such as $2^k = 8, 16, 32$, the prescaling and divisor multiple formation can be obtained by addition of a few terms. Note that prescaled division does not provide the remainder corresponding to the final quotient for the original division problem, but can provide its positive, zero, or negative status.

Prescaled division with remainder In this variation of prescaled division the (unscaled) dividend is divided by a scaled divisor exploiting the efficiency of digit determination. A final digit selection step operates similarly to short reciprocal division to "descale" the divisor and obtain the appropriate quotient–remainder pair.

5.2.2 Iterative refinement algorithms

The methodology for iterative refinement can also be expressed by a general paradigm utilizing the following definitions. For an approximation $\tilde{z}$ to z of relative error ε,

$$\tilde{z} = z(1+\varepsilon),$$

let $(1+\varepsilon)$ be termed the *relative error factor*.

Let the corresponding term $(1-\varepsilon)$ be called the *complementary error factor*. Note that multiplying the approximation $\tilde{z}$ by the complementary error factor yields a much improved approximation to z,

$$z' = \tilde{z}(1-\varepsilon) = z(1-\varepsilon^2).$$

Note also that it is sufficient to determine a good approximation to the relative error factor $(1+\varepsilon)$ to obtain a much improved approximation to z. The *iterative refinement paradigm* itemizes the key steps sequentially or concurrently employed during each iteration:

(i) approximate relative error factor determination;
(ii) complementary error correction factor formation;
(iii) approximate result refinement.

Newton–Raphson division The result of the iterative refinement is to be an approximate reciprocal of the divisor with relative error guaranteed sufficiently small. In a preprocessing step an initial approximate reciprocal ρ_0 of the divisor y is determined. During each iteration the product $y\rho_i$ is determined (completing (i)), and its complement $(2-y\rho_i)$ formed (completing (ii)). The product $\rho_{i+1} = \rho_i(2-y\rho_i)$ then provides the refined approximate reciprocal (completing (iii)).

A postprocessing transformation step to obtain the quotient is performed by suitably rounding the product of the dividend x and a final reciprocal approximation ρ_n, $q = R(x \times \rho_n)$.

If desired the corresponding remainder can be computed from $r = q \times y - x$; this step potentially requires more hardware.

Convergence division The result of the iterative refinement is an approximate quotient q with relative error guaranteed sufficiently small. In a preprocessing step an approximate reciprocal ρ of the divisor y is determined and multiplied by the dividend x to produce an initial approximate quotient $q_1 = x \times \rho$, and concurrently with the divisor y to provide an initial relative error factor $1 - \varepsilon = y\rho$.

During each iteration a complementary error correction factor $2 - (1 - \varepsilon^{2^i}) = (1 + \varepsilon^{2^i})$ and a new relative error factor $1 - \varepsilon^{2^{i+1}} = (1 + \varepsilon^{2^i})(1 - \varepsilon^{2^i})$ are found (completing (ii) and (i)). The product $q_{i+1} = q_i(1 + \varepsilon^{2^{i+1}})$ then provides the refined approximate quotient (completing (iii)).

In a postprocessing step a quotient is obtained by suitably rounding the final approximate quotient $q = R(q_n)$ to a last place consistent with the known bound on the relative error of q_n.

The corresponding remainder is optionally computed from $r = q \times y - x$; this step potentially requires more hardware.

Postscaled division This method defers the initial scaling by a divisor reciprocal approximation and replaces it by a postscaling by a power of the divisor reciprocal.

The divisor is partitioned to form $y = d - \delta$, where d is a leading part of size appropriate for a table index and $\delta/d = \varepsilon$ is small. During each iteration a complementary divisor factor is formed, for concurrent multiplications of both the numerator term, initially x, and the denominator term, initially $y = d - \delta$. The complementary factors $(d + \delta)$, $(d^2 + \delta^2)$, $(d^4 + \delta^4)$, ..., yield denominator products $(d^2 - \delta^2)$, $(d^4 - \delta^4)$, ..., $(d^{2^i} - \delta^{2^i})$, where $(1/d)^{2^i}$ is provided by table look-up when the implicit relative error $(\delta/d)^{2^i}$ is sufficiently small.

The reciprocal power table look-up of $(1/d)^{2^i}$ can be overlapped with the iterative multiplies, removing table look-up latency from the critical path. This procedure avoids the preprocessing table look-up latency of the other iterative refinement division algorithms. The table look-up reciprocal power output must be provided to full precision width, although the index size can be comparable to the other iterative refinement direct lookups.

5.2.3 Resource requirements

The computing requirements of the eight division algorithms summarized here vary over the spectrum of the trade-off between computation speed and hardware size. It is convenient to measure speed in terms of the number of quotient bits per iteration for a constant-time iteration. Hardware size may be compared for these algorithms by noting the number of adders employed, with a $(k \times n)$ bit multiplier assumed equivalent to k n-bit adders.

The slower restoring, non-restoring, and SRT algorithms utilize only an adder as the principal hardware resource. The short reciprocal and prescaled division algorithms require a short-by-long, e.g., $(k \times n)$-bit, multiplier. The hardware size of a short-by-long multiplier scales down from that of a full (e.g., $(n \times n)$-bit) multiplier roughly by the short/long ratio. The short side should not be chosen too small otherwise the method would reduce to a few cascaded steps of SRT division. It should not be chosen too large otherwise the preprocessing steps of short reciprocal and prescaled division would approach that of the full division problem.

For 64-bit division, choosing a radix between 2^{11} and 2^{17}, with a multiplier of aspect ratio between 6 to 1 and 4 to 1, means that the short reciprocal and prescaled division algorithms fill the performance gap between the traditional digit-serial methods and the accelerated iterative refinement methods.

The Newton–Raphson, and convergence division algorithms generally assume a full long-by-long multiplier with a long result.

Note that if the option to also find the corresponding remainder is needed, the full double-length product of the long-by-long result is implicitly required to correctly obtain the remainder.

By increasing the number of adders from one to n (in the long-by-long multiplier) we reduce the time from $O(n)$ iterations for the elementary digit-serial algorithms to $O(\log n)$ iterations for the iterative refinement algorithms. This is close to an equal trade-off of hardware with time. The short reciprocal and prescaled algorithms fill in the trade-off curve. They exhibit an essentially linear decrease in time with increase in hardware (assuming constant-time multiplication) until the short reciprocal determination time dominates, to merge with the iterative refinement schemes in effectiveness.

5.2.4 Reciprocal look-up algorithms

Look-up tables (LUTs). Reciprocal LUT procedures can be summarized with reference to a normalized binary divisor $y = 1.b_1b_2 \cdots b_{p-1}$, where substrings of leading bits provide indices to one or more LUTs whose outputs are summed to provide the approximate reciprocal.

Direct The leading i-bit string $b_1b_2 \cdots b_i$ provides the index to a table of size $j2^i$ bits giving the leading j bits of the normalized reciprocal approximation. Typically $j = i + g$, where the LUT directly provides about i-bit accuracy with $0 \le g \le 2$ denoting a number of guard bits.

Bipartite The leading i-bit string $b_1b_2 \cdots b_i$ provides the index giving a $j \approx \frac{3}{2}i$ bit string output from the primary table. The next $\frac{1}{2}i$ bits $b_{i+1}b_{i+2} \cdots b_{\frac{3}{2}i}$ along with the leading $\frac{1}{2}i$ bits $b_1b_2 \cdots b_{i/2}$ form an i-bit index to an interpolating table, essentially providing a correcting term added to the lower $\frac{1}{2}i$ bits of the primary output to provide an approximate reciprocal of about $\frac{3}{2}i$-bit accuracy. The bipartite process may be considered a direct look-up of a redundant binary (borrow-save or carry-save) represented value.

Multipartite The leading i-bit string $b_1b_2 \cdots b_i$ provides the index giving a $j \approx 2i$ bit string output from the primary table. The two or more interpolating tables each employ additional substrings of bits from the next i bits $b_{i+1}b_{i+2} \cdots b_{2i}$, each substring paired with sufficient initial bits to provide a succession of correcting terms from the successive LUTs with the sum providing an approximate reciprocal of accuracy approaching $2i$ bits.

Multiplicative interpolation reciprocal look-up. The multiplicative interpolation reciprocal procedures employ a leading i-bit string $b_1b_2\cdots b_i$ obtained from the normalized binary divisor $y = 1.b_1b_2\cdots b_{p-1}$ as the index to a coefficient LUT providing two or more coefficients of a polynomial reciprocal approximation. The argument of the polynomial is determined by the trailing bit string $b_{i+1}b_{i+2}\cdots b_{p-1}$, with an argument value implicitly scaled to be less than 2^{-i}.

Linear interpolation For linear interpolation the coefficient table provides a leading $2i$-bit reciprocal value and an i-bit linear term coefficient giving the slope of the reciprocal function for that index.

The i-bit slope term is multiplied by the i-bit proportional part string $b_{i+1}b_{i+2}\cdots b_{2i}$ with the leading i bits of the product forming an interpolating term added to the constant term providing an approximate reciprocal of $2i$-bit accuracy.

Quadratic interpolation The quadratic interpolation coefficient table provides (i) a leading $3i$-bit reciprocal value, (ii) a $2i$-bit linear (slope) coefficient to be multiplied by the $2i$-bit proportional part string $b_{i+1}b_{i+2}\cdots b_{3i}$, and (iii) an i-bit quadratic coefficient.

The i-bit string $b_{i+1}b_{i+2}\cdots b_{2i}$ can provide an index to a squarer LUT providing an i-bit argument square value to be multiplied by the i-bit quadratic coefficient. The two multiplies can be formed concurrently so the quadratic reciprocal approximation of $3i$-bit accuracy can be formed with latency given by the i-bit look-up followed by the small precision multiply latency.

5.3 Quotients and remainders

Division is the reverse problem to multiplication. Multiplication is closed for reals, rationals, integers, and radix-β number types in the sense that the product of two numbers of a given type is a unique number of that same type. Such a convenient characterization of the result of division is possible for reals and rationals, but not, in general, for integer and radix-β number types.

The complete specification of the result of division for integer number types, and in general radix-β numbers, involves determination of a quotient, remainder pair of the corresponding type. All such quotient remainder pairs must satisfy the fundamental *division invariant* equation:

$$\text{dividend} = \text{quotient} \times \text{divisor} + \text{remainder}. \qquad (5.3.1)$$

5.3.1 Integer quotient, remainder pairs

Integer division is a mapping **div**: $\mathbb{Z} \times \mathbb{Z} \to \mathbb{Z} \times \mathbb{Z}$ with the following traditional definition.

Definition 5.3.1 *Given an integer dividend x and an integer divisor $y \neq 0$, the integer quotient q and remainder r are the unique pair satisfying*

$$x = qy + r$$

with

$$|r| \leq |y| - 1,$$

where any non-zero remainder must have the same sign as the dividend.

The result of integer division then comprises a unique quotient and unique remainder, where either may be specified as a separate and uniquely defined function.

Lemma 5.3.2 *The quotient and remainder functions for integer division of x by $y \neq 0$ are*

$$q(x, y) = \begin{cases} \left\lfloor \dfrac{x}{y} \right\rfloor & \text{if } x/y \geq 0, \\ \left\lceil \dfrac{x}{y} \right\rceil & \text{if } x/y < 0, \end{cases}$$

$$r(x, y) = x - yq(x, y). \tag{5.3.2}$$

Although the quotient function $q(x, y) = q(v)$ depends only on the rational value $x/y = v \in \mathbb{Q}$, the remainder function depends on the dividend, divisor pair (x, y) in several ways not evident from the rational value. The remainder shares the same greatest common divisor with the dividend and divisor, and has the same sign as the dividend, both properties which are not computable from the rational value, as illustrated in the following table.

Dividend x	Divisor y	Rational $x/y \in \mathbb{Q}$	Quotient $q(x, y)$	Remainder $r(x, y)$	$\gcd(x, y)$ $=\gcd(y, r)$	Division Invariant $x = qy + r$
8	6	$1.33\cdots$	1	2	2	$8 = 1 \times 6 + 2$
4	3	$1.33\cdots$	1	1	1	$4 = 1 \times 3 + 1$
-7	4	-1.75	-1	-3	1	$-7 = -1 \times 4 + (-3)$
7	-4	-1.75	-1	3	1	$7 = -1 \times (-4) + 3$
7	4	1.75	1	3	1	$7 = 1 \times (4) + 3$
-7	-4	1.75	1	-3	1	$-7 = 1 \times (-4) + (-3)$

The examples show that -7 divided by 4 does not yield the same remainder as 7 divided by -4. It follows that in integer division one may not preprocess the operation by attaching the quotient sign to the dividend and hence treating the divisor as unsigned. It is possible to compute unsigned $|x|$ divided by $|y|$ and then attach the signs to $|q|, |r|$ yielding $q = (-1)^{s(x)+s(y)} \times |q|$, and $r = (-1)^{s(x)} \times |r|$.

The quotient for integer division is obtained by effectively rounding the rational $x/y = v \in \mathbb{Q}$ towards zero. Alternatively v could be rounded away from zero, yielding a complementary quotient, remainder pair.

Definition 5.3.3 *The complementary quotient q^* and complementary remainder r^* for integer division of x by $y \neq 0$ are given by*

$$q^*(x, y) = \begin{cases} \left\lceil \dfrac{x}{y} \right\rceil & \text{if } x/y \geq 0, \\ \left\lfloor \dfrac{x}{y} \right\rfloor & \text{if } x/y < 0, \end{cases}$$

$$r^*(x, y) = x - yq^*(x, y). \tag{5.3.3}$$

Observation 5.3.4 *The remainder and complementary remainder are either both zero, or otherwise $|r^*(x, y)| = |y| - |r(x, y)|$ with $r^*(x, y)$ having a sign opposite to that of $r(x, y)$.*

An alternative precise definition for integer division and the remainder function was employed in the ANSI/IEEE 754 floating-point standard. The remainder function employed there always chooses between (q, r) and (q^*, r^*) so as to obtain a remainder of minimum magnitude of at most $|y/2|$. The quotient is effectively a rounding of x/y to "nearest, ties to even."

Definition 5.3.5 *The integer quotient and remainder functions for IEEE integer division of x by $y \neq 0$ are:*

(i) for $x/y + \frac{1}{2}$ not an integer,

$$q_{ieee}(x, y) = \left\lfloor \frac{x}{y} + \frac{1}{2} \right\rfloor$$

(ii) for $x/y + \frac{1}{2}$ an integer, $x/y + \frac{1}{2} = k$,

$$q_{ieee}(x, y) = \begin{cases} \left\lfloor \frac{x}{y} \right\rfloor = k - 1 & \text{if } k \text{ is odd}, \\ \left\lceil \frac{x}{y} \right\rceil = k & \text{if } k \text{ is even}, \end{cases}$$

and

$$r_{ieee}(x, y) = x - y\, q_{ieee}(x, y).$$

The characterization of unique integer quotient, remainder pairs (q, r) as the result of integer division extends to the case where the dividend and divisor are arbitrary rational fractions $x = i/j$ and $y = k/l$. We assume $x/y = il/jk$ using divisor jk to establish uniqueness of the remainder, where the quotient is determined simply by rounding the rational x/y to the appropriate neighboring integer. If the inputs i, j, k, l have magnitudes at most n, note that the integers q, r can have magnitudes as large as n^2 and $n^2 - n$, respectively. The characterization of unique integer division for $x, y \in \mathbb{Q}_\beta$ also provides the unique integer quotient by rounding the rational value x/y. For $x = i\beta^j$ and $y = k\beta^\ell \neq 0$, the remainder can be made unique by assuming common factors of β are scaled out, so

$x/y = i\beta^{j-l}/k$ with integer dividend $i\beta^{j-l}$ for $j - l \geq 0$, and $x/y = i/k\beta^{l-j}$ with integer divisor $k\beta^{l-j}$ for $j - l < 0$. Even if i and k are bounded, the quotient for integer division with $x = i\beta^j$ and $y = k\beta^l \neq 0$ can be very large. However, the remainder in this case satisfies a useful bound.

Lemma 5.3.6 *Let $x = i\beta^j$ and $y = k\beta^l \neq 0$. Then $|r(x, y)| \leq \max\{|i|, |k|\}$.*

Proof For $j - l > 0$, integer division of $i\beta^{j-l}$ by k yields a remainder of magnitude less than $|k|$. For $j - l < 0$, integer division of i by $k\beta^{l-j}$ yields a remainder of magnitude $|i|$ or smaller. □

Note that the complementary quotient, remainder pair (q^*, r^*) may have r^* very large independent of the bound on i and k in Lemma 5.3.6, but the remainder $r_{ieee}(i\beta^j, k\beta^l)$ will satisfy the bound of Lemma 5.3.6. The fact that the remainder for radix-β division satisfies the bound of $\max\{|i|, |k|\}$ is important for implementing the IEEE floating-point standard which requires such an integer remainder function for floating-point arguments.

5.3.2 Radix-β quotient, remainder pairs

Consider division procedures where the integers x, y are given as standard radix-β digit strings $x = d_{n-1}d_{n-2}\cdots d_0$ and $y = y_{m-1}y_{m-2}\cdots y_0$ yielding integer valued digit strings for the quotient, remainder pair (q, r). By successively scaling down the divisor when $y_0 = 0$ or alternatively scaling up the dividend by factors of β, the resulting integer quotient sequence defined in this manner effectively appends an additional low-order digit for each quotient of the sequence.

For a decimal example initiated by $x = 882$, $y = 7500$, we successively obtain $q(882, 7500) = 0$, $q(882, 750) = 1$, $q(882, 75) = 11$, $q(8820, 75) = 117$, $q(88200, 75) = 1176\cdots$. The quotients will grow in size indefinitely, while the corresponding remainders will, in general, all remain bounded in $[0, r - 1]$, and will eventually cycle.

The family of such (q, r) pairs and their complements (q^*, r^*) can be characterized by introducing radix-β division as a one-to-many mapping β-div: $\mathbb{Q}_\beta \times \mathbb{Q}_\beta \to \mathbb{Q}_\beta \times \mathbb{Q}_\beta$. Here and in the following subsections, for brevity we shall by a small abuse of notation utilize $x, y, q, r \in \mathbb{Q}_\beta$ to apply to particular factorizations, i.e., ulp(q) denotes ulp$(i\beta^j) = \beta^j$, identifying a particular last place index j that will parameterize the family of radix-β (q, r) pairs.

Definition 5.3.7 *For the dividend $x \in \mathbb{Q}_\beta$ and divisor $y \in \mathbb{Q}_\beta$, $y \neq 0$, a* radix-β quotient, remainder pair (q, r) *is any pair $q, r \in \mathbb{Q}_\beta$ of numbers satisfying*

$$x = q \times y + r \tag{5.3.4}$$

with

$$\left|\frac{r}{y}\right| = \left|\frac{x}{y} - q\right| < \text{ulp}(q). \tag{5.3.5}$$

From (5.3.4) we note that $\text{ulp}(r) = \min\{\text{ulp}(x), \text{ulp}(q)\text{ulp}(y)\}$. The bound (5.3.5) expresses the condition that the quotient $q = i\beta^j$ must differ in its last place by less than one unit from the rational x/y, or equivalently $|r| < |y|\beta^j$.

The parameters ulp(q) and remainder sign provide for a convenient tabulation of a family of quotient, remainder pairs.

Example 5.3.1 For the decimal dividend 8.82 and divisor 75, the decimal quotient, remainder pairs and their complements for quotient units in the last place from 10^0 down to 10^{-5} are

ulp(q)	$(q, r \geq 0)$	$(q^*, r^* < 0)$
1	(0, 8.82)	(1, −66.18)
.1	(.1, 1.32)	(.2, −6.18)
.01	(.11, 0.57)	(0.12, −0.18)
.001	(.117, 0.045)	(0.118, −0.030)
.0001	(.1176, 0)	NA
.00001	(.11760, 0)	NA

Note that the sequences of quotients q and q^* for successive last places 10^l form the "best radix-10 approximations" to $v = 882/7500$ as discussed in Section 1.9.1. □

As suggested by the preceding example, the following confirms that (q, r) pairs will be unique if both ulp(quotient) $= \beta^\ell$ and the sign of any non zero remainder are specified.

Lemma 5.3.8 *Given the dividend x, divisor $y \neq 0$ and* ulp(quotient) $= \beta^\ell$, *there is either:*

(i) exactly one quotient, remainder pair (q, r) with ulp(q) $= \beta^\ell$ *and a zero remainder,*

or

(ii) exactly two distinct quotient, remainder pairs (q, r), (q^*, r^*) with ulp(q) = ulp(q^*) $= \beta^\ell$ *and r, r^* of opposite signs.*

In the latter case we further obtain

$$\begin{aligned} q^* - q &= \beta^\ell, \\ r^* - r &= -y\beta^\ell. \end{aligned} \tag{5.3.6}$$

Proof For uniqueness here we must allow the zero values $0 \cdot \beta^i$ for any i. Then for $x = 0 \cdot \beta^i$, the unique result $q = 0 \cdot \beta^i$ and $r = 0 \cdot \beta^j$ with

$\beta^j = \min\{\beta^i, \text{ulp}(y)\beta^\ell\}$ is obtained. For $x \neq 0$, define q by

$$q = \left\lfloor \frac{x}{y\beta^\ell} \right\rfloor \beta^\ell \leq \frac{x}{y}.$$

If $q = x/y$, then $r = x - qy = 0$ and $q^- = q - \text{ulp}(q)$ as well as $q^+ = q + \text{ulp}(q)$ will violate (5.3.3), hence there is exactly one quotient, remainder pair when $r = 0$.

If $q < x/y$ with $r = x - qy$, define

$$q^* = \left\lceil \frac{x}{y\beta^\ell} \right\rceil \beta^\ell > \frac{x}{y}.$$

then $q^* - q = \beta^\ell = \text{ulp}(q)$. Letting $r^* = x - q^*y$ we have $r^* - r = (q - q^*)y$, thus (5.3.6) holds. □

The properly signed quotient, remainder pair for any ulp(quotient) $= \beta^\ell$ either has remainder zero, or a remainder whose sign agrees with that of the dividend. The goal of radix-β division can be formulated as follows.

Definition 5.3.9 *The* radix-β division problem *is:*

Given: *A dividend* $x \in \mathbb{Q}_\beta$, *a divisor* $y \in \mathbb{Q}_\beta$, $y \neq 0$, *and* $\text{ulp}(quotient) = \beta^\ell$.
Find: *The quotient, remainder pair* $q, r \in \mathbb{Q}_\beta$ *with* $\text{ulp}(q) = \beta^\ell$, *satisfying* $\text{sgn}(r) = \text{sgn}(x)$ *for* $r \neq 0$, *and* $|r| < |y|\,\beta^\ell$.

5.3.3 Converting between radix-β quotient, remainder pairs

The foundations for three types of conversion operations between members of the family of radix-β quotient, remainder pairs are given below. These results will be useful in the design of algorithms to solve the division problem.

(i) Precision expansion from a quotient, remainder pair (q, r) with $\text{ulp}(q) = \beta^\ell$ to a successor quotient, remainder pair (q', r') with $\text{ulp}(q') = \beta^{\ell-1}$,
(ii) Complementing a quotient, remainder pair,
(iii) Conversion from a quotient, remainder pair for "unsigned" radix-β division of $|x|$ by $|y|$ to a quotient, remainder pair for properly signed radix-β division of x by y.

The precision expansion operation allows a non-deterministic choice of either of two successors that are complements of each other. The non-determinism allows the operation to be implemented more efficiently as we show in the next subsection. The complement operation provides a way of making a terminal successor value unique.

(i) Precision extension from a quotient with $\mathrm{ulp}(q) = \beta^\ell$ *to a successor quotient* q' *with* $\mathrm{ulp}(q') = \beta^{\ell-1}$

For any radix-β quotient, remainder pair (q, r) for radix-β division of x by $y \neq 0$, the expression $r/\mathrm{ulp}(q)$ is the *normalized remainder* corresponding to q. From (5.3.3) it follows that $|r/\mathrm{ulp}(q)| < |y|$, so the normalized remainder $r/\mathrm{ulp}(q)$ has range $-|y| < r/\mathrm{ulp}(q) < |y|$.

Theorem 5.3.10 *Let* (q, r) *be a quotient, remainder pair for radix-*β *division of* x *by* $y \neq 0$ *with* $\mathrm{ulp}(q) = \beta^\ell$. *Let* q' *and* r' *be defined from* (q, r) *and a digit* d *by*

$$q' = q + d\beta^{\ell-1} \quad (\mathrm{ulp}(q') = \beta^{\ell-1}), \tag{5.3.7}$$

$$\frac{r'}{\mathrm{ulp}(q')} = \frac{r}{\mathrm{ulp}(q)}\beta - d\,y \tag{5.3.8}$$

with d *chosen so that* $\left|r'/\mathrm{ulp}(q')\right| < |y|$. *Then* (q', r') *is a quotient, remainder pair with* $\mathrm{ulp}(q') = \beta^{\ell-1}$.

Proof From (5.3.7) we have $\mathrm{ulp}(q') = \beta^{\ell-1}$ and

$$q'y = qy + d\beta^{\ell-1}y,$$

and from (5.3.8) we obtain

$$r' = r - d\beta^{\ell-1}y,$$

so $q'y + r' = qy + r = x$ confirming (5.3.2) for (q', r'). By the choice of d, the assumption $\left|r'/\mathrm{ulp}(q')\right| < |y|$ implies $\left|r'/y\right| < \mathrm{ulp}(q')$, confirming (5.3.3), so then (q', r') is a radix-β quotient, remainder pair. □

The relation (5.3.8) is termed the fundamental *division recurrence* confirming d to be a *next quotient digit* with $r'/\mathrm{ulp}(q')$, termed the *successor normalized remainder*, being in the open interval $(-|y|; |y|)$. The relation (5.3.8) is popularly stated in the literature as a relation between remainders, where updating of the *scaled remainder* corresponding to (5.3.8) has the form $r' = r\beta - dy$. The remainders computed during the division recurrence are often termed *partial remainders* to indicate that they are not the final remainders of the division. Our formulation of (5.3.8) makes the normalization of the remainder explicit in conformance with the division invariant equation (5.3.1).

The possible values for the digit d are readily defined using the floor and ceiling of a division operation, as shown in the following theorem, the proof of which is left as an exercise.

Theorem 5.3.11 *Let* q, r *be a quotient, remainder pair for radix-*β *division of* x *by* $y \neq 0$ *with* $\mathrm{ulp}(q) = \beta^\ell$. *Then* q', r' *given by*

$$q' = q + d\beta^{\ell-1}, \tag{5.3.9}$$

$$r' = r - d\beta^{\ell-1}y \tag{5.3.10}$$

will be a radix-β quotient, remainder pair if and only if

$$d \text{ is either } \left\lfloor \frac{r\beta}{y\,\mathrm{ulp}(q)} \right\rfloor \text{ or } \left\lceil \frac{r\beta}{y\,\mathrm{ulp}(q)} \right\rceil.$$

(ii) Complementing a quotient, remainder pair

For the radix-β quotient, remainder pair (q, r) with $r \neq 0$, the *complement operation* yields the quotient, remainder pair (q^*, r^*) satisfying $\mathrm{ulp}(q^*) = \mathrm{ulp}(q)$ with r^* of opposite sign to r. Complementing a (q, r) pair involves incrementing the quotient by $\pm\mathrm{ulp}(q)$ and adding or subtracting $\mathrm{ulp}(q) \times y$ to the remainder, allowing the quotient, remainder pair to be converted to or from the properly signed pair.

Lemma 5.3.12 *Given the quotient, remainder pair (q, r) with $r \neq 0$, the complement (q^*, r^*) is formed as follows:*

$$r^* = \begin{cases} r - \mathrm{ulp}(q) \cdot y & \textit{if divisor and remainder signs agree,} \\ r + \mathrm{ulp}(q) \cdot y & \textit{if divisor and remainder signs disagree,} \end{cases}$$

and

$$q^* = \begin{cases} q + \mathrm{ulp}(q) & \textit{if divisor and remainder signs agree,} \\ q - \mathrm{ulp}(q) & \textit{if divisor and remainder signs disagree.} \end{cases}$$

Proof Assume that the divisor and remainder signs agree, so from (5.3.2),

$$\begin{aligned} x &= qy + r \\ &= (q + \mathrm{ulp}(q))y + (r - \mathrm{ulp}(q)y) \\ &= q^*y + r^*. \end{aligned} \tag{5.3.11}$$

Since $|r| < |y|\,\mathrm{ulp}(q)$ and r and y have the same signs, $|r - \mathrm{ulp}(q)\,y| < |y|\,\mathrm{ulp}(q)$, so q^*, r^* is the complementary quotient, remainder pair (q^*, r^*) by (5.3.3) and (5.3.11). The proof is analogous for the case where divisor and remainder signs disagree, the operation then complementing the complementary pair (q^*, r^*). □

Conversion into the properly signed pair can then be expressed as follows.

Algorithm 5.3.13 (Properly signed quotient, remainder pair, PQRP)

Stimulus: *A quotient, remainder pair (q, r) from division of x by $y \neq 0$ with $x = qy + r$ and $|r| \leq |y|\mathrm{ulp}(q)$.*

Response: *The corresponding properly signed quotient, remainder pair $(\hat{q}, \hat{r})$ with $-|y|\,\mathrm{ulp}(q) < \hat{r} < |y|\,\mathrm{ulp}(q)$ and $\mathrm{sgn}(\hat{r}) = \mathrm{sgn}(x)$ for $\hat{r} \neq 0$.*

Method: **if** $r \neq 0$ **then**
  if $\mathrm{sgn}(r) \neq \mathrm{sgn}(y)$ **then**
    if $(\mathrm{sgn}(r) \neq \mathrm{sgn}(x)) \vee (r = -y\,\mathrm{ulp}(q))$
      then $\hat{r} := r + y\,\mathrm{ulp}(q)$; $\hat{q} := q - \mathrm{ulp}(q)$ **end**

else
  if (sgn(r) $\neq$ sgn(x)) $\vee$ ($r = y\,\text{ulp}(q)$)
    then $\hat{r} := r - y\,\text{ulp}(q)$; $\hat{q} := q + \text{ulp}(q)$ **end**
end;
else $\hat{r} := r$; $\hat{q} := q$ **end**;

(iii) Converting from unsigned to signed radix-β division

Theorem 5.3.14 *Given $x, y \in \mathbb{Q}_\beta$, $y \neq 0$, with radix-β quotient, remainder pair (q, r), let*

$$x = (-1)^{s(x)}|x|, \quad y = (-1)^{s(y)}|y| \neq 0, \tag{5.3.12}$$

$$q = (-1)^{s(x)+s(y)}|q|, \quad r = (-1)^{s(x)}|r|. \tag{5.3.13}$$

Then $(|q|, |r|)$ is a properly signed quotient, remainder pair for division of $|x|$ by $|y|$ if and only if (q, r) is a properly signed quotient remainder pair for division of x by y.

The proof of this is left as an exercise.

Problems and exercises

5.3.1 Give an example of bounded single-digit decimal integer inputs $x = i10^j$, $y = k10^l$ with $1 \leq i, k \leq 9$, $0 \leq j, l \leq 9$, where $r^*(x, y)$ is larger than 9. How large can the complementary remainder $r^*(x, y)$ be for these single digit inputs.

5.3.2 For integer division of the binary integer inputs $x = i2^j \in \mathbb{Q}_2$, $y = k2^l \in \mathbb{Q}_2$, with $1 \leq i, k \leq n$ and $0 \leq j, l \leq e_{max}$, give bounds on the size of $q_{ieee}(x, y)$ and $r_{ieee}(x, y)$.

5.3.3 Prove Theorem 5.3.11.

5.3.4 Prove Theorem 5.3.14.

5.4 Deterministic digit-serial division

A deterministic digit-serial division method is characterized by specification of a quotient digit selection function $d = f(g, r, y)$ which recursively from a quotient, remainder pair (q, r) of radix-β division of x by $y \neq 0$ yields a next quotient remainder pair (q', r') by

$$q' = q + d\,\text{ulp}(q'), \quad \text{where} \quad \text{ulp}(q') = \beta^{-1}\text{ulp}(q), \tag{5.4.1}$$

$$r' = r - d\,\text{ulp}(q')\,y. \tag{5.4.2}$$

When the scaling employs iteratively normalized remainders the latter is

$$\frac{r'}{\mathrm{ulp}(q')} = \frac{r}{\mathrm{ulp}(q)}\beta - d \times y. \tag{5.4.3}$$

A first quotient, remainder pair to initialize the recursion typically has $r = x$ with $q = 0 \times \mathrm{ulp}(q)$, where $\mathrm{ulp}(q)$ must be explicitly determined to satisfy the normalized remainder bound $|r/\mathrm{ulp}(q)| < |y|$. That is, $\mathrm{ulp}(q)$ should be sufficiently large that $|x| < |y|\mathrm{ulp}(q)$.

Restoring division, non-restoring division, and binary SRT division are three well known digit-serial methods analyzed in this section. Algorithms for these methods are given corresponding to implementations where the quotient digit selection and remainder update functions can be implemented concurrently with hardware comprising simply a carry-completion adder, a shifter, and a two-state negative/non-negative sign on leading bit detector. No table look-up or hardware multiplier is required and the radix representations employed are non-redundant.

The digit-serial division algorithms presented are each analyzed for execution time by measuring the number of adds per digit of quotient determined, where in this section a carry-completion add is implied to insure a non-redundant representation of results. Times for all three methods are linear in adds per digit, allowing an $n \log n$ time implementation of n-digit radix-β division if a log time carry-completion adder is employed. Execution times for binary versions range from a worst case of a high of 2 adds per bit for restoring binary to a low of 1/2 add per bit in the average case for binary SRT.

Robertson diagrams are defined and presented for each method as illustrations of the quotient digit selection function and the corresponding normalized remainder recurrence equation (5.4.3).

5.4.1 Restoring division

Restoring division yields a sequence of proper quotient, remainder pairs. It is most convenient to describe the digit selection function for division of $|x|$ by $|y| \neq 0$, handling the appropriate attachment of signs as described by Theorem 5.3.14. The quotient digit selection function for radix-β restoring division is then

$$d(q, r, y) = \left\lfloor \frac{r\beta}{|y|\mathrm{ulp}(q)} \right\rfloor. \tag{5.4.4}$$

Starting with $r = |x|$ and the appropriate $\mathrm{ulp}(q)$, the recurrence (5.3.8) insures the normalized remainder is always in the range $[0; |y|)$, so the set of allowed quotient digits in (5.4.4) is $\{0, 1, \ldots, \beta - 1\}$.

For integer radix-β restoring division the sequence of quotient, remainder pairs has finite length terminating with the desired integer quotient and integer remainder. The mechanics of radix-β restoring division are conveniently described for the case of integer division. Integer restoring division is a shift-and-subtract iterative procedure. For division of a positive integer divisor by a positive integer dividend, the divisor is initially shifted left zero or more places to align the leading non-zero digits of the divisor and dividend. The shifted divisor is repeatedly subtracted from the dividend until the result becomes negative, then added back in (restored), yielding the first remainder. The net count of subtractions is the leading quotient digit, which may be zero, but the shifting has defined ulp(q). This one-digit quotient and corresponding remainder is the first quotient, remainder pair. The divisor is then shifted one place to the right and the procedure repeated with the remainder playing the role of the dividend. Successive quotient digits and quotient, remainder pairs are determined until the divisor cannot be shifted any further to the right without being to the right of its original position. The sequence of quotient, remainder pairs is thus determined with the final pair being the result of the integer division problem, as defined in Section 5.5.

Example 5.4.1 (Decimal integer restoring division) For the five-digit dividend 35822 and three-digit divisor 273 the divisor is initially shifted two places to the left, then recursively three quotient digits $d_2(q), d_1(q), d_0(q)$ and three corresponding quotient, remainder pairs are determined as illustrated.

35822		
−273	align left determining ulp(q) $= 10^2$ and subtract	
8522		$d_2(q) = 1$
−273	subtract	
−18778	result negative!	
+273	add	
8522	determines ($q = 1 \times 10^2, r = 8522$)	
−273	shift and subtract	
5792		
−273	subtract	
3062		$d_1(q) = 3$
−273	subtract	
332		
−273	subtract	
−2398	result negative!	
+273	add	
332	determines ($q = 13 \times 10^1, r = 332$)	
−273	shift and subtract	
59		$d_0(q) = 1$
−273	subtract	
−214		
+273	add	
59	determines ($q = 131, r = 59$)	

The quotient $q = \sum_{i=0}^{2} d_i(q)10^i = 131$ with remainder 59 satisfies the invariant equation with all quantities integers,

$$35822 = 131 \times 273 + 59,$$

and is the desired integer division result. □

Algorithm 5.4.1 (Integer radix-β restoring division)

Stimulus: *An integer radix-β dividend $N = (-1)^{s(N)}|N|$ with $|N| < \beta^m$ and an integer radix-β divisor $D = (-1)^{s(D)}|D|$ with $0 < |D| < \beta^n$, $n \geq m$.*

Response: *The quotient sign and digit sequence $s(q)$, $d_{m-n}, d_{m-n-1}, \ldots, d_0$ and remainder r where $q_k = (-1)^{s(g)} \sum_{i=k}^{m-k} d_i \beta^i$ is the radix-β quotient of the proper quotient, remainder pair of last place k for $0 \leq k \leq m-n$, and where (q_0, r) is the integer quotient, remainder pair.*

Method: $r := |N|$; $s(q) = (s(D) + s(N)) \bmod 2$

for $k := m - n$ **downto** *0* **do**

$d := 0$;

L1: **while** $r \geq 0$ **do** $d := d + 1; r := r - |D| \times \beta^k$ **end**;

{*subtract shifted divisor from remainder until negative*}

L2: $d_k := d - 1$; $r := r + |D| \times \beta^k$ {*restore*}

end;

$r := (-1)^{s(D)} \times r$

The time required to execute steps L1 and L2 is dominated by the time required to perform the carry-completion addition/subtractions updating the remainder. The number of such addition/subtractions required to determine a quotient digit of value d is data-dependent and equal to $d + 2$ for $0 \leq d \leq \beta - 1$. The running time of Algorithm 5.4.1 is then measured in terms of the number of such carry-completion additions with worst-case and average-case assessments given in view of the data dependence of the result.

For average-case time we assume all radix-β digit values are equally likely, which is quite reasonable in practice for all but the leading one or two digits of the quotient.

Observation 5.4.2 *The number of carry-completion additions per radix-β quotient digit (adds per digit) in Algorithm 5.4.1 is:*

worst case: $\beta + 1$ adds per digit,
average case: $(\beta + 3)/2$ adds per digit.

Noting that a quotient digit for $\beta = 2^m$ generates m quotient bits; a more useful comparison of higher radices is given in adds per bit.

Observation 5.4.3 *The number of carry-completion additions per quotient bit in Algorithm 5.4.1 with* $\beta = 2^m$ *is:*

> *worse case:* $(2^m + 1)/m$ *adds per bit,*
> *average case:* $(2^m + 3)/2m$ *adds per bit.*

Two questions immediately arise with regard to improving the restoring division algorithm.

(i) Why not avoid the restoring addition in step L2 by saving the previous as well as the current remainder?

The answer here calls for a decision based on alternative resources, specifically saving a value versus performing another addition. The preferred choice could be technology-dependent and so is designated an implementation option.

(ii) Why not limit the number of subtractions to at most $\beta - 1$ in step L1?

For general β this requirement would impose an additional test internal to the loop in step L1. This would cause further implementation complexity and computation time in signaling digit selection completion in step L1. For $\beta = 2$ an alternative simple test for digit selection completion is available in conjunction with limiting the maximum digit value to $\beta - 1 = 1$. This option leads to a distinctive version of the restoring division procedure termed binary restoring division.

Algorithm 5.4.4 (Integer binary restoring division)

Stimulus: *An integer dividend* $N = (-1)^{s(N)}|N|$ *with* $|N| < 2^m$, *and an integer divisor* $D = (-1)^{s(D)}|D|$ *with* $0 < |D| < 2^n$, $n \le m$.

Response: *The quotient sign and bit sequence* $s(q)$, $b_{m-n}, b_{m-n-1}, \ldots, b_0$ *and integer remainder* r *where* $q_k = (-1)^{s(q)} \sum_{i=k}^{m-n} b_i 2^i$ *is the binary quotient of the proper quotient, remainder pair of last place* k *for* $0 \le k \le m - n$ *with* q_0 *the integer quotient.*

Method: $r := |N|;\ s(q) := (s(D) + s(N)) \bmod 2$

for $k := m - n$ **down to** *0* **do**

L1: $b_k := 1;\ r := r - |D| \times 2^k$ {*complete for bit* = 1}

L2: **if** $r < 0$ **then** $b_k := 0;\ r := r + |D| \times 2^k$ {*restore for q bit* = 0}

end*;*

$r := (-1)^{s(D)} \times r$

Observation 5.4.5 *The number of carry-completion additions per quotient bit in Algorithm 5.4.4 is:*

> *worst case: 2 adds per bit,*
> *average case: 3/2 adds per bit.*

Because much more efficient algorithms exist, the restoring algorithms are never used in practice. We shall thus leave them here, except for illustrating the quotient digit selection by a diagram, which will turn out to be very useful later.

5.4.2 Robertson diagrams

The quotient digit selection function $d(q, r, y)$ characterizing a deterministic digit-serial division method for division of x $(= qy + r)$ by $y \neq 0$ is for fixed y generally piecewise constant as a function of the normalized remainder $r/\mathrm{ulp}(q)$. This means that the function giving the next normalized remainder

$$\frac{r'}{\mathrm{ulp}(q')} = \beta\frac{r}{\mathrm{ulp}(q)} - d(q, r, y)y \tag{5.4.5}$$

may for fixed y be graphed in terms of the normalized remainder $r/\mathrm{ulp}(q)$ as a piecewise linear function being linear over each interval of $r/\mathrm{ulp}(q)$ for which the digit selection function is constant.

Definition 5.4.6 *A* Robertson diagram *for a deterministic digit-serial division method for division of x by $y \neq 0$, is the graph of a piecewise linear function for the next normalized remainder $r'/\mathrm{ulp}(q')$, versus the current normalized remainder $r/\mathrm{ulp}(q)$. The function is linear over intervals of $r/\mathrm{ulp}(q)$, and is bounded between certain multiples of $|y|$ for which $d(q, r, y)$ is constant. The diagram specifies both $d(q, r, y)$ and $r'/\mathrm{ulp}(q')$ observing the following conventions in the graph:*

(i) the horizontal axis variable is $\beta\, r/\mathrm{ulp}(q)$ with the scale graduated in units of $|y|$ from $-|y|\,\beta$ to $|y|\,\beta$,
(ii) the vertical axis variable is $r'/\mathrm{ulp}(q')$ with the scale from $-|y|$ to $|y|$,
(iii) the pieces of the piecewise linear function are labelled with the digit value $d = d(q, r, y)$ indicating the constant digit value realized over the indicated interval of $r'/\mathrm{ulp}(q')$ with open ends of the intervals denoted by circles.

Figure 5.4.1 is the Robertson diagram for integer binary restoring division, note that in this particular case remainders are always non-negative, and digits $d \in \{0, 1, \ldots, \beta - 1\}$.

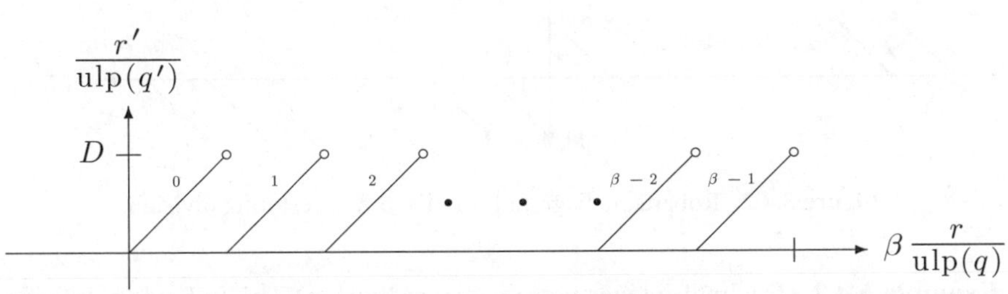

Figure 5.4.1. Robertson diagram for integer restoring division of $N \geq 0$ by $D > 0$.

Full information about a deterministic digit-serial algorithm is given by the Robertson diagram. The piecewise constant digit selection function is superimposed on the graph of the next normalized remainder function by simply labeling the pieces. Each allowed digit value occurs on a unique single piece. The scaling of the normalized remainder in units of $|y|$ handles the dependence of $d(q, r, y)$ on y, and employing $\beta r/\mathrm{ulp}(q)$ as the horizontal axis variable insures that all slopes will be 45°.

The Robertson diagram has an extension to describe the non-deterministic choices available with redundant quotient digits and this subject will be considered in Section 5.5 in connection with higher-radix SRT division.

5.4.3 Non-restoring division

Non-restoring division differs from restoring division in that the restoring step L2 of Algorithm 5.4.1 is not performed. Instead this method proceeds with the negative remainder to the next quotient digit selection stage, then adds the shifted (positive) divisor into the remainder a sufficient number d of times to make the remainder non-negative. This determines the next quotient digit to be $-d$, and the method proceeds alternately generating positive and negative quotient digits from the set $\{-\beta, -(\beta-1), \ldots, -2, -1, 1, 2, \ldots, \beta-1, \beta\}$. Note that the value zero is not in the digit set. The quotient digit selection function formally defining radix-β non-restoring division of $|x|$ by $|y| \neq 0$ is

$$d(q, r, y) = \begin{cases} \left\lfloor \dfrac{r\beta}{|y|\mathrm{ulp}(q)} \right\rfloor + 1 & \text{for } r \geq 0, \\ \left\lfloor \dfrac{r\beta}{|y|\mathrm{ulp}(q)} \right\rfloor & \text{for } r < 0, \end{cases} \tag{5.4.6}$$

where the normalized remainder falls in the range $-|y| \leq r/\mathrm{ulp}(q) < |y|$. Figure 5.4.2 shows the corresponding Robertson diagram.

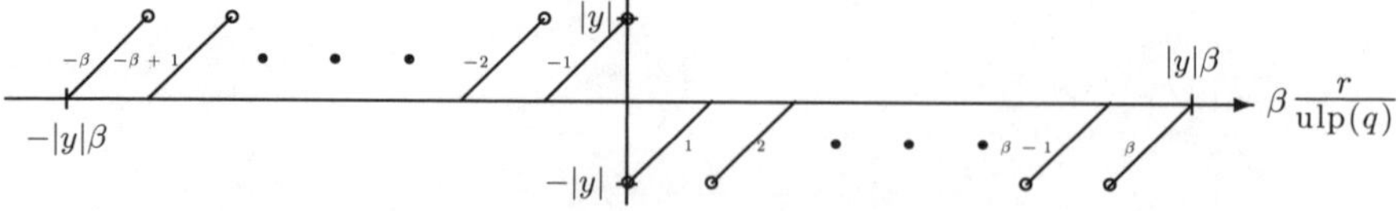

Figure 5.4.2. Robertson diagram for radix-β non-restoring division.

Example 5.4.2 (Decimal integer non-restoring division) The following division of 35822 by 273 illustrates the result of the non-restoring method which may be compared with the previous example in which the same division was done by the restoring method.

$$\begin{array}{rll}
35822 & & \\
-273 & & \\ \hline
8522 & & \left.\right\} d_2 = 2 \\
-273 & & \\ \hline
-18778 & \text{negative! determines } (q = 2 \times 10^2, r = -18778) & \\
+273 & \text{shift and add} & \\ \hline
-16048 & & \\
+273 & \text{add} & \left.\right\} d_1 = -7 \\ \hline
\vdots & & \\ \hline
-3062 & & \\
+273 & \text{add} & \\ \hline
332 & \text{non negative! determines } (q = 2\bar{7} \times 10^1, r = 332) & \\
-273 & & \\ \hline
59 & & \left.\right\} d_2 = 2 \\
-273 & & \\ \hline
-214 & \text{negative! determines } (q = 2\bar{7}2 \times 10^0, r = -214) &
\end{array}$$

There are two problems with the final radix-β quotient, remainder pair. First, the digits are not in the conventional range $\{0, 1, \ldots, 9\}$, the quotient is in a redundant representation. For digit set conversion we note that since digit signs alternate, every pair of digit values $1 \leq d_i \leq 10, -10 \leq d_{i-1} \leq -1$ may be converted simply to $0 \leq d'_i = d_i - 1 \leq 9, 0 \leq d'_{i-1} = d_{i-1} + 10 \leq 9$ without changing the value of q, so ($q = 132 \times 10^0, r = -214$) is determined.

The second problem, which still remains, is that we have not obtained the properly signed quotient, remainder pair as required for integer division. By Lemma 5.3.12 we can simply complement the remainder, obtaining $q = 132 - 1 = 131$ and $r = -214 + 273 = 59$ as the integer quotient and integer remainder results. □

Note from the example that the quotient digit set conversion and final complementation are the same:

(i) a positive digit is reduced by 1,
(ii) a negative digit is increased by β.

These two steps can be incorporated in the non-restoring division method for any non-negative dividend, in effect realizing an on-the-fly conversion (Theorem 2.3.3). Only the final remainder must then be complemented as necessary to obtain the result of division of $|x|$ by $|y|$. The following realizes a signed version of non-restoring division by attaching the sign using Theorem 5.3.14.

Algorithm 5.4.7 (Sign magnitude radix-β non-restoring division)

Stimulus: *A dividend* $x = (-1)^{s(x)} \times |x|$ *with* $|x| < 1$, *and divisor* $y = (-1)^{s(y)} \times |y|$ *with* $1 \le |y| < \beta$, *and* $\text{ulp}(q) = \beta^{\ell}$ *where* $\ell \le -1$.

Response: *The proper quotient, remainder pair* (q, r) *for division of* x *by* y *with* q *in the form* $q = (-1)^{s(q)} \times (0.d_{-1}d_{-2} \cdots d_{\ell})$, $0 \le d_i \le \beta - 1$ *for* $\ell \le i \le -1$.

Method: $s(q) := (s(x) + s(y)) \bmod 2$;
$r := |x|$; $k := -1$;
while $r \neq 0$ **and** $k \ge \ell$ **do** $d_k := 0$;
  if $r > 0$ **then**
    while $r > 0$ **do** $r := r - |y| \times \beta^k$; $d_k := d_k + 1$; **end**;
  else
    while $r < 0$ **do** $r := r + |y| \times \beta^k$; $d_k := d_k - 1$; **end**;
  $k := k - 1$;
end;
if $r < 0$ **then** $r := r + |y| \times \beta^{\ell}$;
$r = (-1)^{s(y)} \times r$;

It follows that in non-restoring division the number of carry-completion additions in determining d_k is $|d_k|$, $1 \le |d_k| \le \beta$, i.e., one fewer add per digit than for restoring division.

The digit values $-\beta$ and β can be disallowed in an alternative form of non-restoring division. The process for $\beta \ge 3$ is not simple to describe in general. For $\beta = 2$ the alternative reduces to a special case worthy of independent investigation, where the average and worst cases are each one add/subtract per bit. Note that the updating of the remainder in this algorithm provides the *scaled remainder* which is only normailzed at the very end of the algorithm.

Algorithm 5.4.8 (2's complement, non-restoring, fixed-point binary division)

Stimulus: *A dividend* x *and divisor* y *in* n*-bit, 2's complement fixed-point representation,* $-1 \le x < 1$ *and* $-1 \le y < 1$.

Response: *An overflow signal* $OF \equiv (y = 0) \vee (x/y < -1) \vee (x/y \ge 1)$.
If $\neg\, OF$ *then also the proper quotient, remainder pair* (q', r') *where* $x = q' \cdot y + r' \cdot 2^{-(n-1)}$, *with* q' *and* r' *in 2's complement representation, and remainder* r' *satisfying* $-|y| < r' < |y|$ *and* $\text{sgn}(r') = \text{sgn}(x)$.

Method: $r_0 := x$; $i := 0$; $OF := (y = 0)$;
while $(i < n - 1) \wedge (\neg\, OF)$ **then**
  if $\text{sgn}(r_i) \neq \text{sgn}(y)$ **then** $b_i := 0$; $r_{i+1} := 2 \times r_i + y$;
    else $b_i := 1$; $r_{i+1} := 2 \times r_i - y$;
  end;
  $i := i + 1$;
  $OF := (r_i < -1) \vee (r_i \ge 1)$; {*2's complement overflow test*}

```
L: end;
   if ¬ OF then
     r := r_{n-1};
     q := -1 + Σ_{j=0}^{n-2} b_j 2^{-j} + 2^{-(n-1)};
     call PQRP(x, y, q, r, q', r'); {convert (q, r) into proper
                                          (q', r')-pair}
     OF := (q' < -1) ∨ (q' ≥ 1); {2's complement overflow test}
   end;
```

Where: *The invariant* $x = r_i 2^{-i} + y(-1 + \sum_{j=0}^{i-1} b_j 2^{-j} + 2^{-i})$ *holds at label L. Note that the overflow tests can be performed as bitwise tests, as can the testing of signs. The function* sgn(i) *returns* 0 *for* $i \geq 0$ *and* 1 *for* $i < 0$ *(the 2's complement "sign-bit"). The computation of* q *can alternatively be realized by a simple updating during the loop, and so can the remainder* r.

Observe that although the quotient digits b'_i actually generated are in the set $\{-1, 1\}$, they are here implicitly being converted into the set $\{0, 1\}$ as follows. Let $b'_i = 2b_{i-1} - 1 \in \{-1, 1\}$, $i = 1, 2, \ldots, n-1$ be the digits which should have been generated, where $b_i \in \{0, 1\}, i = 0, 1, \ldots, n-2$ are the actual bits produced. Then the quotient q can be found as:

$$\begin{aligned} q = \sum_{i=1}^{n-1} b'_i 2^{-i} &= \sum_{i=1}^{n-1} (2b_{i-1} - 1) 2^{-i} \\ &= \sum_{j=0}^{n-2} (2b_j - 1) 2^{-j-1} \\ &= \sum_{j=0}^{n-1} b_j 2^{-j} - \sum_{k=1}^{n} 2^{-k} \\ &= -1 + \sum_{j=0}^{n-2} b_j 2^{-j} + 2^{-(n-1)}. \end{aligned}$$

Also note that the quotient produced is in a non-redundant representation, whether we consider the digit set $\{0, 1\}$ or the digit set $\{-1, 1\}$. The computations must be performed using carry-completing additions, in order to be able to test the sign of r_i and for overflow during the loop. Figure 5.4.3 shows the Robertson diagram for the quotient selection.

Theorem 5.4.9 *For* x *and* y *in* n*-bit, 2's complement, fixed-point representation, where* $-1 \leq x < 1$ *and* $-1 \leq y < 1$*, Algorithm 5.4.8 correctly computes a proper quotient, remainder pair* (q', r') *satisfying* $x = q'y + r'2^{-(n-1)}$ *and*

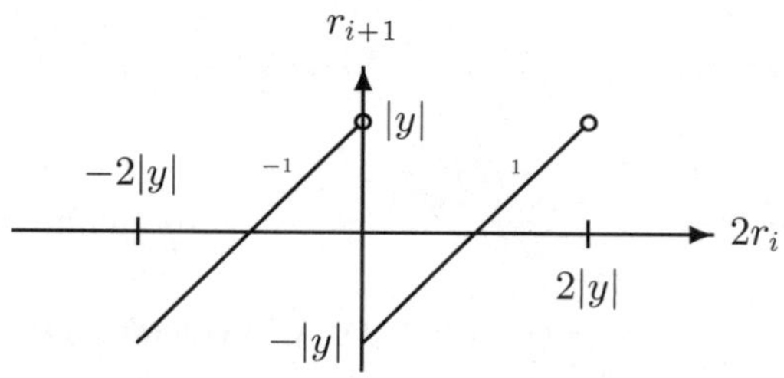

Figure 5.4.3. Robertson diagram for radix-2, non-restoring division.

$-|y| < r' < |y|$ *with* $\text{sgn}(r') = \text{sgn}(x)$, *provided that the overflow signal*

$$OF \equiv (y = 0) \vee \left(\frac{x}{y} < -1\right) \vee \left(\frac{x}{y} \geq 1\right).$$

does not evaluate to true.

Proof First assume that $x/y > 1$ and $y > 0$, then $x > y > 0$, so

$$r_0 = x = y + \varepsilon, \quad \text{with} \quad \varepsilon \geq 2^{-n+1},$$

which implies

$$r_i = y + 2^i\varepsilon, \quad \text{for } i = 0, 1, \dots, n-1,$$

so $r_{n-1} = y + 2^{n-1}\varepsilon \geq y + 1 > 1$, and thus OF will become true before termination of the loop. Similarly for $x/y > 1,\ y < 0$ we find $r_i = y - 2^i\varepsilon$ for some $\varepsilon \geq 2^{-n+1}$, so here also OF becomes true. And the same kind of arguments also cover the cases where $x/y < -1$ for $y > 0$ as well as $y < 0$. If $x/y = 1$ so that $x = y$, then $q_i = 1$ for all $i = 0, 1, \dots, n-1$, and thus the overflow test on q' will yield OF true. Thus we have shown

$$(y = 0) \vee \left(\frac{x}{y} < -1\right) \vee \left(\frac{x}{y} \geq 1\right) \Rightarrow OF.$$

To prove the implication in the other direction we will show that

$$(y \neq 0) \wedge \left(-1 \leq \frac{x}{y} < 1\right) \Rightarrow -1 \leq r_i < 1 \quad \text{for} \quad i = 0, 1, \dots, n-1. \tag{5.4.7}$$

Checking the cases $x < 0$ and $x \geq 0$ for $y \neq 0$ we find for $i = 0, 1, \dots, n-1$

$$\begin{aligned} y > 0 &: -y \leq r_i < y \Rightarrow -y \leq r_{i+1} < y, \\ y < 0 &: y < r_i \leq -y \Rightarrow y < r_{i+1} \leq -y, \end{aligned} \tag{5.4.8}$$

so for $-1 < y < 1$ we have $-1 < r_{i+1} < 1$ for $i = 0, 1, \dots, n-1$. What remains in proving (5.4.7) is the case $y = -1$. From (5.4.8) we only know that $-1 < r_i \leq 1$. So assume there is an i such that $r_i = 1$, then from $r_{i-1} = (r_i + 1)/2$ we find that $r_i = 1$ implies $r_0 = 1$, contradicting the assumption on $r_0 = x \neq 1$.

Finally, we have to prove the invariant. On exiting of the first execution of the loop (i.e. at L) we have $r_1 = 2r_0 - (2b_0 - 1)y$ and $i = 1$, from which it follows that

$$x = r_0 = r_1 2^{-1} + (-1 + b_0 + 2^{-1})y.$$

Then for the general induction step assume that

$$x = r_i 2^{-i} + \left(-1 + \sum_{j=0}^{i-1} b_j 2^{-j} + 2^{-i} \right) y$$

holds. From the updating of the remainder we have $r_i = (r_{i+1} + (2b_i - 1)y)/2$, hence

$$x = r_{i+1} 2^{-i-1} + \left(-1 + \sum_{j=0}^{i} b_j 2^{-j} + 2^{-i-1} \right) y,$$

which by induction proves the invariant. The correctness of the algorithm then follows from the invariant for $i = n - 1$, and Algorithm 5.3.13 (PQRP) for selecting the proper sign. □

For completeness and because 2's complement integers are widely used, we shall conclude the discussion of non-restoring division with corresponding algorithms for this case. First, notice that the dividend x in Algorithm 5.4.8 could be a double-precision fixed-point number in 2's complement. Then the remainders r_i would gradually introduce bits of lower precision when forming $r_{i+1} = 2r_i \pm y$. The necessary hardware resources for the algorithm are equivalent to the hardware used for iterative multiplication, e.g., as shown in Figure 4.2.2, employing a double-length accumulator with an adder/subtractor, whose other operand is in a single-length register. For division the double-length accumulator can initially contain the full $2n$-bit dividend, while the rightmost part is used to store the bits of the quotient, using the space gradually made available while the remainder is shifted left. The setup for division is shown in Figure 5.4.4.

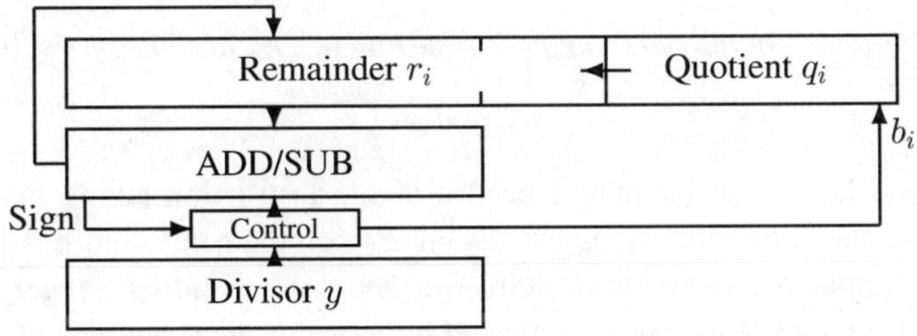

Figure 5.4.4. Hardware structure for binary non-restoring division.

Algorithm 5.4.8 has been described with the initial contents of the accumulator (dividend x), and the other operand (divisor y) as fixed-point numbers with the binary points located at the left end of the registers. First observe that it does not make a difference for the algorithm whether we consider the binary point to be located anywhere else within the accumulator; in particular it could be placed at the right-hand end of the adder (note, not the end of the accumulator), and thus the contents interpreted as integers.

Hence we may also think of it dividing an integer $\hat{x}$ by another integer $\hat{y}$, where

$$\hat{x} = 2^{n-1}x \quad \text{and} \quad \hat{y} = 2^{n-1}y.$$

Since the quotient, remainder pair (q', r') obtained by the algorithm satisfies $x = q'y + r'2^{-(n-1)}$, we can then also interpret the result as

$$\hat{x} = q'\hat{y} + r',$$

where $-1 \leq q' < 1$ and $-1 \leq r' < 1$. However, multiplying both sides by 2^{n-1} we find

$$\tilde{x} = 2^{n-1}\hat{x} = \hat{q}\,\hat{y} + \hat{r},$$

with $-2^{n-1} \leq \hat{q} < 2^{n-1}$ and $-2^{n-1} \leq \hat{r} < 2^{n-1}$ now being n-bit integers like $\hat{y}$, but $\tilde{x}$ being a double-length integer $-2^{2n-1} \leq \tilde{x} < 2^{2n-1}$. This is the type of divide instruction that was often found in early computers, employing the same kind of hardware as needed for multiplication, delivering the result of dividing a double-length fixed-point number or integer, by a single-length fixed-point or integer, as a quotient, remainder pair in the corresponding interpretation.

Corollary 5.4.10 (2's complement, non-restoring, $2n$- by n-bit division) *Provided the overflow flag OF does not return true, Algorithm 5.4.8 correctly computes in 2's complement the quotient, remainder pair (q', r') in n-bit representation, given dividend x in $2n$-bit and divisor y in n-bit representation:*

- *If operands are in fixed-point $-1 \leq x < 1$ and $-1 \leq y < 1$, then the resulting (q', r') are in the same fixed-point representation, satisfying*

$$x = q'y + r'2^{-(n-1)}.$$

- *If operands are integers, then the resulting (q', r') are integers satisfying*

$$x = q'y + r'.$$

Nowadays high-level languages need a divide instruction taking two integer operands of the same precision (usually in 2's complement), and possibly also a separate remainder instruction, delivering the corresponding integer quotient or remainder result. This can be achieved by an equivalent, but actually slightly simpler algorithm.

Algorithm 5.4.11 (2's complement, non-restoring integer division)

Stimulus: *A dividend x and divisor y in n-bit, 2's complement integer representation,* $-2^{n-1} \le x < 2^{n-1}$ *and* $-2^{n-1} \le y < 2^{n-1}$.

Response: *An overflow signal* $OF \equiv (y = 0) \vee (x/y = 2^{n-1})$.
If $\neg OF$, *then also the proper quotient, remainder pair* (q', r'), *where* $x = q' \cdot y + r'$, *with* q' *and* r' *in 2's complement representation, with remainder* r' *satisfying* $-|y| < r' < |y|$ *and if* $r' \neq 0$ *then* $\text{sgn}(r') = \text{sgn}(x)$.

Method: $y := y \times 2^n$; $r := x$; $q := -1$;
for $i := 0$ **to** $n - 1$ **do**
  if $\text{sgn}(r) \neq \text{sgn}(y)$ **then**
    $q := 2 \times q$; $r := 2 \times r + y$;
  else
    $q := 2 \times (q + 1)$; $r := 2 \times r - y$;
  end;
L: end;
$q := q + 1$;
if $r \neq 0$ **then**
  call *PQRP*(x, y, q, r, q', r'); {*Convert* (q, r) *into a proper* (q', r')*-pair*}
  $OF := (q' < -2^{n-1}) \vee (q' \ge 2^{n-1})$; {*2's complement overflow test*}
  $r := r * 2^{-n}$;
end;

Where: *The invariant* $x = ((q + 1)y + r)2^{-i}$ *holds at label L. The overflow test can be performed as a bitwise test, and so can the testing of signs, where the function* $\text{sgn}(i)$ *returns* 0 *for* $i \ge 0$ *and* 1 *for* $i < 0$ *(the 2's complement "sign-bit").*

Note that since there are fewer overflow situations, this algorithm is somewhat simpler than Algorithm 5.4.8, avoiding the initial test for division by zero, and the overflow test during the loop. The zero-divide situation is caught here in the overflow test on q', and so also is the only other possible overflow situation, the division of -2^{n-1} by -1 yielding 2^{n-1}, which is not representable in n-bit 2's complement. We leave the proof of the correctness of the algorithm as an exercise.

5.4.4 Binary SRT division

This method is a particular example of a class of division algorithms developed independently in the late 1950s by Sweeney, Robertson, and Tocher, from whose initials the name derives. While we will discuss the more general SRT methods in the next section, we will here deal with the radix-2 version, as this turns out to be very closely related to the non-restoring algorithm described above.

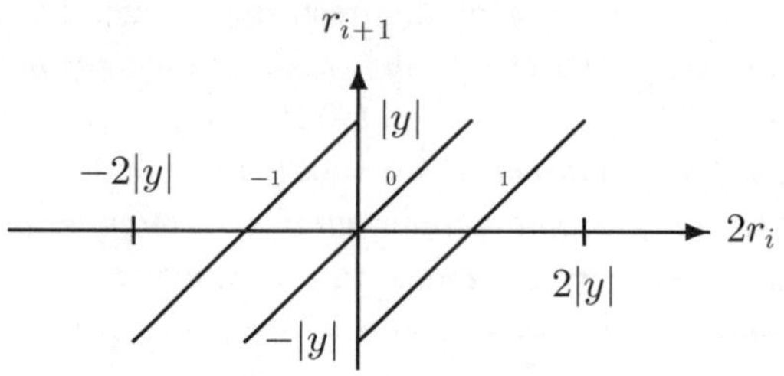

Figure 5.4.5. Robertson diagram for general radix-2 SRT division.

The idea here was to allow the digit value zero, since a step consisting only of a shift can be performed much faster than a shift followed by a carry-completing add/subtract. Considering the Robertson diagram in Figure 5.4.5 with zero also as a permissible digit, it is evident that there now are intervals around zero where there is a choice of next quotient digit $b_{i+1} \in \{-1, 0, 1\}$.

To take advantage of the faster iteration step for quotient digit $b_{i+1} = 0$ one should select the zero value in the widest possible interval $-|y| \le 2r_i \le |y|$. But another choice could be to simplify the quotient selection, avoiding the full-precision comparison of $2r_i$ against the divisor y. Assume that y is normalized, like in floating-point arithmetic where the divisor significand part may be restricted such that $-1 \le y < -\frac{1}{2}$ or $\frac{1}{2} \le y < 1$. In the IEEE-754 floating-point standard, using binary sign-magnitude representation, the interval is $1 \le y < 2$. Then $2r_i$ can be compared with the minimal absolute values of y (i.e., constants), e.g., in the former case in 2's complement arithmetic

$$b_{i+1} = \begin{cases} -1 & \text{if } \; 2r_i < -\frac{1}{2}, \\ 0 & \text{if } \; -\frac{1}{2} \le 2r_i < \frac{1}{2}, \\ 1 & \text{if } \; \frac{1}{2} \le 2r_i, \end{cases}$$

chosen such that the comparisons are easily performed, using only two or three leading bits of $2r_i$ by Observation 3.10.4. Let $2r_i$ have the 2's complement representation $a_1 a_0 . a_{-1} \cdots$, where a_1 is the "sign-bit," then the quotient selection can be performed as

$$\begin{aligned} b_{i+1} = \textbf{if} \quad & (a_1 a_0 = 01) \vee (a_1 a_0 a_{-1} = 001) \;\; \textbf{then} \;\; 1 \\ \textbf{else if} \; & (a_1 a_0 a_{-1} = 000) \vee (a_1 a_0 a_{-1} = 111) \;\; \textbf{then} \;\; 0 \\ & \textbf{else} \;\; -1 \end{aligned}$$

The quotient digits are in the redundant set $\{-1, 0, 1\}$, and the quotient value can then be determined by a sequential conversion process, starting at the least-significant end, but only after all digits have been determined. However, more efficiently, "on-the-fly" conversion (Theorem 2.3.3) can be used in parallel with the remainder updating, so that the quotient value is available immediately after the last digit has been determined.

Observe that the quotient selection in the binary SRT algorithm described above is still deterministic, based on remainders in non-redundant representation. Since there are intervals in Figure 5.4.5 where two quotient values can be selected, it is possible to base the digit selection on a few leading digits of a redundant representation of the remainder. Thus it is not necessary to use carry-completing adders for the implementation of radix-2 SRT division, but we shall cover this aspect in more detail in the next section, dealing with the more general class of SRT algorithms.

Problems and exercises

5.4.1 Prove the correctness of Algorithm 5.4.11. Hint: Use induction with the following expressions, $q_i = 2(q_{i-1} + b_i)$ and $r_i = 2r_{i-1} - (2b_i - 1)y$, where $b_i \in \{0, 1\}$.

5.5 SRT division

This class of division algorithms is characterized by the extended use of redundant representations for the quotient, as well as for the remainder. As noted in the previous section, if there are overlaps between digit selection regions in the Robertson diagram, there is a choice between two digit values. Hence even if the information on the remainder is incomplete (representing an uncertainty interval), it may be possible to choose one of the two alternative quotient digit values. By allowing such a relaxed quotient digit determination, it is possible to base the quotient digit selection on leading digits of the divisor and of the remainder in a redundant representation.

An approximate quotient digit value can also be obtained by multiplying the remainder by some approximation to the reciprocal of the divisor. For higher quotient radices, the use of a rectangular multiplier is also needed to form partial products by multiplying the multiplicand by a digit value. Alternatively such a multiplier can also be used for scaling the dividend and divisor such that the divisor is close to unity, thus simplifying the quotient digit selection. In the next section we shall look at such very high-radix methods, exemplified by short reciprocal and prescaled division.

For radix 4 with the minimally redundant digit set $\{-2, -1, 0, 1, 2\}$ the products can be formed directly by shifts. For radix 4 with digit set $\{-3, -2, -1, 0, 1, 2, 3\}$, for radix 8 and possibly for radix 16, small precomputed tables of multiples may be employed. The SRT methods often use table look-up to determine the next digit as a function of leading digits of the divisor and of the remainder. If implemented by standard tables, the size of these grows exponentially in the number of digits (bits) needed to determine the quotient digit, however, if implemented by optimized

PLAs[1] the growth is more moderate. Anyway, these methods are in practice limited to radix values 4, 8, and 16, the latter ones often being realized by combining two steps of lower radix in one machine cycle.

5.5.1 Fundamentals of SRT division

SRT is not really a specific kind of division, rather it is a class of division methods, characterized by the following:

- The divisor is normalized.
- A redundant symmetric quotient digit set is used.
- Quotient digits are selected by a few leading digits of remainder and divisor.
- The remainders may be in a redundant representation.

For simplicity of notation in the following discussion, we use β to denote the radix (here always of the form $\beta = 2, 4, 8, \ldots$) to avoid the radix power. Let $D = \{-a, \ldots, 0, \ldots, a\}$ be the quotient digit set with $\beta/2 \le a \le \beta - 1$, and define the *redundancy factor*[2]

$$\mu = \frac{a}{\beta - 1} \quad \text{with} \quad \tfrac{1}{2} < \mu \le 1 \tag{5.5.1}$$

where $\mu = 1$ corresponds to a maximally redundant digit set.

Let x be the dividend, y the positive divisor, r_i the remainder (with $r_0 = x$), and d_i the digit selected in the ith step. The purpose of the *digit selection function*, $\sigma(r_i, y)$, is to select the next quotient digit d_{i+1}, while keeping the new remainder $r_{i+1} = \beta r_i - d_{i+1} y$ bounded, say with bounds r_{min} and r_{max},

$$r_{min} \le r_{i+1} \le r_{max}. \tag{5.5.2}$$

Let the selection interval $[L_d, U_d]$ be the interval for βr_i, $\beta r_{min} \le \beta r_i \le \beta r_{max}$, for which it is possible to choose $d_{i+1} = d$ while keeping the updated remainder $r_{i+1} = \beta r_i - dy$ bounded by (5.5.2), i.e., for $L_d \le \beta r_i \le U_d$

$$r_{min} \le r_{i+1} = \beta r_i - dy \le r_{max}$$

must hold, corresponding to the Robertson diagram in Figure 5.5.1.

From the diagram it can be seen that

$$L_d = d\,y + r_{min} \quad \text{and} \quad U_d = d\,y + r_{max}; \tag{5.5.3}$$

in particular (5.5.3) must hold for $d_{i+1} = \pm a$, the extremal digit values, hence

$$L_{-a} = -a\,y + r_{min} \quad \text{and} \quad U_a = a\,y + r_{max}.$$

[1] PLA: programmed logic array.

[2] To be distinguished from the "redundancy index," denoting the number of "redundant digits," $|D| - \beta$.

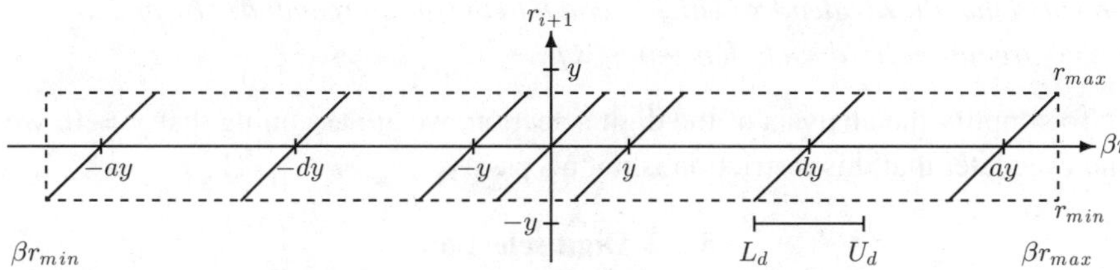

Figure 5.5.1. Robertson diagram for SRT division.

But as seen from Figure 5.5.1, $\beta r_{min} = L_{-a}$ and $\beta r_{max} = U_a$, hence it follows that

$$r_{min} = -\mu\, y \quad \text{and} \quad r_{max} = \mu\, y$$

and from (5.5.3) we find

$$L_d = (d - \mu)y \quad \text{and} \quad U_d = (d + \mu)y. \tag{5.5.4}$$

To insure that at least one digit value can be chosen for any r_i, every value of βr_i must fall in at least one digit selection interval, i.e., it is necessary that

$$U_{d-1} \geq L_d,$$

hence by (5.5.4) we must require that $(d - 1 + \mu)y \geq (d - \mu)y$, or $\mu \geq \frac{1}{2}$, which is always satisfied since $a \geq \beta/2$. Actually by (5.5.1) recalling that we require $y > 0$

$$U_{d-1} - L_d = (2\mu - 1)y > 0, \tag{5.5.5}$$

thus consecutive selection intervals overlap, so that there are values of r_i for which, in general, there is a choice between two digit values (and possibly three for $\mu = 1$).

In summary, provided that $-\mu\, y \leq r_i \leq \mu\, y$, then the selection function can deliver at least one digit value d_{i+1} such that the new remainder satisfies $-\mu\, y \leq r_{i+1} \leq \mu\, y$. However, for $i = 0$, the dividend is used as the first remainder, $r_0 = x$, thus we must require that x and/or y are normalized such that $-\mu\, y \leq x \leq \mu\, y$, or $-\mu \leq x/y \leq \mu$. Any scaling applied for this normalization must then be used to correct the final quotient, remainder pair.

Observation 5.5.1 *With quotient radix $\beta \geq 2$, quotient digit set $\{-a, \ldots, 0, \ldots, a\}$, and $\mu = a/(\beta - 1)$, there exists a digit selection function $\sigma(r_i, y)$ that delivers a next quotient digit d_{i+1} such that the next remainder*

$$r_{i+1} = \beta r_i - d_{i+1} y \quad \textit{for} \quad i = 0, 1, \ldots$$

satisfies

$$-\mu\, y \leq r_{i+1} \leq \mu\, y,$$

provided that the dividend x, which is equal to the initial remainder r_0, *and divisor* $y > 0$ *are normalized such that* $-\mu \le x/y \le \mu$.

To simplify the analysis of the digit selection, we are assuming that $y > 0$; we shall see later that this restriction is not necessary.

5.5.2 Digit selection

Due to the overlap $U_{d-1} - L_d = (2\mu - 1)y > 0$ it is not necessary to know the exact value of the remainder r_i to be able to select a correct next digit d_{i+1}. The digit selection intervals are conveniently illustrated in a *P-D diagram*[3] or *Taylor diagram* as in Figure 5.5.2, showing the intervals as functions of the divisor y, assumed normalized $\frac{1}{2} \le y < 1$, and the shifted remainder βr_i. We shall here assume that the remainder updating and digit selection takes place in binary arithmetic.

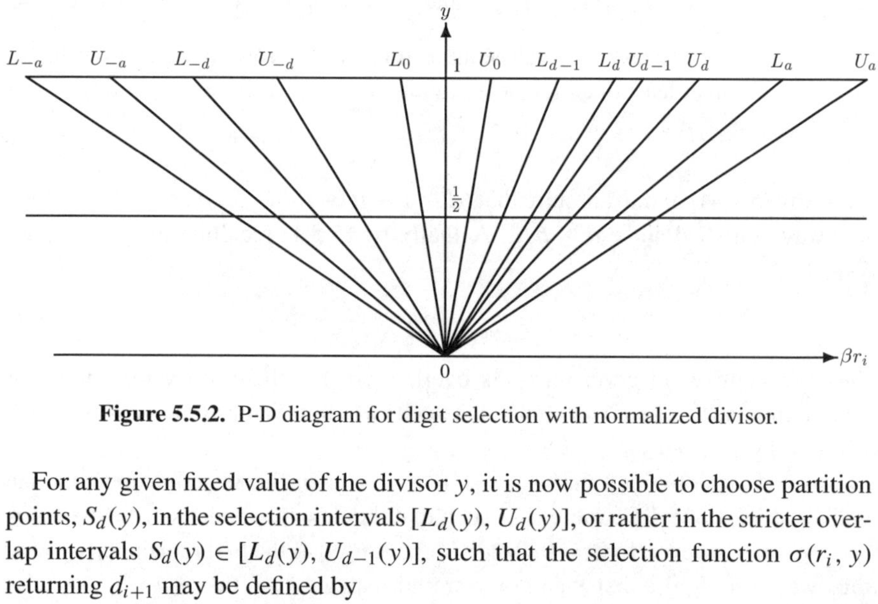

Figure 5.5.2. P-D diagram for digit selection with normalized divisor.

For any given fixed value of the divisor y, it is now possible to choose partition points, $S_d(y)$, in the selection intervals $[L_d(y), U_d(y)]$, or rather in the stricter overlap intervals $S_d(y) \in [L_d(y), U_{d-1}(y)]$, such that the selection function $\sigma(r_i, y)$ returning d_{i+1} may be defined by

$$\begin{aligned} \beta r_i \ge 0 : &\quad S_d(y) \le \beta r_i < S_{d+1}(y) \quad \Rightarrow \quad d_{i+1} = d, \\ \beta r_i < 0 : &\; -S_{d+1}(y) \le \beta r_i < -S_d(y) \quad \Rightarrow \quad d_{i+1} = -d, \end{aligned} \tag{5.5.6}$$

using the symmetry around the y-axis, allowing us to restrict the analysis to $d \ge 0$.

To simplify the following discussion, we shall often assume that y is fixed and drop the argument y in the notation of selection intervals $[L_d(y), U_d(y)]$ and partition points $S_d(y)$. However, these can be pictured as functions of y as in Figure 5.5.3.

[3] From Partial remainder, Divisor diagram, where "partial remainder" is a traditional notation.

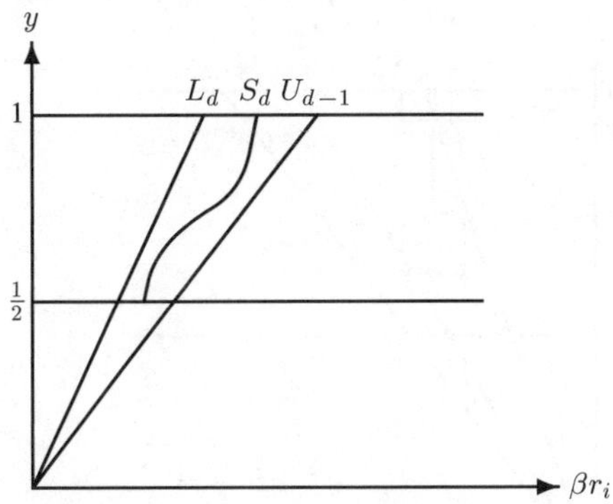

Figure 5.5.3. $S_d(y)$ as a function of y.

Due to the overlap between selection intervals, the partition points can be chosen such that it is sufficient to check a few of the leading digits of the (possibly redundant) value of βr_i. Let $\widehat{\beta r}_i$ denote a truncated value of βr_i, and $\text{ulp}(\widehat{\beta r}_i)$ denote the unit in the last place of the truncated value, say $\text{ulp}(\widehat{\beta r}_i) = 2^{-t}$. Define the truncation error, ϵ_r, by

$$\beta r_i = \widehat{\beta r}_i + \epsilon_r,$$

then for various binary representations of r_i we have:

2's complement:	0	$\leq$	ϵ_r	$<$	$\text{ulp}(\widehat{\beta r}_i)$,
2's complement carry-save:	0	$\leq$	ϵ_r	$<$	$2\text{ulp}(\widehat{\beta r}_i)$,
borrow-save:	$-\text{ulp}(\widehat{\beta r}_i)$	$<$	ϵ_r	$<$	$\text{ulp}(\widehat{\beta r}_i)$,

as illustrated in Figure 5.5.4, where τ is the number of integer bits.

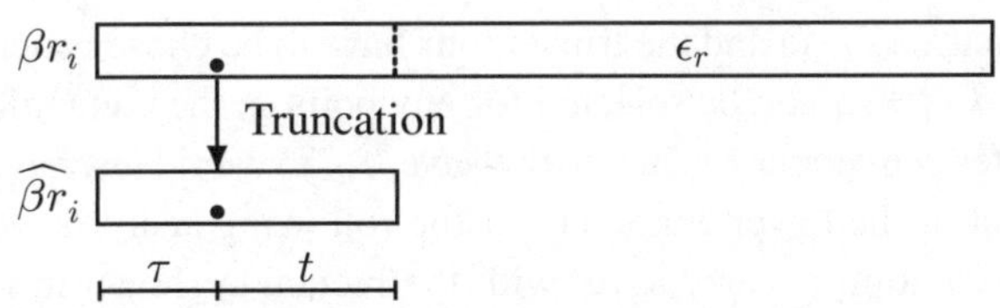

Figure 5.5.4. Truncation of shifted remainder.

It is essential to note that truncation is assumed to take place on the redundant representation, before conversion into non-redundant representation. Thus it is not necessary to convert the full-length remainder into a non-redundant representation.

Similarly, considering only a few leading digits of the divisor, let $\widehat{y}$ denote the truncated value of y with $\text{ulp}(\widehat{y}) = 2^{-u}$ for $u \geq 1$, and truncation error δ defined by $y = \widehat{y} + \delta$, where $0 \leq \delta < \text{ulp}(\widehat{y})$ for y in 2's complement. The function $\widehat{S_d}(y) = S_d(y)$ now becomes a stepwise function, delimiting rectangles within

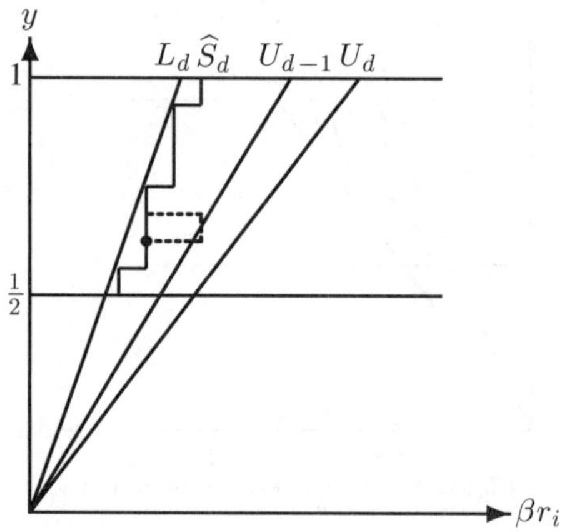

Figure 5.5.5. $\widehat{S}_d(y)$ as a function of truncated divisor $\widehat{y}$.

which a particular quotient digit can be chosen, as depicted in Figure 5.5.5. We shall assume that $\widehat{S}_d(y)$ is chosen to be as far to the left as possible.

The stepwise function $\widehat{S}_d(y)$ is now determined by a set of constants $\widehat{S}_d(\widehat{y})$, corresponding to the various truncated values $\widehat{y}$. These constants are assumed to be specified to the same accuracy as $\widehat{\beta r_i}$ (i.e., $\text{ulp}(\widehat{\beta r_i}) = 2^{-t}$). For fixed $\widehat{y} = k2^{-u}$, where $\text{ulp}(\widehat{y}) = 2^{-u}$, $\widehat{S}_d(\widehat{y})$ can then be written as an integer multiple $s_{d,k}2^{-t}$ of $\text{ulp}(\widehat{\beta r_i}) = 2^{-t}$.

The dotted rectangle in Figure 5.5.5 located with lower left-hand corner at $(\widehat{S}_d(\widehat{y}), \widehat{y})$, for y in non-redundant 2's complement and βr_i redundant carry-save 2's complement, shows the set of points $(\beta r_i, y)$ satisfying

$$\widehat{S}_d(\widehat{y}) \leq \beta r_i < \widehat{S}_d(\widehat{y}) + 2\text{ulp}(\widehat{\beta r_i}) \text{ and } \widehat{y} \leq y < \widehat{y} + \text{ulp}(\widehat{y}), \tag{5.5.7}$$

where the point $(\widehat{S}_d(\widehat{y}), \widehat{y})$ and the truncations have to be chosen such that the next quotient digit $d_{i+1} = d$ can be selected for any point in the rectangle (5.5.7).

For the shifted remainder βr_i in borrow-save, $\widehat{S}_d(\widehat{y})$ would just have to be chosen as the midpoint of the lower edge, but for the following analysis we will assume that the representation is carry-save, with the rectangle shown in more detail in Figure 5.5.6, where the modifications for borrow-save are trivial.

Using (5.5.4) for $d > 0$ the rectangle has to be to the right of the line $L_d(y) = (d - \mu)y$, yielding the following condition on the upper left-hand corner:

$$L_d(\widehat{y} + \text{ulp}(\widehat{y})) = (d - \mu)(\widehat{y} + \text{ulp}(\widehat{y})) \leq \widehat{S}_d(\widehat{y}). \tag{5.5.8}$$

These rectangles overlap, since they are of width $2\text{ulp}(\widehat{\beta r_i})$, but are positioned at a horizontal spacing of $\text{ulp}(\widehat{\beta r_i})$. Since $\widehat{S}_d(\widehat{y})$ is chosen as small as possible, for any point in the (dotted) rectangle overlapping from the left, the digit value $d - 1$ must be chosen. Thus the midpoint of the bottom edge must be to the left of the

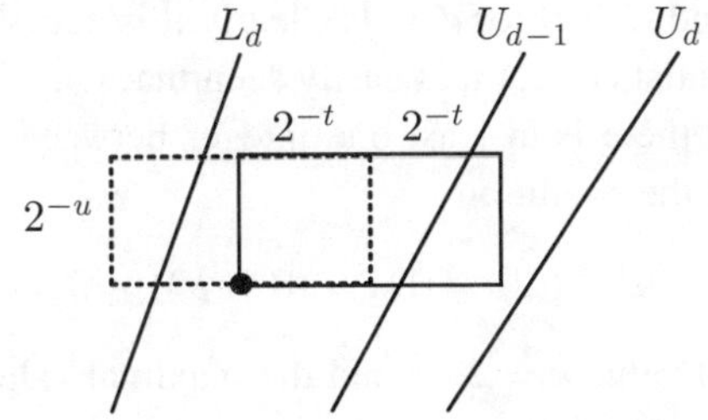

Figure 5.5.6. Overlapping uncertainty rectangles.

line $U_{d-1}(y)$, yielding this additional condition:

$$\widehat{S}_d(\widehat{y}) + \text{ulp}(\widehat{\beta r_i}) \leq U_{d-1}(\widehat{y}) = (d - 1 + \mu)\widehat{y}. \tag{5.5.9}$$

But the lower right corner must also be to the left of the line $U_d(y)$, hence we must also require

$$\widehat{S}_d(\widehat{y}) + 2\text{ulp}(\widehat{\beta r_i}) \leq U_d(\widehat{y}) = (d + \mu)\widehat{y}. \tag{5.5.10}$$

It is easy to see that for $t \geq 1$ (which is always the case), the upper bound on $\widehat{S}_d(\widehat{y})$ obtained from (5.5.9) is smaller than or equal to the bound found from (5.5.10). Thus combining conditions (5.5.7) and (5.5.9) to determine the size and position of the rectangles we must require

$$(d - \mu)(\widehat{y} + \text{ulp}(\widehat{y})) \leq \widehat{S}_d(\widehat{y}) \leq (d - 1 + \mu)\widehat{y} - \text{ulp}(\widehat{\beta r_i}). \tag{5.5.11}$$

But $\widehat{S}_d(\widehat{y})$ has to be an integer multiple of $\text{ulp}(\widehat{\beta r_i}) = 2^{-t}$, hence defining $\widehat{S}_d(\widehat{y}) = s_{d,k}2^{-t}$ we must require

$$\lceil 2^{t-u}(d - \mu)(k + 1)\rceil = s_{d,k} \leq \lfloor 2^{t-u}(d - 1 + \mu)k - 1\rfloor \text{ for } d > 0 \tag{5.5.12}$$

using $\text{ulp}(\widehat{y}) = 2^{-u}$ and defining $\widehat{y} = k2^{-u}$, for integer k, $2^{u-1} \leq k < 2^u$.

Recall that we required $\widehat{\beta r_i} \geq 0$ and thus it should also be possible to choose $d = 0$, but obviously then $\widehat{S}_0(\widehat{y}) = 0$ for all $\widehat{y}$, or $s_{0,k} = 0$ for all k, $2^{u-1} \leq k < 2^u$. Note that the rightmost bounds on the uncertainty rectangles for $d = 0$ are implicitly chosen by the choice of $\widehat{S}_d(\widehat{y})$ for $d = 1$.

Without restrictions on t, u, and μ, there is only the integer term -1 which can be moved in and out of the floor and ceiling functions, but reorganizing terms condition (5.5.12) for $d > 0$ can then be written as

$$\lceil 2^{t-u}(d - \mu)k + 2^{t-u}(d - \mu) + 1\rceil \leq \lfloor 2^{t-u}(d - \mu)k + 2^{t-u}(2\mu - 1)k\rfloor, \tag{5.5.13}$$

where the ceiling and floor expressions are linear functions of k:

$$\lceil Ak + B\rceil \leq \lfloor (A + C)k\rfloor, \tag{5.5.14}$$

with $A \geq 0$, $B \geq 1$ and $C > 0$ for $d \geq 1$. Clearly it is necessary that $Ck \geq B$ for this condition to be satisfied, but it is easily seen that $Ck - B \geq 1$ is a sufficient condition, since then there is at least one integer between the ceiling and floor expressions. Hence if the condition

$$2^{t-u}\left((2\mu - 1)k - (d - \mu)\right) \geq 2$$

holds for the minimal value $k = 2^{u-1}$ and the maximal value $d = a$, then this is sufficient for (5.5.13) to hold.

Thus the stronger condition derived from $Ck - B \geq 1$

$$2^{-t} \leq \left((\mu - \tfrac{1}{2}) - (a - \mu)2^{-u}\right)/2 \tag{5.5.15}$$

may be used to find values of t, $2^{-t} = \mathrm{ulp}(\widehat{\beta r_i})$ and u, $2^{-u} = \mathrm{ulp}(\widehat{y})$, for which (5.5.13) is satisfied for all $d \in \{1, \ldots, a\}$ and all k such that $2^{u-1} \leq k \leq 2^u - 1$, i.e., $\frac{1}{2} \leq \widehat{y} < 1$. Note, however, that the solutions to (5.5.15) need not be optimal in the sense that t is minimal, since it might be sufficient to require $Ck \geq B$, i.e., only to require

$$2^{-t} \leq \left((\mu - \tfrac{1}{2}) - (a - \mu)2^{-u}\right), \tag{5.5.16}$$

which would allow a solution for t which is one smaller than the solution to (5.5.15). We shall return below to the choice between the two conditions.

Obviously, the right-hand sides of both (5.5.15) and (5.5.16) must be strictly positive for solutions to exist for t, hence we want a u satisfying

$$2^{-u} < \frac{\mu - \frac{1}{2}}{a - \mu}, \tag{5.5.17}$$

provided that $a > \mu$, or $\beta > 2$, since $\beta = 2$ is the only case where $\mu = a (= 1)$. As we shall see in Example 5.5.3, the case $\beta = 2$ can be handled separately. Hence (5.5.17) is a sufficient condition on u for all $\beta > 2$, $d \in \{1, \ldots, a\}$ and all k such that $2^{u-1} \leq k \leq 2^u - 1$.

Returning to (5.5.14) and the choice between (5.5.15) and (5.5.16) to determine a minimal value of t, we need the following lemma.

Lemma 5.5.2 *Given constants $A \geq 0$, $C > 0$, then there exist integers $k_0 \leq k_1$ such that the inequality*

$$\lceil Ak + B \rceil \leq \lfloor (A + C)k \rfloor \tag{5.5.18}$$

holds for all $k \geq k_0$, provided that

(i) *$\lceil Ak + B \rceil = \lfloor (A + C)k \rfloor$ for $k_0 \leq k \leq k_1 - 1$, and*
(ii) *$\lceil Ak_1 + B \rceil < \lfloor (A + C)k_1 \rfloor$.*

Proof Define $\Delta(x) = \lfloor (A + C)x \rfloor - \lceil Ax + B \rceil$. Since $\Delta(x) \leq (A + C)x - (Ax + B) = Cx - B$ and $\Delta(x)$ is integral, it follows that $\Delta(x) \leq \lfloor Cx - B \rfloor$. If, on the other hand, $\lfloor Cx - B \rfloor = n$, then $\Delta(x) \geq n - 1$ (there are at least $n - 1$ integers between $Ax + B$ and $(A + C)x$), thus

$$\lfloor Cx - B \rfloor - 1 \leq \Delta(x) \leq \lfloor Cx - B \rfloor. \tag{5.5.19}$$

Hence $\Delta(x) - 1 \leq \lfloor Cx - B \rfloor - 1 \leq \lfloor Cy - B \rfloor - 1 \leq \Delta(y)$ for $y \geq x$, thus

$$y \geq x \Rightarrow \Delta(y) \geq \Delta(x) - 1. \tag{5.5.20}$$

$\Delta(x)$ may not always increase with x as shown in the following figure:

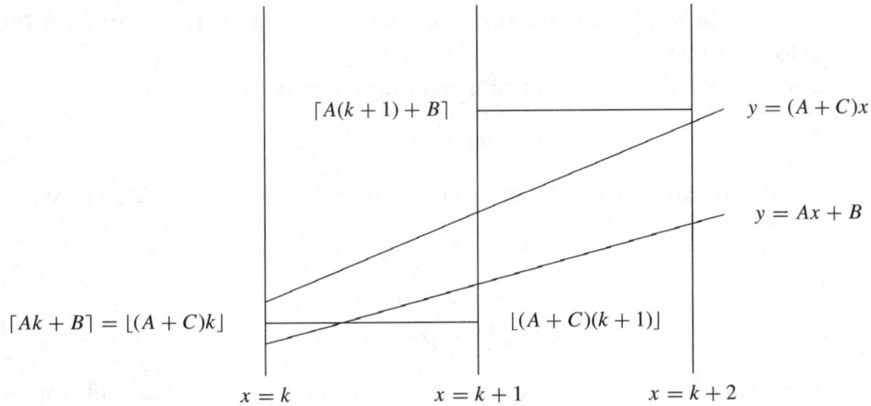

where $\Delta(k) = 0$, but $\Delta(k + 1) = \Delta(k + 2) = -1$.

Since $C > 0$, by (5.5.19) there exists a minimal k_0 such that $\Delta(k) \geq 0$ for $k \geq k_0$ until eventually there is a minimal $k_1 \geq k_0$ such that $\Delta(k_1) \geq 1$. By (5.5.20) the lemma then has been proved. □

We can now combine the previous discussion with the lemma, into the following result.

Theorem 5.5.3 (SRT digit selection constants) *For radix-β SRT division for $\beta > 2$ with digit set $D = \{-a, \ldots, a\}$, $\mu = a/(\beta - 1)$ and remainders r_i in carry-save representation, the selection constants $\widehat{S_d}(\widehat{y}) = s_{d,k}2^{-t}$ can be determined for $1 \leq d \leq a$ and $\widehat{y} = k \cdot \mathrm{ulp}(\widehat{y})$ as*

$$s_{d,k} = \left\lceil 2^{t-u}(d - \mu)(k + 1) \right\rceil$$

for $k = 2^{u-1}, \ldots, 2^u - 1$, using truncation parameters $\mathrm{ulp}(\widehat{S_d}(\widehat{y})) = \mathrm{ulp}(\widehat{\beta r_i}) = 2^{-t}$ and $\mathrm{ulp}(\widehat{y}) = 2^{-u}$, where u has to satisfy

$$2^{-u} < \frac{\mu - \frac{1}{2}}{a - \mu}.$$

To determine t for given u, let t_0 be the smallest t satisfying

$$2^{-t} \leq (\mu - \tfrac{1}{2}) - (a - \mu)2^{-u},$$

and let

$$\Delta(u,t,k) = \left\lfloor 2^{t-u}(a-1+\mu)k - 1 \right\rfloor - \left\lceil 2^{t-u}(a-\mu)(k+1) \right\rceil,$$

then

$$t = \begin{cases} t_0 & \textbf{if } \Delta(u,t_0,2^{u-1}) \geq 1, \\ t_0 & \textbf{if } \Delta(u,t_0,k) = 0, \textbf{ for } k = 2^{u-1},\ldots,k_1-1, \textbf{ and } \Delta(u,t_0,k_1) \geq 1, \\ t_0+1 & \textbf{otherwise}. \end{cases}$$

Proof The expression for $s_{d,k}$ is from (5.5.12), and the condition on u from (5.5.17) was shown, by using $\max d = a$, and $\min k = 2^{u-1}$, to be sufficient for $d > 0$ for all values of t.

Rewriting (5.5.12) we found that u and t must satisfy the condition

$$\lceil Ak + B \rceil \leq \lfloor (A+C)k \rfloor, \tag{5.5.21}$$

or equivalently $\Delta(u,t,k) \geq 0$, for $d = a$ and for all k, $2^{u-1} \leq k < 2^u$, where

$$\begin{aligned} A &= 2^{t-u}(d-\mu) \geq 0, \\ B &= 2^{t-u}(d-\mu) + 1, \\ C &= 2^{t-u}(2\mu - 1) > 0. \end{aligned}$$

Using Lemma 5.5.2 the two first choices for t imply that (5.5.12) holds for all $k \geq k_0 = 2^{u-1}$. As we saw before $Ck - B \geq 0$ for all d, $1 \leq d \leq a$ translates into the condition

$$2^{-t_0} \leq (\mu - \tfrac{1}{2}) - (a-\mu)2^{-u}.$$

If the conditions for the first two choices for t fail, then the stronger condition

$$2^{-t} \leq \left((\mu - \tfrac{1}{2}) - (a-\mu)2^{-u}\right)/2,$$

corresponding to $t = t_0 + 1$, implies that $Ck - B \geq 1$ for $k = k_0 = 2^{u-1}$ and $d = a$, and again by the lemma this implies that condition (5.5.12) holds for all $k \geq k_0$, and then also for all d, $1 \leq d \leq a$. □

Example 5.5.1 (Minimally redundant radix-4 SRT) Let $\beta = 4$ with minimally redundant digit set $D = \{-2,-1,0,1,2\}$, hence $a = 2$ and $\mu = \frac{2}{3}$, and from (5.5.17) we derive $u \geq 4$. Choosing $u = 4$ by Theorem 5.5.3, $t = t_0$ has to be the smallest solution to

$$\begin{aligned} 2^{-t} &\leq (\mu - \tfrac{1}{2}) - (a-\mu)2^{-u} \\ &= (\tfrac{2}{3} - \tfrac{1}{2}) - (2 - \tfrac{2}{3})2^{-4} \\ &= \tfrac{1}{6} - \tfrac{1}{12} = \tfrac{1}{12}, \end{aligned}$$

hence $t_0 = 4$. For $k = 2^{u-1} = 8, 9, 10$, $\Delta(4,4,k) = 0$ but $\Delta(4,4,11) = 1$. Thus by the second choice for t in Theorem 5.5.3, $t = t_0 = 4$. Hence we can compute the values of the constants $\widehat{S}(\widehat{y}) = s_{d,k}2^{-t}$ for $d > 0$ as

$$s_{d,k} = \lceil 2^{t-u}(d-\mu)(k+1) \rceil = \lceil (d-\tfrac{2}{3})(k+1) \rceil ,$$

resulting in the following table of comparison constants:

$k = 16\widehat{y}$	8	9	10	11	12	13	14	15
$16\widehat{S}_1$	3	4	4	4	5	5	5	6
$16\widehat{S}_2$	12	14	15	16	18	19	20	22

Utilizing the definitions (5.5.4) of the functions $L_d(y)$ and $U_d(y)$ we have

	d				
	-2	-1	0	1	2
$L_d(y)$	$\frac{-8}{3}y$	$\frac{-5}{3}y$	$\frac{-2}{3}y$	$\frac{1}{3}y$	$\frac{4}{3}y$
$U_d(y)$	$\frac{-4}{3}y$	$\frac{-1}{3}y$	$\frac{2}{3}y$	$\frac{5}{3}y$	$\frac{8}{3}y$

which together with the table for $\widehat{S}_d$ yields the P-D diagram for the first quadrant shown in Figure 5.5.7. In practice the left part of the diagram (for $\widehat{\beta r_i} < 0$) is not needed as we shall see later, when utilizing symmetries in the uncertainty rectangles. □

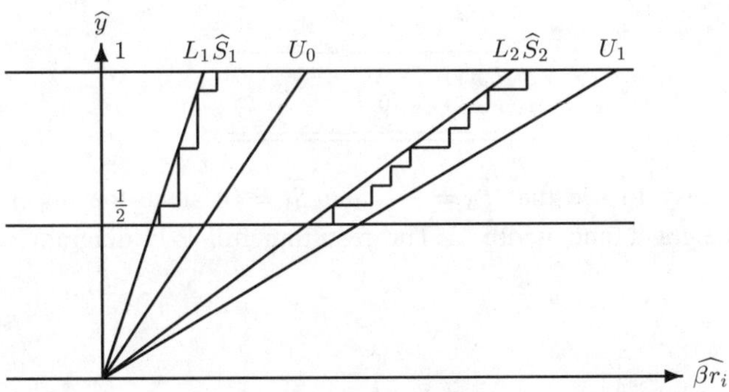

Figure 5.5.7. P-D diagram for minimally redundant radix-4 SRT.

Example 5.5.2 Continuing the previous example, if instead of choosing the minimal $u = 4$ the value $u = 5$ is chosen, we find

$$\begin{aligned} 2^{-t} &\le (\mu - \tfrac{1}{2}) - (a-\mu)2^{-u} \\ &= (\tfrac{2}{3} - \tfrac{1}{2}) - (2 - \tfrac{2}{3})2^{-5} \\ &= \tfrac{1}{6} - \tfrac{1}{24} = \tfrac{1}{8}, \end{aligned}$$

having the minimal solution $t = t_0 = 3$. However, for $k = 2^{u-1} = 16$ we find for $\Delta(u, t_0, k)$

$$\begin{aligned}\Delta(5,3,16) &= \lfloor 2^{-2}\cdot\tfrac{5}{3}\cdot 16 - 1\rfloor - \lceil 2^{-2}\cdot\tfrac{4}{3}\cdot 17\rceil\\ &= \lfloor\tfrac{17}{3}\rfloor - \lceil\tfrac{17}{3}\rceil\\ &= 5 - 6 = -1,\end{aligned}$$

thus by Theorem 5.5.3 it is necessary to increase $t = t_0 + 1 = 4$, hence $(u, t) = (5, 4)$ is also a valid pair of truncation parameters, but obviously is not as good as $(4, 4)$ found in the previous example. As a check we find $\Delta(5, 4, 16) = \lfloor\frac{37}{3}\rfloor - \lceil\frac{34}{3}\rceil = 12 - 12 = 0$ and $\Delta(5, 4, 17) = \lfloor\frac{79}{6}\rfloor - \lceil\frac{36}{3}\rceil = 13 - 12 = 1$. □

Example 5.5.3 (Radix-2 SRT) For $\beta = 2$ and digit set $\{-1, 0, 1\}$ the condition on u in Theorem 5.5.3 cannot be used since $a = \mu = 1$. However, for $u = t = 1$ and $k = 1$, corresponding to $2^{u-1} = 1 = k < 2^u = 2$, it is easily seen that (5.5.12) is satisfied for $d = 1$.

Thus only one bit of the fraction part of y is needed for $\widehat{y}$, and this bit is always 1 since $\frac{1}{2} \le y < 1$, implying that the quotient selection is independent of y. Note that $\mathrm{ulp}(\widehat{\beta r_i}) = 2^{-1} = \frac{1}{2}$. From (5.5.4) the lower and upper bounds are then found to be

	d		
	-1	0	1
$L_d(y)$	$-2y$	$-y$	0
$U_d(y)$	0	y	$2y$

and it now easy to see that $\widehat{S}_0 = -\frac{1}{2}$ and $\widehat{S}_1 = 0$, since the uncertainty rectangle has height $\frac{1}{2}$ and width 1. The resulting full P-D diagram is shown in Figure 5.5.8.

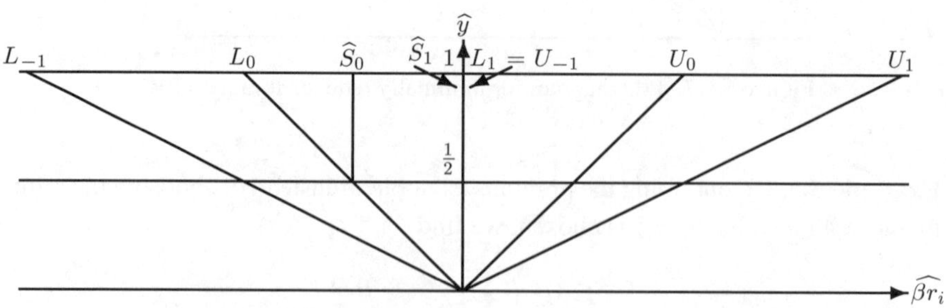

Figure 5.5.8. P-D diagram for radix-2 SRT with redundant remainder.

To determine the number of leading (integer) digits needed to represent βr_i, recall that $r_{min} = -\mu y \leq r_i \leq \mu y = r_{max}$, hence $|\beta r_i| < 2$, thus two integer digits are sufficient to represent the shifted remainder in non-redundant binary. But assuming βr_i is in 2's complement, carry-save representation, then it turns out that three integer digits will be needed. Consider that truncation to $\text{ulp}(\widehat{\beta r}_i) = \frac{1}{2}$ in $\beta r_i = \widehat{\beta r}_i + \varepsilon$ can discard as much as $\varepsilon = 1 - 2^{-n}$, where $2^{-n} = \text{ulp}(\beta r_i)$. Since βr_i can be as small as $\beta r_i = -2 + \delta$, where $\delta \geq \text{ulp}(\beta r_i)$, it is certainly possible that $\delta < \varepsilon$ so that

$$\widehat{\beta r}_i = \beta r_i - \varepsilon = -2 + \delta - \varepsilon < -2.$$

Thus $\widehat{\beta r}_i$ can take any value in the interval between $-\frac{5}{2}$ and $\frac{3}{2}$ in steps of $\frac{1}{2}$, since also $\beta r_i < 2$. Thus three integer bits and one fractional bit are needed to represent $\widehat{\beta r}_i$ when it is converted to non-redundant 2's complement form, and these four bits are then necessary and sufficient for the digit selection. □

Two leading guard digits are needed when updating the remainder, cf, Theorem 3.8.9, and these have to be considered when shifting the remainder to form the new βr_{i+1}. As noted above, three integer digits are needed in the radix-2 case to represent $\widehat{\beta r}_i$ in 2's complement carry-save. The arithmetic of the (redundant) remainder updating then has to be performed with four integer digits; since the new remainder is bounded by $-1 < -y \leq r_{i+1} \leq y < 1$, it will then have three guard digits. After a "brute force" shifting and discarding the most-significant digit, βr_{i+1} will still have the two necessary leading guard digits. After truncation for the next digit selection, these guard-digits are sufficient for the conversion of $\widehat{\beta r}_{i+1}$ to non-redundant representation with three integer bits. This immediately generalizes to higher radices:

Observation 5.5.4 *Remainder updating in radix-2^m SRT division can be performed in carry-save arithmetic with two leading guard digits beyond those needed to represent βr_i in non-redundant 2's complement representation. After remainder updating and left-shifting m positions, the new remainder βr_{i+1} will again contain two guard digits.*

By Theorem 3.8.4 an almost identical result holds when the arithmetic is borrow-save.

Recall that for radix 2 the digit selection is independent of the divisor, simplifying the selection process. For higher values of the radix it is similarly possible to simplify the digit selection if it is known that the divisor is suitably restricted, or after it is made so by an appropriate scaling. By *prescaling* both the dividend and the divisor, it follows that the quotient will remain unchanged, but the remainder will have to be postscaled if the actual value of it is needed. However, for rounding purposes such a postscaling is unnecessary.

If a (rectangular) multiplier is available it may be possible to prescale the operands such that the scaled divisor is so close to unity that the digit selection can be performed by simply rounding a suitable leading prefix of the remainder.

However, a simple prescaling may be obtained by some suitably chosen shift and add operations, allowing the divisor to be brought into an interval where selection is independent of the actual divisor value. Since we are here assuming that the divisor is in the interval $\left(\frac{1}{2}; 1\right]$ the scaled divisor should be at or just below 1, or just above $\frac{1}{2}$.

For minimally redundant radix 4 this would imply a prescaling of the divisor to within one sixteenth of a unit; however, as we shall see below, extending the digit range allowing the digits ± 3 will relax the requirement for prescaling, by doubling the interval. The calculation of the non-trivial multiple $3y$ may be performed in parallel with the prescaling of the operands, before the iterative digit selections take place, now only depending on the value of the remainder.

Example 5.5.4 (Maximally redundant radix-4 SRT) Let $\beta = 4$ with maximally redundant digit set $D = \{-3, -2, -1, 0, 1, 2, 3\}$, hence $a = 3$ and $\mu = 1$, and from (5.5.17) we derive $u \geq 3$. Choosing $u = 3$ by Theorem 5.5.3, $t = t_0$ has to be the smallest solution to

$$2^{-t} \leq (\mu - \tfrac{1}{2}) - (a - \mu)2^{-u} \ = \ \tfrac{1}{4},$$

hence $t_0 = 2$. For $k = 2^{u-1} = 4$ and $k = 5$, $\Delta(3, 2, k) = 0$ but $\Delta(3, 2, 6) = 1$. Thus by the second choice for t in Theorem 5.5.3, $t = t_0 = 2$. Hence we can compute the values of the constants $\widehat{S}(\widehat{y}) = s_{d,k} 2^{-u}$ for $d > 0$ as

$$s_{d,k} = \left\lceil 2^{t-u}(d - \mu)(k + 1) \right\rceil = \left\lceil \tfrac{1}{2}(d - 1)(k + 1) \right\rceil,$$

resulting in the following table of comparison constants:

$k = 8\widehat{y}$	4	5	6	7
$8\widehat{S}_1$	0	0	0	0
$8\widehat{S}_2$	3	3	4	4
$8\widehat{S}_3$	5	6	7	8

Utilizing the definitions (5.5.4) of the functions $L_d(y)$ and $U_d(y)$, $d \geq 0$ we find

	d			
	0	1	2	3
$L_d(y)$	$-y$	0	y	$2y$
$U_d(y)$	y	$2y$	$3y$	$4y$

yielding the following P-D diagram:

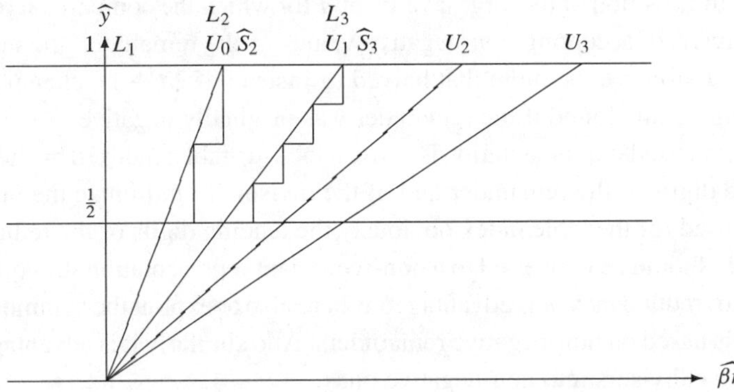

□

Example 5.5.5 (Prescaled maximally redundant radix-4 SRT) The factors needed to prescale dividend, divisor pair (x, y) into (x', y') such that y' belongs to the interval $\left(\frac{1}{2}; \frac{5}{8}\right]$ are listed below, together with the equivalent scaling to obtain the divisor multiple $3y'$:

$y \in$	Factor for x', y'	Factor for $3y'$
$\left(\frac{7}{8}; 1\right]$	$\frac{1}{2} + \frac{1}{8} = \frac{5}{8}$	$2 - \frac{1}{8} = \frac{15}{8}$
$\left(\frac{3}{4}; \frac{7}{8}\right]$	$\frac{1}{2} + \frac{1}{4} - \frac{1}{16} = \frac{11}{16}$	$2 + \frac{1}{16} = \frac{33}{16}$
$\left(\frac{5}{8}; \frac{3}{4}\right]$	$\frac{1}{2} + \frac{1}{4} + \frac{1}{16} = \frac{13}{16}$	$2 + \frac{1}{2} - \frac{1}{16} = \frac{39}{16}$
$\left(\frac{1}{2}; \frac{5}{8}\right]$	1	$2 + 1 = 3$

Note that y' and $3y'$ can be obtained by an 3-to-2 reduction followed by a carry-completing addition, whereas x' may be delivered in redundant form, as it takes the role of the initial remainder.

Due to the prescaling of y the digit selection can now be based exclusively on the remainder, and since it is in redundant form, the selection may even be based directly on the components of that representation, without the need for a conversion.

As seen in the previous example, $\text{ulp}(\widehat{\beta r_i}) = 2^{-t}$ with $t = 2$, thus only two fractional digits of the redundant remainder are needed. From $|r_i| < \mu y$ it follows that $-\frac{5}{2} < \beta r_i < \frac{5}{2}$, and since truncation can discard at most $\frac{1}{2} - 2^{-n}$, we find that $\widehat{\beta r_i}$ can vary between $-\frac{11}{4}$ and $\frac{9}{4}$ in steps of $\frac{1}{4}$, thus three integer carry-save digits are sufficient to represent the (signed) $\widehat{\beta r_i}$. When exploiting the sign symmetry (see below), only two integer bits of the carry and the save parts are needed, and hence an eight-bit look-up table is sufficient to determine the next digit. □

5.5.3 Exploiting symmetries

A simple way of implementing the digit selection function of (5.5.7) is to compare the value of $\widehat{\beta r_i}$ with selection bounds $\widehat{S}_d(\widehat{y})$ corresponding to the staircase function. This can be performed by a set of comparators, using a priority encoder to determine the position of the largest value of d for which the comparison returns true. By exclusively accepting non-negative values of the remainder, the number of such comparators can be more that halved (a instead of $2a + 1$), changing the sign of the digit value found if the remainder was originally negative.

Another often used implementation is to use a look-up table, indexed by the leading truncated digits of the remainder and of the divisor. To minimize the number of bits to be used for the table index obviously the leading digits of the redundant remainder βr_i should be converted to a non-redundant representation. Beyond this assimilation of redundancy, it is advantageous here also to exploit the symmetry, so that look-up is based on non-negative remainders. And similarly it is advantageous to map negative divisors into non-negative ones.

However, it is not quite trivial to see how to change the sign of a negative remainder, since we have to make sure that the "uncertainty" rectangle is mapped correctly from the left half-plane onto the right half-plane of Figure 5.5.5. We will start with the case of remainders in 2's complement carry-save representation, as illustrated in Figure 5.5.9.

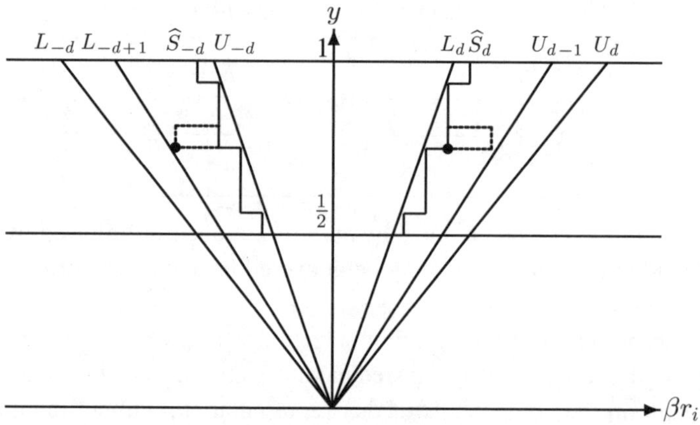

Figure 5.5.9. Mapping negative remainders to positive.

Observe that the left rectangle is determined by the point $(\widehat{\beta r}_i, \widehat{y})$ in its lower *left-hand* corner, assuming $y > 0$. The symmetric point given by $(-\widehat{\beta r}_i, \widehat{y})$ is the lower *right-hand* corner of the symmetric rectangle to the right. But we need the coordinates of the lower *left-hand* corner of it, which is located at the point $(-\widehat{\beta r}_i - 2\mathrm{ulp}(\widehat{\beta r}_i), \widehat{y})$.

Fortunately, there is a very simple way of performing this mapping when $\widehat{\beta r}_i$ is represented in 2's complement carry-save. Recall that negation of a 2's complement carry-save number is performed by inverting the bit-patterns of the "save" part as well as the "carry" part, and adding a unit in the least-significant position of *both*. Hence by just inverting all the bits of the encoding of $\widehat{\beta r}_i$ (i.e., forming the 1's complement, and *not* adding a unit to both), we obtain precisely the effect of subtracting the correction $2\text{ulp}(\widehat{\beta r_i})$, and thus the correct mapping of the left rectangle into the right.

By analogy, it is also possible to map the rectangle corresponding to a negative divisor $\widehat{y}$ from the lower half-plane to the upper one by simply inverting the bits of $\widehat{y}$, which implies that there is no need to perform a time consuming negation of the 2's complement divisor y initially in the division algorithm. If $y < 0$, then the bits of $\widehat{y}$ just have to be inverted.

If the remainder is represented in borrow-save, the situation is also very simple. Here the rectangle of uncertainty is determined by the midpoint of the lower edge of the rectangle. This implies that the mapping is obtained by negating $\widehat{\beta r_i}$, which is the trivial constant-time operation of simply interchanging the negatively and positively weighted bit-vectors of the borrow-save encoding.

Observation 5.5.5 *For the purpose of SRT quotient digit selection, truncated negative remainders and/or divisors can be mapped in constant time into the first quadrant by:*

> **2's complement***: complement all bits of the encoding bit-vector(s)*
> **borrow-save***: interchange the positively and negatively weighted bit-vectors,*

thereby insuring that the "uncertainty rectangles" are mapped into their non-negative equivalents. The selected quotient digit d_{i+1} changes its sign according to the mappings performed, while the updating of the shifted remainder remains $r_{i+1} = \beta r_i - d_{i+1}y$, employing the signed divisor y and remainder r_i.

Of course, to know whether to invert or not, it is necessary to know the sign of $\widehat{\beta r_i}$, but $\widehat{\beta r_i}$ also has to be converted into a non-redundant representation (be compressed). Actually, it is easily seen that a mapping into sign-magnitude representation is what is needed, providing the magnitude part as the result of the mapping. In Section 3.6 it was shown how a 2's complement carry-save represented number can be converted into a sign-magnitude represented one.

However, since the word-length is fairly small, at the expense of some extra area it may be faster to add the carry and the save parts of $\widehat{\beta r_i}$ in parallel, as well as adding the inverted bit-patterns. The sign of the former addition then determines which of the two sums is to be the final (non-negative) compressed result, to be used as index to a table look-up or as input to some other quotient determination logic. Hence the additional delay of the mapping is just that of a selector following the adders, i.e., a constant-time overhead, as shown in Figure 5.5.10. Observe that the adders need to be wider than the wanted result to allow correct sign determination in the 2's complement addition of the carry and the save parts. By Theorem 3.8.9 it is known that two leading guard digits are necessary and sufficient.

Similarly, for borrow-save represented remainders, it is sufficient with two parallel subtractions, followed by a selection of the non-negative result, to realize the compression as well as the mapping.

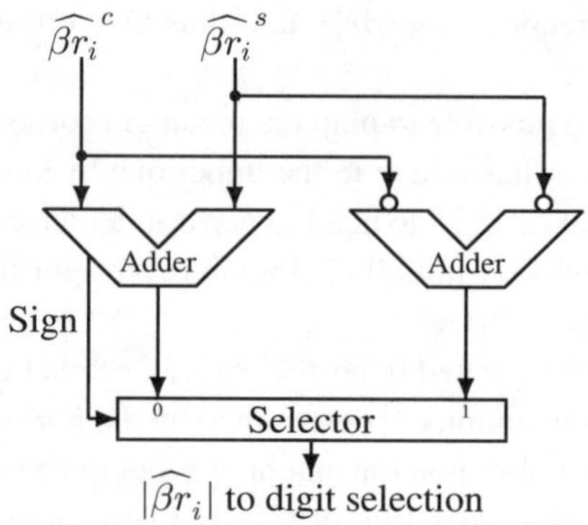

Figure 5.5.10. Mapping into non-negative remainders.

5.5.4 Digit selection by direct comparison

A simple way of implementing the digit selection function of (5.5.7) is in parallel to compare the value of $\widehat{\beta r}_i$ with selection bounds $\widehat{S}_d(\widehat{y})$, corresponding to the different values of d. This can be performed by a set of comparators as sketched in Figure 5.5.11, where the priority encoder determines the position of the largest value of d for which the comparison returns true. We will now assume that negative remainders and divisors have been mapped into non-negative ones, hence only positive digit values are considered. The constant value 1 supplied corresponds to $|\widehat{\beta r}_i| < \widehat{S}_1(|\widehat{y}|)$, where $d_{i+1} = 0$ has to be chosen. The digit value $|d_{i+1}|$ then later has to be supplied with the appropriate sign.

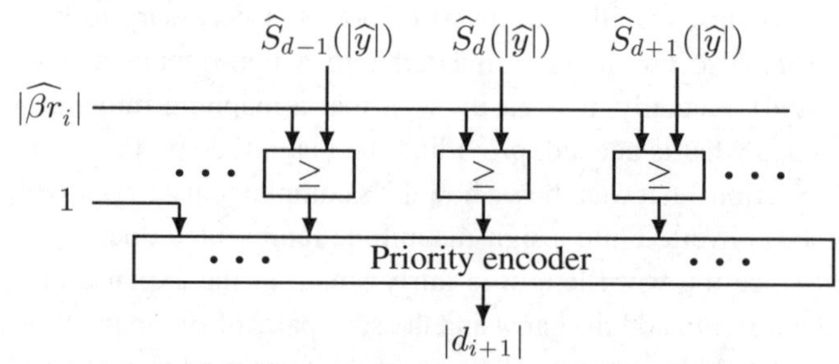

Figure 5.5.11. Quotient determination by direct comparison.

The bit-width of these comparators is determined as $\delta = \tau + t$ by the truncation as illustrated in Figure 5.5.4, where τ can be bounded as

$$\tau = \max\left(\lceil \log_2(|\beta r_i|)\rceil\right) \le \lceil \log_2(\beta r_{max})\rceil \le \lceil \log_2(\beta\mu)\rceil, \qquad (5.5.22)$$

since $r_{max} = \mu y$ and $y < 1$. For example, for $\beta = 4$ and a minimally redundant quotient digit set with $\mu = \frac{2}{3}$ we may use $\tau = \lceil \log_2(\frac{8}{3})\rceil = 2$, and thus $\delta = 6$. Recall that a comparator is simpler than a subtractor as only the carry-out is needed.

Let us conclude this method of quotient determination with some rough estimates of the complexity of the necessary hardware as a function of the quotient radix β. The number of comparators equals the number of non-negative digits in the quotient digit set less 1, which is $a - 1$, which is essentially proportional to β. The bit-width of the comparators is $\delta = \tau + t$, which is proportional to $\log_2(\beta)$, as can be seen from (5.5.16) and (5.5.22). Hence the size (area) of the collection of comparators is $\Theta(\beta \log \beta)$, which also covers the size of the encoder.

Each comparator needs a constant $\widehat{S}_d(|\widehat{y}|)$, which could be obtained from a table whose size is the number of $|\widehat{y}|$ values, 2^{u-1}, and is thus proportional to β, times the bit-width of the constants, hence the total size of the tables in bits is $\Theta(\beta^2 \log \beta)$. For practical implementations, note that $|\widehat{y}|$ is assumed normalized, $\frac{1}{2} \le |\widehat{y}| < 1$, hence the leading bit is always 1, and need not be used to address the tables. Also note that the tables need only to be accessed once, as $|\widehat{y}|$ remains constant during the algorithm, thus the table look-up is not part of the critical path of the loop.

5.5.5 Digit selection by table look-up

The most frequently employed implementation of the quotient digit selection uses leading digits of $|\widehat{y}|$ and the compressed $|\widehat{\beta r_i}|$ concatenated as index (address) to a table, directly providing the quotient digit. This was the method employed in the famous Intel division bug, where some table entries were incorrect in the initial Pentium chips, using the minimally redundant radix-4 SRT division algorithm in the floating-point divide instruction. The errors were introduced during the circuit design of the chip, when table entries were transfered from a file into a design tool generating the hardware table.

Table 5.5.1 lists valid truncations for SRT digit selection tables for radices 2, 4, and 8 and digit sets $\{-a, \ldots, a\}$, based on divisor y in 2's complement with $\mathrm{ulp}(\widehat{y}) = \Delta = 2^{-u}$ and shifted remainder in carry-save or borrow-save encoding, but assimilated into non-redundant representation. It is assumed that divisor and remainder have been mapped into the first quadrant, i.e., they are non-negative. The truncations are specified by the following expanded set of parameters:

$$
\begin{aligned}
\tau &= \text{\# integer bits in assimilated } |\widehat{\beta r_i}|, \text{ including 2's complement sign,}\\
p &= \begin{cases} -\log_2(\mathrm{ulp}(\widehat{\beta r_i^c})) & \text{for remainder in carry-save,}\\ -\log_2(\mathrm{ulp}(\widehat{\beta r_i^p})) & \text{for remainder in borrow-save,} \end{cases}\\
r &= \begin{cases} -\log_2(\mathrm{ulp}(\widehat{\beta r_i^s})) & \text{for remainder in carry-save,}\\ -\log_2(\mathrm{ulp}(\widehat{\beta r_i^n})) & \text{for remainder in borrow-save,} \end{cases}\\
f &= \text{\# fractional bits used for remainder, } f \le \max(p, r),\\
u &= \text{\# fractional bits used for divisor, } u = -\log_2(\mathrm{ulp}(\widehat{y})),\\
w &= \text{total \# bits required for table index, } w = f + (\tau - 1) + (u - 1),
\end{aligned}
$$

Table 5.5.1. *SRT table look-up parameters for redundant remainders radix 2, 4, and 8*

β	a	τ	p	r	f	u	w
2	1	3	1	1	1	1	3
4	2	3	4	4	4	4	9
4	3	4	2	2	2	3	7
8	4	4	5	5	5	8	15
8	4	4	6	5	6	7	15
8	4	4	6	6	5	7	14
8	4	4	8	6	8	6	16
8	4	4	7	7	7	6	15
8	5	4	4	3	4	7	13
8	5	4	5	3	5	6	13
8	5	4	4	4	4	6	12
8	5	4	4	4	3	7	12
8	5	4	5	4	5	5	12
8	5	4	5	5	4	5	11
8	6	4	4	2	4	7	13
8	6	4	5	2	5	6	13
8	6	4	3	3	3	6	11
8	6	4	4	3	4	5	11
8	6	4	4	3	4	5	11
8	6	4	4	4	3	5	10
8	6	4	4	4	2	6	10
8	6	4	9	5	8	4	14
8	7	5	2	2	2	7	12
8	7	5	3	2	3	5	11
8	7	5	3	3	3	4	10
8	7	5	3	3	2	5	10

while in the previous analysis it was assumed that $f = p = r = \text{ulp}(\widehat{\beta r_i}) = t$. With these parameters the truncation errors are $0 \le \epsilon_r < 2^{-p} + 2^{-r}$ for carry-save, and $-2^{-r} < \epsilon_r < 2^{-p}$ for borrow-save remainders before compression to non-redundant representation, thus possibly reducing the width of the uncertainty rectangle. The parameter f specifies the number of fractional bits from the compressed $\widehat{\beta r_i}$ to be used for the table index. The total number of bits is then $w = f + (\tau - 1) + (u - 1)$, where the leading bit from y is not needed since it is always 1, nor is the sign-bit from $|\widehat{\beta r_i}|$. Note that the size of the table is 2^w by 2 to 4 bits for these values of the radix, where the encoding of the quotient digit can be chosen to aid the formation of divisor multiples. It has been found experimentally that when the table is optimized as a PLA, its area grows as $\Theta(\beta^2)$.

5.5.6 Architectures for SRT division

A generic architecture of an implementation of a radix-2^k SRT algorithm may be as illustrated in Figure 5.5.12, assuming that remainder updating takes place in 2's complement carry-save representation (in the carry-save adder (CSA)).

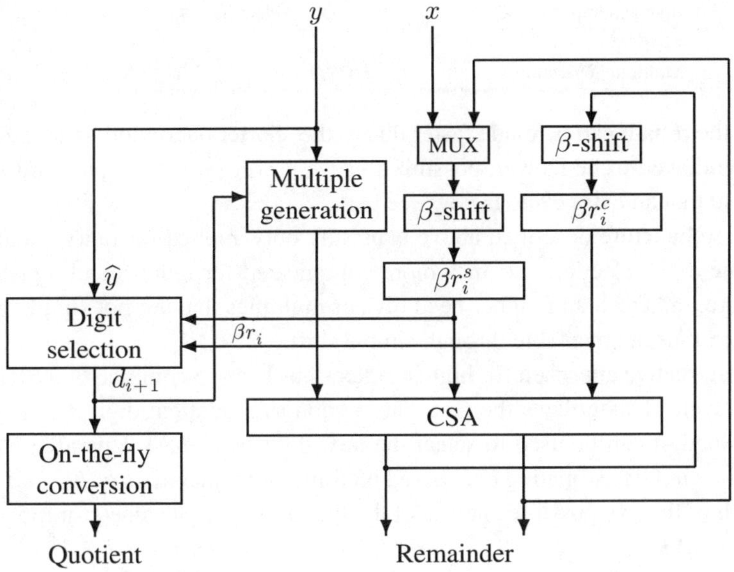

Figure 5.5.12. SRT architecture.

The MUX[4] is used to initialize the βr_i^s buffer, where simultaneously the βr_i^c buffer is reset. The β-shift boxes perform a trivial shift of k positions, corresponding to the assumption that $\beta = 2^k$. Digit selection can be based on either parallel comparisons or table look-up, the controlling input being a suitable number of leading bits from divisor $|\widehat{y}|$ and shifted remainder $|\widehat{\beta r_i}|$. The selected next quotient digit d_{i+1} is fed into the multiple-generation box, generating the product $d_{i+1}y$, to be subtracted from the shifted remainder, thus forming the next remainder. We are not here considering the additional logic needed to form the final quotient and remainder, nor possible rounding.

The critical path is through the digit-selector, the multiple-generation and finally through the CS-adder back to the βr_i-buffers. Sketching this on a time line we have:

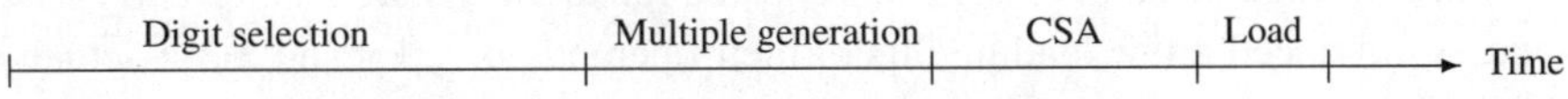

where the digit selection is the most time consuming operation. If a buffer is introduced between the digit selection and the multiple generation, a retiming is introduced such that the digit used is the one produced in the previous cycle. Then there are two parallel paths, one producing the next quotient digit based on the most-significant digits, and another producing the remainder:

[4] MUX: multiplexor, i.e., a selector.

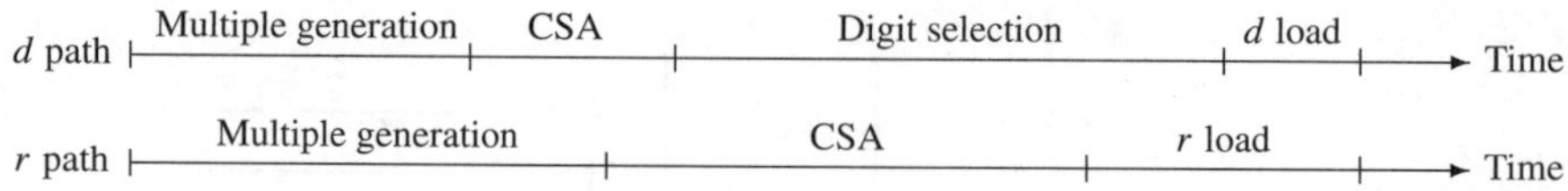

where the d path can be made faster due to the shorter data-width, and the r path can be allowed to be slower, possibly to save power on the longer word-length. But note that an extra cycle is now needed.

The architecture described above is usually only applied for radix 2 and radix 4, as the digit selection becomes more complicated for radix 8 and higher. Furthermore, radices 8 and higher need divisor multiples that are not simple powers of 2, and thus not available through simple shifts.

An alternative approach for higher radices has been to combine two SRT steps in one cycle. This utilizes the fact that as soon as a quotient digit d_{i+1} has been determined, it can be used to select the next digit from $2a + 1$ parallel quotient selection circuits estimating d_{i+2} based on truncated remainders of $\beta r_i - d\,y$, one for each of the five possible choices of the first digit d, as sketched for radix-16 in Figure 5.5.13.

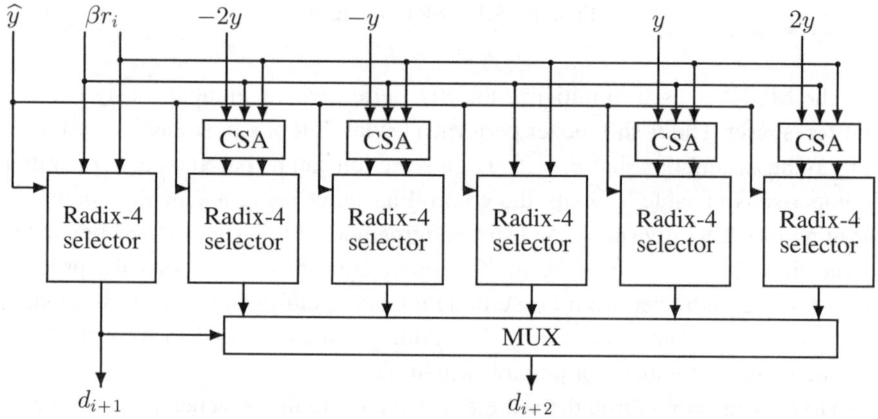

Figure 5.5.13. Radix-16 digit selector made by combining two radix-4 selecton

Since the digit selectors just need truncated remainders, note that the carry-save adders only need a few leading bits of their operands βr_i, $\pm y$ and $\pm 2y$; actually for radix 4, they need two more bits than the left-most selector determining d_{i+1}.

The architecture of the rest of the radix-16 divider now comprises two multiple-generators (MGs) and carry-save adders in sequence, controlled by respectively d_{i+1} and d_{i+2}. The following timing diagram shows the overlapping of operations (not drawn to scale):

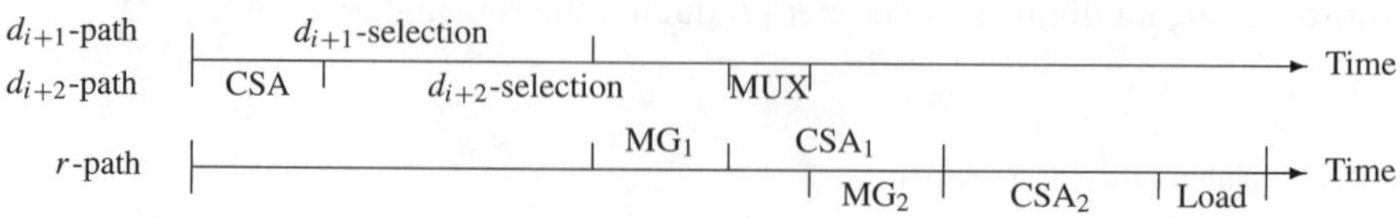

Hence some overlap is obtained, but a detailed design and timing analysis is required to evaluate this solution, compared with two consecutive radix-4 designs, in particular considering that the above radix-16 design requires five radix-4 digit selectors. Generating any of the multiples $3y$, $5y$, and $7y$ in redundant form and subtracting such a redundant multiple from the shifted remainder requires substituting a 4-to-2 adder for the usual 3-to-2 carry-save adder. These multiples could also be generated in non-redundant form, before the digit generating loop is entered, adding some additional clock cycles to the total execution time. If such multiples are generated, then prescaling the divisor should also be considered, thereby simplifying the digit selectors.

Alternatively, a radix-16 digit selector based on direct comparison exploiting sign symmetry, would need seven parallel comparators when the minimally redundant digit set is employed. The comparators need to be slightly wider than for the radix-4 case, but would only be slightly slower.

Problems and exercises

5.5.1 Verify (5.5.3) and (5.5.4), in particular consider why it is not necessary to use bounds r_{min}, r_{max}, L_d, and U_d that depend on the index i.

5.5.2 Show that $L_{-d} = -U_d$ holds in general, and that for maximally redundant quotient digit sets $U_{d-1} = L_{d+1}$ holds.

5.5.3 Work out the details of comparing a borrow-save represented remainder $\widehat{\beta r_i}$ against modified bounds $\widehat{S}_d(\widehat{y}) + \text{ulp}(\widehat{\beta r_i})$.

5.5.4 Check some of the truncation parameters sets for which $f = p = r$ from Table 5.5.1 against the necessary condition (5.5.12).

5.5.5 Design a radix-4 SRT divider by combining two radix-2 quotient digit selections in one clock cycle, and compare it with the prescaled design in Example 5.5.5.

5.6 Multiplicative high-radix division

Digit serial division can be extended to much higher radices when employing a rectangular aspect ratio multiplier. Multiplicative division is particularly efficient for a high radix β^k with a $(k + g) \times p$ digit multiplier where the aspect ratio of the full multiplicand size p to the smaller multiplier size k is in the range $2 \leq p/k \leq 10$, and g is some small number of *guard digits*. Compared to a full $p \times p$ digit multiplier, a large aspect ratio such as $p/k \geq 6$ reduces the hardware size and increases the hardware speed of each $(k + g) \times p$ digit multiplication at the cost of requiring $\lceil p/k \rceil$ multiplications to produce the equivalent of a $p \times p$ digit multiplication. Note that when a multiplier is available to form divisor multiples on demand, there is no need to restrict the digit set to the minimally redundant set. We thus assume that the digit set is maximally redundant, possibly even allowing the first digit to take the value β^k, as occasionally allowed in other non-restoring division algorithms.

Multiplicative digit serial division with a $(k + g) \times p$ digit rectangular multiplier is applicable to radix-β integer, fixed- and floating-point division with the following results:

- For p by p digit *short reciprocal division* we show that $2\lceil p/k \rceil$ multiplications are sufficient to determine the p digit quotient and p digit remainder. A $2p$ by p digit division result can be obtained with at most two additional multiplications.
- For p by p digit *prescaled division* we show that $\lceil p/k \rceil + 2$ multiplications are sufficient to determine the p-digit quotient and a two-bit "round-sticky" remainder state.
- For $2p$ by p digit *prescaled division with remainder* we show a third multiplicative integer division method applicable to multipliers with a relatively large aspect ratio where $\lceil p/k \rceil + 3$ multiplications are sufficient to provide the q, r pair.

Given the resulting quotient and remainder, q, r, of $2p$ by p digit integer division of x by y, note that verification of the identity $x = qy + r$ would itself require $\lceil p/k \rceil$ multiplications with a $(k + g) \times p$ digit multiplier. Thus these high radix-β^k digit serial division procedures make very efficient use of the rectangular multiplier resource.

5.6.1 Short reciprocal division

The digit selection step of digit serial division involves estimation of the remainder *divided* by the divisor. In "short reciprocal" division this operation is realized by forming the *product* of the remainder and an approximate reciprocal of the divisor. In order to obtain the maximum benefit from use of a $(k + g) \times p$ digit multiplier, the approximate reciprocal should have an accuracy corresponding to the shorter size $k + g$ of the rectangular multiplier.

For a normalized decimal fraction such as the six-digit value 0.784731, an approximate reciprocal with an accuracy of three digits is obtained by choosing either of the three-digit values bracketing the reciprocal, where here $1.27 \leq \frac{1}{0.784731} < 1.28$. With the approximate reciprocal 1.28 we note $1.28 \times 0.784731 = 1.00445568$, with accuracy better than one part in 10^2. Without loss of generality, for illustrating decimal division it is convenient to have both the p-digit divisor and the $(k + g)$-digit approximate reciprocal expressed as integers rather than fractions. This is obtained by requiring their product to differ by less than one part in β^k from the value β^{p+k}.

The "short reciprocal division" procedure is similar to hand computed long division, and will be illustrated by a decimal example employing a (3×6)-digit integer multiplier where the quotient is developed digit serially in the higher radix $10^2 = 100$.

Example 5.6.1 (Short reciprocal division)
Problem Find the six-digit integer quotient q and six-digit integer non-negative remainder r with $0 \le r \le y - 1$ such that $x = q \times y + r$, where

$$\begin{aligned} x &= 365748000000 \quad \text{(12-digit dividend)} \\ y &= 784731 \qquad\quad\;\; \text{(6-digit divisor)} \end{aligned}$$

Solution The result is determined digit serially as three radix-100 quotient digits in the redundant digit range $[-100, 100]$, and then converted to standard decimal as follows. Note that the arithmetic employed here is sign-magnitude decimal.
Step 1 Determine the three-digit short reciprocal

$$\rho = \left\lceil \frac{10^8}{784731} \right\rceil = 128.$$

Step 2 Perform digit-serial division radix 100. Digit selection is obtained iteratively by multiplying the three-digit short reciprocal by the six-digit remainder, with the initial quotient digit obtained by multiplying the short reciprocal by the leading six digits of the dividend, and then selecting the signed leading three digits of the nine-digit product as the next redundant radix-100 digit. The digit selection is shown in bold face in the following, and each high radix digit is expressed as a two-digit decimal.

$$\begin{array}{r|l l l}
 & \;\;46\;61\;\overline{19}. & & \\
78\,47\,31 & 36\,57\,48\,00\,00\,00. & & \\
128\times & 36\,57\,48 & = \mathbf{46}\,81\,57\,44 & \text{first digit selection} \\
46 \times 78\,47\,31 = & 36\,09\,76\,26 & & \\
\hline
128\times & \;\;47\,71\,74 & = \mathbf{61}\,07\,82\,72 & \text{second digit selection} \\
61 \times 78\,47\,31 = & \;\;47\,86\,85\,91 & & \\
\hline
128\times & \;\;-15\,11\,91 & = -\mathbf{19}\,35\,24\,48 & \text{third digit selection} \\
-19 \times 78\,47\,31 = & \;\;-14\,90\,98\,89 & & \\
\hline
 & \;\;-20\,92\,11 & &
\end{array}$$

Step 3 Determine the non-negative remainder: $r = 784731 - 209211 = 575520$. Decrement and convert the redundant quotient to standard decimal form:

$$q = 46\ 61\ \overline{19}_{100} - 1 = 466080.$$

The digit selection procedure illustrated in Step 2 is justified as follows. In the first digit selection we seek to estimate a leading radix-100 digit (two decimal digits) of $\frac{365748}{784731}$. Employing the exact divisor reciprocal $\frac{1}{784731} = 127.4322\cdots \times 10^{-4}$, note that $(127.4322\cdots \times 10^{-4}) \times 365748 = 46.60807\cdots$, so "46" should be our leading radix-100 digit. Multiplying by the short reciprocal yields $128 \times 365748 = 046815744$, where the comparably normalized value is larger by less than 1% than the infinitely precise quotient. We select "46" knowing that it is either the correct radix-100 digit (leading to a positive or zero remainder) or possibly one unit too high. The latter situation occurs in estimating the second redundant radix-100 digit. Multiplying the exact divisor reciprocal by the first remainder generates the rest of the infinitely precise quotient $(127.4322\cdots \times 10^{-4}) \times 477174 = 608073.3\cdots$. Multiplication by the short reciprocal yields $128 \times 477174 = 061078272$, leading to the selection of "61" as the second digit generating a *negative* remainder. The fact that the short reciprocal exceeds the true reciprocal by less than 1% insures that the magnitude of the next remainder will always be less than the divisor. □

In practice an integer division implementation would be applied to binary unsigned or 2's complement operands and binary arithmetic with redundant remainders. A formalization of radix-β short reciprocal division incorporating redundancy is summarized in the following results. We will assume that the digit selection by rounding employs some guard digits to reduce the error bounds, but note that the error may be strictly greater than $\frac{1}{2}$. Hence the choice of digit is redundant, since there may be more than one integer value in the uncertainty interval.

Definition 5.6.1 *For the p-digit, normalized divisor* $y = y_0.y_1y_2\cdots y_{p-1} > 0$, *with* $1 \leq y < \beta$, *the k-digit* radix-β short reciprocal ρ, *for* $1 \leq k < p$, *is given by a nearest integer rounding*[5] *as*

$$\rho = \left\lceil \frac{\beta^{k+g}}{y} \right\rfloor \beta^{-g},$$

where $g \geq 0$ *defines a number of fractional* guard digits.

Lemma 5.6.2 *The short reciprocal* ρ *satisfies* $\beta^{k-1} < \rho \leq \beta^k$, *with equality* $\rho = \beta^k$ *occurring if* $y = 1$. *The approximation error* ε *defined by* $\beta^k/y = \rho + \varepsilon$ *satisfies* $|\varepsilon| \leq \frac{1}{2}\beta^{-g}$, *and* $\rho y = \beta^k - \varepsilon y$, *so* $|\rho y - \beta^k| < \frac{1}{2}\beta^{1-g}$.

Proof

$$\rho\beta^g = \left\lceil \frac{\beta^{k+g}}{y} \right\rfloor \leq \beta^{k+g} + \tfrac{1}{2}.$$

Since $\rho\beta^g$ is integral, $\rho \leq \beta^k$, with equality if $y = 1$. The lower bound follows similarly. From $\rho y = \beta^k - \varepsilon y$, it is seen that $|\rho y - \beta^k| < \frac{1}{2}\beta^{1-g}$ since $y < \beta$. □

Given the divisor y, the short reciprocal ρ can be determined by a table look-up procedure. The short reciprocal division algorithm then determines $n = \lceil p/k \rceil$

[5] In the following we will use the notation $\lceil \cdot \rfloor$ to denote "round to nearest," without concern for the choice in the case of a tie.

radix-β^k digits in n passes through a loop, determining a high radix digit d_i, updating the remainder $r_{i+1} = \beta^k r_i - d_i y$ and the quotient $q_{i+1} = q_i + \beta^{-k} d_i$, for $i = 0, 1, \ldots, n-1$, performing two $(k+g) \times p$ multiplications per pass. Initially, $q_0 = 0$ and $r_0 = x$, where the dividend x is assumed bounded so that $|x| < y$, and with normalized divisor $1 \le y < \beta$.

The digit d_i is determined from r_i as $d_i = \lceil \rho r_i \rfloor = \rho r_i - \delta_i$. Assuming that the product ρr_i is in a redundant representation produced by a $(k+g) \times p$ multiplier, without converting the full product to a non-redundant representation, it is necessary to include some redundant, fractional guard digits in the rounding process forming d_i. Note that the rounding to obtain d_i is not necessarily unique since δ_i may be strictly greater than $\frac{1}{2}$. Assuming that β is even, the value of ρr_i may have a fractional part of h guard digits $.d_1 d_2 \cdots d_h$ of the form $.(\beta/2)0 \cdots 0$, in which case the rounding can go either up or down. With $h > 1$ redundant guard digits, and $|d_j| \le \beta - 1$ for $j > h$, the upper bound on $|\delta_i|$ is then

$$|\delta_i| < \tfrac{1}{2} + \beta^{-h}. \tag{5.6.1}$$

Theorem 5.6.3 (Short reciprocal division) *For a p-digit divisor $1 \le y < \beta$ and dividend $|x| \le y$, the short reciprocal division procedure determines high radix-β^k digits d_i, satisfying $|d_1| \le \beta^k$ and $|d_i| \le \beta^k - 1$ for $i = 2, \ldots, n$, where $n = \lceil p/k \rceil$.*

With the short reciprocal ρ given by Definition 5.6.1 using g guard digits, the digits can be chosen serially as

$$d_i = \lceil \rho r_i \rfloor,$$

from redundantly represented remainders, using h guard digits for the rounding. The remainders are computed by the recursion

$$r_{i+1} = \beta^k r_i - d_i y.$$

Then with quotient $q = \sum_1^n d_i \beta^{-ki}$ and remainder $r = r_{n+1}$, the quotient, remainder pair (q, r) satisfies $x = yq + r\beta^{-nk}$, provided that $g = 2$, and for $\beta = 2$, $h = 2$ and $k \ge 3$. For $\beta \ge 4$, $h = 1$ and $k \ge 1$ is sufficient.

Observation 5.6.4 *For p by p-digit short reciprocal division the procedure needs $2\lceil p/k \rceil$ $((k+g) \times p)$-digit multiplications.*

Observation 5.6.5 *The short reciprocal division procedure can be employed to perform 2p by p division in a high radix β^k with $k \ge 2$, employing at most $(2\lceil p/k \rceil + 2)$ $((k+g) \times p)$-digit multiplications for $\beta \ge 3$, and at most $(2\lceil p/k \rceil)$ $((k+g) \times p)$-bit multiplications for binary division.*

The proofs of Theorem 5.6.3 and the observations on efficiency are left as problems. The results on efficiency do not include the resources needed to find the k-digit short reciprocal. The value $\rho = \lceil \beta^{k+g}/y \rfloor \beta^{-g}$ may involve a table look-up procedure supplemented by additions or multiplications when the short reciprocal is needed to be of greater precision than the equivalent of about 12 bits. The determination of approximate reciprocals is discussed in Section 5.8.

It is instructive to compare the digit selections here and in the SRT methods of the previous section. In the SRT methods, a high radix digit value d can be chosen if the shifted remainder $\beta^k r_i$ satisfies

$$L_d = (d - \rho)y \le \beta^k r_i \le (d + \rho)y = U_d$$

or equivalently $d - \rho \le (1/y)\beta^k r_i \le d + \rho$, essentially checking whether $(1/y)\beta^k r_i$ is sufficiently close to the integer d, using comparisons or table look-up. The short reciprocal method just chooses the digit directly as an integer close to the product $(1/y)\beta^k r_i$.

5.6.2 Prescaled division

For $(p \times p)$-digit prescaled division in the high radix β^k, we start with the radix-β $(2p \times p)$-digit input normalization that guarantees a p-digit quotient. The divisor y and dividend x are then each multiplied (prescaled) by a k-digit radix-β short reciprocal ρ of the divisor, yielding a $((p + k) \times (p + k))$-digit division problem with the identical rational valued quotient

$$\frac{x\rho}{y\rho} = \frac{x}{y} = q + \frac{r}{y},$$

and scaled remainder $r\rho$, where $(x\rho) = q(y\rho) + r\rho$. To allow a selection of standard directed and/or nearest roundings, we shall also obtain the values of a two-bit indicator for the unscaled remainder r.

Definition 5.6.6 *For (q, r) a quotient, remainder pair for radix-β division of x by y, the two-bit round-sticky state of the remainder r is given by*

$$\text{round} = \begin{cases} 0 & \textit{if} \quad 0 \le |r/y| < \frac{1}{2}, \\ 1 & \textit{if} \quad \frac{1}{2} \le |r/y| < 1, \end{cases}$$

$$\text{sticky} = \begin{cases} 0 & \textit{if} \qquad |r/y| = 0 \textit{ or } \frac{1}{2}, \\ 1 & \textit{otherwise.} \end{cases}$$

Observation 5.6.7 *The round–sticky state of the remainder is equal to the round–sticky state of the scaled remainder.*

The prescaled division procedure is illustrated by the previous decimal division example of $x = 365748 \times 10^6$ divided by $y = 784731$, employing a (3×6)-digit decimal multiplier, where three quotient digits are to be determined in the high radix $10^2 = 100$. The arithmetic is again for convenience decimal, sign-magnitude, with rounding-towards-zero.

Example 5.6.2 (Prescaled division)
Problem (6×6)-digit decimal division:

$$\begin{aligned} x &= 365748 \times 10^6 && (\textit{six-digit dividend}),\\ y &= 784731 && (\textit{six-digit divisor}),\\ \beta &= 100 && (\textit{two-digit high radix}). \end{aligned}$$

The integer normalization employed here is $10^5 \leq y < 10^6$ and $y10^5 \leq x \leq y10^6 - 1$, which yields a rational valued quotient $10^5 \leq \frac{x}{y} \leq 10^6$.
Step 1 Determine the three-digit short reciprocal

$$\rho = \left\lceil \frac{10^8}{784731} \right\rceil = 128$$

and its "tail"

$$t = y\rho - 10^8 = 445568.$$

Step 2 The divisor and dividend are both multiplied by the short reciprocal (prescaling) to obtain a (9×9)-digit decimal division problem with the same quotient where the nine-digit scaled divisor identifies the tail:

$$\begin{aligned} x\rho &= 46815744 \times 10^6 && (\text{scaled dividend}),\\ y\rho &= 100445568 && (\text{scaled divisor}),\\ t\ &= 445568 && (\text{tail of scaled divisor}). \end{aligned}$$

Since $y\rho$ has leading digits differing from unity by less than 1%, the leading decimal digit tuple "46" of the scaled dividend $x\rho$ can be selected as the the leading radix-100 digit of the quotient.
Step 3 Perform digit-serial division radix 100, with scaled remainder computations comprising multiplications by the tail with the next digit selection obtained directly from each scaled remainder.

	46 61 $\overline{19}$.	
1 00 44 55 68	**46** 81 57 44 00 00 00.	first digit selection
$46 \times$ 44 55 68	(46) 20 49 61 28	
	61 07 82 72	second digit selection
$61 \times$ 44 55 68	(61)27 17 96 48	
	$-$**19** 35 24 48	third digit selection
$-19 \times$ 44 55 68	$-$(19)08 46 57 92	
	$-$26 77 90 08	scaled remainder (fourth digit selection)

Step 4 Adjust the quotient based on the scaled remainder and determine the round-sticky remainder state. Since the "scaled remainder" -26779008 is negative and less than half the scaled divisor 100445568, we decrement the quotient's last place

and obtain the integer quotient and remainder round-sticky state

$$q = 466080 \text{ with } \tfrac{1}{2} < r/y < 1,$$

satisfying $x/y = 466080 + r/y$ with positive scaled remainder $r\rho = 73666560$. □

The digit selection procedure in prescaled division is justified as follows. Since the scaled divisor $y\rho$ has the leading digit string "100," the leading digit string "46" of the scaled dividend is either the correct leading digit or one unit greater. Selecting high radix digit "46" thus guarantees that the absolute value of the next scaled remainder will be less than the scaled divisor. The remainder computation is simplified by implicitly deleting "46" from the scaled remainder and multiplying digit "46" by only the p-digit tail t to update the remainder. This simplification allows the remainder computation to be performed by the $((k+g) \times p)$-digit multiplier, with the next scaled remainder explicitly determining the next high radix digit of the quotient.

In the prescaled division example here the scaled dividend and the next two remainders were, respectively, **46** 81 57 44, **61** 07 82 72, and −**19** 35 24 48. Note that these values are identical to the "side computation" values for digit selection in the preceding short reciprocal division example applied to the same divisor and dividend input.

Let us now assume that the divisor $y = y_0.y_1y_2 \cdots y_{p-1}$ satisfies $1 \leq y < \beta$, and dividend x satisfies $0 \leq x < y$, then $0 \leq q < 1$ and $0 \leq r/y < 1$ in the prescaled problem

$$\frac{x\rho}{y\rho} = \frac{x}{y} = q + \frac{r}{y}.$$

Then note that:

- The short reciprocal ρ, as given by Definition 5.6.1, satisfies $\beta^{k-1} < \rho \leq \beta^k$.
- The p-digit "tail" t satisfies $t = y\rho - \beta^k$ with $|t| \leq \frac{1}{2}\beta^{1-g}$, where g is the number of guard digits used in the rounding of ρ.
- The $(p+k)$-digit *scaled divisor* $y\rho$ satisfies $y\rho = \beta^k(1+\varepsilon)$, where $|\varepsilon| < (\beta/2)\beta^{-(k+g)}$.
- The $(p+k)$-digit *scaled dividend* $x\rho$ is bounded by $|x|\rho < (1 + (\beta/2)\beta^{-(k+g)})\beta^k$ and has its leading k-digit rounded integer part $\lceil x\rho \rfloor \leq \beta^k$, determining the first high radix digit of the quotient in radix β^k.
- The remainder computation can be performed using a $((k+g) \times p)$-digit multiplier with the tail t being the p-digit argument.

Algorithm 5.6.8 (Prescaled division)
Stimulus: *A high radix β^k, a p-digit normalized divisor y with $\beta^{-1} \leq y \leq 1$, a p-digit dividend x with $0 \leq x < y$, and $g \geq 1$, the number of guard digits.*

Response: *A p-digit quotient q with* $0 \le q < 1$ *and a remainder state rstate denoting either (zero)* $r/y = 0$*, (mid)* $r/y = \frac{1}{2}$*, (low)* $0 < r/y < \frac{1}{2}$*, or (high)* $\frac{1}{2} < r/y < 1$*, where q, y, x, and* r/y *satisfy* $x/y = q + r/y$*.*

Method: $q := 0;$
$\rho := \lceil \beta^{k+g}/y \rfloor \beta^{-g}$; {*short reciprocal*}
$r = x\rho$; {*scaled dividend*}
$t = y\rho - \beta^k$; {*tail of scaled divisor*}
$n := \lceil p/k \rceil$; {*number of high radix digits in q*}
for $i := 1$ **to** n **do**
  $d_i := \lceil r \rfloor$; {*digit selection by rounding, using h guard digits*}
  $r := \beta^k(r - d_i)$; {*digit deletion*}
  $r := r - d_i t$; {*remainder determination*}
  $q := q + d_i \beta^{-ki}$ {*quotient accumulation (or on-the fly conversion)*}
end;
if $r < 0$ **then** $q := q - \beta^{-kn}$; $r := r + \beta^k + t$ **end**;
if $r = 0$ **then** *rstate* := *zero*
  elseif $2r = \beta^k + t$ **then** *rstate* := *mid*
  elseif $2r < \beta^k + t$ **then** *rstate* := *low*
  else *rstate* := *high*
end;

In the Prescaled Division Algorithm (Algorithm 5.6.8) the remainder updating is performed in two steps, first deleting the high radix digit from the shifted remainder, such that the next remainder can then be determined by a $((k + g) \times p)$-digit multiplication, demonstrating the usefulness of the p-digit tail t.

Theorem 5.6.9 *The Prescaled Division Algorithm correctly performs p by p digit division in radix* β^k*, with* $k \ge 2$*, employing* $\lceil p/k \rceil + 2$ *multiplications that may be performed by a* $((k + g) \times p)$*-digit multiplier.*

For the key steps of a proof, consider that the remainder updating here is

$$r_{i+1} = \beta^k(r_i - d_i) - d_i t = \beta^k r_i - d_i(\beta^k + t), = \beta^k r_i - d_i y\rho,$$

starting with dividend $x\rho = r_0$ and using divisor $y\rho = \beta^k + t$. Thus the remainders computed here are just scaled versions of the remainders in the short reciprocal division procedure, and hence are identical to the expressions ρr_i used in that algorithm for determining the next digit.

Since a single $((k + g) \times p)$-digit multiplication determines both the next remainder and the next digit selection for the quotient, the total number of multiplications is $\lceil p/k \rceil$ for the remainder updates. In addition there are two multiplications for prescaling the divisor and dividend.

It is instructive to consider the exact integer plus fractional part result of the Prescaled Division Algorithm as shown in Example 5.6.2. At termination we have $\frac{46815744\times 10^6}{10445568} = 466080 + \frac{73666560}{100445568}$. If we were to continue quotient digit selection into the fractional part of the quotient, we would note the next quotient digit must be 73. With the original dividend and divisor normalized as fractions we can thus obtain $\frac{0.365748}{0.784731} = .46608073 + \varepsilon,\ |\varepsilon| < 10^{-8}$. Hence without any more multiplications we can obtain both the correct quotient, remainder pair with a six-digit quotient, and an approximate eight-digit quotient that differs from the infinitely precise quotient by less than one part in 10^8.

Observation 5.6.10 *The Prescaled Division Algorithm implicitly also determines a $(p+k)$-digit quotient accurate to within a unit in the last place.*

The dual result feature of prescaled division is very significant for an implementation where the microcoded division hardware is expected to support both precisely rounded p-digit division and a transcendental function library. The option of obtaining approximate quotients with k extra digits of accuracy, while employing the same hardware resources, can provide the needed intermediate guard digits in transcendental function transformations to support transcendental function output accuracy of less than a unit in the p'th place.

5.6.3 Prescaled division with remainder

Prescaling the divisor and dividend by a k-digit short reciprocal ρ preserves the result of $(p \times p)$-digit division since $x\rho/y\rho = x/y = q + r/y$. However, while the prescaled division procedure yields an identical quotient, the remainder is transformed to a $(p+k)$-digit "scaled remainder" $r\rho$, specifically $x\rho = q(y\rho) + r\rho$, with $0 \le |r\rho| < y\rho$. Scaling the values of the previous integer division problem with $x = 365748000000$ divided by $y = 784731$ using the scale factor $\rho = 128$ and proceeding analogously to the $(p \times p)$-digit prescaled algorithm we could obtain $q = 466081$ and $r\rho = -26779008$. This would leave us with another division problem: "descale" the scaled remainder and determine that $r = \frac{-26779008}{128} = -209211$.

For integer division it is important to obtain the integer remainder. Such an operation is even required in the IEEE floating-point standard. The following shows that the prescaling technique can be altered to support obtaining the integer remainder.

In the *prescaled division with remainder* procedure we avoid the need for remainder descaling, employing instead a divisor descaling step. This division procedure operates in two phases. In the first phase the original (unscaled) dividend x is divided by a scaled divisor $y\rho$, obtaining a "descaled" quotient q' and *oversized remainder* r':

$$x = q'(y\rho) + r' \text{ with } 0 \le |r'| < y\rho. \tag{5.6.2}$$

This phase efficiently generates $(\lceil p/k \rceil - 1)$ high radix digits of the quotient q', using only one $(k \times p)$-digit multiplication per high radix digit. The divisor is then "descaled" by shifting the scale factor to the quotient so that $x = (q'\,\rho)y + r'$ and

$$\frac{x}{y} = q'\,\rho + \frac{r'}{y} = q'' + \frac{r'}{y}, \text{ with } 0 \le \left|\frac{r'}{y}\right| < \rho.$$

In the second phase we continue with division of the oversized remainder r' by the original divisor y, using the short reciprocal division procedure to determine a correction to the quotient $q'' = q'\,\rho$, and obtain the final quotient q and remainder r such that $x = q \times y + r$ with $0 \le |r| < y$.

This two-phase multiplicative pre- and post-scaling division procedure effectively employs the scaled divisor $y\rho$ as a "catalyst." While $y\rho$ is employed as the divisor in the first phase resulting in $(\lceil p/k \rceil - 1)$ quotient digit selections, the term $y\rho$ disappears and the final digit selection phase employs the original divisor y. The following illustrates the two phase algorithm applied to the previous (12×6)-digit decimal example, again using decimal sign-magnitude integer arithmetic.

Example 5.6.3 (Prescaled division with remainder)
Problem For the (12×6)-digit decimal integer division arguments

$$\begin{aligned} x &= 365748375204 && \text{(12 digit dividend)} \\ y &= 784731 && \text{(6 digit divisor)} \end{aligned}$$

find the six-digit integer quotient q and non-negative remainder r with $0 \le r \le y - 1$ such that $x = q \times y + r$.
Step 1 Determine the three-digit short reciprocal:

$$\rho = \left\lceil \frac{10^8}{784731} \right\rceil = 128.$$

Step 2 Determine the scaled divisor and tail:

$$y\rho = 784731 \times 128 = 100445568$$

$$t = 445568$$

Step 3 Perform digit-serial division of the unscaled dividend by the the scaled divisor determining $\lceil p/k \rceil - 1$ digits in the high radix 100. Note that digit selection is obtained directly from each remainder with the remainder computations comprising multiplications by the residual.

$$\begin{array}{rrl} & 36\,41. & \\ \cline{2-2} 1\,00\,44\,55\,68 \,| & \mathbf{36}\ 57\,48\,37\,52\,04. & \text{first digit selection} \\ 36 \times 44\,55\,68 & (36)\,16\,04\,04\,48 & \\ \cline{2-2} & \mathbf{41}\ 44\,33\,04 & \text{second digit selection} \\ 41 \times 44\,55\,68 & (41)18\,26\,82\,88 & \\ \cline{2-2} & 26\,06\,21\,16 & \end{array}$$

Step 4 Perform divisor descaling by quotient scaling:

$$\begin{aligned} 365748\,375204 &= 3641 \times 100445568 + 26062116 & \\ &= 3641 \times (128 \times 784731) + 26062116 & \text{divisor descaling} \\ &= (3641 \times 128) \times 784731 + 26062116 & \text{quotient scaling} \\ &= 466048 \times 784731 + 26062116 & \end{aligned}$$

i.e., forming $q'' = q'\rho = 3641 \times 128 = 46\,60\,48$.
Step 5 Reformat digit serial division and determine a low-order digit increment by one step of the short reciprocal division procedure applied to the "oversized" remainder:

$$\begin{array}{rrll} 128\times & 26\,06\,21\,16 & = \mathbf{33}\,35\,95\,08\,48 & \text{digit selection} \\ 33 \times 78\,47\,31 & 25\,89\,61\,23 & & \text{remainder adjustment} \\ \cline{2-2} & 16\,59\,93 & & \end{array}$$

$$\begin{array}{rrll} & +33. & & \\ & 46\,60\,48. & = 46\,60\,81 & \text{quotient adjustment} \\ \cline{2-2} 78\,47\,31 \,| & 36\,57\,48\,37\,52\,04. & & \end{array}$$

Step 6 Determine (confirm) the positive remainder and provide the quotient in non-redundant form:

$$q = 46\,60\,81, \quad r = 16\,59\,93, \quad 0 \leq r < 78\,47\,31. \qquad \square$$

The digit selection procedure for the first phase of prescaled division with remainder utilized in Step 3 of the example is identical to digit selection in $(p \times p)$-digit prescaled division. It follows that $\lceil p/k \rceil - 1$ high radix digits are determined at a cost of one $((k+g) \times p)$-digit multiplication per digit. The remainder $r' = 26062116$ in Step 5 in general satisfies $|r'| < \beta^{p+k} + t$, so multiplication by the k-digit short reciprocal determines a $(p+2k)$-digit product with leading k-digit tuple magnitude no greater than the short reciprocal. Here we obtain $33 \leq 128$.

The digit selection in Step 5 can be performed as a $((k+g) \times p)$-digit multiplication by deleting the least significant k digits, in the form $\rho \lfloor r'/\beta^k \rfloor = 128 \times 260621 = \underline{33}359488$. Since the augmented digit selected in Step 5 is in the range $[-\rho, \rho]$, the remainder adjustment can be realized by a $((k+g) \times p)$-digit multiplier. In Step 6 the quotient is augmented or decremented as needed to update the remainder to the range $0 \leq r < y$.

Algorithm 5.6.11 (Prescaled division with remainder)

Stimulus: *A high radix β^k with $k \geq 2$, a precision p with $p \geq k + g$, a p-digit divisor y with $1 \leq y < \beta$, a $2p$-digit dividend x with $0 \leq x < y$, together with g, h specifying the number of guard digits.*

Response: *A quotient q with $0 \leq q < 1$, a remainder r with $0 \leq r < y$, where q, r satisfy $x = qy + r$.*

Method: $q := 0;$
$\rho := \lceil \beta^{k+g}/y \rfloor \beta^{-g}$; {*short reciprocal with g guard digits*}
$r = x$; {*initialize remainder*}
$t = y\rho - \beta^k$; {*tail of scaled dividend*}
$n := \lceil p/k \rceil$; {*number of high radix digits in q*}
for $i := 1$ **to** $n - 1$ **do**
  $d_i := \lceil r \rfloor$; {*digit selection by rounding using h guard digits*}
  $r := \beta^k(r - d_i)$; {*leading digit deletion*}
  $r := r - d_i t$; {*remainder determination*}
  $q := q + d_i \beta^{-ik}$; {*quotient accumulation*}
end;
$q := q\rho$; {*quotient scaling*}
$d_n := \lceil \rho r \rfloor$; {*digit augmentation selection by rounding using h guard digits*}
$r := r - d_n y$; {*remainder adjustment*}
$q := q + d_n \beta^{-kn}$; {*quotient adjustment*}
while $r < 0$ **do** $q := q - \beta^{-kn}$; $r := r + y$;
while $r \geq y$ **do** $q := q + \beta^{-kn}$; $r := r - y$;

Observation 5.6.12 *Algorithm 5.6.11 for prescaled division with remainder performs $(2p \times p)$-digit radix-β division in the high radix β^k, with $k \geq 2$, employing $(\lceil p/k)\rceil + 3)$ $((k + g) \times p)$-digit multiplications.*

In summary the $(2p \times p)$-digit prescaled division with remainder algorithm prescales only the initial divisor, and then postscales the resulting "descaled" quotient and *oversized remainder* of this modified division problem. The postscaling of the oversized remainder is employed to find a single high-radix digit to adjust the quotient and compute an appropriately bounded remainder. A formalization of this postscaling step for determining the final high-radix digit for quotient, remainder pair (q, r) determination is provided in the following theorem whose proof is left as an exercise.

Theorem 5.6.13 *For $2 \leq k \leq p - 1$, assume there is:*

- *a p-digit divisor y with $1 \leq y < \beta$,*
- *a $2p$-digit dividend x with $0 \leq x < y$,*
- *a k-digit short reciprocal ρ and tail t satisfying $\rho y = \beta^k + t$ with $0 \leq t < y$,*

- *a* $(\lceil p/k \rceil - 1)$*-digit "descaled" quotient* q' *and oversized remainder* r' *satisfying*

$$x = q'(\rho y) + r' \text{ with } 0 \le r' < \rho y.$$

If $0 \le r' < y$, *then the p-digit quotient* $q = q'\rho$ *and remainder* $r = r'$ *satisfy* $x = qy + r$ *with* $0 \le r < y$. *Else if* $y \le r' < \rho y$, *then the quotient q and remainder r given by*

$$q = q'\rho + \left\lceil \frac{r'\rho - t}{\beta^k} \right\rfloor \beta^{-kn}, \tag{5.6.3}$$
$$r = r' - \left\lceil \frac{r'\rho - t}{\beta^k} \right\rfloor y,$$

satisfy $x = qy + r$ *with* $-y \le r < y$.

Equation (5.6.3) for quotient adjustment includes a $((k+g) \times (p+k))$-digit multiplication for the term $r'\rho$. Algorithm 5.6.11 provides a simplified alternative quotient adjustment employing a $((k+g) \times p)$-digit multiplication for the term $\lceil r'\rho \rfloor$. The quotient adjustment in the algorithm may yield a quotient, remainder pair with a remainder somewhat larger in magnitude than the divisor, thus requiring up to two unit adjustments to the quotient of Algorithm 5.6.11. Further aspects of quotient adjustment are considered in the problems.

5.6.4 Efficiency of multiplicative high radix division

In evaluating the efficiencies of the short reciprocal and two prescaled division algorithms of this section it is important to identify the numbers of multiplications that are dependent and independent in each case. The serial digit selection portion of each algorithm exclusively involves dependent multiplications. The scaling operations provide some opportunities for concurrent parallel/pipelined multiplications.

In $(p \times p)$-digit prescaled division the prescaling of divisor and dividend are independent. In $(2p \times p)$-digit prescaled division with remainder, the scaling of the "descaled" quotient and oversized remainder output from phase one are independent, and the two multiplcations may be executed concurrently. Thus the time critical number of "multiply latencies" is one less than the total number of multiplies performed for both prescaled division algorithms. Table 5.6.1 summarizes the number of multiplies (multiply latencies) required to provide the division operation results for the three algorithms in this section when the results require from two to six high radix quotient digits to be serially generated.

Prescaled $(p \times p)$-digit division, Algorithm 5.6.8, is the most efficient. Providing the remainder by Algorithm 5.6.11 requires only one more $((k+g) \times p)$-digit multiplication. The short reciprocal division algorithm is only competitive for division using multipliers with aspect ratio $\lceil p/k \rceil = 2$.

Regarding the overall efficiency of a division operation implementation, smaller aspect ratios, such as 2 : 1 or 3 : 1, require larger hardware multipliers with more latency. Larger aspect ratio multipliers have smaller areas but entail more multiplications . Also consider that the additional effort to obtain the short reciprocal is larger when the radix is larger, and any multiplies or equivalent table look-up delay needed to obtain the short reciprocal are not reflected in the counts in Table 5.6.1. When the multiplier aspect ratio satisfies $\lceil p/k \rceil = 2$, it may also be important to consider the iterative division algorithms of the next section when determining which division algorithm should be implemented given a particular hardware multiplier.

Table 5.6.1. *The number of* $((k+g) \times p)$*-digit multiplies (multiply latencies) required for three high-radix division algorithms when* $\lceil p/k \rceil = 2, 3, \ldots, 6$ *high-radix quotient digits must be generated*

	Number of high radix digits required				
Division algorithm	2	3	4	5	6
Short reciprocal p by p	4(4)	6(6)	8(8)	10(10)	12(12)
Prescaled p by p	4(3)	5(4)	6(5)	7(6)	8(7)
Prescaled $2p$ by p with remainder	5(4)	6(5)	7(6)	8(7)	9(8)

Problems and exercises

5.6.1 Prove Theorem 5.6.3.

5.6.2 Discuss and justify Observation 5.6.5.

5.6.3 Prove that the round-sticky state of the the scaled remainder is the same as the round-sticky state of the remainder.

5.6.4 Discuss simplification of the specification of the remainder state for $(p \times p)$-bit binary division.

5.6.5 Discuss simplifications of Algorithms 5.6.8 and 5.6.11 for binary implementation of (64 × 64)-bit division, assuming a $((16 + g_1) \times (64 + g_2))$-bit multiplier, where g_1 and g_2 are the respective numbers of guard bits to be chosen in the range $0 \leq g_1, g_2 \leq 3$ to allow a redundant choice of digits by rounding.

5.6.6 Determine the maximum number of unit quotient adjustments required in Algorithm 5.6.11. Provide examples showing these maximum numbers of adjustments may be needed.

5.6.7 Prove Theorem 5.6.13.

5.7 Multiplicative iterative refinement division

Let z_0 be an approximation of z of relative error ε, so then $z_0 = z(1-\varepsilon)$. The expression $(1-\varepsilon)$ is termed the *relative error factor*. Multiplication of z_0 by the *complementary error factor* $(1+\varepsilon) = 2-(1-\varepsilon)$ yields a much improved approximation: $z_1 = z_0(1+\varepsilon) = z(1-\varepsilon^2)$. The process can be iterated with the ith result refinement yielding $z_i = z(1-\varepsilon^{2^i})$. The *Iterative Refinement Paradigm* employs three fundamental steps for each iteration:

- relative error factor determination,
- complementary error factor formation,
- result refinement.

Iterative refinement division algorithm implementations typically employ only one–three iterations and incorporate pre- and/or post-iterative argument processing phases:

- *Preiterative phase* A *seed* value for the function of the iterative refinement phase is determined, typically comprising a table lookup procedure and/or operand scaling,
- *Postiterative phase* The iterative refinement result is multiplied (postscaled) by a term yielding a final quotient approximation.

A high-precision approximate quotient for x divided by y can be obtained by applying iterative refinement either to a divisor reciprocal approximation $\rho_0 = (1/y)(1-\varepsilon)$ or directly to a quotient approximation $q_0 = (x/y)(1-\varepsilon)$. A limit on the convergence of ρ_i to $1/y$ or of q_i to x/y is dictated by the precision of the rounded operations generally employed in implementations of iterative refinement division algorithms.

Without loss of generality, for radix division the exponent scaling of the p-digit divisor $y \in \mathbb{Q}_\beta$ and dividend $x \in \mathbb{Q}_\beta$ can be adjusted to provide a *$(p+g)$-digit approximate quotient* $q + t \approx x/y$ comprising a p-digit integer part q comparable to the quotient of $(p \times p)$-digit division described in Section 5.2 with $\beta^{p-1} \le q < \beta^p$, and a *guard term* $t = i\beta^{-g}$. We focus on approximations with $|i| < \beta^g$, where the fraction t approximates the remainder fraction r/y.

Definition 5.7.1 *The $(p \times p)$-digit radix-β* quarter-ulp division *problem is:*

> Given *A p-digit divisor $y \in \mathbb{Q}_\beta$ and a p-digit dividend $x \in \mathbb{Q}_\beta$, normalized relative to y so that $y\beta^{p-1} \le x < y\beta^p$.*
> Find *A p-digit integer quotient $q \in \mathbb{Q}_\beta$ and a g-digit guard $t = i\beta^{-g}$ with $|i| < \beta^{-g}$ such that*
>
> $$|x/y - (q+t)| < \frac{1}{4}, \tag{5.7.1}$$
>
> *where $q + t$ is termed a quarter-ulp quotient.*

Observation 5.7.2 *A quarter-ulp quotient $q + t$ defines the half-ulp open interval*

$$\left(q + t - \frac{1}{4}, q + t + \frac{1}{4}\right)$$

that must contain x/y and at most one of the three rounding boundary values $q - \frac{1}{2}, q, q + \frac{1}{2}$.

A quarter-ulp quotient and corresponding rounding boundary value are sufficient and convenient for describing implementations requiring a precisely rounded p-digit quotient by procedures described in Section 7.5. Iterative refinement division implementations are typically employed when the hardware resource is a full $((p + g) \times (p + g))$-digit multiplier. Here p is the precision of the $(p \times p)$-digit approximate division problem, and g is a small number, typically $1 \leq g \leq 6$, of multiplier guard digits provided to control accumulated approximation error and support generation of a quarter-ulp quotient for input to a precise rounding operation.

Three iterative refinement division algorithms are described in this section with a performance measured in terms of *total multiplications* and *time latency* as a function of: (i) the number of iterative refinements (#iterations), (ii) the $((p + g) \times (p + g))$-digit *multiplier latency* (L_M), and (iii) the *table look-up latency* (L_T):

- The fundamental Newton–Raphson divisor reciprocal refinement procedure provides a *Newton–Raphson Division* Algorithm utilizing:
  - total multiplications: 2 × (#iterations) + 1,
  - time latency: [2 × (#iterations) + 1] $\times L_M + L_T$.
- Goldschmidt's *Convergence Division* Algorithm employs quotient iterative refinement which allows the steps of relative error factor determination and result refinement to be performed concurrently, utilizing:
  - total multiplications: 2 × (#iterations) + 1,
  - time latency: [(#iterations) + 1] $\times L_M + L_T$.
- The *Postscaled Division* Algorithm postpones divisor reciprocal function scaling so that it is a post-iterative phase rather than a preiterative phase operation removing table look-up from the critical path and thereby utilizing:
  - total multiplications: 2 × (#iterations),
  - time latency: [(#iterations) + 1] $\times L_M$.

The three iterative refinement division algorithms provide basic choices that can be fine tuned to the architecture of the multiplier and supporting hardware. The examples in this section will be given in decimal arithmetic.

5.7.1 Newton–Raphson division

Newton–Raphson iterative refinement division has three phases, the first two of which involve only the divisor:

- Preiterative phase: A seed reciprocal value $\rho_0 \in \mathbb{Q}_\beta$ of the divisor $y \in \mathbb{Q}_\beta$ is determined, typically by a table lookup procedure.
- Iterative refinement phase: A predetermined number $k \geq 1$ of iterative refinements are applied to the seed reciprocal refining $\rho_0 \approx 1/y$ to a highly accurate approximation $\rho_k \approx 1/y$.
- Postiterative phase: The reciprocal approximation ρ_k is multiplied by the dividend to obtain a comparably accurate quotient approximation.

The accuracy and efficiency of this division method depends primarily on the properties of the divisor reciprocal refinement process, derived from a general Newton–Raphson function approximation algorithm.

Definition 5.7.3 *For a p-digit normalized divisor $y \in \mathbb{Q}_\beta$ with $1/\beta \leq y < 1$ and seed reciprocal $\rho_0 =$ with $1 \leq \rho_0 \leq \min\left\{\beta, \frac{3}{2}(1/y)\right\}$, the* Newton–Raphson reciprocal refinement *formula for $i \geq 1$ is*

$$\rho_i = \rho_{i-1}(2 - y\rho_{i-1}).$$

Lemma 5.7.4 *Let $1/\beta \leq y < 1$ be a p-digit normalized divisor and ρ_0 a seed divisor reciprocal approximation satisfying $1 \leq \rho_0 \leq \min\left\{\beta, \frac{3}{2}(1/y)\right\}$. Then the kth reciprocal refinement satisfies*

$$\rho_k = \rho_0 \prod_{j=0}^{k-1}(1 + \varepsilon^{2^j}) = \frac{1}{y}(1 - \varepsilon^{2^k}),$$

where $\varepsilon = ((1/y) - \rho_0)/(1/y) = 1 - y\rho_0$ is the relative error of the divisor seed reciprocal approximation ρ_0. Furthermore, for $k \geq 1$, ρ_k is a subestimate of $1/y$ with relative error ε^{2^k}.

Proof With $y\rho_0 = 1 - \varepsilon$ given, and noting that $(1 - \varepsilon)\prod_{j=0}^{k-1}(1 + \varepsilon^{2^j}) = 1 - \varepsilon^{2^k}$, it is readily shown by induction that $\rho_k = \rho_0 \prod_{j=0}^{k-1}(1 + \varepsilon^{2^j})$. Since $\rho_0 = (1 - \varepsilon)/y$, it follows that $\rho_k = (1/y)(1 - \varepsilon^{2^k})$. The refinements $\rho_1, \rho_2, \ldots, \rho_k$ are subestimates monotonically approaching $1/y$ from below. □

The key to the efficiency and accuracy of Newton–Raphson reciprocal refinement is that the relative error factor of the seed divisor reciprocal value can be determined by a single multiplication.

Observation 5.7.5 *Let y be a p-digit divisor with $1/\beta \leq y < 1$ and let $\rho_0 = i\beta^j$ be an approximate divisor reciprocal satisfying $1 \leq \rho_0 \leq \min\left\{\beta, \frac{3}{2}(1/y)\right\}$ with $j < p$. Then the three steps of the iterative refinement paradigm for divisor reciprocal refinement can be computed as follows:*

- *The relative error factor is determined by a multiplication,*

$$1 - \varepsilon = y\rho_0.$$

- *The complementary error factor is determined by a 2's complement operation,*

$$1 + \varepsilon = 2 - y\rho_0.$$

- *Reciprocal refinement is determined by a multiplication,*

$$\rho_1 = \rho_0(1 + \varepsilon).$$

For a divisor normalization $1/\beta \le y < 1$, the constraint $1 \le \rho_0 \le \min\left\{\beta, \frac{3}{2}(1/y)\right\}$ insures that the relative error of the seed reciprocal falls in the range $\left[-(\beta - 1)/\beta, \frac{1}{2}\right]$. The reciprocal refinement process is robust with ρ_k approaching $1/y$, even when the seed reciprocal is chosen as the constant $\rho_0 = 1$ for any normalized divisor $1/\beta \le y < 1$.

Example 5.7.1 (Newton–Raphson reciprocal refinement) Let $\rho_0 = 1$ for the decimal divisor $y = .7$. Thus ρ_0 approximates $1/y$ with relative error $(\rho_0 - (1/y))/(1/y) = y\rho_0 - 1 = -.3$. The results of each step of the iterative refinement paradigm for four iterations determining $\rho_4 = 1.428\,5714\,2242\,1897$ are tabulated below. The digits of ρ_i that agree with the exact reciprocal $\frac{1}{.7} = 1.428\,5714\,2857\ldots$ are underlined in the table, illustrating that the number of accurate digits in the reciprocal essentially doubles with each iteration.

i	Relative error factor $y\rho_{i-1}$	Complementary error factor $2 - y\rho_{i-1}$	ith result refinement $\rho_i = \rho_{i-1}(2 - y\rho_{i-1})$	Relative error for $\rho_i \approx \frac{1}{y}$
0	–	–	1	$-.3$
1	.7	1.3	1.3	$.9 \times 10^{-1}$
2	.91	1.09	1.417	$.81 \times 10^{-2}$
3	.9919	1.0081	1.4284777	$.6561 \times 10^{-4}$
4	.99993439	1.00006561	1.428571422421897	$.43046721 \times 10^{-8}$

□

In practical computation terms the recurrence $\rho_k = \rho_{k-1}(2 - y\rho_{k-1})$ exhibits a trade-off that is critical for comparing this method with the convergence and postscaled methods of the next sections:

- all multiplications are *dependent* with the kth reciprocal refinement requiring a total of $2k$ back-to-back multiplications;
- all reciprocal approximation roundings are *independent* in the sense that intermediate reciprocal approximations may be rounded to precisions related to their relative error levels with little impact on the final approximation accuracy.

With reference to Example 5.7.1 let $\rho_2 = 1.417$ be rounded to $\rho_2' = 1.42$. Then $\rho_3' = 1.42(2 - 0.7 \times 1.42)$ can be rounded to $\underline{1.4285}$, and $\rho_4' = 1.4285 \times (2 - .7 \times 1.4285)$ can be rounded to $\underline{1.42857142}$. These iterated adaptively rounded reciprocal refinements $\rho_2' = 1.42$, $\rho_3' = 1.4285$ and $\rho_4' = 1.42857142$ approximate $1/y$ with relative errors $.6 \times 10^{-2}$, $.5 \times 10^{-4}$, $.6 \times 10^{-8}$, respectively. Furthermore, when ρ_{i-1} is rounded to a j-digit value to reflect its accuracy, the divisor y need only have its leading $2j$ digits employed in the Newton–Raphson recurrence to obtain the rounded $2j$-digit accuracy of the next iterate ρ_i.

Example 5.7.2 (Newton–Raphson division)
Problem For the (6×6)-digit decimal approximate division arguments:

$$x = 365748 \quad \text{(six-digit dividend)} \quad \text{and} \quad y = .784731 \quad \text{(six-digit divisor)}$$

find a six-digit integer quotient q with $10^5 \le q < 10^6$ and a three-digit decimal guard $t = 0.d_1, d_2, d_3$ such that $|x/y - (q + t)| < \frac{1}{4}$.

Solution

Iteration count: $k = 2$,
Seed reciprocal: $1.28(\approx \frac{1}{.78})$,
(Reference value: $x/y = 466080.7334\cdots$).

i	Divisor y (rounded)	Relative error factor $y\rho_{i-1}$ (rounded)	Complementary error factor $2 - y\rho_{i-1}$	Approximate reciprocal $\rho_i = \rho_{i-1}(2 - y\rho_{i-1})$ (rounded)	Relative error for $\rho_i \approx \frac{1}{y}$ (rounded)	Approximate quotient $q_i = x\rho_i$ (rounded)
0	.78			1.28	$.45 \times 10^{-2}$	$\underline{468}157.$
1	.7847	1.0045	.9955	1.2742	$-.96 \times 10^{-4}$	$\underline{4660}36.$
2	.7847 3100	0.9999 0424	1.0000 9576	1.27432202	$-.69 \times 10^{-8}$	$\underline{466080.73}0$

□

In the table note that two digits of y were employed in obtaining *the seed reciprocal* ρ_0 having two-fractional-digit accuracy. Then four digits of y were employed in determining ρ_1 to four-fractional-digit accuracy, and all six digits of y were utilized only in the final iteration determining ρ_2 to some eight fractional digits of accuracy. The values of $q_i = x\rho_i$ for $i = 0, 1, 2$ are provided to illustrate the approach to the final quotient approximation although only $q_2 = x \times \rho_2$ would be computed as the postiterative phase of Newton–Raphson division.

Observation 5.7.6 *Let $1 \le y < \beta$ be a normalized divisor and let $1/\beta \le \rho < 1$ be a divisor reciprocal approximation with relative error $\varepsilon = \rho y - 1$ satisfying $|\rho y - 1| < \beta^{-j}$ for $j \ge 1$. Let y^* be the round-down value of y truncated at fixed-point position $2j + 1$ for $\beta \ge 3$ or, $2j + 2$ for $\beta = 2$. Then*

$$\rho' = [\rho\,[2 - \rho y^*]]$$

determines a $2j+2$ *digit reciprocal approximation with relative error satisfying* $|\rho' y - 1| < \beta^{-2j}$*, where*

- $[2 - \rho y^*]$ *denotes round-up of* $2 - \rho y^*$ *at fixed-point position* $\begin{cases} 2j+1 & \text{for } \beta \geq 3, \\ 2j+2 & \text{for } \beta = 2, \end{cases}$
- $[\rho[2 - \rho y^*]]$ *denotes round-up of* $\rho[2 - \rho y^*]$ *at fixed-point position* $2j+2$.

Proof From Lemma 5.7.4 note that $\rho(2 - \rho y)$ is a subestimate of $1/y$ with relative error less than β^{-2j}. The three roundings incorporated in (5.7.5) each serve to make the resulting ρ' larger than $\rho(2 - \rho y)$. For $\beta \geq 3$ truncating y increases the subestimate $\rho(2 - \rho y)$ by a relative amount essentially bounded by $(1/y\beta)\beta^{-2k}$, and the successive round-ups further increase the result by relative amounts essentially bounded by $(1/\beta)\beta^{-2k}$ and $(y/\beta^2)\beta^{-2k}$ respectively. Since $((1/y\beta) + (1/\beta) + (y/\beta^2)) < 1$ for $1 \leq y < \beta$ with $\beta \geq 3$ the result follows for radices greater than or equal to 3. A similar argument follows for $\beta = 2$ using roundings at position $2j+2$ in all three cases. □

The employment of adaptive roundings and variable precision multiplications illustrated in Example 5.7.2 requires detailed error analysis that is not our purpose here. A practical consideration is the observation that the whole reciprocal refinement sequence can be computed employing a 2 : 1 aspect ratio rectangular multiplier.

Accelerated Newton–Raphson reciprocal refinement. The Newton–Raphson reciprocal recurrence can be reformulated as a pair of recurrences with the number of dependent multiplications essentially cut in half.

Observation 5.7.7 *Let* $y \in \mathbb{Q}_\beta$ *be a p-digit divisor with* $1/\beta \leq y < 1$*. Let* ρ_0 *be a seed reciprocal satisfying* $1 \leq \rho_0 \leq \min\{\beta, \frac{3}{2}(1/y)\}$*, with* $y\rho_0$ *the initial relative error factor. Then for* $i \geq 1$*, the pair* $(\rho_i, y\rho_i)$ *can be computed from the pair* $(\rho_{i-1}, y\rho_{i-1})$ *by the recurrences*

$$\rho_i = \rho_{i-1} \times (2 - y\rho_{i-1}),$$

$$y\rho_i = y\rho_{i-1} \times (2 - y\rho_{i-1}).$$

The iterative computation of $(\rho_{i-1}, y\rho_{i-1})$ is conveniently visualized as operations performed on the fraction $1/y$ by concurrent scalings of the numerator and denominator terms by successive complementary error factors:

$$\frac{1}{y} = \frac{1 \times \rho_0}{y \times \rho_0} = \frac{\rho_0 \times (2 - y\rho_0)}{y\rho_0 \times (2 - y\rho_0)} = \frac{\rho_1 \times (2 - y\rho_1)}{y\rho_1 \times (2 - y\rho_1)} = \cdots = \frac{\rho_k}{y\rho_k} \qquad (5.7.2)$$

From Lemma 5.7.4 note that the denominator term $y\rho_k = 1 - \varepsilon^{2^k}$ is rapidly approaching unity with increasing k, and the numerator term $\rho_k = (1/y)(1 - \varepsilon^{2^k})$

is rapidly approaching the reciprocal $1/y$ with increasing k. Regarding implementation note that the accelerated Newton–Raphson method requires all intermediate multiplications after the initial seed reciprocal scaling to be implemented with precision comparable to the final result precision, demanding a full $((p+g)\times(p+g))$-digit multiplier.

In (5.7.2) observe that if the numerator term were initiated with the dividend x, then the limit of the numerator term would be a quotient approximation $x\rho_k = (x/y)(1-\varepsilon^{2^k})$. This reorganization of the Newton–Raphson recurrence with initial multiplication by the dividend x in the numerator term is the basis for Goldschmidt's Convergence Division Algorithm, which essentially halves the number of dependent multiplications compared to Newton–Raphson division.

5.7.2 Convergence division

Convergence division is conveniently described by a series of scaling multiplications concurrently and independently applied to a *numerator term* initiated by the dividend x and a *denominator term* initiated by the divisor y, with the rational value of the fraction x/y providing the infinitely precise quotient for algebraically quantifying approximation error.

In a preprocessing phase of convergence division a table look-up generated seed divisor reciprocal approximation ρ is determined and employed to scale the numerator term yielding an initial *quotient approximation* $q_0 = x\times\rho$. A concurrent scaling of the denominator term provides the corresponding *relative error factor* $(1-\varepsilon) = y\times\rho$, with dual operations illustrated as operations on the fraction x/y. If this preprocessing phase is performed with exact multiplications the result is identical to the *prescaling step* of p by p digit prescaled division as described in Section 5.6.2:

$$\frac{x}{y} = \frac{x\times\rho}{y\times\rho} = \frac{q_0}{1-\varepsilon} \qquad \text{(prescaling).}$$

The iterative core of convergence division then performs independent multiplications of the numerator term and of the denominator term by the successive complementary error factors $(1+\varepsilon)$, $(1+\varepsilon^2)$, $(1+\varepsilon^4)$, ... yielding

$$\frac{x}{y} = \frac{q_0\times(1+\varepsilon)}{(1-\varepsilon)\times(1+\varepsilon)} = \frac{q_1}{(1-\varepsilon^2)} \qquad \text{(iteration 1),}$$

$$\frac{x}{y} = \frac{q_1\times(1+\varepsilon^2)}{(1-\varepsilon^2)\times(1+\varepsilon^2)} = \frac{q_2}{(1-\varepsilon^4)} \qquad \text{(iteration 2),}$$

$$\frac{x}{y} = \frac{q_2\times(1+\varepsilon^4)}{(1-\varepsilon^4)\times(1+\varepsilon^4)} = \frac{q_3}{(1-\varepsilon^8)} \qquad \text{(iteration 3).}$$

With infinitely precise arithmetic the ith iteration yields $q_i = (x/y)(1-\varepsilon^{2^i})$ and the "quadratic convergence" rate of the approach to the infinitely precise quotient can be precisely measured. With rounded multiplications implicitly introducing a second source of error the approximation becomes $q_i = (x/y)(1 - \varepsilon^{2^i})(1 + \delta\beta^{-(p+g-1)})$, where δ can be bounded in terms of the number of rounded operations needed to compute q_i. The accumulated rounding errors limit quotient refinement to an approximate precision somewhat less than the guarded precision $(p + g)$ of the rounded multiplications. In practice the seed reciprocal accuracy is typically chosen as about $(p + g)/8$ or $(p + g)/4$, so that three or perhaps just two core iterations are required yielding relatively few roundings in the computation. The number of guard digits is selected to insure that the product of the accumulated round off error factor $1 + \delta\beta^{-(p+g-1)}$ and the relative error factor $1 - \varepsilon^{(2^i)}$ does not exceed a prespecified quotient approximation error factor bound.

The following example illustrates: (i) the general algorithm engineering of jointly prespecifying the look-up table accuracy, number of refinements, and guard digit precision according to the size of the problem, and (ii) a specific numeric walkthrough employing the same input as in the p by p digit prescaled division in Example 5.7.2.

Example 5.7.3 (Convergence division)
Problem For the (6 × 6)-digit decimal approximate division arguments:

$$x = 365748 \quad (\textit{six-digit dividend}) \quad \text{and} \quad y = .784731 \quad (\textit{six-digit divisor})$$

find a six-digit integer quotient q with $10^5 \leq q < 10^6$ and a three-digit decimal guard $t = 0.d_1d_2d_3$ such that $|(x/y) - (q + t)| < \frac{1}{4}$.
Solution The result employs a divisor reciprocal look-up procedure with relative error bound $|\varepsilon| < .021$ and two iterative refinements according to the follow steps illustrated in Figure 5.7.1

Step 1: Determine from a look-up table the three-digit reciprocal $\rho = \mathit{round}\left(\frac{1}{.78}\right) = 1.28$.
Step 2: Prescale the denominator and numerator terms by the seed reciprocal ρ.
Step 3: Perform two iterations of complementary error factor scaling.
Step 4: Output the approximate quotient as comprised of a $(p = 6)$-digit integer $q = 466080$ and a $(g = 3)$-digit fraction $t = .732$ satisfying $|(x/y) - (q + t)| < \frac{1}{4}$. □

The purpose of the algorithm engineering is to set an application specific balance between hardware resource cost and total latency reduction. With a target precision of .25 in the sixth place and a guard precision of $g = 3$ digits, a two-iteration implementation and table precision target relative error bound $|\varepsilon| < .021$ is chosen. For divisor reciprocal approximation, a table precision error bound of $\varepsilon < .021$ can be realized for divisors in the range $.25 \leq y < 1$ with a two-digit look-up table constructed by rounding a two-digit rounded divisor to three places. For divisors with $.1 \leq y < .25$, a three-digit look-up could be used to achieve $\varepsilon < .021$. Step 1 of the numeric walkthrough illustrates a rounded "two-digit input

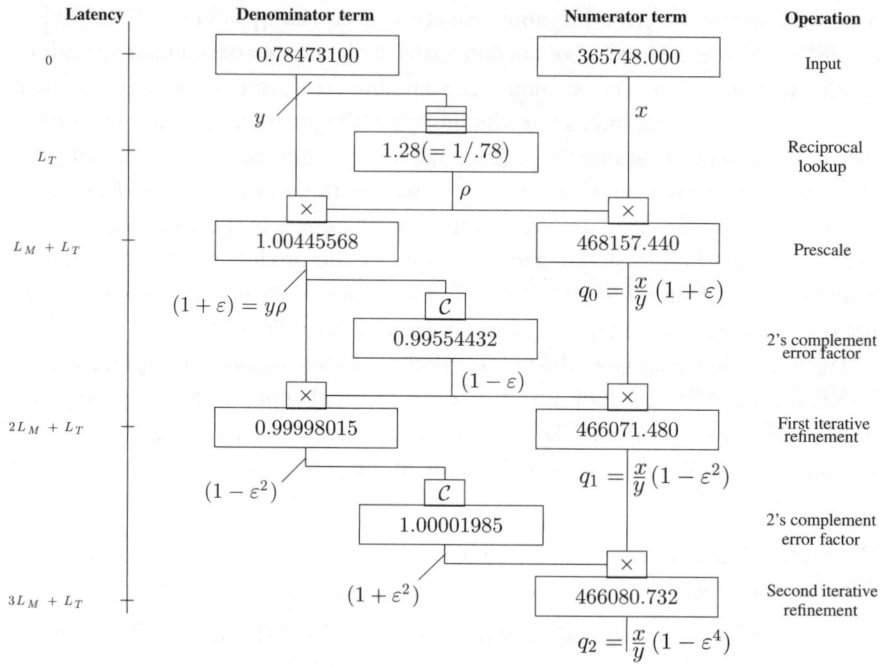

Figure 5.7.1. Convergence division walkthrough of $x = 365748$ divided by $y = .784731$.

with rounded three-digit output" case appropriate to $y = .784731$. Further options on reciprocal table construction and compression are discussed in Section 5.8. It follows that $\varepsilon^4 < .2 \times 10^{-6}$, so confining accumulated roundoff error to below 5×10^{-8} satisfies the total error bound on the approximate quotient.

All four steps of the numeric walkthrough of convergence division are illustrated in Figure 5.7.1, where C denotes 2's complement, $\times$ denotes multiplication, and the horizontal-lined box denotes table look-up. The multiplications applied independently to the denominator and numerator terms are shown in separate columns with the total latency accumulated tabulated down the left axis. Both the relative error factor and the complementary error factor are shown with a fixed-point format $d_0.d_1d_2 \cdots d_{p+g-1}$. The complementation operations shown in the middle column can be accomplished with virtually no latency. The total latency comprises the table lookup latency L_T and three multiply latencies L_M, reflecting the longer chain of dependent multiplications applied to the numerator term.

A multiplication giving the final relative error factor $(1 - \varepsilon^4)$ is optional. Alternatively, $(1 - \varepsilon^4)$ can be estimated using a small table indexed by a couple of leading digits of ε. In this example $(1 - \varepsilon^4) = 1.0000\,0000$ to rounding precision. In conjunction with the guarded quotient fraction value $t = .732$ and accumulated roundoff error bound $(1 \pm 5 \times 10^{-8})$ for the five rounded multiplications, it may further be concluded for this problem that $\frac{1}{2} < x/y - q < 1$. Extensions for completing quotient rounding are discussed in Section 7.5.

Algorithm 5.7.8 (Convergence division)

Stimulus: *A radix β, precision p, guard digit count g, iteration count k, divisor reciprocal lookup function $recip(\cdot)$, along with a p-digit fractional divisor y normalized so that $1/\beta \leq y < 1$, and a p-digit dividend x normalized so that $y\beta^{p-1} \leq x < y\beta^p$.*

Response: *A p-digit integer quotient q and g-digit fractional part t such that $|x/y - (q+t)| < \frac{1}{4}$.*

Method: $\rho := recip(y)$; {*divisor reciprocal lookup*}
$y := y \times \rho$; {*scale denominator toward unity*}
$q := x \times \rho$; {*scale numerator toward quotient*}
for $i := 1$ **to** $k - 1$ **do**
  $\eta := 2 - y$; {*2's complement error factor*}
  $y := y \times \eta$; {*scale denominator toward unity*}
  $q := q \times \eta$; {*scale numerator toward quotient*}
end;
$\eta := 2 - y$; {*2's complement error factor*}
$q := q \times \eta$; {*scale numerator toward quotient*}

The correctness of Algorithm 5.7.8 relies on the parameters β, p, g, k, and the reciprocal look-up function relative error bound ε being chosen so that the relation $|x/y - (q+t)| < \frac{1}{4}$ is satisfied. A rough estimate of accumulated quotient roundoff error is $(k+1)\beta^{-(p+g-1)}$, since the accumulated roundoff error in $1 - \varepsilon^{2^i}$ mostly cancels in computing $1 - \varepsilon^{2^{i+1}}$. Thus each complementary error factor inherits essentially only a single roundoff error, bounded by about $\frac{1}{2}\beta^{-(p+g-1)}$. This estimate of roundoff error can be employed in conjunction with the theoretical error bound $1 - \varepsilon^{2^k}$ to establish the number of guard digits, with a detailed error analysis supplementing this argument to prove correctness. The more complex error analysis for convergence division compared to Newton–Raphson division is the price for the return of reduced latency.

Lemma 5.7.9 *The Convergence Division Algorithm with k iterations employs $(2k+1)\,((p+g) \times (p+g))$-digit multiplications and has a total latency of $L_T + (k+1)L_M$.*

Note that if a full $((p+g) \times (p+g))$-digit multiplier is employed, convergence division has a clear advantage in total latency over the $L_T + (2k+1)L_M$ latency of Newton–Raphson division. If a rectangular multiplier with a $2:1$ aspect ratio is employed, the latencies of the two algorithms are more competitive and the more robust error control of Newton–Raphson division may be preferable giving a lower guard digit count but one that is sufficient to insure the correctness of the implementation.

Implementation of convergence division employing a multicycle pipelined multiplier increases the latency by one cycle and is facilitated by scheduling each denominator term multiplication prior to the corresponding numerator term multiplication.

5.7.3 Postscaled division

Postscaled division applies a series of dependent multiplications to a *numerator term* initiated by the dividend $x \in \mathbb{Q}_\beta$, and concurrently to a *denominator term* initiated by the divisor $y \in \mathbb{Q}_\beta$. The p-digit divisor y has the *normalized divisor partition* $y = d - \delta$ with d a rounded j-digit leading part and $-\delta$ a $(p - j)$-digit rounded-off low-order tail. An implementation of postscaled division with a fixed number of iterations k and partitioned divisor $y = d - \delta$ will be shown to yield an approximate quotient with relative error less than $\beta^{-(j-1)2^k}$. For $k = 2$, about $\frac{1}{4}$ of the divisor digits should be in the leading part and $\frac{3}{4}$ of the divisor digits in the rounded-off tail to obtain an approximate quotient of accuracy comparable to the number of divisor digits.

For the divisor $y = d - \delta$ and any $i \geq 0$, the sum $(d^{2^i} + \delta^{2^i})$ is termed the ith *order sum-of-squares* of the divisor partition. Note that the 0th order sum $(d + \delta)$ can be obtained simply by a low-order part complement of the divisor, with successors obtained iteratively by a product $(d^{(2^i)} - \delta^{(2^i)})(d^{(2^i)} + \delta^{(2^i)}) = d^{(2^{i+1})} - \delta^{(2^{i+1})}$ followed by an appropriate low-order part complement.

Postscaled division has no prescaling and is initiated by performing independent multiplications of the numerator term and the denominator term by successive ith order sum-of-squares terms yielding

$$\frac{x}{y} = \frac{x \times (d + \delta)}{y \times (d + \delta)} = \frac{x(d + \delta)}{d^2 - \delta^2} \qquad \text{(iteration 1)},$$

$$\frac{x}{y} = \frac{x(d + \delta) \times (d^2 + \delta^2)}{(d^2 - \delta^2) \times (d^2 + \delta^2)} = \frac{x(d + \delta) \times (d^2 + \delta^2)}{d^4 - \delta^4} \qquad \text{(iteration 2)}.$$

For the kth iteration a multiplication need only be employed in the numerator term yielding $x \prod_{i=0}^{k-1} (d^{2^k} + \delta^{2^k})$.

A *postprocessing* phase of postscaled division performs a final multiplication of the numerator term by a table look-up approximate reciprocal function value $(1/d)^{2^k}$, yielding the kth *order postscaled approximate quotient*

$$q_k = x \prod_{i=0}^{k-1} (d^{2^i} + \delta^{2^i}) \left(\frac{1}{d}\right)^{2^k} \qquad \text{for } k \geq 1.$$

The accuracy of this approximate quotient is readily determined.

Lemma 5.7.10 *For the dividend x and partitioned divisor $y = d - \delta$, the kth order postscaled approximate quotient q_k is a subestimate of x/y with a relative error factor given by*

$$q_k = \frac{x}{y}\left(1 - \left(\frac{\delta}{d}\right)^{2^k}\right),$$

where absolute error in the quotient approximation is

$$q_k - \frac{x}{y} = -\frac{x}{y}\left(\frac{\delta}{d}\right)^{2^k}.$$

For d a binary round-to-nearest leading part of y, the relative error in q_k is at most 2^{-j2^k}.

Proof Noting that $(d - \delta)\prod_{i=0}^{k-1}(d^{2^i} + \delta^{2^i}) = d^{2^k} - \delta^{2^k}$, it follows that

$$q_k = \frac{d - \delta}{y} \times \frac{x\prod_{i=0}^{k-1}(d^{2^i} + \delta^{2^i})}{d^{2^k}} = \frac{x}{y}\left(1 - \left(\frac{\delta}{d}\right)^{2^k}\right),$$

and the other results follow. □

For a (6×6)-digit decimal example with dividend $x = 36574800$ and divisor partition $y = 78 - (-.4731)$, by Lemma 5.7.10 the approximate quotient and its absolute error for $k = 1$ are

$$q_1 = \frac{36574800 \times (78 - .4731)}{78^2} = \underline{4660}63.5868\cdots,$$

$$q_1 - \frac{x}{y} = -\frac{365748000}{78.4731}\left(\frac{.4731}{78}\right)^2 = -17.1466\cdots.$$

For $k = 2$ we find

$$q_2 = \frac{365748000 \times (78 - .4731) \times (78^2 - .4731^2)}{78^4} = \underline{466080.73}28\cdots,$$

$$q_2 - \frac{x}{y} = -\frac{365748000}{78.4731}\left(\frac{.4731}{78}\right)^4 = -.00063\cdots.$$

Algorithm engineering of postscaled division involves determining an iteration count, divisor partition, and reciprocal function look-up specification so as to bound absolute error somewhat below the quarter-ulp bound, with choice of a guard digit count so the combined computation error remains bounded below a quarter ulp.

Example 5.7.4 (Postscaled division)
Problem
Size: (6×6)-digit decimal approximate division.
Target result: Quotient accurate to one quarter of a unit in the sixth place.

Algorithm engineering
Iteration count: $k = 2$.
Guard digit count: $g = 3$.
Divisor partition:
$y = d - \delta$ with $(\delta/d)^4 < 1.6 \times 10^{-7}$, or $|\delta/d| \leq .02$,

- For $25 \leq y < 100$, let d be a rounded integer part of y with $|\delta| < \frac{1}{2}$.
- For $10 \leq y < 25$, let d be a rounded quarter integer part of y with $|\delta| < \frac{1}{8}$.

Divisor reciprocal function table:

recip(d) is provided by tables with 136 possible values of d providing a nine-digit rounded approximation of $\frac{1}{d^4}$.

Normalization:

$x = 36574800$ and $y = 78.4731$, so $d = 78$, $-\delta = 0.4731$.
$10 \leq y < 100$, $y10^5 \leq x < y10^6$, so the approximate quotient q_2 has a $(p = 6)$-digit integer part q and a $(g = 3)$-digit fraction tail t.

Numeric walkthrough: The steps of the computation are illustrated in Figure 5.7.2 with roundings to $p + g = 9$ decimal places, where C denotes the appropriate low order part complement (fractional complement by this scaling).

Algorithm engineering summary: For this (6×6)-digit postscaled division example, a table look-up precision of about $\frac{1}{4}$ of the six-digit goal seemed reasonable so the iteration count was set to $k = 2$. Partitioning the divisor $y = d - \delta$ so that $|\delta/d| \leq .02$ provided the exact error bound $(\delta/d)^4 < 1.6 \times 10^{-7}$, with 0.9×10^{-7} being the slack allowed for accumulated rounding errors. Setting $g = 3$ was then sufficient as there are just five roundoff errors in the computation. Note that the divisor reciprocal table with index provided by a two-digit rounded d for $25 \leq d \leq 100$ with $|\delta| \leq \frac{1}{2}$ achieves $|\delta/d| \leq .02$ for that range of divisors with 76 table entries. An alternative partition for $10 \leq d \leq 25$ should be used to guarantee $|\delta/d| \leq .02$ for that portion of the table (see Section 5.8 for more details). Note that the table entries should be computed at the target precision of $p + g = 9$ digits to control the roundoff accumulation. The walkthrough illustrated in Figure 5.7.2 shows the step-by-step, nine-digit rounded computation of $q_2 = x \times (d + \delta) \times (d^2 + \delta^2) \times 1/d^4$ which in this case differed by less than two units in the ninth place from $x/y = 466080.7334\cdots$. □

Algorithm 5.7.11 (Postscaled division)
Stimulus: *A radix-β, precision p, guard digit count g, iteration count k, divisor reciprocal look-up function* recip(·)*, a normalized partitioned divisor $y = d - \delta$ with j-digit rounded integer part $\beta^{j-1} \leq d \leq \beta^j$ and a*

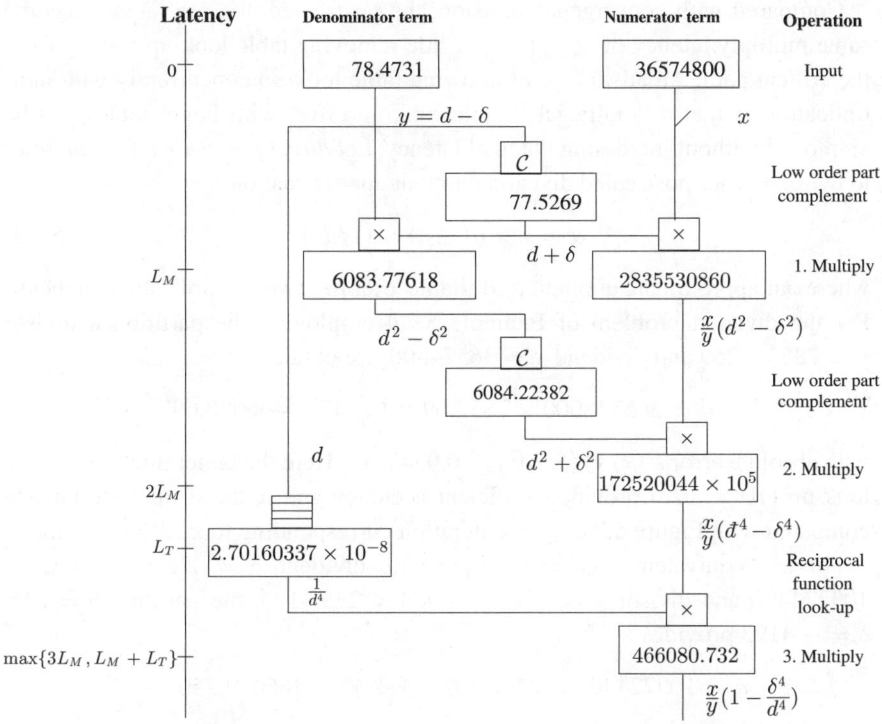

Figure 5.7.2. Postscaled division walkthrough of $x = 36574800$ divided by $y = 78.4731$.

$(p - j)$-digit fractional tail $|\delta| \leq \frac{1}{2}$, and a p-digit dividend x normalized so that $y\beta^{p-1} \leq x < y\beta^p$.

Response: *A p-digit integer quotient q and g-digit fractional tail t such that*

$$|x/y - (q + t)| < \tfrac{1}{4}.$$

Method: $\rho := \text{recip}(d^{2^k})$; {*divisor reciprocal lookup*}
for $i := 1$ **to** k **do**
  $y^\star := 2\left\lfloor y + \frac{1}{2} \right\rfloor - y$; {*low order part complement*}
  $x := x \times y^\star$ {*scale numerator towards qd^{2^i}*}
  $y := y \times y^\star$ {*scale denominator towards d^{2^i}*}
end;
$q := x \times \rho$ {*scale numerator toward quotient*}

Observation 5.7.12 *The Postscaled Division Algorithm (Algorithm 5.7.11) with iteration count k determines a p-digit quarter-ulp quotient approximation with $2k$ total multiplies and a time latency of $(k + 1)$ multiply latencies.*

Compared with convergence division, the postscaled division latency has the same multiply latency of $(k+1)L_M$, while removing table look-up latency from the critical path. An advantage of allowing table look-up concurrently with multiplication is that a "multicycle" table look-up effort with larger tables can be employed without increasing the total latency. Let *direct postscaled division* refer to the first order postscaled division quotient approximation

$$q = x \times (d + \delta) \times \left(1/d^2\right), \tag{5.7.3}$$

where our approximate quotient is available after just two multiplication latencies. For the division problem of Example 5.7.4 employing the partitioned divisor $y = 785 - .269$ and dividend $x = 36574800$, we obtain

$$q = 365748000 \times 785.269 \times \left(\tfrac{1}{785}\right)^2 = 466080.697$$

with absolute error $-(x/y)\left(\delta^2/d^2\right) = 0.0547\cdots$. Here the larger three-digit table look-up (10 × size) provided sufficient accuracy and reduced the time latency compared with Figure 5.7.2 by one iteration corresponding to a 33% reduction.

For an "equivalent" scaled example with dividend $x = (.7) \times 36574800 = 109724400$ and divisor $y = (.7) \times 784.731 = 235.4193$, the partition $d = 235$, $\delta = -.4193$ provides

$$q = 109724400 \times 234.5807 \times \left(\tfrac{1}{235}\right)^2 = 466079.250\cdots,$$

with absolute error $-(x/y)\left(\delta^2/d^2\right) = 1.484\cdots$, which is not sufficient for our quarter-ulp quotient approximation target. Thus direct postscaled division would require a larger index size for a direct table look-up of $\left(1/d^2\right)$ to insure our quarter-ulp quotient result. An alternative reciprocal look-up procedure is to employ a two-term "bipartite" look-up process as described in Section 5.8 with the approximate quotient obtained according to the following lemma.

Lemma 5.7.13 *For a partitioned divisor $y = d - \delta$ and dividend x, the* bipartite reciprocal *direct postscaled approximate quotient*

$$q = x \times (d + \delta) \times \left(\frac{1}{d^2} + \frac{\delta^2}{d^4}\right) \tag{5.7.4}$$

satisfies $q = (x/y)\left(1 - \delta^4/d^4\right)$ with absolute error $-(x/y)(\delta/d)^4$.

Proof

$$q = \frac{x}{y}(d - \delta)(d + \delta)\left(\frac{1}{d^2} + \frac{\delta^2}{d^4}\right) = \frac{x}{y}\left(1 - \frac{\delta^4}{d^4}\right). \qquad \square$$

For a practical application of the bipartite reciprocal quotient equation (5.7.4) it is reasonable to assume a divisor partition where $(\delta/d)^4$ is much smaller than is needed for our quarter-ulp bound. The effort then focuses on approximating the secondary lookup term δ^2/d^4 with sufficient accuracy to obtain the quarter-ulp quotient accuracy bound. For our preceding scaled example, note that $\delta^2/d^4 = \frac{(-.4193)^2}{(235)^4} = .0000576 \times 10^{-6}$, with $1/d^2 = 18.1077411 \times 10^{-6}$. If we can simply determine that $.5 \times 10^{-10} \leq \delta^2/d^4 \leq .6 \times 10^{-10}$, then q as given by (5.7.3) with $d = 235$ and $\delta = -.4193$ satisfies $466080.537 < q < 466080.796$, which is sufficient for our bound.

A two-term bipartite look-up process can also be employed to enhance the accuracy of a two-iteration postscaled division algorithm. There are two options for enhancing the accuracy as given in the following observation.

Observation 5.7.14 *For a partitioned divisor $y = d - \delta$ and dividend x, a* bipartite reciprocal, *two-iteration postscaled quotient is given either by*

$$q_2 = x \times (d + \delta) \times (d^2 + \delta^2) \times \left(\frac{1}{d^4} + \frac{\delta^4}{d^8}\right),$$

with relative error factor $q_2 = (x/y)\left(1 - \delta^8/d^8\right)$, or by

$$q_2' = x \times (d + \delta) \times \frac{1}{d^4} \times \left(d^2 + \delta^2 + \frac{\delta^4}{d^6}\right),$$

with relative error factor $q_2' = (x/y)\left(1 - \delta^6/d^6\right)$.

5.7.4 Efficiency of iterative refinement division

Let j be the number of leading digits of the divisor comprising an index to a direct look-up table providing a divisor reciprocal function seed value. Iterative refinement division algorithms effectively double the precision of the approximation each iteration, with precisions of roughly $2j$, $4j$, and $8j$ achieved after the first, second, and third iterations, respectively. Essentially $\lceil \log_2((p + g)/2j) \rceil$ iterations then suffice for the iterative refinement phase to obtain the desired $(p + g)$-digit approximate result. Larger values of the table look-up index size j serve to decrease the number of iterations at the cost of an increase in the table look-up latency with exponential growth in table size. Even when hardware tables may be compressed so that size is not the primary constraint, the trade-off between the table (decompression) look-up latency and the multiplier latency measured in some comparable unit such as logic levels or machine cycles needs to be considered in comparing various iterative refinement division algorithms.

The performance of iterative refinement division is summarized by giving both the *total number of multiplications* employed (regarding operation throughput and hardware power) and the total *time latency* (regarding hardware stall) to obtain the $(p + g)$-digit approximate quotient in terms of the *number of iterations*. The time latency is determined both by inherent dependences within the table look-up procedure and by the stream of dependent multiplications within and between iterations that must be scheduled back-to-back, limiting the ability of parallel and pipelined architectures to achieve any further speed-up of these algorithms. If hardware simplicity reduces the opportunities for parallel operations, the total number of multiplications is the dominant measure for comparison and all three procedures are roughly comparable. Newton–Raphson division is a likely choice in this case due to its simplicity and more robust error accumulation control. For a pipelined multiplier where a modest latency table look-up allows three iterations to be sufficient for target approximate quotient accuracy, convergence division reduces latency to essentially $4L_M$ compared to Newton–Raphson's $7L_M$ latency, a reduction of over 40%. If a more costly table look-up with L_T essentially equal to L_M is available that can reduce the needed iteration count from three to two, the postscaled algorithm can approach a 25% further reduction in latency over the convergence algorithm. The latencies for the three algorithms then have the ratio of about 6 : 4 : 3. The preceeding subsections provided the essential features of each algorithm and illustrated the method on a common example problem. The most important features were described, with additional features and opportunities for enhancement left to the problems.

Problems and exercises

5.7.1 Prove the following identity extending Lemma 5.7.4 to provide a cubic reciprocal refinement procedure:

$$\rho_k = \rho_0 \prod_{j=0}^{k-1} (1 + \varepsilon^{3^j} + (\varepsilon^{3^j})^2) = \frac{1}{y}(1 - \varepsilon^{3^k}).$$

Then show a cubic refinement recurrence is given by:

$$\rho_i = \rho_{i-1}(2 - y\rho_{i-1} + (1 - y\rho_{i-1})^2)$$

5.7.2 Generate a table similar to that of Example 5.7.1 showing three iterations of cubic refinement, where the "cubic complementary error factors" given by $(2 - y\rho_{i-1} + (1 - y\rho_{i-1})^2)$ for $i = 1$, 2, and 3, are 1.39, 1.027, and 1.000019683 respectively, and where the resulting relative error is less than 10^{-14}.

5.7.3 For the six-digit decimal division of $x = 365748$ by $y = 0.784731$ illustrated in Figure 5.7.1 the prescaling step yields $\varepsilon = \rho y - 1 = 0.445568 \times 10^{-2}$. Consider enhancing the complementary factor $(1 - \varepsilon) = 0.99554432$ by using the factor $(1 - \varepsilon + \varepsilon^2)$, where ε^2 is obtained from a low-precision approximation of ε^2 added to $(1 - \varepsilon)$, e.g., $\varepsilon^2 = 0.44^2 \times 10^{-4}$ or $\varepsilon^2 = 0.45^2 \times 10^{-4}$.

Determine how much precision you would need in estimating ε^2 to obtain an acceptable six-digit approximate quotient using the enhanced factor $(1 - \varepsilon + \varepsilon^2)$ to avoid the final iterative refinement shown in Figure 5.7.1. Also consider the resource needed to estimate ε^2 by a table look-up and/or by the simplified squaring operations described in Section 4.7.

5.7.4 Additive prescaling: Let the binary divisor y have the alternative normalization $3/2 \leq y \leq 3$. Show that the reciprocal $1/3 \leq 1/y \leq 2/3$ has a radix-4 representation $1/y = 0.2d_2d_3\ldots$ where $d_i \in \{-2, -1, 0, 1, 2\}$ for $i \geq 2$. Then show that a three-digit radix-4 prescale value $\rho = 0.2.d_2d_3$ with $d_2, d_3 \in \{-2, -1, 0, 1, 2\}$ can be determined from at most seven leading bits of y so that a prescaled divisor can be given by a three term sum $\rho y = y^{2-1} + d_2y^{2-4} + d_3y^{2-6}$ that satisfies $|\rho y - 1| \leq 2^{-5}$.

5.7.5 For normalized binary division of the dividend x by the divisor y let four factor multiplicative division refer to determination of the approximate quotient $q_2 = x(d+\delta)(d^2+\delta^2)(1/d_4)$. Then for $y = 1.b_1b_2\ldots b_{\rho-1}$, let $y = d - \delta$, where

$$d = 1.b_1b_2\ldots b_k + 2^{-(k+1)},$$
$$\delta = (1 - b_{k+1}.b_{k+2}\ldots b_{\rho-1})2^{-(k+1)}.$$

Discuss the table sizes and squaring operation resources needed so that the approximate quotient q_2 has precision of about 32 bits. Hint: note that $(d+\delta)$ is a low-order part complement and that d^2 and $1/\mathrm{d}_4$ may be determined by table look-up with a k-bit index. Then, δ^2 may be found either by table look-up or by an approximate squaring circuit as described in Section 4.7.

5.8 Table look-up support for reciprocals

The divisor reciprocal operation is a special case of division where fast generation of a low-precision binary approximate reciprocal has many applications. For a normalized binary divisor $y \in \mathbb{Q}_2$, $y \in [1; 2)$, let an *approximate reciprocal* denote a function $\mathrm{recip}(y) \in \mathbb{Q}_2$ with range $y \in [\frac{1}{2}; 1]$ satisfying a specified absolute (or relative) error bound, such as $|1/y - \mathrm{recip}(y)| < 2^{-j}$ for all $y \in [1; 2)$. The bound partitions the bits of $\mathrm{recip}(y) = 0.1a_2a_3\cdots a_ja_{j+1}\cdots a_{j+g}$ into a j-bit leading part and a trailing part here having g *guard bits*.

Binary approximate reciprocal functions have numerous applications in arithmetic unit implementations. These applications and the hardware unit destination of the result serve to classify approximate reciprocals into three types as shown in Table 5.8.1.

Table 5.8.1. *Classification of approximate reciprocals by their intended application and associated destination unit*

Type	Destination	Application
Seed reciprocal	Large multiplier	Iterative division (Newton–Raphson and convergence)
Short reciprocal	Rectangular multiplier (short input)	High-radix division (prescaled and short reciprocal)
Reciprocal function value	Single precision word	Reciprocal instruction

Table 5.8.2. *Classification of approximate reciprocal algorithms by the hardware resources utilized. Typical precisions obtained from comparable sized total tables are included for each algorithm class*

	Approximate Reciprocal Algorithms		
	Direct look-up	Bipartite/multipartite look-up	Linear/quadratic interpolation
Hardware resources	Table	Tables, adder	Tables, multiplier
Table index size	8–12 bits	7–11 bits	7–9 bits
Total table size	$\frac{1}{4}$–6 k-bytes	$\frac{1}{4}$–8 K-bytes	1–6 K-bytes
Precision obtained	8–12 bits	10–22 bits	16–30 bits

Both seed- and short-type reciprocals generated within an ALU can be forwarded in redundant binary form to an appropriate multiplier recoder. A seed reciprocal can typically retain guard bits due to the large size of the multiplier required for iterative division implementations. The short reciprocal destined for the short size input of a rectangular multiplier should have the approximation bound comparable to the short size. Therefore any incidental guard bits generated by table-assisted approximate reciprocal computation should be appropriately rounded off as part of the short reciprocal computation process.

For the type of binary approximate reciprocal destined for a memory word under user control, it is desirable that the behavior of the real-valued reciprocal function be modeled as well as efficient computation allows. This requires investigation of the monotonicity as well as accuracy of the resulting reciprocal function.

There are numerous algorithms for generating approximate reciprocals employing look-up tables. These algorithms utilize tables dedicated to reciprocal generation and also employ "shared" adders and multipliers otherwise available within the ALU. The algorithms can be classified by the supplemental support provided by adders and (small) multipliers as shown in Table 5.8.2.

The binary direct and multipartite approximate reciprocal algorithms of Table 5.8.2 each evaluate an expression having one-to-four terms of decreasing significance given by $\text{recip}(y) = \text{LUT}_1(y) + \text{LUT}_2(y)2^{-e_2} + \text{LUT}_3(y)2^{-e_3} + \text{LUT}_4(y)2^{-e_4}$. The multiplicative interpolations utilize look-up table values as operands for one or more multiplicative operations. Evaluation of a multiterm expression for recip(y) has a first phase of concurrent table look-ups. An accumulate or multiply accumulate phase follows generating a redundant reciprocal. A final phase involves either rounding, recoding, or a $2-1$ addition depending on the reciprocal type. The recip(y) output is thus generated by a single pass through this "reciprocal unit". In a pipelined ALU this allows single-cycle throughput with typical latencies of one-to-three cycles for seed and short reciprocals, and two-to-four cycles for a reciprocal function value output in storage format.

Table 5.8.2 also provides a preview summarizing the range of reciprocal precisions obtainable employing comparable ranges of total table sizes.

5.8.1 Direct table look-up

The error in a direct table look-up reciprocal value arises from two distinct identifiable sources where either may be dominant. The *input discretization* component of the total error derives from the necessity to pick a single table entry for an index designating an interval of input arguments. Determination of the real-valued output, minimizing the error over the input interval, provides a first step in determining a table entry for this index.

A second *output discretization* (rounding) error for an arbitrary radix-β direct look-up table occurs, since the table output must have some fixed number of digits, deviating by some rounding from the error-minimizing real approximation. Either the input discretization error or the output rounding error may be dominant depending on the application.

Our focus on normalized binary inputs $y \in [1;2)$ and outputs $\mathrm{recip}(y) \in [\frac{1}{2}, 1]$ minimizes the radix-dependent "range size" component of these input/output discretizations. We are also able conveniently to characterize table values minimizing maximum relative error, which is of greatest importance for seed values of relatively small precision. The relative error for a reciprocal function approximation can be expressed in terms of a product of argument and result.

$$\left| \frac{1/y - \mathrm{recip}(y)}{1/y} \right| = |1 - y\, \mathrm{recip}(y)|. \tag{5.8.1}$$

Thus for binary inputs and outputs, the relative error can be expressed as an exact binary value.

Direct table look-up provides the greatest control over the behavior of a reciprocal approximation. The joint effects of both error sources may be combined, allowing a preferred table entry to be chosen, satisfying min–max error bounds *and* seamlessly preserving monotonicity.

In practice binary direct table look-up is generally restricted to tables having from 2^6 to 2^{12} entries, due primarily to table size limits and secondarily to a bound on table look-up time. For direct table look-up we determine entries minimizing the maximum relative error. We also identify input arguments that realize the maximum relative error for each table entry, termed *extreme point* inputs for the reciprocal approximation. Extreme points for the whole table, and local extreme points for each table entry, provide good input values for test sets for empirical validation of division algorithm implementations using the table.

Determination of the entries in a direct look-up table is guided by the following general result on reciprocal approximation for an arbitrary input interval.

Lemma 5.8.1 (Best reciprocal approximation by reals) *For any divisor* y *in the interval* $[y_{min}; y_{max}]$ *with* $0 < y_{min} \leq y_{max}$*, the interval midpoint reciprocal* $2/(y_{min} + y_{max})$ *satisfies*

$$\left|\frac{1}{y} - \frac{2}{y_{min} + y_{max}}\right| \leq \frac{y_{max} - y_{min}}{y_{max} + y_{min}}. \tag{5.8.2}$$

Furthermore $\text{recip}(y) = 2/(y_{min} + y_{max})$ *is the reciprocal approximation minimizing the maximum relative error over* $[y_{min}, y_{max}]$*, with both interval end points* $\{y_{min}, y_{max}\}$ *being extreme points realizing the maximum relative error* $(y_{max} - y_{min})/(y_{max} + y_{min})$.

Proof The relative errors for approximating the end points $\{y_{min}, y_{max}\}$, where

$$\frac{1}{y_{max}} \leq \frac{1}{y} \leq \frac{1}{y_{min}},$$

are

$$\left|1 - y_{min}\frac{2}{y_{min} + y_{max}}\right| = \frac{y_{max} - y_{min}}{y_{min} + y_{max}} \quad \text{and}$$

$$\left|1 - y_{max}\frac{2}{y_{min} + y_{max}}\right| = \frac{y_{max} - y_{min}}{y_{min} + y_{max}}.$$

Thus $\text{recip}(y) = 2/(y_{min} + y_{max})$ is the value for which the relative errors at the endpoints are identical. Any other choice for $\text{recip}(y)$ would result in one of these end points having larger relative error. □

Definition 5.8.2 *A look-up table is a one-to-one mapping* $\text{LUT} : \mathbb{N} \to \mathbb{M}$ *of a discrete set of index values* $\mathbb{N}$ *to a set of real valued outputs* $\mathbb{M}$*. A binary lookup table* $\text{LUT} : \mathbb{Q}_2 \to \mathbb{Q}_2$ *has binary inputs typically each representing an input region for a function and binary outputs typically representing approximations of that function.*

We shall often use the notation $\text{LUT}(\cdot)$ informally where the argument implicitly references the index, possibly also denoting the function being approximated as well as the index. The context should be sufficient to convey the specific use implied.

In practice a look-up table is typically stored in a ROM, with an i-bit index employed to retrieve an arbitrary j-bit output as the look-up value.

Observation 5.8.3 *For* y *in the normalized binary interval* $[1; 2)$*, the* n*-entry approximate reciprocal table minimizing the maximum relative error is obtained by partitioning the interval* $[1; 2)$ *into* n *subintervals* $[2^{k/n}; 2^{(k+1)n})$ *for* $k = 0, 1, \ldots, n - 1$*, of* equal relative *size.*

Note that applying Lemma 5.8.1 separately to determine recip(y) for each of these n subintervals results in the same maximum relative error $(2^{1/n} - 1)/(2^{1/n} + 1)$ being obtained for each subinterval. Considering that

$$\frac{2^{1/n} - 1}{2^{1/n} + 1} = \frac{\ln 2}{2n} + O\left(\frac{1}{n^2}\right),$$

we obtain a limit on the "min–max" relative error accuracy achievable from "best" discretization of an input interval.

Employing interval boundaries $\{2^{k/n}\}$ to partition p-bit divisors to obtain a table index is costly to implement. Potential table accuracy must be sacrificed in binary direct look-up table design for ease of determining the table index and corresponding interval partition. For a normalized binary divisor $y = 1.b_1 b_2 \cdots b_{p-1}$, let $y_i = y_i(y) = 1.b_1 b_2 \cdots b_i$. The interval $[y_i; y_i + \beta^{-i})$ can utilize the i-bit string $b_1 b_2 \cdots b_i$ as an index to address a look-up table. The 2^i entries are the table values chosen according to Lemma 5.8.1 as the midpoint reciprocal $\text{recip}(y) = 1/(y_i + 2^{-(i+1)})$.

Observation 5.8.4 *For a normalized binary divisor* $y = 1.b_1 b_2 \cdots b_{p-1}$*, the (rational) reciprocal approximation* $\text{recip}(y) = 1/(y_i + 2^{-(i+1)})$ *determined by the leading part* $y_i(y) = 1.b_1 b_2 \cdots b_i$ *has the extreme point* y_i *for the interval* $[y_i, y_i + 2^{-i})$ *yielding relative error* $(1/(y_i + 2^{-(i+1)}))2^{-(i+1)}$*. The extreme point relative errors over the* 2^i *intervals range from a high of* $1/(2^{i+1} - 1)$ *for the extreme point* $y = y_i = 1$*, down to* $1/(2^{i+2} - 1)$ *for the extreme point* $y = y_i = 2(1 - 2^{-(i+1)})$.

For binary direct look-up reciprocal tables the output is also preferably limited to a reciprocal approximation having a fixed number of bits for all entries in the table (where the argument $y = 1$ is treated as an exceptional case). This output discretizised (rounding) error compounds the overall reciprocal approximation error of the direct look-up table and requires separate investigation.

For determining rounded binary table entries here and in subsequent multipart tables it is convenient to employ notation for fixed-point roundings defined on the full unit interval. For $z \in [0; 1]$, the *fixed-point round-down* $\text{RD}_j(z)$, *round-up* $\text{RN}_j(z)$, and *round-to-nearest (midpoint down)* $\text{RN}_j(z)$ roundings each determine either the j-bit unnormalized binary value $k/2^j = 0.a_1 a_2 \cdots a_j$ with $0 \le k \le 2^j - 1$, or the "special case value" $k/2^j = 1.00 \cdots 0$ with $k = 2^j$. Specifically we have

$$\text{RN}_j(z) = \max_k \left\{ \frac{k}{2^j} \,\middle|\, \frac{k}{2^j} \le z + \frac{1}{2} \right\},$$

with similar expressions for $\text{RU}_j(z)$ and $\text{RD}_j(z)$. For normalized input $z \in [\frac{1}{2}, 1)$, the output is normalized, e.g., $\text{RN}_j(z) = 0.1a_2 a_3 \cdots a_j$ or $\text{RN}_j(z) = 1$.

For the interval $[y_i; y_i + 2^{-i})$, it follows from Observation 5.8.4 that selecting the rounded output $\text{recip}(y) = \text{LUT}(b_1 b_2 \cdots b_i) = \text{RD}_{j+1}(1/(y_i + 2^{-(i+1)}) = 0.1a_2 a_3 \cdots a_{j+1}$ results in the reciprocal range upper bound $1/y_i$ generating the maximum relative error. Employing (5.8.1) this input interval would have the extreme point y_i realizing the maximum relative error $|1 - y_i \text{RD}(1/(y_i + 2^{-(i+1)}))|$. Alternatively selecting $\text{recip}(y) = \text{LUT}(b_1 b_2 \cdots b_i) = \text{RU}_{j+1}(1/(y_i + 2^{-(i+1)})$ means $y_i + 2^{-i}$ is the extreme point with the corresponding maximum relative error $|1 - (y_i + 2^{-i})\text{RU}_{j+1}(1/(y_i + 2^{-(i+1)}))|$. The preferred choice is the rounding direction yielding the minimum of these two maximum relative errors. For the case of binary input intervals $[y_i, y_i + 2^{-i})$, there is a simply determined choice.

Theorem 5.8.5 (Best i-bits-in, j-bits-out reciprocal approximation) *For the normalized binary divisor $y = 1.b_1 b_2 \cdots$, let $y_i = 1.b_1 b_2 \cdots b_i$ with $i, j \geq 2$. Then for the interval $[y_i; y_i + 2^{-i})$ the reciprocal approximation*

$$\text{recip}(y) = \text{RN}_{j+1}\left(\frac{1}{y_i + 2^{-(i+1)}}\right),$$

determined by rounding the midpoint reciprocal $1/(y_i + 2^{-(i+1)})$ to the nearest fixed point $(j+1)$-bit value, minimizes the maximum relative error for $y \in [y_i; y_i + 2^{-i})$. Furthermore, if the round-to-nearest direction of $\text{RN}_{j+1}(1/(y_i + 2^{-(i+1)}))$ is down, then the input extreme point is y_i yielding the maximum relative error $1 - y_i \text{RN}_{j+1}(1/(y_i + 2^{-(i+1)}))$. If the round-to-nearest direction is up, then $(y_i + 2^{-i})$ is an extreme point with corresponding maximum relative error $(y_i + 2^{-i})\text{RN}_{j+1}(1/(y_i + 2^{-(i+1)})) - 1$.

Proof For the interval $[y_i; y_i + 2^{-i}) = [k/2^i; (k+1)/2^i)$, the midpoint reciprocal is $2^{i+1}/(2k+1)$. Consider first that this midpoint reciprocal falls strictly within the upper half of a fixed-point $(j+1)$-bit output interval, so then

$$\frac{n + \frac{1}{2}}{2^{j+1}} < \frac{2^{i+1}}{2k+1} < \frac{n+1}{2^{j+1}}.$$

Then for some integer m

$$0 < \frac{2^{i+1}}{2k+1} - \frac{2n+1}{2^{j+2}} = \frac{2^{i+j+3} - (2k+1)(2n+1)}{(2k+1)2^{j+2}} = \frac{2m+1}{(2k+1)2^{j+2}}.$$

Now $m \geq 0$ since $(2k+1)(2n+1)$ is an odd integer. After rearrangement we have $m = 2^{i+j+2} - (kn + (k+1)(n+1))$. Now if we were to round opposite to nearest and chose the table entry as the round-down value of $2^{i+1}/(2k+1)$, the maximum relative error would be $1 - y_i \text{RD}_{j+1}(1/(y_i + 2^{-(i+1)}))$ $= (2^{i+j+1} - kn)/2^{i+j+1}$. But this error is always greater than or equal to the round up choice maximum relative error $(y_i + 2^{-i})\text{RU}_{j+1}(1/(y_i + 2^{-(i+1)})) - 1 =$

$((k+1)(n+1) - 2^{i+j+1})/2^{i+j+1}$ since

$$(2^{i+j+1} - kn) - ((k+1)(n+1) - 2^{i+j+1}) = m \geq 0.$$

Thus $y_i + 2^{-i}$ is always an input extreme point when $(n + \frac{1}{2})/2^{i+1} < 1/(y_i + 2^{-(i+1)}) < n/2^{j+1}$, with corresponding maximum relative error $(y_i + 2^{-i})\mathrm{RN}_{j+1}(1/(y_i + 2^{-(i+1)})) - 1$. It is straight forward to show that y_i is the unique extreme point when $n/2^{j+1} \leq 1/(y_i + 2^{-(i+1)}) \leq (n + \frac{1}{2})/2^{j+1}$, with $\mathrm{recip}(y) = \mathrm{RN}_{j+1}(1/(y_i + 2^{-(i+1)})) = n/2^{j+1}$ having the maximum relative error $1 - y_i \mathrm{RN}_{j+1}(1/(y_i + 2^{-(i+1)}))$. □

From the proof of Theorem 5.8.5 we note that the input interval midpoint reciprocal $1/(y_i + 2^{-(i+1)})$ must always fall at least $1/(y_i 2^{(i+1)} + 1)$ ulps away from an output midpoint. Proof of this and the following is left to the problems.

Observation 5.8.6 *For the normalized binary divisor $y \in [1; 2)$ and any $i, j \geq 2$, let $y_i = 1.b_1 b_2 \cdots b_i$. Then the maximum absolute rounding error of an input interval midpoint reciprocal is strictly less than one-half* ulp, *specifically*

$$\left| \frac{1}{y_i + 2^{-(i+1)}} - \mathrm{RN}_{j+1}\left(\frac{1}{y_i + 2^{-(i+1)}}\right) \right| \leq \left(\frac{1}{2} - \frac{1}{y_i 2^{(i+1)} + 1}\right) \text{ulps.}$$

Definition 5.8.7 *An i-bits-in, j-bits-out reciprocal table for the normalized divisor $y = 1.b_1 b_2 \cdots$ is a direct look-up table over the interval $y \in [1; 2)$ of 2^i entries. Each entry is determined by the i leading fraction bits $y_i(y) = 1.b_1 b_2 \cdots b_i$ of y, where the index $b_1 b_2 \cdots b_i$ selects the j-bit output string $\mathrm{LUT}(b_1 b_2 \cdots b_i) = a_2 a_3 \cdots a_{j+1}$ for the reciprocal approximation* $\mathrm{recip}(y) = 0.1 a_2 a_3 \cdots a_{j+1}$, *with* $\mathrm{recip}(y) = 1$ *a special case.*

Note that for binary reciprocal tables the bit counts i and j refer to bit-string lengths excluding the assumed leading bit. Thus the index has length i bits and the stored output word size is j bits, with a *table size* $2^i\ j$ bits.

Definition 5.8.8 *The* precision *(in bits) of an i-bits-in, j-bits-out reciprocal table denotes the negative of the radix 2 logarithm of the maximum relative error in approximating $1/y$ for $y \in [1; 2)$. Furthermore, a* maximum precision *direct look-up reciprocal table is an i-bits-in, j-bits-out reciprocal table, where each entry minimizes the maximum relative error for that entry.*

The precision of a maximum precision direct look-up reciprocal table, denoted by $\mathrm{prec}(i, j)$, is then determined from

$$\mathrm{prec}(i, j) = -\log_2 \left(\max_{y_i} \left\{ \min_{\delta=0,1} \left\{ \left| 1 - (y_i + \delta 2^{-i})\, \mathrm{RN}_{j+1}\left(\frac{1}{y_i + 2^{-(i+1)}}\right) \right| \right\} \right\} \right).$$

Maximum precision i-bits-in, j-bits-out reciprocal tables are uniquely determined by Theorem 5.8.5, and the entries and precisions of such tables for moderate size

Table 5.8.3. *The precisions of maximum precision i-bits-in, j-bits-out direct look-up reciprocal tables for* $6 \le i, j \le 12$

i	j = 6	7	8	9	10	11	12
6	**6.476**	6.790	6.907	6.950	7.000	7.000	7.000
7	6.790	**7.484**	7.775	7.888	7.948	7.976	8.000
8	6.907	7.775	**8.453**	8.719	8.886	8.944	8.974
9	6.950	7.888	8.719	**9.430**	9.725	9.852	9.942
10	7.000	7.948	8.886	9.725	**10.443**	10.693	10.858
11	7.000	7.976	8.944	9.582	10.693	**11.429**	11.701
12	7.000	8.000	8.974	9.942	10.858	11.701	**12.428**

i, j can be exhaustively computed. Such precisions are reported for $6 \le i, j \le 12$ in Table 5.8.3.

Note that for $j \gg i$, the total error reflects essentially just the input discretization component. Applying Lemma 5.8.1 to the input interval $[1; 1 + 2^{-i})$ yields a midpoint reciprocal approximation determining the maximum relative error to be $1/(2^{i+1} + 1)$. Thus the limiting table precision is essentially $i + 1$ bits, as seen in Table 5.8.3 for $j \gg i$. For $i \gg j$, the total error is dominated by the output discretization (rounding) component. The table value from Theorem 5.8.5 is then essentially $\mathrm{RN}_{j+1}(1/y)$ with maximum relative error near $1/2^{j+1}$ for inputs y near 2. Thus the table precision for $i \gg j$ is essentially $(j + 1)$ bits, as also seen in Table 5.8.3.

Observation 5.8.9 *The precisions of maximum precision i-bits-in, j-bits-out and j-bits-in, i-bits-out reciprocal tables are identical.*

The symmetry in the look-up table precisions is unexpected due to the lack of symmetry in the corresponding look-up table size. The proof is left as an exercise.

The precision of a maximum precision i-bits-in, j-bits-out reciprocal table is then essentially determined by the minimum of i and j. In particular, an i-bits-in, i-bits-out reciprocal table discretizes both the input range and output domain into the same number, 2^i, of subintervals, and is generally said to provide roughly i-bit accuracy. The highlighted values on the diagonal in Table 5.8.3 show that a precision of over $i + 0.4$ bits can be obtained from the maximum precision tables obtained by Theorem 5.8.5 for $6 \le i \le 12$.

Guard bits and table size. It is convenient to parameterize binary direct look-up reciprocal tables as i-bits-in, $(i + g)$-bits-out reciprocal tables with $g \ge 0$ denoting the number of output *guard* bits. The following result confirms that only a few guard bits are needed to approach the $(i + 1)$-bit precision bound, and that this number of guard bits is determined independent of the size i of the input index.

Lemma 5.8.10 (Guard Bit Lemma) *The precision of a maximum precision i-bits-in, $(i+g)$-bits-out reciprocal table for any $i \geq 2$, with $g \geq 0$ guard bits, is at least $i + 1 - \log_2(1 + 1/2^{g+1})$.*

Table 5.8.4 provides exhaustively computed maximum table precisions for i-bits-in, j-bits-out reciprocal tables for $i \in \{8, 12, 16\}$ and $0 \leq g \leq 4$, along with the maximum table precision lower bound valid for all $i \geq 2$ for $0 \leq g \leq 4$ as given by Lemma 5.8.10.

Table 5.8.4. *Minimum precisions of reciprocal tables with different numbers of guard bits*

	g				
i	0	1	2	3	4
8	8.453	8.719	8.886	8.944	8.974
12	12.428	12.687	12.844	12.918	12.963
16	16.418	16.679	16.833	16.914	16.956
Lower bound (all i)	$i + .415$	$i + .678$	$i + .830$	$i + .912$	$i + .955$

The use of a few guard bits is efficient in increasing reciprocal table precision without incurring excessive cost in the growth of table size. Without guard bits an increase by one bit of precision requires more than doubling the i-bits-in, i-bits-out table size. For example, with $i = 12$ the table size is $2^{12} \times (12)$ bits or 6 Kbytes. For $i = 13$ the table size is $2^{13} \times (13)$ bits or 13 Kbytes, an increase that adds another 117% to the table size to obtain one more bit of precision.

From our exhaustive computations and Lemma 5.8.11 below it is seen that the use of just three output guard bits results in an increase of one-half bit in table precision. For the $(i = 12)$-bit case the size of the 12-bits-in, 15-bits-out reciprocal table is $2^{12} \times (15)$ bits or $7\frac{1}{2}$ Kbytes, so only a 25% increase in table size is needed to achieve a half-unit increase in precision. As a rule of thumb to enhance direct look-up table precision (for seed reciprocals) it is desirable to employ i-bits-in, $(i+g)$-bits-out reciprocal tables using two–four guard bits.

Implementations of Newton–Raphson and convergence division algorithms as discussed in Section 5.7 typically include a full-by-full or half-by-full multiplier, with ample room for any guard bits in a seed reciprocal provided by a direct look-up table. For the short reciprocal and prescaled division algorithms it is sometimes desirable to have near maximum accuracy "to the last place" in the reciprocal approximation, as measured by the smaller size of a short-by-long multiplier. For this case the following lemma on the absolute error bound achievable from an $(i+g)$-bits-in, i-bits-out reciprocal table, having $g \geq 0$ *input* guard bits, and determined by Theorem 5.8.5 is applicable.

Lemma 5.8.11 (Input Guard Bit Lemma) *For $i \geq 2,\ g \geq 0$, a maximum precision $(i+g)$-bits-in, i-bits-out reciprocal table has a maximum relative error less than $(\frac{1}{2} + 2^{-g})2^{-i+1}$.*

It follows from Lemma 5.8.11 that one input guard bit is sufficient to insure that the result $\text{recip}(1/y) = 0.1b_2b_3 \cdots b_{i+1}$ will be accurate to a unit in the last place. For the short reciprocal and prescaled algorithms as described in Section 5.6, the table output can have each entry incremented by a unit to provide a short reciprocal that is an upper bound, larger by at most 2 ulps (with $\text{recip}(1/y) = 1.00$ treated as a special case).

5.8.2 Ulp accurate and monotonic reciprocal approximations

For applications as seed reciprocals the extra guard digits obtained in multiterm approximations can be retained, such that the associated functional approximation relative or absolute error bound reflects the total approximation error. For application as a short reciprocal or as a reciprocal function instruction output, essentially all the guard bits must be rounded off. This rather severe output discretization affects both the accuracy and the step-by-step discrete function behavior regarding preservation of monotonicity.

For normalized binary reciprocal function instructions and short reciprocals, the number of distinct input points or intervals (respectively) is comparable to the number of distinct values in the output range. Specifically for i-bits-in, i-bits-out binary reciprocal tables there are 2^i input intervals and the output range has size $2^i + 1$. A normalized i-bit binary reciprocal function instruction has 2^{i-1} input points and an output range of 2^{i-1} values. Approximations that minimize error do not yield a one-to-one mapping into the output range in either case.

Such approximate reciprocal mappings must have some distinct table entries that yield the same output either by direct look-up or as the result of a rounded multiterm computation. For these approximations the error in ulps of output and step-by-step differences in ulps of output per input ulp are important measures of the quality of function approximation behavior. For $y \in [1; 2)$ the continuous reciprocal function $1/y \in (\frac{1}{2}; 1]$ is monotonically strictly decreasing from 1 down to $\frac{1}{2}$ at a rate that is four times greater at 1 than it is on approaching $\frac{1}{2}$. It is important to determine if severely discretizised approximate reciprocals have strictly monotonic decreasing values starting from unity and at least monotonic (i.e., monotonically non-increasing) behavior for $1/y$ approaching one half.

Observation 5.8.12 *For an i-bits-in, i-bits-out reciprocal table, the sequence of table entries given by $\text{RN}_{i+1}(1/(y_i + 2^{-(i+1)}))$ is monotonically strictly decreasing with steps of one or two* ulps *for $y_i \in [1; \sqrt{2 + 2^{-2(i+1)}})$ and monotonically non-increasing with steps of zero or one* ulps *for $y_i \in (\sqrt{2 + 2^{-2(i+1)}}; 2)$.*

Proof The sequence of midpoint reciprocals $1/(y_i + 2^{-(i+1)})$ has differences

$$\frac{1}{y_i - 2^{-(i+1)}} - \frac{1}{y_i + 2^{-(i+1)}} = \frac{2}{y_i^2 - 2^{-2(i+1)}} 2^{-(i+1)}.$$

Thus the midpoint reciprocals (before roundings) drop by amounts of $\frac{2}{y_i^2 - 2^{-2(i+1)}}$ ulps. For $y_i \in [1 + 2^{-i}; \sqrt{2 + 2^{-2(i+1)}})$ the decrements satisfy $1 < 2/(y_i^2 - 2^{-2(i+1)}) < 2$. Therefore the round-to-nearest values of the two successive values must decrease by one or two ulps. For $y_i \in (\sqrt{2 + 2^{-2(i+1)}}; 2 - 2^{-i}]$, we have $y_i^2 \in (2 + 2^{-2(i+1)}; 4)$, so $\frac{1}{2} < 2/(y_i^2 - 2^{-2(i+1)}) < 1$. Therefore successive round-to-nearest values differ by zero or one ulps. □

Figure 5.8.1(a) illustrates the four-bits-in, four-bits-out reciprocal table as a step function graph with 2^i steps for $y \in [1; 2)$. The graph effectively plots the round-to-nearest input midpoint reciprocal function $\text{RN}_5(32/(33 + 2k))$ for $k = 0, 1, \ldots, 15$.

The use of one output guard bit doubles the size of the output values. This sharpens the approximation of $1/y$ at each of the steps, and increases the granularity of step sizes between steps. Figure 5.8.1(b) shows the four-bits-in, five-bits-out reciprocal table step function graph. The use of a single guard bit here has visibly smoothed the output. The 16 visibly distinct downward steps serve to model the reciprocal function derivative $(d/dy)(1/y) = -1/y^2$.

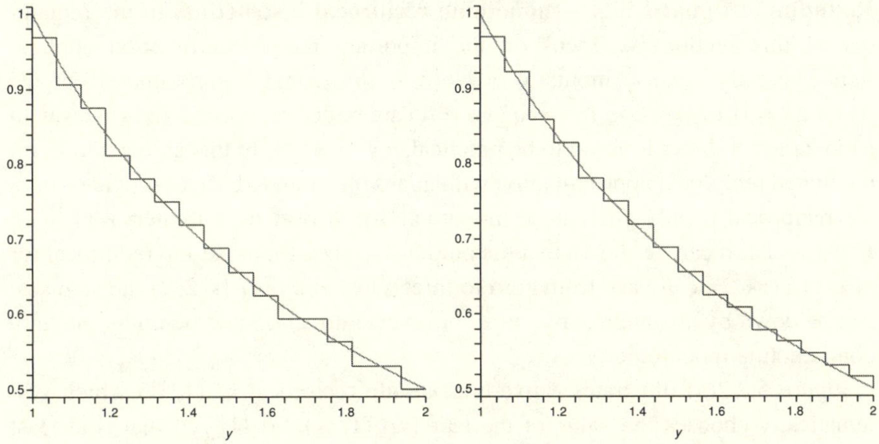

(a) (b)

Figure 5.8.1. Graphs of: (a) four-bits-in, four-bits-out, and (b) four-bits-in, five-bits-out reciprocal approximations.

Observation 5.8.13 *For an i-bits-in, $(i + g)$-bits-out reciprocal table with $g \geq 1$, the sequence of table entries is monotonically strictly decreasing. In particular the i-bits-in, $(i + 1)$-bits-out table has successive entries decreasing by one–four units as measured in the last place position $2^{-(i+2)}$. Furthermore, the transitions* *in step size are smooth in that they pass from four or three units to three or two units to two or one units as y_i increases over* $[1; 2)$.

The reciprocal approximation $\text{recip}(y) = 0.1a_2a_3 \cdots a_j$ is termed a j-bit *one-ulp reciprocal* when $|1/y - \text{recip}(y)| < 2^{-j}$ for all normalized binary divisors $y \in [1; 2)$, and similarly is a j-bit $\frac{3}{4}$*-ulp reciprocal* when $|1/y - \text{recip}(y)| < \frac{3}{4}2^{-j}$.

Observation 5.8.14 *A j-bit one-*ulp *reciprocal* recip(y) *is either the round-up or round-down value of* $1/y$ *for all normalized binary divisors* $y \in [1; 2)$. *That is,*

$$\text{recip}(y) \in \left\{ \text{RD}_j\left(\frac{1}{y}\right), \text{RU}_j\left(\frac{1}{y}\right) \right\} \quad \textit{for all } y \in [1; 2).$$

Note that a one-ulp reciprocal is efficiently computable by first obtaining a multiple term reciprocal approximation $\text{recip}_a(y) = 0.1a_2a_3 \cdots a_ja_{j+1} \cdots a_{j+g}$, with g guard bits $a_{j+1}a_{j+2} \cdots a_{j+g}$, that satisfies $|1/y - \text{recip}_a(y)| < \frac{1}{2}2^{-j}$. Then the guard bits are rounded off to obtain the j-bit one-ulp reciprocal $\text{recip}(y) = \text{RN}_j(\text{recip}_a(y))$. Such one-ulp reciprocals have applications as short reciprocals in high radix division algorithms and as the approximate reciprocal function value for a reciprocal instruction implementation. For implementation of a one-ulp reciprocal function as a reciprocal instruction, it is also desirable to investigate the monotonicity properties of such an approximate function.

Rounding off guard bits – monotonic reciprocal instruction. In the remainder of this section we focus on the important reciprocal function application where the (exact) inputs $y_i = 1.b_1b_2 \cdots b_{i-1}$, and (approximate) outputs $\text{recip}(y_i) = 0.1a_2a_3 \cdots a_i$ (or $\text{recip}(y_i) = 1$) are both i-bit normalized values with i too large for direct look-up to be practical, e.g., $i = 24$. In this case a multiterm computed reciprocal approximation with guard bits rounded off, to provide a one-ulp reciprocal is only guaranteed monotonic for y_i over the subinterval $[1; \sqrt{2})$. In particular, it can be shown that the output step size, for a one-ulp reciprocal for y_i over $[1; \sqrt{2})$, can vary from zero to three ulps, and over $[\sqrt{2}; 2)$ the step size can be down by as much as two ulps, or reverse direction and be up by one ulp, contradicting monotonicity.

Figure 5.8.2(a) illustrates a five-bit, one-ulp reciprocal $rf_5(1/y_5)$ which systematically chooses the value of the pair $\{\text{RD}_5(1/y_5), \text{RU}_5(1/y_5)\}$ that is at least one half-ulp away from $1/y_5$, i.e., where $\frac{1}{2}2^{-5} < |1/y_5 - rf_5(1/y)| < 2^{-5}$ for $y_5 \in (1; 2)$. The step function graph in Figure 5.8.2(a) clearly illustrates that such a perverse, one-ulp reciprocal can have exaggerated variability in step size over $[1; \sqrt{2})$ and be non-monotonic to the extent of virtual oscillation over $[\sqrt{2}; 2)$.

Note that computing $\text{recip}_a(y) = 0.1a_2a_3 \cdots a_ia_{i+1} \cdots$ satisfying $|1/y - \text{recip}_a(y)| < \frac{1}{4}2^{-i}$ results in $\text{recip}(y_i) = \text{RN}_i(\text{recip}_a(y))$ being a $\frac{3}{4}$-ulp reciprocal. Figure 5.8.2(b) illustrates for $i = 5$ a five-bit $\frac{3}{4}$-ulp reciprocal function $\text{recip}(y_5)$

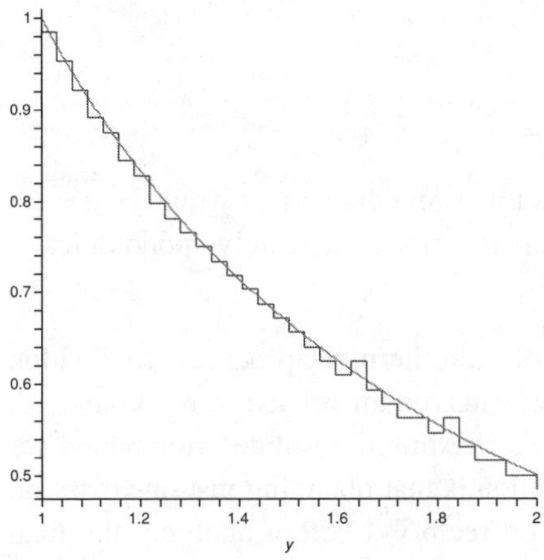

(a)

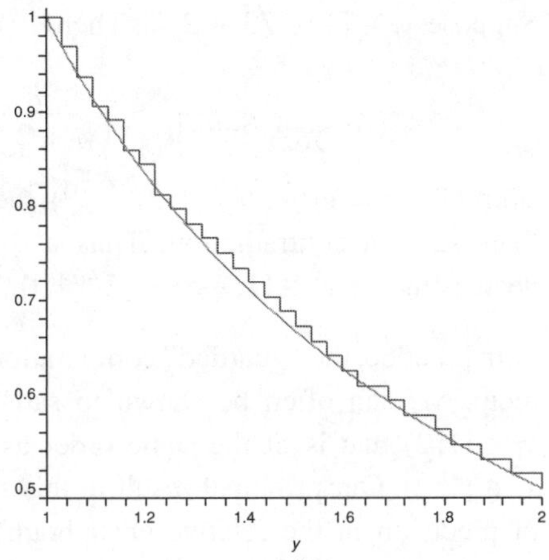

(b)

Figure 5.8.2. (a) A five-bit non-monotonic "round-farthest," one-ulp approximation, and (b) a $\frac{3}{4}$-ulp monotonic approximation.

that chooses the farthest away of $\{\text{RD}_5(1/y_5), \text{RU}_i(1/y_5)\}$ whenever the farthest yields $|1/y_5 - \text{recip}(y_5)| < \frac{3}{4}2^{-5}$, and otherwise chooses the unique one satisfying the $\frac{3}{4}$-ulp bound.

Lemma 5.8.15 *For a normalized i-bit divisor* $y_i = 1.b_1b_2\cdots b_{i-1}$, *an i-bit* $\frac{3}{4}$-ulp *reciprocal function is monotonic over the interval* $[1;2)$, *and strictly monotonic over the portion* $[1; \frac{2}{3}\sqrt{3} - 2^{-(i+1)})$.

Proof For consecutive exact inputs $y_i, y_i + 2^{-(i-1)} \in [1;2]$ the consecutive reciprocals decrease by

$$\frac{1}{y_i} - \frac{1}{y_i + 2^{-(i-1)}} = \frac{1}{y_i(y_i + 2^{-(i-1)})}2^{-(i-1)} > \frac{1}{2}2^{-i}.$$

Thus exact outputs decrease by at least one-half ulp of output. Suppose

$$\text{RD}_i\left(\frac{1}{y_i}\right) < \text{RU}_i\left(\frac{1}{y_i + 2^{-(i-1)}}\right).$$

Then the sum of the rounding errors satisfies

$$\left(\frac{1}{y_i} - \text{RD}_i\left(\frac{1}{y_i}\right)\right) + \left(\text{RU}_i\left(\frac{1}{y_i + 2^{-(i-1)}}\right) - \frac{1}{y_i + 2^{-(i-1)}}\right)$$
$$= \left(\frac{1}{y_i} - \frac{1}{y_i + 2^{-(i-1)}}\right) + \left(\text{RU}_i\left(\frac{1}{y_i + 2^{-(i-1)}}\right) - \text{RD}_i\left(\frac{1}{y_i}\right)\right) > \frac{3}{2}2^{-i}.$$

Hence at least one of the rounding errors would be greater than $\frac{3}{4}$ulp, a contradiction. Thus $\text{RD}_i(1/y_i) \geq \text{RU}_i(1/(y_i + 2)^{-(i-1)})$ holds for all $y_i \in [1;2)$, and a $\frac{3}{4}$-ulp reciprocal function is monotonic (i.e., monotonically non increasing) over $[1;2)$.

Suppose $y_i \in [1; \frac{2}{3}\sqrt{3} - 2^{-i})$. Then

$$\frac{1}{y_i} - \frac{1}{y_i + 2^{-(i-1)}} > \frac{1}{(y_i + 2^{-i})^2} 2^{-(i-1)} > \frac{3}{4} 2^{-(i-1)} = \frac{3}{2} 2^{-i}.$$

If $\text{RD}_i(1/y_i) = \text{RU}_i(1/(y_i + 2)^{-(i-1)})$, the sum of rounding errors would be greater than $\frac{3}{2}2^{-i}$, a contradiction. Thus a $\frac{3}{4}$-ulp reciprocal is strictly monotonically decreasing for $y_i \in [1; \frac{2}{3}\sqrt{3} - 2^{-(i+1)})$. □

In practice, the "guarded" computation of a multiterm reciprocal approximation $\text{recip}_a(y_i)$ can often be shown to satisfy a maximum relative error bound for $y_i \in [1; 2)$ that is of the same order as the maximum absolute error bound for $y_i \in [1; 2)$. Our principal result in this section is that obtaining just one extra bit of precision in the relative error bound on $\text{recip}_a(y)$ before applying the final rounding is sufficient to yield monotonicity.

Theorem 5.8.16 (Monotonicity Theorem) *For a normalized i-bit divisor $y_i = 1.b_1 b_2 \cdots b_{i-1}$, let $\text{recip}_a(y_i) = 0.1a_2a_3 \cdots a_i a_{i+1} \cdots$ be a reciprocal approximation with relative error strictly less than $\frac{1}{2}2^{-i}$ for all $y_i \in [1; 2 - 2^{-(i-1)}]$. Then $\text{recip}(y_i) = \text{RN}_i(\text{recip}_a(y_i))$ is a monotonic, one-ulp reciprocal function over $[1; 2 - 2^{-(i-1)}]$.*

Proof A reciprocal approximation $\text{recip}_a(y)$ with relative error strictly less than $\frac{1}{2}2^{-i}$ satisfies $|(1/y_i) - \text{recip}_a(y_i)| < (1/2y_i)2^{-i}$. So then $\text{RN}_i(\text{recip}_a(y_i))$ is a one-ulp reciprocal for $y_i \in [1; 2 - 2^{-(i-1)}]$ satisfying $|(1/y_i) - \text{RN}_i(\text{recip}_a(y_i))| < (\frac{1}{2} + (1/2y_i))2^{-i}$. (Note that this bound scales down towards $\frac{3}{4}$-ulp as $y \to 2$). For successive i-bit normalized divisors y_i, $y_i + 2^{-(i-1)} \in [1; 2]$, the difference of their reciprocals satisfies

$$\frac{1}{y_i} - \frac{1}{y_i + 2^{-(i-1)}} = \frac{2}{y_i(y_i + 2^{-(i-1)})} 2^{-i} \geq \frac{1}{y_i} 2^{-i}.$$

Assume that

$$\text{RN}_i(\text{recip}_a(y_i)) = \text{RD}_i\left(\frac{1}{y_i}\right) < \text{RU}_i\left(\frac{1}{y_i + 2^{-(i-1)}}\right) = \text{RN}_i(\text{recip}_a(y_i + 2^{-(i-1)})).$$

Then the sum of these successive reciprocal rounding errors satisfies

$$\left(\frac{1}{y_i} - \text{RN}_i(\text{recip}_a(y_i)\right) + \left(\text{RN}_i(\text{recip}_a(y_i + 2^{-(i-1)})) - \frac{1}{y_i + 2^{-(i-1)}}\right)$$
$$\geq \left(\frac{1}{y_i} - \frac{1}{y_i + 2^{-(i-1)}}\right) + 2^{-i} \geq \left(1 + \frac{1}{y_i}\right) 2^{-i}.$$

So at least one of these rounding errors is greater than or equal to $(\frac{1}{2} + (1/2y_i))2^{-i}$, a contradiction. Thus $\text{RN}_i(\text{recip}_a(y_i)) \geq \text{RN}_i(\text{recip}_a(y_i + 2^{-(i-1)}))$, and $\text{recip}(y_i) = \text{RN}_i(\text{recip}_a(y_i))$ is monotonic for $y_i \in [1; 2 - 2^{-(i-1)}]$. □

5.8.3 Bipartite tables

The bipartite table look-up process for determining an approximate reciprocal of a normalized binary divisor $y = 1.b_1b_2 \cdots b_{p-1}$, comprises the use of two distinct binary direct look-up tables with table indices of comparable size. These tables are concurrently addressed by distinct substrings of divisor bits, with each table fashioned to provide a distinct part of a carry-save or borrow-save representation of the approximate reciprocal. Specifically, our bipartite reciprocal approximations are of the form

$$\text{recip}(y) = \text{LUT}_1(y_{2k}) + \text{LUT}_2(y)2^{-i}.$$

With $y_{2k} = 1.b_1b_2 \cdots b_{2k}$, the primary approximation $\text{LUT}_1(y_{2k}) \in [\frac{1}{2}; 1]$ is determined by the $2k$ bit index $b_1b_2 \cdots b_{2k}$. The secondary approximation term is determined by some leading bits and some supplementary trailing bits $b_{2k+1}b_{2k+2} \cdots$. The approximation may be fashioned so that $\text{LUT}_2(y)$ is exclusively positive or negative, with magnitudes less than a unit, or sign-symmetric with magnitude less that half a unit. Partitioning the operand y into three equal k-bit parts, Figure 5.8.3 illustrates the look-up process.

The compelling advantage of the two-table bipartite look-up process is that it provides a simple procedure to achieve essentially $1\frac{1}{2}$ times the precision, at a cost of less than twice the table size, compared to a single direct look-up table. For use

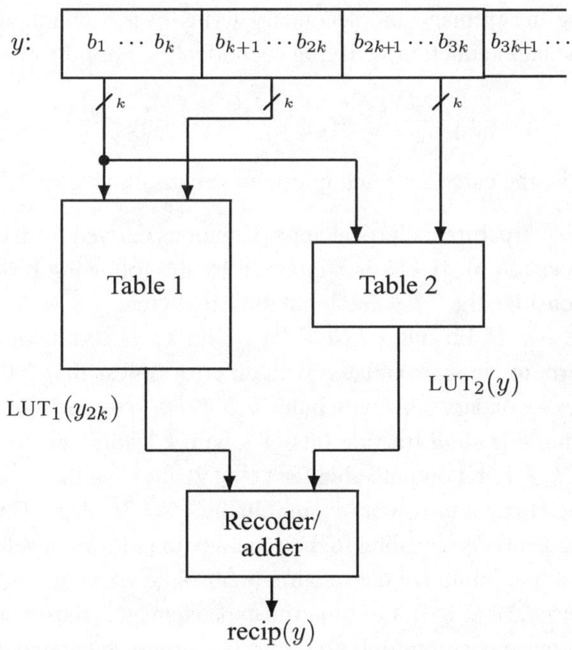

Figure 5.8.3. The bipartite table look-up method.

as a seed or short reciprocal in application to division algorithms the redundant reciprocal approximation may be sent directly to an appropriate multiplier recoder. For reciprocal function output in standard binary form, the two outputs require a supplementary addition.

Determination of the entries in a bipartite look-up table pair is guided by the following exact expansions particular to the reciprocal function:

Theorem 5.8.17 (Bipartite reciprocal identities) *For the normalized binary divisor partition* $y = y_i + f2^{-i}$ *where*

$$y_i = 1.b_1 b_2 \cdots b_i,$$
$$f = 0.b_{i+1} b_{i+2} \cdots b_{p-1},$$

the reciprocal $1/y \in \left(\frac{1}{2}; 1\right]$ *can be expanded to the sum of a primary term, determined by* y_i*, and a secondary term of magnitude less than* 2^{-i}*, according to any of the following*

$$\frac{1}{y} = \frac{1}{y_i} - \frac{f}{yy_i}2^{-i} \quad \textit{(borrow-save expansion)} \tag{5.8.3}$$

$$\frac{1}{y} = \frac{1}{y_i + 2^{-i}} + \frac{1-f}{y(y_i + 2^{-i})}2^{-i} \quad \textit{(carry-save expansion)} \tag{5.8.4}$$

$$\frac{1}{y} = \frac{1}{y_i + 2^{-(i+1)}} + \frac{\frac{1}{2} - f}{y(y_i + 2^{-(i+1)})}2^{-i} \quad \textit{(midpoint expansion).} \tag{5.8.5}$$

Proof Putting the primary and secondary terms over a common denominator yields an immediate reduction verifying each identity. For borrow save

$$\frac{y}{yy_i} - \frac{f}{yy_i}2^{-i} = \frac{y - f2^{-i}}{yy_i} = \frac{y_i}{yy_i} = \frac{1}{y},$$

and similarly for the carry-save and midpoint expansion identities. □

The claim that bipartite reciprocal approximations derived from (5.8.3)–(5.8.5) can have a precision of $\frac{3}{2}i$ bits is supported by the following observations. Let $i = 2k$ and consider the $\frac{3}{2}i = 3k$ input bits in Figure 5.8.3. A primary table employing the $i = 2k$ bit index $b_1 b_2 \cdots b_{2k}$ with $\frac{3}{2}i = 3k$ bits of output allows the primary term to be approximated with an error of less than half a unit in the $3k + 1$ place. A secondary table with index $b_1 b_2 \cdots b_k \circ b_{2k+1} b_{2k+2} \cdots b_{3k}$ formed by concatenating k leading fraction bits of y with k leading bits of f. This table can provide a $(k + 1)$-bit output value for $\text{LUT}_2(y)$ allowing the secondary term to be approximated to near the order of a unit in the $\frac{3}{2}i = 3k$ place. These arguments will now be made precise, leading to the specification of formulas for direct look-up table entries that minimize the maximum absolute errors in each of the terms in expansions (5.8.3)–(5.8.5). For bipartite expansions it is most convenient to fix a common last place position for both terms and minimize the maximum absolute error contributed by each term.

Note that the primary terms in each of the identities have exact inputs. Their evaluation can provide entries to i-bits-in, j-bits-out direct look-up tables with an absolute error for each entry bounded by $2^{-(j+2)}$, due only to rounding the exact output to the output table size. For example, for the borrow-save expansion with $i = 2k$, let the output size be $3k + g$ bits, where $g \geq 0$ is a small number of guard bits. Excluding the special case $y_i = 1$, the primary term approximation in the $2k$-bits-in, $(3k + g)$-bits-out table is

$$\text{LUT}_1(y_{2k}) = \text{RN}_{3k+g+1}\left(\frac{1}{y_{2k}}\right) = 0.1a_2a_3\cdots a_{3k+g+1}. \qquad (5.8.6)$$

The primary table size for $\text{LUT}_1(y_{2k})$ is $2^{2k}(3k + g)$ bits with maximum absolute error $2^{-(3k+g+2)}$.

The secondary term approximation $\text{LUT}_2(y) \approx -f/yy_{2k}$ for the borrow-save reciprocal expansion (5.8.3) with $i = 2k$ will be determined from the k leading fraction bits of y, where $y_k = 1.b_1b_2\cdots b_k$ along with the k leading bits of f with $f_k = .b_{2k+1}b_{2k+2}\cdots b_{3k}$. Thus we have $\text{LUT}_2(y) = \text{LUT}_2(y_k, f_k)$. In terms of the arguments (y_k, f_k) we note the following bounds on each of the factors of the secondary term f/yy_{2k}.

$$\begin{aligned} f_k \leq f &< f_k + 2^{-k}, \\ y_k \leq y_{2k} &< y_k + 2^{-k} - 2^{-2k}, \\ y_k + f_k2^{-2k} \leq y &< y_k + 2^{-k} - 2^{-2k} + f_k2^{-2k} + 2^{-3k}. \end{aligned}$$

Lemma 5.8.18 *For the divisor $y = 1.b_1b_2\cdots$ with $2k \geq 4$, let*

$$\begin{aligned} y_k &= 1.b_1b_2\cdots b_k, \\ y_{2k} &= 1.b_1b_2\cdots b_{2k}, \\ f &= .b_{2k+1}b_{2k+2}\cdots, \\ f_k &= .b_{2k+1}b_{2k+2}\cdots b_{3k}. \end{aligned}$$

Then the borrow-save expansion secondary term satisfies the following bounds where f/yy_{2k} can be arbitrarily close to either bound:

$$\frac{f_k}{(y_k + 2^{-k})(y_k + 2^{-k} - 2^{-2k} + f_k2^{-2k} + 2^{-3k})} < \frac{f}{yy_{2k}} < \frac{f_k + 2^{-k}}{y_k(y_k + f_k2^{-k})}. \qquad (5.8.7)$$

Let $m(y_k, f_k)$ be the midpoint of the interval determined by (5.8.7). The value $\text{m}(y_k, f_k)$ minimizes the maximum absolute error for approximation of f/yy_{2k} in each of the separate regions determined by each index $b_1b_2\cdots b_k \circ b_{2k+1}b_{2k+2}\cdots b_k$. The maximum error over all the regions will occur for the argument pair $y_k = 1,\ f_k = 0$ with index $0^{[k]}$ leading to the following result.

Corollary 5.8.19 *Let $m(y_k, f_k)$ be the midpoint of the interval determined by (5.8.7). Then*

$$\left|\frac{f}{yy_{2k}} - m(y_k, f_k)\right| < \tfrac{3}{2}2^{-3k}.$$

The secondary term in our bipartite borrow-save approximation is then determined by rounding $m(y_k, f_k)$ to the last place position $k + g + 1$:

$$\text{LUT}_2(y_k, f_k) = -\text{RN}_{k+g+1}(m(y_k, f_k)). \tag{5.8.8}$$

Including the two rounding errors we further obtain the following from Corollary 5.8.19.

Corollary 5.8.20 *The borrow-save bipartite reciprocal approximation for the normalized divisor $y = 1.b_1b_2\cdots$ is given by*

$$\text{recip}(y) = \text{RN}_{3k+g+1}\left(\frac{1}{y_{2k}}\right) - \text{RN}_{k+g+1}(\text{m}(y_k, f_k))2^{-2k} \tag{5.8.9}$$

and satisfies the absolute error bound

$$\left|\frac{1}{y} - \text{recip}(y)\right| < \left(\frac{3}{2} + \frac{1}{2^{g+1}}\right)2^{-3k}.$$

For $g = 0$ the maximum error is then $2^{-(3k-1)}$ with total table size $2^{2k}(4k + 1)$. With just a few guard bits we approach the case for $g \to \infty$, where $\text{recip}(y) = 1/y_{2k} - \text{m}(y_k, f_k)2^{-2k}$ with $|1/y - \text{recip}(y)| < \frac{3}{4}2^{-(3k-1)}$. It can be shown that the maximum relative error for such a bipartite table occurs for $y \to 1$, so the precision is at least $(3k - 1)$-bits.

In practice, bipartite tables are found most effective for total index lengths $9 \leq 3k \leq 18$, where each part has between three and six bits. Considering variable sized parts the preferred partitions of index parts are $k|k|k$, $(k + 1)|k|k$, and $(k + 1)|k|(k + 1)$.

For the carry-save reciprocal identity

$$\frac{1}{y} = \frac{1}{y_i + 2^{-i}} + \frac{1 - f}{y(y_i + 2^{-i})}2^{-i}$$

we obtain

$$\text{recip}(y) = \text{RN}_{3k+g+1}\left(\frac{1}{y_i+2^{-i}}\right) + \text{RN}_{k+g+1}(\text{m}'(y_k, f_k))2^{-2k}. \tag{5.8.10}$$

In (5.8.10) the term $\text{m}'(y_k, f_k)$ denotes the midpoint of the tightly bounded range of values for $(1 - f)/y(y_i + 2^{-i})$, which can be determined in the same manner as for the borrow-save secondary term in Lemma 5.8.18. Note that (5.8.9) and (5.8.10) provide the respective borrow-save and carry-save redundant representations directly from the two tables.

Exploiting symmetry in bipartite tables. The midpoint expansion (5.8.5) allows for the design of a sign symmetric, bipartite table process providing one additional bit of accuracy. For the symmetric case a low-order part of the input bits and the secondary term approximation are subject to conditional complementation.

For the bipartite midpoint identity with $i = 2k$

$$\frac{1}{y} = \frac{1}{y_{2k} + 2^{-(2k+1)}} + \frac{1-2f}{y(y_{2k} + 2^{-(2k+1)})} 2^{-(2k+1)}.$$

Then let $(1 - 2f) = (-1)^{b_{2k+1}} f^*$, where

$$f^* = .b^*_{2k+2}b^*_{2k+3}\cdots = \begin{cases} .\overline{b}_{2k+2}\overline{b}_{2k+3}\cdots & \text{for } b_{2k+1} = 0, \\ .b_{2k+2}b_{2k+3}\cdots & \text{for } b_{2k+1} = 1 \end{cases}$$

with $\overline{b}_j$ the complement of bit b_j. Then with $f^*_k = .b^*_{2k+2}b^*_{2k+3}\cdots b^*_{3k+1}$ determined by the leading k bits of f^*, we obtain

$$\frac{f^*_k}{(y_k + 2^{-k})^2} < \frac{f^*}{y(y_{2k} + 2^{-(2k+1)})} < \frac{f^*_k + 2^{-k}}{y_k^2}. \tag{5.8.11}$$

Let $\mathrm{m}^*(y_k, f^*_k)$ be the midpoint of the interval bounding $f^*/y(y_{2k} + 2^{-(2k+1)})$ in (5.8.11). Then the symmetric bipartite expansion for the normalized divisor $y = 1.b_1b_2\cdots$ is

$$\begin{aligned}\mathrm{recip}(y) = {} & \mathrm{RN}_{3k+g+2}\left(\frac{1}{y_{2k} + 2^{-(2k+1)}}\right) \\ & +(-1)^{b_{2k+1}}(\mathrm{RD}_{k+g+1}(\mathrm{m}^*(y_k, f^*_k)) + 2^{-(k+g+2)})2^{-(2k+1)} \end{aligned} \tag{5.8.12}$$

The secondary term can be determined by the $2k$-bit index $b_1b_2\cdots b_k \circ b^*_{2k+2}b^*_{2k+3}\cdots b^*_{3k+1}$ employing only a conditional 1's complement to the lower input bits $b_{2k+2}b_{2k+3}\cdots b_{3k+1}$. The output multiplication by $(-1)^{b_{2k+1}}$ is also computable as a conditional 1's complement, since we employed the secondary term rounding to the nearest odd $(k + g + 2)$-bit value $\mathrm{RD}_{k+g+1}(\cdot) + 2^{-k+g+2} = 0.a_1a_2\cdots a_{k+g+1}1$.

In practical applications, the reciprocal approximation (5.8.12) is appropriate for generating seed and short reciprocals for normalized binary inputs $y_i = 1.b_1b_2\cdots b_{p-1}$, where $p \gg 3k$. In this case f^* practically fills the interval $[f^*_k; f^*_k + 2^{-k})$, justifying $\mathrm{m}^*(y_{2k}, f^*_k)$ being the midpoint of the interval determined by (5.8.11) with f^* obtained as noted using a conditional 1's complement.

When the approximate reciprocal type is a reciprocal function defined on "exact" input points, the midpoint expansion (5.8.5) can be modified to yield a symmetric secondary term.

Symmetric bipartite reciprocal functions. For the normalized p-bit divisor $y = 1.b_1b_2\cdots b_{p-1}$, the secondary part of the partition $y = y_i + f2^{-i}$ has $f = 0.b_{i+1}b_{i+2}\cdots b_{p-1}$ with $f \in [0; 1 - 2^{-(p-i-1)}]$. The secondary part can be centered by subtracting $(\frac{1}{2} - 2^{-(p-i)})$ and adding the same to the primary part. The *symmetric divisor partition* for the p-bit normalized divisor $y = 1.b_1b_2\cdots b_{p-1}$ is then

$$y = y_i - 2^{-p} + 2^{-(i+1)} + (f + 2^{-(p-i)} - \tfrac{1}{2})2^{-i}. \qquad (5.8.13)$$

From (5.8.13) for any precision p the *symmetric bipartite identity* for $1/y$ is

$$\frac{1}{y} = \frac{1}{y_i + 2^{-(i+1)} - 2^{-p}} + \frac{1 - 2(f + 2^{-(p-i)})}{y(y_i + 2^{-(i+1)} - 2^{-p})}2^{-(i+1)}. \qquad (5.8.14)$$

Then the *symmetric bipartite reciprocal function* for the $(3k+2)$-bit normalized binary divisor $y = 1.b_1b_2\cdots b_{3k-1}$ is determined from (5.8.14) with $i = 2k$ and $p = 3k+2$ by

$$\frac{1}{y} = \frac{1}{y_{2k}2^{-(2k+1)} - 2^{-p}} + (-1)^{b_{2k+1}}\frac{f^*}{y(y_{2k} + 2^{-(2k+1)} - 2^{-p})}2^{-(2k+1)}. \qquad (5.8.15)$$

Here f^* is determined so that $1 - 2(f + 2^{-(k+2)}) = (-1)^{b_{2k+1}} f^*$, where

$$f^* = .b^*_{2k+2}b^*_{2k+3}\cdots b^*_{3k+1}1 = \begin{cases} .\bar{b}_{2k+2}\bar{b}_{2k+3}\cdots\bar{b}_{3k+1}1 & \text{for} \quad b_{2k+1} = 0, \\ .b_{2k+2}b_{2k+3}\cdots b_{3k+1}1 & \text{for} \quad b_{2k+1} = 1. \end{cases}$$

This allows the bounds

$$\frac{f^*}{(y_k + 2^{-k})^2} < \frac{f^*}{y(y_{2k} + 2^{-(2k+1)} - 2^{-p})} < \frac{f^*}{y_k^2}, \qquad (5.8.16)$$

where the interval midpoint $\mathrm{m}^*(y_k, f^*)$ from (5.8.16) yields an approximate reciprocal function in the same manner as (5.8.12). The centering of the secondary part in (5.8.13) and (5.8.14) thus provides for a sharper result than (5.8.12) since f^* is now exact, and shares the practical convenience of determining f^* by a 1's complement.

Minimizing the double rounding penalty in bipartite approximation. The bipartite reciprocal identities of Theorem 5.8.17 can be expressed as follows:

$$\frac{1}{y} = \mathrm{LUT}_1(y_i) + \mathrm{LUT}_2(y)2^{-i} + \delta 2^{-\frac{3}{2}i}. \qquad (5.8.17)$$

The input discretization error that occurs from using an $(i+g)$-bit index for the secondary term in (5.8.17) is reflected in the bound on $|\delta|$ in the approximation error $\delta 2^{-\frac{3}{2}i}$. From the nature of the secondary terms in the bipartite identities of Theorem 5.8.17 it is clear that we can obtain $|\delta| < \frac{1}{2}$ or $|\delta| < \frac{1}{4}$ by employing more than i bits in the secondary term index.

Let the normalized divisor $y = 1.b_1b_2\cdots b_{j-1}$ be an exact j-bit divisor with $j = \lfloor\frac{3}{2}i\rfloor + 1$. Then the bipartite reciprocal approximation process with primary term recip(y_i) can be modified to yield a j-bit one-ulp monotonic reciprocal recip(y) $= 0.1a_2a_3 \cdots a_j$ (or recip(y) $= 1$). The process is modified by allowing the secondary term to have $i+1$ or $i+2$ bits in the index and by employing one or two output guard bits in both the primary and secondary term tables that are rounded off in obtaining the final one-ulp monotonic reciprocal recip(y). The process is described implicitly in the following theorem.

Theorem 5.8.21 (Bipartite Monotonicity Theorem) *Let the normalized divisor* $y = 1.b_1b_2 \cdots b_{j-1}$ *be an exact* j*-bit divisor with* $j = \lfloor \frac{3}{2}i \rfloor + 1$*. Let* y *have the divisor partition* $y = y_i + f2^{-i}$*, where* $f = 0.b_{i+1}b_{i-2} \cdots b_{\lfloor \frac{3}{2}i \rfloor}$ *is an exact* $\lfloor \frac{1}{2}i \rfloor$*-bit fraction. Let* $1/y$ *satisfy the identity*

$$\frac{1}{y} = \text{LUT}_1(y_i) + \text{LUT}_2(y)2^{-i} + \frac{\delta}{y^2}2^{-(j+1)} \qquad (5.8.18)$$

with $|\delta| < 1/4y^2$*, then the reciprocal approximation*

$$\text{recip}(y) = \text{RN}_j(\text{RN}_{j+2}(\text{LUT}_1(y_i)) + \text{RD}_{j+2}(\text{LUT}_2(y)2^{-i}))$$

is a j*-bit one-ulp monotonic reciprocal function for* $y \in [1;2)$.

A key step in proving the theorem is provided by the following observation.

Observation 5.8.22 *Let* $u, v \geq 0$, $0 < u + v \leq 1$ *be binary numbers. Then*

$$u + v - \text{RN}_j(\text{RN}_{j+2}(u) + \text{RD}_{j+2}(v)) \leq \tfrac{5}{8}2^{-j}.$$

Multipartite tables. The bipartite table look-up process for determining an approximate reciprocal can be expanded to a tripartite or multipartite process. The result is then the sum of three or more terms obtained from three of more table look-ups indexed by comparably sized indices. Tripartite tables, in principle, should achieve about $1\frac{3}{4}$ times the precision and cost about 3 times the table size as a single direct lookup table.

In practice, multipartite tables are arguably most effective for tables with total input index lengths and resulting output approximation precisions in the range 15–24 bits. This range can be covered employing three or four term sums with primary table indices bounded by 11 bits. These practical bounds keep the total table size moderate. They also allow table look-up and subsequent addition time to be kept small.

For practical table indices of size at most 11 bits, the marginal improvement in tripartite and quadripartite table approximations for each additional part is only two or three bits per part. For these index ranges the multipartite process is conveniently visualized by recognizing the divisor partition as a partial recoding operation.

Exploiting recoding in multipartite tables.

Definition 5.8.23 *For $f = 0.b_1b_2\cdots$ and $k \geq 1$, a k-digit* partial recoding *(Booth radix 4) of $(f - \frac{1}{2})$ denotes the expansion*

$$f - \frac{1}{2} = d_1 2^{-2} + d_2 2^{-4} + d_3 2^{-6} + \cdots + d_k^{-2k} + t2^{-(2k+1)} \tag{5.8.19}$$

with the tail *t satisfying $t \in [-1; 1)$ and $d_i \in \{-2, -1, 0, 1, 2\}$ for $1 \leq i \leq k$. Furthermore, for the $(j+1)$-bit fraction $f = 0.b_1b_2\cdots b_j1$ with $f \in [2^{-(j+1)}; 1 - 2^{-(j+1)}]$ and $j \geq 2k+3$, the tail t in (5.8.19) satisfies $t \in [-(1 - 2^{-(j-2k)}); (1 - 2^{-(j-2k)})]$.*

Note that the condition on the range of the tail $t \in [-1; 1)$ for $f \in [0; 1)$ and $t \in [-(1 - 2^{-(j-2k)}); (1 - 2^{-(j-2k)})]$ for $f \in [2^{-(j+1)}; 1 - 2^{-(j+1)}]$ makes the expansion (5.8.19) unique. In practice the digits d_i are determined from the bit triples $b_{2i-1}b_{2i}b_{2i+1}$ concurrently as in standard Booth recodings. The tail $t = .b^*_{2k+2}b^*_{2k+3}\cdots$ (or $t = .b^*_{2k+2}b^*_{2k+3}\cdots b^*_j1$) is determined from conditionally complementing the bits $b_{2k+2}b_{2k+3}\cdots$ depending on bit b_{2k+1} as described for symmetric bipartite expansions. The notion of partial recoding is extendable to Booth radix-8 recodings in the obvious way.

The divisor (input) partition for the bipartite midpoint expansion (5.8.5) is $y = (y_i + 2^{-i+1}) + (f - \frac{1}{2})2^{-i}$. This provides the basis for multipartite (output) expansions by partial recoding of the secondary term of (5.8.5).

Observation 5.8.24 *Let the normalized binary divisor $y \in [1; 2)$ have the recoded tripartite partition $y = (y_i + 2^{-(i+1)}) + d2^{-(i+2)} + t2^{-(i+3)}$ with $y_i = 1.b_1b_2\cdots b_i$, $d \in \{-2, -1, 0, 1, 2\}$, and $-1 \leq t < 1$. Then*

$$\frac{1}{y} = \frac{1}{y_i + 2^{-(i+1)}} - \frac{d}{y(y_i + 2^{-(i+1)})}2^{-(i+2)} - \frac{t}{y(y_i + 2^{-(i+1)})}2^{-(i+3)}. \tag{5.8.20}$$

Proof The result is obtained by substituting the partial recoding $(f - \frac{1}{2}) = d2^{-2} + t2^{-3}$ into the bipartite midpoint expansion (5.8.5). □

From (5.8.20) with $i = 2k$ we obtain a recoded tripartite expansion for use as a seed or short reciprocal:

$$\begin{aligned}\text{recip}(y) = {} & \text{RN}_{3k+g+2}\left(\frac{1}{y_{2k} + 2^{-(2k+1)}}\right) \\ & - d\left(\text{RD}_{k+g+1}\left(\frac{1}{(y_{2k} + 2^{-(2k+1)})^2}\right) + 2^{-(k+g+2)}\right)2^{-(2k+2)} \\ & + (-1)^{b_{2k+3}}\text{LUT}_3(b_1b_2\cdots b_k \circ b^*_{2k+4}b^*_{2k+5}\cdots b^*_{3k+3})2^{-(2k+3)} \end{aligned} \tag{5.8.21}$$

The $2k$-bit index $b_1b_2\cdots b_{2k}$ can be used to retrieve both

a $(3k+2+g)$-bit output for $\text{RN}_{3k+2+g}(\frac{1}{y_{2k}+2^{-(2k+1)}})$ and
a $(k+1+g)$-bit output for $\text{RN}_{k+g+1}(\frac{1}{(y_i+2^{-(i+1)})^2})$.

The second output can be conditionally complemented and/or shifted to determine $\mathrm{LUT}_2(y)$ as an approximation of $-(d/y(y_i+2^{-(i+1)}))$ satisfying

$$\left|\mathrm{LUT}_2(y) - \left(-\frac{d}{y(y_i+2^{-(i+1)})}\right)\right| 2^{-(2k+2)} < 2^{-(4k+2)} + 2^{-(3k+3+g)}.$$

The approximation for the *terminal term* $\mathrm{LUT}_3(y) \approx -t/y(y_i+2^{-(i+1)})$ is handled as for the secondary term in the symmetric bipartite expansion employing the $2k$ bit string $b_1b_2\cdots b_k \circ b^*_{2k+4}b^*_{2k+5}\cdots b^*_{3k+3}$ as the index to a separate terminal term table. The recoded tripartite expansion here employs an intermediate Booth 4-digit in the tripartite divisor partial recoding to obtain a two-bit enhancement of the precision of the result compared to a symmetric bipartite reciprocal approximation. A recoded four-part divisor partition including two intermediate Booth 8-digits is described in the problems.

5.8.4 Linear and quadratic interpolation

The three bipartite reciprocal identities of Theorem 5.8.17 each have the reciprocal $1/y$ as a factor in the secondary terms. This recursive property of the identities (5.8.3)–(5.8.5) allows any one of them to be substituted in the secondary term of another. Three such expansions are of particular interest for investigating linear interpolation with a closed form residual term.

Lemma 5.8.25 (Linear reciprocal interpolation identities) *Let the normalized p-bit binary divisor have the partition $y = y_i + f2^{-i}$, where $y_i = 1.b_1b_2\cdots b_i$ and $f = 0.b_{i+1}b_{i+2}\cdots$. Then the reciprocal $1/y \in (\frac{1}{2};1]$ can be expanded to a piecewise linear function with both coefficients determined by y_i, according to any of the following:*

$$\frac{1}{y} = \frac{1}{y_i} - \frac{f}{y_i(y_i+2^{-i})}2^{-i} - \frac{1}{y}\frac{4f(1-f)}{y_i(y_i+2^{-i})}2^{-2(i+1)}; \qquad (5.8.22)$$

$$\frac{1}{y} = \frac{1}{y_i+2^{-i}} + \frac{1-f}{y_i(y_i+2^{-i})}2^{-i} - \frac{1}{y}\frac{4f(1-f)}{y_i(y_i+2^{-i})}2^{-2(i+1)}; \qquad (5.8.23)$$

$$\frac{1}{y} = \frac{1}{y_i+2^{-(i+1)}} - \frac{2f-1}{(y_i+2^{-(i+1)})^2}2^{-(i+1)} + \frac{1}{y}\frac{(2f-1)^2}{(y_i+2^{-(i+1)})^2}2^{-2(i+1)}. \qquad (5.8.24)$$

Proof From Theorem 5.8.17, substituting (5.8.4) for $1/y$ into (5.8.3) yields the identity (5.8.22). Alternatively, substituting (5.8.3) into (5.8.4) yields (5.8.23), and substituting (5.8.5) into itself yields (5.8.24). □

Identity (5.8.22) interpolates over a one-ulp interval down from the base-point $1/y_i$, and (5.8.23) similarly interpolates up from $1/(y_i+2^{-i})$ with the same slope as (5.8.22). Identity (5.8.24) interpolates out from the midpoint by one half-ulp in each direction. The slope in (5.8.24) differs from the prior slopes only in a lower order term since

$$\frac{1}{(y_i + 2^{-(i+1)})^2} = \frac{1}{y_i(y_i + 2^{-i}) + 2^{-2(i+1)}}.$$

The residuals are closely related since the denominators are equal (to a smaller-order term) and the numerators are complementary, i.e., $(2f - 1)^2 = 1 - (4f(1 - f))$.

The constant-term coefficients can be slightly adjusted to center the residual terms in each of the identities of Lemma 5.8.25. Thus the piecewise linear functions determined by each of the identities will have for the interval $y_i \le y \le y_i + 2^{-i}$ a maximum absolute error bound of order $(1/(y_i + 2^{-(i+1)})^3)2^{-(2i+3)}$ and maximum relative error bound of $(1/(y_i + 2^{-i+1})^2)2^{-(2i+3)}$. Considering the dependence of the error on f, it follows that the overall maximum absolute and relative error bounds are realized at the endpoints and midpoint of the first interval where $1 \le y < 1 + 2^{-i}$.

Lemma 5.8.26 *A piecewise linear function* recip(y) *can be determined by each of the identities (5.8.22)–(5.8.24) using an i-bits-in, $2(i + 1)$-bits-out table for the constant-term coefficient and an i-bits-in, $(i + 3)$-bits-out table for the linear-term coefficient. Each approximation satisfies*

- $|1/y - \text{recip}(y)| < 2^{-2(i+1)}$;
- recip(y) *has precision at least* $2i + 1$ *bits;*
- recip(y) *employs tables of size totalling* $2^i(3i + 5)$ *bits.*

Proof For (5.8.22) the increment $(1/y_i(y_i + 2^{-i})(y_i + 2^{-(i+1)}))2^{-(2i+3)}$ may be added to the constant term to center the residual term and bound the residual term absolute error by $\frac{1}{2}2^{-2(i+1)}$. The constant term value

$$\text{RN}_{2i+3}\left(\frac{1}{y_i} + \frac{1}{y_i(y_i + 2^{-i})(y_i + 2^{-(i+1)})}\right) = 0.1a_2a_3 \cdots a_{2i+3}$$

can be provided by an i-bits-in, $2(i + 1)$-bits-out look-up table contributing absolute error at most $\frac{1}{4}2^{-2(i+1)}$. The linear term coefficient $\text{RN}_{i+3}(1/y_i(y_i + 2^{-i})) = 0.a'_1a'_2 \cdots a'_{i+3}$ can be provided by an i-bits-in, $(i + 3)$-bits-out table contributing absolute error in the linear term of at most $\frac{1}{4}2^{-2(i+1)}$. Thus the piecewise linear reciprocal approximation

$$\text{recip}(y) = \text{RN}_{2i+3}\left(\frac{1}{y_i} + \frac{1}{y_i(y_i + 2^{-i})(y_i + 2^{-(i+1)})}\right) - \text{RN}_{i+3}\left(\frac{1}{y_i(y_i + 2^{-i})}\right) f2^i$$

satisfies $|1/y - \text{recip}(y)| < 2^{-2(i+1)}$, and the conditions of the lemma are satisfied. The arguments for (5.8.22) and (5.8.24) are similar, where the linear coefficient table for (5.8.24) can have one fewer output bit. □

Exploiting linear interpolation reformulation. The piecewise linear reciprocal approximations determined by the constant and linear terms of identities (5.8.22) and (5.8.24) may be reformulated to reduce the required total table size. Our first reformulation is obtained by placing the constant and linear terms in (5.8.22) over a common denominator.

Lemma 5.8.27 (Multiplicative reciprocal identity) *For the normalized binary divisor* $y = y_i + f2^{-i}$

$$\frac{1}{y} = \frac{y_i + (1-f)2^{-i}}{y_i(y_i + 2^{-i})} - \frac{4f(1-f)}{yy_i(y_i + 2^{-i})}2^{-2(i+1)}. \tag{5.8.25}$$

The *modified divisor* $\hat{y}(i) = y_i + (1-f)2^{-i} = 1.b_1b_2\cdots b_i\overline{b}_{i+1}\overline{b}_{i+2}\cdots$ in (5.8.25) is readily obtained in an implementation by complementing the low order part f of the input partition. Then

$$\text{recip}(y) = \hat{y}(i)\,\text{RD}_{2i+3}\left(\frac{1}{y_i(y_i + 2^{-i})}\right) \tag{5.8.26}$$

can be obtained using an i-bits-in, $(2i+3)$-bits-out table. The approximation error satisfies $-2^{-2(i+1)} < 1/y - \text{recip}(y) < -2^{-2(i+3)}$. Considering the residual term reduction as y increases from 1 to 2 it follows that the magnitude of the relative error is bounded by $2^{-2(i+1)}$, so the purely multiplicative approximation (5.8.16) has a precision of $2(i+1)$ bits.

Thus the multiplicative reciprocal approximation (5.8.26) achieves a maximum error bound and precision at least as good as the traditional two-table multiply-add interpolations of Lemma 5.8.26. The table size of $2^i(2i+3)$ bits is just $\frac{2}{3}$ the size of traditional implementation. In practice, the major cost of the multiplicative reciprocal approximation is that it requires a $((2i+g)\times(2i+3))$-bit multiplier, about four times the size of the $((i+3)\times(i+g))$-bit multiplier needed for the traditional multiply-add formulation.

A second reformulated linear interpolation is available for reducing table size by merging reciprocal identities (5.8.23) and (5.8.24) so they share a common constant term as the basis of interpolations in different directions.

Theorem 5.8.28 (Bidirectional linear reciprocal interpolation identity) *For the binary divisor* $y = y_i + f2^{-i}$ *with* $y_{i-1} = 1.b_1b_2\cdots b_{i_1}$ *and* $y_i = 1.b_1b_2\cdots b_i$,

$$\frac{1}{y} = \frac{1}{y_{i-1} + 2^{-i}} + \frac{(1 - b_i - f)}{y_i(y_i + 2^{-i})}2^{-i} - \frac{1}{y}\frac{4f(1-f)}{y_i(y_i + 2^{-i})}2^{-2(i+1)}. \tag{5.8.27}$$

Proof For the case $b_i = 0$, we have $y_{i-1} = y_i$ and (5.8.27) is identical to (5.8.24). For $b_i = 1$, (5.8.27) is identical to (5.8.23). □

The identity (5.8.27) provides a fourth reformulation of the same piecewise linear reciprocal approximation

$$\text{recip}(y_i, f) = \frac{1}{y_{i-1} + 2^{-i}} + \frac{(1 - b_i - f)}{y_i(y_i + 2^{-i})} 2^{-i}. \qquad (5.8.28)$$

Bidirectional linear interpolation employing (5.8.27) has its base points located two ulps apart with interpolation both up and down over ulp intervals from each base point. The interpolation direction is implicitly determined by bit b_i, and separate slopes are determined for interpolation over each ulp interval. Bidirectional linear interpolation requires storage of only half as many base points as linear interpolation employing formulation (5.8.12), at the cost of complementing the multiplier input (for $b_i = 0$) or output (for $b_i = 1$) depending on bit b_i.

Observation 5.8.29 (Bidirectional linear interpolation) *For the normalized binary divisor $y = y_i + f2^{-i}$ the binary linear reciprocal interpolation formulation*

$$\begin{aligned}\text{recip}(y) = {} & \mathrm{RN}_{2i+3}\left(\frac{1}{y_{i-1} + 2^{-i}} - \frac{1}{y_{i-1}(y_{i-1} + 2^{-i})(y_{i-1} + 2^{-(i+1)})}\right) \\ & + \mathrm{RN}_{i+3}\left(\frac{1}{y_i(y_i + 2^{-i})}\right)(1 - b_i - f)2^{-i}\end{aligned}$$

using an $(i-1)$-bits-in, $2(i+1)$-bits-out table for the constant term coefficient and an i-bits-in, $i+3$-bits-out table for the linear term coefficient satisfies

- $|(1/y) - \text{recip}(y)| < 2^{-2(i+1)}$;
- $\text{recip}(y)$ *has precision* $2i + 1$ *bits*;
- $\text{recip}(y)$ *employs tables of size totalling* $2^i(2i + 4\frac{1}{2})$ *bits.*

Proof The arguments are the same as for the proof of Lemma 5.8.26, noting here that the primary table size is one half of the one employed there. □

The table size of $2^i(2i + 4\frac{1}{2})$ bits for the bidirectional $\text{recip}(y)$ formulation is again just two-thirds the size of the traditional implementations considered in Lemma 5.8.26, and can be supported by an $((i + 3) \times (i + g))$-bit multiplier as in that case.

Exploiting concurrent multiplication. For implementations where a partitioned multiplier is available to support two concurrent, distinct, relatively small multiplications, reciprocal approximations of precision three times the table index size are feasible. The method is derived from a quadratic reciprocal identity.

Lemma 5.8.30 (Quadratic reciprocal interpolation identity) *Let the normalized binary divisor $y \in [1; 2)$ have the divisor partition $y = y_i + f2^{-i}$. Then the* *reciprocal can be expanded to a piecewise quadratic function with coefficients determined by* y_i*:*

$$\frac{1}{y} = \frac{1}{y_{i-1}+2^{-i}} - \frac{(1-b_i-f)}{y_i(y_i+2^{-i})}2^{-i} - \frac{4f(1-f)}{y_i(y_i+2^{-i})(y_i+2^{-(i+1)})}2^{-2(i+1)}$$
$$- \frac{8f(1-f)(1-2f)}{yy_i(y_i+2^{-i})(y_i+2^{-(i+1)})}2^{-(3i+4)}, \qquad (5.8.29)$$

where the residual coefficient satisfies

$$\left|\frac{8f(1-f)(1-2f)}{yy_i(y_i+2^{-i})(y_i+2^{-i-1})}\right| < |8f(1-f)(1-2f)| < 0.77.$$

Proof The result is obtained by substituting the bipartite midpoint expansion identity (5.8.5) into the bidirectional linear reciprocal identity (5.8.16). The bound on the residual is obtained from taking the derivative of $f(1-f)(1-2f)$ to find the maximum. □

Since the residual is zero at the interval end points, a continuous piecewise quadratic reciprocal approximation is determined from the three leading terms.

Corollary 5.8.31 (Quadratic reciprocal approximation) *Let the normalized binary divisor* $y \in [1;2)$ *have the divisor partitions* $y = y_i + f2^{-i} = y_{i-1} + (b_i + f)2^{-i}$. *Then the piecewise quadratic reciprocal approximation*

$$\text{recip}_a(y) = \frac{1}{y_{i-1}+2^{-i}} - \frac{(1-b_i-f)}{y_i(y_i+2^{-i})}2^{-i} - \frac{4f(1-f)}{y_i(y_i+2^{-i})(y_i+2^{-(i+1)})}2^{-2(i+1)}$$

is continuous with absolute error and relative error over $[1;2)$ *satisfying*

$$\left|\frac{1}{y} - \text{recip}_a(y)\right| \le \left|1 - y\,\text{recip}_a(y)\right| < \max_{0\le f\le 1}\{8f(1-f)(1-2f)\}2^{-(3i+4)}$$
$$< 0.77\, 2^{-(3i+4)}.$$

The identity (5.8.29) has utilized all three of the recursive bipartite identities (5.8.3)–(5.8.5) to obtain a closed approximation with a precision more than one bit greater than that obtained by simply employing (5.8.5) recursively for quadratic expansion about the midpoint $y_i + 2^{-(i+1)}$.

The coefficients of the constant, linear, and quadratic terms in (5.8.29) can all be obtained by table look-up. The argument, $4f(1-f)$, of the quadratic term can also be obtained by table look-up to an accuracy approaching the size of the residual term in (5.8.29). Let the divisor partition be extended to a three-part partition $y = y_i + f2^{-i} = y_i + f_{i+1}2^{-i} + t2^{-(2i+2)}$. Here we use the $(i+1)$-bit midpoint fraction and terminal part

$$f_{i+1} = .b_{i+1}b_{i+2}\cdots b_{2i+1} + 2^{-(2i+2)},$$
$$t = b_{2i+2} \cdot b_{2i+3} \cdots - 1 \tag{5.8.30}$$

with $|t| \leq 1$. Then

$$\begin{aligned} f(1-f) &= (f_{i+1} + t2^{-(i+2)})(1 - f_{i+1} - t2^{-(i+2)}) \\ &= f_{i+1}(1 - f_{i+1}) + (1 - 2f_{i+1})t2^{-(i+2)} - t^2 2^{-(2i+4)}. \end{aligned}$$

The quadratic argument $4f_{i+1}(1 - f_{i+1})$ may be obtained by table look-up. The symmetry in $f_{i+1}(1 - f_{i+1})$ can be exploited using bit b_{i+1} to conditionally complement bits $b_{i+2}b_{i+3}\cdots b_{2i+1}$ employing a 1's complement for the centered f_{i+1}. Thus $\text{LUT}_3(f_{i+1}) = \pm 4f_{i+1}(1 - f_{i+1})$ can be obtained using an i-bit index.

Observation 5.8.32 *Let the normalized binary divisor $y \in [1; 2)$ have the divisor partitions $y = y_i + f2^{-i} = y_i + f_{i-1}2^{-i} + t2^{-(2i+2)}$ specified in (5.8.30). Let*

$$\begin{aligned} \text{recip}(y) = {} & \text{LUT}_0(y_{i-1}) + \text{LUT}_1(y_i) \times (1 - b_i - f)2^{-i} \\ & - \text{LUT}_2(y_i) \times \text{LUT}_3(f_{i+1})2^{-(2i+2)} + \delta 2^{-(3i+1)}, \end{aligned} \tag{5.8.31}$$

where LUT_1, LUT_2 and $\text{LUT}_3(f_{i+1}) = \pm 4f_{i+1}(1 - f_{i+1})$ are indexed by an (i-bit) index and LUT_0 employs an $(i-1)$-bit index with values determined by (5.8.21). Then with all table outputs rounded to yield terms with last place position $(3i + 2 + g)$, the absolute and the relative errors satisfy

$$\left|\frac{1}{y} - \text{recip}(y)\right| < |1 - y\,\text{recip}(y)| < \left(\frac{5}{8} + \delta'(g)\right)2^{-(3i+1)},$$

where $\delta'(g)$ may be made arbitrarily small by choosing a sufficient number of guard bits g.

In practice, (5.8.31) for recip(y) can be implemented by four concurrent table look-ups followed by two concurrent multiplications. Employing a partitionable multiplier allowing a $((2i + 2 + g) \times (2i + 2 + g))$-bit multiplication along with an $((i + g) \times (i + g))$-bit multiplication, the value of recip(y) can be determined by a single pass through the multiplier as shown in Figure 5.8.4. The outputs of the independent multiplications are passed to the adder in redundant form in the design of Figure 5.8.4, so that at most a single carry-completion adder is required for final non-redundant output.

In applications employing a pipelined multiplier and tables indexed by eight bits (seven bits for LUT_0) a 24-bit reciprocal, accurate to less than a unit in the last place, can be obtained with single cycle throughput by this process with total table size less than 2 Kbytes.

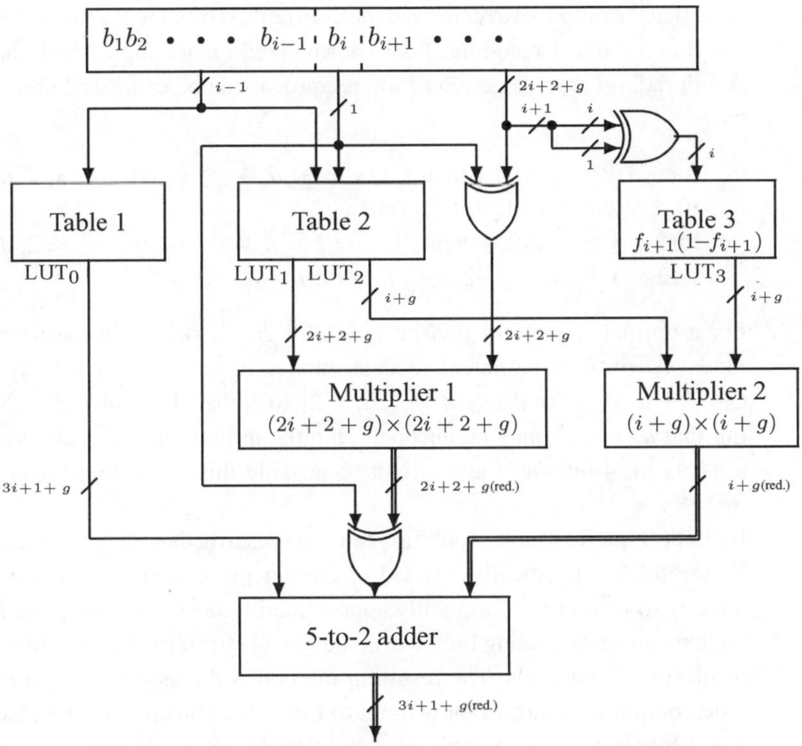

Figure 5.8.4. Quadratic reciprocal interpolation circuit.

Problems and exercises

5.8.1 Determine the eight entries for a 3-bits-in, 3-bits out maximum precision direct look-up reciprocal table and tabulate the relative error bounds for each entry. Confirm that the worst-case error yields a precision guarantee of 3.540 bits for this table. Then if the output is extended to 4-bits-out, show that precision of 4.0 bits can be guaranteed.

5.8.2 Prove Observation 5.8.6, and provide an argument to justify the comment in the paragraph before Observation 5.8.6.

5.8.3 Prove Observation 5.8.9, which verifies the symmetry illustrated in Table 5.8.3.

5.8.4 Explain why an $(i+1)$-bits-in, i-bits-out reciprocal table is sufficient to yield a one-ulp monotonic reciprocal function for any $i \geq 2$.

5.8.5 For the $2^{23} = 8,388,608$ single precision values in the interval $[1, 2)$, determine the percentage of values that an $(i+1)$-bits-in, i-bits-out reciprocal table effectively rounds up, rounds down, and rounds to nearest for $i = 5, 6, 7, 8, 9, 10$.

5.8.6 Note that the output of a reciprocal look-up table is often sent to a multiplier for Booth radix-4 recoding. For a normalized binary input show that a Booth radix-4 representation of the reciprocal can be expressed uniquely by:

(a) for $3/2 \leq y < 2$, then $1/y = \sum d_i\, 4^{-i}$ where $d_i = 2, d_2 \in \{0, 1, 2\}, d_i \in \{-1, 0, 1, 2\}$ for $i \geq 3$,
(b) for $1 \leq y < 3/2$, then $1/y = 2\sum d_i\, 4^{-1}$ where $d_i = 2, d_2 \in \{-2, -1, 0\}, d_i \in \{-2, -1, 0, 1\}$ for $i \geq 3$.

5.8.7 For a normalized binary divisor $y = 1.1\, b_2 b_3 \ldots$ show that the four-bit index $b_2 b_3 b_4 b_5$ is sufficient to determine a prescale factor $\rho = 1/2 + d_2 4^{-2} + d_3 4^{-3}$ with $d_2, d_3 \in \{-1, 0, 1, 2\}$ such that $|1 - y\rho| \leq 2^{-5}$. Noting that d_2 and d_3 may be encoded employing two bits for each, design a 4-bits in, 4-bits-out logic circuit to provide this prescaling factor for $3/2 \leq y < 2$.

5.8.8 Indirect bipartite table look-up can be described with reference to Figure 5.8.3. Specifically, instead of having the k leading bits of y go directly to $\text{LUT}_2(y)$ for implicitly approximating the $(1/y)^2$ term, the first table may use $2k$ leading bits to evaluate the $(1/y)^2$ term and partition the result into 2^k intervals. The resulting interval is then signaled by a k-bit index output from LUT_1 that is input to LUT_2. Specifically, for the case of $k = 3$ bits, determine an eight-part partition of the $(1/y)^2$ term range that provides the optimum precision enhancement for the resulting bipartite approximation result.

5.8.9 Develop a piecewise cubic reciprocal interpolation identity similar to the quadratic reciprocal interpolation identity of Lemma 5.8.30, and provide a bound on the corresponding residual coefficient.

5.8.10 Illustrate and discuss a cubic reciprocal interpolation circuit extending the design of Figure 5.8.4. Specifically, incorporate a radix-4 squaring circuit such as described in Section 4.7.3 to replace Table 3 of Figure 5.8.4. Then obtain the cubic term employing separate table look-ups of a coefficient received as a third output of Table 2 and a "fraction cubed" factor received as output of a revised Table 3.

5.9 Notes on the literature

Being the most complicated and time consuming operation, division has been extensively studied in the computing literature. There is a constant stream of conference presentations and journal papers on the subject, often also covering the closely related problem of square root extraction. All textbooks on computer arithmetic cover alternative algorithms and implementations for division, digit-serial as well as iterative, but so far only one monograph exists,

[EL94] by Ercegovac and Lang, covering exclusively digit-serial methods. Oberman and Flynn [OF97] provide a comprehensive survey of division algorithms and their implementations. Soderquist and Leeser in [SL96] discuss trade-offs in divide and square root implementations, in particular with respect to pipelining.

Just as for the other basic arithmetic operations, [BGvN46] discusses the implementation of division, briefly starting with the restoring and then in more detail the non-restoring algorithm for 2's complement representation, including the use of the quotient digit set $\{-1, 1\}$ and its implicit conversion to 2's complement. It points out how the same register structure as needed for multiplication (see Figure 4.5.1) can be used for division, when a left-shift capability is added, and observes how this division algorithm in a natural way deals with negative operands, as opposed to multiplication where the 2's complement representation creates a certain complexity.

In 1961 Wilson and Ledley [WL61] introduced the notion of "shifting over zeroes" in non-restoring division. By normalizing the remainder after each remainder update, employing the quotient digit set $\{-1, 0, 1\}$, they provided what is now termed a form of radix-2 SRT-type division. They related this process to the recoding of multipliers into optimal, but not necessarily non-adjacent (or canonical) form, as do MacSorley [Mac61] and Metze [Met62].

An SRT-type method was first described for the radix-2 case by Nadler [Nad56], and then more generally independently by Sweeney in an internal IBM note, Robertson [Rob58] and Tocher [Toc58]. The standard reference for the introduction of SRT methods is [Rob58], which also introduced "Robertson diagrams". The SRT algorithms were given that name by Freiman, based on the initials of these three authors, in a paper analyzing this class of algorithms [Fre61]. Atkins [Atk68] extended the SRT algorithms, basing the quotient digit selection on estimates of the divisor and remainder, the latter being in redundant representation. It is in this interpretation that SRT methods have become widely used and discussed extensively in the literature, e.g., in [Rob70, Atk75, Tan78, Par87, WHAY87, EL88, Fan89, EL94, OF97, ALMN02, Kor05].

In [Tay85] Taylor investigated alternative ways of implementing radix-16 SRT division using "overlapped quotient selection stages" (see Figure 5.5.13), using radix-2 and radix-4 building blocks.

Traditionally optimal (minimal) values of the SRT truncation parameters have been found by searching for the largest possible "uncertainty rectangle" that fits into the P-D diagram [WH86, BW95]. The bound (5.5.16) on $\mathrm{ulp}(\widehat{y}) = 2^{-u}$ can be found in [CM90] and [Par01] has bounds on $\mathrm{ulp}(\widehat{y})$ as well as $\mathrm{ulp}(\beta r_i) = 2^{-t}$. Theorem 5.5.3 to determine the truncation parameter t given u satisfying (5.5.16) and Observation 5.5.5 on mapping truncated remainders and divisors into non-negative values is from [Kor05] by Kornerup. Table 5.5.1 is a modified copy (with permission) from [Jen98] by Jensen.

SRT division methods were popular for commodity floating-point coprocessors through the 1980s as they only employed an adder. Radix-4 SRT division was employed in the Intel Pentium™ microprocessors introduced in 1993.

The short reciprocal division procedure developed by Briggs and Matula [BM93] was implemented in the Cyrix83D87 floating point coprocessor in 1989 employing a rectangular multiplier [BM91].

The idea of prescaling was introduced by Svoboda in [Svo63], and further developed by Krishnamurthy in [Kri70]. Multiplicative prescaled division in the high radix $\beta = 2048$ was employed in a version of the Cyrix 387 class floating-point coprocessor introduced in 1991. The procedure described in [BM95] employs a (13×68)-bit short-by-long rectangular multiplier. Burgess in [Bur94] employs a simple additive prescaling in conjunction with a maximally redundant radix-4 SRT-type division, which influenced the single cycle radix-16 division procedure employed in the Intel Penryn™ series.

Multiplicative and additive integer prescaled division was described in [MFF03]. Ercegovac Lang, and Montuschi discuss selection by rounding after prescaling in [ELM94].

Multiplier-based, iterative algorithms for obtaining the reciprocal were first described in detail by Ferrari in [Fer67], referencing Wallace [Wal64], who in turn attributes the idea to Wheeler in [WWG51]. Determination of reciprocals along with other unary functions such as square root and root-reciprocal has been pursued by many researchers, e.g. [Tak98, ELMT00, PB02]. In a master's thesis, Goldschmidt developed the multiplicative convergence division procedure [Gol64], with an acceleration procedure given in [EIM+00]. In [Fly70], Flynn analyzed various divisions by functional iteration. Postscaled division procedures where the table look-up reciprocal function is obtained concurrently with a first multiplication of the dividend are introduced by Matula and Iordache in [Mat01, MI04].

Oberman [Obe99] describe an implementation of Goldschmidt's multiplicative division algorithm employed in the AMD Athlon™, where the pipelined iterative-dependent multiplications of the division algorithm were interleaved with independent multiplications to enhance throughput. This procedure was also employed in the IBM P6™ processor [TSSK07]. An alternative prescaling and series approximation method was previously employed by Agarwal, Gustavson, and Schmookler for the Power3™ processor [AGS99], exploiting the versatility of the fused multiply-add operation.

Testing the results of a precisely rounded division implementation has been investigated by Kahan and others employing a variety of number theoretic techniques [Tan89, Par00, MM01, AAM+05].

Look-up tables for divisor reciprocals have been investigated by many researchers, since they provide a key step in limiting the number of iterations required for Newton–Raphson and convergence division. The accuracy of

i-bits-in, j-bits-out direct look-up tables is investigated in [DM94]. Bipartite tables were introduced by DasSarma and Matula in [DM95]. Further investigations on bipartite and multipartite table assisted look-up occur in [HT95, IM99, Sei99, SS99, dDT05, KM05, TSSK07]. Extensions of linear interpolation utilizing the fused multiply-add operation were introduced by Ito, Takagi, and Yajima in [ITY97, Tak98]. Quadratic interpolation table look-up procedures using concurrent multiplications were developed for a Lawrence Livermore National Laboratory supercomputer by Farmwald in [Far81] and are further discussed in [CWC01, WS05].

References

[AAM+05] M. Aharoni, S. Asaf, R. Maharik, Solving constraints on the invisible bits of the intermediate result for floating-point verification. In *Proc. 17th IEEE Symposium on Computer Arithmetic*, pages 76–83. IEEE Computer Society, 2005.

[AGS99] R. C. Agarwal, F. G. Gustavson, and M. S. Schmookler. Series approximation methods for divide and square root in the Power3 Processor. In *Proc. 14th IEEE Symposium on Computer Arithmetic*, pages 116–123. IEEE Computer Society, 1999.

[ALMN02] E. Antelo, T. Lang, P. Montuschi, and A. Nannarelli. Fast radix-4 retimed division with selection by comparisons. In *Proc. Application-Specific Systems, Architectures and Procdessors (ASAP2002)*, pages 185–196. IEEE, 2002.

[Atk68] D. E. Atkins. Higher-radix division using estimates of the divisor and partial remainders. *IEEE Trans. Computers*, C-17:925–934, 1968. Reprinted in [Swa80].

[Atk75] D. E. Atkins. Higher radix, non-restoring division: history and recent developments. In *Proc. 3rd IEEE Symposium on Computer Arithmetic*, pages 158–167. IEEE Computer Society, 1975.

[BGvN46] A. Burks, H. H. Goldstine, and J. von Neumann. *Preliminary Discussion of the Logic Design of an Electronic Computing Instrument.* Technical report, Institute for Advanced Study, Princeton, 1946. Reprinted in C. G. Bell, *Computer Structures, Readings and Examples*, McGraw-Hill, 1971.

[BM91] W. S. Briggs and D. W. Matula. Method and Apparatus for Performing the Division Function Using a Rectangular Aspect Ratio Multiplier. US Patent No. 5,046,038, 1991.

[BM93] W. S. Briggs and D. W. Matula. A 17×69 bit multiply and add unit with redundant binary feedback and single cycle latency. In *Proc. 11th IEEE Symposium on Computer Arithmetic*, pages 163–170. IEEE Computer Society, 1993.

[BM95] W. S. Briggs and D. W. Matula. Method and Apparatus for Performing Prescaled Division. US Patent No. 5,475,630, 1995.

[Bur94] N. Burgess. Prescaled maximally-redundant radix-4 SRT divider. *IEE Electron. Lett.*, 30(23):1926–1928, November 1994.

[BW95] N. Burgess and T. Williams. Choices of operand truncation in the SRT division algorithm. *IEEE Trans. Computers*, 44(7):933–938, July 1995.

[CM90] L. Ciminiera and P. Montuschi. Higher radix square rooting. *IEEE Trans. Computers*, 39(10):1220–1231, October 1990.

[CWC01] J. Cao, B. W. Y. Wei, and J. Cheng. High-performance architecture for elementary function generation. In *Proc. 15th IEEE Symposium on Computer Arithmetic*, pages 136–144. IEEE Computer Society, 2001.

[dDT05] F. de Dinechin and A. Tisserand. Multipartite table methods. *IEEE Trans. Computers*, 54(3):319–330, 2005.

[DM94] D. DasSarma and D. W. Matula. Measuring the accuracy of ROM reciprocal tables. *IEEE Trans. Computers*, 43(8):932–940, August 1994.

[DM95] D. DasSarma and D. W. Matula. Faithful bipartite ROM reciprocal tables. In *Proc. 12th IEEE Symposium on Computer Arithmetic*, pages 17–28. IEEE Computer Society, 1995.

[EIM+00] M. D. Ercegovac, L. Imbert, D. W. Matula, J.-M. Muller, and G. Wei. Improving Goldschmidt division, square root, and square root reciprocal. *IEEE Trans. Computers*, 49(7):759–763, July 2000.

[EL88] M. D. Ercegovac and T. Lang. *Radix-4 Division with Scaling and Quotient Prediction*. Technical report, CSD-880049 University of California, Los Angeles, CA 90024-1596, July 1988.

[EL94] M. D. Ercegovac and T. Lang. *Division and Square Root: Digit-Recurrence Algorithms and Implementations*. Kluwer Academic Publishers, 1994.

[ELM94] M. D. Ercegovac, T. Lang, and P. Montuschi. Very-high radix division with prescaling and selection by rounding. *IEEE Trans. Computers*, 43(8):909–918, 1994.

[ELMT00] M. D. Ercegovac, T. Lang, J.-M. Muller, and A. Tisserand. Reciprocation, square root, inverse square root, and some elementary functions using small multipliers. *IEEE Trans. on Computers*, 49(7):628–637, July 2000.

[Fan89] J. Fandrianto. Algorithms for high speed shared radix 8 division and radix 8 square root. In *Proc. 9th IEEE Symposium on Computer Arithmetic*, pages 68–75. IEEE Computer Society, 1989.

[Far81] P. M. Farmwald. High bandwidth evaluation of elementary functions. In *Proc. 5th IEEE Symposium on Computer Arithmetic*, pages 139–142. IEEE Computer Society, 1981.

[Fer67] D. Ferrari. A division method using a parallel multiplier. *IEEE Trans. Electronic Computers*, EC-16:224–226, 1967. Reprinted in [Swa80].

[Fly70] M. J. Flynn. On division by functional iteration. *IEEE Trans. Computers*, C-19:702–706, 1970. Reprinted in [Swa80].

[Fre61] C. V. Freiman. Statistical analysis of certain binary division algorithms. *Proce. IRE*, pages 91–103, January 1961.

[Gol64] R. E. Goldschmidt. Applications of division by convergence. Master's thesis, MIT, June 1964.

[HT95] H. Hassler and N. Takagi. Function evaluation by table look-up and addition. In *Proc. 12th IEEE Symposium on Computer Arithmetic*, pages 10–16. IEEE Computer Society, 1995.

[IM99] C. Iordache and D. W. Matula. Analysis of reciprocal and square root reciprocal instructions in the AMD K6-2 implementation of 3DNow. *Electron. Notes Theor. Computer Sci.*, 24, 1999.

[ITY97] M. Ito, N. Takagi, and S. Yajima. Efficient initial approximation for multiplicative division and square root by a multiplication with operand modification. *IEEE Trans. Computers*, 46(4):495–498, 1997.

[Jen98] T. A. Jensen. Alternative implementations of SRT division and square root algorithms. Master's thesis, Odense University, July 1998.

[KM05] P. Kornerup and D. W. Matula. Single precision reciprocals by multipartite table look-up. In *Proc. 17th IEEE Symposium on Computer Arithmetic*, pages 240–248. IEEE Computer Society, 2005.

[Kor05] P. Kornerup. Digit selection for SRT division and square root. *IEEE Trans. Computers*, 54(3):294–303, March 2005.

[Kri70] E. V. Krishnamurthy. On range-transformation techniques for division. *IEEE Trans. Computers*, C-19:157–159, February 1970.

[Mac61] O. L. MacSorley. High-speed arithmetic in binary computers. *Proc. IRE*, 49:67–91, January 1961. Reprinted in [Swa80].

[Mat01] D. W. Matula. Improved table lookup algorithms for postscaled division. In *Proc. 15th IEEE Symposium on Computer Arithmetic*, pages 101–108. IEEE Computer Society, 2001.

[Met62] G. Metze. A class of binary divisions yielding minimally represented quotients. *IRE Trans. Electronic Computers*, pages 761–764, 1962.

[MFF03] D. W. Matula and A. Fit-Florea. Prescaled integer division. In *Proc. 16th IEEE Symposium on Computer Arithmetic*, pages 63–68. IEEE Computer Society, 2003.

[MI04] D. W. Matula and C. S. Iordache. Method and Apparatus for Performing Division and Square Root Functions Using a Multiplier and a Multipartite Table. US Patent No. 6,782,405 B1, 2004.

[MM01] L. D. McFearin and D. W. Matula. Generation and analysis of hard to round cases for binary floating point division. In *Proc. 15th IEEE Symposium on Computer Arithmetic*, pages 119–135. IEEE Computer Society, 2001.

[Nad56] M. Nadler. A high speed electronic arithmetic unit for automatic computing machines. *Acta Tech. (Prague)*, 6:464–478, 1956.

[Obe99] S. F. Oberman. Floating point division and square root algorithms and implimentation in the AMD-K7 microprocessor. In *Proc. 14th IEEE Symposium on Computer Arithmetic*, pages 106–115. IEEE Computer Society, 1999.

[OF97] S. F. Oberman and M. J. Flynn. Division algorithms and implementations. *IEEE Trans. Computers*, 46(8):833–854, August 1997.

[Par87] B. Parhami. On the complexity of table lookup for iterative division. *IEEE Trans. Computers*, C-36:1233–1236, 1987.

[Par00] M. Parks. Number theoretic test generation for directed rounding. *IEEE Trans. Computers*, 49(7):651–658, July 2000.

[Par01] B. Parhami. Precision requirements for quotient digit selection in high-radix division. In *Proc. 35-th Asilomar Conference on Circuits, Systems and Computers*, pages 1670–1673. IEEE Press, 2001.

[PB02] J. A Pineiro and J. D. Bruguera. High-speed double-precision computation of reciprocal, division, square root and inverse square root. *IEEE Trans. Computers*, 51(12):1377–1388, 2002.

[Rob58] J. E. Robertson. A new class of digital division methods. *IRE Trans. Electronic Computers*, EC-7:218–222, 1958. Reprinted in [Swa80].

[Rob70] J. E. Robertson. The correspondance between methods of digital division and multiplier recoding procedures. *IEEE Trans. Computers*, C-19:692–701, August 1970.

[Sei99] P.-M. Seidel. High-speed redundant reciprocal approximation. *INTEGRATION, the VLSI Journal*, 28:1–12, 1999.

[SL96] P. Soderquist and M. Leeser. Area and performance tradeoffs in floating-point divide and square-root implementations. *ACM Computing Surveys*, 28(3):518–564, September 1996.

[SS99] M. Schulte and J. Stine. Approximating elementary functions with symmetric bipartite tables. *IEEE Trans. Computers*, 48(8):842–847, 1999.

[Svo63] A. Svoboda. An algorithm for division. *Information Processing Machines*, 9:25–32, 1963. Reprinted in [Swa80].

[Swa80] E. E. Swartzlander, editor. *Computer Arithmetic,* volume I. Dowden, Hutchinson and Ross, Inc., 1980. Reprinted by IEEE Computer Society Press, 1990.

[Tak98] N. Takagi. Powering by a table look-up and a multiplication with operand modification. *IEEE Trans. Computers*, 47(11):1216–1222, 1998.

[Tan78] K. G. Tan. The theory and implementations of high-radix division. In *Proc. 4th IEEE Symposium on Computer Arithmetic*, pages 154–163. IEEE Computer Society, 1978.

[Tan89] P. T. P. Tang. Testing Computer Arithmetic by Elementary Number Theory. Preprint MCS-P84-0889, Mathematics and Computer Science Division, Argonne National Laboratory, August 1989.

[Tay85] G. S. Taylor. Radix 16 SRT dividers with overlapped quotient selection stages. In *Proc. 7th IEEE Symposium on Computer Arithmetic*, pages 64–71. IEEE Computer Society, 1985.

[Toc58] K. D. Tocher. Techniques of multiplication and division for automatic binary computers. *Q. J. Mech. Appl. Math.*, 11:364–384, 1958.

[TSSK07] S. D. Trong, M. Schmookler, E. M. Schwarz, and M. Kroener. P6 binary floating-point unit. In *Proc. 18th IEEE Symposium on Computer Arithmetic*, pages 77–86. IEEE Computer Society, 2007.

[Wal64] C. S. Wallace. A suggestion for a fast multiplier. *IEEE Trans. Electronic Computers*, EC-13:14–17, 1964. Reprinted in [Swa80].

[WH86] T. E. Williams and M. Horowitz. *SRT Division Diagrams and Their Usage in Designing Custom Integrated Circuits for Division.* Technical Report CSL-TR-87-326, Stanford University, 1986.

[WHAY87] T. E. Williams, M. Horowitz, R. L. Alverson, and T. S. Yang. A self-timed chip for division. In Paul Losleben, editor, *Advanced Research in VLSI. Proceedings of the 1987 Stanford Conference*, pages 75–95. The MIT Press, 1987.

[WL61] J. B. Wilson and R. S. Ledley. An algorithm for rapid binary division. *IRE Trans. Electronic Computers*, EC-10:662–670, 1961.

[WS05] E. G. Walters and M. J. Schulte. Efficient function approximation using truncated multipliers and squarers. In *Proc. 17th IEEE Symposium on Computer Arithmetic*, pages 232–239. IEEE Computer Society, 2005.

[WWG51] M. W. Wilkes, D. J. Wheeler, and S. Gill. *Preparation of Programs for an Electronic Digital Computer*. Addison-Wesley, 1951.

6

Square root

6.1 Introduction

Radix square root pertains to the unary square root operation where the radicand and result are in $\mathbb{Q}_\beta$. Square root is the inverse of the squaring operation discussed in Section 4.7. The precision hierarchy of square root results for the operand $x \in \mathbb{Q}_\beta$ is classified in Table 6.1.1.

Until the appearance of the IEEE floating-point standard, the square root operation was just considered a mathematical function, typically provided by a software implementation in the mathematical function library. Its frequency of use, together with the possibility of efficiently providing precise rounding as for the other basic arithmetic operations, convinced the defining committee that it should also be standardized since the square root operation was simple to provide – if not in hardware, then at least in software. This gradually convinced CPU manufacturers that square root should be implemented as an instruction directly in hardware, supported by the fact that it can be implemented sharing hardware with the divide instruction.

Owing to the resemblance between division and square root, the methods described here will follow closely the methods described in the previous chapter. Thus we shall here show corresponding digit-serial as well as iterative

Table 6.1.1. *Classification of square root types for radicands in* $\mathbb{Q}_\beta$

Result name	Type	Result set
Infinitely precise root	Algebraic number	Unique value
Root, remainder	Exact radix-β pair	
Root and remainder sign	One-ulp interval	Discrete family parameterized by root last place
One-ulp root	Two-ulp interval	
Approximate root	Interval	Family parameterized by error bounds

Table 6.1.2. *Classification of square root algorithms*

Digit Serial Algorithms		
Deterministic	Non-deterministic	Iterative refinement algorithms
Restoring (school method)	SRT	Newton–Raphson square root
Binary non-restoring	Short reciprocal	Newton–Raphson root-reciprocal
	Prescaled	Convergence

algorithms for square root, and also implementations where division and square root share resources. These square root algorithms are conveniently classified in Table 6.1.2.

Section 6.2 provides a brief formalization of radix square root, introducing radix-β root, remainder pairs and one-ulp root, tail partitions as a basis for characterizing intermediate and final results of the digit serial algorithms of Sections 6.3 and 6.4. The restoring "school method" should be familiar in decimal, and serves to motivate all the digit serial algorithms.

Section 6.3 covers restoring, non-restoring, and SRT algorithms appropriate for serial implementation employing only an adder. The SRT discussion presents details on integration with SRT division. The short reciprocal and prescaled algorithms covered in Section 6.4 are appropriate for implementation employing a short-by-long rectangular multiplier. The short reciprocal scales each successive remainder by the precomputed short reciprocal to obtain a digit in the high radix 2^k (typically $k \geq 8$), with interleaved remainder updating, where both steps serially employ the short-by-long multiplier. The prescaled algorithm scales the radicand to a small region ($1 \pm 2^{-k}, k \geq 8$) near unity, so a scaled root can be obtained digit serially from leading parts of successive scaled remainders. The digits of the scaled remainder can be serially postscaled to obtain partial square root terms that can be accumulated concurrently with each scaled remainder update. This procedure is attractive for a pipelined rectangular multiplier implementation.

Section 6.5 covers several iterative refinement algorithms appropriate for implementations employing a full multiplier. With a full multiplier the typical method for square root function software implementation has been to employ Newton–Raphson square root, which from an initial approximation essentially doubles the number of accurate digits each iteration. The method is classic dating back to the Babylonians, and its history is discussed further in the bibliography. Provided that a good approximation can be found by table look-up, very few iterations are needed to obtain the precisions required for the IEEE standard output family. For very fast arithmetic, it is now fairly common to use iterative methods in pipelined ALU architectures, supported by fast multiplier hardware.

Newton–Raphson iterative square root is conceptually very simple but is not well suited for a hardware implementation as it requires iterative divisions. Thus most implementations use a reformulated root-reciprocal algorithm, requiring only multiplications, assumed to be efficient in modern hardware, with a final multiplication by the radicand. Both Newton–Raphson algorithms are discussed in Section 6.5 along with the convergence variation. The convergence square root algorithm simultaneously develops the root as well as the root-reciprocal employing concurrent multiplications suitable for a piplelined full multiplier.

6.2 Roots and remainders

The *square root* operation is denoted by $\sqrt{x}$, where the operator symbol "$\sqrt{\ }$" is termed the *radical*, the argument $x \geq 0$ is termed the *radicand*, and the result is termed the *root*. Square root is a unary operation inverse to the operation of *square* ($x^2 = x \times x$), i.e., square is the unary multiplication operation with both arguments identical. Squaring is a very frequent operation in practice and is investigated in Section 4.7. Square root is sufficiently important to a variety of practical applications that it is typically considered worthy of implementation as a primitive arithmetic operation.

For computation hosted in radix arithmetic, square root can often be implemented at a performance level and resource cost comparable to that of the division operation. Hardware joint implementations of division and square root using similar algorithms can benefit from sharing of resources. The IEEE floating-point standard has mandated that square root be supported with the same precise roundings as the dyadic operations $(+, -, \times, /)$, so radix square root is an essential arithmetic operation for practical radix computer arithmetic.

For the one-(half-)ulp root partition $\sqrt{x} = q + t$ with leading part $q = i\beta^{\ell} \in \mathbb{Q}_\beta$, q is termed a *one-ulp root* of the radicand x if the rounded off tail t satisfies $|t| = |\sqrt{x} - q| < \beta^{l}$, and a *half-ulp root* if $|t| \leq \frac{1}{2}\beta^{\ell}$. If the corresponding $\mathrm{sgn}(t)$ is known, then $(q, \mathrm{sgn}(t)) \in \mathbb{Q}_\beta \times \{-1, 0, 1\}$ is a one-(half-)ulp *directed root* of x. Except for the relatively infrequent instances where the radicand is the square of a radix-β number, e.g., $x = i^2\beta^{2\ell}$, the tail t is not in $\mathbb{Q}_\beta$ even though the radicand $x \in \mathbb{Q}_\beta$. In practice, it is more effective to use a remainder function as a continuation function for the tail in developing the algorithms for radix square root. The fundamental *square root invariant* equation

$$\text{radicand} = \text{root} \times \text{root} + \text{remainder}$$

provides the foundation for the square root operation and illustrates the similarity with the division operation and the division invariant equation

$$\text{dividend} = \text{quotient} \times \text{divisor} + \text{remainder}.$$

It is instructive to compare the square root invariant and division invariant equations. Note that the root plays the role of both the quotient and the divisor of the division operation. If we have a radix-β root denoted root_1, then the expression

$$\frac{\text{radicand}}{\text{root}_1} = \text{root}_1 + \frac{\text{remainder}}{\text{root}_1} \approx \text{root}_2$$

provides the basis for a recurrence where the term $\text{remainder}/\text{root}_1$ can be utilized to improve the precision of the radix-β root. The $\text{radicand}/\text{root}_1$ expression is analogous to the division expression for a radix-β quotient

$$\frac{\text{dividend}}{\text{divisor}} = \text{quotient}_1 + \frac{\text{remainder}_1}{\text{divisor}} \approx \text{quotient}_2$$

Following this analogy we shall use the notation $\rho \approx 1/\text{root}_1$ to here denote a radix-β short root-reciprocal, since it serves the same remainder scaling role in digit serial square root as the radix-β divisor short reciprocal $\rho = 1/\text{divisor}$ in division. For radix-β square root, this expression can be developed into an iterative algorithm since both the remainder and approximate roots can be developed in appropriate radix-β representations.

Formally for the radicand $x \geq 0$, a radix-β one-(half-)ulp root, remainder pair (q, r) comprises a one-(half-)ulp root $q = i\beta^\ell \in \mathbb{Q}_\beta$ with $q \geq 0$ satisfying

$$x = q^2 + r, \tag{6.2.1}$$

where $r = x - q^2$ is the *remainder* and $t = \sqrt{x} - q$ is the *rounded off tail.*

Observation 6.2.1 *For a one-ulp root, remainder pair (q, r) of the radicand $x \in \mathbb{Q}_\beta$, the remainder is related to the tail by the identity,*

$$r = t(2q + t). \tag{6.2.2}$$

Proof Note that $\sqrt{x} + q = 2q + t$, so then $r = (x - q^2) = (\sqrt{x} - q)(\sqrt{x} + q) = t(2q + t)$. □

Observation 6.2.2 *For $x = i\beta^j > 0$ and last place ℓ, there is a unique one-ulp root, remainder pair $(q, r) \in \mathbb{Q}_\beta \times \mathbb{Q}_\beta$ with $q = k\beta^\ell \geq 0$ and $r = m\beta^n \geq 0$ with $n = min(j, 2\ell)$.*

For last place ℓ, the radicand x satisfies $i\beta^\ell \leq x < (i+1)\beta^\ell$, and so has a unique *one-ulp-floor* root, remainder pair $(i\beta^\ell, r)$ with $r \geq 0$, and a unique *one-ulp-ceiling* pair $((i+1)\beta^\ell, r^*)$ with $r^* < 0$ for $r \neq 0$, otherwise $(i\beta^\ell, 0)$, that are *complementary* one-ulp pairs.

Observation 6.2.3 *For the radicand $x \geq 0$, and one-ulp root, remainder pair (q, r) with last place ℓ, the complementary one-ulp pair $(\bar{q}, \bar{r})$ is given by*

$$\bar{q} = q + \text{sgn}(r)\beta^\ell, \tag{6.2.3}$$

$$\bar{r} = r - (2q + \text{sgn}(r)\beta^\ell)\text{sgn}(r)\beta^\ell, \tag{6.2.4}$$

where $\bar{t} = t - \text{sgn}(r)\beta^\ell$ is the complementary rounded off tail.

Proof Let $s = \operatorname{sgn}(r)$. Inserting $\bar{t} = t - s\beta^\ell$ and $\bar{q} = q + s\beta^\ell$ in $\bar{r} = \bar{t}(2\bar{q} + \bar{t})$ we obtain

$$\begin{aligned} \bar{r} &= (t - s\beta^\ell)(2(q + s\beta^\ell) + t - s\beta^\ell) && (6.2.5)\\ &= (t - s\beta^\ell)(2q + t + s\beta^\ell) && (6.2.6)\\ &= (2qt + t^2) - 2qs\beta^\ell + s^2\beta^{2\ell} && (6.2.7)\\ &= r - (2q + s\beta^\ell)s\beta^\ell, && (6.2.8)\end{aligned}$$

verifying the expression for $\bar{r}$, so that $(\bar{q}, \bar{r})$ is the complementary one-ulp root, remainder pair. □

Problems and exercises

6.2.1 Let $q \in \mathbb{Q}_\beta$ be a one-ulp root of the real-valued radicand $x \geq 0$ and let the real-valued remainder $r = x - q^2$. If x is of numeric type transcendental, irrational, non-radix-β rational, or a radix-β number, then discuss whether the remainder r is respectively of numeric type transcendental, irrational, non radix-β rational or radix-β.

6.2.2 Let $q \in \mathbb{Q}_\beta$ be a one-ulp root of $x = i\beta^j \in \mathbb{Q}_\beta$ with β prime. Show that the tail $t = \sqrt{x} - q = k\beta^\ell \in \mathbb{Q}_\beta$ iff $i = k^2$ is a perfect square, and $j = 2\ell$ is even, so $t = k\beta^\ell$.

6.2.3 Provide an example to show that the condition of Problem 6.2.2 fails if β is a composite radix.

6.2.4 For a root-reciprocal approximation $\rho \approx 1/\sqrt{x}$, show that $\operatorname{sgn}(\rho - 1/\sqrt{x}) = \operatorname{sgn}(\rho^2 x - 1)$. Then discuss how you might provide a half-ulp directed root-reciprocal to support implementation of a precisely rounded root-reciprocal operation.

6.3 Digit-serial square root

The process of obtaining leading parts of the tail to append to a one-ulp root relies on the relation between the remainder and corresponding tail.

Lemma 6.3.1 *The remainder $r = x - q^2$ and tail $t = \sqrt{x} - q$ corresponding to the one-ulp root $q = i\beta^\ell$ of $x > 0$ have* $\operatorname{sgn}(r) = \operatorname{sgn}(t)$ *and satisfy*

$$\frac{r}{2\sqrt{x}} \leq t = \frac{r}{2q + t} \leq \frac{r}{2q} = t\left(1 + \frac{t}{2q}\right), \tag{6.3.1}$$

$$|r| \leq (2q + \operatorname{sgn}(r)\beta^\ell)\beta^\ell. \tag{6.3.2}$$

Proof Note that $r = x - q^2 = (\sqrt{x} - q)(\sqrt{x} + q) = t(2q + t)$, and $2\sqrt{x} = 2q + 2t$, from which the equalities in (6.3.1) follow. The bound is easily seen to hold for both positive and negative t and r. The bound (6.3.2) follows from (6.2.4) since $\mathrm{sgn}(r) = -\mathrm{sgn}(\bar{r})$. □

For $q = i\beta^\ell$ with $i \gg 1$, then $|t| < \beta^\ell \ll q$, and it follows that $r/2q$ is a very good approximation of t. This suggests that $2q$ may play the same role as the divisor y in adapting digit-serial divide algorithms to digit-serial square root algorithms. Furthermore, t is very close to the midpoint of the interval $[r/2\sqrt{x}; r/2q]$, so $r/2q$ and $r/2\sqrt{x}$ are comparably good approximations to t, located on opposite sides of t.

We shall now consider precision extension from $\mathrm{ulp}(q) = \beta^\ell$ to the successor q' with $\mathrm{ulp}(q') = \beta^{\ell-1}$, i.e., the root and remainder updating.

Theorem 6.3.2 *Let (q, r) be a one-ulp root, remainder pair for the radicand x with $q = i\beta^\ell$. Let the successors q' and r' be defined from q, r and a selected digit d by*

$$q' = q + d\beta^{\ell-1}, \tag{6.3.3}$$
$$r' = r - (2q + d\beta^{\ell-1})d\beta^{\ell-1}, \tag{6.3.4}$$

where the digit d is chosen from the maximally redundant digit set $\{-(\beta - 1), \ldots, \beta - 1\}$ so that $|r'| < (2q' + \mathrm{sgn}(r')\beta^{\ell-1})\beta^{\ell-1}$. Then (q', r') is a one-ulp root, remainder pair satisfying $|\sqrt{x} - q'| < \beta^{\ell-1}$.

Proof From (6.3.3) and (6.3.4) we have

$$\begin{aligned} q'q' &= (q + d\beta^{\ell-1})(q + d\beta^{\ell-1}) \\ &= q^2 + (2q + d\beta^{\ell-1})d\beta^{\ell-1} \\ &= x - r + (2q + d\beta^{\ell-1})d\beta^{\ell-1} \\ &= x - (r - (2q + d\beta^{\ell-1})d\beta^{\ell-1}) \\ &= x - r', \end{aligned}$$

and by the condition on d and Lemma 6.3.1, (q', r') is a one-ulp root, remainder pair for x. However, it remains to be proven that such a digit d exists. Define $r(d) = r - (2q + d\beta^{\ell-1})d\beta^{\ell-1}$ and $q(d) = q + d\beta^{\ell-1}$, and assume first that $r > 0$. Then $r(0) = r > 0$ and by Lemma 6.3.1, $r(\beta) = r - (2q + \beta^\ell)\beta^\ell \le 0$, hence there must be a value of d, $0 \le d \le \beta - 1$, such that

$$r(d) \ge 0 > r(d + 1)$$

with $r(d) - r(d + 1) = (2q + \beta^{\ell-1})\beta^{\ell-1}$, and $q(d + 1) - q(d) = \beta^{\ell-1}$. Thus for $r(d) > 0$, $(q', r') = (q(d), r(d))$ and $(r(d + 1), q(d + 1))$ form complementary root, remainder pairs for last place $\ell - 1$. If $r(d) = 0$, then d is the digit to be chosen. The equivalent results for $r = r(0) < 0$ with $r(-\beta) = r + (2q - \beta^\ell)\beta^\ell \ge 0$ follow similarly, and finally if $r(0) = 0$, then $d = 0$ is to be chosen. □

Note the similarity with the updating of remainders in division algorithms, where here the term $(2q + d\beta^{\ell-1})$ plays the same role as the divisor in division. Conforming to popular practice in the following, we shall also develop algorithms using shifted remainders, $\hat{r}_i = r_i\beta^i$, where then the quotient q_i and shifted remainder $\hat{r}_i$ have the same last place position. Note the relationship to the correspondingly shifted tails $\hat{t}_i = t_i\beta^i$, as given by $\hat{r}_i = (x - q_i^2)\beta^i = (2q_i + \hat{t}_i\beta^{-i})\hat{t}_i = \hat{t}_i(\sqrt{x} + q_i)$.

Corollary 6.3.3 *Let $(q_i, \hat{r}_i)$ be a one-ulp root, shifted-remainder pair for the radicand x with $\beta^{-2} \le x < 1$ and $\text{ulp}(q_i) = \beta^{-i}$. Let q_{i+1} and $\hat{r}_{i+1}$ be defined from $q_i, \hat{r}_i$ and a digit d_{i+1} by*

$$\begin{aligned} q_{i+1} &= q_i + d_{i+1}\beta^{-i-1}, \\ \hat{r}_{i+1} &= \beta\hat{r}_i - (2q_i + d_{i+1}\beta^{-i-1})d_{i+1}, \end{aligned}$$

where d_{i+1} is chosen so that $|\hat{r}_{i+1}| \le 2q_{i+1} + \text{sgn}(\hat{r}_{i+1})\beta^{-i-1}$. Then $(q_{i+1}, \hat{r}_{i+1})$ is a root, shifted-remainder pair satisfying $x = q_{i+1}^2 + \hat{r}_{i+1}\beta^{-i-1}$, and $\text{ulp}(\hat{r}_{i+1}) = \text{ulp}(q_{i+1}) = \beta^{-i-1}$.

Proof Following the proof of Theorem 6.3.2:

$$\begin{aligned} q_{i+1}^2 &= (q_i + d_{i+1}\beta^{-i-1})(q_i + d_{i+1}\beta^{-i-1}) \\ &= x - \hat{r}_i\beta^{-i} + (2q_i + d_{i+1}\beta^{-i-1})d_{i+1}\beta^{-i-1} \\ &= x - (\hat{r}_i\beta - (2q_i + d_{i+1}\beta^{-i-1})d_{i+1})\beta^{-i-1} \\ &= x - \hat{r}_{i+1}\beta^{-i-1}, \end{aligned}$$

hence $(q_{i+1}, \hat{r}_{i+1})$ is a root, shifted-remainder for x with $\text{ulp}(q_{i+1}) = \text{ulp}(\hat{r}_{i+1}) = \beta^{-i-1}$. It also follows that d_{i+1} can be chosen such that $|\hat{r}_{i+1}| \le 2q_{i+1} + \text{sgn}(\hat{r}_{i+1})\beta^{-i-1}$. □

6.3.1 Restoring and non-restoring square root

Theorem 6.3.2 leads immediately to a square root algorithm which was taught in schools until around 1970, when electronic calculators became common. For decimal arithmetic the factor $2\beta = 20$ was the so-called *"mysterious factor 20."* It uses a slightly rearranged version of (6.3.4)

$$r' = r - (2\beta q + d\,\text{ulp}(q))d \cdot \beta^{\ell-2},$$

thus for $\beta = 10$, $r' = r - (20q + d\,\text{ulp}(q))d \cdot 10^{\ell-2}$. The algorithm turns out to be the equivalent of restoring division, as commonly used for hand calculations.

Example 6.3.1 Let us illustrate the *school method* in decimal by an example, finding approximations to the square root of the number 23456.78. We start by pairing digits on both sides of the decimal point, and subtracting the largest integer square d^2 smaller than or equal to the value of the leading pair, recording d as the first digit of the result. Subsequent digits d are then chosen as the maximal digits satisfying

$$(20q + d\,\text{ulp}(q))d\,10^{\ell-2} \leq r.$$

Of course, when performing the algorithm by hand a guess is made and checked, instead of successively trying with $0, 1, 2, \ldots,$ as specified for the restoring type of algorithm. Noting that $\text{ulp}(r) = \text{ulp}^2(q)$, the calculations can then be displayed as:

	1 5 3 . 1 d	
	$\sqrt{}$ 02 34 56 . 78 00	
1	1	ulp(q)=100
	1 34	
25	1 25	ulp(q)=10
	9 56	
303	9 09	ulp(q)=1
	47 . 78	
3061	30 . 61	ulp(q)=0.1
	17 . 17 00	
3062d	xx .xx xx	ulp(q)=0.01

the last step illustrating how a digit d is chosen. Here $d = 5$ turns out to be the maximal usable value, since $(20 \times 1531 + 5) \times 5 = 30625 \times 5 = 15\,31\,25$, but $(20 \times 1531 + 6) \times 6 = 18\,37\,56$. The guess of the digit 5 can be determined simply by a "trial division" of 171700 by the "trial divisor" $20q \approx 30620$ or alternatively, by division of $17.17\,00$ by $2q = 306.2$ yielding $0.05607\cdots$, which by (6.3.1) corresponds to calculating an approximation to the tail $r/2q \approx t$, from which the digit 5 also can be found. Note that we then actually have

$$q' \approx q + \frac{r}{2q}$$

as a very good approximation of the square root. In this case we have $q + r/2q = 153.15607\cdots$, satisfying $153.15607^2 = 23456.781777\cdots$. As the number of digits determined becomes larger, the digit to be chosen has less and less influence on the "divisor" $(2\beta q + d\beta^{\ell})$, and the more the algorithm resembles an ordinary division. If terminating with the remainder 17.17 we find that $23456.78 = 153.1^2 + 17.17$, thus satisfying $x = q^2 + r$. □

The school method can be illustrated by the simple process of algebraically extending $x - q^2$ to $x - (q + d)^2$ as shown symbolically without explicit normalization in Figure 6.3.1.

Based on Theorem 6.3.2 and in particular Corollary 6.3.3, it is straightforward to develop a binary non-restoring square root algorithm, in analogy with Algorithm 5.4.8 for binary non-restoring division, but now explicitly using the

$$\begin{array}{lll} & \;\;q \quad + \quad d & \\ & \sqrt{x} & \\ & \;\;q^2 & \\ \hline & \;\;x - q^2 & \text{first remainder} \\ (2q+d) & \;\;d(2q+d) & \\ \hline & \;\;x - (q^2 + 2qd + d^2) & \text{second remainder} \\ & \;\;[= x - (q+d)^2] & \text{“completing the square”} \end{array}$$

Figure 6.3.1. The "school method" of square root interpreted as "completing the square".

digit set $\{-1, 1\}$, and effectively developing a shifted remainder with deterministic digit selection. In the following the radicand is simply assumed to be in the unit interval.

Algorithm 6.3.4 (Binary non-restoring square root)

Stimulus: *A non-negative radicand* x, $0 \le x < 1$, *and integer* $n > 0$.

Response: *An* n*-bit square root* q *and shifted remainder* $\hat{r}$ *satisfying* $x = q^2 + \hat{r}2^{-n}$ *and* $|\sqrt{x} - q| < 2^{-n} = \text{ulp}(\hat{r}) = \text{ulp}(q)$.
If $x \neq 0$ *then* $-2q < \hat{r} - 2^{-n} < 2q$, *and for* $x = 0$, $q = \hat{r} = 0$.

Method: $q_0 := 0$; $\hat{r}_0 := x$;
for $i := 1$ **to** n **do**
  if $\hat{r}_{i-1} = 0$ **then**
    $\hat{r} = 0$; $q = q_{i-1}$; **return**;
  if $\hat{r}_{i-1} < 0$ **then**
    $\hat{r}_i := 2\hat{r}_{i-1} + (2q_{i-1} - 2^{-i})$; $q_i := q_{i-1} - 2^{-i}$;
  else
    $\hat{r}_i := 2\hat{r}_{i-1} - (2q_{i-1} + 2^{-i})$; $q_i := q_{i-1} + 2^{-i}$;
L: **end**;
end;
$\hat{r} := \hat{r}_n$; $q := q_n$;

Where: *The invariant* $(x = q_i^2 + \hat{r}_i 2^{-i}) \wedge (-2q_i < \hat{r}_i - 2^{-i} < 2q_i)$, *holds at label L.*

First note for an implementation that the updating of q_i can never cause a carry ripple beyond one position. Obviously there is no carry ripple when 2^{-i} is added. Otherwise since $x \ge 0$, $q_1 = 2^{-1}$, assume that the last bit of q_{i-1}, $b_{-(i-1)} = 1$. Hence if $q_i = q_{i-1} - 2^{-i}$ it follows that q_i will have bits $b'_{-(i-1)} = 0$ and $b'_{-i} = 1$. Similarly in forming $2q_{i-1} \pm 2^{-i}$ there can never be a carry ripple beyond the terminal one in q_{i-1}.

Proof (of invariant) The updating of q_i and $\hat{r}_i$ consists of choosing a digit $d_i \in \{-1, 1\}$, depending on the sign of $\hat{r}_{i-1}$, and then forming

$$q_i = q_{i-1} + d_i 2^{-i}$$

and remainder

$$\begin{aligned}\hat{r}_i &= 2\hat{r}_{i-1} - d_i(2q_{i-1} + d_i 2^{-i}) \\ &= 2\hat{r}_{i-1} - d_i(2q_i - d_i 2^{-i}),\end{aligned} \tag{6.3.5}$$

by rewriting q_{i-1}. Note that the left part of the invariant

$$(x = q_i^2 + \hat{r}_i 2^{-i}) \wedge (-2q_i < \hat{r}_i - 2^{-i} < 2q_i)$$

trivially holds before the loop is entered ($q_0 = 0$ and $\hat{r}_0 = x$). By induction, assuming it holds for $i-1$, we obtain by inserting $\hat{r}_{i-1} = \frac{1}{2}(\hat{r}_i + d_i(2q_i - d_i 2^{-i}))$ and $q_i = q_{i-1} + d_i 2^{-i}$

$$\begin{aligned}x &= q_{i-1}^2 + \hat{r}_{i-1}2^{-i} \\ &= (q_i - d_i 2^{-i})^2 + (\hat{r}_i + d_i(2q_i - d_i 2^{-i}))2^{-i} \\ &= q_i^2 + \hat{r}_i 2^{-i},\end{aligned}$$

thus the left part of the invariant holds.

The right part is satisfied for $i = 1$ since $\hat{r}_1 = 2x - \frac{1}{2}$ and $q_1 = 2^{-1}$, so assume it holds for $i-1$. As $\hat{r}_{i-1} > 0 \Rightarrow d_i = -1$ using (6.3.5) and $q_{i-1} = q_i - 2^{-i}$, by rewriting the invariant for $i-1$ in the case $\hat{r}_{i-1} > 0$ we have

$$\begin{aligned} 0 &< \hat{r}_{i-1} < 2q_{i-1} + 2^{-(i-1)} \\ -(2q_{i-1} + 2^{-i}) &< \hat{r}_i < 2(2q_{i-1} + 2^{-i+1}) - (2q_{i-1} + 2^{-i}) \\ -2q_{i-1} - 2^{-i} &< \hat{r}_i < 2q_{i-1} + 3 \cdot 2^{-i} \\ -2q_i + 2^{-i} &< \hat{r}_i < 2q_i + 2^{-i},\end{aligned}$$

and similarly for $\hat{r}_{i-1} < 0 \Rightarrow d_i = 1$, thus the right part holds for $\hat{r}_i \neq 0$. It now remains to be shown that $-2q_i + 2^{-i} < 0$, or $q_i > 2^{-i-1}$, for $\hat{r}_i = 0$. But from $q_1 = 2^{-1}$ it follows that $q_i \geq 2^{-i}$, since the smallest value is obtained by the digit sequence $1\bar{1}\cdots\bar{1}$. □

Theorem 6.3.5 *Given a radicand x, $0 \leq x < 1$, and* ulp$(q) = 2^{-n}$, *Algorithm 6.3.4 computes a root, remainder pair $(q, \hat{r}2^{-n})$ satisfying $x = q^2 + \hat{r}2^{-n}$ with $|\sqrt{x} - q| < 2^{-n}$.*

Proof First notice that the remainder $\hat{r}$ computed in the algorithm has been shifted n positions, compared to the remainder used in Lemma 6.3.1. For $x > 0$ from the invariant for $i = n$ we have on exit $x = q^2 + \hat{r}2^{-n}$ and $-2q < \hat{r} - 2^{-n} < 2q$, which by the lemma proves the response in this case. For $x = 0$, $q = \hat{r} = 0$ is trivially true. □

6.3.2 SRT square root

Like SRT division, the SRT square root algorithms form a class characterized by the following:

- The radicand is normalized.
- A redundant symmetric quotient digit set is used.
- Root digits are selected by a few leading digits of remainder and root.
- The remainders may be in a redundant representation.

Note that here we assume normalization to be "relaxed," since preferably the scaling factor applied for normalization should be an even power of the radix used in the representation of the radicand x and in the arithmetic. Normally this will mean that a binary representation and binary arithmetic are used, and that the high radix used for representing the quotient is of the form 2^k with $k = 2$ or 3. For simplicity of notation in discussing SRT square root, we use β to denote the high radix ($\beta = 4, 8, \ldots$) to avoid having the radix power in all the exponents. We will initially assume that normalization is to the interval $\frac{1}{4} \leq x < 1$.

Let β be the quotient radix and $D = \{-a, \ldots, 0, \ldots, a\}$ the quotient digit set with maximum digit magnitude a satisfying $\beta/2 \leq a \leq \beta - 1$, and *redundancy factor*

$$\mu = \frac{a}{\beta - 1} \quad \text{with} \quad \tfrac{1}{2} < \mu \leq 1. \tag{6.3.6}$$

This insures that each successive quotient will be a μ-ulp quotient with a maximally redundant digit set yielding one-ulp quotients.

Let x be the normalized radicand and (q_i, r_i) the root, shifted-remainder pair such that $x = q_i^2 + r_i\beta^{-i}$, and let d_{i+1} be the digit selected in the $(i + 1)$th step. The purpose of the *digit selection function*, $\sigma(r_i, y)$, is to select the next root digit $d = d_{i+1}$, while keeping the new remainder $r_{i+1} = \beta r_i - (2q_i + d\beta^{-i-1})d$ bounded. Let the shifted remainder bounds be r_i^{min} and r_i^{max}, such that

$$r_i^{min} \leq r_i \leq r_i^{max}, \tag{6.3.7}$$

where we shall see that the bounds here turn out to depend on the iteration index i.

Let $[L_d(i), U_d(i)]$ be the selection interval of βr_i for which we can select the digit $d_{i+1} = d$, while keeping the shifted remainder r_{i+1} bounded:

$$\begin{gathered} \beta r_i \in [L_d(i), U_d(i)] \\ \Downarrow \\ r_{i+1}^{min} \leq r_{i+1} = \beta r_i - (2q_i + d\beta^{-i-1})d \leq r_{i+1}^{max}, \end{gathered}$$

and update the root by $q_{i+1} = q_i + d_{i+1}\beta^{-i-1}$, using Corollary 6.3.3. Hence we have the bounds

$$L_d(i) = r_{i+1}^{min} + 2q_id + d^2\beta^{-i-1} \leq \beta r_i \leq r_{i+1}^{max} + 2q_id + d^2\beta^{-i-1} = U_d(i). \tag{6.3.8}$$

With $d = a$, the maximal digit value, we have for βr_i maximal, $\beta r_i = \beta r_i^{max}$,

$$r_{i+1}^{max} = \beta r_i^{max} - (2q_i + a\beta^{-i-1})a \tag{6.3.9}$$

whose solution is $r_i^{max} = 2\mu q_i + \mu^2\beta^{-i}$, which can be checked by insertion. Similarly, the following expression for r_i^{min} can be found, such that the remainder r_i for all i must satisfy

$$r_i^{min} \;=\; -2\mu q_i + \mu^2\beta^{-i} \;\le\; r_i \;\le\; 2\mu q_i + \mu^2\beta^{-i} \;=\; r_i^{max}. \tag{6.3.10}$$

To find the overlap between selection intervals we have using (6.3.8) that

$$\begin{aligned} L_d(i) &= r_{i+1}^{min} + 2q_i d + d^2\beta^{-i-1} = 2q_i(d-\mu) + (d-\mu)^2\beta^{-i-1}, \\ U_d(i) &= r_{i+1}^{max} + 2q_i d + d^2\beta^{-i-1} = 2q_i(d+\mu) + (d+\mu)^2\beta^{-i-1}, \end{aligned} \tag{6.3.11}$$

where it is necessary for the digit selection that each value of βr_i falls in at least one selection interval, i.e., $U_{d-1}(i) \ge L_d(i)$, or

$$U_{d-1}(i) - L_d(i) = (2\mu - 1)\left(2q_i + (2d-1)\beta^{-i-1}\right) \ge 0. \tag{6.3.12}$$

Note that this inequality depends not only on d as in SRT division, but also on the iteration index $i = 1, 2, \ldots$. Since $2\mu - 1 > 0$, for (6.3.12) to hold for all $d \in \{-a, \ldots, a\}$, i.e., for all $d \ge -a = -\mu(\beta - 1)$, and since

$$1 - 2d \le 1 + 2\mu(\beta - 1) = 2\mu\beta - (2\mu - 1) \le 2\mu\beta,$$

in order to insure an overlap of selection intervals it is required that

$$q_i \ge \mu\beta^{-i}. \tag{6.3.13}$$

Since $|t_i| = |\sqrt{x} - q_i| < \mu\,\mathrm{ulp}(q_i) = \mu\,\beta^{-i}$ and $\frac{1}{2} \le \sqrt{x} < 1$, condition (6.3.13) will be satisfied for all $i \ge i'$ for some $i' \ge 1$. Since $\mu \le 1$ it is sufficient to require $q_i \ge \beta^{-i}$, in particular it is possible to use $q_0 = d_0 = 1$ as the initial value. This is necessary whenever $\mu < 1$, since the maximally obtainable value of $q_n = \sum_1^n d_i\beta^{-i}$ without a β^0 term is

$$q_n \le \sum_1^n a\beta^{-i} < \frac{a}{\beta - 1} = \mu.$$

For maximally redundant digit sets where $\mu = 1$, it is feasible to start with $d_0 = 0$ and $d_1 = \lceil \beta/2 \rceil$ yielding $1 > q_1 \ge \frac{1}{2} \ge \beta^{-1}$.

6.3.3 Combining SRT square root with division

Looking for a square root algorithm that is as close as possible to SRT division, we want the selection intervals $[L_d(i), U_{d-1}(i)] = [L_d^i(2q_i), U_{d-1}^i(2q_i)]$ to be independent of i, and as functions of $2q_i$ to coincide with the bounds for division as functions of the divisor y, at least for $i \ge i'$ for some $i' \ge 1$. Since the bounds for root extraction are

$$\begin{aligned} U^i_{d-1}(2q_i) &= 2q_i(d-1+\mu) + (d-1+\mu)^2\beta^{-(i+1)}, \\ L^i_d(2q_i) &= 2q_i(d-\mu) + (d-\mu)^2\beta^{-(i+1)}, \end{aligned} \tag{6.3.14}$$

where the second term in $U^i_{d-1}(2q_i)$ is non-negative and small for large i, this term may be discarded, thus yielding the smaller upper bound

$$U^*_{d-1}(2q_i) = 2q_i(d-1+\mu), \tag{6.3.15}$$

which is now identical to the upper bound $U_{d-1}(y) = y(d-1+\mu)$ for division, when y is substituted by $2q_i$.

However, where the range of the divisor y is the interval $\left[\frac{1}{2};1\right)$, this is unfortunately not identical to the range of $2q_i \approx 2\sqrt{x} \in [1;2)$, using the previously assumed normalization of the radicand $x \in \left[\frac{1}{4};1\right)$. But changing the normalization of the radicand x such that $x \in \left[\frac{1}{16};\frac{1}{4}\right)$, implies $2q_i \approx 2\sqrt{x} \in \left[\frac{1}{2};1\right)$, which now coincides with the interval of divisor y.

Below we shall see how to determine an index i' such that an overlap of selection intervals is insured for all $i \geq i'$. Recalling that for SRT division it is assumed that dividend x and divisor y are normalized such that $-\mu \leq x/y \leq \mu$, both an approximate quotient q_i and an approximate root q_i can be written as $q_i = \sum_1^i d_i\beta^i$, i.e., as a proper fraction.

Rewriting the second of equations (6.3.14) as

$$L^i_d(2q_i) = \left(2q_i + (d-\mu)\beta^{-(i+1)}\right)(d-\mu),$$

we notice that this can be considered a value of the bound for division $L_d(z) = z(d-\mu)$, where the argument $z = 2q_i + \delta^d_i$ is a perturbed version of the argument y used for division, the perturbation being $\delta^d_i = (d-\mu)\beta^{-(i+1)}$. As for division we shall again assume that $\beta r_i > 0$, handling negative remainders by symmetry. Observing then that

$$0 \leq d-\mu \leq a-\mu \quad \text{for} \quad 1 \leq d \leq a,$$

the perturbation δ^d_i can be made arbitrarily small by requiring $i \geq i'$ for some sufficiently large i'. Since the line $L_d(z)$ is to be the left boundary for the uncertainty rectangles, the (non-negative) perturbation can be compensated for by increasing the height of these rectangles by the maximal perturbation.

We shall now turn our attention to the selection function, in the form of determining the step-wise functions $S_d(z)$, where $z = 2q_i$ here takes the place of the divisor y in division. In particular we will first attempt to determine a value i', such that the perturbations $\max_d(\delta^d_i)$ are sufficiently small for $i \geq i'$, using the same truncation parameters u and t as for division. Recalling (5.5.11) for choosing the step function $\widehat{S_d}(\widehat{z})$ for division, but now adding the perturbation in the left bound we get the condition:

$$(d-\mu)(\widehat{z}+\mathrm{ulp}(\widehat{z})+\delta_i^d) \le \widehat{S}_d(\widehat{z}) \le (d-1+\mu)\widehat{z}-\mathrm{ulp}(\widehat{\beta r_i}), \tag{6.3.16}$$

where $\widehat{z}$ is now either $\widehat{y}$ or $\widehat{2q_i}$.

As for division, $\widehat{S}_d(\widehat{z})$ has to be an integer multiple of $\mathrm{ulp}(\widehat{\beta r_i}) = 2^{-t}$, hence defining $\widehat{S}_d(\widehat{z}) = s_{d,k}2^{-t}$ we must require

$$\lceil 2^{t-u}(d-\mu)(k+1+2^u\delta_i^d)\rceil = s_{d,k} \le \lfloor 2^{t-u}(d-1+\mu)k-1\rfloor \tag{6.3.17}$$

using $\mathrm{ulp}(\widehat{z}) = 2^{-u}$, and defining $\widehat{z} = k2^{-u}$, for integer k, $2^{u-1} \le k < 2^u$, assuming that the range of z is the half-open interval $\frac{1}{2} \le z < 1$.

Now we want to determine a minimal bound $\varepsilon > 0$ as a function of i', such that

$$\max_d(\delta_i^d) = (a-\mu)\beta^{-(i+1)} \le \varepsilon = (a-\mu)\beta^{-(i'+1)} \quad \text{for} \quad i \ge i', \tag{6.3.18}$$

preferably without changing the values of the discretization parameters u and t.

However, this will not be possible in general. Let $\Delta(u,t,k,\varepsilon)$ be defined as in Theorem 5.5.3, but modified with the inclusion of a perturbation ε:

$$\Delta(u,t,k,\varepsilon) = \lfloor 2^{t-u}(a-1+\mu)k-1\rfloor - \lceil 2^{t-u}(a-\mu)(k+1+2^u\varepsilon)\rceil. \tag{6.3.19}$$

Consider the case of radix-4 SRT as in Example 5.5.1, where $\Delta(4,4,k,0) \ge 0$ for $8 \le k \le 15$, but $\Delta(4,4,8,\varepsilon) \le -1$ for any $\varepsilon > 0$, since $2^{t-u}(a-\mu)(k+1)$ happens to be integral. Thus (6.3.17) cannot be satisfied and it will be necessary to increase u and/or t.

Inserting the bound ε from (6.3.18) in (6.3.19), changing the function Δ to be a function of i' instead of ε we obtain

$$\begin{aligned}\Delta'(u,t,k,i') = {} & \lfloor 2^{t-u}(a-1+\mu)k-1\rfloor \\ & - \left\lceil 2^{t-u}(a-\mu)\left(k+1+2^u(a-\mu)\beta^{-(i'+1)}\right)\right\rceil,\end{aligned} \tag{6.3.20}$$

whose values must be non-negative for all values of k, $2^{u-1} \le k \le 2^u - 1$ for a chosen set of parameters u, t, and i'.

Theorem 6.3.6 *The digit selection of radix-($\beta \ge 2$) SRT square root can for $i \ge i'$ be implemented using the same algorithms as SRT division, when i' is chosen such that*

$$\Delta'(u,t,2^{u-1},i') \ge 1$$

or

$$\left(\Delta'(u,t,k,i') = 0 \ \textit{for} \ k = 2^{u-1},\ldots,k_1-1, \ \textit{and} \ \Delta'(u,t,k_1,i') \ge 1\right)$$

for suitable values of u and t. These values must satisfy $u \ge u'$ and $t \ge t'$, where u' and t' are the values determined for division by Theorem 5.5.3. It is assumed *that the divisor y is normalized to the interval* $\left[\frac{1}{2};1\right)$ *and the radicand x to the interval* $\left[\frac{1}{16};\frac{1}{4}\right)$.

Proof By Theorem 5.5.3, the choices of u and t insure that digit selection for division is possible for $\beta > 2$; what remains is to show that they are also valid for digit selection for square root when $i \geq i'$ in these cases. The condition of the theorem on u, t, and i' implies by Lemma 5.5.2 that $\Delta'(u, t, k, i') \geq 0$ for all k, $2^{u-1} \leq k \leq 2^u - 1$, such that condition (6.3.17), or equivalently (6.3.16) holds for all digits $d > 0$. But the latter condition on the step function $S(z)$ defined by the points $\widehat{S}_d(\widehat{z})$ is equivalent to requiring that the "height extended" uncertainty rectangles for all k, d, δ_i^d and $i \geq i'$ lie between the lines $L_d(z)$ and $U_{d-1}^*(z)$, both as defined for division.

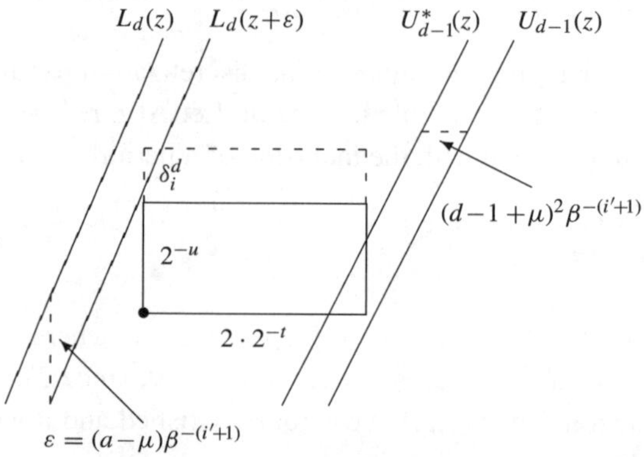

This follows from the equivalent of (5.5.8), specifying that the upper left-hand corner must be to the right of the line $L_d(z)$, as specified by

$$L_d(\widehat{z} + \text{ulp}(\widehat{z}) + \varepsilon) = (d - \mu)(\widehat{z} + \text{ulp}(\widehat{z}) + \varepsilon) \leq \widehat{S}_d(\widehat{z}),$$

and similarly that the midpoint of the lower edge of the rectangle must be to the left of the line $U_{d-1}^*(z)$:

$$\widehat{S}(\widehat{z}) + \text{ulp}(\widehat{\beta r_i}) \leq U_{d-1}^*(\widehat{z}) = (d - 1 + \mu))\widehat{z}.$$

These bounds are equivalent to (6.3.16), thus the uncertainty rectangles, for division as well as for square root, for all digits $d > 0$ and $2^{u-1} \leq k < 2^u$, for $i \geq i'$ are properly located.

What remains is to handle the case in which $\beta = 2$, where, since $\mu = 1$, it is easily seen that the perturbations $\delta_i^d \leq 0$. For $u = t = 1$, checking the cases by insertion, it is found that (6.3.17) is satisfied for all values of $d \in \{-1, 0, 1\}$ and $k = 1$ for all $i \geq 0$. But (6.3.17) in this case is simply a rewriting of the condition that $\Delta'(1, 1, 1, 0) \geq 0$. □

Values of Δ' can thus be used to verify parameter choices, e.g., for $\beta = 4$, minimally redundant, where we know that $u = t = 4$ is sufficient for division, but that u and/or t must be increased. Trying $u = 4$, $t = 5$, and $i' = 3$, we find

$$\{\Delta'(4, 5, k, 3) \mid k = 8..15\} = \{0, 2, 2, 2, 4, 4, 4, 6\},$$

but also for $i' = 2$

$$\{\Delta'(4, 5, k, 2) \mid k = 8..15\} = \{0, 1, 1, 2, 3, 3, 4, 5\},$$

both satisfying Theorem 6.3.6. Choosing instead $u = 5$ and $t = 4$ turns out also to be possible:

$$\{\Delta'(5, 4, k, 2) \mid k = 16..31\} = \{0, 0, 0, 0, 0, 0, 1, 1, 1, 1, 1, 1, 2, 2, 2, 2\}.$$

As illustrations some examples are shown in Table 6.3.1, including the ones found above. The smallest possible values of u, t and i' have been chosen, allowing for non-negative values of $\Delta'(u, t, k, i')$ for all k such that $2^{u-1} \leq k \leq 2^u - 1$.

Table 6.3.1. *Parameters for some combined SRT divide and square root algorithms*

β	a	μ	u	t	i'
2	1	1	1	1	0
4	2	$\frac{2}{3}$	4	5	2
4	2	$\frac{2}{3}$	5	4	2
4	3	1	4	4	1
8	4	$\frac{4}{7}$	7	6	3
8	7	1	5	4	2

However, recall that it is not necessary to check all values of k, it is sufficient to check a few initial values, e.g., for $\beta = 8$, minimally redundant

$$\{\Delta'(7, 6, k, 3) \mid k = 64..127\} = \{1, 1, 0, 1, 1, 1, 2, 1, \ldots, 5, 5, 4, 5, 5, 5, 6, 5\}$$

the first value $\Delta'(7, 6, 64, 3) = 1$ is sufficient to insure that the remaining values are non-negative.

Finally, note that while y is a constant in division, normally given in non-redundant representation, for square root q_i is computed for each step. Hence q_i is likely to be in a redundant representation, but can be converted to a non-redundant representation "on-the-fly," and then truncated to accuracy $\text{ulp}(\widehat{2q}_i) = 2^{-u}$ for digit selection.

It is interesting to see whether the truncation parameters t and u can be chosen such that some constant value, $z = 2q$, found by table look-up in non-redundant form, can be used as an approximation of $\widehat{2q}_i$ during the iterative phase of the square root algorithm, i.e., for $i > i'$, i' suitably chosen.

Assuming that the quotient radix is of the form $\beta = 2^m$, where $m \geq 1$, having determined $2q_{i'}$ (and implicitly digits $d_1, \ldots, d_{i'}$) by table look-up, then

$\text{ulp}(2q_{i'}) = 2^{-mi'+1}$. From the proof of Theorem 6.3.6 it is easy to see that the vertical location and size of an uncertainty rectangle can be determined by a modified condition (6.3.20) on Δ', to remain fixed for $i > i'$ at $\widehat{z} = \widehat{2q} = \widehat{2q}_{i'}$ with $\text{ulp}(\widehat{2q}) = 2^{-mi'+1}$. Since

$$|2q_i - \widehat{2q}| \le 2(|q_i - \sqrt{x}| + |\sqrt{x} - \widehat{q}|) \le 2\,\text{ulp}(\widehat{2q}) = 2^{-mi'+2},$$

it is possible to choose u, t, and i' such that all subsequent points $(2q_i + \delta_i^d, \beta r_i)$ for $i > i'$ are inside such suitably located uncertainty rectangles. We shall, however, not pursue this possibility further, but instead see that it is possible to avoid the initial table look-up phase in the frequently used case of $\beta = 4$.

Example 6.3.2 *(Combined divide and square root for minimally redundant radix 4)* Let $\beta = 4$ with minimally redundant digit set $D = \{-2, -1, 0, 1, 2\}$. From Table 6.3.1 we choose $u = 4$, $t = 5$, and $i' = 2$. We can now compute the values of the constants $\widehat{S}(\widehat{y}) = s_{d,k}2^{-t}$ for $d > 0$ as

$$s_{d,k} = \left\lceil 2^{t-u}(d - \mu)(k + 1 + 2^u\varepsilon)\right\rceil = \left\lceil 2(d - \tfrac{2}{3})(k + \tfrac{4}{3})\right\rceil,$$

resulting in the following table of comparison constants valid for square root digit selection for $i \ge 2$ (and for all $i \ge 1$ for division), assuming z is in the interval $[\frac{1}{2}; 1)$:

$k = 16\widehat{z}$	8	9	10	11	12	13	14	15
$32\widehat{S}_1$	7	7	8	9	9	10	11	11
$32\widehat{S}_2$	25	28	31	33	36	39	41	44

Utilizing that the definitions of the functions $L_d(z)$ and $U_d^*(z)$ are the same as for division, combining with the table of $\widehat{S}_d$, yields the P-D diagram for the first quadrant as shown in Figure 6.3.2.

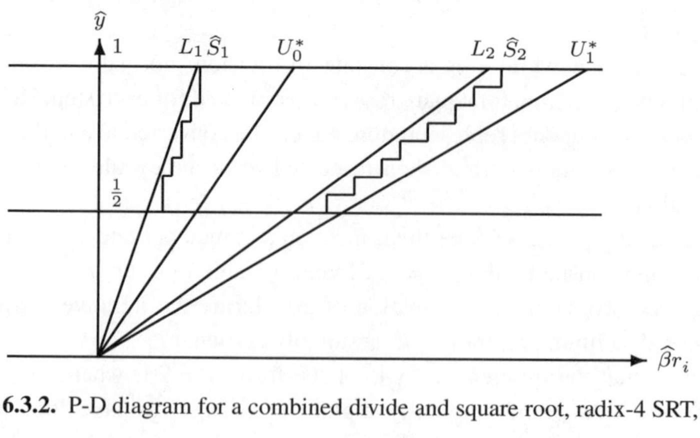

Figure 6.3.2. P-D diagram for a combined divide and square root, radix-4 SRT, $i > 1$.

Since $q_i = \sum_1^i d_i 2^{-i} \approx \sqrt{x} \in \left[\frac{1}{4}; \frac{1}{2}\right)$ and $i' = 2$, it is sufficient initially to determine d_1, before the general SRT algorithm can be applied. We now want to see if it is possible to find common selection intervals for d_1, valid for division as well as square root.

For the square root of the radicand x, there are two possibilities for the choice of the first digit, $d_1 \in \{1, 2\}$, as seen from the following based on $r_0 = x$:

$d_1 = 1$: here $q_1 = \frac{1}{4}$, thus $r_1 = 4x - \frac{1}{4}$, and the bounds $r_1^{min} \leq 4x - \frac{1}{4} \leq r_1^{max}$ are satisfied for $\frac{1}{144} \leq x \leq \frac{25}{144}$;

$d_1 = 2$: here $q_1 = \frac{1}{2}$, thus $r_1 = 4x - 1$, and the bounds $r_1^{min} \leq 4x - 1 \leq r_1^{max}$ are satisfied for $\frac{16}{144} \leq x \leq \frac{56}{144}$,

using the remainder bounds $r_i^{min} \leq r_i \leq r^{max}$ from (6.3.10). Note that the whole range $\frac{1}{16} \leq x < \frac{1}{4}$ is covered, with an overlap interval $\left[\frac{16}{144}; \frac{25}{144}\right]$ of width $\frac{1}{16}$, where a choice of d_1 is possible. Translating these bounds on x into bounds on $\beta r_0 = 4x$ we find

$$d_1 = 1 \quad \text{can be chosen for} \quad \frac{1}{4} \leq \beta r_0 \leq \frac{25}{36}$$

$$d_1 = 2 \quad \text{can be chosen for} \quad \frac{16}{36} \leq \beta r_0 \leq 1,$$

with an overlap of width $\frac{1}{4}$ between these intervals.

Note that $q_0 = 0$, hence the value of $\widehat{2q_0} = 0$ cannot be used for digit selection together with $\widehat{\beta r_0} = \widehat{4x}$, since $\widehat{z} = \widehat{2q_0}$ is supposed to be in the interval $\left[\frac{1}{2}; 1\right)$. But we are free to choose any suitable value for $\widehat{2q_0}$ (or k) in the comparisons or table look-up to determine $d_1 \in \{1, 2\}$; we may even translate or scale the initial remainder βr_0, as long as we insure that the standard digit selection chooses the proper value of d_1 according to the selection intervals just found.

It is thus possible just to use $2\widehat{\beta r_0} = 2 * \widehat{4x}$ instead of $\widehat{4x}$ for the initial digit selection. Recall that the entries in the table of selection constants are lower bounds for the digit selections, since $2 * \frac{16}{36} = \frac{32}{36} < \frac{31}{32}$, then $\frac{31}{32}$ can be used as a lower bound for selecting $d_1 = 2$, choosing $k = 10$ for $i = 1$. The corresponding lower bound for selecting $d_1 = 1$ is then $2 \times \frac{1}{4} = \frac{16}{32}$, with an upper bound way beyond the lower bound for choosing $d_1 = 2$. Note that multiplication by the factor 2 is permissible, since the uncertainty rectangle of βr_0 here only has half the width because $\widehat{\beta r_0} = \widehat{4x}$ is in non-redundant representation, thus it has only half the error as values of $\widehat{\beta r_i}$ used in later iterations.

Finally, there is another minor problem with the digit selection for square root determination. If the initial digit $d_1 = 2$ is chosen, then $q_1 = \frac{1}{2}$ which is not in the interval $\left[\frac{1}{4}; \frac{1}{2}\right)$, and there are no selection constants corresponding to the value $k = 16$ in the table above. Of course, entries for $k = 16$ (and other larger values of k) can easily be added, but for a table look-up implementation this would imply that one additional bit would be needed in the address k for the look-up. But since the entries for $k = 16$ would be 12 and 47, it is easily seen from the P-D diagram in Figure 6.3.2 that there is plenty of "room" to change the values for $k = 15$ from 11 respectively 44 to the values 12 respectively 47, and just use the selection constants for $k = 15$ instead of the missing for $k = 16$. Thus the following table of selection constants is valid for radix-4 division and square root for all $i \geq 1$:

$k = 16\hat{z}$	8	9	10	11	12	13	14	15,16
$32\widehat{S}_1$	7	7	8	9	9	10	11	12
$32\widehat{S}_2$	25	28	31	33	36	39	41	47

using $k = 10$ and $2\widehat{\beta r}_0 = 2 * \widehat{4x}$ instead of $\widehat{4x}$ for the initial square root digit selection. □

Problems and exercises

6.3.1 Develop an algorithm for binary restoring square root based on Theorem 6.3.2.

6.3.2 Check the solution to (6.3.9) by insertion.

6.3.3 Develop the details of a radix-2 SRT square root algorithm, reusing information from Example 5.5.3.

6.3.4 In the SRT square root algorithms, $2q_i$ replaces the divisor y in SRT division. As discussed on page 414 it is possible to determine an index i' such that a constant value $2q$ can be used instead of $\widehat{2q}_i$ for $i \geq i'$. For minimally redundant, radix 4 determine this value, and discuss the implications for an initial table look-up of it, as well as the corresponding initial root digits.

6.4 Multiplicative high-radix square root

The traditional "school method" of digit-serial high-radix square root can be extended to a high *root radix* β^k employing a rectangular multiplier. For high-radix square root the radix β is the host radix of the arithmetic unit. The high radix β^k with $k \geq 2$ is the radix of the digit-serial procedure. Consider the radicand x with root, remainder pair (q, r) and tail t satisfying $x = q^2 + r = (q + t)^2$. In the restoring (hand computation) method a next root-digit $d = \lceil t\beta \rfloor$ is obtained by estimating $r/2q$ corresponding to the upper bound of the range $r/2\sqrt{x} \leq t \leq r/2q$ from Lemma 6.3.1. In the multiplicative non-restoring *short reciprocal square root* procedure of this section, the successor high-radix root-digit is obtained, with reference to the lower bound from Lemma 6.3.1, by multiplying half the normalized remainder $r/2$ by a precomputed *short root-reciprocal* ρ, satisfying $|\rho - 1/\sqrt{x}| < \beta^{-k}$. Obtaining root-digit selection by rounding a fixed size, short-by-long product $d = \lceil \frac{r}{2}\rho\beta^k \rfloor \approx \lceil t\beta^k \rfloor$ is sufficient for digit-serial development of the square root, and avoids the variable precision division suggested by the approximation $t \approx r/2q$. The successor root, remainder pair (q', r') with $q' = q + d\beta^{-k}$ and $r' = (x - (q')^2) = r - (2q + d)d\beta^{-k}$ is then obtained employing a short-by-long multiply-subtract operation for remainder update.

Example 6.4.1 Three (2×7)-digit decimal multiply-subtract remainder updates are employed in computing the six-digit decimal square root $\sqrt{.651795}$ as a three-root-digit result in the high radix $10^2 = 100$. For convenience in decimal digit selection, note that we shall use a table look-up three-digit decimal short half-root reciprocal $\rho/2 = .619$ obtained from $1/(2 \times \sqrt{.651795}) = 0.6193\cdots$, and rounded root $q_1 = \mathrm{R}(\sqrt{.651795}) = .80$. Given the initial root, remainder pair $(.80, 1.1795 \times 10^{-2})$, then $d_2 = \lceil .619 \times 1.1795 \times 10^2 \rfloor = \lceil \mathbf{73}.01105 \rfloor = 73$ and $d_3 = \lceil .619 \times 0.6171 \times 10^2 \rfloor = \lceil \mathbf{38}.19849 \rfloor = 38$, showing that short-by-long multiplies are sufficient to obtain the second and third high radix root-digits in this case:

$$
\begin{array}{rrl}
 & .80 \quad 73 \quad 38 & \\
 & \sqrt{.6517\ 9500\ 0000} & \\
(0 + .80) \times .80 = & .6400 \qquad\qquad\quad & \\
\hline
.619\times & 117\ 9500 \qquad & (= \mathbf{73}.0\,..)\ \text{second digit selection} \\
(16000 + 73) \times 73 = & 117\ 3329 \qquad & \\
\hline
.619\times & 6171\ 0000 & (= \mathbf{38}.1\,..)\ \text{third digit selection} \\
(1614600 + 38) \times 38 = & 6135\ 6244 & \\
\hline
 & 35\ 3756 & \text{third remainder}
\end{array}
$$

□

Continuing the algorithm yields $d_4 = \lceil .619 \times .353756 \times 10^2 \rfloor = \lceil 21.89 \cdots \rfloor = 22$ as the fourth root-digit. This will result in a negative fourth remainder, implying as expected that a redundant root-digit set is necessary to allow for a non-restoring algorithm. Computing the fourth remainder will also require a larger precision multiplication.

We assumed the first digit to be given by other means, e.g., a table look-up. Actually, since the "short half-root reciprocal" is an approximate value of $\frac{1}{2\sqrt{.651795}}$ the digit could be chosen as the round-to-nearest value of $.619 \times (2 \times 65.1795) = 80.69\cdots$. This would provide 81 as the first root-digit, implying that the next root-digit would become negative. Thus the first digit could be chosen by the same iterative procedure seeding the algorithm with twice the radicand as an initial remainder. Since the short root reciprocal is likely to be determined by a table look-up, the first root-digit and its square might as well be looked up in parallel to expedite the computation by one iteration.

The distinctions between the short reciprocal square root procedure and short reciprocal division are summarized as follows.

- *Digit selection* For square root an approximation of $1/2\sqrt{x}$ is employed to scale the remainder to select the next root-digit in contrast to an approximation of $1/y$ for division. Since $1/2\sqrt{x} = \lim_{t\to 0}(1/(2q + t))$ is itself an approximation to the (variable) term $1/(2q + t)$ required for the identity $(1/(2q + t)) \times r = t$, more analysis is needed for the square root-digit selection procedure to insure that all digit selections are confined to an acceptable redundant root-digit set, to establish the non-restoring feature for short reciprocal square root.
- *Remainder update* For square root the factor $2q + d\beta^{-k}$ must be updated after each root-digit selection and grows from the initial "short size" corresponding to $\ell = 0$, to a precision greater than p digits when the root q has p or more digits. Thus obtaining a square root of precision more than one guard high-radix root-digit greater than the larger dimension of a supporting rectangular multiplier will require a multiple precision multiplication for remainder update. In contrast for division an arbitrarily high precision quotient can be obtained digit serially employing a single fixed precision short-by-long multiply of quotient digit by divisor for remainder update after each quotient digit selection.

In this section we also investigate an extension of short reciprocal square root termed "prescaled square root." This extension is initiated by precomputing a short radicand-reciprocal prescaling factor $\rho^2 \approx 1/x$ along with a high-precision postscaling-multiplicand $z \approx 1/\rho$ satisfying $|z - 1/\rho| < \beta^{-nk}$ for a target $p = nk$ digit radix-β square root. The radicand x is then prescaled by ρ^2, with the short root reciprocal algorithm applied to the scaled radicand $\rho^2 \times x \approx 1$. For radicands near unity, digit selection is simplified to selecting one half the shifted remainder (avoiding multiplication). The trade-off cost is a multiplication of each resulting scaled quotient digit by the nk-digit radix-β postscaling multiplicand z employed to yield partial square root terms accumulated to provide a value for $\sqrt{x}$.

Computationally, we show that a $((k + 2) \times (nk + 1))$-digit radix-$\beta$ multiplier can be employed to support digit selection (or digit postscaling) as well as remainder updating for determining an n-digit high-radix square root, which then provides a result of precision $p = nk$ digits in the host radix β. Methods for obtaining either an n-digit radix-β^k root, remainder pair or (more simply) just an n digit one-ulp square root are described with computational latencies determined from the following.

- The *short reciprocal square root* algorithm allows an n-digit high radix-β^k root, remainder pair to be determined employing $2n - 1$ dependent multiplications.
- The *prescaled square root* algorithm allows an n-digit high-radix-β^k one-ulp square root to be determined employing $(2n + 1)$ multiplications where

at most $(n+1)$ are dependent. The method can be extended to also provide the corresponding remainder employing a total of at most $(3n-1)$ multiplications with the same latency of $n+1$ dependent multiplications.

6.4.1 Short reciprocal square root

Short reciprocal square root is a high-radix digit-serial square root procedure employing a redundant digit set to provide a non-restoring algorithm. With a host radix β and radicand x normalized at the granularity of the host radix so that $1/\beta^2 \le x < 1$, the square root $\sqrt{x}$ will be a normalized fraction $1/\beta \le \sqrt{x} < 1$, which we intend to develop digit serially in the high radix β^k with $k \ge 2$. Let $q_i = d_1\beta^{-k} + d_2\beta^{-2k} + \cdots + d_i\beta^{-ik}$ denote an i-digit high-radix one-ulp square root with $\beta^{k-1} \le d_1 \le \beta^k$, $|d_j| \le \beta^k - 1$, for $2 \le j \le i$. Let the *ith tail* $t_i = (\sqrt{x} - q_i)$ satisfy $|t_i| < (1-\beta^{-k})\beta^{-ik}$ with corresponding *ith remainder* $r_i = (x - q_i^2) = (2q_i + t_i)t_i = t_i(\sqrt{x} + q_i)$.

Ideally the $(i+1)$th high-radix digit of $\sqrt{x}$ would be the round-to-nearest value $\lceil t_i\beta^{(i+1)k} \rfloor$ yielding $|t_{i+1}| \le \frac{1}{2}\beta^{-ik}$. In short reciprocal square root we shall instead select the $(i+1)$th high-radix digit by $d_{i+1} = \lceil \rho\beta^k \mathrm{R}(r_i) \rfloor = \lceil \rho\beta^k r_i + \delta \rfloor$, where δ represents the effect of using a rounded value $\mathrm{R}(x)$ instead of r_i. Here $t_{i+1} = t_i\beta^k - d_{i+1}$ must satisfy $|t_{i+1}| < 1 - 1/\beta^k$. This requires us to determine bounds on three independent approximations so as to jointly satisfy $|t_i\beta^k - (\rho\beta^k r_i + \delta)| < \frac{1}{2} - 1/\beta^k$. Specifically we must bound

(i) $\left|t_i\beta^k - (\beta^k/2\sqrt{x})r_i\right|$: this is the "remainder scaling factor" error
(ii) $\left|(\rho/2) - (\beta^k/2\sqrt{x})\right|$: this is the discretization error ε in determining the short half root reciprocal $\rho/2 \approx 1/2\sqrt{x}$ by table lookup assisted methods based on a leading part of x;
(iii) $|\delta|$: this is the error tolerance in using an approximate remainder $\mathrm{R}(x)$ for determining d_{i+1}, where only a leading digit portion of a redundant representation of r_i is employed.

Definition 6.4.1 *For the radicand x with $1/\beta^2 \le x < 1$, let $\rho/2$ be an η-ulp short half root reciprocal for the radicand $x \in \mathbb{Q}_\beta$ satisfying*

$$\frac{\rho}{2} = \frac{1}{2\sqrt{x}}(1 + \varepsilon\beta^{-k}) \textit{ with } |\varepsilon| < \eta. \tag{6.4.1}$$

Note that $1 < \rho/2 \le \beta$.

Determining required accuracy of $\rho/2$. Here $\rho/2$ must be determined using a rounded value $\mathrm{R}(x)$, so to obtain $|\varepsilon| \le \eta$ for some suitable value of η, it is sufficient to require

$$|x - \mathrm{R}(x)| \le \frac{2\eta}{\beta^k}\mathrm{R}(x) \text{ for all } \frac{1}{\beta^2} \le x < 1,$$

or $|x - \mathrm{R}(x)| \le 2\eta\beta^{-(k+2)}$. Let $\mathrm{R}(x) = \lceil \beta^{k+g} x \rfloor / \beta^{k+g}$ be a $(k+g)$-digit nearest rounding of x, employing g guard digits, then it follows that $|\varepsilon| < \eta = 1/4\beta^{g-2}$ will be sufficient.

The digit recurrence loop. It follows that $\sqrt{x}$ may be developed digit serially for all $i \ge 2$, with recursively determined maximally redundant digits in the high radix β^k, obtained by rounding (with some tolerance) the product of the ith remainder and the $(k+g)$-digit short half-root reciprocal $\rho/2$, chosen with $g = 2$ so that $\eta = \frac{1}{4}$, provided we have suitably initiated the recursion.

To obtain proper starting values, let $\mathrm{R}(x) = \lceil \beta^{k+2} x \rfloor / \beta^{k+2}$ be a $(k+2)$-digit rounding of x to be used for a table look-up to determine both

$$\frac{\rho}{2} = \left\lceil \frac{\beta^{k+2}}{2\sqrt{\mathrm{R}(x)}} \right\rfloor \beta^{-k+2} \quad \text{and} \quad d_1 = \left\lceil \beta^k \sqrt{\mathrm{R}(x)} \right\rfloor,$$

from which it can be noted that $\beta^{k-1} \le d_1 \le \beta^k$, and q_1, r_1 can be defined as

$$q_1 = d_1 \beta^{-k} \quad \text{and} \quad r_1 = (x - q_1^2).$$

It is now possible to bound t_1 by

$$\begin{aligned} |t_1| &= |\sqrt{x} - q_1| = |\sqrt{x} - d_1\beta^{-k}| \\ &\le \left|\sqrt{x} - \sqrt{\mathrm{R}(x)}\right| + \left|\sqrt{\mathrm{R}(x)} - \lceil \sqrt{\mathrm{R}(x)} \rfloor\right| \le \left(\frac{1}{4\beta^{k+1}} + \frac{1}{2}\right)\beta^{-k} \le \frac{17}{32}\beta^{-k}, \end{aligned} \tag{6.4.2}$$

since

$$\left|\sqrt{x} - \sqrt{\mathrm{R}(x)}\right| = \frac{|x - \mathrm{R}(x)|}{\sqrt{x} + \sqrt{\mathrm{R}(x)}} \le \frac{\frac{1}{2}\beta^{-(k+2)}}{2} = \frac{1}{4\beta^{k+2}}.$$

Lemma 6.4.2 *Given the radicand x with $1/\beta^2 \le x < 1$, and an i-digit root radix β^k, $k \ge 2$, one-ulp root, scaled-remainder pair (q_i, r_i) with ith tail $t_i = (\sqrt{x} - q_i)$ satisfying $|t_i|\beta^{ik} < 1 - 1/\beta^k$ and $r_i = t_i(\sqrt{x} + q_i)$ for $i \ge 1$, then t_i and r_i satisfy the identity*

$$t_i - \frac{1}{2\sqrt{x}} r_i = \frac{t_i^2}{2\sqrt{x}} \qquad \text{for } i \ge 1, \tag{6.4.3}$$

and inequalities

$$0 \le t_i\beta^k - \frac{\beta^k}{2\sqrt{x}} r_i < \frac{1}{2\beta^{k-1}} \qquad \text{for } i \ge 2, \tag{6.4.4}$$

$$\frac{|r_i|}{2\sqrt{x}} < \beta^{-k} \quad \text{for } i \ge 1. \tag{6.4.5}$$

Proof Note that

$$t_i - \frac{r_i}{2\sqrt{x}} = \frac{t_i 2\sqrt{x} - t_i(\sqrt{x}+q_i)}{2\sqrt{x}} = t_i \frac{\sqrt{x}-q_i}{2\sqrt{x}} = \frac{t_i^2}{2\sqrt{x}},$$

proving (6.4.3), from which (6.4.4) follows easily. But also note that $r_i/2\sqrt{x} = \left(1 - t_i/2\sqrt{x}\right) t_i$ so that $r_i/2\sqrt{x} < 1$ for r_i and t_i both non-negative and $i \geq 1$. From (6.4.3) and (6.4.4) for r_i and t_i negative and $i \geq 2$

$$\frac{-r_i\beta^{-ik}}{2\sqrt{x}} < -t_i\beta^{-ik} + \frac{1}{2\beta^{2k-1}} < 1 - \frac{1}{\beta^k} + \frac{1}{2\beta^{2k-1}} < 1,$$

and for $i = 1$ from (6.4.3)

$$\frac{-r_1}{2\sqrt{x}} < -t_1 + \frac{1}{2\sqrt{x}\beta^k} < \frac{17}{32} + \frac{1}{2\beta^{k-1}} < 1,$$

proving (6.4.5) for $i \geq 1$. □

Inequality (6.4.4) provides the basis for determining the discretization error that can be allowed in a table look-up error for an approximation of $1/2\sqrt{x}$. The first and major result only applies for $i \geq 2$, implying that, in general, there is a problem with determining d_2 as illustrated in the following decimal example.

Example 6.4.2 Let $\beta = 10$ and $k = 4$, then for $x = 0.0408625$, using two guard digits in determining $\mathrm{R}(x)$, the following values are found: $\rho/2 = 2.47346$, $d_1 = 2021$, $r_1 = 0.180900 \times 10^{-4}$, $t_1 = 0.447501 \times 10^{-4}$, so that $t_1 10^8 = 4475.01$, whereas $(\rho/2)r_1 10^8 = 4474.49$ is obtained yielding $d_2 = 4474$, and resulting in $t_2 = 1.011734 \times 10^{-8}$. Owing to d_2 being too small, the algorithm fails as indicated by the following choice of $d_3 = 10117 > 10^4$. Had the value $d_2 = 4475$ been chosen, then $d_3 = 117$ would have been the result. □

For values of $\beta > 2$ it is, of course, possible to determine suitable values of d_2 by table look-up, along with d_1 and $\rho/2$, by precalculating and checking the values for each entry during the table construction. Fortunately, for the common binary case $\beta = 2$, the second digit may be computed by the same expression as the subsequent digits.

More importantly note that while $|d_i| \leq \beta^k - 1$ for $i > 1$, d_1 may take the value β^k, which may not be convenient for an implementation. It will happen when x is very close to 1, in which case the next non-zero digit must be negative, as x is strictly smaller than 1, and thus the remainder r_1 must be negative. However, it is possible in the table look-up determining d_1, in this situation to force its value to be $\beta^k - 1$, thereby leaving a positive remainder r_1, forcing the next digit to be positive. The problem may reoccur, hence it will be necessary in the loop to check for $d_{i+1} = \beta^k$ and possibly force $d_{i+1} = \beta^k - 1$.

Binary short reciprocal square root. For a binary-based root radix 2^k a short reciprocal needs to be selected with a relative error factor $(1 + \varepsilon 2^{-k})$ having $|\varepsilon| < \eta = \frac{1}{2} - 1/2^{k-1} - |\delta|$ for a suitable δ. For any $k \geq 8$, this requires only that $\eta < \frac{1}{2} - \frac{1}{128} - |\delta|$, that is, just slightly less than $\frac{1}{2}$ for the sum $(\eta + |\delta|)$. For radicands x satisfying $\frac{1}{4} \leq x < 1$, we use separate tables for the two binades $[\frac{1}{4}; \frac{1}{2})$ and $[\frac{1}{2}; 1)$. Employing the terminology for direct binary table look-up used in Section 5.8, we provide a brief overview of table size for selecting a short root reciprocal that allows a useful rounding tolerance.

Short root reciprocal table.

- Radicand range $\frac{1}{2} \leq x < 1$: A $(k+1)$-bits-in, $(k+1)$-bits-out look-up table has an input relative error bound of $(1/8x)2^{-k}$ which is reduced to about $(1/16x)2^{-k}$ for the function $1/2\sqrt{x}$, along with an additional output rounding relative error bound of $(\sqrt{x}/4)2^{-k}$ for determining $2^k \rho/2 = 1\,b'_{k-2}\,b'_{k-3} \cdots b'_0\,.\,b'_{-1}b'_{-2}$, yielding a total relative error bound $\eta = \frac{5}{16}2^{-k}$ (approached as $x \to 1$), allowing $\delta = \frac{23}{128}$ for any $k \geq 8$.
- Radicand range $\frac{1}{4} \leq x < \frac{1}{2}$: A k-bits-in, $(k+1)$-bits-out look-up table similarly yields a bound of about $(1/16x)2^{-k}$ for the function $1/2\sqrt{x}$, along with an additional output rounding relative error bound limited to $(\sqrt{x}/4)2^{-k}$ yielding a total relative error bound $\frac{3}{8}2^{-k}$ (approached as $x \to \frac{1}{4}$) allowing $\delta = \frac{15}{128}$ for any $k \geq 8$. With a $(k+1)$-bits-in, $(k+1)$-bits-out table the total relative error bound is reduced to $\frac{1}{4}2^{-k}$, allowing $\delta = \frac{31}{128}$ for any $k \geq 8$.

Leading root digit table.

- Separate $(k+1)$-bits-in, $(k-1)$-bits-out tables for x in each binade $\frac{1}{4} \leq x < \frac{1}{2}$ and $\frac{1}{2} \leq x < 1$ yield an input relative error bound of $\frac{1}{4}2^{-k}$ (approached as $x \to \frac{1}{4}$, respectively $x \to \frac{1}{2}$). This relative error is reduced to about $\frac{1}{8}2^{-k}$ for the function $\sqrt{x}$ which then contributes an absolute error bound limited to $\frac{1}{8}2^{-k}$, so the rounding determining d_1 for the table gives a tail t_1 with $|t_1|$ essentially bounded by $\frac{5}{8}2^{-k}$ (approached as $x \to 1$).

6.4.2 Prescaled square root

The idea of this algorithm is to prescale a radicand x by a *short radicand-reciprocal* prescaling factor $\rho^2 = \mathrm{R}(1/x)$ so that application of the short reciprocal square root procedure to the *scaled radicand* $\rho^2 x$ allows high-radix digit selection from essentially just shifting and rounding the remainder. Specifically, if the scaled radicand $\rho^2 x$ is sufficiently close to unity that $|\sqrt{\rho^2 x} - 1| < \frac{2}{5}\beta^{-k}$, then the following initial high-radix selections are:

- first root-digit: $d_1 = \beta^k$,
- first partial root: $q_1 = 1.0$,
- first tail bound: $|t_1| < \frac{5}{8}\beta^{-k}$,
- first remainder: $r_1 = \rho^2 x - 1$.

It is interesting to note that the first remainder $r_1 = \rho^2 x - 1$ is the relative error $r_1 = \varepsilon\beta^{-k}$ in specifying the short-radicand reciprocal $\rho^2 = (1/x)(1 + \varepsilon\beta^{-k})$.

Each root radix digit d_i can then be concurrently "postscaled" by an independent multiplication by a *postscaling multiplicand* $z = \mathrm{R}(1/\rho)$ predetermined to an accuracy of $nk + 1$ radix-β digits. Note that if the short radicand reciprocal $\rho^2 = \mathrm{R}(1/x)$ is determined by direct table look-up from the equivalent of 6–12 bits of the normalized radicand x, then the corresponding postscaling factor z can be given as a companion table entry as a half-ulp $(nk + 1)$-digit radix-β rounded value $\mathrm{RN}(1/\rho)$. The postscaling products $d_i z\beta^i$ provide partial square roots that are accumulated to obtain the root $q = \sum_{i=1}^{n} d_i z\beta^{-nk}$ in the same manner as digit serial multiplication.

Example 6.4.3 Let the six-digit decimal (three-digit high-radix 10^2) radicand $x = .651795$ be prescaled by a short reciprocal factor $\rho^2 = 1.54 = \frac{1}{.651795}(1 + \varepsilon 10^{-2})$ with $\varepsilon = .37643$ giving the scaled radicand $\rho^2 x = 1.0037\,6430$. The second and each subsequent high-radix digit selection for $\sqrt{\rho^2 x}$ is available by rounding one-half the shifted scaled remainder's integer portion (where one-half of an odd integer rounds away from zero).

$$
\begin{array}{rlll}
 & 1.00 \quad 19 \quad \overline{20} & & \\
 & \sqrt{1.0037 \quad 6430 \quad 0000} & & \text{scaled radicand} \\
 & 1.0000 & & \\
\frac{1}{2}\times & \underline{37}. \quad 6430 & & (= 19)\ \text{second digit selection} \\
([2.00]19) \times 19 = & -[38.] \quad 0361 & & \\
\frac{1}{2}\times & -\underline{39}.31 \quad 0000 & & (= \overline{20})\ \text{third digit selection} \\
([2.00]3780) \times (-20) = & [40.]07 \quad 5600 & & \\
 & \underline{76}. \quad 5600 & & (= 38)\ \text{guard digit selection}
\end{array}
$$

The radicand reciprocal $\rho^2 = 1.54$ has the associated seven-digit decimal postscaling multiplicand $z = \mathrm{RN}(1/\rho) = \left\lceil \frac{10^7}{\sqrt{1.54}} \right\rfloor 10^{-7} = \lceil 8058229.6\cdots \rfloor 10^{-7} = 0.8058\,230$ allowing concurrent digit-serial accumulation of the root $q_n(x) = \sum_1^n z d_i \beta^{-nk}$, employing only a short-by-long multiplier.

$$
\begin{array}{rll}
 & 0.\underline{80}\ 58\ 23\ 00\cdots & q_1(x) = \mathrm{RN}(1/\rho)\ \text{(first partial root)} \\
19 \times 10^{-4} \times (.8058\,230) & 0.00\ 15\ 31\ 06\cdots & \text{second partial root} \\
 & 0.\underline{80\ 73}\ 54\ 06\cdots & q_2(x) \\
-20 \times 10^{-6} \times (.8058\,230) & -0.00\ 00\ 16\ 11\cdots & \text{third partial root} \\
 & 0.\underline{80\ 73\ 37}\ 95\cdots & q_3(x) \\
38 \times 10^{-8} \times (.8058\,230) & 0.00\ 00\ 00\ 30\cdots & \text{guard partial root} \\
 & 0.\underline{80\ 73\ 38}\,(25)\cdots & q_4(x)
\end{array}
$$

Then $q = \lceil 10^6 q_4(x) \rfloor 10^{-6} = 0.807338$ is a one-ulp square root of .651795. Note that the postscaling can also be postponed to the end when all digits have been found, and then performed with a single multiplication, but then this must be a full, long-by-long multiplication. Depending on the architecture, this may or may not be faster than performing a number of independent short-by-long rectangular multiplications, pipelined with the derivation of the digits, as performed above. □

In the following we shall restrict the discussion to the case $\beta = 2$.

Definition 6.4.3 *For the high radix 2^k with $k \geq 2$, precision nk with $n \geq 2$, and radicand x with $\frac{1}{4} \leq x < 1$, let the $(k+1)$-bit* short radicand reciprocal ρ^2 *with $1 \leq \rho^2 \leq 4$ and corresponding $(nk+1)$-bit (long) normalized postscaling multiplicand z with $\frac{1}{2} \leq z \leq 1$ satisfy*

$$\rho^2 = \frac{1}{x}\left(1 + \varepsilon 2^{-k}\right) \text{ with } |\varepsilon| < \frac{5}{8},$$

$$z = \left\lceil \frac{2^{nk+1}}{\rho} \right\rceil 2^{-(nk+1)}.$$

Setting the specific relative error bound $|\varepsilon| < \frac{3}{4}$ with specific sizes of $k+1$ digits for ρ^2 and $nk+1$ digits for z simplifies the presentation without placing significant additional resource requirements on an implementation. It follows that $|\sqrt{\rho^2 x} - 1| < \frac{5}{16}2^{-k}$, allowing the initial values for (d_1, q_1, r_1) previously noted.

In practice, it is convenient to have both ρ^2 and z provided by a direct look-up table based on a rounded off leading part index R(x). For example, from Theorem 5.8.5, a k-bits-in, k-bits-out reciprocal table provides a $(k+1)$-bit value for ρ^2 with relative error less than $\frac{3}{4}2^{-k}$ for any $k \geq 3$, so a table of 2^k entries for ρ^2 and z is sufficient for each input binade $\frac{1}{4} \leq x < \frac{1}{2}$ and $\frac{1}{2} \leq x < 1$.

The prescaled short reciprocal (one-ulp) square root algorithm is characterized by the recursive computation of the 4-tuples $(d_i, q_i, r_i,\ q_i(x))$ satisfying the bound $|\sqrt{x} - q_i(x)| < 2^{-ik}$ with terms defined as follows: $q_i(\rho^2 x) = d_1 2^{-k} + d_2 2^{-2k} + \cdots + d_i 2^{-ik}$ denotes an ik-bit one-ulp *scaled square root* of the *scaled radicand* $\rho^2 x$ for $1 \leq i \leq n+1$, where $d_1 = 2^k$ and $|d_j| \leq 2^k - 1$ for $2 \leq j \leq n+1$, with ith scaled remainder $r_i = (\rho^2 x - q_i^2)2^{ik}$ and ith scaled tail $t_i = (\sqrt{\rho^2 x} - q_i)2^{-ik}$ for $1 \leq i \leq n$; $q_i(x) = \sum_{j=1}^{i} z d_j 2^{-kj}$ denotes the ith partial sum of the postscaled digit terms for $1 \leq i \leq n+1$, with $q_{n+1}(x)$ including the contribution from the guard digit d_{i+1} available from the nth scaled remainder r_n. The procedure is initiated for $i = 1$ with $d_1 = 2^k,\ q_1(\rho^2 x) = 1,\ q_1(x) = z$, and $r_1 = (\rho^2 x - 1)2^k$.

Theorem 6.4.4 *Given a radicand x with $\frac{1}{2^2} \leq x < 1$, a high radix 2^k with $k \geq 6$, the radicand reciprocal ρ^2 and postscaling multiplicand z with $d_1 = 2^k,\ q_1 = 1,\ r_1 = (\rho^2 x - 1) = \varepsilon$ with $|\varepsilon| < \frac{3}{4}$, and $q_1(x) = f$, the* 4*-tuples $(d_{i+1}, q_{i+1}, r_{i+1}, q_{i+1}(x))$ may be determined digit serially with $|d_{i+1}| \leq 2^k - 1$ and with the ith tail $t_{i+1} = t_i 2^k - d_{i+1}\, 2^{-ik}$ satisfying the invariant $|t_{i+1}|2^{ik} < 1 - \frac{1}{2^k}$ for $i = 1, 2, \ldots, n$, by*

- $d_{i+1} = \lceil r_i/2 \rfloor$,
- $q_{i+1} = q_i + d_{i+1}2^{-(i+1)k}$,
- $r_{i+1} = r_i 2^k - \left(2q_i + d_{i+1}2^{-(i+1)k}\right) d_{i+1}$,
- $q_{i+1}(x) = q_i(x) + z \times d_{i+1}2^{-(i+1)k}$.

Furthermore,

$$|\sqrt{x} - q_i(x)| < \begin{cases} 2^{-ik} & \textit{for } 1 \le i \le n-1, \\ \frac{3}{2}2^{-nk} & i = n, \\ \frac{1}{2}2^{-nk} & i = n+1. \end{cases} \tag{6.4.6}$$

Proof It is sufficient to show that remainder scaling by $\frac{1}{2}$ satisfies the defining constraint (6.4.1), so that the validity of the digit-serial loop for $d_{i+1}, q_{i+1}, r_{i+1}$ for all $i \ge 1$, where the invariant $|t_{i+1}|2^{ik} < 1 - 1/2^k$, can be shown to hold for $i \ge 2$ and $k \ge 2$, and for $i = 1$ when $k = 1$. Then $|\varepsilon| = |t_1|2^{-k} = |\varepsilon/2| + (\varepsilon^2/8)2^{-k} + (\varepsilon^3/8)2^{-2k} < \frac{3}{8} + \frac{1}{8}2^{-k}$, satisfying the constraint of (6.4.1) for $2^k \ge 64$.

With $\sqrt{\rho^2 x}$ determined through the guard digit d_{n+1}, we have $\sqrt{\rho^2 x} = q_i + t_i 2^{-ki}$ and $q_i(x) = zq_i$ for $1 \le i \le n+1$. So then

$$\sqrt{x} - q_i(x) = \left(\frac{1}{\sqrt{\rho^2}} - z\right) q_i + \frac{1}{\sqrt{\rho^2}} t_i \text{ for } 1 \le i \le n+1.$$

Since $|1/\sqrt{\rho^2} - z| \le \frac{1}{4}2^{-nk}$ from the definition of z,

$$|\sqrt{x} - q_i(x)| \le \tfrac{1}{4}2^{-nk} + \left(1 - \tfrac{1}{2}\right) 2^{-ki} \text{ for } 1 \le i \le n+1,$$

and the bounds (6.4.6) follow. □

Given the successor digit selection $d_{i+1} = \lceil 2^{k-1} r_i \rfloor$, the 4-tuple $(d_{i+1}, q_{i+1}, r_{i+1}, q_{i+1}(x))$ is computed employing two independent multiplications, with $q = \lceil 2^{nk} \sum_1^{n+1} d_i z \rfloor 2^{-nk}$ satisfying $|\sqrt{x} - q| < 2^{-nk}$. Furthermore, for reasonable high radices such as $2^6 = 64$, there is considerable latitude in the rounding determining the high-radix digits, e.g., $|d_{i+1} - 2^{k-1} r_i| < \frac{1}{2} + \frac{3}{64}$ is sufficient. This allows for rounding d_i from leading bits of the redundant remainder for all $i \ge 1$.

Regarding computational latency, each short-by-long digit postscaling multiplication $d_i \times z$ can be performed concurrently with the short-by-long remainder update multiplication $d_i \times (2q_{i-1} + d_i)2^{-(i+1)k}$ employed to determine r_i and d_{i+1}. Thus the total multiply latency is $n+1$ dependent short-by-long multiplications to obtain $q_{n+1}(x)$. It is usually sufficient only to compute $q_n(x)$ when our target precision of p bits corresponds to less than n high-radix digits. In particular, let $q(x)$ have $\mathrm{ulp}(q(x)) = 2^{-p}$ satisfying $|q_n(x) - q(x)| \le 2^{-p-1}$. Then $|\sqrt{x} - q(x)| < (\frac{1}{2} + 3/2^{nk-p+1})2^{-p}$, so $q(x)$ is a one-ulp square root of x for any $p \le nk - 1$, and similarly $q(x)$ is a one-ulp square root whenever $p \le nk - 2$. As seen in the example the successive values of $q_i(x)$ serve to identify an additional (two-ulp accurate) high-radix digit of $\sqrt{x}$ with each postscaled digit augmentation.

The prescaled short radicand-reciprocal procedure can be extended from a one-ulp square root algorithm to an algorithm providing a root, remainder pair for $\sqrt{x}$ with additional concurrent computations that essentially delay the final result by at most one more multiply latency. The *prescaled short Radicand-Reciprocal Algorithm* for a $(p = nk)$-bit *square root and remainder pair* $q_n,\ r_n$ can be characterized by the recursive computation of the 4-tuple $(d_i(x),\ \mathrm{tail}_i(x),\ q_i(x),\ r_i(x))$ for $i = 1, 2, \ldots, n$. We leave the details of this extended algorithm to an exercise.

Problems and exercises

6.4.1 For normalized operands $1/\beta^2 \le x < 1$ show that the remainder and tail satisfy $(2/\beta)|t| \le |r| \le 2|t|$.

6.4.2 Show that (6.4.3) also holds for $i = 1$ using the bound $t_1 \le \frac{5}{8}\beta^{-k}$.

6.4.3 Can we develop prescaled binary square root if we assume the radicand has the standard normalization $1 \le x < 2$?

6.4.4 Determine the size of direct binary look-up tables needed to obtain prescaling, postscaling pairs (ρ^2, z) for $(p = 53)$-bit precision using the root radix 2^k for $k = 6, 7, 8, 9, 10, 11$. Note that one "short" value for ρ^2 may be stored with two "long" values for $z \approx 1/\rho$.

6.4.5 Are there any practical alternatives to direct table look-up for obtaining the prescaling and postscaling pairs (ρ^2, z) for prescaled square root? Investigate, in particular, the root radix case 2^{18} for precision $p = 53$ bits and radix 2^{19} for precision $p = 113$ bits.

6.4.6 Work out the details for developing the remainder along with the postscaled root.

6.5 Iterative refinement square root

The Newton–Raphson process provides analytic iterative refinement formulas for both the *root* $\sqrt{x}$ and the *root reciprocal* $1/\sqrt{x}$. Throughout this section we shall use $q_i = \sqrt{x}(1 + \varepsilon_i)$ to denote an *ith iterative root approximation* with implicitly defined relative error ε_i, and $\rho_i = (1/\sqrt{x})(1 + \varepsilon_i)$ to denote an *ith iterative root-reciprocal approximation* with its implicit relative error similarly denoted by ε_i.

When the radicand x is a radix-β number in the range $1/\beta^2 \le x < 1$, and the ith iterative root approximation $q_i = \sqrt{x}(1 + \varepsilon_i)$ has a relative error satisfying $|\varepsilon_i| < \frac{1}{2}\beta^{\ell}$, then the final root approximation may be given as a one-ulp root of x by $q = \lceil q_i \beta^{-\ell} \rfloor \beta^{\ell}$ using Lemma 1.9.4. The further computation needed to determine a directed one-(half-)ulp root or a root, remainder pair from q and x for all last places $p \geq \ell$ is investigated in the final subsection. The first three subsections are each devoted to a particular iterative refinement procedure characterized as follows:

- *Newton–Raphson square root*

$$q_{i+1} = \frac{1}{2}\left(q_i + \frac{x}{q_i}\right). \tag{6.5.1}$$

- *Newton–Raphson root reciprocal*

$$\rho_{i+1} = \rho_i \left(\tfrac{1}{2}(3 - \rho_i^2 x)\right). \tag{6.5.2}$$

- *Convergence (Goldschmidt)*

$$\begin{aligned} q_{i+1} &= q_i \left(\tfrac{1}{2}(3 - q_i \rho_i)\right), \\ \rho_{i+1} &= \rho_i \left(\tfrac{1}{2}(3 - q_i \rho_i)\right), \end{aligned} \tag{6.5.3}$$

  maintaining the invariant ratio $q_{i+1}/\rho_{i+1} = x$.

Algebraic arguments are used in this section to confirm the respective limits $q_i \to \sqrt{x}$, $\rho_i \to 1/\sqrt{x}$, with the rates of approach being $\varepsilon_{i+1} = \frac{1}{2}\varepsilon_i^2 + O(\varepsilon_i^3)$ for the root refinement (6.5.1), and the slightly slower rate $\varepsilon_{i+1} = -\frac{3}{2}\varepsilon_i^2 + O(\varepsilon_i^3)$ for the root-reciprocal and convergence refinements.

The Newton–Raphson square root refinement (6.5.1) is shown to have an alternative "remainder form" more convenient for radix arithmetic. The remainder form employs interleaved dual refinements of the root and root reciprocal with the division operation replaced by a root-reciprocal multiplication. All three refinements are investigated with regard to their demands on the multiplier resource. In particular, the opportunities for sharing resources in a pipelined multiplier or a large partitionable multiplier are investigated and summarized by separate enumeration of the total multiplications and the dependent multiplication chain length. The multiplication size is considered with a summary count of supplemental multiplications required if the resource is only a half-by-full multiplier.

Table 6.5.1 summarizes the independent versus dependent multiplication sequencing issues relevant to time latency for determining a final square root result. Note, in particular, that each of the refinements can be organized so that only two dependent multiplications are required per iteration. Furthermore, the concurrent independent multiplications can be scheduled in pairs sharing a common operand. Regarding multiplication size, the main result is that if only a half-by-full multiplier is employed, each refinement method can still be implemented involving at most one additional multiplication (independent of the number of iterations) with no increase in the chain of dependent multiplications.

Table 6.5.1. *Multiplication summary for the iterative refinement algorithms*

Refinement algorithm	Multiplications per iteration: Total	Multiplications per iteration: Dependent	Supplemental half-by-full multiplications
Newton–Raphson square root (remainder form)	4	2	0
Newton–Raphson root reciprocal	3	2	0 (root reciprocal)
Convergence	3	2	1 (root) in the last iteration

6.5.1 Newton–Raphson square root

The analytic Newton–Raphson refinement formula for a square root function approximation $q_i = \sqrt{x}(1+\varepsilon_i)$ is $q_{i+1} = \frac{1}{2}(q_i + x/q_i)$. The term $x/q_i = \sqrt{x}(1+\varepsilon_i)^{-1}$ constitutes a *geometric mean complement* of q_i in the sense that the exact root $\sqrt{x}$ is the geometric mean of the terms q_i and x/q_i. The error of the refined approximation $q_{i+1} = \sqrt{x}(1+\varepsilon_{i+1})$ is determined by the extent to which the arithmetic mean approximates the geometric mean.

Theorem 6.5.1 (Newton–Raphson square root) *Given a radicand $x > 0$ and an initial root approximation $q_1 = \sqrt{x}(1+\varepsilon_1)$ with $|\varepsilon_1| < \frac{1}{2}$, let the sequence of root approximations $q_i = \sqrt{x}(1+\varepsilon_i)$ be determined by the iterative refinements*

$$q_{i+1} = \frac{1}{2}\left(q_i + \frac{x}{q_i}\right) \text{ for all } i \geq 1. \tag{6.5.4}$$

Then $\lim_{i\to\infty} q_i = \sqrt{x}$, with the relative errors decreasing at the quadratic rate

$$\varepsilon_{i+1} = \frac{1}{2}\varepsilon_i^2 - \frac{\varepsilon_i^3}{2(1+\varepsilon_i)} \text{ for all } i \geq 2. \tag{6.5.5}$$

Proof Note that

$$q_{i+1} = \frac{1}{2}\left(q_i + \frac{x}{q_i}\right) = \frac{\sqrt{x}}{2}\left(1+\varepsilon_i + \frac{1}{1+\varepsilon_i}\right) = \sqrt{x}\left(1 + \frac{\varepsilon_i^2}{2} - \frac{\varepsilon_i^3}{2(1+\varepsilon_i)}\right),$$

yielding (6.5.5). Furthermore,

$$|\varepsilon_{i+1}| < |\varepsilon_i|\left(\frac{|\varepsilon_i|}{2} + \frac{|\varepsilon_i^2|}{2(1-|\varepsilon_i|)}\right) < \frac{1}{2}|\varepsilon_i| \text{ for } |\varepsilon_i| < \frac{1}{2},$$

confirming that $\lim_{i\to\infty} q_i = \sqrt{x}$. □

Since ε_{i+1} behaves as $\frac{1}{2}\varepsilon_i^2$ for sufficiently small ε_i, it follows that successive refinements converge to $\sqrt{x}$ from above, with the number of *bits-of-accuracy*, as measured by $-\log_2|\varepsilon_i|$, exhibiting a rate of "doubling plus one" each iteration. These observations can be employed in determining a format and rounding direction for obtaining a sufficient leading part of q_{i+1}. The iterative refinement formula is robust in that if q_{i+1} is to be determined with a format size (and accuracy) of

j digits, then only a j-digit half-ulp leading part (x_{i+1}) of the radicand need be employed to determine a j-digit geometric mean complement x_{i+1}/q_i from which to obtain q_{i+1}. The iterations are conveniently illustrated by the triples $(x_i, x_i/q_{i-1}, q_i)$ presented in a common length format that doubles each iteration with a (rounded down) guard digit attached for x_i/q_{i-1} and q_i until the guard digit becomes sufficiently accurate to jump the result by an extra digit.

Table 6.5.2 shows the triples for three iterative refinements $q_i = \frac{1}{2}(q_{i-1} + x_i/q_{i-1}) = \sqrt{\pi}(1+\varepsilon_i)$ for $i = 2, 3, 4$ starting with a one-digit half-ulp root $q_1 = 2$. The refinements are shown with formats of (2 + guard), (5 + guard), and (10 + guard), employing half-ulp leading parts of π of size 3-, 6-, and 11-digits respectively. The directed roundings for x_i/q_{i-1} and q_i increase the rate of bits-of-accuracy improvement to better than "doubling plus one" per iteration in this example.

Table 6.5.2. *Three Newton–Raphson iterative refinements of the square root of π*

i	x_i	x_i/q_{i-1}	q_i	ε_i
1	3.	–	2.	$2^{-2.96}$
2	3.1(4)	1.5(7)	1.7(8)	$2^{-7.88}$
3	3.1415(9)	1.7649(3)	1.7724(6)	$2^{-18.1}$
4	3.1415 92653(6)	1.7724 47701 (8)	1.7724 53850(9)	2^{-38}

Note that each division operation, e.g., $x_3/q_2 = \frac{3.14159}{1.78} = 1.76493\cdots$ in the example in Table 6.5.1, must be executed to *twice* the number of digits (i.e., twice the bits-of-accuracy equivalent) of the square root approximation employed as the divisor. This level of division accuracy precludes the use of substituting multiplication by a sufficiently refined root reciprocal. The refinement formula (6.5.4) is better suited to computation implemented in rational rather than radix arithmetic.

There is a reformulation of the refinement formula (6.5.4) utilizing the remainder $r = x - q^2$ that allows multiplication by a sufficiently refined root-reciprocal approximation $\rho \approx 1/\sqrt{x}$ to replace the division operation. This reformulation exhibits a fundamental relationship between the digit-serial *school method* covered in Section 6.3 and the analytic Newton–Raphson refinement formula.

Theorem 6.5.2 *Let $q > 0$ be a one-ulp radix-β root of the radicand $x > 0$, where q has an associated remainder r satisfying $r = x - q^2$. Then the school method square root approximation $q^* = q + r/2q$ equals the Newton–Raphson refined square root approximation $q' = \frac{1}{2}(q + x/q)$,* i.e.,

$$q + \frac{r}{2q} = \frac{1}{2}\left(q + \frac{x}{q}\right).$$

Proof

$$q' = \frac{1}{2}\left(q + \frac{x}{q}\right) = \frac{1}{2}\left(2q + \frac{x - q^2}{q}\right) = q + \frac{r}{2q} = q^*. \qquad \square$$

Definition 6.5.3 *The* remainder form *of the Newton–Raphson square root refinement formula for the radicand $x > 0$ is*

$$q_{i+1} = q_i + \frac{1}{2q_i}\left(x - q_i^2\right), \tag{6.5.6}$$

where $r = x - q_i^2$ is the corresponding remainder of q_i.

The refinement formula (6.5.6) is more amenable to computation hosted in radix arithmetic. In particular, there is a fully multiplicative implementation employing concurrent "interleaved" refinements of a root reciprocal $\rho \approx 1/\sqrt{x} \approx 1/q$ to provide a factor, replacing the division operation by a multiplication operation.

Lemma 6.5.4 (Newton–Raphson square root: remainder form) *Given a radicand $x > 0$, an initial root approximation $q_1 = \sqrt{x}(1 + \varepsilon_1)$ satisfying $|\varepsilon_1| < \frac{1}{2}$, and an initial root-reciprocal approximation $\rho_1 = (1/\sqrt{x})(1 + \delta_1)$ satisfying $|\delta_1| < \frac{1}{2}$, let the sequence of root and root-reciprocal approximations $q_i = \sqrt{x}(1 + \varepsilon_i)$, $\rho_i = (1/\sqrt{x})(1 + \delta_i)$ be given by the interleaved iterative refinements for $i \geq 1$*

$$q_{i+1} = q_i + \frac{\rho_i}{2}(x - q_i^2), \tag{6.5.7}$$

$$\rho_{i+1} = 2\rho_i - \rho_i^2 q_{i+1} \tag{6.5.8}$$

Then $\lim_{i\to\infty} q_i = \sqrt{x}$ and $\lim_{i\to\infty} \rho_i = 1/\sqrt{x}$, where the relative errors decrease at a quadratic rate governed by

$$\varepsilon_{i+1} = -\varepsilon_i\left(\delta_i + \tfrac{1}{2}\varepsilon_i\right) - \tfrac{1}{2}\varepsilon_i^2\delta_i, \tag{6.5.9}$$

$$\delta_{i+1} = -\left(\delta_i^2 - \varepsilon_i\delta_i - \tfrac{1}{2}\varepsilon_i^2\right) + c(\max\{|\varepsilon_i|, |\delta_i|\})^3,\ |c| < 3. \tag{6.5.10}$$

Proof Substituting $q_i\sqrt{x}(1 + \varepsilon_i)$ and $\rho_i = (1/\sqrt{x})(1 + \delta_i)$ into (6.5.7) yields $q_{i+1} = \sqrt{x}(1 + \varepsilon_i + \frac{1}{2}(1 + \delta_i)(1 - (1 + \varepsilon_i)^2))$, so $-(\varepsilon_i\delta_i + \frac{1}{2}\varepsilon_i^2) - \frac{1}{2}\varepsilon_i^2\delta_i$. Similar substitutions into (6.5.8) using this expression for ε_{i+1} confirm (6.5.10). $\square$

Note that both $|\varepsilon_i|$ and $|\delta_i|$ decrease at a quadratic rate bounded by $\frac{3}{2}\max\{|\varepsilon_{i-1}|, |\delta_{i-1}|\}^2$, which nearly doubles the number of bits-of-accuracy with each refinement, as measured by $-\log_2(\max\{|\varepsilon_i|, |\delta_i|\})$. The square root approximation can experience a jump in refinement accuracy if $\delta_i \approx -\varepsilon_i/2$. The root-reciprocal refinement is governed by $\frac{1}{2}\delta_i^2 < |\delta_{i+1}| < \frac{3}{2}\delta_i^2$, ignoring lower-order terms. Thus the interleaved dual refinement's decreasing relative error rate is limited to the rate for the root-reciprocal refinement. Two procedures for *enforcing* comparable initial relative errors lead to sharper distinct results for q_2 and ρ_2. Determining q_1 from ρ_1 by employing a multiplication $q_1 = x\rho_1$ enforces $\delta_1 = \varepsilon_1$.

Selecting ρ_1 as essentially the inverse of q_1 by direct table look-up of both from a leading part of x enforces the "opposite" extreme $\delta_1 = -\varepsilon_1/(1+\varepsilon_1)$.

Corollary 6.5.5 *In the dual iterative refinements (6.5.7) and (6.5.8) for root and root-reciprocal approximations,*

(i) if $\rho_i = 1/q_i$*, then* $\delta_i = -\varepsilon_i/(1+\varepsilon_i)$*, and*

$$\varepsilon_{i+1} = \frac{1}{2}\frac{\varepsilon_i^2}{(1+\varepsilon_i)^2},$$

$$\delta_{i+1} = -\frac{3}{2}\frac{\varepsilon_i^2}{(1+\varepsilon_i)^2} + \frac{\varepsilon_i^3}{2(1+\varepsilon_i)^3};$$

(ii) if $\delta_i = \varepsilon_i$*, then*

$$\varepsilon_{i+1} = -\frac{3}{2}\varepsilon_i^2 - \frac{1}{2}\varepsilon_i^3,$$

$$\delta_{i+1} = \frac{1}{2}\varepsilon_i^2 + c(\varepsilon_i)^3, \quad |c| < 3.$$

The dual iterative refinements (6.5.7) and (6.5.8) are robust in that if q_1 and ρ_1 are given to a format (and bits-of-accuracy equivalent) of j digits, then a $2j$-digit half-ulp leading part, x_2, of the radicand is sufficient for determining q_2 and ρ_2 in $2j$-digit leading part formats with nearly $2j$-digit leading part accuracy.

Table 6.5.3 gives the results of three such dual iterative refinements for $x = \pi$ starting with the half-ulp leading part $q_1 = 2$ and enforcing $\rho_1 = 1/q_1 = .5$. In this case ε_i remains less than δ_i in magnitude and of opposite sign for $i \geq 2$, with both decreasing at a rate essentially governed by $|\delta_{i+1}| \approx |\delta_i|^2$, as seen in the third iteration of the example. The vertical bars illustrate where the nearest leading parts of q_i and ρ_i would be one-ulp leading parts in these cases.

Table 6.5.3. *Three dual iterative refinements of the square root and root reciprocal of* π*, showing the triples* (x_i, q_i, ρ_i) *and the errors*

i	x_i	q_i	ρ_i	ε_i	δ_i
1	3.	2.	.5	$2^{-2.96}$	$-2^{-3.13}$
2	3.1	1.8|	.5|5	$2^{-5.98}$	$-2^{-5.32}$
3	3.142	1.773|	.563|7	$2^{-11.7}$	$-2^{-10.2}$
4	3.141 5927	1.772 454|2	.5641 89|04	$2^{-22.3}$	$-2^{-20.0}$

For each triple (x_i, q_i, ρ_i), we must employ a multiplication to compute each of the squares q_i^2 and ρ_i^2 to $j2^{i+1}$ digits with a total of four multiplications per "dual refinement" iteration. The size of the multiplications and scheduling of concurrent multiplications for three dual iterative refinements are given in Table 6.5.4. The initial terms q_1, ρ_1, and q_1^2 are assumed available in $j\bar{v}$, j- and $2j$-digit formats respectively, from table look-up using a j-digit directed one-ulp leading part of x

Table 6.5.4. *Multiplication sizes and scheduling summary for three interleaved dual iterative refinements*

Time latency	Terms computed	Multiplication sizes	Shared operand	Total multiplies
L_T	q_1, ρ_1, q_1^2			
$L_M + L_T$	q_2, ρ_1^2	$2j \times j,\ j \times j$	ρ_1	2
$2L_M + L_T$	ρ_2, q_2^2	$2j \times 2j,\ 2j \times 2j$	q_2	4
$3L_M + L_T$	q_3, ρ_2^2	$4j \times 2j,\ 2j \times 2j$	ρ_2	6
$4L_M + L_T$	ρ_3, q_3^2	$4j \times 4j,\ 4j \times 4j$	q_3	8
$5L_M + L_T$	q_4	$8j \times 4j$		9

as an index. The formats show the potential for realizing a final "full precision" result employing only half-by-full multiplications.

Observation 6.5.6 *A total of* $4(k-1)+1$ *half-by-full multiplications is sufficient to determine a kth refined square root* $q_{k+1} = \sqrt{x}(1+\varepsilon_{k+1})$ *having bits-of-accuracy nearly* 2^k *times the initial measure of bits-of-accuracy with latency* $(2k-1)L_{HM} + L_T$*, where* L_{HM} *is the latency of a half-by-full multiplier and* L_T *is the table lookup latency.*

6.5.2 Newton–Raphson root-reciprocal

The analytic Newton–Raphson refinement formula for a root-reciprocal function approximation $\rho = (1/\sqrt{x})(1+\varepsilon)$ is

$$\rho' = \frac{\rho}{2}(3 - \rho^2 x). \tag{6.5.11}$$

Corresponding to the implicit relative error factor $(1+\varepsilon)$ inherent in the approximation ρ, the explicitly computable scale factor $\omega = \frac{1}{2}(3-\rho^2 x)$ satisfies $\omega = (1 - \varepsilon - \varepsilon^2/2)$ and constitutes a *complementary error factor* of $(1+\varepsilon)$. The reduction in relative error from successive iterative refinements drives $\rho \to 1/\sqrt{x}$ at a quadratic rate somewhat slower than for the root refinement of Theorem 6.5.1.

Theorem 6.5.7 (Newton–Raphson root reciprocal) *Given a radicand* $x > 0$ *and an initial root-reciprocal approximation* $\rho_1 = (1/\sqrt{x})(1+\varepsilon_1)$ *with* $|\varepsilon_1| < \frac{1}{2}$*, let the sequence* $\rho_i = (1/\sqrt{x})(1+\varepsilon_i)$ *of root-reciprocal approximations be determined by the iterative refinement*

$$\rho_{i+1} = \rho_i \left(\tfrac{1}{2}(3 - \rho_i^2 x)\right) \quad \textit{for all } i \geq 1. \tag{6.5.12}$$

Then $\lim_{i\to\infty} \rho_i = 1/\sqrt{x}$*, with the relative errors decreasing at the quadratic rate*

$$\varepsilon_{i+1} = -\tfrac{3}{2}\varepsilon_i^2 - \tfrac{1}{2}\varepsilon_i^3 \quad \textit{for all } i \geq 2. \tag{6.5.13}$$

Proof The ith scale factor $\omega_i = \frac{1}{2}(3 - \rho_i^2 x) = (1 - \varepsilon_i - \varepsilon_i^2/2)$, so then

$$\rho_{i+1} = \rho_i \omega_i = \frac{1}{\sqrt{x}}(1+\varepsilon_i)\left(1 - \varepsilon_i - \frac{\varepsilon_i^2}{2}\right) = \frac{1}{\sqrt{x}}\left(1 - \frac{3\varepsilon_i^2}{2} - \frac{\varepsilon_i^3}{2}\right)$$

yielding (6.5.13). Furthermore,

$$|\varepsilon_{i+1}| \le |\varepsilon_i|\left(\frac{3|\varepsilon_i|}{2} + \frac{\varepsilon_i^2}{2}\right) < \frac{7}{8}|\varepsilon_i| \text{ for } |\varepsilon_i| < \frac{1}{2},$$

confirming that $\lim_{i\to\infty} \rho_i = 1/\sqrt{x}$. □

Since ε_{i+1} behaves as $-\frac{3}{2}\varepsilon_i^2$ for sufficiently small ε_i, it follows that successive refinements converge to $1/\sqrt{x}$ from below, with the bits-of-accuracy as measured by $-\log_2|\varepsilon_i|$ exhibiting a rate of "doubling minus 0.585" each iteration.

The iterative computation of ρ_i for $2 \le i \le k+1$ can be followed by a final multiplication by the radicand to determine a square root approximation $q = \rho_{k+1}x = \sqrt{x}(1+\varepsilon_{k+1})$. This final multiplication can be integrated into the final iterative refinement yielding $q_k = (\rho_k x)(\frac{1}{2}(3 - \rho_k(\rho_k x)))$. Thus a total of $3k$ multiplications is sufficient to obtain the desired root approximation.

Multiplication size and latency for Newton–Raphson root reciprocal.
Size and format for x, ρ_i: The recurrence for Newton–Raphson root reciprocal is robust in the sense that each multiplication can be of a reduced size reflecting the relative error anticipated in each product term. With a j-digit initial root approximation determined from a j-digit leading part of the radicand x, a Newton–Raphson root-reciprocal algorithm can be characterized by the computation of a series of pairs (x_i, ρ_i) of $j2^{i-1}$-digit radix-β leading parts, with $\rho_{i+1} = \rho_i(\frac{1}{2}(3 - \rho_i^2 x_{i+1}))$ rounded up in the direction towards $1/\sqrt{x_{i+1}}$.

Table 6.5.5 illustrates the computation of (x_i, ρ_i) in this manner for determining the root reciprocal of π with three iterative refinements applied to the initial one-digit half-ulp leading part $\rho_1 = (1/\sqrt{\pi})(1+\varepsilon_1) = 0.6$. Noting that $1/\sqrt{\pi}$ has the half-ulp directed leading part $0.5641\ 8958^+$, the values of ρ_2 and ρ_3 are each one-ulp leading parts of $1/\pi$, with ρ_4 yielding a seven-digit half-ulp leading part, following quite well the "doubling minus .585" estimated bits-of-accuracy measure.

Table 6.5.5. *Three iterative refinements of the Newton–Raphson root-reciprocal of π*

i	x_i	ρ_i	ε_i
1	3.	0.6|	$2^{-3.98}$
2	3.1	0.57|	$2^{-6.60}$
3	3.142	0.5641|	$-2^{12.62}$
4	3.141 5927	0.5641 895|6	$-2^{24.51}$

The formats illustrate the potential for implementing the algorithm employing a half-by-full multiplier.

Observation 6.5.8 *A total of* $3k + 1$ *half-by-full multiplications is sufficient to determine a full precision root approximation* $q_k = \sqrt{x}(1 + \varepsilon_{k+1})$.

Proof The final iteration contains the only products needed for full precision. In modified form for a root approximation, note that $q_k = [\rho_k x] \times [\frac{1}{2}(3 - \rho_k(\rho_k x))]$. Only the multiplication by terms in the square brackets requires a full-by-full multiplication, equivalently two half-by-full multiplications sharing a common "full sized" operand. □

Latency: The iterative refinement formula $\rho_{i+1} = \rho_i(\frac{1}{2}(3 - \rho_i^2 x_{i+1}))$ can be equivalently expressed by distributing ρ_i over the sum yielding $\rho_{i+1} = \frac{3}{2}\rho_i - (\rho_i^2) \times (\rho_i x_{i+1})$. Thus the root-reciprocal recurrence can be implemented with two dependent multiplications per iteration, with an additional dependent multiplication employed to determine the root q_k. Let L_T denote the table look-up latency, with L_M and L_{HM} denoting the latency of a full-by-full and a half-by-full multiplication, respectively.

Observation 6.5.9 *The Newton–Raphson root-reciprocal algorithm can be implemented employing k root-reciprocal refinements to obtain a final square root approximation* $q_k = \sqrt{x}(1 + \varepsilon_{k+1})$ *with latency* $(2k + 1)L_M + L_T$, *where concurrent multiplications occur in pairs sharing a common operand. Using a half-by-full multiplier the latency is* $(2k + 1)L_{HM} + L_T$.

6.5.3 Convergence square root

Convergence square root is conveniently described by a series of scaling multiplications by complementary error factors that are concurrently and independently applied to numerator and denominator terms. In a *prescaling* step a table look-up seed root-reciprocal approximation establishes the *first denominator term* $\rho_1 = (1/\sqrt{x})(1 + \varepsilon_1)$ and is employed to scale the radicand to initiate the *first numerator term* $q_1 = x\rho_1 = \sqrt{x}(1 + \varepsilon_1)$ yielding

$$\frac{x}{1} = \frac{x \times \rho_1}{1 \times \rho_1} = \frac{q_1}{\rho_1} = \frac{\sqrt{x}(1 + \varepsilon_1)}{(1/\sqrt{x})(1 + \varepsilon_1)} \text{ (prescaling).}$$

The common relative error factor $(1 + \varepsilon_1)$ implicit in both the first root approximation q_1 and the first root-reciprocal approximation ρ_1 has the same explicitly computable *complementary error factor* $\omega_1 = \frac{1}{2}(3 - q_1\rho_1) = 1 - \varepsilon_1 - \frac{1}{2}\varepsilon_1^2$ as employed for the Newton–Raphson root-reciprocal refinement (6.5.12). Scaling the first numerator and denominator terms by ω_1 yields a second pair (q_2, ρ_2) of root and root-reciprocal approximations preserving the ratio $q_2/\rho_2 = x$. The iterative core of convergence square root involves scaling the ith numerator term $q_i = \sqrt{x}(1 + \varepsilon_i)$ and the ith denominator term $\rho_i = (1/\sqrt{x})(1 + \varepsilon_i)$ by the ith *complementary error factor* $\omega_i \equiv (1 - \varepsilon_i - \frac{1}{2}\varepsilon_i^2) = \frac{1}{2}(3 - q_i\rho_i)$ yielding the

$(i+1)$th pair (q_{i+1}, ρ_{i+1}):

$$\frac{x}{1} = \frac{q_i}{\rho_i} \times \frac{\omega_i}{\omega_i} = \frac{\sqrt{x}(1 - \frac{3}{2}\varepsilon_i^2 - \frac{1}{2}\varepsilon_i^3)}{(1/\sqrt{x})(1 - \frac{3}{2}\varepsilon_i^2 - \frac{1}{2}\varepsilon_i^3)}$$
$$= \frac{\sqrt{x}(1 + \varepsilon_{i+1})}{(1/\sqrt{x})(1 + \varepsilon_{i+1})} = \frac{q_{i+1}}{\rho_{i+1}} \quad \textit{for } i \geq 1,$$

satisfying the *invariant* $q_{i+1}/\rho_{i+1} = x$. The *convergence square root algorithm* can be simply characterized as the iterative computation of the triples $(q_i, \overline{q_i}, \omega_i)$.

Theorem 6.5.10 (Convergence square root) *Given a radicand $x > 0$ and a root-reciprocal approximation $\rho_1 = (1/\sqrt{x})(1 + \varepsilon)$ with $|\varepsilon| < \frac{1}{2}$, let the triples (q_i, ρ_i, ω_i) be given iteratively by*

$$(q_1, \rho_1, \omega_1) = \left(x\rho_1, \rho_1, \tfrac{1}{2}(3 - x\rho^2)\right), \tag{6.5.14}$$
$$(q_i, \rho_i, \omega_i) = \left(q_{i-1}\omega_{i-1}, \rho_{i-1}\omega_{i-1}, \tfrac{1}{2}(3 - q_i\rho_i)\right) \quad \textit{for } i \geq 2. \tag{6.5.15}$$

Then $\lim_{i\to\infty} q_i = \sqrt{x}$ *and* $\lim_{i\to\infty} \rho_i = 1/\sqrt{x}$*, with q_i, ρ_i jointly approaching the limits satisfying the invariant $q_i/\rho_i = x$ for all $i \geq 1$. Furthermore, the relative error ε_i implicitly defined by $q_i = \sqrt{x}(1 + \varepsilon_i)$ for all $i \geq 1$ decreases at the quadratic rate*

$$\varepsilon_{i+1} = -\tfrac{3}{2}\varepsilon_i^2 - \tfrac{1}{2}\varepsilon_i^3 \quad \textit{for } i \geq 1. \tag{6.5.16}$$

Proof Initially $q_1/\rho_1 = x\rho/\rho = x$, and iteratively $q_i/\rho_i = q_{i-1}\omega_{i-1}/\rho_{i-1}\omega_{i-1} = q_{i-1}/\rho_{i-1}$, so $q_i/\rho_i = x$ is invariant by induction for all $i \geq 1$. To obtain the quadratic refinement rate for relative error note that $q_{i+1} = q_i\omega_i$ with $q_i = \sqrt{x}(1 + \varepsilon_i)$ and $\omega_i = \frac{1}{2}(3 - q_i\rho_i) = 1 - \varepsilon_i - \frac{1}{2}\varepsilon_i^2$, so $q_{i+1} = \sqrt{x}(1 - \frac{3}{2}\varepsilon_i^2 - \frac{1}{2}\varepsilon_i^3)$, yielding (6.5.16). □

The convergence square root algorithm is fundamentally a reformulation of the Newton–Raphson root-reciprocal algorithm fashioned to emphasize concurrent computation. It follows that the quadratic refinement rate for convergence square root (6.5.16) is the same as for the Newton–Raphson root-reciprocal (6.5.13) with the growth rate of bits-of-accuracy also being "doubling minus .585" for each iteration, with both approximations q_i, ρ_i converging from below.

Table 6.5.6 illustrates the computation of (q_i, ρ_i, ω_i) for determining the root and root reciprocal of $x = \pi$ with three iterative refinements applied to the initial one-digit one-ulp root reciprocal $\rho_1 = 0.5$. The computed values of q_i and ρ_i are each partitioned by a vertical bar indicating where a *rounded-up* leading part would be a one-ulp value of $\sqrt{x}$ and $1/\sqrt{x}$, respectively.

Multiplication size, format and latency for convergence square root.
Size and format for (q_i, ρ_i, ω_i): The recurrence for convergence square root is not robust. The accuracy of q_i and ρ_i must be maintained at the target accuracy

of the final result as q_i comprises "low-order part" information needed from the original radicand x, and ρ_i must be kept at a comparable accuracy to maintain the invariant $q_i/\rho_i = x$ at the target level of precision. For the p-digit radix-β radicand x satisfying $1 \le x < \beta^2$, the root and root-reciprocal leading parts should use $(p+g)$-digit formats

$$\rho_i = 0.d_1 d_2 \cdots d_{p+g},$$

$$q_i = d_0.d_1 d_2 \cdots d_{p+g-1},$$

to reflect the original accuracy of the radicand x needed for the target result accuracy. The number of guard digits g should be chosen to control the accumulated round-off error inherited by the final root approximation (e.g., $g = 3$ bits or 1 decimal digit is sufficient for three–four iterations).

The ith complementary error factor $\omega_i = (1 - \varepsilon_i - \frac{1}{2}\varepsilon_i^2)$ used as a scale factor may be rounded to a leading part of accuracy corresponding to the bound on ε_i^2, i.e., to the anticipated bits-of-accuracy equivalent of the next root and root-reciprocal approximations q_{i+1} and ρ_{i+1}. If ρ_1 is obtained by table look-up to j-digit one-ulp leading part size and accuracy, then it is sufficient for ω_i to be rounded up into the format $\omega_i = d_0.d_1 d_2 \cdots d_{j2^i}$. The bounded size of the scale factors ω_i result in ρ_i having a limited number of non-zero digits, effectively ρ_i is a $j(2^i - 1)$-digit value which does not exceed the target size $p + g$ until the final iteration as illustrated in Table 6.5.6. This results in all multiplications having relatively small arguments except for the scalings of q_i, since ω_i need only be a $(j2^i + 1)$-digit leading part. The "order" of sizes of the multiplications for three iterations of convergence square root is included in Table 6.5.7.

Table 6.5.6. *The triples (q_i, ρ_i, ω_i) for three iterative refinements of convergence square root for the radicand $x = \pi$ with the seed root reciprocal $\rho_1 = .5$*

i	q_i	ρ_i	ω_i	ε_i
1	1.|5707 963(3)	0.5|	1.1(1)	$-2^{-3.14}$
2	1.7|435 839(3)	0.55|5	1.016(2)	$-2^{-5.94}$
3	1.771|8 299(9)	0.563|9 91	1.0003 519(2)	$-2^{-11.5}$
4	1.7724 53|5(3)	0.5641 894|(8)	–	$-2^{-22.4}$

Latency: The computation of the ith triple (q_i, ρ_i, ω_i) involves a concurrent pair of "scaling" multiplies $[q_{i-1} \times \omega_{i-1}$ and $\rho_{i-1} \times \omega_{i-1}]$ employing a common argument ω_{i-1}, followed by a dependent multiplication $q_i \times \rho_i$ needed to obtain ω_i. With the initial direct table look-up used to provide both ρ_1 and ρ_1^2, a first pair of concurrent multiplies $[\rho_1 \times x$ and $\rho_1^2 \times x]$ employing a common argument x allows the first triple (q_1, ρ_1, ω_1) to be available after a time equal to the *table look-up latency* L_T plus the *multiplier latency* L_M.

The earliest times at which the individual terms q_i, ρ_i, and ω_i become available measured in time latency steps $(jL_M + L_T)$ respecting the dependent

Table 6.5.7. *Latency scheduling of multiplications for three iterative refinements of convergence square root*

Time latency	Terms computed	Multiplication sizes	Shared operand	Total multiplies
L_T	ρ_1, ρ_1^2	–	–	–
$L_M + L_T$	q_1, ω_1	$8j \times j,\ 8j \times 2j$	x	2
$2L_M + L_T$	q_2, ρ_2	$8j \times 2j,\ 8j \times 2j$	ω_1	4
$3L_M + L_T$	ω_2	$4j \times 4j$	–	5
$4L_M + L_T$	q_3, ρ_3	$8j \times 4j,\ 8j \times 4j$	ω_2	7
$5L_M + L_T$	ω_3	$8j \times 8j$		8
$6L_M + L_T$	q_4	$4j \times 4j$		9

multiplications are shown for terms of the triples (q_i, ρ_i, ω_i) for $i = 1, 2, 3$ and for q_4 in Table 6.5.7. The formats and multiplication size orders in Table 6.5.7 illustrate the potential for implementing the algorithm using a half-by-full multiplier.

Observation 6.5.11 *The ith triple (q_i, ρ_i, ω_i) of convergence square root can be computed with latency $(2(i-1)+1)L_M + L_T$ employing a total of $(3i-1)$ multiplies for all $i \geq 2$, where all concurrent multiplications are paired with one operand in common. For determining a final kth root approximation by the convergence square root algorithm, all but one of the $3(k-1)$ total multiplies can be performed as half-by-full multiplies. Furthermore, the exceptional product $(q_{k-1}) \times (\rho_{k-1})$ needed to determine $\omega_{k-1} = 1 + \beta^{-\lceil \frac{p}{2} \rceil}(d_0.d_1 d_2 \cdots d_{\lceil \frac{p}{2} \rceil + 1})$ can be obtained by two concurrent half-by-full multiplies sharing a common argument.*

Proof The formats for q_i, ρ_i, and ω_i confirm that the scaling multiplications $q_{i+1} = (q_i) \times (\omega_i)$ and $\rho_{i+1} = (\rho_i) \times (\omega_i)$ are half-by-full multiplications for $1 \leq i \leq k-2$, and $q_k = q_{k-1}\omega_{k-1} = q_{k-1} + q_{k-1}\beta^{-\lceil \frac{p}{2} \rceil}(d_0'.d_1'd_2' \cdots d'_{\lceil \frac{p}{2} \rceil + 1})$ comprises a half-by-full multiplication and an addition. The multiplication $(q_i) \times (\rho_i)$ needed to obtain $\omega_i = \frac{1}{2}(3 - q_i \rho_i)$ can be obtained using only a $(\lceil p/2 \rceil + 2)$-digit leading part of q_i (and/or ρ_i) for $1 \leq i \leq k-2$ since ω_i will be obtained in a $(\lceil p/2 \rceil + 2)$-digit leading part format. □

It follows from Observation 6.5.11 that a final kth root approximation of full target precision can be computed by the convergence square root algorithm with latency $2(k-1)L_{HM} + L_T$ employing $3(k-1)+1$ half-by-full multiplications, where L_{HM} is the latency of a half-by-full multiplier. The first and last pair of concurrent multiplications share a common long operand and the $k-2$ intermediate pairs of multiplications share a common short operand.

6.5.4 Exact and directed one-ulp roots

For efficient radix representation of an iteratively refined approximate square root, it is typically desired that the final output be given as a directed one-ulp root. When both the radicand and one-ulp root are p-digit radix-β numbers, the final output may be required in the more complete form of the unique half-ulp radix-β root, remainder pair, where the remainder is also a p-digit radix-β number. The computation of either sgn(tail) or the remainder may be simplified to avoid the full cost of direct evaluation of the remainder from the defining equation $r \equiv x - q^2$.

We limit our investigation of obtaining the directed one-ulp root to the binary case. Given a $(p+g)$-bit one-ulp root q, the determination of the exact bit (is $\text{sgn}(\sqrt{x} - q) = 0$?) of the p-bit radicand x can be reduced to checking whether there are a sufficient number of low-order zeroes in q, since an exact root of a p-bit radicand can have no more than $\lceil p/2 \rceil$ of its leading bits non-zero.

Lemma 6.5.12 *For the p-bit radicand* $x = 0.1b_2b_3 \cdots b_p$ *satisfying*

$$\frac{1}{2} + \frac{1}{2^p} \leq x \leq 1 - \frac{1}{2^p},$$

let $q = i2^{-(p+1)}$ *be a* $(p+1)$*-bit one-ulp root of* x. *Then* $q = \sqrt{x}$ *if and only if* q *has an irreducible radix-*β *factorization* $q = i2^{-j}$ *with* $j \leq \lfloor p/2 \rfloor$. *For the p-bit radicand* $x = 0.01b_3b_4 \cdots b_{p+1}$ *satisfying*

$$\frac{1}{4} + \frac{1}{2^{p+1}} \leq x \leq \frac{1}{2} - \frac{1}{2^{p+1}},$$

let $q = i2^{-(p+2)}$ *be a* $(p+2)$*-bit one-ulp root of* x. *Then* $q = \sqrt{x}$ *if and only if* q *has an irreducible radix-*β *factorization* $q = i2^{-j}$ *with* $j \leq \lceil p/2 \rceil$.

Proof For x in the binade

$$\frac{1}{2} + \frac{1}{2^p} \leq x \leq 1 - \frac{1}{2^p},$$

any $(p+1)$-bit one-ulp root must fall in the internal $\sqrt{2}/2 < q \leq 1 - 1/2^{p+1}$. Suppose the irreducible factorization $q = i2^{-j}$ has $j \geq \lfloor p/2 \rfloor + 1$. Then $\frac{1}{2} < q^2 = i^2 2^{-2j} < 1$ with $2^j \geq p+1$, so q cannot be an exact root of any p-bit radicand. If otherwise $j \leq \lfloor p/2 \rfloor$, then $q = i2^{-j}$ is an exact root of the p-bit radicand q^2. To complete the proof for $1/2 + 1/2^p \leq x \leq 1 - 1/2^p$, it is sufficient to show that the interval of real values for which $q = i2^{-j}$ is a one-ulp $(p+1)$-bit root contains only one p-bit radicand.

Now y has a $(p+1)$-bit one-ulp root $q = i2^{-j}$ with $j \leq \lfloor p/2 \rfloor$ if and only if $(i2^{-j} - 2^{-(p+1)}) < \sqrt{y} < (i2^{-j} + 2^{-(p+1)})$. Then $y < (i2^{-j} + 2^{-(p+1)})^2 < q^2 + (1 - 1/2^{p+1})2^{-p} + 2^{-2(p+1)} < q^2 + 2^{-p}$, and similarly $y > (i2^{-j} - 2^{-(p+1)})^2 > q^2 - 2^{-p}$. Since x has q as a $(p+1)$-bit one-ulp root, it follows that $x = q^2$. □

The test for an exact p-bit root of a p-bit radicand has a practical interpretation in terms of verifying a string of low order zero bits in the one-ulp root.

Corollary 6.5.13 (Exact p-bit root test) *For precision* $p \geq 2$ *and guard bits* $g \geq 1$, *let* $\text{R}(\sqrt{(x)}) = 0.1b_2b_3 \cdots b_{p+g}$ *be a one-ulp* $(p+g)$*-bit root of the p-bit* *radicand x where* $\frac{1}{4} \le x \le 1 - 1/2^p$. *Then* $0.1b_2b_3 \cdots b_{p+g} = \sqrt{x}$ *if and only if*

$$b_{i+1}b_{i+2} \cdots b_{p+g} = 0^{p+g-i}, \quad \text{where} \begin{cases} i = \lfloor p/2 \rfloor & \text{and } g \ge 1 \text{ for } x \ge \frac{1}{2}, \\ i = \lceil p/2 \rceil & \text{and } g \ge 2 \text{ for } x < \frac{1}{2}. \end{cases} \tag{6.5.17}$$

Table 6.5.8 illustrates the application of the exact root test of Corollary 6.5.13 to five 9-bit radicands. The need for testing a sufficient number of guard bits is seen in the first and fourth cases. In particular, note that $\sqrt{0.10101\ 0011} = .1101\ 0^5 10 \cdots$, so the nine-bit root prefix $.1101\, 0^5$ is not sufficient by itself to confirm or deny an exact root. Similarly the root prefix $.10101\, 0^5$ needs another guard bit to resolve the exact root status. When $q \ne \sqrt{x}$, then q is one of two distinct $(p+g)$-bit one-ulp roots of x. If there are at least two guard bits for q, then it is often possible to identify the unique half-ulp p-bit root and whether it is a subestimate or superestimate of $\sqrt{x}$ by simple properties of the guard bits.

Table 6.5.8. *Application of exact root test*

Nine-bit radicand	$(9+g)$-bit root prefix	g	Exact root test result
0.10101 0011	$0.1101\ 0^5$	0	Uncertain
0.10101 0010	$0.1101\ 0^6$	1	Exact
0.10110 1101		1	Not exact ($b_5 \ne 0$)
0.01101 11010	$0.10101\ 0^5$	1	Uncertain
0.01101 11011	$0.10101\ 0^6$	2	Exact

Lemma 6.5.14 *For precision* $p \ge 2$ *and guard bits* $g \ge 2$, *let* $0.1b_2b_3 \cdots b_{p+g}$ *be a* $(p+g)$*-bit root of the p-bit radicand x, where* $\frac{1}{4} \le x \le 1 - 1/2^p$. *Then if* $b_{p+2}b_{p+3} \cdots b_{p+g} \ne 0^{[g-1]}$, *the p-bit half-ulp directed root is given by*

$$q = 0.1b_2b_3 \cdots b_p + b_{p+1}2^{-p} \text{ with } \operatorname{sgn}(\sqrt{x} - q) = (-1)^{b_{p+1}}.$$

The identification of the p-bit half-ulp directed root with its direction towards $\sqrt{x}$ provides sufficient information to determine the rounded root by any of the required rounding modes and smaller precisions of the IEEE floating-point standard (see Chapter 7). As discussed for quotient rounding in Section 7.5, obtaining some 5–10 accurate guard bits may well be possible in the large majority of cases without adding too much to the size of a hardware implementation. This is especially true for the Newton–Raphson and convergence square root refinements where a good estimate of the final relative error of the particular output (rather than the worst case) is available as a concurrently provided by-product of the final iteration. With conformation of from five up to ten accurate guard bits, the need for further computation to determine the half-ulp directed root can be reduced to an event occurring in between 5% and a fraction of 1% of all inputs since only the all-zero string of guard bits will require more computation in view of Lemma 6.5.14 and Corollary 6.5.13.

For efficient remainder determination, a p-digit one-ulp root, remainder pair (q, r) is more conveniently described employing an integer scaling for the radicand, root, and remainder. The following results are applicable to any *even* radix $\beta \geq 2$. Let the p-digit *integer* radicand x have the factorization $x = k\beta^{p-1}$, with $\beta^{2(p-1)} \leq x \leq \beta^{2p} - 1$. A p-digit one-ulp root of x is then an integer i, with $\beta^{p-1} \leq i \leq \beta^p - 1$, and $x = i^2 + r$ with $-2(i-1) \leq r \leq 2i$. Since x is a p-digit integer, $k = d_p d_{p-1} \cdots d_0$ is a p-digit integer with $d_0 = 0$ whenever $x \geq 2^{2p-1}$. The computation of the remainder r given x and i is simplified by partitioning the digit string $i = d_{p-1}d_{p-2}\cdots d_0$ between $d_{\lceil p/2\rceil}$ and $d_{\lceil p/2\rceil+1}$ expressing i as the sum of a $(\lfloor p/2\rfloor - 1)$-digit high order part $h\beta^{\lceil \frac{p}{2}\rceil+1} = d_{p-1}d_{p-2}\cdots d_{\lceil p/2\rceil+1}0^{\lceil \frac{p}{2}\rceil+1}$ and a $(\lceil p/2\rceil + 1)$-digit low-order part $t = d_{\lceil p/2\rceil}d_{\lceil p/2\rceil-1}\cdots d_0$.

Theorem 6.5.15 *Let $i = h\beta^{\lceil p/2\rceil+1} + t$ with $\beta^{\lceil p/2\rceil-2} \leq h \leq \beta^{\lceil p/2\rceil-1} - 1$ and $0 \leq t \leq \beta^{\lceil p/2\rceil+1} - 1$, be a p-digit one-ulp integer root of the p-digit radicand $x = k\beta^{p-1}$, where $\beta^{2(p-1)} \leq x \leq \beta^{2p} - 1$, with $\beta \geq 2$ being an even radix. Then the integer remainder $r = x - i^2$ can be determined employing at most three of the p digits of x and a half-by-full $((\lceil p/2\rceil + 1) \times (p+1))$-digit integer multiplication to determine the non-negative remainder residue*

$$r^+ = \left| |k|_{4\beta}\beta^{p-1} - (2I\beta^{\lceil \frac{p}{2}\rceil+1} + t)t \right|_{4\beta^p}, \tag{6.5.18}$$

where then

$$r = \begin{cases} r^+ & \text{for } r^+ < 2\beta^p, \\ r^+ - 4\beta^p & \text{for } r^+ \geq 2\beta^p. \end{cases}$$

Proof Since $\sqrt{x} + i < 2\beta^p$, then $|r| = |\sqrt{x} - i|(\sqrt{x} + i) \leq |2|\beta^p - 1||$. Hence $r = |r|_{4\beta^p}$ for $0 \leq |r|_{4\beta^p} \leq 2\beta^p - 1$, and $r = 4\beta^p - |r|_{4\beta^p}$ for $2\beta^p \leq |r|_{4\beta^p}| \leq 4\beta^p - 1$. For any even radix $\beta \geq 2$, $|i^2|_{4\beta^p} = |(2I\beta^{\lceil \frac{p}{2}\rceil+1} + t)t|_{4\beta^p}$, and $|x|_{4\beta^p} = |k|_{4\beta}\beta^{p-1}$. □

Note that $|k|_{4\beta}$ is determined by at most three of the low-order digits of the p-digit x for any even radix β, and by at most two digits when $\beta^{2p-1} \leq x \leq \beta^{2p} - 1$. If a one-ulp $(p+g)$-digit root has a guard digit string $d_{p+1}d_{p+2}\cdots d_{p+g}$ that does not equal either zero or $\frac{1}{2}$ ulp, then the unique half-ulp p-digit root can be identified along with its direction towards the precise root, i.e., the sign of the corresponding p-digit remainder is then also determined. With this information the p-digit remainder determination can be reduced to a half-by-full integer multiplication modulo β^p, which is independent of the p-digit radicand x when $\beta^{2p-1} \leq x \leq \beta^{2p} - 1$.

Observation 6.5.16 *Let i be the p-digit half-ulp integer root of the p-digit radicand x, where $\beta^{2p-1} \leq x \leq \beta^{2p} - 1$, and assume the direction from* *i to $\sqrt{x}$ is known. Then with $i = d_{p-1}d_{p-2}\cdots d_0 = h\beta^{\lceil p/2\rceil} + t$, where $h = d_{p-1}d_{p-2}\cdots d_{\lceil \frac{p}{2}\rceil}\beta^{\lceil p/2\rceil}$, and $t = d_{\lceil p/2\rceil -1}d_{\lceil p/2\rceil -2}\cdots d_0$, the remainder $r = x - i^2$ is given by*

$$r = \begin{cases} |-(2h\beta^{\lceil p/2\rceil} + t)t|_{\beta^p} & \text{for } i < \sqrt{x}, \\ -|(2h\beta^{\lceil p/2\rceil} + t)t|_{\beta^p} & \text{for } i > \sqrt{x}. \end{cases} \tag{6.5.19}$$

If the guard digit string $d_{p+1}d_{p+2}\cdots d_{p+g}$ with $g \geq 1$ is all zeroes, then $i = 0.d_1d_2\cdots d_p$ is a half-ulp root with indeterminate sign, and if the guard digit string equals $\frac{1}{2}$ ulp, then i is a one-ulp root with a positive sign. It follows that with just one guard digit in any even radix, the remainder may be calculated modulo $2\beta^p$. The multiplication size and digits of x needed to compute the remainder by Theorem 6.5.15 and the extensions discussed here are summarized in the following.

Observation 6.5.17 *Let i be a p-digit integer one-ulp root of a p-digit integer radicand x with $\beta^{2(p-1)} \leq x \leq \beta^{2p} - 1$, and $\beta \geq 2$ an even radix. Then a sufficiently large modulus m for computing the remainder is:*

- *$m = \beta^p$ if i is known to be the directed half-ulp integer root;*
- *$m = 2\beta^p$ if either i is the half-ulp integer root, or if $\mathrm{sgn}(\sqrt{x} - i)$ is known;*
- *$m = 4\beta^p$ if i is known only to be a one-ulp root.*

The number of low-order digits of x needed to determine $|x|_m$ for $\beta^{2p-2} \leq x \leq \beta^{2p-1} - 1$ is then 1, 2, or 3 for $m = \beta^p, 2\beta^p$, or $4\beta^p$ respectively, and for $\beta^{2p-1} \leq x \leq \beta^{2p} - 1$, the number is 0, 1, or 2 respectively. The multiplication size obtained using the identity $|i^2|_m = |2(i-t)t|_m$ requires t to include $\lceil p/2\rceil$, $\lceil (p+1)/2\rceil$, or $\lceil (p+2)/2\rceil$ low order digits of i for $m = \beta^p, 2\beta^p$, or $4\beta^p$ respectively.

By using a reasonable number of guard digits so that say $\beta^g = 100$, it is likely that the half-ulp status of i *and* its direction to $\sqrt{x}$ can be determined in some 98% or more of all cases. In this large majority of cases the computation of the unique half-ulp root, remainder pair involves simply a $(\lceil p/2\rceil \times p)$-digit multiplication and uses at most the odd or even status of the lowest digit (pth digit) of x.

Table 6.5.9 provides a number of examples of the size of the modular computation needed to determine the remainder for various precisions, numbers of guard digits and radices 2 and 10 according to Observation 6.5.17. The remainder computation for the half-ulp root $i = 807338$ with positive remainder is obtained using (6.5.19), $r = |-(614\,338)338|_{10^6} = |-646\,244|_{10^6} = 353756$. For the half-ulp root 270 967 with negative remainder $|73\,4231 \times 10^5|_{10^6} = 100\,000$, the remainder is $r = |-(540\,967)967 - 100\,000|_{10^6} = -|115089 - 100\,000|10^6 = -15089$. When a guard digit string is all-zero yielding the one-ulp root i and positive remainder case, and if the computed r then satisfies $r \geq i + 1$, the pair $(i+1, 2(i+1) - r)$ is the half-ulp root, remainder pair.

Table 6.5.9. *Reduced modulus and half-by-full multiplication size for computing a half-ulp root, remainder pair from a $(p+g)$-digit one-ulp root*

x	p	g	$(p+g)$-digit one-ulp root	p-digit root i	Ulp bound	sgn(r)	Modulus m	$\|x\|_m$	Mult. size
651795×10^6	6	0	807339	807339	1	$\pm$	4×10^6	95×10^6	4×7
651795×10^6	6	1	807338.2	807338	$\frac{1}{2}$	$+$	10^6	0	3×6
734231×10^5	6	1	270967.0	270967	$\frac{1}{2}$	$\pm$	2×10^6	31×10^5	4×7
734231×10^5	6	2	270967.97	270967	$\frac{1}{2}$		10^6	1×10^5	3×6
98765×10^5	5	0	99381	99381	1	$\pm$	4×10^5	65×10^5	4×6
98765×10^5	5	1	99380.5	99380	1	$+$	2×10^5	1×10^5	3×6
98765×10^5	5	2	99380.58	99381	$\frac{1}{2}$		10^5	0	3×6
13579×10^4	5	1	11652.9	11653	$\frac{1}{2}$		10^5	9×10^4	3×5
$100\,0000 \times 2^7$	7	2	101 101010	1011010	1	$+$	2×2^7	0	4×8
$111\,1111 \times 2^6$	7	2	101 1010.00	101 1010	$\frac{1}{2}$	$\pm$	2×2^7	$11_2 \times 2^6$	4×8
$111\,1111 \times 2^6$	7	0	101 1010	101 1010	1	$\pm$	4×2^7	$111_2 \times 2^6$	5×8

Problems and exercises

6.5.1 In the remainder form of Newton–Raphson square root with decreasing relative error rate given by (6.5.9) and (6.5.10), consider the ratio $\varepsilon_{i+1}/\delta_{i+1}$ for (a) $\varepsilon_i = \delta_i$, (b) $\varepsilon_i = -3\delta_i$, (c) $\varepsilon_i = -\delta_i$, and (iv) $\varepsilon_i = -\delta_i/2$.

6.5.2 Is there a limiting ratio of $\varepsilon_{i+1}/\delta_{i+1}$ under the recurrences (6.5.9) and (6.5.10), or is the behavior "chaotic." If $\varepsilon_i = O(\delta_i^2)$, does ε_i stay of smaller order compared to δ_i as $i \to \infty$?

6.5.3 Discuss the implementation of convergence square root in a pipelined multiplier allowing a dependent multiplication to be initiated after a four-cycle delay. Give reasonable total cycle counts for obtaining the final root approximation q_k for $k = 3, 4$.

6.5.4 Repeat problem 6.5.3 assuming a half-by-full multiplier with a three-cycle delay.

6.5.5 Repeat problem 6.5.3 for the case where the multiplier can be partitioned to allow two half-by-full multiplies to be computed in parallel.

6.5.6 Suppose linear interpolation is employed to obtain a more accurate initial root-reciprocal approximation ρ for convergence square root that allows our target accuracy for the square root to be achieved with one fewer iterative refinement. Discuss the effect on overall time latency, total multiplications, and other aspects related to a practical implementation.

6.5.7 Let the radicand x be an n-digit radix-β number and let $\rho_1 = 0.d_1d_2\cdots d_j$ be a one-ulp j-digit radix-β seed root-reciprocal of x. Then what is the maximum size of each radix-β term of the 3-tuple (q_i, ρ_i, ω_i) of convergence square root in terms of (n, j, i). Discuss the growth rate of the size of q_i and ρ_i versus the growth in the number of bits-of-accuracy (in equivalent radix-β digits) of q_i and ρ_i.

6.5.8 Discuss comparative advantages for implementation of a square root operation employing:

(a) Newton–Raphson square root (remainder form),
(b) Newton–Raphson root-reciprocal,
(c) convergence square root.

6.5.9 For a radicand x with $1 \leq x < \beta^2$, a j-digit radix-β one-ulp root-reciprocal $\rho = i\beta^{-j} = 0.d_1d_2 \cdots d_j$ satisfies $|1/\sqrt{x} - \rho| < 1/\beta^j$. Describe how to obtain a one-ulp root reciprocal by iterative refinement.

6.5.10 Let ρ be a one-ulp n-digit radix-β root-reciprocal of the n-digit radix-β radicand x with $1 \leq x < \beta^2$. Discuss the multiplication resources needed to determine the sign of $(1/\sqrt{x} - \rho)$ in terms of (a) total multiplications, (b) dependent multiplications, and (c) the minimum size of each multiplication.

6.5.11 Which radix-β radicands x with $1 \leq x < \beta^2$ have an exact radix-β root-reciprocal for:

(a) $\beta = 2,\ 8$, and 10,
(b) β a prime,
(c) β a prime power, or composite.

6.6 Notes on the literature

Before 1980 the literature on square root algorithms and implementations was limited to a few classic papers. With the introduction of the IEEE floating-point standard a large volume of literature emerged on square root algorithms suitable for hardware implementation. Digit-serial examples include [OE82, MC89, EL89, EL90, CM90], and also the book [EL94] by Ercegovac and Lang.

The combination of division and square root SRT was first discussed by Taylor in [Tay81], where a radix-4 SRT division shares hardware with a radix-2 restoring square root algorithm. In [ZG87] Zurawski and Gosling discuss an SRT-like shared implementation of division and square root (and multiply). Fandrianto describes the combination of radix-4 SRT division and square root in [Fan87] (and radix-8 in [Fan89]), using a shared prediction PLA. To reduce the size of the PLA he transforms the remainder into sign magnitude, and discusses reducing the PLA size for higher-radix algorithms by restricting the range of the divisor, using prescaling of dividend and divisor. Other references to sharing algorithms are [Has90, SL96, LA03, LM95]

In [EL90] Ercegovac and Lang discuss the possibility of avoiding an initial PLA for radix-4 SRT square root, cf. Example 6.3.2. Ciminiera and Montuschi in

[CM90] investigate when a constant value of the root approximation, determined by the initial table look-up, can be used during the second, iterative phase of SRT square root. As for division, the determination of optimal truncation parameters is from [Kor05], which also discusses shared division and square root, including avoiding an initial PLA.

The short reciprocal square root algorithm discussed in [BM93] was introduced by Matula and implemented by Briggs, Brightman and Matula [BBM91] in the Cyrix 83D87 coprocessor. The procedure employs two Newton–Raphson root-reciprocal iterations to determine a 19-bit short root-reciprocal which is employed to scale successive remainders to determine successive high radix digits in the radix 2^{17}, using a (17×69)-bit multiply-add unit.

In [LM92] Lang and Montuschi describe a prescaled square root procedure with prescaling by a short scaling factor a^2 and postscaling by division by the divisor a, where a is an 11-bit value for their high radix 256 example. In [LM95] the same authors describe a combined prescaled divide and square root algorithm.

Newton–Raphson iteration is a well known and useful technique for finding zeroes of functions. It was first introduced by Newton around 1669 [New71], to solve polynomial equations (without explicit use of the derivative), and generalized by Raphson a few years later. Actually, the algorithm applied for square root extraction goes back much further. Al-Khwarizmi mentions the method in his arithmetic book [DDP86]. Moreover, it was already used by Heron of Alexandria, and seems to have been known by the Babylonians 2000 years before Heron [FR98]. More recent iterative algorithms can be found in [RGK72, Maj85, IM99, KM97]. In [Obe99] Oberman describes an implementation of Goldschmidt's multiplicative division and square root algorithms that were employed in the AMD Athlon™ microprocessors. The algorithms employed scalings with reciprocals and root reciprocals accurate to about 15 bits so that only two iterations were needed to obtain double precision quarter-ulp accurate approximations.

An alternative prescaling and series approximation method was employed for the Power3™ processor square root operation as described in [AGS99].

Choosing good/optimal starting values for square root iterative algorithms has also been a subject of research, e.g., as found in [Eve63, Fik66, Wil70, MM91, KM06]. The bipartite look-up tables employed by Oberman for seed values in [Obe99] used 11-bit indices to obtain the 15-bit root-reciprocals following the bipartite look-up procedure introduced by DasSarma and Matula in [DM95]. The exact square root test was described by Cristina S. Iordache and Matula in [IM99], along with a procedure for determining a precisely rounded root-reciprocal.

References

[AGS99] R. C. Agarwal, F. G. Gustavson, and M. S. Schmookler. Series approximation methods for divide and square root in the Power3 Processor. In *Proc. 14th IEEE*

Symposium on Computer Arithmetic, pages 116–123. IEEE Computer Society, 1999.

[BBM91] W. S. Briggs, T. B. Brightman, and D. W. Matula. Method and Apparatus for Performing the Square Root Function Using a Rectangular Aspect Ratio Multiplier. US Patent No. 5,060,182, 1991.

[BM93] W. S. Briggs and D. W. Matula. A 17 × 69 bit multiply and add unit with redundant binary feedback and single cycle latency. In *Proc. 11th IEEE Symposium on Computer Arithmetic*, pages 163–170. IEEE Computer Society, 1993.

[CM90] L. Ciminiera and P. Montuschi. Higher radix square rooting. *IEEE Trans. Computers*, 39(10):1220–1231, October 1990.

[DDP86] A. Dahan-Dalmedico and J. Peiffer. *Histoire des Mathématiques*. Editions du Seuil, 1986.

[DM95] D. DasSarma and D. W. Matula. Faithful bipartite ROM reciprocal tables. In *Proc. 12th IEEE Symposium on Computer Arithmetic*, pages 17–28. IEEE Computer Society, 1995.

[EL89] M. D. Ercegovac and T. Lang. On-the-fly rounding for division and square root. In *Proc. 9th IEEE Symposium on Computer Arithmetic*, pages 169–173. IEEE Computer Society, 1989.

[EL90] M. D. Ercegovac and T. Lang. Radix-4 square root without initial PLA. *IEEE Trans. Computers*, C-39(9):1016–1024, August 1990.

[EL94] M. D. Ercegovac and T. Lang. *Division and Square Root: Digit-Recurrence Algorithms and Implementations*. Kluwer Academic Publishers, 1994.

[Eve63] J. Eve. Starting approximations for the iterative calculation of square roots. *Computer J.*, 6:274–276, Oct. 1963.

[Fan87] J. Fandrianto. Algorithms for high speed shared radix 4 division and radix 4 square root. In *Proc. 8th IEEE Symposium on Computer Arithmetic*, pages 73–79. IEEE Computer Society, 1987.

[Fan89] J. Fandrianto. Algorithms for high speed shared radix 8 division and radix 8 square root. In *Proc. 9th IEEE Symposium on Computer Arithmetic*, pages 68–75. IEEE Computer Society, 1989.

[Fik66] C. T. Fike. Starting approximations for square root calculation on IBM System/360. *Commun. ACM*, 9(4):297–299, April 1966.

[FR98] D. Fowler and E. Robson. Square root approximations in Old Babylonian mathematics: YBC 7289 in context. *Hist. Math.*, 25:366–378, 1998.

[Has90] R. Hashemian. Square rooting algorithms for integer and floating-point numbers. *IEEE Trans. Computers*, C-39(9):1025–1029, August 1990.

[IM99] C. Iordache and D. W. Matula. On infinitely precise rounding for division, square root, reciprocal, and square root reciprocal. In *Proc. 14th IEEE Symposium on Computer Arithmetic*, pages 233–240. IEEE Computer Society, 1999.

[KM97] A. H. Karp and P. Markstein. High-precision division and square root. *(ACM) Trans. Mathematical Software*, 23(4):561–589, Dec 1997.

[KM06] P. Kornerup and J.-M. Muller. Choosing starting values for certain Newton–Raphson iterations. *Theor. Comput. Sci.*, 351:101–110, 2006.

[Kor05] P. Kornerup. Digit selection for SRT division and square root. *IEEE Trans. Computers*, 54(3):294–303, March 2005.

[LA03] T. Lang and E. Antelo. Radix-4 reciprocal square-root and its combination with division and square root. *IEEE Trans. Computers*, 52(9):1100–1114, 2003.

[LM92] T. Lang and P. Montuschi. Higher radix square root with prescaling. *IEEE Trans. Computers*, 41(8):996–1009, August 1992.

[LM95] T. Lang and P. Montuschi. Very-high radix combined division and square root with prescaling and selection by rounding. In *Proc. 12th IEEE Symposium on Computer Arithmetic*. IEEE Computer Society, 1995.

[Maj85] S. Majerski. Square-rooting algorithms for high-speed digital circuit. *IEEE Trans. Computers*, C-34(8):724–733, August 1985.

[MC89] P. Montuschi and L. Cinimera. On the efficient implementation of higher radix square root algorithms. In *Proc. 9th IEEE Symposium on Computer Arithmetic*, pages 154–161. IEEE Computer Society, 1989.

[MM91] P. Montuschi and M. Mezzalama. Optimal absolute error starting values for Newton–Raphson calculation of square root. *Computing*, 46:67–86, 1991.

[New71] I. Newton. *Methodus Fluxionem et Serierum Infinitarum*. 1664-1671.

[Obe99] S. F. Oberman. Floating point division and square root algorithms and implementation in the AMD-K7 microprocessor. In *Proc. 14th IEEE Symposium on Computer Arithmetic*, pages 106–115. IEEE Computer Society, 1999.

[OE82] V. G. Oklobdzija and M. D. Ercegovac. An on-line square root algorithm. *IEEE Trans. Computers*, C-31(1):70–75, January 1982.

[RGK72] C. V. Ramamoorthy, J. R. Goodman, and K. H. Kim. Some properties of iterative square-rooting methods using high-speed multiplication. *IEEE Trans. Computers*, C-21:837–847, 1972.

[SL96] P. Soderquist and M. Leeser. Area and performance tradeoffs in floating-point divide and square-root implementations. *ACM Computing Surveys*, 28(3):518–564, September 1996.

[Tay81] G. S. Taylor. Compatible hardware for division and square root. In *Proc. 5th IEEE Symposium on Computer Arithmetic*, pages 127–134. IEEE Computer Society, 1981.

[Wil70] M. W. Wilson. Optimal starting approximations for generating square root for slow or no divide. *Commun. ACM*, 13(9):559–560, September 1970.

[ZG87] J. H. P. Zurawski and J. B. Gosling. Design of a high-speed square root multiply and divide unit. IEEE Trans. Computers, C36: 13–23, 1987.

7

Floating-point number systems

7.1 Introduction

Floating-point number systems evolved from the process of expressing numbers in "scientific decimal notation," e.g., 2.734×10^6 and -5.23×10^{-4}. In practice, the display of a value in scientific notation generally employed the length (precision) of the display as a measure of the accuracy of the display. Values such as $\alpha = 6.0247 \times 10^{23}$ for Avogadro's number or $\pi = 3.14159 \times 10^0$ for the natural constant π were subject to various interpretations regarding the final digit with respect to the correspondence to the presumed infinitely precise specification of the real number for which the scientific notation was an approximation. Terms such as *correct, rounded*, and *significant* were often explicitly or implicitly implied with the display.

Specifically, the digits were termed correct if they agreed with the initial digits of an infinitely precise specification, such as 3.1415 or 3.141592 for π. This interpretation is now popularly termed the *round-towards-zero* result. The notions of rounded and significant digits did not convey such an unambiguous interpretation. Saying that 2.734×10^{-6} was a rounded value of x did not necessarily mean that $|x - 2.734 \times 10^{-6}|$ was guaranteed to be at most $.5 \times 10^{-9}$. More particularly, disregarding inherent measurement error, the rounding of the exact "midpoint" value $x = 2.7345 \times 10^{-6}$ to four places was subject to different interpretations. Most ambiguous was the statement that the digits 6.0247×10^{23} were significant in representing a number such as Avogadro's constant. This probably meant only that the relative error was less than one-half part in 6024 but not necessarily less than one-half part in 60247. The fifth digit "7" here serves as a "guard digit" to roughly center the "four significant digit" relative error. The procedures for computation with rules for significant digit arithmetic provided only a crude approximation to the well founded notion of interval arithmetic. For example, note that an interval representation such as

$\alpha = (6.0247 \pm 0.0002) \times 10^{23}$ is an unambiguous approximation for Avogadro's number.

While the familiarity of scientific notation served to advance the cause of performing floating-point arithmetic as approximate real arithmetic on a computer, the ambiguity of the rules for rounding input or the intermediate results of significant digit computations gave little guidance for designing a "natural" floating-point arithmetic.

The seminal 1946 paper by Burks, Goldstine, and von Neumann [BGvN46] outlined many of the fundamental properties of computers as we know them today, including the use of 2's complement representation, and presented algorithms for the basic arithmetic operations. However, discussing floating-point representations, as being proposed for other contemporary computers, they concluded: "It would therefore seem to us not at all clear whether the modest advantages of a floating binary point offset the loss of memory capacity and the increased complexity of the arithmetic and control units." As an alternative, they pointed to the possibility of using multiple word representations. Time and technology advances should prove them wrong on this point.

Early floating-point arithmetic implementations were provided in software packages using hardware integer arithmetic for underlying support. Multiple precision integer arithmetic could be employed if needed to support variable precision floating-point representations. Such software packages had the potential to be updated to include improved procedures for interpreting the results of floating-point computation without expensive hardware arithmetic modifications. This provided some degree of assurance for computation within an unnatural and difficult host system.

The demand for faster floating-point computation in the 1960s and thereafter led most commercial machine designers to incorporate floating-point arithmetic in hardware as a competitive necessity. By this time the storage of data as well as instructions was restricted to formats in fixed-size words varying between 32 and 64 bits, with widely used machine implementations employing different word sizes such as 32, 36, 48, 60, and 64. The lack of well defined floating-point arithmetic rules afforded the hardware designer leeway in such areas as allocating bits between the exponent and significand fields, choice of radix, method of rounding, and handling of exceptions. In virtually all commercial machines the accuracy needs of scientific computation and the advantages of hardware versus software implementation speed meant that only one fixed combination of precision and exponent range was available to host all floating-point computation.

The closure of floating-point arithmetic by rules for rounding and/or exception handling over a finite set of values characterized by fixed-precision floating-point factorizations, were at that time not well appreciated by hardware designers. In many hardware implementations the rules themselves were implicitly set or altered as by-products of decisions made to increase the speed of floating-point computation. In such environments portable mathematical software became practically impossible to develop for the various host floating-point systems. To make an unfortunate situation worse, the detailed implicit rules for the resulting floating-point arithmetic to the last bit were not documented as part of the typical hardware description available to users. The inevitable consequence was that by the 1970s *floating-point computation for scientific computation had been reduced to an empirical science*.

Towards the end of the 1970s the Institute of Electrical and Electronic Engineers (IEEE) set up a committee to propose a binary floating-point arithmetic standard for the evolving generation of microprocessors. The purpose was that the microprocessor-based computers could share a common floating-point interface to software. This committee accumulated the best ideas available on floating-point number systems and arithmetic design from the viewpoint of both arithmetic computation stability and the efficiency of suitable hardware implementation, culminating in the IEEE-754 standard [IEE85]. This work effectively established the foundations for reproducible computations in floating-point number systems across different platforms, and more generally for computations in a dual precision hierarchy of single- (32-bit) and double-precision (64-bit) floating-point number systems. The impact of this standard has been tremendous, being observed by most computer manufacturers since then, except for a few notable exceptions in some supercomputers and signal processors. Recognizing the need for higher precision (128-bit) and other revisions, a new committee produced an expanded version of this standard, which was approved in 2008 [IEE08].

This chapter will first discuss floating-point number representation in fairly general terms, considered as triples (sign, exponent, significand), describing in factored form $(-1)^{sign} \times \beta^{exponent} \times$ significand, as signed, scaled numbers, without regard to specific encodings of these components, but with restricted significand range and sign represented as a factor (a sign-magnitude representation). The discussion will not be tied exclusively to the IEEE standard, but is to a large extent flavored by the ideas influencing the careful design of the standard.

Sections 7.2 and 7.3 cover general aspects of floating-point representations, the distribution of floating-point numbers, and the rounding into such floating-point systems. Beyond discussing properties of roundings, the specifics of some particular roundings (rounding modes) are introduced. A concept of *rounding equivalent intervals (eqv-intervals)* is defined as a partition over the reals, over which a *precise rounding* is constant. An eqv-interval is essentially an interval determined by two neighboring representable values and their midpoint, together with what is generally known as the *round* and *sticky* bits, providing sufficient information for determining the result of a rounding, when applied to any value in that interval. Determination of such finitely representable eqv-intervals thus allows (the defining property) the result of a precise rounding applied to an eqv-interval to be the very same as if the rounding had been applied to any exact real value in that interval, even though that exact value may not be representable. This discussion will also deal with a generalization to intermediate, redundantly represented eqv-intervals.

Section 7.4 discusses the implementation of addition and multiplication, both operations characterized by the property that a result of either of these, when applied to finitely represented operands, is itself finitely representable. Addition, or rather subtraction of "close" operands, presents some special problems treated in some detail. Section 7.5 then deals with quotient and square root rounding, where the result, in general, is not finitely representable. Finally, the rest of the chapter is devoted to the specifics of the revised IEEE standard [IEE08], its features and the hierarchy of finite precision number systems, defined by their encoding of finite values, as well as their extensions to make them closed systems under the set of required arithmetic operations.

7.2 Floating-point factorization and normalization

To avoid many irrelevant details we shall in the following often omit specifics of particular representations of floating-point numbers and their encodings, as employed in actual *floating-point unit* (FPU) implementations.

7.2.1 Floating-point number factorization

A *floating-point number* is a real number with a radix-β normalized factorization (*radix-β factorization*)

$$v = (-1)^s \beta^e f, \tag{7.2.1}$$

where for $v \neq 0$:

- $(-1)^s$ is an optional *sign factor* determined by a *sign bit*, $s \in \{0, 1\}$;
- β^e is a *scale factor* determined by the radix β and an integer *exponent* e;
- f is a *significand* with a restricted *normalization range* $0 < f_{min} \leq f \leq f_{max}$.

A radix-β factorization is not unique. For $\beta = 10$ we have $-5\frac{3}{4} = (-1)^1\, 10^1\, 0.575 = (-1)^1\, 10^0\, 5.75 = (-1)^1\, 10^{-2}\, 575$, and similarly in binary $101.1101 = (-1)^0\, 2^2\, 1.011101 = (-1)^0\, 2^{-4}\, 1011101$. It has been customary in the past to obtain uniqueness in representation by requiring the significand to have its radix point in some specific position. However, to simplify the following discussion, without any loss of generality, we will assume it is normalized as specified in the IEEE standard and is consistent with scientific notation, such that $f \in [1; \beta)$.

Historically some floating-point number representations have used alternative normalization such as a fraction normalized significand $f \in [1/\beta; 1)$, or a full precision integer significand $f \in [\beta^{p-1}; \beta^p - 1)$. Other representations of significands such as 1's and 2's complement have also been employed, obviously then without an explicit sign factor.

We shall assume that a non-zero *normalized* floating-point number has the *standard* significand range $f \in [1; \beta)$, with a normalized radix-β factorization $v = (-1)^s \beta^e f$. With implicit reference to this unique normalized radix-β factorization it is then meaningful for every non-zero real v to refer to the *sign* $s(v)$, *exponent* $e(v)$, and *significand* $f(v)$ as well defined properties of the radix-β number v. The triple $(s(v), e(v), f(v))$ effectively represents the floating-point number $v = (-1)^s \beta^e f$ by this one-to-one correspondence. The value $v = 0$ has no such normalized factorization but still is typically included in floating-point number systems as an exceptional value with a reserved encoding.

For convenience of exposition for a general radix we shall employ the normalized factorization $v = (-1)^s \beta^e f$ when only the values of factors are relevant, not their encodings. The digital significand expression $v = (-1)^s \beta^e d_0.d_1 d_2 \cdots$ will be employed when properties of the digits are relevant. In this chapter we shall use the digit string without $\| \cdot \|_\beta$ in displaying floating-point factorizations, as is customary in scientific notation. Note that the normalized radix-β factorization $v = (-1)^{s(v)} \beta^{e(v)} d_0(v).d_1(v) d_2(v) \cdots$ for non-zero v has each digit $d_i(v)$ uniquely determined for all $i \geq 0$, $\beta \geq 2$ only when the digit set employed is non-redundant, implicitly the standard digit set $\{0, 1, \ldots, \beta - 1\}$ unless otherwise specified.

For binary floating-point numbers the interval $[2^e; 2^{e+1})$ is termed a *binade*, with the exponent $e(v)$ designating the particular binade for $|v| \neq 0$. The *principal binade* for normalization is $[1; 2)$. This normalization is particularly convenient when the significand is represented by a finite or infinite precision standard bit-string $f = 1.b_1 b_2 \cdots b_{p-1} b_p \cdots$. For normalized binary floating-point numbers the significand's *fractional part* bit-string $b_1 b_2 \cdots$ is then sufficient to characterize the significand.

The decimal interval $[10^e; 10^{e+1})$ is termed a *decade*, with $[1; 10)$ the *principal* decade for decimal floating-point normalization. The decimal significand $f = d_0.d_1 d_2 \cdots$ has the non-zero leading digit satisfying $1 \leq d_0 \leq 9$. The term *scientific notation* corresponds to the normalized decimal factorization $v = (-1)^s 10^e d_0.d_1 d_2 \cdots$. Scientific notation highlights the leading digit as an integer with the less-significant digits forming a fractional tail. The exponent and leading digit provide a concise reference to the order of magnitude of the number, with the fractional tail yielding greater refinement of the value, as needed for the application. Historically scientific notation was widely used as a basis for hand and software based scientific computation and clearly influenced the formalization of general floating-point number systems as a foundation for the implementation of a hardware floating-point unit (FPU).

Although scientific notation can be given a firm number theoretic foundation for any radix, its particular radix-dependent properties relegated its status in classical mathematics to be most often just a non-deterministic approximation. Our approach here is to introduce a minimal number theoretic foundation.

7.2.2 Finite precision floating-point number systems

When discussing digit-string representations of significands, it is obvious that, for a unique normalization of the value, it is necessary that the digit set employed is non-redundant; we will insure this is the case here by using the standard digit set $\{0, 1, \ldots, \beta - 1\}$.

A *p-digit* normalized floating-point number has the significand range $f \in [1; \beta - \beta^{-(p-1)}]$ where $f \in \{i/\beta^{p-1} \mid \beta^{p-1} \le i \le \beta^p - 1\}$. Notationally the factorization is given employing a fixed-point string representation of the significand, where low-order zeroes are explicitly displayed to characterize the value of p:

$$v = \begin{cases} (-1)^s \beta^e\, d_0.d_1d_2 \cdots d_{p-1} & \text{for} \quad \beta \ge 3, \\ (-1)^s 2^e\, 1.b_1b_2 \cdots b_{p-1} & \text{for} \quad \beta = 2. \end{cases} \tag{7.2.2}$$

The unique (sign, exponent, significand) triple representation $s(v)$, $e(v)$, $f(v)$ for $v \neq 0$ then has $f(v) = d_0.d_1d_2 \cdots d_{p-1}$ with $1 \le d_0 \le \beta - 1$, and $0 \le d_i \le \beta - 1$ for $1 \le i \le p - 1$. The p-digit normalized floating-point radix-β numbers form a conveniently characterized subset of the radix-β numbers defined in Chapter 1 by $\mathbb{Q}_\beta = \{k\beta^\ell | k, \ell \in \mathbb{Z}\}$, disregarding for the moment the usual finiteness of the set of exponent values in practical floating-point systems.

Definition 7.2.1 *For the radix $\beta \ge 2$, the p-digit floating-point numbers are given by the set $\mathbb{Q}_\beta^p = \{k\beta^\ell | k, \ell \in \mathbb{Z}, |k| \le \beta^p - 1\}$, where p is termed the precision of $\mathbb{Q}_\beta^p$.*

Observation 7.2.2 *For $v \in \mathbb{Q}_\beta^p$, $v \neq 0$ has the unique integer normalized factorization $v = k\beta^\ell$, where the integer significand k satisfies $\beta^{p-1} \le |k| \le \beta^p - 1$, and $\ell = e - (p - 1)$ is the least-significant position. Furthermore, e is the index m of the most-significant digit position in the radix polynomial for v,*

$$v = (-1)^s(d_m\beta^m + d_{m-1} + \cdots + d_\ell\beta^\ell) = (-1)^s\beta^e d_0.d_1d_2 \cdots d_{p-1}$$

with $e = m$ being alternative designations of the order of that polynomial.

Proof For $v = (-1)^s\beta^e\, d_0.d_1d_2 \cdots d_{p-1} = (-1)^s\beta^e f$, note that $k = (-1)^s f\beta^{p-1}$ is an integer satisfying $\beta^{p-1} \le |k| \le \beta^p - 1$. Thus $v = (-1)^s\beta^e f = k\beta^{e-(p-1)} = k\beta^\ell$, and the result follows. □

For $v \in \mathbb{Q}_\beta^p$ we shall employ the term "normalized factorization" for both the *integer significand* factorization $v = k\beta^\ell$ with $\beta^{p-1} \le |k| < \beta^p$ and the scientific standard significand factorization $v = (-1)^s\beta^e f$ with $f \in [1, \beta)$, with the context and/or the placing of the radix point in the digit string clarifying the choice. We shall reserve the term "exponent" for the power "e" of the radix in the scientific factorization and "last position" ℓ for the power of the radix in the integer factorization, with $e = \ell + (p - 1)$ for normalized factorizations. This exponent adjusting correspondence between scientific notation and the normalized integer significand factorization, provides a foundation for the notion of "significant low-order zeroes" displayed in the fixed-point string. The number theoretic integer significand factorization for $v \in \mathbb{Q}_\beta^p$ is useful in proving results such as the following regarding exact radix conversion (see the proof of Theorem 1.3.5).

Observation 7.2.3 *If β divides γ, then any p-digit normalized radix-β floating-point number $v = (-1)^s \beta^e f$ is a $(p + |e| - 1)$-digit normalized radix-γ floating-point number $v = (-1)^s \gamma^{e'} f'$. Furthermore, if $e \geq p - 1$, then v is an integer and is a $\lceil \log_\gamma \beta^{e+1} \rceil$-digit number.*

By Observation 7.2.3 we note that any p-bit normalized floating-point number $v = (-1)^s 2^e f$ is a $(p + |e| - 1)$-digit normalized decimal floating-point number. Furthermore, if $e \geq p - 1$, then v is a $\lceil \log_{10} 2^{e+1} \rceil$-digit decimal integer. For example, $1.101011_2 \times 2^{-4} = 1.03515625_{10} \times 10^{-1}$, and $1.10101 \times 2^{10} = 1.696 \times 10^3 = 1696$.

Despite the convenient number theoretic characterization in Observation 7.2.2, it should be emphasized that the set $\mathbb{Q}_\beta^p$ of p-digit radix-β floating-point numbers is far less "natural" than the axiomatic integer, rational, and real number systems. The radix-β numbers $\mathbb{Q}_\beta$ rely only on one arbitrary parameter β. In contrast the p-digit radix-β floating-point numbers $\mathbb{Q}_\beta^p$ rely on two parameters, the radix β and the precision p. If the range of integer exponent values is also bounded to obtain a finite set, the range parameters also become parameters of the system. For digit-string representation, the digit set also becomes a parameter of the system.

Employing intermediate redundant representations. When implementing basic floating-point arithmetic operations and during composite calculations like inner product or evaluation of standard functions, it may be advantageous to employ intermediate redundant representations. The most obvious example is the use of such during processing of the significand values.

Formally we may introduce *floating-point polynomials* for $P(v) \in \mathcal{P}[\beta, D]$ as

$$P(v) = (-1)^{s(v)}[\beta]^{e(v)} \sum_{i=0}^{p-1} d_i[\beta]^{-i}, \tag{7.2.3}$$

denoting a factorization with the sign factor, the monomial scale factor, and the significand radix polynomial, each separately determined. Here the digit set D may or may not be redundant; the former case allows us to distinguish different representations of (possibly) the same value of the significand. This factorization again designates the exponent $e(v)$ as the order of the polynomial, assuming of course that $d_0 \neq 0$.

Although expressed as if (7.2.3) is normalized, it need not be so due to redundancy of the digit set. For example, it may contain non-zero leading guard digits representing a zero-valued leading part. It may also contain trailing guard digits needed for a precise rounding, to be discussed in the next section.

It should be emphasized that the exponent function also allows redundant integer representations of exponents, which can be employed to describe intermediate floating-point factorizations within an FPU.

It should be noted and appreciated that for radix $\beta = 2$, there is no alternative non-redundant digit set to consider other than $\{0, 1\}$. This is one of the many features compelling us to select binary floating point as the most natural of floating-point systems. The representation-dependent preferences are independent from, but consistent with, the preference for binary floating point based on current machine architectural considerations.

Bounding the exponent range. Finite floating-point factorizations will generally employ a bounded exponent with a range consisting of a contiguous set of integers $e_{min} \leq e \leq e_{max}$. However, an unbounded exponent range may be conveniently utilized when analyzing the numerical accuracy of algorithms, leaving the boundedness question to a separate analysis. Of course, when implementing standard floating-point operations on machine encodings of floating-point values the exponent range will necessarily be bounded, where normally the exponent range is "almost symmetric", i.e., $-e_{min}$ is close to e_{max}.

The actual encoding of the exponent must allow for positive as well as negative values, and traditionally 2's complement or biased integer representations have been employed.

In machine encodings for storage representation of floating-point values, the field allocated for the exponent value may employ certain bit patterns to represent special floating-point values, such as infinities or not-a-number. Thus introducing special bit patterns in the allocated field to be interpreted as "escape codes" allows not only for different interpretations of the field, but also for a different encoding of the whole floating-point encoding, e.g., for it to be interpreted as an unnormalized representation.

Examples of finite floating-point number systems having finite range and exceptional floating-point values are provided by the IEEE standard floating-point number systems which are analyzed in detail in Section 7.6.

7.2.3 Distribution of finite precision floating-point numbers

The distance between successive p-digit radix-β floating-point numbers is termed an *ulp*. For $v = (-1)^s \beta^e d_0.d_1 d_2 \cdots d_{p-1}$, the *ulp function* is defined by $\mathrm{ulp}(v) = \beta^{e-(p-1)} = \beta^{\ell}$. The uniform spacing between p-digit values is convenient for implementing and analyzing roundings of reals to p-digit values, provided that the boundary values $|v| = \beta^e$, where the ulp function jumps by a factor of β, can be efficiently identified.

Finite floating-point number systems contain a subset of radix-β numbers obtained by bounding the exponent range as well as the number of digits in the factorization. The positive normalized three-bit floating-point numbers $v = 2^e\, 1.b_1 b_2$ for $-2 \leq e \leq 2$ are illustrated in Figure 7.2.1 along with the upper binade boundary $v = 2^3 1.00$.

The 21 values $v = 2^e\, 1.b_1 b_2$ partition the interval $[\frac{1}{4}; 8]$ into 20 one-ulp *gaps* where the gap size is constant over each binade $[2^e; 2^{e+1})$, and is doubled in passing from one binade $[2^e, 2^{e+1})$ to the next binade $[2^{e+1}; 2^{e+2})$. In Figure 7.2.1 the gap size is 16 times as large approaching the upper bound $v \to 8$, as approaching the lower bound $v \to \frac{1}{4}$.

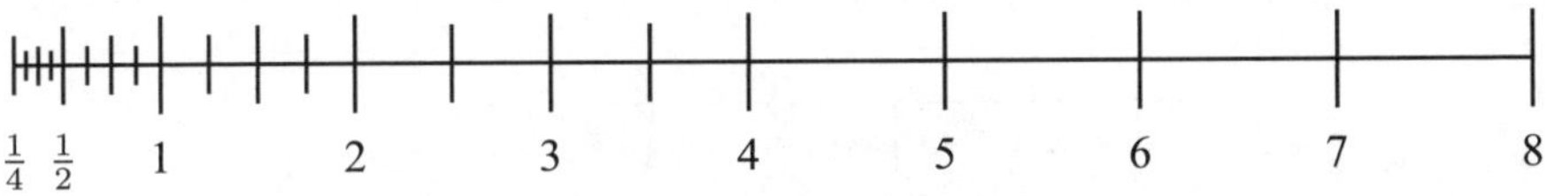

Figure 7.2.1. Tickmarks showing $v = 2^e\, 1.b_1 b_2$ over the interval $[\frac{1}{4}; 8]$.

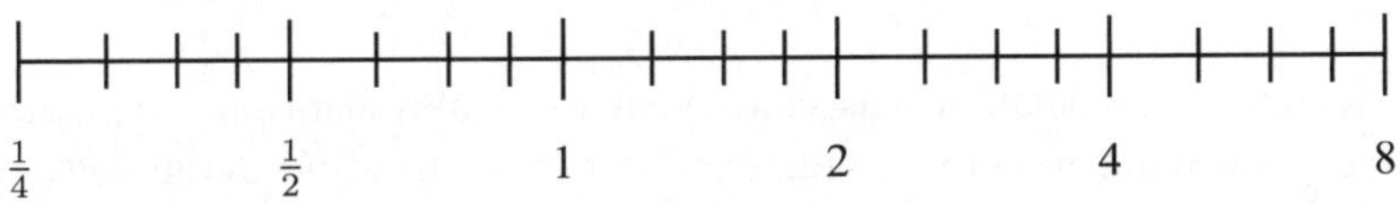

Figure 7.2.2. Tickmarks showing $v = 2^e\, 1.b_1 b_2$ over the interval $[\frac{1}{4}; 8]$ on a log scale.

A better understanding of the distribution of floating-point numbers over a large range incorporating many binades is then provided by graphing the floating-point values as tickmarks on a log scale. Figure 7.2.2 illustrates the normalized three-bit floating-point numbers $v = 2^e(1.b_1 b_2)$ over the five-binade interval $v \in \left[\frac{1}{4}; 8\right]$. The periodicity is evident on the log scale so that the distribution for a floating-point number system with a wide exponent range $|e| \leq n$ is visualized in terms of the periodic extension to $2n$ periods.

The value of the ulp function over the full range is most readily visualized as a step function plotted on a log–log scale. Figure 7.2.3 illustrates the ulp function behavior for $v = 2^e\, 1.b_1 b_2$ over the five-binade interval $[\frac{1}{4}; 8)$.

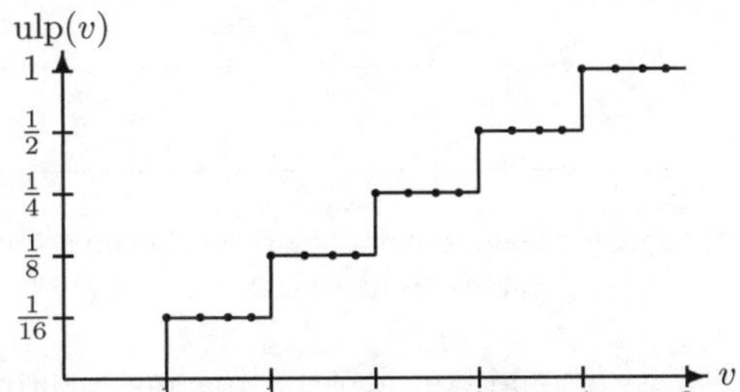

Figure 7.2.3. Graph of $\mathrm{ulp}(v)$ for $v = 2^e\, 1.b_1 b_2$ over the interval $\left[\frac{1}{4}; 8\right)$.

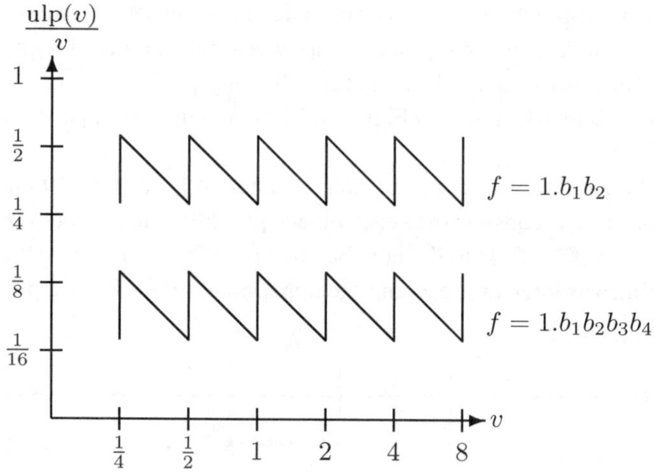

Figure 7.2.4. Graph of the relative gap $(\mathrm{ulp}(v))/v$ for $v \in \left[\frac{1}{4}; 8\right)$ with significands $f = 1.b_1b_2$ and $f = 1.b_1b_2b_3b_4$.

In scientific computation it is convenient to employ hardware implemented floating-point numbers with a wide exponent range to avoid the tedious chore of scaling values to an integer range by user controlled software. For scientific computations employing floating-point arithmetic over a wide range it is best to consider measurement and computation errors on a relative scale.

Figure 7.2.4 shows the graph of the *relative gap* between floating-point numbers at v, defined by $(\mathrm{ulp}(v))/v$ for all real v over the floating-point range, plotted on a log–log scale.

Collectively, Figures 7.2.1–7.2.4 illustrate that we need to interpret floating-point number systems differently at the micro and the macro range. At the micro level we specifically imply intervals $[a; b]$, where a and b fall in either the same or adjacent binades (decades). In this case the gap between floating-point values is constant if a and b are in the same binade (decade), or differs by a factor of the radix if they are in adjacent binades (decades) as illustrated in Figure 7.2.5.

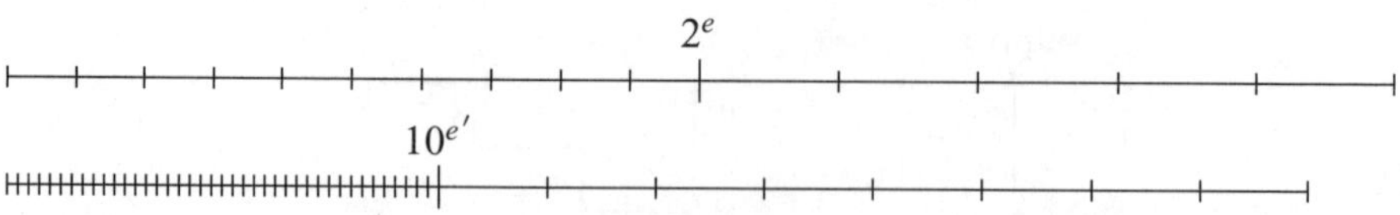

Figure 7.2.5. Micro range: floating-point binary and decimal values in adjacent binades (decades).

At the macro level we should consider the spacing as providing essentially bounded relative spacing over the whole exponent range with periodic wobbles as is evident in the sawtooth plot of Figure 7.2.6.

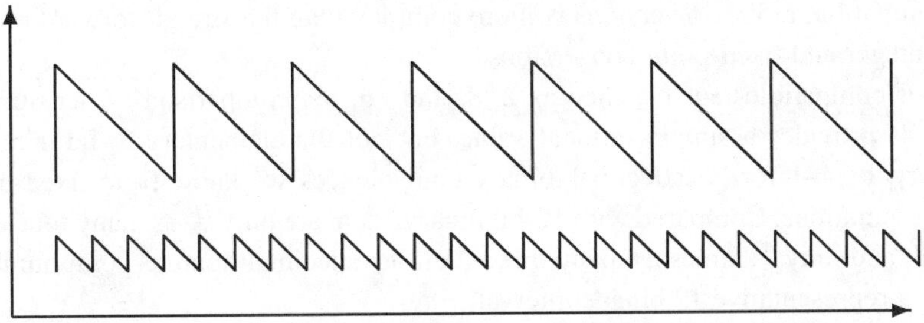

Figure 7.2.6. Macro range: precision wobble for binary and decimal floating point.

Referring to $\log_\beta(\text{ulp}(v)/v)$ as the "effective precision" of v, the term *precision wobble* then refers to this periodic behavior illustrating that the relative accuracy of p-digit radix-β approximations just above a boundary β^e is not much better than a $(p-1)$-digit approximation just below that boundary. This effective loss of accuracy is more critical for large radices, particularly for systems with smaller precisions.

7.2.4 Floating-point base conversion and equivalent digits

In every binade of values from $\mathbb{Q}_2^p$ there are 2^{p-1} distinct values, as is evident from the normalization having a leading unit. Similarly, there are $9 \times 10^{p-1}$ distinct decimal values in each decade of $\mathbb{Q}_{10}^p$. With reference to Figure 7.2.1, note that there are 21 three-bit binary numbers over the interval $[\frac{1}{4}; 8]$, but this interval contains only 15 one-digit decimal numbers. Inspection of larger intervals confirms that there are *more three-bit floating-point binary numbers than one-digit decimal floating-point numbers*. In contrast to integer representation where each decimal digit is "equivalent" to $\log_2 10 = 3.32$ bits, the notion of an "equivalent number" of digits must be modified for floating point.

Observation 7.2.4 *For the p-digit radix-β numbers, $\mathbb{Q}_\beta^p$, and the q-digit radix-γ numbers, $\mathbb{Q}_\gamma^q$, the ratio of the densities of representable values is given by*

$$\lim_{m\to\infty} \frac{\left|\left\{w \middle| w \in \mathbb{Q}_\gamma^q, \frac{1}{m} \le |w| \le m\right\}\right|}{\left|\left\{v \middle| v \in \mathbb{Q}_\beta^p, \frac{1}{m} \le |v| \le m\right\}\right|} = \frac{(\gamma-1)\gamma^{q-1}}{(\beta-1)\beta^{p-1}} \log_\gamma \beta, \tag{7.2.4}$$

where as usual $|\cdot|$ on a set denotes the cardinality of the set.

For the well known problem traditionally termed "floating-point base conversion" it is desired to determine an equivalent number of digits in the radix of the target system, and this number can be formalized by requiring, that the target system has essentially the same density of representable values over a large range as the source system. The equivalent number of digits for both compatible and incompatible radix conversions is more complex than the straightforward results on integer and fixed-point conversions.

For compatible radices, such as 2, 8, and 16, the comparison of densities in (7.2.4) provides a simple rational value, but not the elementary 3–1 for octal–binary, or 4–1 for hexadecimal–binary equivalences we know from fixed-point representations. Compared with 12-bit binary, there are only $\frac{7}{12}$ as many four-digit octal, and only $\frac{15}{32}$ times as many three-digit hexadecimal floating-point numbers over a representative 12 binade interval.

In general, to find an "equivalent digit formula" for floating-point radix conversion for arbitrary radices $2 \leq \beta < \gamma$, it is sufficient to equate the right-hand side of (7.2.4) to unity, and solve for q in terms of p.

Observation 7.2.5 *The number of digits q in a radix-γ system providing the same density of floating-point numbers as a p-digit radix-β system is*

$$q = p \log_\gamma \beta + \log_\gamma \left(\frac{\gamma(\beta - 1)}{\beta(\gamma - 1)} \right) - \log_\gamma \log_\gamma \beta.$$

Thus the "equivalent digit formula" for binary–decimal floating-point radix conversion is

$$\#\text{bits} = 3.322 \times (\#\text{decimal digits}) - 0.884.$$

This loss of almost one bit in representation density explains the fact that there are more three-bit (binary) than one-digit (decimal) numbers, as noted with reference to Figure 7.2.1. Furthermore, the loss for the compatible radices 8 and 16 compared with binary is also about one bit in each case:

$$\#\text{bits} = 3 \times (\#\text{octal digits}) - 0.778,$$
$$\#\text{bits} = 4 \times (\#\text{hexadecimal digits}) - 1.093.$$

The fact that the density of a three-digit hexadecimal is only $\frac{45}{56} \sim 80\%$ that of the four-digit octal floating-point numbers does not reveal the full story. By plotting the gap functions we note that there are regions, such as the interval [64; 256), where three-digit hexadecimal is twice as dense as four-digit octal, and other regions, such as [16; 64), where three-digit octal is four times as dense as three-digit hexadecimal.

Perhaps most surprising, as is evident from a plot of the relative gap functions, is the fact that four-digit decimal is more dense than four-digit hexadecimal floating point over some intervals, such as [0.625; 1), even though the overall density is only 17.6%.

To obtain that the target system must be more dense throughout any large range, it is essentially necessary to give up a digit in the target system, and to insure a uniformly sparser target system, a digit of the source system must be sacrificed, as summarized in the following

Theorem 7.2.6 (Floating-Point Base Conversion Theorem) *For incompatible radices β and γ, the gap sizes* $\mathrm{ulp}(v)$ *for* $v \in \mathbb{Q}_\beta^p$ *and* $\mathrm{ulp}(w)$ *for* $w \in \mathbb{Q}_\gamma^q$ *with* $w \le v < w'$ *satisfy:*

- $\mathrm{ulp}(w) < \mathrm{ulp}(v)$ *for all* $v \in \mathbb{Q}_\beta^p$ *iff* $\gamma^{q-1} \ge \beta^p - 1$,
- $\mathrm{ulp}(w) \ge \mathrm{ulp}(v)$ *for all* $v \in \mathbb{Q}_\beta^p$ *iff* $\beta^{p-1} \ge \gamma^q - 1$.

The proof relies on a theorem of Kronecker, stating that the integer multiples of an irrational modulo unity are dense in the unit interval, and the details are left as an exercise.

Problems and exercises

7.2.1 Determine the equivalent number of octal digits for a p-digit hexadecimal floating-point number system, showing it is less than $\frac{4}{3}p$.

7.2.2 Plot the relative gap functions $\mathrm{ulp}(v)/v$ over the interval $[\frac{1}{1000}; 1000)$ for four-digit decimal and four-digit hexadecimal floating-point number systems. Which subintervals of $[\frac{1}{1000}; 1000)$ have the least and greatest densities of four-digit decimal compared to four-digit hexadecimal floating-point numbers?

7.2.3 Which subintervals of $[\frac{1}{1000}; 1000)$ have the least and greatest densities of seven-digit decimal compared with:

- six-digit hexadecimal floating-point numbers,
- 24-bit binary floating-point numbers?

7.2.4 Prove Theorem 7.2.6, the Base Conversion Theorem.

7.2.5 Plot the relative gap functions for the three-digit hexadecimal and the four-digit octal floating-point number systems for the interval $[1; 4096)$. Why is this a "representative interval" for these functions?

7.3 Floating-point roundings

In Section 1.9 we considered the series of "best radix approximations" to a particular real value x, where the series was indexed by decreasing "last place positions" ℓ in the approximations of x. Two types of series were defined based on generating a monotonic approximation sequence or a sequence in which each approximation has minimum error for that last place position. Our focus here is an orthogonal viewpoint in that we look at issues related to mapping of the set of reals into a particular finite precision system $\mathbb{Q}_\beta^p$, i.e., p is fixed. In this sense the roundings are viewed as mappings of the reals into $\mathbb{Q}_\beta^p$. Again we have options in defining the mapping based on observing direction or minimizing error when we have a choice. We also consider the roundings inverse mappings as mappings from $\mathbb{Q}_\beta^p$ into real intervals, in particular the one-ulp width intervals represented by the rounded value.

7.3.1 Precise roundings

Precise roundings always choose a value minimizing error subject only to the option that the error may or may not have a fixed sign, i.e., providing precise nearest or directed roundings. Precise roundings into the p-digit radix-β floating-point numbers $\mathbb{Q}_\beta^p$ are aided by the following terms for successive values and midpoints between successive values in $\mathbb{Q}_\beta^p$:

- $v' = \min\{w|w > v, w \in \mathbb{Q}_\beta^p$ for $v \in \mathbb{Q}_\beta^p, v \neq 0\}$, and v' is the *one-ulp successor* of v in $\mathbb{Q}_\beta^p$;
- v, v' is an *adjacent pair* of $\mathbb{Q}_\beta^p$ with $v' - v = \text{ulp}(v)$;
- v'' is the one-ulp successor of v' and v, v', v'' is an *adjacent triple* of $\mathbb{Q}_\beta^p$;
- $v_{mid} = (v + v')/2$ is the *midpoint* between v and v';
- v'_{mid} is the *midpoint* between v' and v''.

In this section we restrict β to even values such as 2 and 10, where the midpoint operation then yields $v_{mid} \in \mathbb{Q}_\beta^{(p+1)}$. Specifically, v_{mid} has a significand with an additional low-order digit $d_p = \beta/2$.

Adapting the definition of a *rounding* from Section 1.9 as a mapping from the set of reals into an arbitrary set of representable numbers we have the following definition for roundings into $\mathbb{Q}_\beta^p$.

Definition 7.3.1 *A* precise rounding *is a mapping* $\text{R} : \mathbb{R} \to \mathbb{Q}_\beta^p$ *satisfying the following properties:*

monotonic: $x < y \Rightarrow \text{R}(x) \leq \text{R}(y)$;
fixed points: $x \in \mathbb{Q}_\beta^p \Rightarrow \text{R}(x) = x$;
precise: $v \in \mathbb{Q}_\beta^p, 0 < v < x < y < v_{mid}$ *OR* $v_{mid} < x < y < v' \Rightarrow \text{R}(x) = \text{R}(y)$.

Any precise rounding is then a uniquely defined function from the reals onto $\mathbb{Q}_\beta^p$ defined parametrically in terms of (p, β). The mapping $\text{R} : \mathbb{R} \to \mathbb{Q}_\beta^p$ is a *directed precise rounding* if the open ulp interval $(v; v')$ rounds to just one value, v or v', with that value chosen consistently for all adjacent pairs $v, v' \in \mathbb{Q}_\beta^p$ based only on the order $v < v'$ and the sign of v. There are then four distinct directed roundings onto $\mathbb{Q}_\beta^p$ that may be distinguished simply by each of four rounding directions: towards minus infinity, towards plus infinity, towards zero, and away from zero (towards unsigned infinity).

Definition 7.3.2 *The directed roundings* RD, RU, RZ, RA : $\mathbb{R} \to \mathbb{Q}_\beta^p$ *are given for all real x with v, v' denoting adjacent radix-β numbers in $\mathbb{Q}_\beta^p$ by:*

(i) RD: *round-down(-towards-minus-infinity)*

$$\text{RD}(0) = 0,$$
$$\text{RD}(x) = v \text{ for } v \le x < v',\ v \in \mathbb{Q}_\beta^p;$$

(ii) RU: *round-up(-towards-plus-infinity)*

$$\text{RU}(0) = 0,$$
$$\text{RU}(x) = v' \text{ for } v < x \le v',\ v' \in \mathbb{Q}_\beta^p;$$

(iii) RZ: *round-towards-zero*

$$\text{RZ}(0) = 0,$$
$$\text{RZ}(x) = \begin{cases} v & \text{for } 0 < v \le x < v',\ v \in \mathbb{Q}_\beta^p, \\ v' & \text{for } v < x \le v' < 0,\ v' \in \mathbb{Q}_\beta^p; \end{cases}$$

(iv) RA: *round-away-from-zero*

$$\text{RA}(0) = 0,$$
$$\text{RA}(x) = \begin{cases} v' & \text{for } 0 < v \le x < v',\ v' \in \mathbb{Q}_\beta^p, \\ v & \text{for } v < x \le v' < 0,\ v \in \mathbb{Q}_\beta^p. \end{cases}$$

It is easily seen that the roundings then satisfy Definition 7.3.1. Regarding the sign of x, note that the "magnitude" directed roundings RZ and RA also satisfy the antisymmetry property $\text{R}(-x) = -\text{R}(x)$, while the "algebraically" directed roundings RU and RD satisfy $\text{RU}(-x) = -\text{RD}(x)$.

Observation 7.3.3 *Given the radix-β factorization $v = (-1)^s \beta^e f \neq 0$, the algebraically directed roundings* RU, RD *are given in terms of the magnitude directed roundings* RZ, RA *applied to $|v| = \beta^e f$ by*

$$\text{RU}((-1)^s \beta^e f) = \begin{cases} (-1)^s \text{RA}(\beta^e f) & \text{for } s = 0, \\ (-1)^s \text{RZ}(\beta^e f) & \text{for } s = 1, \end{cases}$$

$$\text{RD}((-1)^s \beta^e f) = \begin{cases} (-1)^s \text{RZ}(\beta^e f) & \text{for } s = 0, \\ (-1)^s \text{RA}(\beta^e f) & \text{for } s = 1. \end{cases}$$

A *nearest precise rounding* $\text{R} : \mathbb{R} \to \mathbb{Q}_\beta^p$ maps every real x into $\mathbb{Q}_\beta^p$ so as to minimize the difference between x and its rounded value, i.e., $|x - \text{R}(x)| = \min\{|x - v| \mid v \in \mathbb{Q}_\beta^p\}$. Nearest roundings are unique except for the rounding direction of the midpoint $v_{mid} = (v + v')/2$ for every adjacent pair $v, v' \in \mathbb{Q}_\beta^p$. Nearest roundings may be made unique simply by rounding all midpoints by a specified directed rounding such as RZ or RA.

However, when nearest roundings are intended to avoid statistical bias in the rounding direction, an unbiased rounding direction rule is needed for midpoints. For guidance, consider that the round-to-nearest-even rounding of the reals to the integers $\text{RN} : \mathbb{R} \to \mathbb{Z}$ maps midpoints $2k \pm \frac{1}{2}$ to their nearest even integer value $\text{RN}_e(2k \pm \frac{1}{2}) = 2k$. A round-to-nearest-even mapping from the reals into $\mathbb{Q}_\beta^p$ can be created by appropriately extending the notion of parity to elements of $\mathbb{Q}_\beta^p$.

Definition 7.3.4 *The parity of v in $\mathbb{Q}_\beta^p$ is the parity of the integer significand in the unique normalized factorization $v = (-1)^s k \beta^\ell$ with $\beta^{p-1} \le k \le \beta^p - 1$.*

Definition 7.3.5 (RN$_e$: round-to-nearest-even) *For $\beta \geq 2$ and $p \geq 2$ or $\beta \geq 3$ and $p = 1$, the rounding $\text{RN}_e : \mathbb{R} \to \mathbb{Q}^p_\beta$ is given for all real x with v, v', v'' denoting adjacent triples in $\mathbb{Q}^p_\beta$ by,*

$$\text{RN}_e(0) = 0,$$
$$\text{RN}_e(x) = v' \ \textit{for} \ v_{mid} < x < v'_{mid},$$
$$\text{RN}_e(v_{mid}) = \begin{cases} v & \textit{for} \quad v \textit{ of even parity}, \\ v' & \textit{for} \quad v' \textit{ of even parity}. \end{cases}$$

The RN$_e$ rounding is unbiased in the sense that one half of the $\beta^{p-1}(\beta - 1)$ p-digit midpoints v_{mid} in the interval $[\beta^j; \beta^{j+1}]$ round to the larger value v' and one half to the smaller value v. Round-to-nearest (even) is the required nearest rounding in the IEEE standard for binary floating-point arithmetic.

Amongst the four directed roundings RU, RD, RZ, RA and the nearest rounding RN$_e$ only the last effectively depends on the factorization of $v \in \mathbb{Q}^p_\beta$ for its definition. For the critical binary ($\beta = 2$) and decimal ($\beta = 10$) cases, the following shows a helpful simplification for algorithmic implementation of RN$_e$.

Observation 7.3.6 *For even β and $p \geq 2$, let the non-zero $v \in \mathbb{Q}^p_\beta$ be given by the unique normalized radix-β factorization $v = (-1)^s \beta^e d_0.d_1 d_2 \cdots d_{p-1}$. Further, let v' be the successor of v in $\mathbb{Q}^p_\beta$. Then the round-to-nearest-even rounding $\text{RN}_e(v_{mid})$ is determined by*

$$\text{RN}_e\,(v_{mid}) = \begin{cases} v & \textit{if} \quad d_{p-1} \textit{ is even}, \\ v' & \textit{if} \quad d_{p-1} \textit{ is odd}. \end{cases} \tag{7.3.1}$$

The consequence of Observation 7.3.6 is that the parity of the low-order digit is sufficient to resolve the representation dependence for even bases such as binary and decimal. It should be noted here that a round-to-nearest-odd can be defined for even bases and $p \geq 2$ simply by reversing the choices in (7.3.1). Such a rounding has been described in the literature. For an odd radix, parity determination effectively requires inspection of the whole significand. Thus an implementation of RN$_e$ for an odd radix could be more costly in algorithmic or hardware resources than for an even radix.

Traditionally another (but biased) round-to-nearest has been used and is still being used in financial calculations, where the tie situation at the midpoint is resolved by "rounding up" as it is usually denoted. But "rounding up" in a sign-magnitude system actually means "away from zero," hence a more precise term is *round-to-nearest-away*, RN$_a$, implicitly understanding "-from-zero." This is a rounding mode selectable for decimal arithmetic in the IEEE standard (see Section 7.6.2).

Definition 7.3.7 (RN$_a$: round-to-nearest-away) *Let $\beta \geq 2$ and $p \geq 1$; the nearest rounding $\text{RN}_a : \mathbb{R} \to \mathbb{Q}^p_\beta$ is given for all real x with v, v', v'' denoting adjacent* *triples in* $\mathbb{Q}_\beta^p$ *by*

$$\text{RN}_a(0) = 0,$$

$$\text{RN}_a(x) = \begin{cases} v' & \textit{for} \quad v_{mid} < x < v'_{mid}, \\ v' & \textit{for} \quad x = v_{mid} > 0, \\ v & \textit{for} \quad x = v_{mid} < 0. \end{cases}$$

The result of a floating-point addition $v + w$ computed by rounding-to-nearest-away is more likely to be biased than might be expected, considering that real-valued computation with continuously valued operands would almost never show a bias with such a final rounding. Choosing p-digit binary numbers v, w uniformly with $1 \leq v, w < 2$, note that if v and w have different parities, then $v + w$ will be a midpoint u of some ulp interval (u, u') of $\mathbb{Q}_2^p$. Thus in about 25% of the cases, $\text{RN}_a(v + w)$ will be greater than $\text{RN}_e(v + w)$. For p-digit decimal addition with v, w chosen uniformly with $\frac{1}{2} \leq v, w < 1$, the low-order digit of the sum will be a 5 in 8% of the cases, and the different nearest roundings would still differ in about 4% of all cases. This round-up bias should be recognized as an issue in financial accounting employing floating-point arithmetic.

In practice, a *precise rounded arithmetic over* $\mathbb{Q}_\beta^p$ is a set of closed rounded arithmetic operations $\otimes_\text{R} : \mathbb{Q}_\beta^p \times \mathbb{Q}_\beta^p \to \mathbb{Q}_\beta^p$, where $\otimes_\text{R}(v, w) = \text{R}(v \otimes w)$ has the domain $v, w \in \mathbb{Q}_\beta^p$, and $\otimes$ is any of the arithmetic operations addition (+), multiplication ($\times$), or division (/) with $w \neq 0$, and where $\text{R}(\cdot)$ is a particular precise rounding here chosen from RD, RU, RZ, RA, RN_e, RN_a.

In summary, the investigation leading to the determination of which rounded arithmetic should be implemented requires an assessment of the efficiencies of the algorithms and architectures with respect to the necessity of qualitative properties. These comparisons of efficiencies and qualitative properties are affected by the radix β of the host as seen by the choices made in the IEEE standard.

7.3.2 One-ulp roundings and tails

A *one-ulp rounding* of $y \in \mathbb{R}$ denotes a mapping $\text{R} : \mathbb{R} \to \mathbb{Q}_\beta^p$, where $\text{R}(y) = i\beta^\ell$ satisfies $|y - \text{R}(y)| < \beta^\ell$. Thus a one-ulp rounding is non-deterministic in the limited sense that it always returns one of at most two values: $\text{RD}(y)$ and $\text{RU}(y)$. A *half-ulp rounding* $\text{R}(y) = i\beta^\ell$ satisfies $|y - \text{R}(y)| \leq \frac{1}{2}\beta^\ell$, and so is equivalent to $\text{RN}(y)$ unless y is a midpoint v_{mid} in $\mathbb{Q}_\beta^p$, in which case the result is either v or v'.

Algorithms for arithmetic operations other than $\{\pm, \times\}$ and most functions typically provide an intermediate result in the non-deterministic form of a one-ulp rounded value $\text{R}(y)$ rather than the exact value y, and further operation-dependent steps are required to obtain a precise rounding. Precise roundings specify unique results that are desirable for portability between different implementations and are required by the IEEE floating-point standard for all the primitive operations.

A one-ulp rounding partitions $y \in \mathbb{R}$ into the sum $y = \mathrm{R}(y) + t$, where t is the *rounded-off tail* so that if $\mathrm{R}(y) = i\beta^\ell$, then $|t| < \beta^\ell$. More generally, if y is developed with the radix-β redundant digit set $D = \{-a, -a+1, \ldots, a\}$ having redundancy factor $\mu = a/(\beta - 1)$ with $\frac{1}{2} < \mu \leq 1$, then a *μ-ulp rounding* $\mathrm{R}(y) = i\beta^\ell$ satisfies $|\mathrm{R}(y) - y| = |t| < \mu\beta^\ell$ for all $y \in \mathbb{R}$.

Observation 7.3.8 *Let* $\mathrm{R}(y) = y - t$ *be a μ-ulp rounding employing a minimally redundant digit set. Then* $\mathrm{R}(y)$ *is the unique round-to-nearest value whenever the leading digit of the tail t is not a maximum magnitude digit* $\pm\beta/2$.

To obtain a precise rounding into $\mathbb{Q}_\beta^p$ from a preliminary procedure for developing any one-ulp rounding, it is sufficient to obtain a μ-ulp rounded value with one guard digit, i.e., $\mathrm{R}(y) \in \mathbb{Q}_\beta^{p+1}$, and the sign of the rounded-off tail $t = y - \mathrm{R}(y)$, where $|t| < \beta^{\ell-1}$. This procedure is conveniently formalized by investigating the inverse of the precise rounding.

7.3.3 Inverses of the rounding mappings

A precise rounding partitions the reals into a set of disjoint intervals corresponding to the sets of reals that map into each particular $v \in \mathbb{Q}_\beta^p$.

Observation 7.3.9 *The inverse precise roundings* R^{-1} *of* $\mathbb{Q}_\beta^p$ *for* $\beta \geq 2$, $p \geq 1$, *with* $\mathrm{R} \in \{\mathrm{RD}, \mathrm{RU}, \mathrm{RZ}, \mathrm{RA}, \mathrm{RN}_e, \mathrm{RN}_a\}$ *all have* $\mathrm{R}^{-1}(0) = [0]$, *and otherwise for the adjacent triple* $v, v', v'' \in \mathbb{Q}_\beta^p$, $v < v' < v''$, *determine intervals relative to the middle member* v' *by:*

- $\mathrm{RU}^{-1}(v') = (v; v']$;
- $\mathrm{RD}^{-1}(v') = [v'; v'')$;
- $\mathrm{RZ}^{-1}(v') = \begin{cases} [v'; v'') & \textit{for} \quad v' > 0, \\ (v; v'] & \textit{for} \quad v' < 0, \end{cases}$
- $\mathrm{RA}^{-1}(v') = \begin{cases} (v; v'] & \textit{for} \quad v' > 0, \\ [v'; v'') & \textit{for} \quad v' < 0; \end{cases}$
- $\mathrm{RN}_e^{-1}(v') = \begin{cases} [v_{mid}; v'_{mid}] & \textit{for} \quad v' \textit{ of even parity in } \mathbb{Q}_\beta^p, \\ (v_{mid}; v'_{mid}) & \textit{for} \quad v' \textit{ of odd parity in } \mathbb{Q}_\beta^p; \end{cases}$
- $\mathrm{RN}_a^{-1}(v') = \begin{cases} [v_{mid}; v'_{mid}) & \textit{for} \quad v' > 0, \\ (v_{mid}; v'_{mid}] & \textit{for} \quad v' < 0. \end{cases}$

All of the inverse mapping intervals $\mathrm{R}^{-1}(v')$, in Observation 7.3.9 have $\mathrm{ulp}(v')$ width whenever $v' \neq \beta^e$, and contain v' and no other element from $\mathbb{Q}_\beta^p$. All of these ulp intervals are in turn composed from certain fundamental partial-ulp subintervals determined by the members of $\mathbb{Q}_\beta^p$ and the midpoints of adjacent pairs in $\mathbb{Q}_\beta^p$.

Definition 7.3.10 *The set of* rounding-equivalent intervals *(eqv-intervals) for the radix* $\beta \geq 2$ *and p-digit numbers* $\mathbb{Q}_\beta^p$, $p \geq 1$, *is given by*

$$\{[0]\} \cup \{[v], (v; v_{mid}), [v_{mid}], (v_{mid}; v') \mid v, v' \textit{ adjacent in } \mathbb{Q}_\beta^p\}.$$

Definition 7.3.11 *Any precise rounding onto* $\mathbb{Q}_\beta^p$ *is constant over every eqv-interval.*

The eqv-intervals form a disjoint cover of the reals and partition the reals into "point"-type and "half-ulp-open"-type intervals. The seven eqv-intervals covering the two-ulp-wide interval $(v; v'')$ forming the inverse of a one-ulp rounding to v' are illustrated in Figure 7.3.1.

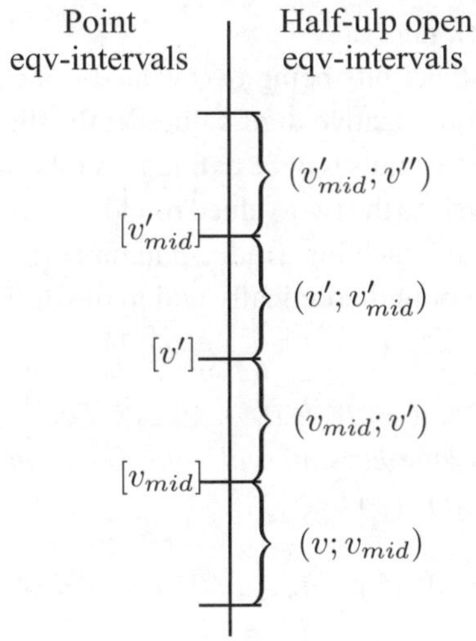

Figure 7.3.1. The seven eqv-intervals of a one-ulp rounding to v'.

From the figure note that each of the inverse rounding intervals in Observation 7.3.9 is composed of two half-ulp open eqv-intervals with a common boundary point, and from one to three point-type eqv-intervals. Any directed or nearest rounding onto $\mathbb{Q}_\beta^p$ will be a precise rounding onto $\mathbb{Q}_\beta^p$ for any $\beta \geq 2$, $p \geq 2$.

The significance of eqv-intervals is that it is not necessary to develop the exact result of a radix-β arithmetic operation $v \odot w$ for $v, w \in \mathbb{Q}_\beta^p$, in order to determine the result of the precisely rounded operation $\mathrm{R}(v \odot w) \in \mathbb{Q}_\beta^p$. It is sufficient to determine the specific eqv-interval containing $v \otimes w$. Then the rounding can be completed by applying R to any member of the eqv-interval, all the members being *rounding equivalent*.

The eqv-intervals are conveniently denoted by appending a two-bit subscript to each $v \in \mathbb{Q}_\beta^p$, $v \neq 0$, where $r \in \{0, 1\}$ is termed the *round bit*, and $z \in \{0, 1\}$ is termed the *sticky* bit. Specifically the four eqv-intervals v_{rz} comprising the ulp interval $[v; v')$ are

$$\begin{aligned} v_{00} &= [v], \\ v_{01} &= (v; v_{mid}), \\ v_{10} &= [v_{mid}], \\ v_{11} &= (v_{mid}; v'). \end{aligned}$$

Conventional binary floating-point terminology for the round bit r simply refers to the value of the bit b_p, indicating that the value to be rounded is at or above the midpoint, and the sticky bit z denotes for $z = 0$ that the "tail" satisfies $\|b_{p+1}b_{p+2}\cdots\|_2 > 0$. This obviously generalizes to higher values of the radix β and standard digit sets of the form $\{0, 1, \ldots, \beta - 1\}$. Regarding intervals, the sticky bit z denotes for $z = 0$ that v_{r0} is a single point-type interval, and for $z = 1$ that $v_{r,1}$ is a half-ulp open interval.

Note that round and sticky bits being two-valued is a consequence of the digit set exclusively having non-negative digit values. If the digit set contains negative digits then round and sticky digits will be defined over the signed bit set $\{-1, 0, 1\}$. First we restrict the analysis to the two-valued round and sticky bits. Characterizing the sticky bit z as a type bit indicating a radix-β number ($z = 0$) or an open half-ulp interval ($z = 1$) is useful both theoretically and in the design of algorithms.

Lemma 7.3.12 *For $\beta \geq 2$, $p \geq 2$, let $v_{rz(p)}$ be any eqv-interval with $v = \beta^e d_0.d_1 \cdots d_{p-1} \in \mathbb{Q}_\beta^p$. Let $w = \beta^e d_0.d_1 \cdots d_{p-2} \in \mathbb{Q}_\beta^{p-1}$. Then $v_{rz(p)} \subseteq w_{rz(p-1)}$, where $w_{rz(p-1)}$ is the unique eqv-interval with round and sticky bits $rz(p-1)$ determined from d_{p-1} and $rz(p)$ by*

$$rz(p-1) = \begin{cases} 00 & \text{for } d_{p-1} = 0,\ rz(p) = 00, \\ 01 & \text{for } \begin{cases} d_{p-1} = 0,\ rz(p) \neq 00, \\ 1 \leq d_{p-1} \leq \beta/2 - 1 \end{cases} \\ 10 & \text{for } d_{p-1} = \beta/2,\ rz(p) = 00, \\ 11 & \text{for } \begin{cases} d_{p-1} = \beta/2,\ rz(p) \neq 00, \\ d_{p-1} > \beta/2. \end{cases} \end{cases} \tag{7.3.2}$$

Proof Let w' denote the successor of w in $\mathbb{Q}_\beta^{p-1}$, with $w' - w = \text{ulp}(w) = \beta\text{ulp}(v)$. It follows that $v = w + (d_{p-1}/\beta)\text{ulp}(w)$, $v' = w + ((d_{p-1} + 1)/\beta)\text{ulp}(w)$, and $v_{mid} = w + ((d_{p-1} + \frac{1}{2})/\beta)\text{ulp}(w)$. The comparisons of these three boundary values of eqv-intervals with the real values w, $w' = w + \text{ulp}(w)$, and $w_{mid} = w + \frac{1}{2}\text{ulp}(w)$, are then sufficient to verify the round and sticky update formulas for $rz(p-1)$ in (7.3.2). □

The nesting of eqv-intervals as noted in Lemma 7.3.12 is illustrated in Figure 7.3.2.

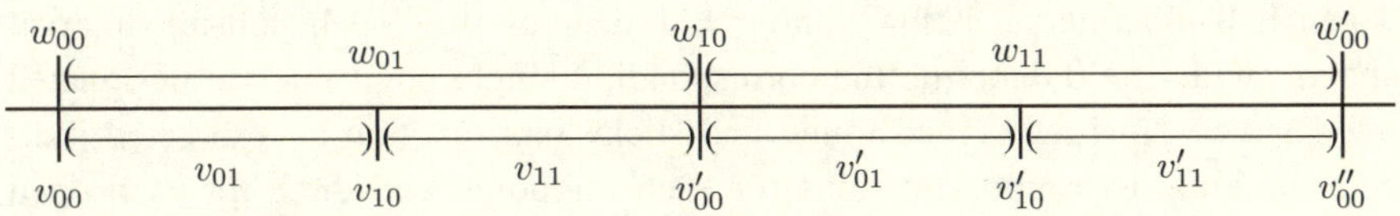

Figure 7.3.2. Partition of an ulp interval $[w; w']$ into five $(p-1)$-bit eqv-intervals, with a nested subpartition into nine p-bit eqv-intervals.

In general, the p-digit eqv-intervals form a partition of the reals that is a subpartition of the partition formed by the q-digit eqv-intervals for $q < p$. *Precision conversion* is the process of finding the q-digit eqv-interval $v_{rz(q)}$ for $w \in \mathbb{Q}_\beta^q$ containing a given eqv-interval $v_{rz(p)}$, which can be done by extending the lemma to inspection of tail sequences $d_q d_{q-1} \cdots d_{p-1}$ along with $rz(p)$.

Given the p-digit eqv-interval $v_{rz(p)}$, a precise rounding $\text{R} : \mathbb{R} \to \mathbb{Q}_\beta^q$ to any lower precision $2 \leq q \leq p$ is then obtained by performing precision conversion to obtain $w_{rz(q)}$, followed by evaluating $\text{R}(x) = u \in \mathbb{Q}_\beta^q$ for any $x \in w_{rz(q)}$ by any rounding mode $\text{R} \in \{\text{RD}, \text{RU}, \text{RZ}, \text{RA}, \text{RN}_e, \text{RN}_a\}$. Note that rounding does not change the sign, and the exponent changes by at most a unit carry, i.e., $e(u) = e(v) + c$ with $c \in \{0, 1\}$.

A reduced precision rounding can thus be expressed as an algorithm at the digit level, accepting as input the sign bit and significand of v, the round and sticky bits rz, and the precision and rounding modes. The output is simply expressed by the significand of u and an *exponent carry bit* c. In a hardware implementation with separate significand and exponent units, the exponent unit could either add c after its computation is completed, or use it to select between precomputed values of e and $e + 1$. For binary significands, such reduced precision roundings can be given as bit level operations.

To provide further options in algorithm design it is useful to allow limited redundancy in the representation of the p-digit, radix-β eqv-intervals, in particular of the significands. The use of redundant digit sets with negative digits suggests that signed round and sticky bits should be introduced.

For $v \in \mathbb{Q}_\beta^p$, the eqv-intervals can be denoted by a *signed round bit* $r \in \{-1, 0, 1\}$ and a *signed sticky bit* $z \in \{-1, 0, 1\}$. Notationally, for v, v' an adjacent pair, the four eqv-intervals covering $[v, v')$ have the redundant representations

$$
\begin{aligned}
[v] &= v_{00}, \\
(v, v_{mid}) &= v_{01} = v_{1\bar{1}} = v'_{\bar{1}\bar{1}}, \\
[v_{mid}] &= v_{10} = v'_{\bar{1}0}, \\
(v_{mid}, v') &= v_{11} = v'_{\bar{1}1} = v'_{0\bar{1}}.
\end{aligned}
$$

Note that for $z = 0$ the signed round digit denotes the "midpoint below" for $r = -1$ and the "midpoint above" for $r = 1$. Similarly $z = -1$ denotes the "open half-ulp interval below" and $z = 1$ denotes the "open half-ulp interval above," with $z = 0$ denoting the corresponding single point interval designated by v and r. Thus the signed round and sticky bits function as a directed indicator yielding two representations for each midpoint, and three for each open half-ulp interval. Single point intervals for $v \in \mathbb{Q}_\beta^p$ still have the unique interval designation v_{00}.

For carry-save representations it is necessary to allow the round digit $r \in \{0, 1, 2, 3\}$, with the sticky bit $z \in \{0, 1\}$, since the "tail" is non-negative. As usual $r = 1$ represents the "midpoint above" (one half-ulp above), the extended value $r = 2$ represents the successor v' (one ulp), and $r = 3$ represents three half-ulps above, with $z = 0$ indicating a point eqv-interval and $z = 1$ specifying a half-ulp open eqv-interval. Note that for rounding purposes there is then essentially no difference between the ordinary binary and carry-save representations, except that for $r \geq 2$ the "base point" is shifted.

We may thus summarize the previous discussion in the following definition.

Definition 7.3.13 A normalized rounding eqv-interval *is an eqv-interval of the form v_{rz}, where the significand of v may be in redundant representation, normalized such that it is in $[1; \beta)$, and rz are (possibly signed or extended) round and sticky digits.*

Determining the signed/extended round and sticky digits is, in general, a non-trivial problem. Let us first consider the case of determining the round and sticky digits from a carry-save represented "tail," $d_p d_{p+1} \cdots d_q$ with digits in $\{0, 1, 2\}$, where the tail represents a value in the interval $2^{-p}\left[0; 2(2 - 2^{p-q})\right)$, the upper bound obtained for a string of 2s.

By Observation 2.5.11, it is possible (with $k = 1$) by applying the PQ-mapping to convert the carry-save representation into borrow-save, with a tail now strictly bounded in absolute value by one ulp. By the same observation a borrow-save representation may be transformed by a PN-mapping into another borrow-save representation, again where the tail is strictly bounded in absolute value by one ulp. Thus in both cases the problem of determining the round and sticky digits is reduced to the case of a borrow-save represented tail, whose absolute value is strictly bounded by one ulp.

Observation 7.3.14 *For any precise rounding mode* R, *a rounding equivalent interval, denoted by v_{rz}, is equivalent to the rounded representation of x, $\text{R}(x) \in \mathbb{Q}_\beta^p$, for any $x \in v_{rz}$.*

Hence v_{rz} may be used instead of $\text{R}(x)$ wherever the rounded representation is used in rounding mode R. In binary we may write $x = v_{rz}$, denoted by appending the rz bits forming the $(p + 2)$-bit significand $b_0.b_1 b_2 \cdots b_{p-1} rz$ for x. This intermediate redundant representation of the value $\text{R}(x)$ can be used as an operand in further computations, thus avoiding the "expensive" log-time rounding delay.

Additionally, as opposed to the rounded value, note that the representation as an eqv-interval also allows conversion to reduced precision by truncation of the significand, and appropriate modifications of the round and sticky bits.

It is important here to notice that keeping a value represented as an eqv-interval as long as possible is very useful. There are situations where a value is generated to some precision p, but later needed at precision p', where $p' < p$. Here a *double rounding* should be avoided, since if a value is first rounded to some precision, and next again rounded to a lower precision, the result will not, in general, be the same as when the rounding is performed in one step from the original to the lowest precision.

In floating-point arithmetic ideally it is required that the result of an arithmetic operation is the precisely rounded value of the exact result of that operation. Thus double rounding is not permitted and the eqv-interval representation is a convenient way of representing a result until the actual rounding based on a specific rounding mode is to take place.

In particular this problem may occur if the system supports so-called *subnormal* or *denormalized* numbers, used to allow a "gradual underflow" when a value cannot be represented with the minimal value of the exponent, and consequently is allowed to be represented with an unnormalized significand, as in the IEEE standard (see Section 7.6). Hence the actual rounding should not be performed before it is known if such an underflow situation has been identified. Performing first one precision reduction to an eqv-interval and then in the exceptional case of gradual underflow performing another precision reduction is then a design option which may be exploited.

Problems and exercises

7.3.1 Show that for odd radices and fixed $v \in \mathbb{Q}_\beta^p$, the parity of $v \in \mathbb{Q}_\beta^{p+1}$ is the same as the parity of $v \in \mathbb{Q}_\beta^p$.

7.3.2 Let $\beta \geq 2$ and $p \geq 2$, or $\beta = 2k + 1 \geq 3$ and $p = 1$. Show that adjacent pair $v, v' \in \mathbb{Q}_\beta^p$ have opposite parities in $\mathbb{Q}_\beta^p$.

7.3.3 Prove that for p-digit decimal addition with $v, w \in \mathbb{Q}_{10}^p$ chosen uniformly in $\frac{1}{2} \leq v, w < 1$, that the roundings RN_a and RN_e of the result into $\mathbb{Q}_{10}^p$ will differ in about 5% of the cases.

7.3.4 For $v, w, x, y \in \mathbb{Q}_2^p$ chosen uniformly over the binade $[1; 2)$, determine the distribution of $\mathrm{RN}_a(\mathrm{RN}_a(v + w) + \mathrm{RN}_a(x + y)) - \mathrm{RN}_e(\mathrm{RN}_e(v + w) + \mathrm{RN}_e(x + y))$. Can the difference be as large as three ulps?

7.3.5 Discuss the widths of the various ulp intervals containing $v = \beta^e$, i.e., the significand is unity.

7.3.6 Employing signed round and sticky bits, discuss whether or not the exponent carry should also be signed, i.e., $c \in \{-1, 0, 1\}$. Hint: let $v = \beta^e$.

7.3.7 Discuss how many carry-save digits of the tail $d_p d_{p+1} \cdots$ are needed to determine the extended round digit $r \in \{0, 1, 2, 3\}$, and sticky bit $z \in \{0, 1\}$.

7.3.8 What is the range of the normalized rounding eqv-intervals v_{rz} for rz signed and/or extended and v normalized to the interval $[1; \beta)$?

7.3.9 Provide instances in binary and decimal of the undesired "double roundings" that we wish to avoid when employing eqv-intervals.

7.3.10 Discuss why and how it is useful to obtain a guard bit along with the round and sticky bits if a redundant binary arithmetic result is prenormalized only to fall in the two binade interval $1 \le v \otimes w < 4$.

7.4 Rounded binary sum and product implementation

For addition, subtraction, multiplication, and fused multiply-add of floating-point numbers it is possible to deliver the exact result as an intermediate value in $\mathbb{Q}_\beta$ given in radix polynomial representation on which the rounding algorithm can be applied. Recall that it is advantageous to leave intermediate results in redundant representation until the rounding is to take place, hence this will be our objective. In the following we shall assume that the operands are provided with significands represented with the same number of significant digits (same precision), if necessary with shorter operands extended with trailing zeroes.

Restricting our discussion here to binary floating-point, we are assuming operands have unsigned significands in the interval $[1; 2)$, hence the resulting significand in addition and multiplication is in $[1; 4)$. In subtraction a cancellation of leading digits may occur, requiring a normalization, a log-time operation, and the result may even be negative, but fortunately negation on redundant representations can be performed in constant time. As we shall see in the following there are arguments for preferring borrow-save over 2's complement carry-save representations, noting that these are equivalent as it is possible to convert either one into the other in constant time.

When the exponent values differ significantly there is obviously no need to generate the exact representation of the sum or difference of the two aligned operands. It is better to generate the intermediate result directly in the form of a suitably truncated significand value in redundant form with extended round and sticky digits (an eqv-interval). We shall assume that rounding of a resulting eqv-interval may take place to any precision lower than or equal to that of the most precise operand.

For a precise rounding the multiplier must deliver the double-length product of the two significands, and we shall assume that it is delivered in redundant form. Note that it is possible with additional circuitry in log time, in parallel with the summation of partial products, to calculate the sticky digit based on the least-significant half.

For a *fused multiply-add operation*, $a \times b + c$ with a single rounding, it is required to add the full-length product to the additive operand after alignment. It is here assumed that the additive operand is extended with zero-valued digits to the same double length. But note that the product is assumed to be available in redundant representation, hence the addition (effective add/subtract operation) will have to employ redundant adders of type npn or pnp to produce the redundant significand of the result as an eqv-interval. Depending on the situation the round and sticky digits may originate from the redundant product, or from the non-redundant additive operand.

We shall thus require that the intermediate result to be rounded is available as a *quasi-normalized* eqv-interval formally defined as follows.

Definition 7.4.1 *A binary quasi-normalized rounding interval is an extended eqv-interval of the form v_{rz}, where the significand of v_{rz} is the significand of $v = v_{00}$ in redundant representation, normalized such that it is in the two-binade interval* $[1; 4)$, *and r and z are (possibly signed) round and sticky digits.*

Note that the bounds on the significand of v_{rz} are those directly satisfied when adding or multiplying unsigned significands. In subtraction with cancellation, a normalization may be necessary, requiring at least logarithmic time.

Lemma 7.4.2 *When the significand is in borrow-save representation, without loss of generality we may assume that a quasi-normalized v_{rz} has the following factorization:*

$$v = (-1)^s\, 2^e\, \|1b_0.b_{-1} \cdots b_{-(p-1)}\|_2$$

with $b_i \in \{-1, 0, 1\}$ for $i \leq 0$ and $b_0 = -1 \Rightarrow \|b_{-1}b_{-2} \cdots b_{-p+1}\|_2 \geq 0$, together with the round and the sticky digits in $\{-1, 0, 1\}$. For v in carry-save representation

$$v = (-1)^s\, 2^e\, \|d_0.d_{-1} \cdots d_{-(p-1)}\|_2 \text{ with } d_0 \in \{1, 2\} \text{ and } d_i \in \{0, 1, 2, 3\} \text{ for } i < 0,$$

with the round digit in $\{0, 1, 2, 3\}$ and the sticky bit in $\{0, 1\}$.

Proof The factorization for the last case follows trivially from Theorem 3.8.3, possibly after recoding of leading guard digits for unsigned carry-save representation.

For the borrow-save representation consider the encoding of the significand as the difference between the values of two bit vectors, and recode it into carry-save by changing the negatively weighted vector into its complement, and add a carry-in. Then rewrite the integer part of that prefix according to the result above, and convert the result back into borrow-save by applying the Q-mapping of Section 2.5, which is a constant-time transformation. □

As we shall see, to determine a more specific range of quasi-normalized borrow-save represented values, it is often sufficient to consider only the integer part $\cdots b_3b_2b_1b_0$ of the representation of an intermediate result. Here the leading guard digits are b_3b_2, since the value is in the interval $[1;4)$; it is thus not necessary to map the representation into the form of the lemma.

7.4.1 Determining quasi-normalized rounding intervals

First notice that with the most common floating-point representations (based on encodings like the IEEE 754 standard), it is possible to distinguish operands of value zero from those with non-zero values based on special encodings of the exponent field. This implies that the result of any of the standard binary operations other than the fused multiply-add can be determined immediately if one or both operands are zero. For fused multiply-add if either the multiplier or the multiplicand is zero, the result is immediate. *In the following we will assume that all operands are non-zero.*

When implementing addition and subtraction of floating-point operands, what is essential is the effective operation to be performed, as determined by the signs of the operands. In the following let the two operands be $x = (-1)^s\, 2^e\, \|b_0.b_{-1}\cdots b_{-(p-1)}\|_2$ and $y = (-1)^{s'}\, 2^{e'}\, \|b'_0.b'_{-1}\cdots b'_{-(p-1)}\|_2$, where without loss of generality we assume that $e \geq e'$.

Addition and subtraction in carry-save representation. For effective addition ($s = s'$) there are three different situations depending on the extent of "overlap" after alignment of the significands. Let us first assume that the result is wanted in carry-save representation. In the case of $e = e'$ the quasi-normalized rounding equivalent interval for their sum w_{rz} is trivially obtained in constant time with r and z both zero. For $e > e' \geq e - p$ there is an overlap (including the "touch" situation $e - e' = p$), again the significand of their sum w is trivially found, adding only the "overlapping parts," simply by pairing them to be interpreted as a carry-save representation of the sum with exponent $\max(e, e')$. Then the round bit of the result is $r = b'_{(e-e')-p}$, and the sticky bit is $z = \bigvee_{(e-e')-p-1}^{-(p-1)} b'_i$, the logical OR of the "tail". Finally when $e - e' > p$, then $w = v$ with round bit zero and sticky bit 1. Note that it is never necessary to employ registers of width greater than p plus possibly a few guard bits.

Turning to the more complicated case of effective subtraction, without loss of generality let us assume that operand significands v and v' have been ordered such that subtraction is to take place as $v - v'2^{e'-e}$ with exponents $e \geq e'$. Cancellation of many leading digits can then only occur if $e - e' \leq 1$, since for $e - e' \geq 2$ the difference of aligned significands is then bounded away from zero, i.e., $\frac{3}{4} \leq v - v'2^{e'-e} < \frac{7}{4}$. Pipelined implementations of subtraction distinguish these two situations (see Section 7.4.3), the case $e - e' \leq 1$ performed in the *near path* and

$e - e' \geq 2$ in the *far path* also handles effective addition, due to a significant difference in the sequence of operations to be performed.

For the last case, subtraction with $e - e' \geq 2$ in carry-save is performed by adding the aligned and inverted subtrahend with a unit carry-in. Note that it is always the subtrahend that has to be aligned, insuring a positive result. Due to the above mentioned bound on the resulting significand, at most a single left-shift is needed for quasi-normalization of the result. Hence such a shift may unconditionally be performed safely before the addition, such that the exponent becomes $\max(e, e') - 1$.

The resulting significand part can then be obtained by pairing the sign extended, inverted "overlapping" part of the subtrahend with the augend. The determination of extended round and sticky digits then has to be done from the inverted "tail" part together with the carry-in. Adding the carry-in to the tail may create a carry-out, but this can be stopped by allowing the most-significant position to take the value 2. Since the digit in this position is to be used as the extended round digit, this is feasible.

It is possible to determine the possibility of a carry into the most-significant position of the tail by checking for all ones in the remaining positions. Since these bits are the inverses of the bits of the original subtrahend, this test becomes the very same as that which has to be performed to determine the sticky bit in the case of addition described above. Hence the extended round and sticky digits in the case of subtraction in the far path may be determined by some simple modifications of those of addition.

Continuing with subtraction in the case $e - e' \leq 1$ (in the near path), note that alignment here is a constant-time operation, and that the result is always exactly representable in precision $p + 1$, hence the sticky digit is always zero. Note that for $e = e'$ the result may be negative. For both cases of $e - e' \leq 1$ a normalization may be needed, requiring first the determination of the number of leading zeroes. As described in Section 3.10 a bit pattern having the same number of leading zeroes as, or one fewer than the non-redundant representation of the value can be obtained in constant time from the borrow-save represented difference. From this value the approximate number of normalization shifts can be obtained, employing the function $\mathrm{lzd}(\cdot)$, also determining the sign and signaling if the value is zero. By Observation 3.10.14 it is possible, in parallel with the determination of the approximate number of shifts needed, to obtain information on whether an additional shift is needed to get proper normalization. If the result is non-zero, the necessary shifting can be performed with a normalized, but still redundantly represented result, and the exponent can be appropriately adjusted.

Determination of the shift amount and the normalization are both log-time operations, but zero detection and sign determination are implicitly obtained. If the sign has come out negative, then the necessary sign inversion can be obtained in constant time when combined with conversion to a carry-save represented quasi-normalized rounding interval.

Addition and subtraction in borrow-save representation. Effective addition is here realized by applying the Q-mapping (as defined in Section 2.5) to the leading part of the paired aligned operands, performing a conversion from carry-save to borrow-save in constant time. For effective subtraction with borrow-save encoding of the result, the (possibly unnormalized) significand is trivially obtained by pairing the aligned operands. As there are only small differences between handling subtraction in the case $e - e' \geq 2$, and handling effective addition, these can both be executed in the far path. Again it is sufficient to use registers of width p plus possibly a few extra bits.

The case of subtraction in the near path when $e - e' \leq 1$ is handled as described above since the difference here is in borrow-save. And a possible final sign inversion is trivially obtained in constant time by swapping the encoding bit patterns.

From the above, and also the discussion on normalization in Section 3.10, note that the borrow-save representation seems more convenient than carry-save for the implementation of floating-point addition and subtraction. To summarize let us be more specific about the details of the significand determination when employing borrow-save representation, subtraction being considered a special case of addition.

Observation 7.4.3 *For addition in borrow-save arithmetic, let the sign and exponent of the operands be s, e, respectively s', e'. Then the result can be delivered as an eqv-interval w_{rz} with a quasi-normalized significand as follows:*

$s = s'$ *The resulting significand is obtained as the Q-mapping applied to the overlapping parts of the aligned significands, these comprising a carry-save encoding of the sum. The exponent of the result is* $\max(e, e')$. *The round and sticky digits are both zero for $e = e'$, otherwise they are determined from the (non-negative) tail of the smaller operand.*

$s = -s'$ *This case is split in two, depending on the exponent difference:*

$|e - e'| \leq 1$ *The result is obtained by left-normalizing the aligned significands, these being considered a borrow-save encoding of the difference. The exponent is accordingly adjusted and the significand sign defined as found by the leading zeroes determination. The sticky digit is always zero, and for $e = e'$ the round digit is also zero, otherwise if no normalization shift was performed, the round digit is found from the single-bit tail with the appropriately chosen sign. Note that in this case the significand may be delivered correctly normalized to the interval* $[1; 2)$.

$|e - e'| \geq 2$ *Left-shift both operands one position and define the revised exponent to be* $\max(e, e') - 1$. *The resulting significand is then obtained by considering the overlapping parts a borrow-save encoding of the result, with the round and the sticky digit determined from the remaining tail of the minuend or subtrahend after the left-shift.*

Subtraction is obtained as addition with the sign s' inverted.

Observation 7.4.4 *Addition and subtraction of two binary, precision p, floating-point numbers can be performed with the result in general represented as a redundant quasi-normalized rounding equivalent interval w_{rz}. Log-time operations are necessary to perform alignment, normalization, sign detection, and sticky digit determination; however, all of these will not always be needed. It is a design issue how these operations can be scheduled in a (pipelined) implementation of a floating-point adder.*

Multiplication. Turning to multiplication of two binary p-bit significands, we assume that the multiplier delivers the double-length result in a redundant representation. The unit can, in parallel with the accumulation of partial products, also deliver the round and sticky digits based on the least-significant digits of the product. If the representation of the product is borrow-save, then the tail $t^* = \|0.d_{-p-1} \cdots d_{-2p+2}\|_2$ is bounded $|t^*| < 1$. It is then trivial to compose the result as a quasi-normalized rounding interval w_{rz}, the round digit being $r = d_{-p}$, and the sticky digit $z = \text{sgn}(t)$. Even if the multiplier is tree structured, this sign can be determined in parallel in log time.

The situation is slightly more complicated if the carry-save representation is employed, since in this case the bounds on the tail are $0 \leq t^* < 2$. Thus in some way possible carries must be brought forward, such that a sticky bit in $\{0, 1\}$ can be delivered. This can be achieved by applying Q-mapping, so that the representation is converted into borrow-save. But then, of course, the multiplier might as well be designed to deliver the product directly in borrow-save.

Observation 7.4.5 (Sticky digit determination) *Except in the case of subtraction where exponents differ by no more than 1 (the near path), sticky digit determination for an addition, subtraction, or multiplication instruction in general requires a log-time calculation of* $\text{sgn}(t)$ *or equivalent, where t^* is the value of the resulting "tail's" $(p-2)$-digit string $0.d_{p+1}d_{p+2} \cdots d_{2p-2}$. For multiplication this calculation may be performed in parallel with the multiplication itself.*

The fused multiply-add operation. The purpose of providing the fused multiply-add instruction $x = a \times b + c$ is to reduce the error in accumulating products, by insuring that only a single rounding takes place when combining multiplication with an addition. Since the exact result of the multiplication is a double-length significand, assumed here to be in redundant representation, there are some new issues to be dealt with when performing the subsequent addition:

- The operands are of different length (precision).
- One operand may be in a redundant representation.
- The range of results is larger than for ordinary addition.

Dealing with the last issue first, note that the range is $[2;6)$ if the signs agree and an effective addition is performed, in which case a right-shift can safely be performed to yield a quasi-normalized result.

When the signs differ it turns out to be convenient first to reduce the range of the product significand $p \in [1;4)$. Assuming it is represented in borrow-save, let the integer part of the representation be $h = \|\cdots b_3 b_2 b_1 b_0\|_2$ and the fractional part $t^* = \|0.b_{-1}\cdots b_{2p+2}\|_2$. Then if $h = 3$ the product may be right-shifted since $|t^*| < 1$, so that $p = h + t > 2$. Let p' be the possibly transformed product significand, it follows that $p' \in [1;3)$, and let r be the resulting difference between the aligned significands. For exponents e, e' with $|e - e'| \leq 1$ it follows that $r \in (-1;2)$, so a normalization and a sign determination are needed. By checking extreme values for $|e - e'| \geq 2$ it is found that $r \in (\frac{1}{4};3)$, hence let $h = \|\cdots b_3 b_2 b_1 b_0\|_2$ be the value of the integer part of r. Thus if $h \leq 0$ it is safe to left-shift the result by two positions, so that the possibly transformed significand $r' \in [1;4)$ is quasi-normalized. The addition or subtraction can be realized by an array of the 3-to-2, pnp-, respectively npn-, adders described in Section 3.5.

Observation 7.4.6 *For the fused multiply-add operation in borrow-save arithmetic, let the sign and exponent of the product be s, e and those of the addend be s', e'. Then the result can be delivered with a quasi-normalized significand as follows:*

> $s = s'$ *The resulting significand is obtained by adding the aligned significands, with exponent* $\max(e, e') + 1$.
>
> $s = -s'$ *Let h be the value of the integer part of the product significand representation, $h = \|\cdots b_3 b_2 b_1 b_0\|_2$. Then range-reduce the significand by right-shifting it if $h = 3$ while increasing the exponent e by* 1. *Depending on the exponent difference then:*
>
> > $|e - e'| \leq 1$ *The result is the left-normalized difference of the aligned significands, with the exponent accordingly adjusted and the sign determined by the normalization.*
> >
> > $|e - e'| \geq 2$ *Let $h' = \|\cdots b'_3 b_2 b'_1 b'_0\|_2$ be the value of the integer part of the representation of the difference r' of the aligned significands. If $h' \geq 1$, then r' is quasi-normalized, otherwise left-shift r' by two positions and adjust the result exponent accordingly.*

Note that for subtraction when exponents differ by at most 1, the redundant result can be delivered properly normalized, thus in the close path avoiding a subsequent log-time conversion from quasi-normalized form.

Returning to the other two complications in fused multiply-add, the problem of addends having different lengths may simply be handled by zero-extending the additive operand. The effective addition or subtraction can then take place, employing a 3-to-2 constant-time adder, due to the product significand being in redundant representation. Actually for p-bit operands (exercise) a register structure of essentially $3p$-bit (digit) width, and only a p-digit adder is sufficient. A few additional positions may be needed to handle cases where e and e' differ significantly, to separate parts of the result where only the less-significant part is needed to determine the sticky digit.

However, determining (extended) round and sticky digits becomes slightly more complicated, as the tail part may be in redundant form. Since the round and sticky digits depend on the tail, they can only be determined after the normalization shift has been found. But in parallel as discussed in Section 3.10, it is possible by a look-ahead structure to calculate the signs of all the "tail strings" or suffixes, so that after normalization the sticky digit can be found as the sign of such a suffix in a fixed position.

Assume that the difference to be normalized is a p-digit borrow-save fractional number in string notation $x = d_0.d_{-1} \cdots d_{-p+1}$. In parallel with the determination of the approximate number of leading zeroes (say with result k) also calculate all the signs $\sigma_i = \sigma(d_i \cdots d_{-p+1})$ of the suffixes, employing Corollary 3.10.8, these signs forming another digit string $s = \sigma_0.\sigma_{-1} \cdots \sigma_{-p+1}$. Then after left-shifting both digit strings by k positions (quasi-normalizing x), if the sign in position 0 is the opposite of that in position -1 (originally the signs σ_{-k+1} and σ_{-k}); another left-shift is required to obtain proper normalization. Now also the sign of the appropriate tail needed for rounding can be found in a fixed position.

7.4.2 Rounding from quasi-normalized rounding intervals

Before a rounding can take place, it is necessary to normalize the intermediate result, or equivalently to determine its rounding position. Given a quasi-normalized rounding interval v_{rz} this corresponds to determining whether the significand of v is in $[1; 2)$ or in $[2; 4)$. We shall show below that it is sufficient to perform a sign determination operation. Note that converting the significand to non-redundant representation is much more complicated, and is not needed until the actual rounding is subsequently to be performed.

By Lemma 7.4.2, when a carry-save representation is used, it is fairly straightforward to see when a normalization is needed. For borrow-save with $1b_0 = 10$, it is necessary to determine the sign of the fractional part, so that $\text{sgn}(\|b_{-1} \cdots b_{-(p-1)}\|_2) \geq 0$ implies that a right-shift is needed. However, determining whether the prefix is $1b_0 = 10$ requires the elimination of some possibly non-zero leading guard digits. As the lemma does not immediately provide the conditions for normalization, we need to elaborate on it.

Theorem 7.4.7 (Normalization of a quasi-normalized rounding interval) *Determining whether a right-shift normalization of a quasi-normalized rounding interval is needed can be done by evaluating the sign of the fractional part of the significand, and can be performed as a log-time operation in parallel with the sticky digit determination.*

When the significand $\cdots b_3 b_2 b_1 b_0 . b_{-1} \cdots b_{-p+1}$ *is in borrow-save encoding, the value of* $h = \| |b_2| b_1 b_0 \|_2$ *determines whether a normalization is needed such that*

$$h \in \{3, 4\} \Rightarrow \textit{right-shift}$$
$$h = 2 \Rightarrow \textbf{if } \operatorname{sgn}(\|b_{-1} \cdots b_{-(p-1)}\|_2) \geq 0 \textbf{ then } \textit{right-shift},$$

For the significand $d_0 . d_{-1} \cdots d_{-(p-1)}$, $d_0 \in \{1, 2\}$ *in carry-save representation, the value of* d_0 *determines whether a normalization is needed such that*

$$d_0 = 2 \Rightarrow \textit{right-shift}$$
$$d_0 = 1 \Rightarrow \textbf{if } \operatorname{sgn}(\|b_{-1} \cdots b_{-(p-1)}\|_2) > 0 \textbf{ then } \textit{right-shift},$$

where $b_i = d_i - 1 \in \{-1, 0, 1\}$ *for* $i = -1, \ldots, -p+1$.

Proof Assume that the significand is $\cdots b_3 b_2 b_1 b_0 . b_{-1} \cdots b_{-p+1}$ in borrow-save representation with guard digits b_3, b_2, then by Theorem 3.8.4, $b_2 = 0$ implies that the prefix $\|Q\| = 0$ and $b_2 \neq 0$ that $\|Q\| = 1$ since the sign is known to be positive. Thus $|b_2| b_1 b_0 . b_{-1} \cdots b_{-p+1}$ represents the value of a quasi-normalized number, where obviously $b_2 \neq 0 \Rightarrow b_1 \leq 0$. Let $h = \| |b_2| b_1 b_0 \|_2$ be the value of the integer part and $t = \|0.b_{-1} b_{-2} \cdots b_{-p+1}\|_2$ the value of the fractional part. If $h \in \{3, 4\}$, then unconditionally a right-shift is needed to normalize the value, and if $h \in \{0, 1\}$ the value is already normalized. But for $h = 2$ the sign of t determines whether or not to right-shift.

By Lemma 7.4.2 when carry-save is used, then for $d_0 = 2$ unconditionally a right-shift is needed, but for $d_0 = 1$ it is necessary to determine whether $\|0.d_{-1} \cdots d_{-(p-1)}\| \geq 1$, or equivalently (exercise) $\operatorname{sgn}(\|0.b_{-1} \cdots b_{-(p-1)}\|_2) > 0$ with $b_i = d_i - 1 \in \{-1, 0, 1\}$. □

After normalization of a quasi-normalized rounding interval v_{rz}, to obtain a normalized eqv-interval of the form $v'_{r'z'}$ according to Definition 7.3.13, the exponent must be incremented if a right-shift was performed, and the round and sticky digits must be adjusted.

Obviously, the (signed) digit shifted out becomes the new extended round digit r', and the old round digit r is "prepended" to the old sticky digit z to form the new sticky digit $z' = \operatorname{sgn}(2r + z)$. The actual rounding may then be realized in combination with conversion of the redundant significand to non-redundant representation.

Final conversion to non-redundant representation. The rounding of an eqv-interval v_{rz} with the significand in $[1;2)$ may now be realized as an addition with a carry-in, chosen by the rounding rule and the values of the round and sticky digits, as determined by a table look-up. For the RN_e rounding, the choice when $rz = 10$ or $\bar{1}0$ is determined by the value of the least-significant digit $d_{-p+1} \in \{0, 1, 2\}$.

However, there are, in general, now three possible carry values for borrow-save in the set $\{-1, 0, 1\}$ and four for carry-save representations in $\{0, 1, 2, 3\}$. But as discussed in Observation 3.6.7, a carry-look-ahead adder is capable of absorbing such a carry, by combining a carry-in at the bottom of the look-ahead tree with a carry-based selection at the output. If the usual borrow-save and carry-save encodings are employed, then the two encoding bits may trivially be used directly as the two carry-in bits.

Observe that the rounding table look-up may be performed in parallel with the conversion to non-redundant representation, if the carry-completing adder provides two output values for selection by the carry-in bit. Thus no further log-time operations will be necessary.

At extreme values the carry-in may cause an overflow or an underflow, in which case an exponent adjustment (up or down) will be needed. Note that the resulting significand then has "trivial" values like $1.00\cdots0$ (with an incremented exponent), respectively $1.11\cdots1$ (with a decremented exponent).

7.4.3 Implementing floating-point addition and subtraction

There are a number of design options when implementing add, multiply, and (if required) also the combined fused multiply-add instructions. For the following discussion on add and subtract operations, we will first cover the case where two operands are of the same floating-point format, and the result is to be delivered in the same format.

As mentioned before there are some significant differences in the scheduling of the individual steps to be performed, when the effective operation is a subtraction between two very close operands (where cancellation of leading digits may occur), or when it is a more general addition or subtraction. Modern pipelined implementations of add and subtract units distinguish these two situations to reduce the number of pipe stages needed, at the expense of duplicating the adder structure.

Thus a comparison between the operand exponents must be performed. If the exponent difference $|e - e'| \leq 1$ and a subtraction is to be performed, the so-called *near path* is to be taken, otherwise the *far path is used.* In the following we will assume that the borrow-save representation is used, and that both operands are non-zero. Due to the much shorter exponent range we will consider comparison, addition, and subtraction on exponents as constant-time operations.

The near path. The sequential steps to be performed for subtraction when $|e - e'| \leq 1$ are:

Form difference The two operand significands are paired in the proper order and alignment (at most a single shift) to form the borrow-save representation of the result significand as their difference, possibly swapping of operands, and if so a sign flag is set. The exponent is set to $\max(e, e')$.

Determine shift amount The number of left-shifts needed for proper normalization, as well as result sign, is determined by the $\text{lzd}(\cdot)$ algorithm described in Section 3.10.

Normalize The significand is normalized and possibly negated to the interval $[1;2)$, according to Observation 3.10.14. Recall that the sticky bit is zero here, and so is the round bit when $e = e'$. If negation took place the sign flag is reversed. The number of leftshifts performed is subtracted from the exponent, so that the result is now a properly normalized eqv-interval $x = v_{rz}$.

Rounding If the exponent is within the representable range (no underflow) then round $x = v_{rz}$ and convert to non-redundant representation, otherwise form a subnormal number by performing further right-shifting with exponent adjust, rounding, and conversion (with the possibility of underflowing to zero).

Note that the first step, "form difference," is a constant-time operation, whereas the rest are log-time steps. Thus the two first steps may be merged into a single (first) pipe stage, as illustrated in Figure 7.4.1, which shows in simplified form a possible three-stage pipeline implementation.

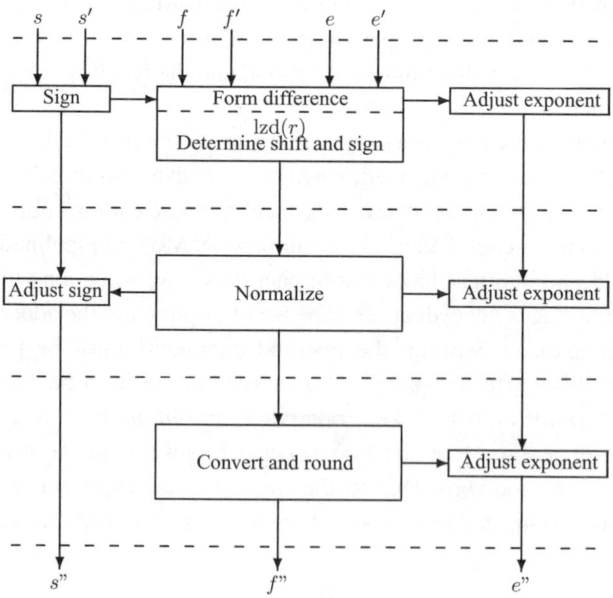

Figure 7.4.1. Near path, subtraction when $|e - e'| \leq 1$.

The far path. When $|e - e'| \geq 2$ for effective addition or subtraction perform the following:

Align operands Based on the exponent difference, shift the smaller operand significand after possibly swapping. The exponent is set to $\max(e, e')$.
Form sum/difference Pair operands to form a borrow-save representation of sum/difference. For addition apply Q-mapping and right-shift the result one position while incrementing the exponent. For subtraction possibly swap operands to insure a positive result, and set a sign flag appropriately, followed by left-shifting the result one position while decrementing the exponent. The significand is now quasi-normalized.
Adjust significand Calculate the sign of the fractional part to determine the need for a right-shift for normalization. In parallel, determine round and sticky bits. The result is a properly normalized eqv-interval $x = v_{rz}$.
Rounding If the exponent is within the representable range (no overflow), then round $x = v_{rz}$ and convert the significand to a non-redundant representation, otherwise map to infinity with the appropriate sign. Note that the carry-in during rounding may cause an overflow situation.

There are again three log-time steps, but here the constant-time step is the second, "form sum/difference," which may be merged with the third step to form a (second) pipe stage, as shown in Figure 7.4.2.

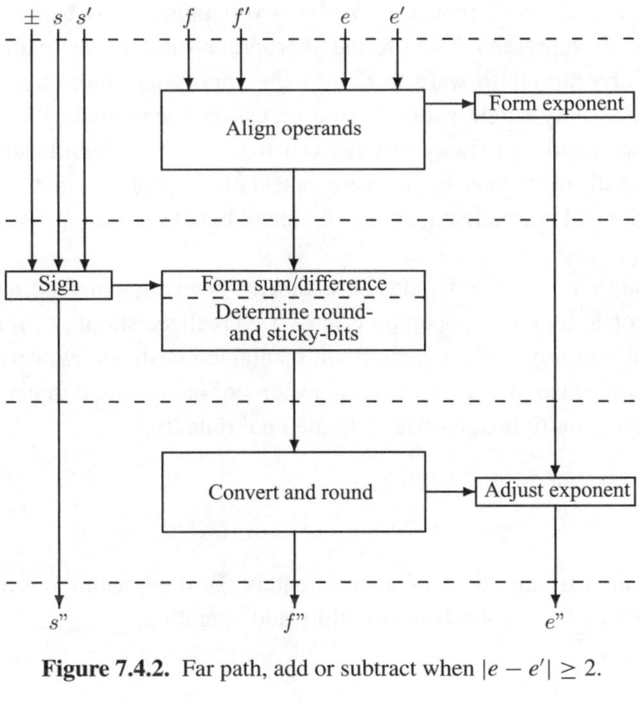

Figure 7.4.2. Far path, add or subtract when $|e - e'| \geq 2$.

Note that in both paths the three stages may be reduced to two. In the near path, conversion to a non-redundant representation may be performed in parallel with the determination of the shift for normalization. The sticky bit is always zero, and rounding is only needed if $|e - e'| = 1$ and there is no cancelation. In the far path, a merge of the last two stages is possible, where rounding is realized by choosing between two outputs of the adder. However, we will not further discuss the details of these possibilities.

Dealing with operands in different formats. When the two operands of an add or subtract operation are not in the same format it is straightforward to convert the operand of less precision to the format of the other operand by supplying the former with trailing zeroes. In fact, if the CPU only has a single floating-point adder, both paths of the adder will most likely be designed for the largest implemented format, and the rounding must then be built to support rounding into possibly several "shorter" formats. However, if the CPU is built with several floating-point adders, capable of operating in parallel, to these may be built to support different formats, and then the "wider" ones must similarly be built for rounding into the shorter formats.

Since a fused multiply/add instruction is defined to deliver the rounded result of the exact sum of the product and the addend, then this is a special case to be considered, in particular if the (exact, double-length) product is delivered in redundant representation. After operand alignment it may be necessary to perform an initial 3-to-2 reduction of the "overlapping" part before delivering the redundantly represented sum to the appropriate (near or far) path. In the far path it is fairly straightforward to deliver the appropriate most-significant part of the sum, as only single shifts will be necessary for normalization, and then to determine round and sticky bits derived from the least-significant part. But in the near path there may be a severe cancelation, implying that leading zero determination and normalization must be able to handle essentially double length operands.

Often when a fused multiply/add instruction has been implemented, addition and multiplication as individual operations have been realized simply by initialization of the fused multiply/add instruction with suitable constants. However, it is, of course, possible also to reuse and combine parts of independent implementations of addition and multiplication with a single final rounding.

Problems and exercises

7.4.1 Explain why an adder of approximately $3p$-digit width is sufficient for implementing a p-bit fused multiply/add operation.

7.4.2 Show that $\|0.d_{-1}\cdots d_{-(p-1)}\|_2 \geq 1$ is equivalent to $\mathrm{sgn}(\|b_{-1}\cdots b_{-(p-1)}\|_2) > 0$ with $b_i = d_i - 1 \in \{-1, 0, 1\}$ as claimed on page 478. If $d_i \in \{0, 1, 2\}$ is given in carry-save encoding, show how b_i can be determined in borrow-save encoding.

7.4.3 Find a constant-time mapping, adding a carry-in $\{-1, 0, 1\}$ to a borrow-save represented number. (Hint: reduce the logic of a 4-to-2 borrow-save adder.)

7.5 Quotient and square root rounding

The procedures for radix division and square root rounding are quite different from those for product and sum rounding. Operands in $\mathbb{Q}_\beta$ typically lead to exact rational quotients and irrational square roots that are not in $\mathbb{Q}_\beta$. The algorithms presented in Chapters 5 and 6 can deliver a result in the form of an arbitrary precision radix-β quotient or root one-ulp approximation, possibly (but not always) together with a remainder defining the residual, representing a *continuation function*, from which an arbitrary number of subsequent digits of one-ulp radix approximations may be determined.

Starting from the position of having a p-digit quotient or root one-ulp radix approximation $q \in \mathbb{Q}_\beta^p$, the goal of precise rounding is then reduced to determining the signed round and sticky bits r, z such that the eqv-interval q_{rz} must contain the infinitely precise quotient or root. Two approaches will be investigated in detail:

- obtaining the round and sticky bits from the sign of the possibly redundant remainder;
- obtaining the round and sticky bits from d_p and the sign of the possibly redundant tail $d_{p+1}d_{p+2}\cdots d_{p'}$ of a p'-digit one-ulp quotient approximation with $p' \gg p$.

Digit-serial methods always provide the value of the remainder, although it will often be in a redundant representation, which may complicate its use. Iterative refinement algorithms do not, in general, deliver a remainder, so it may be necessary to perform a *back-computation* to calculate the remainder from the division invariant equation. Note that the latter methods cannot deliver a quotient value with a remainder having a specific sign.

The use of extra accurate tails is principally of value in iterative refinement implementations where a hierarchy of result precisions may be parametrically determined, e.g., it is shown that a single precision tail may be extracted from a "double-plus-one" precision, best radix approximation to determine the round and sticky bits for the single precision precise rounding.

Another difference for quotient and root rounding into $\mathbb{Q}_\beta^p$ compared to product and sum rounding is that, for dividends, divisors, and radicands given normalized from $\mathbb{Q}_\beta$, the operands can be conveniently prenormalized by at most a one-digit shift so division or square root algorithms can provide quotient or root one-ulp approximations $q \in \mathbb{Q}_\beta^p$ that will be in the normalized interval $[1;\beta)$ with a predetermined last place $\ell = p - 1$.

7.5.1 Prenormalizing rounded quotients and roots

Radix division with the determination of a rounded quotient in $\mathbb{Q}_\beta^p$ is conveniently put in the general setting of rounding a positive rational fraction x/y into $\mathbb{Q}_\beta^p$. Without loss of generality we take *normalized p-digit radix-β division* of x by y to imply the dividend x and the divisor y are arbitrary positive integers, normalized so that $1 \le y \le x \le \beta y$, with $\beta \ge 2$. This normalization insures that the infinitely precise rational-valued quotient x/y will be normalized into the interval $[1;\beta)$, in which then a one-ulp approximation $q \in \mathbb{Q}_\beta^p$ of x/y is termed a *one-ulp quotient* satisfying $|q - x/y| < \beta^{-(p-1)} = \text{ulp}(x/y)$, where $q = i\beta^{-(p-1)}$ has the integer significand i normalized so that $\beta^{p-1} \le i < \beta^p - 1$.

Note that quotient prenormalization is a log-time operation, which generally can be performed concurrently with a division algorithm operating on the significands.

Quotient normalization could be deferred until after a result in the interval $[\frac{1}{\beta};\beta)$ is determined, and handled similarly to the sum and product normalization procedures. We employ prenormalization of the quotient as it provides simpler descriptions of the options in rounding procedures. Avoiding the boundary case $q = 1$ we then have

$$1 \le \frac{i-1}{\beta^{p-1}} < \frac{x}{y} < \frac{i+1}{\beta^{p-1}} \le \beta. \tag{7.5.1}$$

Furthermore, x/y falls in precisely one of seven eqv-intervals q_{rz}, where $r = r(x, y, q)$ and $z = z(x, y, q)$ are signed round and sticky bits as illustrated in Figure 7.5.1. Thus we need to determine r and z to have an intermediate result sufficient to be able to perform any of the precise roundings into $\mathbb{Q}_\beta^p$.

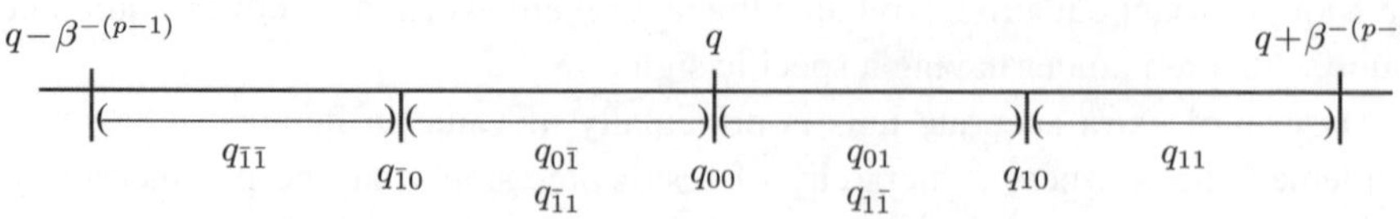

Figure 7.5.1. The seven eqv-intervals determined by the one-ulp quotient q and the signed round and sticky bits r, z.

For square root rounding into $\mathbb{Q}_\beta^p$, let the positive radicand $x = \beta^e f$ be normalized by left-shifting the significand whenever the exponent is odd so that $x = \beta^{2\lfloor e/2 \rfloor} f'$, where f' is then in the interval $[1;\beta^2)$. This insures the significand of the square root will be normalized into $[1;\beta)$ with the square root exponent $\lfloor e/2 \rfloor$.

Note that this prenormalization for square root with an even radix is a *constant-time operation* and an essential part of typical square root algorithms. For example, initial table look-up of an approximate square root in binary can be implemented by an index incorporating the low-order bit of the exponent (for exponent parity) and the high-order bits of the significand.

The importance of these prenormalization steps for division and square root is that the round position d_p of the possibly redundant result significand digit string $d_0.d_1d_2\cdots$ for either the quotient or the root can be known prior to reducing the representation to non-redundant form, allowing digits $d_pd_{p+1}\cdots$ of the tail, and/or the corresponding remainder to be identified for computation of the signed round and sticky bits, concurrent with compressing the leading digits of the significand to non-redundant form.

7.5.2 Quotient rounding using remainder sign

Note that $|(x/y) - q| = |r/y|$, where r is the corresponding remainder satisfying $|r| < \text{ulp}(q)y = \beta^{-(p-1)}y$. In general, we cannot insure that the remainder has a specific sign, and for digit-serial methods with signed, redundant digit sets, there is no point in developing a quotient value with a corresponding remainder of a specific sign, when a rounding has to take place anyway. However, for implementing a precise rounding it is sufficient to know the sign of an appropriate remainder.

Let $q = i/\beta^{p-1}$ be a one-ulp quotient for normalized p-digit radix-β division of x by y. The *remainder sign* of q is the three-valued function[1] $\text{sgn}(r) = \text{sgn}(x - qy) \in \{-1, 0, 1\}$. Note that (i) if $\text{sgn}(r) = -1$, then $(i-1)/\beta^{p-1} < x/y < i/\beta^{p-1}$, and (ii) if $\text{sgn}(r) = 0$, then $i/\beta^{p-1} = x/y$, and (iii) if $\text{sgn}(r) = 1$, then $i/\beta^{p-1} < x/y < (i+1)/\beta^{p-1}$.

Observation 7.5.1 *For $q \in \mathbb{Q}_\beta^p$ a one-ulp quotient for normalized radix division of x by y, the remainder sign is a signed round digit with $x/y \in q_{rz}$ for $r = \text{sgn}(r) = \text{sgn}(x - qy)$. Furthermore, then $q = i\beta^{p-1}$ and r are sufficient to*

[1] Note that the remainder sign, sgn(r), is here defined as a function of x, y, q, and that it takes at least log time to determine sgn(r), even directly from the remainder. In Section 3.10 it was shown how sgn(r) can be determined in a time equivalent to that for a carry look-ahead addition.

determine any directed precise rounding of x/y, where specifically

$$
\begin{aligned}
\mathrm{RU}\left(\frac{x}{y}\right) &= \quad i\beta^{-(p-1)} && \text{if} \quad \mathrm{sgn}(x - qy) \le 0;\\
\mathrm{RU}\left(\frac{x}{y}\right) &= (i+1)\beta^{-(p-1)} && \text{if} \quad \mathrm{sgn}(x - qy) = 1;\\
\mathrm{RD}\left(\frac{x}{y}\right) &= \quad i\beta^{-(p-1)} && \text{if} \quad \mathrm{sgn}(x - qy) \ge 0;\\
\mathrm{RD}\left(\frac{x}{y}\right) &= (i-1)\beta^{-(p-1)} && \text{if} \quad \mathrm{sgn}(x - qy) = -1.
\end{aligned}
$$

For a one-ulp quotient with one guard digit, the remainder sign is sufficient to determine any precise rounding of nearest or directed type.

Lemma 7.5.2 *For normalized radix division of x by y in radix $\beta \ge 2$, let $1 \le q = d_0.d_1d_2\cdots d_p < \beta$ be a $(p+1)$-digit, one-ulp quotient, with remainder sign* $\mathrm{sgn}(r)$. *Then*

- *q is rounding equivalent to x/y when $d_p \notin \{0, \pm\beta/2\}$;*
- *$q + \mathrm{sgn}(r)\beta^{-(p+1)}$ is rounding equivalent to x/y when $d_p \in \{0, \pm\beta/2\}$.*

Proof For $d_p \in \{0, \pm\beta/2\}$, the one-ulp quotient $q = d_0.d_1d_2\cdots d_p$ has the signed round digit $r = 0$ with $z = 0$, $d_p = \pm\beta/2$ and an end point for $d_p = 0$. It is then sufficient for radix rounding to determine whether x/y is larger than, equal to, or smaller than q, since the absolute error is less than $1/\beta^p$. □

For the very important case of a quotient in ordinary or signed-bit binary representation, we immediately have in continuation of the above.

Observation 7.5.3 *For normalized binary division of x by y, let $1 \le q = b_0.b_1\cdots b_p < 2$ be a $(p+1)$-bit, ordinary or signed-bit binary, one-ulp quotient. Then $q + \mathrm{sgn}(r)2^{-(p+1)}$ is rounding equivalent to x/y. Thus b_p forms the round bit and $\mathrm{sgn}(r) \in \{-1, 0, 1\}$ the sticky bit.*

For nearest precise roundings of x/y into $\mathbb{Q}_\beta^p$ with even radix β, it is sufficient to determine a normalized $(p+1)$-digit one-ulp quotient and corresponding remainder sign to obtain a normalized p-digit one-ulp quotient with signed round and sticky digits.

Observation 7.5.4 *Let $q^* = d_0.d_1\cdots d_p$ be a $(p+1)$-digit one-ulp quotient for normalized binary division of x by y, with $d_i \in \{-1, 0, 1\}$ for $0 \le i \le p$, where $q^* \in [1;2)$. Then $q = d_0.d_1\cdots d_{p-1} \in \mathbb{Q}_2^p$ satisfies $q \in [1;2)$, with $x/y \in q_{rz}$ where a signed round digit is $r = d_p$ and the signed sticky digit is*

$z = \mathrm{sgn}(x - q^*y)$. *Specifically for binary round-to-nearest-even we obtain:*

$$\text{RN}_e\left(\frac{x}{y}\right) = \quad i2^{-(p-1)} \quad \textit{if} \;\; d_p = 0 \qquad \textit{or} \;\; d_p = -\mathrm{sgn}(x - q^*y);$$

$$\text{RN}_e\left(\frac{x}{y}\right) = (i + d_p)2^{-(p-1)} \quad \textit{if} \;\; |d_p| = 1 \qquad \textit{and} \;\; d_p = \mathrm{sgn}(x - q^*y);$$

$$\text{RN}_e\left(\frac{x}{y}\right) = (i + d_p)2^{-(p-1)} \quad \textit{if} \;\; |d_p| = |d_{p-1}| \;\; \textit{and} \;\; \mathrm{sgn}(x - q^*y) = 0.$$

For even radices $\beta \geq 4$, it is possible to determine both the signed round and sticky digits from d_p whenever $d_p \notin \{0, \pm\beta/2\}$, and otherwise to compute the sign of the remainder only when $d_p \in \{0, \pm\beta/2\}$.

Often digit-serial methods develop the quotient or root in a redundant digit set, which is likely also to be in a higher radix like β^k, most-significant digit first. In Section 2.3 we saw how it is possible in digitwise, most-significant digit first order, to perform the conversion from a redundant to a non-redundant representation by on-the-fly conversion. It is possible to perform the remainder updating in parallel with the conversion. In Section 7.5.6 we shall show that this type of conversion can be extended to perform the actual rounding by a simple selection, according to the values of the round and sticky digits and the current rounding rule.

7.5.3 Rounding equivalence of extra accurate quotients

Iterative refinement division algorithms have the capability of generating one-ulp quotients $q^* = d_0.d_1d_2 \cdots d_{p-1}d_p \cdots d_{p-1+g}$ with somewhat more than p digits at only a moderate extra cost. Such "extra accurate" one-ulp quotients q^* are said to contain g *guard digits* with $g \geq 1$. We are particularly interested in cases where q^* is rounding equivalent to x/y, i.e., q^* and x/y are in the same eqv-interval for precise roundings to $\mathbb{Q}_\beta^p$. Then $\text{R}(q^*)$ can be implemented in the same manner as for precise roundings of exact sums and products into $\mathbb{Q}_\beta^p$. For normalized p-digit radix-β division of x by y, precise rounding of x/y can be simplified using an extra accurate one-ulp quotient q^* from the following considerations:

- As the number of guard digits grows from $g = 2$ to $g = p$, there is an exponentially decreasing probability that the one-ulp quotient is not rounding equivalent to the infinitely precise quotient, so the need for remainder sign computation significantly decreases.
- When the tail of guard digits satisfies $d_pd_{p+1} \cdots d_{p+g-1} = d_p00 \cdots 0$ with $d_p \in \{0, \beta/2\}$, the amount of further computation to determine the remainder sign can be reduced to a modular $((p - g + 2) \times (p - g + 2))$-digit multiplication and subtraction. Thus a full back-computation of the remainder $x - qy$ may be avoided.

Observation 7.5.5 *Let $q^* = d_0.d_1d_2 \cdots d_{p+g-1}$ be a $(p + g)$-digit one-ulp quotient for normalized p-digit radix-β division of x by y with β even. A sufficient* *condition for q^* to be rounding equivalent to x/y is* either *that $d_i \neq 0$ for some $p+1 \leq i \leq p+g-1$* or *that $d_p \notin \{0, \pm\beta/2\}$. Equivalently, the $(p+g)$-digit one-ulp quotient q^* is rounding equivalent to x/y, except when $q^* \in \mathbb{Q}_\beta^p$, or q^* is a midpoint of $\mathbb{Q}_\beta^p$.*

For a normalized radix-β division algorithm applied to a wide range of inputs x and y, it is reasonable to estimate that β^{-g} is the probability that a string of g guard digits will all be zero. Assume the probability that a $(p+g)$-digit one-ulp quotient will be rounding equivalent to x/y is essentially $1-2\beta^{-g}$. This means e.g., that a quotient generation algorithm generating an one-ulp quotient in decimal with three guard digits, or in binary with ten guard bits, will result in a rounding equivalent quotient without need for remainder sign computation in about 99.8% of all cases. Increasing the number of guard digits to six decimal digits or 20 bits, the probability of needing to compute the remainder sign reduces to about two per million divisions.

For suitably bounded integer divisors and dividends it is possible to prove that a one-ulp quotient with a sufficient number of guard digits is *necessarily rounding equivalent* to the exact quotient, including the confirmation of exact midpoints or membership in $\mathbb{Q}_\beta^p$.

Lemma 7.5.6 *For normalized p'-digit radix-β division of x by y with β even, where $y = y^*\beta^p$ with $\beta^{p'-1} \leq y^* < \beta^{p'}$ and $x \in \mathbb{Q}_\beta^{p+p'}$, let $q^* = d_0.d_1d_2\cdots d_{p+p'}$ be a $(p+p'+1)$-digit one-ulp quotient with $p'+1$ guard digits. Then q^* is rounding equivalent to $x/y \in \mathbb{Q}_\beta^p$.*

Proof q^* will be rounding equivalent to x/y by Observation 7.5.5 unless $d_i = 0$ for $p+1 \leq i \leq p+p'$, with $d_p \in \{0, \pm\beta/2\}$. Now let us assume $q^* = d_0.d_1d_2\cdots d_p = i/\beta^p \in \mathbb{Q}_\beta^p$ (for $d_p = 0$) or is a p-digit midpoint (for $d_p = \pm\beta/2$). For the case $|i/\beta^p - x/y| \neq 0$, we obtain

$$\left|\frac{i}{\beta^p} - \frac{x}{y^*\beta^p}\right| = \frac{|iy^* - x|}{y^*\beta^p} > \frac{1}{\beta^{p+p'}},$$

contradicting the fact that $q^* = d_0.d_1d_2\cdots d_{p+p'}$ is a $(p+p'+1)$-digit one ulp quotient, thus $q^* = x/y$. So in all cases q^* is rounding equivalent to x/y. □

An important consequence of Lemma 7.5.6 is that division in radix $\beta \geq 2$ of a $2p$-digit "double-precision" dividend x by a p-digit "single-precision" divisor y resulting in a $(2p+1)$-digit "double-precision-plus-one" one-ulp quotient q^* that is rounding equivalent to x/y in $\mathbb{Q}_\beta^p$.

7.5.4 Precisely rounded division in $\mathbb{Q}_\beta^p$

In this section we focus on "closed" precisely rounded division over $\mathbb{Q}_\beta^p$. Specifically we investigate features of normalized p-digit radix-β division of x by y, where $x, y \in \mathbb{Q}_\beta^p$ and where x/y is to be precisely rounded into $\mathbb{Q}_\beta^p$.

Theorem 7.5.7 *For normalized p-digit radix-β division of x by y with even β, a $(2p+1)$-digit one-ulp quotient $q = d_0.d_1d_2\cdots d_{2p}$ is rounding equivalent to x/y. Furthermore, $(2p+1)$ is the least possible one-ulp quotient precision for this result in that for $\beta \geq 2$ and any $p \geq 2$ there exists a $(p \times p)$-digit fraction x/y and a $2p$-digit, one-ulp quotient for x/y which is not rounding equivalent to x/y.*

Proof The sufficiency of the $(2p+1)$-digit one-ulp quotient follows from Lemma 7.5.6. To show $2p+1$ is least possible, let $x = \beta^p$ and $y = \beta^p - 1$. Then

$$\frac{x}{y} = \frac{\beta^p}{\beta^p - 1} = 1 + \beta^{-p} + \beta^{-2p} + \cdots,$$

and $q = 1 + \beta^{-p}$ satisfies

$$\frac{x}{y} - q = \beta^{-2p}\frac{\beta^p}{\beta^p - 1} > \beta^{2p}.$$

Thus q is a $2p$-digit one-ulp quotient for x/y that is not rounding equivalent to x/y. □

A very practical use of this result is in the implementation of a hierarchical precision division environment, such as single–double, when the "double" one-ulp approximation is ready for rounding to single precision, and the double-precision rounded result can be obtained by further back-computation of remainder sign.

Regarding the determination of round and sticky digits from a one-ulp quotient with $2 < g \leq p$ guard bits, the following results are useful.

Theorem 7.5.8 *For normalized binary division of x by y and any $p \geq 2$, x/y cannot be a p-bit midpoint, or equivalently if $|b_p| = 1$, then the remainder is non-zero.*

Proof To the contrary assume for some $p \geq 2$ that there is a $(p \times p)$-bit fraction that is a p-bit midpoint. Since the division is normalized, $y \leq 2^p - 1$, there is then an i, $2^{p-1} \leq i \leq 2^p - 1$, such that $x/y = (2i+1)/2^p$. Then $x2^p = (2i+1)y$ yields a contradiction, since $(2i+1)$ is odd and 2^p cannot divide y since $y \leq 2^p - 1$. □

The fact that x/y cannot be a midpoint greatly simplifies the nearest rounding of a quotient, since there is no ambiguity in how to round a midpoint.

Corollary 7.5.9 *For closed rounded division over $\mathbb{Q}_2^p$, any nearest quotient rounding satisfies $|q - x/y| < \frac{1}{2}\text{ulp}(q)$.*

In contrast to binary, the occurrence of exact midpoints is of considerable concern for even radices $\beta \geq 4$. Note that the p-digit divisor range $1 \leq y \leq \beta^p - 1$ includes very small divisors $y = 2, 3, 4, 5, \ldots$ that frequently occur as constants in formulas, rather than as p-digit approximations of a continuous variable. When dividing by an integer of very small size, there is a considerable likelihood of generating an exact quotient. In particular consider the case of division by 2 for an even radix $\beta \geq 4$.

Observation 7.5.10 *For radix $\beta > 2$ and any precision $p \geq 2$, let $y = 2\beta^{p-1}$ with $y + 1 \leq x \leq 2y - 1$. Then the exact quotient $q = x/y = x/2\beta^{p-1}$ is a midpoint when x is odd, and otherwise $q \in \mathbb{Q}_\beta^p$.*

In decimal division by the integer 2 we are then quite likely to have to round a midpoint, so the bias of the IEEE's decimal condition to round-to-nearest-away must be considered relevant.

Returning to binary precise roundings and further consequences of Theorem 7.5.8, it is then clear that if $|x2^p - (2i+1)y| = 1$, we would obtain x, y pairs where the one-ulp quotients would have a maximum length string of zero guard digits following $d_p = \pm 1$, making these x, y pairs the extremal "hard-to-round" cases for binary floating-point division. For those cases where$(2i+1)y \equiv \pm 1 \pmod{2^p}$, such pairs can be found systematically employing the theory of generating multiplicative inverses modulo 2^p. Thus Theorem 7.5.8 along with Theorem 7.5.7 provides the basis for generating suites of test cases for verifying floating-point division implementations.

Importantly, it is sufficient to obtain a $(p+2)$-bit one-ulp quotient to test for an exact quotient by counting low-order zeroes.

Corollary 7.5.11 *Let $q = k/2^{p+1} = 1.b_1b_2 \cdots b_{p+1}$ be a one-ulp $(p+2)$-bit quotient with $p \geq 2$ for normalized $(p \times p)$-bit division of x by y with $2^{p-1} \leq y \leq 2^p - 1$. Then $x/y = k/2^{p+1}$ and $k/2^{p+1} \in \mathbb{Q}_2^p$ if and only if the number of low order zero-bits in k and y is at least $p + 1$.*

Example 7.5.1

(1) (8×8)-bit division, $g = 2$ guard bits:

$x = 231 = 11100111$, $y = 168 = 10101000$, with three low-order zeroes;
$q = 1.011000000 = k/2^9$, with six low-order zeroes, including two guard bits;
low-order zero sum: $6 + 3 \geq p + 1 = 9$;
conclusion: $x/y = q = 1.0110000$ exactly with zero remainder.

(2) (8×8)-bit division, $g = 2$ guard bits:

$x = 298 = 100101010$, $y = 180 = 10110100$, with two low-order zeroes;
$q = 1.101010000 = Q/2^9$, with four low-order zeroes, including two guard bits;
low-order zero sum: $4 + 2 < p + 1 = 9$;
conclusion: $x/y \neq q$. (Note: $x/y = 1.10100111110 \cdots$.) □

Another advantage of computing at least two guard digits for a one-ulp quotient when the radix is even is that it simplifies the computation of the round and sticky digits, even when a back-computation of remainder sign is necessary.

Observation 7.5.12 *Let* $q^* = d_0.d_1d_2 \cdots d_{p-1+g} = i\beta^{-(p-1+g)}$ *be a one-ulp quotient for normalized p-digit radix* β *division of x by y, with* $x, y \in \mathbb{Q}_\beta^p$, $2 \le g \le p$ *and* β *even. Then the round digit is determined from* d_p, *and* $\mathrm{sgn}(r)$ *can be found from y and i independent of x.*

The next theorem provides the details for computing $\mathrm{sgn}(r)$ when $q^* \in \mathbb{Q}_\beta^p$ or is a midpoint of $\mathbb{Q}_\beta^p$.

Modular computation of remainder sign. For the "low probability" cases where a reasonable number of guard digits $2 \le g \le p$ are computed with $d_p \in \{0, \pm\beta/2\}$ and all subsequent digits are zero ($d_{p+1} \cdots d_{p-1+g} = 0$), there are some useful results that can be derived from the knowledge that $x, y \in \mathbb{Q}_\beta^p$.

Theorem 7.5.13 *For even radix* $\beta \ge 2$, *let* $q^* = d_0.d_1d_2 \cdots d_{p-1}00 \cdots 0 = i/2\beta^{p-1}$ *with* $2\beta^{p-1} + 1 \le i \le 2\beta^p - 1$ *be a* $(p+g)$*-digit one-ulp quotient for normalized p-digit radix division of x by y with* $x, y \in \mathbb{Q}_\beta^p$, *where* $2 \le g \le p$. *Then* $\mathrm{sgn}(r) = \mathrm{sgn}(x - qy)$ *is given independent of x by*

$$\mathrm{sgn}(r) = \begin{cases} 1 & \text{for} \quad 2\beta^{p-g} \le |yi|_{4\beta^{p-g}} \le 4\beta^{p-g} - 1, \\ 0 & \text{for} \quad |yi|_{4\beta^{p-g}} = 0, \\ -1 & \text{for} \quad 1 \le |yi|_{4\beta^{p-g}} \le 2\beta^{p-g} - 1, \end{cases}$$

and $q^* + \mathrm{sgn}(r)\beta^{-(p+1)}$ *is rounding equivalent to* x/y. *Furthermore, for* $\beta \ge 4$ *with* $|yi|_{4\beta^{p-g}} = d'_{p-g}d'_{p-g-1} \cdots d'_0 \ne 0$, *the leading digit* d'_{p-g} *is sufficient to determine* $\mathrm{sgn}(r)$.

Proof Since $q = i/2\beta^{p-1}$ is a $(p+g)$-digit one-ulp quotient with $2 \le g \le p$, we obtain $|x/y - i/2\beta^{p-1}| < 1/\beta^{p+g-1}$. Without loss of generality, assume $y < \beta^p$ and let $k = |yi - 2x\beta^{p-1}|$. Then $0 \le k < 2y/\beta^g < 2\beta^{p-g}$, so $0 \le k \le 2\beta^{p-g} - 1$ where $0 \le p - g \le p - 2$.

Assume $x/y \le i/2\beta^{p-1}$, in which case $k = yi - 2x\beta^{p-1} \ge 0$. But then k satisfies $k \equiv yi - 2x\beta^{p-1} \pmod{4\beta^{p-g}}$, or as a standard residue $k = |yi - 2x\beta^{p-1}|_{4\beta^{p-g}} = |yi|_{4\beta^{p-g}}$ since $4\beta^{p-g}$ divides $2x\beta^{p-1}$ ($p - g \le p - 2$). In particular $|yi|_{4\beta^{p-g}} = k = 0$ occurs only for $(x/y) - (i/2\beta^{p-1}) = 0$, and $|yi|_{4\beta^{p-g}} = k$ with $1 \le k \le 2\beta^{p-g} - 1$ for $x/y < i/2\beta^{p-1}$.

Now assume $x/y > i/2\beta^{p-1}$, in which case $yi - 2x\beta^{p-1} = -k$. It follows that $|yi - 2x\beta^{p-1}|_{4\beta^{p-g}} = |yi|_{4\beta^{p-g}} = 4\beta^{p-g} - k$, where then $2\beta^{p-g} < |yi|_{4\beta^{p-g}} \le 4\beta^{p-g} - 1$. The conclusion then follows from Lemma 7.5.2. □

Example 7.5.2 For division in $\mathbb{Q}_{10}^3$ with $g = 2$ guard digits, let $x = 2290$, $y = 308$, with $q^* = 7.4350 = \frac{1487}{200}$, being a one-ulp midpoint. Then

$$|y|_{40} = |308|_{40} = 28,$$
$$|i|_{40} = |1487|_{40} = 7,$$
$$|yi|_{40} = |196|_{40} = 36.$$

Conclusion: $20 \le |yi|_{40} \le 39$ implies $\mathrm{sgn}(r) = 1$ and $x/y > q^*$, with $q^{*\prime} = 7.4351$ then rounding equivalent to $\frac{2290}{308}$. □

For $\beta = 2$, the result of Theorem 7.5.13 simplifies to a $((p - g + 2) \times (p - g + 2))$-bit modular multiplication for remainder sign determination and rounding equivalent quotient specification.

Corollary 7.5.14 *For $\beta = 2$, let*

$$q^* = \overbrace{1.b_1 b_2 \cdots b_p}^{p+1}\overbrace{00\cdots 0}^{g-1} = i/2^p$$

with $2^p + 1 \le i \le 2^{p+1} - 1$, be a $(p + g)$-bit one-ulp quotient with $g \ge 2$ guard bits for normalized p-bit division of x by y with $x, y \in \mathbb{Q}_\beta^p$. Then $q^{*\prime} = q^* + \mathrm{sgn}(r)2^{-(p+1)}$ is rounding equivalent to x/y, where $\mathrm{sgn}(r) = \mathrm{sgn}(x - qy)$ can be determined as:

$$\mathrm{sgn}(r) = \begin{cases} 1 & \text{for} \quad 2^{p-g+1} \le |yi|_{2^{p-g+2}} \le 2^{p-g+2} - 1, \\ 0 & \text{for} \quad |yi|_{2^{p-g+2}} = 0, \\ -1 & \text{for} \quad 1 \le |yi|_{2^{p-g+2}} \le 2^{p-g+1} - 1. \end{cases} \tag{7.5.2}$$

In particular, when $|yi|_{2^{p-g+2}} \neq 0$, the leading bit of $|yi|_{2^{p-g+2}}$, is sufficient to determine $\mathrm{sgn}(r)$. This is a simpler way of determining $\mathrm{sgn}(r)$ than performing the full back-computation, and using the definition $\mathrm{sgn}(r) = \mathrm{sgn}(x - qy)$.

Example 7.5.3

(1) For $p = 12$ let $x = 101\,010\,101\,010$ and $y = 100\,011\,000\,000$. With $g = 2$ guard bits let a $(p + g)$-bit quotient be $q^* = 1.001\,110\,000\,000\,0$ having guard bits $b_{12} = b_{13} = 0$. Hence $i = \lfloor q^* 2^{12} \rfloor = 1\,001\,110\,000\,000$ with $p - g + 2 = 12$, and it is easily seen (here by counting trailing zeroes) that $|yi|_{12} = 0$, thus $\mathrm{sgn}(r) = 0$ and the quotient q^* is exact.

(2) Now decrementing the dividend to $x = 101\,010\,101\,001$, but otherwise using the same parameters, let a one-ulp 14-bit quotient be $q^* = 1.001\,101\,111\,110\,0$ with guard-bits $b_{12} = b_{13} = 0$, so $i = \lfloor q 2^{12} \rfloor = 1\,001\,101\,111\,110$. Then $|yi|_{2^{12}} = 111\,010\,000\,000$ hence $\mathrm{sgn}(r) = 1$, corresponding to $x - qy = 0.000\,110\,000\,000 > 0$. By the corollary we can thus conclude that $q^{*\prime} = 1.001\,101\,111\,11\underline{0}\,\underline{1}$ is rounding equivalent of $x/y = 1.001\,101\,111\,110\,001 \cdots$. The underlined bits of $q^{*\prime}$ are respectively the round bit and the sticky bit. Depending on the rounding mode, the

actual rounded value of q^* will then be

$$\mathrm{R}(q) = \begin{cases} 1.001\,101\,111\,11 & \text{for effective round-down,} \\ 1.001\,101\,111\,11 & \text{for round-to-nearest,} \\ 1.001\,110\,000\,00 & \text{for effective round-up.} \end{cases}$$

□

Observation 7.5.15 *For $\beta = 2$, let $q^* = 1.b_1b_2\cdots b_pb_{p+1}\cdots b_{p+g-1} = i/2^p$, with $2^p + 1 \le i \le 2^{p+1} - 1$, be a $(p+g)$-bit one ulp quotient with $g \ge 2$ guard bits for normalized p-bit division of x by y. Then $q^{*\prime} = 1.b_1b_2\cdots b_pb^*_{p+1}$ is rounding equivalent to x/y with signed sticky bit*

$$b^*_{p+1} = \begin{cases} \mathrm{sgn}(r) & \text{if } b_{p+1}\cdots b_{p+g-1} = 0\cdots 0, \\ \mathrm{sgn}(b_{p+1}\cdots b_{p+g-1}) & \text{otherwise,} \end{cases}$$

where sgn(r) *may be determined by (7.5.2).*

7.5.5 Precisely rounded square root

In the same manner as precise quotient rounding from one-ulp quotient and remainder sign (see Observation 7.5.1 and Lemma 7.5.2), it is sufficient to determine a half-ulp root to obtain precise square root roundings $\mathrm{R}(\sqrt{x})$. Our focus here is then on simplifying the process of obtaining the remainder sign sgn(r), particularly for the case of binary square root.

It follows as a corollary to Lemma 6.5.12 on directed half-ulp roots that if a one-ulp root with some p or $(p+1)$ accurate guard bits is obtained, then the guard bit-string will be all zeroes only when the value q is exact, so the half-ulp directed root is *guaranteed to be obtainable* from such a "double-size" one-ulp root. However, obtaining up to $p+1$ accurate guard bits (i.e., a $2(p+1)$-bit one-ulp root) is probably more costly than computing $r = x - q^2$ directly. For those cases where a reasonable number of accurate guard bits (e.g., $4 \le g \le 12$) results in an all-zero string, it is possible to determine the needed sign of $x - q^2$ by a less costly squaring operation generating only low order bits of the square. Furthermore, as the size of g grows, the number of low-order bits of q^2 that need to be computed decreases by a like amount.

Lemma 7.5.16 *Let $q = 0.1b_2b_3\cdots b_{p+1}0^{g-1} = i2^{-(p+1)}$ be a $(p+g)$-bit one-ulp root of the p-bit radicand x where $\frac{1}{4} \le x = k2^{-(p+1)} \le 1 - 1/2^p$, with $\sqrt{x} \ne q$, and $3 \le g < p$. Then the sign of the remainder is given by*

$$\mathrm{sgn}(r) = \begin{cases} -1 & \text{if} \quad 0 < \left|i^2\right|_{2^{p+1}} < 2^p, \\ 0 & \text{if} \quad \left|i^2\right|_{2^{p+1}} = 0, \\ +1 & \text{if} \quad 2^p < \left|i^2\right|_{2^{p+1}} < 2^{p+1}. \end{cases}$$

Proof By definition i is an integer with $i^2 = q^2 2^{2(p+1)}$ and k is an integer with $k2^{p+1} = x2^{2(p+1)}$, so

$$r2^{2(p+1)} = (x - q^2)2^{2(p+1)} = k2^{p+1} - i^2.$$

Since $\sqrt{x} + q < 2$, then r satisfies $|r| = |x - q^2| = |\sqrt{x} - q|(\sqrt{x} + q) < 2^{-(p+g-1)}$, and $|r2^{2(p+1)}| < 2^{p+3-g}$, hence

$$\left|k2^{p+1} - i^2\right| < 2^{p+3-g} \leq 2^p, \tag{7.5.3}$$

since $g \geq 3$. It is now easy to see that

$$\begin{aligned}
0 < \quad & i^2 - k2^{p+1} && < 2^p && \Rightarrow \operatorname{sgn}(r) = -1,\\
& i^2 = k2^{p+1} && && \Rightarrow \operatorname{sgn}(r) = 0,\\
-2^p < \quad & i^2 - k2^{p+1} && < 0 && \text{or equivalently,}\\
2^p < {} & i^2 - (k-1)2^{p+1} && < 2^{p+1} && \Rightarrow \operatorname{sgn}(r) = 1,
\end{aligned}$$

thus the result is obtained by computing $i^2 \bmod 2^{p+1}$. □

Note that with i^2 having the binary representation $q_{2p+2} \cdots q_{p+1} q_p \cdots q_o$, then $r \leq 0$ iff $q_p = 0$ and $r > 0$ iff $q_p = 1$, hence these tests are single-bit tests. Furthermore, $q_p \cdots q_0 = 0^{p+1} \Rightarrow r = 0$ in conformance with Corollary 6.5.13. Also note that if computations are performed with more than $g = 3$ guard bits, fewer bits of i^2 need to be computed, since actually only $|i^2|_{2^{p+4-g}}$ is needed, cf. (7.5.3).

Note that $|i^2|_{2^j}$ requires squaring of just the low-order $j - 1$ bits of i. Since we need only the highest-order bit of $|i^2|_{2^j}$, a $(j-1)$-bit-in, 1-bit-out table is sufficient for direct look-up. For $j \leq 14$ such a table would have a size of 2^{13} bits or 1 KByte. For values of j in the range 24–30, a special squaring circuit as described in Section 4.7.2 could be employed. For larger precision squares a half-by-full multiplier is sufficient. In particular the computation of $|i^2|_{2^{p+1}}$ can be realized as a half-by-full, $(\lceil (p+1)/2 \rceil \times (p+1))$-bit integer multiplication, employing the following general lemma.

Lemma 7.5.17 (Modular Square Lemma) *For any integer i and modulus* $2^n \geq 2$, *let* $t = |i|_{2^{\lceil n/2 \rceil}}$. *Then*

$$|i^2|_{2^n} = |t(|2i - t|_{2^n})|_{2^n}. \tag{7.5.4}$$

Proof $i^2 = ((i - t) + t)^2 = (i - t)^2 + 2(i - t)t + t^2 = (i - t)^2 + t(2i - t)$. Since $|i - t|_{2^{\lceil \frac{n}{2} \rceil}} = 0$, $|(i - t)^2|_{2^n} = 0$, yielding (7.5.4). □

The preceding results provide the basis for an efficient procedure for determining the p-bit half-ulp directed root $(q, \operatorname{sgn}(\sqrt{x} - q))$ for a p-bit radicand from a $(p + g)$-bit one-ulp root for $g \geq 3$.

Algorithm 7.5.18 (Half-ulp directed root)

Stimulus: *An integer* $p \geq 3$, *a p-bit radicand* x *with* $\frac{1}{4} \leq x \leq 1 - 1/2^{2p}$, *and a* $(p+g)$*-bit, one-ulp root* $q = 0.1b_2 \cdots b_{p+g}$ *computed with* $g \geq 3$ *guard bits,* $g \leq p$.

Response: *A p-bit half-ulp root* q *and* $s = \operatorname{sgn}(\sqrt{x} - q)$.

Method: **if** $(b_{\lceil p/2\rceil+1}b_{\lceil p/2\rceil+2}\cdots b_g = 0^{g-\lceil p/2\rceil})$ **and** $((x \geq \frac{1}{2})$ **or** $((x < \frac{1}{2})$ **and** $(b_{\lfloor p/2\rfloor+1} = 0)))$
then
$s := 0;$
else if $b_{p+2}b_{p+3}\cdots b_{p+g} \neq 0^{g-1}$
then
$q := q + b_{p+1}2^{-p};$
$s := (-1)^{b_{p+1}};$
else
$i := q2^{p+1}; t := i \ \mathbf{mod}\ 2^{\lceil \frac{p}{2}\rceil};$
$b'_p b'_{p-1}\cdots b'_0 := (2i - t)t \ \mathbf{mod}\ 2^{p+1};$
$q := q + b_{p+1}2^{-p};$
$s := (-1)^{b'_p};$
end

Example 7.5.4 For the seven-bit radicand $\frac{1}{2} = .1000\ 000$, let a 13-bit one-ulp root q with $g = 6$ guard bits be obtained for application of Algorithm 7.5.18. Assume an iterative refinement algorithm then provides $q = .1011\ 0101\ 0000\ 0$ as a 13-bit one-ulp root.

Since $b_8 = 1$, q is not an exact root. As $b_9b_{10}\cdots b_{13} = 0^5$, the half-by-full remainder multiplication will be needed. So then $i = q2^8 = 1011\ 0101$, $t = |i|_{2^4} = 0101$, and $|(2i - t)t|_{2^8} = |1011\ 00101 \times 0101|_{2^8} = 1111\ 1001$, with leading bit $b'_7 = 1$. Thus $q := q + 2^{-7} = .1011\ 011$ and $\operatorname{sgn}(\sqrt{x} - q) = \operatorname{sgn}(x - q^2) = -1$ determines the desired half-ulp directed root.

Since the number of guard bits here is quite large, it is possible to reduce the number of bits needed from i^2. It is sufficient to compute $(2i - t)t$ modulo $2^{p+4-g} = 2^5$, with $t = |i|_{2^3} = 101$. Hence we get $|(2i - t)t|_{2^5} = |00101 \times 101|_{2^5} = 11001$, with leading bit $b'_4 = 1$, yielding the same result as above. □

7.5.6 On-the-fly rounding

As described by Theorem 2.3.3, on-the-fly conversion consists of updating several polynomials in parallel, each representing a prefix of the result, corresponding to some possible value of the carry-in, i.e., there is one polynomial for each member in the carry set C. Specifically, if the carry set is $\{0, 1\}$, then there is a polynomial representing a quotient (or root value) q, along with a second polynomial representing $q + \operatorname{ulp}(q)$, which are selectively available when all digits have been determined.

To perform the final rounding it may be necessary to choose between q and $q + \text{ulp}(q)$, and possibly also $q - \text{ulp}(q)$. However, it is also easy to develop the polynomial representing $q - \text{ulp}(q)$, simply by an addition assuming the possibility of a carry of value -1, although this value may never be generated as a carry. Thus all possible rounded values can be available for selection when the last digit has been generated, without the need for a time-consuming addition or subtraction of a unit.

To perform a precise rounding, it is necessary to know the sign of the remainder, which is only available when the last digit is known, possibly even in redundant representation. However, as a digression for certain computations (e.g., in signal processing), it may not always be necessary to perform a precise rounding, and a result can thus be available as soon as the last digit has been found. Because a precise rounding requires knowledge of the round and sticky bits, it may be necessary to use additional hardware and clock cycles to compute the sticky bit to obtain the precise rounding. For such applications one may instead consider whether a "relaxed" rounding using only the round bit would be sufficient.

When the quotient has been developed in a higher radix, a radix conversion also has to take place along with the digit set conversion, e.g., when quotient digits are from a redundant digit set for $\beta = 2^m$, and the result is to be delivered in radix 2. Then the radix conversion may easily be integrated in the on-the-fly conversion, as in the following example.

Example 7.5.5 Assume that the quotient is developed in minimally redundant, radix 4, with digit set $D = \{-2, -1, 0, 1, 2\}$, and the quotient is to be delivered in 2's complement. The conversion mapping $\alpha : (C, D) \to (C, E)$, with $E = \{0, 1, 2, 3\}$ and carry set $C = \{-1, 0\}$, then specifies a digit set conversion in radix 4:

α	-2	-1	0	1	2
-1	$\bar{1}1$	$\bar{1}2$	$\bar{1}3$	00	01
0	$\bar{1}2$	$\bar{1}3$	00	01	02
1	$\bar{1}3$	00	01	02	03

where the leftmost digit is the carry-out, and the rightmost is a digit in E. Note that only carries in $\{-1, 0\}$ are being generated, but the table has been augmented so that it also includes the possibility of receiving a carry-in of 1, which is needed to be able to generate $q + \text{ulp}(q)$ for rounding. This can be rewritten into also performing radix conversion into radix 2:

α	-2	-1	0	1	2
-1	$\bar{1}01$	$\bar{1}10$	$\bar{1}11$	000	001
0	$\bar{1}10$	$\bar{1}11$	000	001	010
1	$\bar{1}11$	000	001	010	011

where the rightmost two bits now form bit pairs of the 2's complement representation.

Assuming that the quotient is of the form $q = d_0.d_1 \cdots d_{p-1}$, reversing the indexing used in Theorem 2.3.3, then updating is based on the rule $i_k^c = i_{k-1}^{c'} + e_k[4]^{-k}$ with $\alpha(c, d) = (c', e)$. From the second table where now $e \in \{00, 01, 10, 11\}$, we can derive the following rules for updating the 2's complement polynomials $\{Q_k^c\}_{c \in C}$ in string representation, when a radix-4 digit d_k of the quotient $q = 0.d_1 d_2 \cdots d_p$ becomes known in the order $k = 1, 2, \ldots, p$:

d_k	Q_k^{-1}	Q_k^0	Q_k^1
-2	$Q_{k-1}^{-1} \circ 01$	$Q_{k-1}^{-1} \circ 10$	$Q_{k-1}^{-1} \circ 11$
-1	$Q_{k-1}^{-1} \circ 10$	$Q_{k-1}^{-1} \circ 11$	$Q_{k-1}^{0} \circ 00$
0	$Q_{k-1}^{-1} \circ 11$	$Q_{k-1}^{0} \circ 00$	$Q_{k-1}^{0} \circ 01$
1	$Q_{k-1}^{0} \circ 00$	$Q_{k-1}^{0} \circ 01$	$Q_{k-1}^{0} \circ 10$
2	$Q_{k-1}^{0} \circ 01$	$Q_{k-1}^{0} \circ 10$	$Q_{k-1}^{0} \circ 11$

in which $\circ$ means string concatenation, and the polynomials have been initialized as

$$Q_0^{-1} = (\bar{1})1, \quad Q_0^0 = (0)0 \quad \text{and} \quad Q_0^1 = (\mathtt{X})\,\mathtt{X},$$

where the bracketed digits as usual are not needed in practice.

It is necessary with a two-digit initialization to be able to form Q_0^{-1} in 2's complement representation, since when $d_1 = -2$ an initialization to $\bar{1}$ appended with 01 would not yield a proper 2's complement form. Observe that the initialization of Q_{-1}^1 is a "don't-care," since the value is never used. Also note that for rounding in this context we are assuming that the quotient is normalized and positive. Since p is often even and we need $p + 1$ bits for rounding, and the most significant radix-4 digit $d_0 = 1$, the algorithm will develop an even number of fractional bits, the last one being the round bit. □

Obtaining a rounded quotient q', applying Lemma 7.5.2 (using target precision $p + 1$), is then just a matter of choosing the appropriate resulting polynomial Q_{p+1}^s, as specified by the least-significant digit d_p and remainder sign sgn(r), or for binary directly computing $q' = q + \text{sgn}(r)\beta^{-p}$, where $q = Q_{p+1}^0$ using Observation 7.5.4.

Problems and exercises

7.5.1 Find the round-to-nearest quotient for normalized three-digit decimal division of a seven-digit dividend by a four-digit divisor in the following cases, where Lemma 7.5.6 dictates that an eight-digit one-ulp quotient must be rounding equivalent to x/y:

(a) $x = 3,017,615$ and $y = 6373 \times 10^2$.
(b) $x = 3,018,089$ and $y = 6374 \times 10^2$.

7.5.2 Let ρ be a one-ulp n-digit radix-β root-reciprocal of the n-digit radix-β radicand x with $1 \leq x < \beta^2$. Discuss the multiplication resources needed to determine the sign of $(1/\sqrt{x} - \rho)$ in terms of: (a) total multiplications, (b) dependent multiplications, and (c) the minimum size of each multiplication.

7.5.3 Discuss pre- and post-processing steps to obtain the leading bit of $|i^2|_{2^j}$ so than an index of less than $j - 1$ bits is sufficient for the direct look-up table. How large a value of j can you obtain with such steps and at most a 1 KByte table?

7.6 The IEEE standard for floating-point systems

When specifying finite integer number systems and arithmetic, the phrase "unsigned n-bit integer arithmetic" quite naturally identifies simultaneously a finite precision, finite range set of representable values, *and* an encoding into bit strings. Furthermore, the results of addition and multiplication for this integer number system are implicitly characterized as forming a closed system modulo the word size 2^n, with only integer division needing clarification as being the quotient $\lfloor x/y \rfloor$.

In contrast the design of a finite floating-point number system and associated arithmetic involves specifying a number of distinct interrelated parameters and sometimes subtle procedures that are not simply reduced to a natural and/or canonical choice set. To settle on a preferred floating-point system, an IEEE standards committee was established with an industrial and academic membership, so as to define a standard for floating-point number systems and arithmetic.

In this section we shall use the term "Standard" to refer to the specifications of the IEEE document "IEEE Standard for Floating-Point Arithmetic" [IEE08], prepared as a revision of the ANSI/IEEE Std 754-1985 standard [IEE85].

Features of the Standard have been chosen to avoid the many pitfalls that have occurred in previous ad hoc floating-point implementations. Our purpose here is to cover the realistic user expectations guiding the specifications of the Standard along with selected tables illustrating the broad scope of what has become known simply as "IEEE standard floating-point arithmetic."

The five-*property paradigm* governing the Standard for a finite floating-point number system and arithmetic is that it should be: closed, deterministic, well behaved, scalable, and efficient.

Closed The set of representable floating-point values should be chosen so that supported operations applied to any floating-point value(s) of the set must return a floating-point value of the set.

Deterministic Every operation or function on floating-point values must return a well defined result. When a result is a numerical approximation, it must be a precisely specified approximation. User selectable options are supported with each choice yielding a unique floating-point value.

Well behaved The set of normalized floating-point values should be chosen to support roughly equivalent relative errors for rounded numeric results and roughly equivalent ranges for multiplicative inverse. The numeric values should provide uniform absolute spacing in the neighborhood of zero. The signs of results indicating approaches to infinity and/or zero from one side should be respected where possible. Real-valued properties of a sequence of results, such as monotonicity, should be obtained where practically tractable. Directed roundings should be provided to facilitate the implementation of interval and multiple precision arithmetic.

Scalable The system should provide user selectable members of a hierarchy of precision and range for numeric values. The system should provide user accessible functions to determine the granularity and range of any floating-point variable defined over the hierarchy.

Efficient The required floating-point values and arithmetic should be determined so as to allow efficient hardware implementation of arithmetic operations, and encodings should provide for concise storage and efficient retrieval of all floating-point values.

7.6.1 Precision and range

Finite floating-point number systems characterize a finite set of real values having finite precision and with magnitudes restricted to a finite range disjoint from zero. Each Standard floating-point number system is in effect a hybrid system providing a finite sequence of normalized p-digit radix-β reals from $\mathbb{Q}_\beta^p$ over the symmetric, signed intervals $(-\beta^{emax+1}; \beta^{emin}]$ and $[-\beta^{emin}; \beta^{emax+1})$, and filling the interval gap $(-\beta^{emin}; \beta^{emin})$ with a fixed-point system of reals including zero.

Definition 7.6.1 *A Standard (i.e., IEEE Standard for Floating-Point arithmetic, [IEE08]) floating-point number system* FPNS$(\beta, p, e_{max}, e_{min}) \in \mathbb{Q}_\beta^p$ *is a finite set of real numbers including $v = 0$ and all reals expressed by a unique* floating-point factorization

$$v = (-1)^s \beta^e d_0.d_1 d_2 \cdots d_{p-1},$$

where

- *s is the* sign bit *with $s \in \{0, 1\}$,*
- *β is the* radix *with $\beta \in \{2, 10\}$,*

- e *is the* exponent *with exponent range* $e_{min} \leq e \leq e_{max}$,
- e_{max} *is the* maximum exponent,
- e_{min} *is the* minimum exponent, *where either* $e_{min} = 1 - e_{max}$ *or* $e_{min} = -e_{max}$,
- p *is the* precision *with* p *denoting the number of digits and* $d_0.d_1d_2 \cdots d_{p-1}$ *is the digit-string of the* significand, *with* $0 \leq d_i \leq \beta - 1$ *for* $0 \leq i \leq p - 1$,

and where

(i) $d_0 \neq 0$, *characterizing the subset of* normalized *floating-point numbers of* FPNS($\beta, p, e_{max}, e_{min}$), *or*
(ii) $d_0 = 0$, $d_i \neq 0$ *for some* $1 \leq i \leq p - 1$, $e = -e_{min}$, *characterizing the subset of* subnormal *floating point numbers of* FPNS($\beta, p, e_{max}, e_{min}$), *or*
(iii) $d_i = 0$ *for* $0 \leq i \leq p - 1$, $e - e_{min}$, *denoting magnitude zero. The real value* $v = 0$ *is neither normal nor subnormal, and has two representations corresponding to* $s = 0$ *and* $s = 1$.

Furthermore for any FPNS($\beta, p, e_{max}, e_{min}$)

- v_{max} *is the maximum value with* $v_{max} = \beta^{e_{max}+1}(1 - \beta^{-p})$;
- v_{min} *is the minimum positive value with* $v_{min} = \beta^{e_{min}-(p-1)}$.

With regard to zero, in the context of finite real operations the sign is redundant and the value is an exact zero; in the context of exceptional results (see Table 7.6.4), or as an exceptional operand the sign is recognized to distinguish between plus and minus zero.

Note that the positive normalized reals of FPNS($\beta, p, e_{max}, e_{min}$) fall in the closed interval between the minimum positive normalized value $v = \beta^{e_{min}}$ and v_{max}, and the positive subnormals fall in the closed interval between v_{min} and $v = \beta^{e_{min}}(1 - \beta^{-p})$, with $\beta^{e_{min}}$ being the *break point* between the subnormals and the normalized floating point reals.

The characterization of each floating-point non-zero real $v \in$ FPNS($\beta, p, e_{max}, e_{min}$) by a unique factorization allows us to write

$$v = (-1)^{s(v)}\beta^{e(v)}d_0(v).d_1(v)d_2(v) \cdots d_{p-1}(v),$$

where the sign bit $s(v)$, exponent $e(v)$, and digit values $d_i(v)$ for $0 \leq i \leq p - 1$ are uniquely determined integer parameters associated with the real number v and are independent of how these values may be encoded in a floating-point word. Unique factorization provides a foundation for investigating these integer-valued parameters associated with every non-zero floating-point real number that is similar in construction to the foundation for referring to the prime factors of an integer without regard to the integer's representation.

Example 7.6.1 (Binary single precision) The Standard's *binary single precision* numbers are characterized by FPNS(2, 24, 127, −126). For this system we have

$v_{max} = 2^{128}(1 - 2^{-24})$, and $v_{min} = 2^{-149}$, with 2^{-126} being the smallest positive normalized value. Note that the positive normalized values cover 254 binades, with $2^{23} \sim 8$ million distinct real values in each binade, determined by a fixed exponent e. The positive subnormal values sparsely cover 23 binades with $(2^{23} - 1) \sim 8$ million distinct values in total. Thus single precision binary has about 254 times as many normalized real values than subnormal real values.

For the real $v = 5\frac{3}{8}$, the unique (normalized) factorization is $5\frac{3}{8} = (-1)^0 2^2 1.01011$, and for the real $v = 5\frac{3}{8} \times 2^{-128}$, the unique single precision (subnormal) factorization is $5\frac{3}{8} \times 2^{-128} = (1)^0 2^{-126} 0.0101011$. Note that single precision binary has only about 4 billion distinct real numbers in the whole number system, so that such a system is inadequate for approximate real scientific computation where several billion operations are performed each second. □

$\mathrm{FPNS}(\beta, p, e_{max}, e_{min})$ is a finite set so that every $v \in \mathrm{FPNS}$, $v \neq v_{max}$, has a *successor* $v' = \min\{u | u > v, u \in \mathrm{FPNS}\}$. Then $\mathrm{ulp}(v) = v' - v$ is defined for all $-v_{max} \leq v < v_{max}$, where we further define $\mathrm{ulp}(v_{max}) = \beta^{e_{max}-(p-1)}$. Thus for $v_{min} \leq v \leq v_{max}$, $\mathrm{ulp}(v)$ is a step function constant over each binade (decade) independent of the significand with $\mathrm{ulp}(v) = \beta^{e(v)-(p-1)}$.

The inclusion of the subnormals and zero in the FPNS set means that the neighborhood of zero effectively constitutes a fixed-point number system with values $v = (-1)^s (d_{e_{min}} \beta^e_{min} + d_{e_{min}-1} \beta^{e_{min}-1} + \cdots + d_{e_{min}-(p-1)} \beta^{e_{min}-(p-1)})$. In contrast to an unbounded range finite precision floating-point number system, the successor function over $\mathrm{FPNS}(\beta, p, e_{max}, e_{min})$ passes smoothly from negative values to 0 to positive values, i.e., $(-2v_{min})' = -v_{min}$, $(-v_{min})' = 0$, $0' = v_{min}$, and $v'_{min} = 2v_{min}$.

Observation 7.6.2 *The unit in the last place at zero is* $\mathrm{ulp}(0) = v_{min} = \beta^{e_{min}-(p-1)}$.

The value $\mathrm{ulp}(0)$ gives the smallest gap between successive real floating-point numbers of $\mathrm{FPNS}(\beta, p, e_{max}, e_{min})$, and it is a fundamental unit in the characterization of the members of $\mathrm{FPNS}(\beta, p, e_{max}, e_{min})$ and has many properties.

Observation 7.6.3 *Every* $v \in \mathrm{FPNS}(\beta, p, e_{max}, e_{min})$ *is an integer multiple of* $\mathrm{ulp}(0) = \beta^{e_{min}-(p-1)}$, *and* $\mathrm{ulp}(v) = 2^{e(v)-e_{min}} \mathrm{ulp}(0)$ *for all* $v_{min} \leq v \leq v_{max}$, *with* $\mathrm{ulp}(v)$ *monotonically non-decreasing over* $0 \leq v \leq v_{max}$.

Observation 7.6.4 *The finite floating-point number system* $\mathrm{FPNS}(\beta, p, e_{max}, e_{min})$ *has a "fixed-point neighborhood of zero" in that the members of* FPNS *in the closed interval* $[-\beta^{e_{min}+1}, \beta^{e_{min}+1}]$ *form a set of* $2\beta^p + 1$ *reals uniformly spaced at distance* $ulp(0) = \beta^{e_{min}-(p-1)}$ *and symmetric about zero. This set includes boundary values* $[-\beta^{e_{min}+1}, 0, \beta^{e_{min}+1}]$, *all subnormal* $v \in$FPNS, *and*

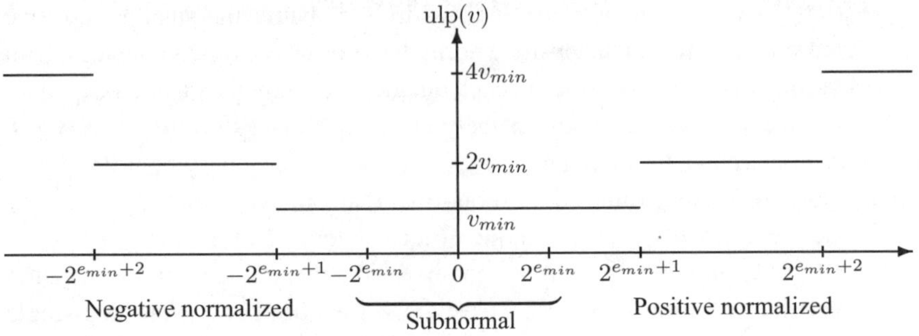

Figure 7.6.1. Gap function $\mathrm{ulp}(v)$ between floating-point real values around zero.

the normalized $v \in$ FPNS *from the smallest magnitude positive and negative binades* (decades for $\beta = 10$).

The value of the gap function in the neighborhood of zero is illustrated in Figure 7.6.1. The neighborhood includes the flat fixed-point neighborhood of zero with granularity $\mathrm{ulp}(v) = v_{min} = \beta^{emin-(p-1)}$ throughout, the neighboring binades where $\mathrm{ulp}(v) = 2v_{min}$, and portions of the binades with granularity $\mathrm{ulp}(v) = 2^2 v_{min}$.

The uniform gaps with $\mathrm{ulp}(v) = \mathrm{ulp}(0) = v_{min}$ between floating-point numbers over the interval $[-\beta^{emin+1}, \beta^{emin+1}]$, afforded by including the subnormal reals and zero in the Standard floating-point number system specification, are of great importance to the developers of mathematical software, where approaches to a limit are often measured by small discrete differences in approximations that in theory approach zero.

For regions away from zero where $\mathrm{ulp}(v) = 2^k v_{min}$ for large k, it is more illustrative to graph the *relative gap* $\mathrm{ulp}(v)/v$.

Observation 7.6.5 *Over the range of positive normalized floating-point numbers the relative gap satisfies*

$$\beta^{-p} < \mathrm{ulp}(v)/v \leq \beta^{p-1} \quad \textit{for} \quad \beta^{emin} \leq v < v_{max}.$$

The relative gap $\mathrm{ulp}(v)/v$ for the binary single-precision system FPNS(2, 24, 127, −126) for $v_{min} \leq v < v_{max}$ is illustrated in Figure 7.6.2. The figure shows the nearly uniform precision with periodic "precision wobble" over the normalized value range $[2^{-126}; 2^{128})$, and the progressive decrease in precision for subnormal values over $[2^{-149}; 2^{-126})$, decreasing towards $\mathrm{ulp}(v_{min})/v_{min} = 1$.

The near-uniform relative gap given by $\mathrm{ulp}(v)/v$ over the much larger normalized range is desirable for controlling accumulated approximate relative error in scientific computation.

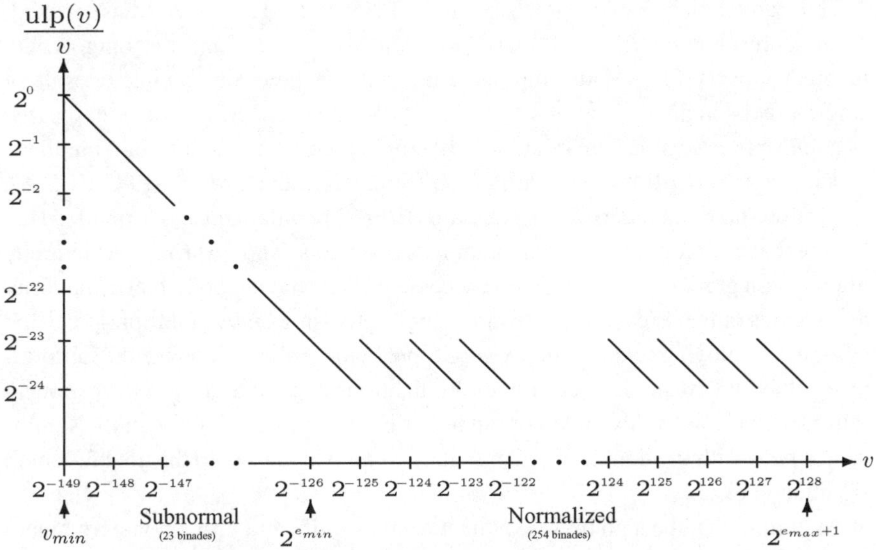

Figure 7.6.2. Relative gap between the Standard's positive single precision floating-point numbers.

It should also be noted that the characterization of the non-zero reals of a binary FPNS by unique factorizations is of considerable practical benefit in providing a more efficient encoding of each binary real number into a bit string of a floating-point word (see Section 7.6.4).

To appreciate the relation between finite range and finite precision for the floating-point numbers FPNS(β, p, e_{max}, e_{min}), it is convenient to specify a minimal fixed-point system containing FPNS.

Observation 7.6.6 *Every* $v \in$ FPNS(β, p, e_{max}, e_{min}) *has a fixed-point representation*

$$v = (-1)^s d_{emax} d_{emax-1} \cdots d_1 d_0 . d_{-1} \cdots d_{emin-(p-1)}, \tag{7.6.1}$$

and $(e_{max} - e_{min} + p)$ *is the minimum width in digits of a signed fixed-point radix-*β *number system containing* FPNS(β, p, e_{max}, e_{min})*. Furthermore, the members of* FPNS *are precisely the values* $v = (-1)^s d_m d_{m-1} \cdots d_l$ *specified by (7.6.1) with* $d_m \neq 0$, $d_l \neq 0$, *and* $m - l \leq p$.

The relationship between precision measured simply as the width in digits of the significand, and the much greater width in digits of the fixed-point value of (7.6.1) is best visualized by expressing the latter as a multiple of the former.

Definition 7.6.7 *The* range width *of* FPNS(β, p, e_{max}, e_{min}) *measured in precisions is given by* $(e_{max} - e_{min} + p)/p$.

Thus any binary single-precision $v \in \text{FPNS}(2, 24, 127, -126)$ has a fixed-point representation $v = (-1)^s b_{127}b_{126}\cdots b_0.b_{-1}\cdots b_{-149}$ and the range width in precisions is $11\frac{13}{24}$. This implies, e.g., that the precisely rounded result of single-precision floating-point inner products that do not overflow can be expressed as a multiple-precision sum of at most 12 single precision floating-point numbers.

The original IEEE binary floating-point standard, Standard 754-1985 [IEE85], and follow-up radix independent standard [IEE87] popularized and formalized the feature that a "floating-point computation environment" should provide a hierarchy of precision granularity levels, with successive levels having both finer granularity and greater range. Prevailing software double-precision and/or multiple-precision systems employing two or more single-precision values allowed the unfortunate behavior that single-precision multiplications could overflow in the exponent range, which was not extended in range in double precision, although the significand's field with the double-word granularity could always hold the double-length significand exactly.

Assuming that the sign bit s, exponent e, and significand f of a non-zero binary floating-point number will be encoded into three fields of a single-length word (e.g., 32 bits for binary single precision), note that the double-length word still needs only a single bit for the sign and only three or four more exponent bits to increase the range width (in bits) by essentially a factor of 8 or 16 respectively. Thus the precision for the binary double-precision numbers can readily be set to more than $2p$, still accommodating an efficient encoding. Table 7.6.1 provides the parameters p, e_{max}, e_{min}, characterizing six recognized format levels of standardized binary systems $\text{FPNS}(\beta, p, e_{max}, e_{min})$ according to the Standard.

Table 7.6.1. *Standard binary hierarchy of floating-point number systems with specifications of precision p, range (e_{min}, e_{max}), range width $(e_{max} - e_{min} + p)/p$, maximum value v_{max} and subnormal fixed-point gap $\text{ulp}(0) = 2^{e_{min}-(p-1)}$ for each format level*

	Half	Single	Single	Double	Double	Quad
Format level	Storage-16	Basic-32	Extended-40	Basic-64	Extended-80	Basic-128
p bits	11	24	≥ 32	53	≥ 64	113
e_{max}	15	127	≥ 1023	1023	≥ 16383	16383
e_{min}	-14	-126	≤ -1022	-1022	≤ -16383	-16382
exponent bias	15	127		1023		16383
Range width	3.6	11.5	65	40	513	291
v_{max}	65504	$10^{38.5}$	10^{308}	10^{308}	10^{4932}	10^{4932}
$\text{ulp}(0) = v_{min}$	0.6×10^{-7}	10^{-45}	10^{-317}	10^{-323}	10^{-4950}	10^{-4965}

The terms "basic" and "extended" refer to format levels that must support arithmetic in an implementation, with the sizes 32, 64, and 128 corresponding to bit-string lengths of the encodings to be discussed in Section 7.6.4. The term "storage" refers to a narrower format level which is not intended for arithmetic but which must support conversions to and from higher format levels within the hierarchy.

The four intermediate levels from basic single to double extended are carried over from the 754-1985 standard [IEE85]. Values for the range width in precisions, the maximum real value v_{max}, and the gap, ulp(0), between the represented real values in the neighborhood of zero are given as approximate decimals for easier interpretation and comparison. The range width and ulp(0) values for the extended precisions correspond to $p = 32$ and $p = 64$ as implemented in the de-facto standard ×86 implementations and their legacy follow-up microprocessor families.

Table 7.6.2 similarly provides the parameters for the Standard's decimal floating-point system hierarchy, with the extended-80 entries for range width, v_{max} and v_{min} corresponding to minimum choices for p, e_{max}, and $-e_{min}$. The systems are chosen to have similar effective maximum precisions as measured by the minimum relative gap sizes but the precision wobble is over three times as large for decimal floating-point. Figure 7.6.3 illustrates an overlay of the relative gap function for the decimal storage-32 system onto the binary single-precision relative gap function from Figure 7.6.2. It should be emphasized that the Standard treats the decimal 32-bit format specification only as a storage format and not a level for which arithmetic should be implemented.

Table 7.6.2. *Standard decimal hierarchy of floating-point number systems with specifications of precision p, range* (e_{min}, e_{max})*, system range-width* $(e_{max} - e_{min} + p)/p$*, maximum value* v_{max}*, and subnormal fixed point gap* $\text{ulp}(0) = 10^{e_{min}-(p-1)}$ *for each level*

	Single	Double	Double	Quad
Format level	Storage-32	Basic-64	Extended-80	Basic-128
p digits	7	16	≥20	34
e_{max}	96	384	≥6144	6144
e_{min}	−95	−383	≤−6143	−6143
Exponent bias	101	398		6176
Range width	28	49	615	362
v_{max}	$10^{97}(1 - 10^{7})$	$10^{385}(1 - 10^{16})$	$10^{6145}(1 - 10^{20})$	$10^{6145}(1 - 10^{34})$
$\text{ulp}(0) = v_{min}$	10^{-101}	10^{-398}	10^{-6162}	10^{-6176}

Precision granularity and range functions over $F(\beta, p, e_{max}, e_{min})$**.** For the development of robust approximate mathematical software for any floating-point number system such as those of the Standard, it is essential to respect the discreteness of the number system. To aid this the Standard requires specific functions to be implemented to access the precision granularity, range, and radix of representation for real values within a floating-point number system hierarchy. These functions provide explicit user access to parameters that might otherwise be hidden from

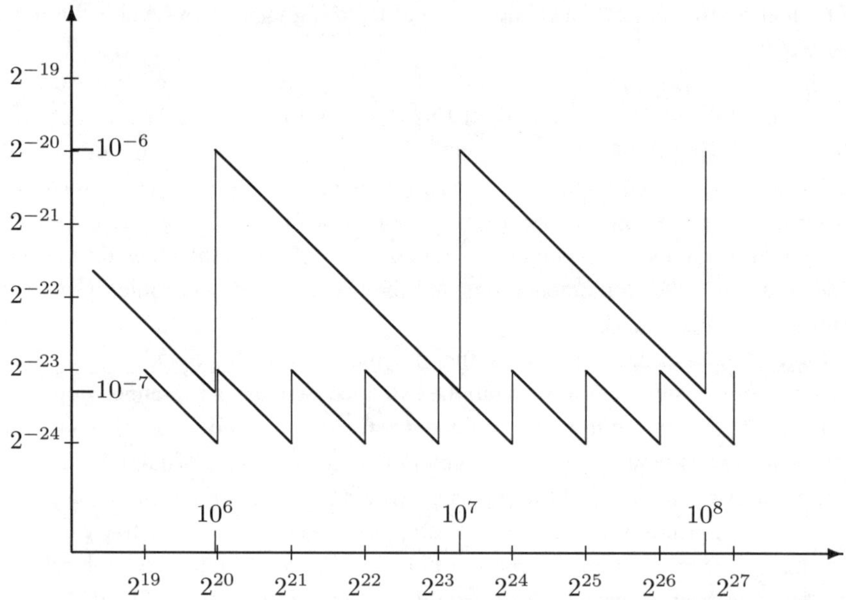

Figure 7.6.3. Relative gap functions for the Standard's binary and decimal 32-bit format levels, showing seven-digit decimal versus 24-bit binary precisions with a common integer interval $[2^{23}; 10^7]$.

Table 7.6.3. *Granularity, range, and radix representation user access functions required by the Standard over each supported system FPNS(β, p, e_{max}, e_{min})*

Category	Required operations	Result range
General functions	successor(v), predecessor(v)	$F(\beta, e_{max}, e_{min}) \cup \{+\infty, -\infty\}$
Range predicates	isNormal(v), isSubnormal(v), isZero(v)	Boolean
Representation components	exponent(v)	Integers $\cup\{-\infty\}$
Representation components	radix (v)	2,10

the user and only available for system implementation. The functions are given in Table 7.6.3, and in many cases can be characterized for $v \in F(\beta, p, e_{max}, e_{min})$ by components of the unique factorization $v = (-1)^{s(v)}\beta^{e(v)}d_0(v).d_1(v)\cdots d_{p-1}(v)$ pertaining to all $v \neq 0$.

For the precision granularity functions:

successor(v) is denoted by v' and defined by

$$v' = \begin{cases} \beta^{e(v)-(p-1)} & \text{for} \quad v \neq 0, v \neq -\beta^i \text{ with } -e_{max}+1 \leq i \leq -e_{min}+1, \\ \beta^{e_{min}-(p-1)} & \text{for} \quad v = 0, \\ \beta^{e(v)-p} & \text{for} \quad v = -\beta^i \text{ with } -e_{max}+1 \leq i \leq -e_{min}+1; \end{cases}$$

predecessor(v) $= -(-v)'$.

For the range predicates:

isNormal(v) is true iff $\beta^{emin} \leq |v| \leq \beta^{emax+1}(1 - \beta^{-p})$,
isSubnormal(v) is true iff $\beta^{emin-(p-1)} \leq |v| \leq \beta^{emin}(1 - \beta^{-(p-1)})$,
isZero(v) is true iff $v = 0$.

For the representation components:

exponent(v) is defined by

$$\text{exponent}(v) = \begin{cases} \lfloor \log_\beta(|v|) \rfloor & \text{for} \quad v \neq 0, \\ -\infty & \text{for} \quad v = 0 \end{cases}$$

and invoking exponent(0) triggers the divide-by-zero exception. Thus the value of exponent(v) is effectively defined by a unique factorization over $\mathbb{Q}_\beta^p$. radix (v) yields the radix $\beta(v) \in \{2, 10\}$ for any $v \in F(\beta, p, e_{max}, e_{min})$, as determined by a data encoding tag or the language data type of the expression v.

In summary, regarding the five-property paradigm governing the Standard's design, the precision and range hierarchies specified in Tables 7.6.1 and 7.6.2 satisfy the "scalability goal" for floating-point number systems and have been carefully chosen for efficient encodings into 16-, 32-, 64-, or 128-bit words as will be shown in Section 7.6.4. The hybrid normalization dovetailing floating-point with a fixed-point neighborhood of zero satisfies the "well behaved" subgoal of near uniform relative gap size over the normalized range with uniform gaps over an interval around zero.

7.6.2 Operations on floating-point numbers

In this section, we focus on floating-point arithmetic, comparison, and representation operations, where the operands are all binary numbers from the binary hierarchy specified by Table 7.6.1, termed $F_2 \in \mathbb{Q}_\beta^p$, or all decimal numbers from the decimal hierarchy specified by Table 7.6.2, termed $F_{10} \in \mathbb{Q}_\beta^p$. The operations fall into three categories classified as follows:

- computational arithmetic operations over F_β,
- selection and comparison operations and functions over $F(\beta, p, e_{max}, e_{min})$,
- base conversion mappings[2] between F_2 and F_{10}.

Computational arithmetic operations over F_β. These operations include the Standard arithmetic operations $+, -, \times, /, \sqrt{}$ associated with real arithmetic. Floating-point arithmetic operations always have an explicitly or implicitly

[2] Here we follow the traditional notation of binary–decimal and decimal–binary conversion, where "base" is synonymous with "radix."

specified destination system $F(\beta, p, e_{max}, e_{min})$. The results may trigger one or more of the five *exception conditions* {`invalid, divide-by-zero, overflow, underflow, inexact`} and lead to any of the five *exceptional floating point values* {$+\infty$, $-\infty$, $+0$, -0, not-a-number (NaN)} as specified in Table 7.6.4.

Table 7.6.4. *Results of the Standard's arithmetic operations defined over each finite floating-point number hierarchy F_β, for $\beta = 2, 10$, giving destination extended value ranges and exception conditions*

Arithmetic Operations on F_β	Extended Value Range of Results	Exceptions
Addition, subtraction, multiplication, fused multiply-add	$F(\beta, p, e_{max}, e_{min}) \cup \{\pm\infty, \pm 0\}$	`inexact over/underflow`
Divide	$F(\beta, p, e_{max}, e_{min}) \cup \{\text{NaN}, \pm\infty, \pm 0\}$	`inexact over/ underflow invalid divide- by-zero`
Remainder	$F(\beta, p, e_{max}, e_{min}) \cup \text{NaN}$	`invalid`
Square root	$F(\beta, p, e_{max}, e_{min}) \cup \{\text{NaN}, +0\}$	`inexact underflow invalid`

The arithmetic operations of Table 7.6.4 are interpreted as being evaluated by a two-stage process:

(i) the operation is to be performed as if by real arithmetic yielding an intermediate finite real number of unbounded range an infinite precision, with exceptions `invalid` or `divide-by-zero` triggered at this stage if the result is undefined over the reals,

(ii) the intermediate finite value is to be rounded into either a floating-point number of the destination system $F(\beta, p, e_{max}, e_{min})$, or one of the extended floating point values $+\infty$, $-\infty$, $+0$, -0. The exceptions that can be triggered at this stage are `inexact, overflow, and underflow.`

The Standard's exceptional valued results have the following meanings.

Definition 7.6.8 *The five floating-point exceptional values resulting from operations on floating point numbers of F_β are:*

- *Not-a-number (*NaN*) denotes that the result was undefined as a real operation on the operands.*
- *Plus infinity ($+\infty$) and minus infinity ($-\infty$) denote that the real result was a finite value larger in magnitude than the destination systems largest magnitude, which by the rounding rule rounded to infinity with the sign of the finite valued result.*

- *Plus zero (+0) and minus zero (−0) denote that the real result was a finite value smaller in magnitude than the destination systems smallest magnitude, which rounded by the rounding rule to zero with the sign of the finite valued result.*

Regarding implementation of the exceptional values $\{+0, -0, +\infty, -\infty, \text{NaN}\}$ note that these five values are to be encoded, stored, and available as arguments of the arithmetic operations with results specified in Section 7.6.3 on closure of operations over $F \cup \{+0, -0, +\infty, -\infty, \text{NaN}\}$. The exceptions triggered are assumed to be recorded by the system as sticky flags independently available to the user by further operations prescribed in the Standard's documentation [IEE08] and beyond the scope of our presentation.

Regarding the first stage of computing an intermediate result, consider the different groupings in Table 7.6.4. The groupings are related both to the exceptions that can occur, and to practical aspects of their implementations.

Addition and multiplication. For $x, y, z, \in F$, the operations addition $(x + y)$, subtraction $(x - y)$, multiplication $(x \times y)$, and fused multiply-add $(x \times y + z)$ yield exact intermediate binary (or decimal) results without exception.

Division, remainder, and square root. For $x, y \in F$ these operations can yield invalid results or real results requiring special intermediate binary (or decimal) results for implementing the subsequent rounding. Specifically **division** (x/y) yields

- NaN triggering the `invalid` exception when both $x, y = 0$.
- $(-1)^{s(x)}\infty$ when $y = 0$ and $x \neq 0$.
- A rational x/y when $y \neq 0$. The intermediate rational value may be indicated by a one-ulp quotient of sufficient accuracy possibly accompanied by a remainder sign.

The remainder function **remainder** (x, y) yields

- NaN triggering the `invalid` exception when $y = 0$.
- The exact binary (or decimal) number $(x - n \times y)$ where n is the nearest integer to (x/y) with n chosen as the even integer if there is a tie. The remainder function is only defined for x and y in the same floating-point system $F(\cdot)$, and the result is exact in that system.

The square root function $\sqrt{\ }$ yields

- NaN triggering the `invalid` exception when $x < 0$.
- The real value $\sqrt{x} \geq 0$ that may be indicated by an intermediate one-ulp square root of sufficient accuracy, possibly augmented by the sign of $(x - z^2)$.

Table 7.6.5. *The Standard's floating-point precise rounding modes*

Rounding types	Precise rounding mode
Nearest	RN_e: round-to-nearest even RN_a: round-to-nearest away
Directed	RU: round up (towards ∞) RD: round down (toward $-\infty$) RZ: round towards zero (preserving sign)

The intermediate real-valued results of the arithmetic operations of Table 7.6.4 have then to be rounded by any of the five user selectable rounding modes enumerated in Table 7.6.5 to provide the floating-point-valued result. Note that the nearest-away mode RN_a is only required for decimal systems, with the other four roundings required for both binary and decimal systems.

The five rounding modes affecting the first group of six computational operations in Table 7.6.4 round the reals either into the bounded finite set $F(\beta, p, e_{max}, e_{min})$ or to ± 0 or $\pm\infty$, and are conveniently defined by three ranges of magnitude.

- *Normalized range* For reals with magnitudes in $[\beta^{e_{min}}, v_{max}]$, the roundings into $F(\beta, p, e_{max}, e_{min})$ are identical to the precise roundings into the unbounded range and precision p set $\mathbb{Q}_\beta^p$ defined in Section 7.6.2.
- *Subnormal and underflow range* For reals over $(-\beta^{e_{min}}, \beta^{e_{min}})$ including the subnormal values $F(\beta, p, e_{max}, e_{min})$ and zero, the roundings conform with fixed-point roundings. Note that any non-zero value from $(-v_{min}, v_{min})$ that rounds to zero triggers an `underflow` exception, and yields a similarly signed zero.
- *Overflow range* For reals with magnitudes in the overflow range (v_{max}, ∞), the roundings yield one of the boundary magnitudes v_{max} or ∞ with the same sign as the real values and *may or may not* trigger an `overflow` exception. The selections are governed by giving preference to the rounding direction for directed roundings and to triggering `overflow` only when the rounding distance is at least an ulp for directed roundings or one half ulp for nearest roundings. Specifically,
    - Directed roundings that round values not in $\mathbb{Q}_\beta^n$ towards zero round all values with magnitudes in (v_{max}, ∞) to v_{max} with the same sign. Values with magnitudes in the ulp interval $(v_{max}, \beta^{e_{max}+1})$ that round to v_{max} do not trigger `overflow`, all others trigger `overflow`. Similarly, directed roundings that round values not in $\mathbb{Q}_\beta^n$ away from zero round all values with magnitudes in (v_{max}, ∞) to ∞ with the same sign and always trigger `overflow`.

- Nearest roundings round magnitudes from the half ulp interval $(v_{max}, v_{max} + \frac{1}{2}\text{ulp}(v_{max}))$ to v_{max} without overflow being triggered, and preserve the sign. Values with magnitudes at least $v_{max} + \frac{1}{2}\text{ulp}(v_{max}) = \beta^{e_{max}+1}(1 - \frac{1}{2}\beta^{p})$ round to ∞ preserving the sign and signaling `overflow`.

Observation 7.6.9 *Real values that round by any of the five rounding modes in Table 7.6.4 to* $\pm v_{max}$ *in* $F(\beta, p, e_{max}, e_{min})$ *that would also round to* $\pm v_{max}$ *in the unbounded range* $\mathbb{Q}_\beta^p$ *never trigger* `overflow`. *Values with magnitudes at least* v_{max} *that round to a value of larger magnitude in* $\mathbb{Q}_\beta^p$ *always trigger* `overflow`.

From Observation 7.6.9 we note that roundings into $v \in F(\beta, p, e_{max}, e_{min})$ that do not trigger overflow have a maximum rounding interval width of ulp(v) whenever the magnitude of v is not a power of β. This is useful for preserving the integrity of measuring rounding error near the upper magnitude bound v_{max}. The choice of roundings to magnitudes v_{max} or infinity for directed roundings preserves the integrity of rounding direction control that is essential for support of interval arithmetic.

Selection and comparison operations and functions over $F(\beta, p, e_{max}, e_{min})$**.** For every supported (non-storage) floating-point number system $F(\beta, p, e_{max}, e_{min})$ implemented according to the Standard, the operations of Table 7.6.6 must be implemented. Except for the comparison predicates the operations are *closed and exact* as real-valued operations over $F(\beta, p, e_{max}, e_{min})$.

Table 7.6.6. *Operations and functions required by the Standard on each supported system* $F(\beta, p, e_{max}, e_{min})$ *having exact results in the same system or as Boolean values*

Category	Required operations
Selection operations	$\min(x, y)$, $\max(x, y)$, minmag(x, y), maxmag(x, y)
Sign operations	$\lvert x\rvert$, negate(x), copysign(x, y) for $y \neq 0$
Integer part functions	$\text{RNI}_a(x)$, $\text{RNI}_e(x)$, $\text{RZI}(x)$, $\text{RUI}(x)$, $\text{RDI}(x)$
Comparison predicate operations	$=, \neq, >, \geq, <, \leq$

The functions in Table 7.6.6 have their usual real-valued meanings with the following extensions:

minmag(x, y) selects $\min(\lvert x\rvert, \lvert y\rvert)$ as the magnitude of the result with the sign of the selected term. If the magnitudes are equal it selects $\min(x, y)$. A similar definition applies to **maxmag**(x, y)
copysign(x, y) is defined only if $y \neq 0$ and returns magnitude $\lvert x\rvert$ with the sign of y for $x \neq 0$.

Note that the minmag and maxmag functions can be useful in the formation of interval arithmetic operations.

The round-to-integer-part functions are simplified by the following important property of floating point number systems.

Observation 7.6.10 *Let* $v \in F(\beta, p, e_{max}, e_{min})$. *Then*

$$v \text{ is an integer for } |v| \geq \beta^{p-1},$$
$$\lfloor v \rfloor, \ \lceil v \rceil \in F(\beta, p, e_{max}, e_{min}) \text{ for } |v| < \beta^{p-1}.$$

Thus all five integer part roundings are the identity whenever $|v| \geq \beta^{p-1}$.

The directed roundings always then choose either $\lfloor v \rfloor$ or $\lceil v \rceil$ in the appropriate direction. The nearest roundings choose $\lfloor v + \frac{1}{2} \rfloor$ whenever $v + \frac{1}{2}$ is not an integer. If $v + \frac{1}{2}$ is an integer, then $\text{RNI}_e(v)$ chooses the nearest even integer and $\text{RNI}_a(v)$ chooses the neighboring integer of largest magnitude. In summary, the integer part functions over $F(\beta, p, e_{max}, e_{min})$ behave as if they were selecting $\lfloor v \rfloor$ or $\lceil v \rceil$ over the reals.

The Standard specifies that all arithmetic operations be performed with operands and (unexceptional) results contained in the same floating-point number hierarchy F, i.e., with representations having the same radix. This allows for more efficient hardware implementation since implicit conversions to different range and precision bounds are common practice in current ALU designs. Thus the fundamental adders and multipliers may be optimized based on a binary architecture or alternatively on a decimal digit encoding architecture determined by the intended user community. The required computations in the alternative radix could then be provided by software, probably taking advantage of efficient integer arithmetic that could have separate optimized hardware, most likely in native binary for maximum range and simplicity of design. In an age of multicore microprocessors it might be practical to provide at least one core optimized for floating-point arithmetic in each radix.

Base conversion mappings between F_2 and F_{10}. The connection between the binary and decimal domains is provided solely by base conversion mappings, having explicitly selected destinations in the alternative radix representation system. The mappings are of the form

$$\text{R} : F(\beta, p, e_{max}, e_{min}) \longrightarrow F(\beta', p', e'_{max}, e'_{min}),$$

where

- the function $\text{R} \in \{\text{RN}_e, \text{RN}_a, \text{RU}, \text{RD}, \text{RZ}\}$ is any of the Standard's five precise rounding modes described (see Table 7.6.5),
- the domain of R is the supported format levels of the radix β hierarchy,
- the range of R is the supported format levels of the radix $\beta' \neq \beta$ hierarchy (including storage formats of each).

Note that the range and domain of R include the storage formats of each.

Observation 7.6.11 *All of the Standard's base conversion mappings are uniquely specified. In particular all of the mappings must be the identity by any rounding mode over the set of values exactly representable in both binary and decimal systems, i.e., over* $F(2, p, e_{max}, e_{min}) \cap F(10, p', e'_{max}, e'_{min})$*. Thus iterated roundings in a fixed directed rounding mode must eventually converge to a member of this subset.*

To visualize a comparison of the effective precision of the binary and decimal n-bit format levels for $n = 32$, 64, and 128, and to determine properties of the conversion mappings it is instructive to consider the binary versus decimal relative gap functions. Figure 7.6.3 provides such a comparison between the 24-bit binary and 7-digit decimal precisions of the respective 32-bit format levels. Note that the relative gap functions coincide over the interval $[2^{23}; 10^7]$, where both floating-point systems yield in common a sequence of integer values.

The periods of precision wobble of the decimal and binary relative gap functions have an irrational ratio. Since $10^6 = 2^{19.93} < 2^{23} < 10^7 = 2^{23.25} < 2^{24}$, there will then be binades where the gaps for 24-bit binary are either smaller, or have a break between smaller and larger, than the 7-digit decimal numbers, and one interval (of each sign) where they are equal.

Observation 7.6.12 *The intersection of the integer binade* $[2^{p-1}; 2^p]$ *and the integer decade* $[10^{p'-1}; 10^p]$ *is the only positive real interval where the p-bit binary and p'-digit decimal floating-point numbers have equal gap functions. In particular, these intersecting intervals are* $[2^{23} = (8, 388, 608); 10^7]$, $[2^{52}, 2^{53}]$, *and* $[2^{112}, 10^{34}]$ *for the Standard's 16-, 32-, and 64-bit format levels.*

Outline of proof Each gap is a power of 2 in binary and is a power of 10 in decimal. The only equal powers are $2^0 = 10^0 = 1$, so we are restricted to integer intervals for equal gap sizes. □

Note than binary–decimal conversion by any rounding mode is an onto mapping over intervals where the binary gaps are smaller than or equal to the decimal gaps and is one-to-one when the binary gaps are greater than or equal to the decimal gaps, with a similar result for decimal–binary conversion.

Observation 7.6.13 *Binary–decimal and decimal–binary base conversions have a piecewise alternation between being one-to-one and onto, with at most one change per decade for the base conversions between binary and decimal n-bit formats by any rounding mode for the Standard's 32-, 64-, and 128-bit format levels.*

7.6.3 Closure

In this section we focus on floating-point extended operations and functions where the arguments are members of the extended set $F_\beta \cup \{+0, -0, +\infty, -\infty\}$, where at least one argument is an exceptional value.

Definition 7.6.14 *A Standard closed floating-point system F^*_β is given by*

$$F^*_\beta = F_\beta \cup \{+0, -0, +\infty, -\infty, \mathrm{NaN}\},$$

where

- $v \in \{+0, -0, +\infty, -\infty, \mathrm{NaN}\}$ *is an* exceptional value *of* F^*_β,
- $v \in F_\beta$ *is a* finite value *of* F^*_β,
- $\beta \in \{2, 10\}$ *is the* radix *of* F^*_β,
- $v \in \{+0, -0\}$ *is a* signed zero *of* F^*_β,
- $v \in \{+\infty, -\infty\}$ *is a* signed infinity *of* F^*_β.

The extension of the domain of definition of the floating-point number system by appending five exceptional values was designed to obtain an algebraically closed floating-point arithmetic. Furthermore, the extension was crafted to provide useful numeric structure for mathematical software development, without seriously impacting run time efficiency of an implementation, compared with operations restricted to the finite floating-point numbers $F_\beta \in \mathbb{Q}^p_\beta$.

The arithmetic structure of F^*_β is crafted to respect the logical distinction between the real value $v = 0 \in F_\beta$, sometimes referred to as the exact zero, and either of the signed zero closure values $v \in \{+0, -0\}$. For purposes of appreciating the arithmetic structure of F^*_β, it is useful to visualize $v \in \{+0, -0\}$ as a very tiny non-zero quantity of known sign, as might be suggested by $\epsilon > 0$ in discussions of practical applications of calculus in science. Similarly, $v \in \{+\infty, -\infty\}$ suggests an unboundedly large quantity of known sign analogous to $1/\epsilon$.

That there is a distinction between the roles of an exact real-valued zero and the signed zero closure values is suggested by considering two levels of interpretation, as illustrated in Figure 7.6.4. The two levels are respected implicitly by the Standard arithmetic defined over F^*_β.

(a) $-\infty <$ ———————— 0 ———————— $< +\infty$

(b) $\{-\infty <$ ———————— $< -0\} \cup \{+0 <$ ———————— $< +\infty\} \cup \mathrm{NaN}$

Figure 7.6.4. (a) The extended reals with exact zero, and (b) the algebraically closed floating-point value system with signed zeroes.

Computational arithmetic operations with an exceptional valued argument. For the arithmetic operations addition, subtraction, multiplication, and division, if one operand is in $\{+0, -0, +\infty, -\infty\}$ and the other is not a NaN, then the operation is interpreted as if the exceptional values are appropriately signed finite real tiny or huge values $\pm\epsilon$, respectively $\pm\infty$. If the result is then a tiny ($\pm\delta$) real or a huge ($\pm 1/\delta$) real the result is an appropriate member of $\{+0, -0, +\infty, -\infty\}$. For $v \in F_\beta$, $v \neq 0$, addition or subtraction of zero in a context where zero is considered exact the result is v. If the result is not tiny or huge with known sign, then the result is a NaN.

Operations yielding a NaN with the invalid exception condition signaled include:

- effective addition of oppositely signed zeros or infinities,
- multiplication of a zero by an infinity,
- division of a zero by a zero or of an infinity by an infinity.

If either argument is a NaN the result is a NaN. For further specifications see [IEE08].

Range predicates and range boundary functions over F_β^*. The Standard's closed floating-point system F_β^* has specifications for range predicates that allow user queries for determining specific closure values of floating-point variables of the system:

isFinite(v) is true if and only if v is zero, subnormal or normal (not infinity or NaN).
isZero(v) is true if and only if $v \in \{+0, -0\}$. **isZero** v is also true for v being a real exact zero.
isInfinity(v) is true if and only if $v \in \{+\infty, -\infty\}$.
isSigned(v) is true if and only if v has a negative sign.
isNaN(v) is true if and only if v has the value NaN.

The range sequencing function values for adjacencies with exceptional values are given by

$$\text{successor}(v) = \begin{cases} +\infty & \text{for } v = +\infty \text{ or } v_{max}, \\ v_{min} & \text{for } v \in \{+0, -0\}, \\ -0 & \text{for } v = -v_{min}, \\ -v_{max} & \text{for } v = -\infty, \\ \text{NaN} & \text{for } v = \text{NaN}, \end{cases}$$

$$\text{predecessor}(v) = \begin{cases} v_{max} & \text{for } v = +\infty, \\ 0 & \text{for } v = v_{min}, \\ -v_{min} & \text{for } v \in \{+0, -0\}, \\ -\infty & \text{for } v = -v_{max} \text{ or } -\infty, \\ \text{NaN} & \text{for } v = \text{NaN}. \end{cases}$$

Observation 7.6.15 *The successor and predecessor are inverse functions for $v \in F_\beta \cup NaN$.*

7.6.4 Floating-point encodings

A binary encoding is a mapping of a numeric or non-numeric value into a bit string. The Standard specifies encodings from the hierarchies of basic and storage floating-point values into bit strings of standard binary power word sizes from 16 to 128 bits. The standardized encodings are intended to insure that basic and storage, binary and decimal, floating-point values may be simply interchanged between conforming platforms. The Standard's extended formats are considered implementation-dependent and their encodings are not specified.

The encodings are generically partitioned into sign, exponent, and significand fields crafted primarily to efficiently encode the floating-point numbers. Figure 7.6.5 illustrates a partition according to the factorization $v = (-1)^s \beta^e d_0.d_1 \cdots d_{p-1}$.

Sign s	Exponent/combination $(e, d_0, \text{exceptions})$	Significand fractional part $(d_1, d_2, \ldots, d_{p-1})$

Figure 7.6.5. Floating-point value encoding partition.

As shown in Figure 7.6.5, for the real non-zero floating-point numbers, the sign field encodes the sign factor's sign bit s, the significand field encodes the significand factor's digit string fractional part $d_1 d_2 \cdots d_{p-1}$, and the exponent field determines an exponent and leading digit pair (e, d_0). The exponent field further serves as a control field with reserved encodings that can change the interpretation of the significand and sign fields to facilitate encoding of the exceptional values $\{-0, +0, -\infty, +\infty, \text{NaN}\}$ and to facilitate the passing of diagnostic information with the NaN value.

The exponent e is encoded as a binary integer in all decimal and binary floating-point number encodings. As shown in Figure 7.6.5, for decimal floating-point numbers the significand can be encoded as a decimal digit string, or an alternative encoding can be used where the significand is a binary integer with the significand field containing all but the four leading bits of the significand.

The following provides an overview of the encodings and summarizes some auxiliary properties obtained by the Standard's artfully crafted encodings.

Sign field. The one-bit sign field is present in all floating-point value encodings. It is recognized as part of the operand or is redundant (ignored) as follows:

- Non-zero finite reals and infinity – recognized.
- NaN – redundant.
- Zero – in the context of finite real operations the sign is redundant and the value is an exact zero; in the context of exceptional results (see Table 7.6.4) or as an exceptional operand the sign is recognized to distinguish between plus and minus zero.

The significand field and exponent field are treated quite differently for binary and decimal encodings and are best traversed separately.

Standard binary floating-point system hierarchy encoding. The binary hierarchy of floating-point bit-string encoding partitions is illustrated in Figure 7.6.6, with the interpretation of the encoding as a floating-point number or exceptional value controlled by the exponent field.

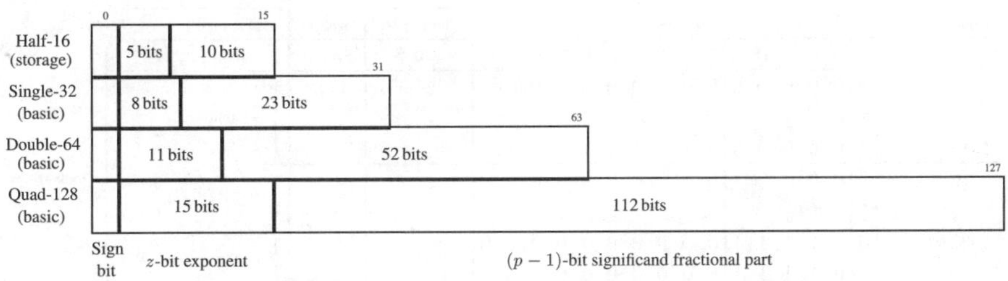

Figure 7.6.6. Binary floating-point hierarchy encodings.

z-bit binary exponent field. For each of the four exponent field widths $z \in \{5, 8, 11, 15\}$:

- All zeroes field: this denotes that the number is subnormal or zero with leading significand bit zero, and with the significand's field storing a fractional part $0.b_1b_2 \cdots b_{p-1}$ scaled by 2^{emin}.
- All ones field: this denotes the encoding of an exceptional value of $\{-\infty, +\infty, \text{NaN}\}$ to be determined jointly with the other fields.
- Not all zeroes or all ones field: this denotes the encoding of a normalized floating-point number with leading significand bit one, and having a biased exponent given by $1 \leq e + bias \leq 2^z - 2$ with bias $2^{z-1} - 1$.

$(p - 1)$-bit binary significand field. For each of the four significand field widths $p - 1 \in \{10, 23, 52, 112\}$:

- For a binary floating-point number the significand field encodes the $(p - 1)$-bit fractional part of the significand's magnitude in a one-to-one manner.
- For the exceptional values $\{-0, +0, -\infty, +\infty\}$ the significand field is all zeroes, providing a unique encoding for these values.
- For a NaN the significand field is non-zero and may be used to store and pass implementation-dependent diagnostic information, termed the "payload."

The exponent and significand fields of the Standard's binary floating-point encodings have been crafted to seamlessly exploit the native ordering and distinctness of bit strings.

Observation 7.6.16 *For* $n = 16, 32, 64, 128$ *the Standard's binary floating-point system n-bit encoding provides an order preserving one-to-one correspondence between the extended real floating-point values from* $+0$ *to* $+\infty$ *and the integers of the closed interval* $[0; 2^{n-1} - 2^{p-1}]$.

The mapping between all positive floating-point values and the integers $[0; 2^{n-1}]$ is illustrated in Figure 7.6.7.

Integer:	Binary:			Floating point values: storage-16	general	value type	
0	0	0 0 0 0 0	0 0 0 0 0 0 0 0 0 0	0	zero	zero	One-to-one encodings of positive reals
1	0	0 0 0 0 0	0 0 0 0 0 0 0 0 0 1	2^{-24}	v_{min}	Subnormals	
⋮	⋮	⋮	⋮	⋮	⋮		
1023	0	0 0 0 0 0	1 1 1 1 1 1 1 1 1 1	$2^{-14}-2^{-24}$	–		
1024	0	0 0 0 0 1	0 0 0 0 0 0 0 0 0 0	2^{-14}	$2^{e_{min}}$	Normalized FP	
⋮	⋮	⋮	⋮	⋮	⋮		
15,360	0	0 1 1 1 1	0 0 0 0 0 0 0 0 0 0	$2^0 = 1$	1		
15,361	0	0 1 1 1 1	0 0 0 0 0 0 0 0 0 1	$1+2^{-10}$	$1+2^{-(p-1)}$		
⋮	⋮	⋮	⋮	⋮	⋮		
31,743	0	1 1 1 1 0	1 1 1 1 1 1 1 1 1 1	$2^{16}-2^5$	v_{max}		
31,744	0	1 1 1 1 1	0 0 0 0 0 0 0 0 0 0	$+\infty$	$+\infty$	Plus infinity	
31,745	0	1 1 1 1 1	0 0 0 0 0 0 0 0 0 1	–	NaN	Redundant NaNs	
⋮	⋮	⋮	⋮	⋮	⋮		
32,767	0	1 1 1 1 1	1 1 1 1 1 1 1 1 1 1	–	NaN		

Figure 7.6.7. Interpretations of the Standard 16-bit storage format word.

Properties of the standard's binary floating-point encodings.

Zero detection The positive floating-point zero is the integer zero, and a signed floating-point zero is equivalent to a signed integer zero in a sign-magnitude integer system.

Comparison Comparisons over the extended reals are equivalent to integer comparisons in a sign-magnitude integer system for all integer magnitudes up to $2^{n-1} - 2^{p-1}$.

Continuity The positive extended real floating-point values correspond to integers of the interval $[0; 2^{n-1} - 2^{p-1}]$. Integers of the interval $[2^{n-1} - 2^{p-1} + 1; 2^{n-1} - 1]$ give redundant representations of a NaN.

The Standard's synergistic binary encodings provide storage efficiency in terms of the proportion of bit patterns that represent distinct real values.

Observation 7.6.17 *For* $n = 16, 32, 64, 128$ *and corresponding values of* $p = 11, 24, 53, 113$, *the Standard's binary floating-point system n-bit encodings give distinct representations to* $2^n - 2^p - 2$ *non-zero real values and four extended real values, with* $2^p - 2$ *redundant representations for a NaN.*

From Observation 7.6.17 we note that over 99.8% of the 32-bit single-precision binary floating-point system bit-string encodings uniquely represent real numbers, growing to over 99.998% of the 128-bit quad precision encodings.

Standard decimal floating-point system hierarchy encodings. The implicit goals for the Standard's encodings of decimal floating-point systems into bit strings are:

- to provide efficient access to the sign, exponent, and significand components of the decimal floating-point factorization;
- to support efficient storage density of decimal values in bit-strings;
- to provide a hierarchy of precision and range levels comparable to the binary floating-point system levels for the same format sizes.

Note that the second goal effectively excludes traditional binary coded decimal (BCD) encoding of a decimal digit-string for the integer significand, since using four bits per decimal digit results in three decimal digits needing twelve bits where ten are sufficient. Thus alternative significand encodings are introduced to achieve more compactness, and more complex interpretation of the controlling exponent/combination field is required for the encoding/decoding mapping which is not a one-to-one correspondence.

The Standard's decimal floating-point encodings are best understood with reference to the decimal floating-point integer significand factorization

$$v = (-1)^s 10^{(\text{integer exponent})} \times (\text{integer significand}).$$

The decimal hierarchy is limited to three levels corresponding to a decimal 32-bit storage format and decimal 64- and 128-bit basic formats with the field partitions given in Figure 7.6.8.

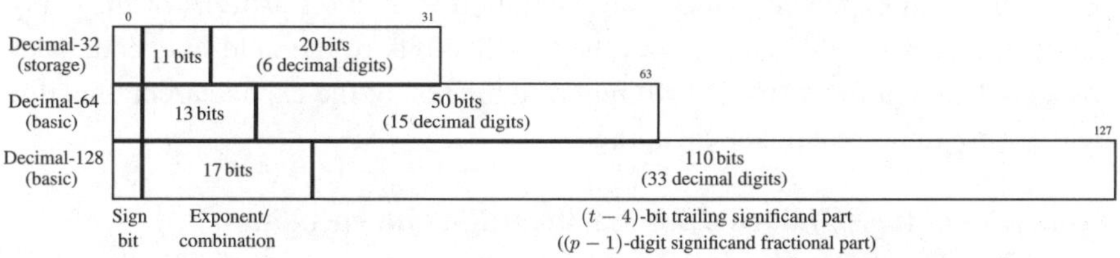

Figure 7.6.8. Decimal floating-point hierarchy encodings.

Two encodings are allowed based on whether the significand is interpreted as a "native" p-digit decimal string or as a binary integer t-bit string where $t = 10(p-1)/3 + 4$ and the range is limited to $[0; 10^p - 1]$. The choice of decimal or binary integer significand format affects the encoding of both the trailing significand field and the exponent/combination field.

p-digit decimal integer significand format, ($p \in \{7, 16, 34\}$). The trailing significand field is subdivided into $(p-1)/3$ 10-bit fields termed *declets*, each declet storing an integer with the range $[0; 999]$ in the densely packed decimal (DPD) encoding. The Standard further specifies mappings between a ten-bit declet and a twelve-bit field hosting three decimal digits, each encoded in BCD in a four-bit field, thus yielding the $(p-1)$-digit least-significant digits $d_1 d_2 \cdots d_{p-1}$. This declet decimal integer encoding is a dense packing of a decimal floating-point value, for interchange between conforming units or platforms with the $4p$-bit decimal string digit encoding available for designing operations within an implementation such as a decimal hardware oriented ALU.

The z-bit *exponent/combination* field with $z \in \{11, 13, 17\}$ jointly encodes:

- the leading decimal significand digit with a range of ten distinct digit values;
- the biased integer exponent with a range of $3 \times 2^{z-5}$ distinct integer values;
- the exceptional values NaN and infinity, with $+\infty$, $-\infty$ determined by the sign field.

t-bit binary integer significand decimal floating-point format. The trailing significand field contains the low-order $t-4$ bits of a t-bit integer significand with $t \in \{24, 54, 114\}$.

The exponent/combination field encodes the ten distinct leading four-bit values 0000, 0001, ..., 1001 as the high four bits of the t-bit integer significand. The range of accepted values of the integer significand is limited to $[0; 10^p - 1]$, noting that $p = \lfloor \log_{10} 2^t \rfloor \in \{7, 16, 34\}$.

The exponent/combination field also encodes $3 \times 2^{z-5}$ biased integer exponent values and exceptional value encodings for NaN and infinity as in the decimal integer significand format.

Note that the encodings of the exponent/combination field for both the decimal and binary integer significand formats interleave the significand leading part values and exponent values to consume $30 \times 2^{z-5}$ bit patterns of the exponent/combination field. This allows the first five bits of the field to encode NaN (by 11111) and infinity (by 11110) ignoring the rest of the exponent/combination field as providing redundant encodings.

Properties of the standard's decimal floating-point encodings.

- The choice of a decimal integer significand format can facilitate a decimal hardware-based ALU implementation. The choice of a binary integer significand format can facilitate a software implementation of decimal floating-point arithmetic.
- The integer significands are not required to be normalized. This introduces some redundancy that can facilitate embedded small integer arithmetic, counting, and small step incrementation without the need for intermediate normalizations or changing the exponent/combination field.

The challenging goal of the decimal floating-point encodings into n-bit strings is to obtain efficient utilization of the n-bit format's capacity to host up to 2^n distinct values, while maintaining straightforward access to the exponent and significand terms, providing a hierarchy of encodings closely resembling the precision and range hierarchy of the Standard's pre-existing binary floating-point n-bit formats.

Table 7.6.7. *Binary versus decimal floating-point hierarchy comparisons of representation densities, exponent and precision ranges*

		n-bit format level					
		32		64		128	
		Binary	Decimal	Binary	Decimal	Binary	Decimal
Real number density		99.8%	80.5%	99.98%	75%	99.99%	65%
Effective exponent	$e_{max}+1$	128	322.2	1024	1278.9	16384	20413
range in bits	e_{min}	−126	−315.6	−1022	1272.3	−16382	−20407
Precision wobble	max	24	23.25	53	53.15	113	112.95
in bits	min	23	19.92	52	49.83	112	109.62

Table 7.6.7 compares the performances of the binary and decimal encodings over the 32-, 64-, and 128-bit format levels employing measures of comparative density, range, and precision:

- Density: the density measure is given by the ratio of the number of distinct reals represented compared to the capacity 2^n. The careful crafting of the exponent/combination field encoding has resulted in about 80.5% of the 32-bit word decimal floating-point system bitstrings representing distinct real numbers, with the distinct reals for the 128-bit words at about 65.0%. This amounts to less than one bit of representation capacity lost even for the 128-bit case.
- Range: the range measure is given by the interval of effective exponent range for normalized floating-point numbers. The maximum is $e_{max}+1$ using a maximum significand to obtain the upper exponent bound with minimum exponent bound e_{min}. The decimal exponents are given as equivalent binary exponents for easy comparison. Note that the decimal systems have somewhat more exponent range at all levels.
- Precision: the precision measure in bits is determined by the range of the radix 2 logarithms of the relative gap function. This varies by one bit in binary but by 3.32 bits in decimal as shown in Table 7.6.7. The precision wobble over the range was illustrated in the relative gap plot of Figure 7.6.3. Although the maximum effective precisions are comparable, the decimal system minimums are 2–3 bits less than the binary system minmums at each level.

In summary, the binary and decimal hierarchies are reasonably comparable at all three format levels, 32-, 64-, and 128-bits. The decimal systems have slightly more range and 2–3 bits less effective precision. This feature is implicitly recognized in that the 32-bit format in decimal is the smallest of the hierarchy and is exclusively a storage format. In binary, the 32-bit format is a basic format where the Standard requires implementation of arithmetic at this level.

7.7 Notes on the literature

Arithmetic in the first fully functional digital computer, the Z-3 built by Konrad Zuse in 1941 using electro-mechanical relays, employed a binary floating-point number representation. However, the work by Zuse was not known outside a few circles in Germany until much later. And as mentioned in the introduction of the chapter, [BGvN46] foresaw the need for a number representation to have "a floating binary (or decimal) point," but dismissed its hardware implementation by referring to "programming it out of operations built into the computer." Many early computers relied on such software implementations built on integer arithmetic, but soon computers designed to support "scientific" computations were supplied with hardware implementations of floating-point operations, so that by 1960 it was quite common. Unfortunately, manufacturers each chose their own way of representing floating-point values, e.g., how the number of bits in a word were allocated to exponent and significand parts, and the way of representing the signs of the significand and exponent. The radix of the representation could be chosen to be higher values like 8 or 16, to allow for a greater range of exponent values for a given allocated number of bits. But this was achieved with the unfortunate side effect that, e.g., normalization in the IBM 360/370 series of computers, using hexadecimal exponents together with a (binary) significand part, could result in up to three leading zeroes and thus the corresponding loss of up to three least-significant bits. There were a few initial efforts to support arithmetic on decimal floating-point representations, but only very lately have these again become of interest, due to the revision [IEE08] of the IEEE standard.

An example of the difficulty in determining equivalent digits for binary–decimal floating-point base conversion was noted in [Gol67]. The formalization of floating-point base conversion of Section 7.2 is based on the work by Matula in [Mat68a, Mat68b, Mat70].

Floating-point addition, being the algorithmically most complicated of the operations, has been the subject of many investigations and numerous patents, see, e.g., [Swe65, VLP89, QF91]. In [Far81] Farmwald suggested splitting the floating-point adder in two paths, depending on the exponent difference, an approach that has been widely followed since [BSBLL99, NMLE00, SE01] and used in many patents.

The proposal of a floating-point fused multiply-add instruction has similarly created many suggestions for its implementation, e.g., [HMC90, BM93, MHR90, LB04, LB05, SST05, QSL08], and this attention has been further emphasized by being requested in the revised standard [IEE08]. In particular, [LB04] by Lang and Bruguera provides a very detailed explanation of a double-precision fused multiply/add implementation, which is further developed in [LB05] into a dual path design of reduced latency. Traditionally when fused multiply/add units were provided, they were also used either as the floating-point adder or as the multiplier by choosing suitable input values, thus excluding the parallel execution of an addition and a multiplication. However, in [QSL08] Quinnell, Swartzlander and Lemonds proposed adding a "bridge" unit, allowing the reuse of existing (and independent) FADD and FMUL components.

The handling of denormalized floating-point numbers is often dealt with in software, however [SST05] by Schwartz, Schmookler, and Tang describes a hardware supported implementation.

In the past there have been several widely used machines implementing floating-point arithmetic that did not result in precise roundings. For example, some have not even implemented such roundings for addition, which is very trivial to do and at little cost. Some supercomputers have sacrificed precise rounding for speed when implementing the divide instruction.

The *double rounding* problem is discussed in [Lee89], where it is shown that a two-bit tag is sufficient to carry rounding information forward from one rounding to a subsequent rounding into a lower precision, in order to avoid the result of the two consecutive roundings being different from that obtained if only a single rounding was performed.

The problem of rounding in general is discussed in [KPS77, Yoh73, QTF91], and plenty of published sources deal with rounding in connection with specific operations, e.g., multiplication in [Ura68, SBH89, ES00], division/square root in [ELM93, LM95, OF96, IM99], and in particular when digits are developed sequentially in [EL89], where the on-the-fly conversion algorithm is proposed for rounding.

Normalization of the significand was from an early stage considered the standard representation. However, it was suggested in [AM58] to represent values in unnormalized form. The idea was that represented digits should reflect "trustworthy information," such that, e.g., input data of low precision, and the possible loss of significant digits in subtraction with cancellation, carried through the calculations thereby could be signaled to the user. This so-called *significance arithmetic* was implemented in the Maniac III computer [Ash62].

Note that while some integer and fixed-point operations deliver the exact "double-length" result and integer division yields an exact "double-length" quotient, remainder pair, this is not the case for typical floating-point operations. In [BWKM91] it was argued that it is possible to implement an "exact" floating-point arithmetic, leaving it up to the user to discard the "less-significant" part of a two-word result if not needed. However, if needed, employing only standard floating-point arithmetic, it is possible to provide the exact sum of two floating-point numbers x and y in the form of $x + y = h + l$ where $h = RN_e(x + y)$ and l are in the same floating-point format as the operands x, y, as discussed by Knuth [Knu98] and Møller [Møl65]. If nothing is known on the signs of x and y, six floating-point add/subtract instructions are needed to determine h and l. Similarly, for multiplication it is possible to deliver h and l such that $x \times y = h + l$. For division, the result may be delivered as a quotient and remainder, and similarly for square root.

With the advent of the first single-chip microprocessors, it was anticipated that the increase in transistor density would soon permit hardware on-chip support of floating-point arithmetic. This led to the formation of the first IEEE committee with representation from industry and academia, with the aim of establishing a standard for the format of floating-point representations and requirements on the supported arithmetic operations.

The overriding goal in the effort to provide the standard was to satisfy the desire of end users to have numeric software that runs on any conforming platform, where it could be reliably expected to provide identical results as specified by the standard. The first draft of the "Proposed Standard for Binary Floating-point Arithmetic" was issued in 1982 and in just a few years became the ANSI standard ANSI/IEEE Std 754-1985 [IEE85], also known as IEC 60559:1989. A companion "A Radix- and Word-length-independent Standard for Floating-point Arithmetic" [IEE87] was issued in 1987. The latter provided some insight into the logic governing the various parameter choices. A proposed major extension of the Standard was issued in 2006 [IEE06] and approved as a new IEEE standard in 2008 [IEE08] and its features are incorporated in our discussion here. Detailed specifications are available in the documents.

For ease of developing robust mathematical software there was considerable support for the range width in significands being quite large, e.g., several orders of magnitude. On the other hand in the late 1970s when the trade-offs between precision and range were debated by the standards committee, it was commonly agreed that single precision p must be as large as possible to avoid excessive accumulated round-off error in typical scientific computations of that era. The resulting agreement was to provide precision rather than range enhancement for the single-precision format and then to provide a more balanced enhancement of both precision and range for the basic double format to accommodate critical large-scale scientific computations. To accommodate the mathematical software community it was strongly recommended that the highest basic format implemented, be it single or double, be enhanced by a companion "extended" format including considerably greater range with just limited additional precision. The double extended format with $p = 64$ encoded in an 80-bit (10 byte) word was incorporated in the popular $\times 86$ microprocessor family. This resulted in the double extended format becoming a de facto standard for hosting the body of existing software, with a range-width over 500 and a fixed-point granularity of at least about one part in 10^{4950} around zero, preserved in later generations as a legacy *de-facto* requirement. The specification of a single extended format, as having minimum bounds on precision and range-width, allowed 32-bit word architectures implementing only basic single and basic double formats to be compliant, since basic double could then be considered a single extended level.

The half-precision storage level and the basic quad level were added in the revised standard [IEE08]. The storage level is relevant to the large data sets of multimedia applications where results to 10 bits (i.e., three decimal digit equivalent) are sufficient and compact storage is desirable. Furthermore, for architectures with 128-bit words, the basic quad word can alternatively store 2 doubles, 4 singles, or 8 half-level floating-point values.

The specification of a hierarchy of decimal floating-point number systems was added by the 754-2008 standard with parameters given in Table 7.6.2. Note that the three specified formats used for computation (non-storage) have decimal precisions of at least 16 digits ($\sim$50 bits) and range-widths of essentially several orders of magnitude. The decimal floating-point systems are largely specified for the application needs of the financial accounting community where legacy pre-computer decimal-based accounting systems were not readily transferred to binary floating-point systems. Due to the more recent specifications for decimal formats there is a renewed interest in the implementation of the basic arithmetic operations performed on decimal encodings, e.g., multiplication employing decimal carry-save addition [VAM07, EHS09].

The single precision binary floating-point numbers $F(2, 24, 127, -126)$ contain only about 2^{32} or 4 billion distinct real values. For microprocessors operating at a gigaflop, it takes just four seconds to perform a floating-point operation on every single precision binary number. For machines operating at the teraflop level of about 2^{40} floating-point operations per second, it takes only an hour to perform a single cycle floating-point operation on every double precision binary floating-point number from the normalized interval $[1; 2)$, with a comparable bound for the basic 64 normalized decimal floating-point numbers.

References

[AM58] R. L. Ashenhurst and N. Metropolis. *Unnormalized Floating Point Arithmetic*. Technical report, Institute for Computer Research, University of Chicago, Chicago, Illinois, 1958.

[Ash62] R. L. Ashenhurst. The Maniac III Arithmetic System. In *Proc. 1962, Spring Joint Computer Conference*, pages 195–202. AFIPS, May 1962.

[BGvN46] A. Burks, H. H. Goldstine, and J. von Neumann. *Preliminary Discussion of the Logic Design of an Electronic Computing Instrument*. Technical report, Institute for Advanced Study, Princeton, 1946. Reprinted in C. G. Bell, *Computer Structures, Readings and Examples*, McGraw-Hill, 1971.

[BM93] W. S. Briggs and D. W. Matula. A 17×69-bit multiply and add unit with redundant binary feed-back and single cycle latency. In *Proc. 11th IEEE Symposium on Computer Arithmetic*, pages 163–170. IEEE Computer Society, 1993.

[BSBLL99] A. Beaumont-Schmith, N. Burgess, S. Lefrere, and C. C. Lim. Reduced latency IEEE floating-point standard adder architectures. In *Proc. 14th IEEE Symposium on Computer Arithmetic*, pages 35–42. IEEE Computer Society, 1999.

[BWKM91] G. Bohlender, W. Walter, P. Kornerup, and D. W. Matula. Semantics for exact floating point operations. In *Proc. 10th IEEE Symposium on Computer Arithmetic*. IEEE Computer Society.

[EHS09] M. Erle, B. J. Hickmann, and M. J. Schulte. Decimal floating-point multiplication. *IEEE Trans. Computers*, 58(7):902–916, July 2009.

[EL89] M. D. Ercegovac and T. Lang. On-the-fly rounding for division and square root. In *Proc. 9th IEEE Symposium on Computer Arithmetic*, pages 169–173. IEEE Computer Society, 1989.

[ELM93] M. D. Ercegovac, T. Lang, and P. Montuschi. Very high radix division with selection by rounding and prescaling. In *Proc. 11th IEEE Symposium on Computer Arithmetic*, pages 112–119. IEEE Computer Society, 1993.

[ES00] G. Even and P.-M. Seidel. A comparison of three rounding algorithms for IEEE floating-point multiplication. *IEEE Trans. Computers*, 49(7):638–650, July 2000.

[Far81] P. M. Farmwald. On the design of high performance digital arithmetic units. Ph.D. thesis, Stanford University, 1981.

[Gol67] I. B. Goldberg. 27 bits is not enough for 8-digit accuracy. *CACM*, 10(2):105–106, February 1967.

[HMC90] E. Hokenek, R. K. Montoye, and P. W. Cook. Second-generation RISC floating point with multiply-add fused. *IEEE J. Sol. State Circuits*, 25(5):1207–1213, 1990.

[IEE85] IEEE. *Std 754-1985 IEEE Standard for Binary Floating-Point Arithmetic*. Standards Committee of The IEEE Computer Society. 345 East 47th Street, New York, NY 10017, USA, 1985.

[IEE87] IEEE. *Std 854-1987 IEEE Standard for Radix-Independent Floating-Point Arithmetic*. Standards Committee of The IEEE Computer Society. 345 East 47th Street, New York, NY 10017, USA, 1987.

[IEE06] IEEE. *DRAFT Standard for Floating-Point Arithmtic P754*. IEEE Standards Activities Department. 445 Hoes Lane, Piscataway, NJ 08854, USA, 2006.

[IEE08] IEEE. *IEEE Std. 754^TM-2008 Standard for Floating-Point Arithmtic*. IEEE, 3 Park Avenue, NY 10016-5997, USA, August 2008.

[IM99] C. Iordache and D. W. Matula. On infinitely precise rounding for division, square root, reciprocal and square root reciprocal. In *Proc. 14th IEEE Symposium on Computer Arithmetic*, pages 233–240. IEEE Computer Society, April 1999.

[Knu98] D. E. Knuth. *Seminumerical Algorithms*, volume 2 of *The Art of Computer Programming*. Addison Wesley, first edition 1969, second edition 1981, third edition, 1998.

[KPS77] D. J. Kuck, D. S. Parker, and A. H. Sameh. Analysis of rounding methods in floating-point arithmetic. *IEEE Trans. Computers*, C-26:643–650, 1977.

[LB04] T. Lang and J. D. Bruguera. Floating-point multiply-add-fused with reduced latency. *IEEE Trans. Computers*, 53(8):988–1003, 2004.

[LB05] T. Lang and J. D. Bruguera. Floating-point fused multiply-add: reduced latency for floating point addition. In *Proc. 17th IEEE Symposium on Computer Arithmetic*, pages 42–51. IEEE Computer Society, 2005.

[Lee89] C. Lee. Multistep gradual rounding. *IEEE Trans. Computers*, 38(4):595–600, April 1989.

[LM95] C. N. Lyu and D. W. Matula. Redundant binary Booth encoding. In *Proc. 12th IEEE Symposium on Computer Arithmetic*. IEEE Computer Society, 1995.

[Mat68a] D. W. Matula. In-and-out conversions. *CACM*, 11(1):47–50, January 1968.

[Mat68b] D. W. Matula. The base conversion theorem. *Proc. Amer. Math. Soc.*, 19(3):716–723, January 1968.

[Mat70] D. W. Matula. A formalization of floating-point numeric base conversion. *IEEE Trans. Computers*, C-19(8):681–692, August 1970.

[MHR90] R. K. Montoye, E. Hokenek, and S. L. Runyon. Design of the IBM RISC System/6000 Floating-Point Execution Unit. *IBM J. Res. Develop.*, 34(1):59–67, January 1990.

[Møl65] O. Møller. Quasi double-precision in floating-point addition. *BIT*, 5(1):37–50, 1965.

[NMLE00] A. Munk Nielsen, D. W. Matula, C. N. Lyu, and G. Even. An IEEE compliant floating-point adder that conforms with the pipelined packet-forwarding paradigm. *IEEE Trans. Computers*, 49(1):33–47, January 2000.

[OF96] S. F. Oberman and M. J. Flynn. *Fast IEEE Rounding for Division by Functional Iteration.* Technical Report CSL-TR-96-700, Stanford University, July 1996.

[QF91] N. Quach and M. J. Flynn. *Design and Implementation of the SNAP Floating Point Adder.* Technical Report CSL-TR-91-501, Stanford University, December 1991.

[QSL08] E. Quinnell, E. E. Swartzlander, and C. Lemonds. Bridge floating-point fused multiply-add design. *IEEE Trans. VLSI Systems*, 16(12):1726–30, December 2008.

[QTF91] N. Quach, N. Takagi, and M. J. Flynn. *On Fast IEEE Rounding*. Technical Report CSL-TR-91-459, Stanford University, Jan. 1991.

[SBH89] M. R. Santoro, G. Bewick, and M. A. Horowitz. Rounding algorithms for IEEE multipliers. In *Proc. 9th IEEE Symposium on Computer Arithmetic*, pages 176–183. IEEE Computer Society, 1989.

[SE01] P.-M. Seidel and G. Even. On the design of fast IEEE floating point adders. In *Proc. 15th IEEE Symposium on Computer Arithmetic*, pages 184–194. IEEE Computer Society, 2001.

[SST05] E. M. Schwartz, M. Schmookler, and S. D. Trong. FPU implementations with denormalized numbers. *IEEE Trans. Computers*, 54(7):825–836, July 2005.

[Swe65] D. W. Sweeney. An analysis of floating-point addition. *IBM Systems J.*, 4:31–42, 1965.

[Ura68] M. Urabe. Roundoff error distribution in fixed-point multiplication and a remark about the rounding rule. *SIAM J. Num. Anal.*, 5(2):202–210, June 1968.

[VAM07] A. Vazquez, E. Antelo, and P. Montuschi. A new family of high performance parallel decimal multipliers. In *Proc. 18th IEEE Symposium on Computer Arithmetic*, pages 195–204. IEEE Computer Society, 2007.

[VLP89] S. Vassiliadis, D. S. Lemon, and M. Putrino. S/370 sign-magnitude floating-point adder. *IEEE J. Sol.-State Circuits*, 24:1062–1070, 1989.

[Yoh73] J. M. Yohe. Roundings in floating-point arithmetic. *IEEE Trans. Computers*, C-22(6):577–586, June 1973.

8

Modular arithmetic and residue number systems

8.1 Introduction

In many applications integer computations are to be performed modulo some given constant. One such area is cryptology, where often multiplications, inversions, and exponentiations are to be performed modulo some very large integer. Hence we shall here investigate algorithms for such operations in their own right, but also because these can be used as primitives for the implementation of *multiple modulus systems*, also denoted *residue number systems* and abbreviated to *RNSs*. Here an integer is represented by a set of residues (the values of the integer modulo a set of given integer moduli, often chosen to be prime numbers).

In such systems computations in a large integer domain can be performed truly in parallel on completely independent processors (often called "channels"), one for each modulus from the set of moduli, and thus operating in a much smaller domain. Due to this independence, additions can be performed "carry-free" in the sense that there is no interaction between the computations of the channels, each of which is operating on integers from a smaller domain. The same applies to multiplication, and as we have pointed out in Chapter 3, such arithmetic is one way to minimize the h_j-separable sets, and thus to decrease the computation time by exploiting parallelism. Addition, subtraction, and multiplication in particular can thus be efficiently implemented, whereas other operations like division and comparisons are much more difficult. The problem with the latter is that these require information about the magnitude of the number represented, information which is distributed in all the residues of the representation. Hence these operations need implicitly to perform a mapping into the value in some kind of weighted number representation, e.g., as can be realized by the Chinese Remainder Theorem. Since mapping between an RNS and an ordinary weighted representation is quite complex, it is only feasible if a fairly extensive computation is to take place so that its cost can be amortized over many operations.

Arithmetic in RNSs has in particular been promoted for applications in DSP, where a dominant part of the computations consists of inner product calculations (multiply-add), used for filtering and other transformations of signals. While DSP applications are usually performed in fairly low precision, and thus only a few moduli are needed, RNS arithmetic can also be used for cryptographic algorithms on very large numbers, whose representation requires a large set of moduli.

Due to the independence of the computations in the different channels in an RNS, it is possible to add redundant channels for the purpose of performing error checking and possibly error correction. However, we shall not pursue this topic here, but refer the reader to the literature on fault tolerant computing.

Another application of a single modulus system is to perform arithmetic in the domain of rational numbers. Fractions from a set with bounded numerators and denominators can be mapped uniquely to and from a bounded integer domain, and arithmetic can be performed on single integer values, instead of operating on the numerator and denominator values of the fractions according to the usual rules for rational arithmetic. Provided that input and output rational values are representable in the domain of the system, the system allows exact computations in the rational domain. Note that division is possible and simple here, being realized as multiplication by the inverse.

In the domain of rationals it is also possible to apply multiple modulus systems, and thus perform arithmetic in parallel on many independent processors, but again comparison and conversion in and out of the representation remains costly.

Finally for completeness we also briefly cover the so-called Hensel codes, based on truncated p-adic number representations, which turn out to be equivalent to the above-mentioned single modulus rational system, based on a modulus of the form p^k for some prime p. Thus they also allow simplified exact computations in a bounded rational domain. However, it turns out that there are certain problems with arithmetic operations in this domain. Due to the above-mentioned equivalence there is no real advantage in using this system.

8.2 Single-modulus integer systems and arithmetic

In their own right, but also as a foundation for the later discussion of multiple-modulus systems we shall first investigate the simpler types of systems based on a single modulus. Formally, for any multiple-modulus system there is an equivalent single-modulus system. As we shall see later, the choice of moduli has a significant impact on the properties of the system as well as on the implementation of its arithmetic.

A single-modulus integer system is defined by a given number $m > 1$, the *system modulus*, together with a complete residue system of integers modulo m.

Traditionally two such complete residue systems are used:

$$\mathbb{Z}_m = \{i \in \mathbb{Z} \mid 0 \le i < m\} \qquad \textit{(the standard system)}$$

and for m odd:

$$\mathbb{S}_m = \left\{ i \in \mathbb{Z} \mid \frac{-m+1}{2} \le i \le \frac{m-1}{2} \right\} \qquad \textit{(the symmetric system).}$$

In *single-modulus integer arithmetic* we map each integer $a \in \mathbb{Z}$ into the chosen system, and perform arithmetic on operands from the system by mapping the results of arithmetic operations back into the system. The mapping employed for the standard system is defined as follows.

Definition 8.2.1 *The mapping* $|\cdot|_m : \mathbb{Z} \to \mathbb{Z}_m$ *is defined by*

$$|a|_m = r \in \mathbb{Z}_m \text{ for } a \in \mathbb{Z}$$

if and only if $a \equiv r \pmod{m}$.

The value of $|a|_m$ is called the *residue* of a modulo m. The mapping for the symmetric system is defined equivalently, defining *symmetric residues* modulo m:

Definition 8.2.2 *The mapping* $\langle\cdot\rangle_m : \mathbb{Z} \to \mathbb{S}_m$ *is defined by*

$$\langle a\rangle_m = r \in \mathbb{S}_m \text{ for } a \in \mathbb{Z}$$

if and only if $a \equiv r \pmod{m}$.

The symmetric system carries information on the sign of a number in a natural way. Hence if the data for a computation as well as its final results are known to be in $\mathbb{S}_m$, then the signs of the results are immediately available. Of course, this requires that m is chosen sufficiently large for the computation performed.

For the standard system it is also possible to associate sign information by interpreting numbers in the range $[0; (m-1)/2]$ as positive, and in $[(m+1)/2; m-1]$ as negative, assuming m is odd. The standard system is more easily dealt with, so we shall not pursue the symmetric system further since it is equivalent to the standard system.

The following results are well known from algebra:

Theorem 8.2.3 *Let* $a, b, m \in \mathbb{Z}$, $m > 1$. *Then*

$$|a+b|_m = ||a|_m + |b|_m|_m = ||a|_m + b|_m = |a + |b|_m|_m$$

and

$$|ab|_m = ||a|_m |b|_m|_m = ||a|_m b|_m = |a|b|_m|_m .$$

The system $(\mathbb{Z}_m, \oplus_m, \otimes_m)$, where $\oplus_m$ and $\otimes_m$ denote integer addition and multiplication modulo m respectively, is a finite commutative ring with identity.

Note that we can then also define subtraction in the obvious way since $m - a$ acts as an *additive inverse* modulo m. The existence of a *multiplicative inverse* is not always guaranteed as also known from algebra.

Theorem 8.2.4 *Let $a \in \mathbb{Z}$. Then there exists a unique integer $b \in \mathbb{Z}$ such that*

$$|ab|_m = 1$$

if and only if $|a|_m \neq 0$ and $\gcd(a, m) = 1$.

We shall denote the multiplicative inverse of a modulo m by $|a^{-1}|_m$, or just a^{-1} if the modulus is understood from the context. If m is a prime, then any a such that $|a|_m \neq 0$ has a multiplicative inverse, and hence $(\mathbb{Z}_m, \oplus_m, \otimes_m)$ constitutes a finite field isomorphic to the Galois field $GF(m)$.

It is essential here to notice that the existence of a multiplicative inverse does not necessarily imply that it is possible to perform division in the ordinary sense in the domain of integers, where for the division of a by b we want an integer quotient q and remainder r such that $a = bq + r$.

Example 8.2.1 With $m = 7$ we find that the multiplicative inverse of $a = 2$ is $a^{-1} = 4$ since $|2 \times 4|_7 = |8|_7 = 1$. Then $|6 \times 2^{-1}|_7 = |6 \times 4|_7 = 3$, which is what we expect since 6 div 2 = 3. But $|3 \times 2^{-1}|_7 = |3 \times 4|_7 = 5$, so since 2 does not divide 3 we cannot expect to get an integral result. However, 5 behaves algebraically as $\frac{3}{2}$ when considered over the field of rationals modulo 7, since $|5 + 5|_7 = |10|_7 = 3$ and $|5 \times 4|_7 = 6$ corresponding to $\frac{3}{2} + \frac{3}{2} = 3$ and $\frac{3}{2} \times 4 = 6$. □

We shall return to residue systems over the rationals in Section 8.6, so here we will restrict considerations to situations where we interpret the system $(\mathbb{Z}_m, \oplus_m, \otimes_m)$ as a representation of the integers $\mathbb{Z}$. Each member a of $\mathbb{Z}_m$ is then a *representation* of any member of the residue class

$$C_a^m = \{i \in \mathbb{Z} \mid |i|_m = a\}, \ a \in \mathbb{Z}_m,$$

which is precisely what is wanted in certain computations (e.g., in cryptology). In other situations $(\mathbb{Z}_m, \oplus_m, \otimes_m)$ is used for computations in the integer domain where m is chosen sufficiently large that it is guaranteed that the (integral) results belong to $\mathbb{Z}_m$. Then a computation can be performed in a finite domain without any rounding errors occurring. Note that intermediate values of the actual computation may "overflow" the system with no effect on the final result, if this is guaranteed to be in $\mathbb{Z}_m$.

Integer division producing a quotient and a remainder only makes sense in modular arithmetic if the dividend and the divisor both have their correct value: if either or both of these are substituted by arbitrary other members of their respective residue classes, the results will be rather unpredictable. We shall thus leave the problem of integer division for now, but return to it in Section 8.5 when dealing with multiple modulus systems usually applied to situations where the "dynamic range" of the system is chosen such that the computation stays within the bounds of the system.

8.2.1 Determining the residue $|a|_m$

By Definition 8.2.1, the value of $|a|_m$ can be found by normal integer division with remainder; however, in general this is too costly and faster methods exist. As usual we assume that a is given in some radix representation, so in particular if m is a power of the radix, the value of $|a|_m$ is trivially found by truncation of the radix representation of a. Note that this also holds for β-complement representations: e.g., let $a = -2^n + a'$ with $0 \le a' < 2^n$ be an n-bit 2's complement negative number, then $|a|_{2^k} = \left||-2^n|_{2^k} + |a'|_{2^k}\right|_{2^k} = |a'|_{2^k}$ for $k < n$.

In some situations it is preferable that the system modulus m is a prime, e.g., if it is required that there is a multiplicative inverse of any non-zero member of $\mathbb{Z}_m$. In cases where any prime of a certain order of magnitude can be used, a good choice may be a number of the form $2^n - 1$ or $2^n + 1$, some of which are prime (e.g., $2^3 - 1 = 7$ or $2^{61} - 1$, which are both examples of *Mersenne primes*). For such moduli there is a simple way to reduce modulo m without implied division by m, and instead use division by 2^n:

Lemma 8.2.5

$$|a|_{m-1} = a \bmod (m-1) = ((a \bmod m) + (a \operatorname{div} m)) \bmod (m-1) \qquad (8.2.1)$$

and[1]

$$|a|_{m+1} = a \bmod (m+1) = ((a \bmod m) - (a \operatorname{div} m)) \bmod (m+1). \qquad (8.2.2)$$

Proof Equation (8.2.1) follows from

$$\begin{aligned} a &= (a \bmod m) + m \cdot (a \operatorname{div} m) \\ &= (a \bmod m) + (a \operatorname{div} m) + (m-1)(a \operatorname{div} m) \end{aligned}$$

and (8.2.2) follows similarly. □

When the modulus m can be factored, reduction can be performed in two steps.

Theorem 8.2.6 *Let $m = pq$ be a composite integer with $p > 1$ and $q > 1$, then*

$$|a|_m = \left|\left\lfloor \frac{a}{p} \right\rfloor\right|_q p + |a|_p.$$

Proof Rewriting a, there exist integers f', f'' such that

$$a = f'p + |a|_p = (f''q + |f'|_q)p + |a|_p = f''\,pq + \left|\left\lfloor \frac{a}{p} \right\rfloor\right|_q p + |a|_p,$$

[1] Note that here mod and div are used as binary operators.

hence $|a|_{pq} \equiv ||\lfloor a/p \rfloor|_q p + |a|_p \pmod{pq}$, and $0 \le ||\lfloor a/p \rfloor|_q p + |a|_p \le (q-1)p + p - 1 = pq - 1$, which proves the theorem. □

This result is particularly useful if say p is a power of the radix, since division and multiplication by p then becomes trivial.

For certain values of the modulus m, there is the possibility of simplifying the reduction modulo m, when operating in binary arithmetic.

Using periodicity properties of $|2^i|_m$. We shall start with some useful definitions.

Definition 8.2.7 *The* period $P(m)$ *of the odd modulus m is the minimal distance between two* 1's *in the sequence of residues $|2^i|_m$, i.e., $P(m) = \min\{i \mid i > 0$ and $|2^i|_m = 1\}$.*

Obviously $P(m)$ always exists, and $0 < P(m) \le m - 1$. Also

$$\left|2^{t \cdot P(m)+i}\right|_m = |2^i|_m \quad \text{for any integers } t, i \ge 0. \tag{8.2.3}$$

Lemma 8.2.8 *If $P(m)$ is the period of m, then for any integer a, $|a|_m = \left||a|_{2^{P(m)}-1}\right|_m$.*

Proof By definition $|2^{P(m)}|_m = 1$, hence

$$\begin{aligned} |a|_m &= \left|\left\lfloor \frac{a}{2^{P(m)}-1} \right\rfloor \left(2^{P(m)} - 1\right) + |a|_{2^{P(m)}-1}\right|_m \\ &= \left||a|_{2^{P(m)}-1}\right|_m . \end{aligned}$$ □

Definition 8.2.9 *The* half-period $HP(m)$ *of the odd modulus m is the minimal distance between a pair of 1 and $m-1$ in the sequence of residues $|2^i|_m$, $i = 0, 1, \ldots$, provided that $m-1$ is in the sequence.*

Note that $m - 1 \equiv -1 \pmod{m}$. When $HP(m)$ exists, then $P(m) = 2HP(m)$. Similarly to the above it can be shown that

$$\left|2^{t \cdot HP(m)+i}\right|_m = (-1)^t \, |2^i|_m \quad \text{for any integer } t \ge 0, \tag{8.2.4}$$

and that the following lemma holds.

Lemma 8.2.10 *If the half-period $HP(m)$ exists, then for any integer a, $|a|_m = \left||a|_{2^{HP(m)}+1}\right|_m$.*

Finally, we may note the following:

- $HP(2^p - 1)$ does not exist, but $P(2^p - 1) = p$;
- $HP(2^p + 1) = p$ and $P(2^p + 1) = 2p$;
- if $HP(m)$ exists, then $\min(P(m), HP(m)) \le (m-1)/2$.

Example 8.2.2 For $m = 7$ we have the following table:

i	0	1	2	3	4	5	6
$\lvert 2^i \rvert_7$	1	2	4	1	2	4	1

so $P(7) = 3$ and $HP(7)$ does not exist. For $m = 17$ we find

i	0	1	2	3	4	5	6	7	8	9	10	11	12	13	14	15	16
$\lvert 2^i \rvert_{17}$	1	2	4	8	16	15	13	9	1	2	4	8	16	15	13	9	1

showing that $P(17) = 8$ and $HP(17) = 4$, satisfying $P(17) = 2HP(17)$. □

Provided that m is odd, and the period ($P(m)$ or $HP(m)$) is small, then a simplification is possible by first reducing modulo a power of 2. This is, of course, particularly useful when the operand a is very large compared with m. In Section 8.4 we shall show an alternative way of exploiting the periodicity, in connection with the parallel modular reductions mapping of an integer into the RNS representation. But note that these methods are only useful for those values of m, with short periods.

If m is even, it is possible to "factor out" a power of 2, by applying Theorem 8.2.6, changing the problem to the reduction of the odd component of $m = 2^k q$ for q odd.

8.2.2 The multiplicative inverse

The multiplicative inverse plays an important role in many algorithms, so let us consider how to determine $\lvert a^{-1} \rvert_m$ given a and m. The classical approach for m prime is to use *Fermat's Little Theorem.*

Theorem 8.2.11 (Fermat) *If p is prime, then*

$$\lvert a^p \rvert_p = \lvert a \rvert_p.$$

Proof The statement is obviously true for $a = 0$ and $a = 1$. Assume it is true for a, then using the binomial expansion

$$\lvert (a+1)^p \rvert_p = \lvert a^p + 1 \rvert_p = \lvert a + 1 \rvert_p,$$

since $\binom{p}{i}$ for $0 < i < p$ is divisible by p when p is a prime, so by induction the theorem holds for $a \geq 0$. The result for $a < 0$ follows similarly. □

Corollary 8.2.12 *If $\lvert a \rvert_p \neq 0$, then $\lvert a^{-1} \rvert_p = \lvert a^{p-2} \rvert_p$ for p prime.*

Proof By Fermat's Little Theorem, $\big\lvert \lvert a^{p-2} \rvert_p \cdot a \big\rvert_p = \lvert a^{p-1} \rvert_p = 1$. □

Notice that Fermat's Little Theorem does not hold for p composite, e.g., $|2^6|_6 = 4 \neq |2|_6$, hence other methods must be used in this case when it is known that $\gcd(a, m) = 1$ and thus $|a^{-1}|_m$ is known to exist.

For the case where the modulus is a power of 2, $m = 2^k$, the multiplicative inverse b of any odd number a, $0 \le a < 2^k$, can be constructed bitwise sequentially, initially noticing that $b_0 = 1$ since $a_0 = 1$. Mimicking the standard iterative algorithm for multiplication of the (known) multiplicand a by the (only partially known) multiplier b, we can start by initializing the accumulated partial products s with the value $s = a$ since $b_0 = 1$. In step i we choose b_i such that when $a2^i$ is added into s, then bit i of s is annihilated, hence b_i is chosen to be equal to bit i of the present value of s. Since $a_0 = 1$, $s + b_i a 2^i$ will then become zero in position i, and $s \bmod 2^{i+1} = 1$. The process can be illustrated as in the following diagram of the multiplication:

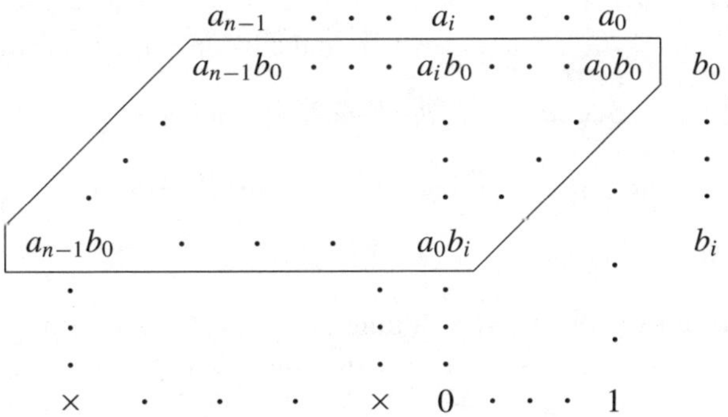

Observe that $s = a \cdot (b \bmod 2^{i+1})$, so that after $k - 1$ steps we have:

$$s \equiv 1 \pmod{2^k} \quad \text{and} \quad s \equiv ab \pmod{2^k},$$

hence $b = |a^{-1}|_{2^k}$. We have thus shown the correctness of the following algorithm, where for convenience s is gradually right-shifted to allow the testing of s to be performed as a parity check.

Algorithm 8.2.13 (Multiplicative inverse modulo 2^n)
Stimulus: *An odd integer a,* $1 \le a < 2^n$ *for* $n \ge 1$.
Response: *An integer b such that* $ab \equiv 1 \pmod{2^n}$.
Method: $s := a;\ b := 1;\ t := 1;$
for $i := 1$ **to** $n - 1$ **do**
  $s := s$ div $2;\ t := t * 2;$
  if odd(s) **then** $s := s + a;\ b := b + t$ **end**
end

The cycle complexity of this algorithm is obviously linear in n. However, there is an algorithm whose complexity is only $O(\log n)$ based on the following lemma.

Lemma 8.2.14 *If* $ab \equiv 1 \pmod{m}$ *such that* $b = |a^{-1}|_m$ *is the inverse of* a *modulo* m*, then*

$$|a^{-1}|_{m^2} = |b(2 - ab)|_{m^2} .$$

Proof Since $a = cm + |a|_m$ for some integer c, we have $a|a^{-1}|_m = dm + 1$ for some integer d, then $2 - a|a^{-1}|_m = -dm + 1$ and hence

$$a\left|a^{-1}\right|_m (2 - a\left|a^{-1}\right|_m) = (dm + 1)(-dm + 1) = -d^2m^2 + 1.$$

With $b = \left|a^{-1}\right|_m$ we then have

$$|a \cdot b(2 - ab)|_{m^2} = 1,$$

which proves the lemma. □

Thus if m is a power of 2, $m = 2^e$, then $m^2 = 2^{2e}$, so if binary arithmetic is used, knowing $|a^{-1}|_{2^e}$ it is fairly easy to find the inverse $|a^{-1}|_{2^{2e}}$ using the lemma.

Theorem 8.2.15 *If* $b = |a^{-1}|_{2^e}$ *is the inverse of* a *modulo* 2^e*, then*

$$\left|a^{-1}\right|_{2^{2e}} = |b(2 - ab)|_{e^{2e}} = b + 2^e|-bk|_{2^e},$$

where k *is defined by* $|ab|_{2^{2e}} = 1 + k \cdot 2^e$.

Proof The result follows by substituting $|ab|_{2^{2e}} = 1 + k \cdot 2^e$ for the term ab in $|b(2 - ab)|_{e^{2e}}$, which can then be rewritten into $|b + 2^e|-bk|_{2^e}|_{2^{2e}} = b + 2^e|-bk|_{2^e}$. □

For example, if $a = 256a_1 + a_0$, then $|a^{-1}|_{256} = |a_0^{-1}|_{256}$ can be found by say an 8-bit table look-up using the least-significant eight bits of a. By Theorem 8.2.15 $|a^{-1}|_{256^2} = |a^{-1}|_{2^{16}}$ is then easily computed by concatenating $|-bk|_{2^e}$ as the high-order part to $b = |a^{-1}|$, and this can then be repeated to find $|a^{-1}|_{2^{32}}$, etc.

Another useful result is specified by the following lemma, allowing the inverse $a^{-1} \bmod b$ to be determined from $b^{-1} \bmod a$.

Lemma 8.2.16 *If* $\gcd(a, b) = 1$ *for non-negative integers* a, b*, then*

$$a^{-1} \bmod b = \frac{1 + b(-b^{-1} \bmod a)}{a}.$$

Proof Let $c = a(a^{-1} \bmod b) + b(b^{-1} \bmod a)$, then $c \equiv 1 \pmod{a}$ and $c \equiv 1 \pmod{b}$ and thus $c \equiv 1 \pmod{ab}$. Since $1 < a + b \le c < 2ab$, then $c = 1 + ab$ and the result follows. □

This lemma is particularly useful if either a or b is a power of the radix of the number representation used, e.g., $\beta = 10$, as in the following example, which also employs Lemma 8.2.14.

Example 8.2.3 Let $a = 10^2 = 100$ and $b = 223$, satisfying $\gcd(100, 223) = 1$. By Lemma 8.2.14 we can find $|223^{-1}|_{100}| = |23^{-1}|_{100}$ from $|3^{-1}|_{10} = 7$, as

$$|223^{-1}|_{100} = |7(2 - 223 \cdot 7)|_{100} = |7(2 - 23 \cdot 7)|_{100} = |-1113|_{100} = 87,$$

and from Lemma 8.2.16

$$100^{-1} \bmod 223 = \frac{1 + 223(-87 \bmod 100)}{100} = \frac{1 + 223 \cdot 13}{100} = \frac{2900}{100} = 29.$$

The check $100 \cdot 29 = 13 \cdot 223 + 1$ implicit in the above confirms the result. □

For the most general case of finding the multiplicative inverse we shall use an extended version of the well known Euclidean Algorithm for calculating $\gcd(a, b)$, but initially let us consider a special sequence of recursively generated integer pairs.

For any sequence of integers $\{q_0, q_1, q_2, \ldots\}$ define for $i = 0, 1, 2, \ldots$

$$\begin{aligned} a_i &= a_{i-2} - q_i a_{i-1}, \\ b_i &= b_{i-2} - q_i b_{i-1} \end{aligned} \tag{8.2.5}$$

with initial values

$$\begin{bmatrix} a_{-2} & b_{-2} \\ a_{-1} & b_{-1} \end{bmatrix} = \begin{bmatrix} a & b \\ c & d \end{bmatrix} \quad \text{(the \textit{seed matrix}).} \tag{8.2.6}$$

Lemma 8.2.17 *If in seed matrix (8.2.6)* $ad \equiv bc \pmod{m}$*, then for* $i \geq -2$, $j \geq -2$

$$a_i b_j \equiv a_j b_i \pmod{m},$$

where $\{(a_i, b_i)\}$ *is the sequence of pairs generated by (8.2.6) using any* $q_0, q_1, q_2, \ldots$.

Proof

$$\begin{aligned} a_i b_{i-1} - a_{i-1} b_i &= (a_{i-2} - q_i a_{i-1}) b_{i-1} - a_{i-1}(b_{i-2} - q_i b_{i-1}) \\ &= (a_{i-2} b_{i-1} - a_{i-1} b_{i-2}) + q_i \cdot 0 \\ &\vdots \\ &= (-1)^i (a_{-2} b_{-1} - a_{-1} b_{-2}) \\ &= (-1)^i (ad - bc) \\ &\equiv 0 \pmod{m}. \end{aligned}$$

Also similarly,

$$\begin{aligned} a_i b_{i-2} - a_{i-2} b_i &= a_i (b_i + q_i b_{i-1}) - (a_i + q_i a_{i-1}) b_i \\ &= q_i (a_i b_{i-1} - a_{i-1} b_i) \\ &\equiv 0 \pmod{m}. \end{aligned}$$

Hence the result follows by repeated applications assuming say $i > j$. □

Now if we choose $q_i = \lfloor a_{i-2}/a_{i-1} \rfloor$ for $a_{i-1} \neq 0$, $i = 0, 1, 2, \ldots,$ then $\{a_0, a_1, \ldots, a_n\}$ is the sequence of partial remainders when the Euclidean Algorithm is applied to (a, c), terminating at step n where $a_n = 0$, $a_{n-1} = \gcd(a, c)$.

Algorithm 8.2.18 *(Extended Euclidean Algorithm, EEA)*
Stimulus: *Any four integers a, b, c, d in a seed matrix*

$$\begin{bmatrix} a_{-2} & b_{-2} \\ a_{-1} & b_{-1} \end{bmatrix} = \begin{bmatrix} a & b \\ c & d \end{bmatrix}$$

Response: *An integer $n \geq 0$ and a sequence (a_i, b_i), $i = 0, 1, \ldots, n$ with $a_i \neq 0$ for $i < n$ and $a_n = 0$. Furthermore $a_{n-1} = \gcd(a, c)$.*
Method: $i := 0$;
while $a_{i-1} \neq 0$ **do**
  $q_i := \lfloor a_{i-2}/a_{i-1} \rfloor$;
  $a_i := a_{i-2} - q_i a_{i-1}$;
  $b_i := b_{i-2} - q_i b_{i-1}$;
  $i := i + 1$ **end**

The correctness of Algorithm 8.2.18 follows from the well-known standard Euclidean Algorithm. Now initialize the algorithm with the seed matrix

$$\begin{bmatrix} a & b \\ c & d \end{bmatrix} = \begin{bmatrix} m & 0 \\ c & 1 \end{bmatrix},$$

where $\gcd(c, m) = 1$, so $ad \equiv bc \pmod m$. Hence Lemma 8.2.17 applies, and in particular for $i = -1$, $j = n - 1$ we obtain for the pair (a_{n-1}, b_{n-1}) found by the EEA Algorithm:

$$cb_{n-1} \equiv a_{n-1} \cdot 1 \pmod m.$$

Since $a_{n-1} = \gcd(m, c) = 1$

$$cb_{n-1} \equiv 1 \pmod m,$$

hence $|b_{n-1}|_m$ is the multiplicative inverse of c modulo m.

We have thus proved the following theorem.

Theorem 8.2.19 (Multiplicative inverse) *When the EEA Algorithm is applied to the seed matrix*

$$\begin{bmatrix} m & 0 \\ c & 1 \end{bmatrix}$$

generating the sequence $\{(a_i, b_i)\}_{i=0,1,\cdots n}$, terminating with $a_n = 0$, then when $a_{n-1} = 1$ $(= \gcd(c, m))$ the multiplicative inverse of c modulo m exists and is found as

$$|c^{-1}|_m = |b_{n-1}|_m.$$

Example 8.2.4 In the system with $m = 39$ we can determine whether a multiplicative inverse of 17 exists and find its value as follows:

i	q_i	a_i	b_i
−2		39	0
−1		17	1
0	2	5	−2
1	3	2	7
2	2	1	−16
3	2	0	39

Since $a_3 = 0$ and $a_2 = 1 = \gcd(39, 17)$ we find $|17^{-1}|_{39} = |-16|_{39} = 23$. □

Changing the lower right-hand entry of the seed matrix from 1 to d, it follows that all b_i generated are being multiplied by d, thus we have the following corollary.

Corollary 8.2.20 (Modular division) *When the EEA Algorithm is applied to the seed matrix*

$$\begin{bmatrix} m & 0 \\ c & d \end{bmatrix}$$

generating the sequence $\{(a_i, b_i)\}_{i=0,1,\cdots n}$*, terminating with* $a_n = 0$*, then when* $a_{n-1} = 1$ $(= \gcd(c, m))$ *the multiplicative inverse of c modulo m exists and*

$$\left|\frac{d}{c}\right|_m = d \times |c^{-1}|_m = |b_{n-1}|_m.$$

Note that modular multiplication of the inverse of c by d in this way may be obtained "for free," cf. Example 8.2.1.

There is also a binary version of the EEA Algorithm, working from right to left. It is based on the following observation. If a is even and b is odd, then $\gcd(a, b) = \gcd(a/2, b)$. If a and b are both even, then it is easy to see that either $a + b$ or $a - b$ is divisible by 4, thus there are two cases:

$(a + b) \bmod 4 = 0$ Here $\gcd(a, b) = \gcd((a + b)/2, b)$, where $(a + b)/2$ is even and $|(a + b)/2| \le \max(|a|, |b|)$.

$(a - b) \bmod 4 = 0$ Here $\gcd(a, b) = \gcd((a - b)/2, b)$, where $(a - b)/2$ is even and $|(a - b)/2| \le \max(|a|, |b|)$.

The following gcd-type algorithm will make progress by choosing to reduce the larger of $|a|$ and $|b|$. It does so by keeping track of the magnitudes of $|a|$ and $|b|$ using a variable δ representing the difference $\alpha - \beta$, where α and β are such that $|a| < 2^\alpha$ and $|b| < 2^\beta$.

Note that the algorithm determines reductions by looking at the least-significant bits only, and performs right-shifts, i.e., that reduction takes place from right to left. This implies that the additions and subtractions may be performed in constant time on redundant representations, and most decisions can be made directly on least-significant digits. There are, however, some tests where "tricks" have to be applied, we shall return to these below.

Algorithm 8.2.21 (Binary EEA for modular division, B-EEA)
Stimulus: *Integers p, q, m in a seed matrix:*

$$\begin{bmatrix} a & c \\ b & d \end{bmatrix} = \begin{bmatrix} m & 0 \\ q & p \end{bmatrix}$$

where the odd modulus m satisfies $2^{n-1} < m < 2^n$ for some $n > 0$, $0 \le p < m$ and $0 < q < m$.
Response: *If* $\gcd(q, m) = 1$ *then* $r = |p/q|_m$ *with* $0 \le r < m$, *otherwise "error".*
Method: $\delta = 0$;

while $a \neq 0$ **do**
  while $a \bmod 2 = 0$ **do**
    $a := a/2;\ c := c/2 \bmod m;\ \delta := \delta - 1;$
  end;
  if $\delta < 0$ **then**
    $t := a;\ a := b;\ b := t;$
    $t := c;\ c := d;\ d := t;$
    $\delta := -\delta;$
  end;
  if $(a + b) \bmod 4 = 0$ **then**
    $a := (a + b)/2;\ c := (c + d)/2 \bmod m;$
  else
    $a := (a - b)/2;\ c := (c - d)/2 \bmod m;$
  end;
end;
if $b = 1$ **then** $r := d$;
elseif $b = -1$ **then** $r := m - d$;
else return *"error"*

Where: *Modular division by 2, $x/2 \bmod m$, returns $x/2$ if x is even, and $(x + m)/2$ if x is odd.*

The counter δ need only be of width $\log \log n$, or can alternatively be implemented by a sign bit and shift register with a single bit in position $i = |\delta|$. The test for $a \neq 0$ can be substituted by a bit test of position zero in a register, initialized to 2^{n+1} (a single bit), and right-shifted whenever a is divided by 2. The correctness of the algorithm follows from Lemma 8.2.17.

8.2.3 Implementation of modular addition and multiplication

To discuss the implementation of modular arithmetic operations it is useful to consider the representation of the integral-valued residues. If the system modulus

m is a power of some integer β, e.g., $m = \beta^n$, then an obvious choice is to represent the residues in the fixed-point radix system, radix β:

$$\mathcal{F}_{0,n-1}[\beta, D] \text{ with } D = \{0, 1, \ldots, \beta - 1\},$$

since the set of representable integer values is then precisely $\{0, 1, \ldots, m - 1\}$.

In this case modular addition is realized as standard radix addition where the carry-out of the most-significant position is just discarded. Since radix multiplication is realized as repeated additions of shifted values of the multiplicand, modular multiplication can be realized by trivial modifications (and simplifications) to standard multipliers, just truncating any digits or carries of weight β^k for $k \geq n$. Notice that accumulation of partial products in a redundant representation (e.g., carry-save for $\beta = 2$) is possible, with a final conversion into non-redundant representation, discarding any carry-out.

This immediately raises the question of whether a redundant digit set could be used more generally for the representation of residues. And by Theorem 8.2.3 it is seen that for any composite expression of additions and multiplications it is not necessary to reduce modulo m after each arithmetic operation. A redundant and basic digit set D necessarily insures that $\mathbb{Z}_m = \{0, 1, \ldots, \beta^n - 1\}$ is a subset of the values which can be represented by $\mathcal{F}_{0,n-1}[\beta, D]$. Thus we may note the following.

Observation 8.2.22 *For any composite computation involving only additions and multiplications modulo $m = \beta^n$, $\beta \geq 2$, it is possible to perform the computations in $\mathcal{F}_{0,n-1}[\beta, D]$, where D is a redundant, basic digit set for radix $\beta > 0$, by just discarding digits of weight greater than or equal to β^n. By a final conversion into $\mathcal{F}_{0,n-1}[\beta, D']$, where D' is the standard digit set $\{0, 1, \ldots, \beta - 1\}$, the correct residue modulo $m = \beta^n$ is obtained.*

From Lemma 8.2.5 it follows that addition modulo $2^n - 1$ can be performed as addition modulo 2^n with an "end-around carry," and similarly for addition modulo $2^n + 1$. Note that in the case of addition of two residues, say $a = |x|_m + |y|_m$, the addition of the term a div m cannot produce a new carry since $a \leq 2m - 2$, and hence results in the proper residue. This term then acts as an "end-around carry" in the sense that a most-significant bit (a carry) is added into the least-significant end of the number.

For multiplication modulo $m = 2^n - 1$, the product of two residues $c = |a|_m |b|_m$ can be split so that

$$||a|_m |b|_m|_m = \left|(c \bmod 2^n) + (c \operatorname{div} 2^n)\right|_m,$$

where the right-hand addition can be performed with "end-around carry" as above. For $m = 2^n + 1$, (8.2.2) applies similarly.

Again notice that a redundant digit set can be employed, e.g., carry-save in the case of $m = 2^n \pm 1$. And for the final conversion from carry-save into non-redundant binary (8.2.1) or (8.2.2) may also be applied, since such a conversion is equivalent to the addition of two residues.

For a binary implementation of addition modulo m in the more general case, assume n is chosen such that $2^{n-1} < m < 2^n$.

Since

$$a + b + (2^n - m) \geq 2^n \Longleftrightarrow a + b \geq m,$$

let $p = 2^n - m$ and compute in parallel:

$$c = |a|_m + |b|_m \qquad \text{and} \qquad c' = |a|_m + |b|_m + p,$$

we then obtain:

$$||a|_m + |b|_m|_m = \begin{cases} c & \text{if } c' \text{ div } 2^n = 0, \\ c' \bmod 2^n & \text{if } c' \text{ div } 2^n = 1. \end{cases}$$

Hence the carry-out of the biased c'-computation can be used to select between c and $c' \bmod 2^n$ from two parallel n-bit carry-completing additions as in the *modular addition select adder* of Figure 8.2.1.

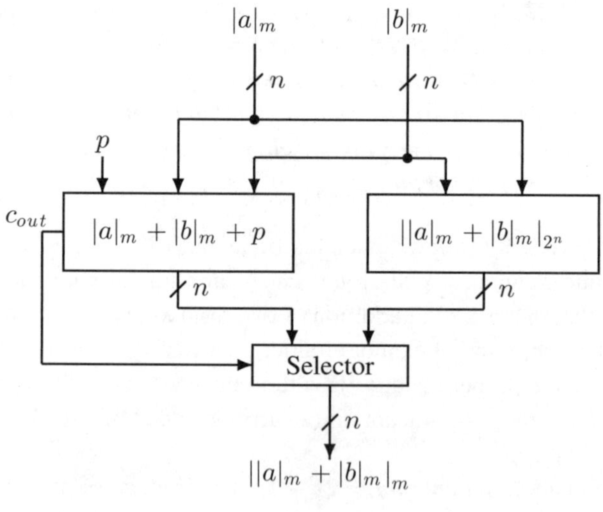

Figure 8.2.1. Modular addition select adder.

Now consider the general case of modular multiplication, where the reduced product $|x|_m = ||a|_m \cdot |b|_m|_m$ is wanted. Assume again that m satisfies $2^{n-1} < m < 2^n$, that $p = 2^n - m$ where $1 \leq p < 2^{n-1}$, and $k = 1 + \lfloor \log_2 p \rfloor$ is such that

$1 \le k \le n-1$. The binary double-length product

$$X = \sum_{i=0}^{2n-1} x_i 2^i = |a|_m \cdot |b|_m$$

can then be partitioned such that

$$X = D\,2^{2n-1-k} + C\,2^n + B\,2^{n-1} + A$$

with

$$A = \sum_0^{n-2} x_i 2^i, \quad B = x_{n-1}, \quad C = \sum_n^{2n-k-2} x_i 2^{i-n}, \quad D = \sum_{2n-k-1}^{2n-1} x_i 2^{i-(2n-k-1)}.$$

Noting that by the definition of p, $|2^n|_m = p$, and we have

$$|X|_m = \left||D\,2^{2n-1-k} + B\,2^{n-1}|_m + C\,p + A\right|_m, \tag{8.2.7}$$

where $A \le 2^{n-1} - 1 = (m+p)/2 - 1$ and $C\,p \le (2^{n-k-1} - 1)\,2^k = 2^{n-1} - 2^k = (m+p)/2 - 2^k$, thus

$$C\,p + A \le m + p - 2^k - 1 = m - (2^{k-1} - p + 1) < m.$$

What remains to be considered in (8.2.7) is the term $|D\,2^{2n-1-k} + B\,2^{n-1}|_m$, where we note that D is a $(k+1)$-bit quantity and B is a single bit, hence that term can be found by a table look-up in a ROM,[2] based on $k+2$ bits of input assuming that k is fairly small. To perform the computation of (8.2.7) a rectangular $(n-k-1)$ by k multiplier is needed to compute Cp, and a carry-save (3-to-2) adder is needed to reduce the three terms to the sum of two, which then can be added modulo m in a modular addition select adder as in Figure 8.2.1.

Note that the output of the ROM is a residue modulo m, and the sum of A and the output of the small multiplier is $Cp + A < m$. Thus the input to the modular addition select adder satisfies its bounds. The total structure of the *modular multiplier* is shown in Figure 8.2.2.

The topmost $n \times n$ multiplier must deliver its result in non-redundant form, but the smaller rectangular multiplier can deliver its result in carry-save form, provided that the carry-save adder is changed into a 4-to-2 redundant adder, thus saving a final carry completion in the small multiplier. Provided that k is quite small the total timing of the modular multiplier is essentially the sum of the time for an ordinary $n \times n$ multiplier plus the time for a carry completing n-bit adder; formally the time is then $O(\log n)$.

[2] ROM: read-only memory.

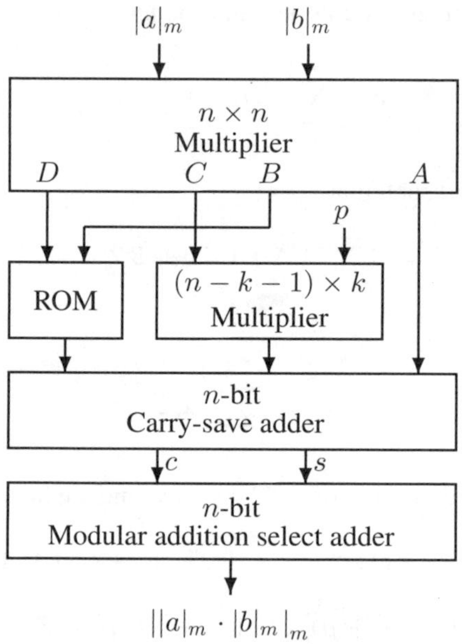

Figure 8.2.2. Modular multiplication.

8.2.4 Multioperand modular addition

The idea of the modular addition select adder can be generalized to multioperand modular addition, where the conditional addition of the term $2^n - m$ is postponed to the next step as described in the following algorithm.

Algorithm 8.2.23 (Multioperand modular addition)
Stimulus: *A modulus* m, $2^{n-1} < m < 2^n$, $p = 2^n - m$, *and a set of* k *residues,* $\{a_i\}_{i=1,2,\ldots,k}$, $0 \le a_i < m$.
Response: *The sum* $S = \left(\sum_{i=1}^{k} a_i\right) \bmod m$.
Method: $c := 0;\ S := p;$
for $i := 1$ **to** k **do**
  if $c = 1$ **then** $S := S + a_i + p$ **else** $S := S + a_i$;
  $c := S \text{ div } 2^n;\ S := S \bmod 2^n;$
end;
if $c = 0$ **then** $S := S - p$ **end**;

The proof of the correctness of the algorithm is based on the fact that by initializing the sum to $p = 2^n - m$, the carry-out signals that the accumulated value exceeds m. Hence m can safely be subtracted, as realized by discarding the carry-out and adding p ($S := S - 2^n + 2^n - m$). Using the definition of p it is easily seen that $S < 2^{n+1}$, so $c \in \{0, 1\}$. The final correction step effectively subtracts the initial bias p

This algorithm has a straightforward hardware implementation where the input of p to a three-input adder/accumulator is conditional on the latched value of the carry-out from the previous iteration. Obviously this accumulation takes time $O(k \log n)$, assuming that a carry-look-ahead adder is used. However, as usual there is no need to use a carry-completing adder for the internal accumulations of the summation. With a slight complication in the selection of the input to the adder we can modify Algorithm 8.2.23 to perform accumulation in, say, carry-save representation. Here it turns out to be possible and necessary to subtract larger multiples of 2^n, for which the following lemma is useful.

Lemma 8.2.24 *For any k the following equivalence holds:*

$$x - k2^n + (k2^n \bmod m) \equiv x \pmod m.$$

This lemma allows us to perform an "approximate" reduction also for larger carry-out.

Algorithm 8.2.25 (Multioperand modular redundant addition)

Stimulus: *A modulus* m, $2^{n-1} < m < 2^n$, $p = 2^n - m = 2^n \bmod m$, $p' = 2^{n+1} \bmod m$ *and a set of* $k \geq 2$ *residues,* $\{a_i\}_{i=1,2,\ldots,k}$, $0 \leq a_i < m$.

Response: *The sum* $S = \left(\sum_{i=1}^{k} a_i\right) \bmod m$.

Method: $c_1 := 0;\ S^c := p;\ S^s := a_1;$
for $i := 2$ **to** k **do**
  $S := S + a_i;$
  $c_2 := S^c \text{ div } 2^n;\ S^c := S^c \bmod 2^n;$
  case (c_1, c_2) **of**
  $00: \quad S := S;$
  $01, 10: S := S + p;$
  $11: \quad S := S + p';$
  end;
  $c_1 := S^c \text{ div } 2^n;\ S^c := S^c \bmod 2^n;$
end;
if $c_1 = 1$ **then** $S := S + p$ **end**;
$S := S^c + S^s$; {*Note: carry-completion addition*}
if $S \geq 2^n$ **then** $S := S - 2^n$ **else** $S := S - p$;

Where: $S = (S^c, S^s)$ *is the carry-save representation of S, and all additions of the form* $S := S + x$ *are 3-to-2 carry-save additions.*

After each carry-save addition it follows that $S^s < 2^n$ and $S^c < 2^{n+1}$, so $c_1, c_2 \in \{0, 1\}$ and the correctness of the algorithm then follows from Lemma 8.2.24. The timing of the complete multioperand addition is then $O(k + \log n)$, which can be reduced further to $O(\log k + \log n)$ if a tree structure of carry-save adders is used for the accumulation.

Each node of such a tree structure receives two operands A and B in carry-save representation, and the node produces a result S (also in carry-save) such that

$$S \equiv A + B \pmod{m}.$$

The computation at each node can then be described as follows.

Algorithm 8.2.26 (4-to-2 redundant modular addition node)

Stimulus: *A modulus* m, $2^{n-1} < m < 2^n$ *and a set of constants* $p_k = k2^n \bmod m$, $k = 0, 1, 2, 3, 4$.
$A = A^s + A^c$, $0 \le A^s < 2^n$ *and* $0 \le A^c < 2^{n+1}$.
$B = B^s + B^c$, $0 \le B^s < 2^n$ *and* $0 \le B^c < 2^{n+1}$.

Response: $S = S^s + S^c$, $0 \le S^s < 2^n$ *and* $0 \le S^c < 2^{n+1}$ *satisfying* $S \equiv A + B \pmod{m}$.

Method:
$c_1 := A^c \text{ div } 2^n;\ A^c := A^c \bmod 2^n;$
$c_2 := B^c \text{ div } 2^n;\ B^c := B^c \bmod 2^n;$
$S := A + B^c;$
$c_3 := S^c \text{ div } 2^n;\ S^c := S^c \bmod 2^n;$
$S := S + B^s;$
$c_4 := S^c \text{ div } 2^n;\ S^c := S^c \bmod 2^n;$
$k := c_1 + c_2 + c_3 + c_4;$
$S := S + p_k$

Where: *All additions of the form* $S := X + x$ *are 3-to-2 carry-save additions.*

Again the correctness follows from Lemma 8.2.24 together with trivial bounds on the operands. Note that in a tree of such nodes the leaf nodes accept four non-redundant operands at the inputs A^s, A^c, B^s, and B^c. Thus here the c_1 and c_2 signals will become zero, but at one leaf node one of these can be forced to a one to supply the initial bias $p = p_1$. The final result can be obtained at the root of the tree after a conversion and modular reduction identical to the one in Algorithm 8.2.25, also removing the initial bias.

In a system where the modulus m is such that either the period $P(m)$ or the half-period $HP(m)$ is reasonably small, another efficient method applies. Let

$$\begin{aligned} a_0 &= b_{0n-1}b_{0n-2}\cdots b_{00}, \\ a_1 &= b_{1n-1}b_{1n-2}\cdots b_{10}, \\ &\ \vdots \\ a_{k-1} &= b_{k-1n-1}b_{k-1n-2}\cdots b_{k-10} \end{aligned}$$

be k residues modulo m to be added, given by their n-bit binary representations. We want to perform a multioperand modular addition of these, and by Lemma 8.2.8

we have for the period given by $P(m)$:

$$
\begin{aligned}
|S|_m &= \left| \sum_{i=0}^{k-1} a_i \right|_m \\
&= \left| \sum_{j=0}^{n-1} |2^j|_m \cdot \left(\sum_{i=0}^{k-1} b_{ij} \right) \right|_m \\
&= \left| \left| \sum_{j=0}^{n-1} |2^{j \bmod P(m)}|_m \cdot \left(\sum_{i=0}^{k-1} b_{ij} \right) \right|_{2^{P(m)}-1} \right|_m ,
\end{aligned}
\tag{8.2.8}
$$

where the inner summations may be performed in a network of full- and half-adders, modulo $2^{P(m)}$ with end-around carry, according to (8.2.1). Note that some of the inner sums share the same weight $|2^t|_m$ due to (8.2.3). Only at the very end will a modulo m reduction be needed. We shall not here go into the details of designing the adder networks. If the periodicity is given by the half-period, $HP(m)$, there is an equivalent algorithm, but based on reducing modulo $2^{HP(m)} + 1$.

8.2.5 ROM-based addition and multiplication

For moderate values of m it is possible to use a table look-up to perform the reduction modulo m. If a carry-completing adder is used, the $(n + 1)$-bit sum of $|a|_m$ and $|b|_m$ can be used as an address into a ROM as in Figure 8.2.3.

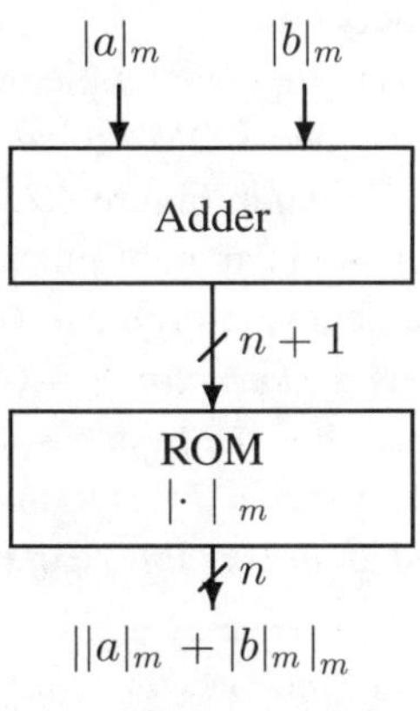

Figure 8.2.3. A ROM-based modular adder.

ROM-based methods may also be used for multiplication when m is of moderate size. One idea is to use the *quarter-square method*, based on the following rewriting of a product:

$$ab = \left(\frac{a+b}{2}\right)^2 - \left(\frac{a-b}{2}\right)^2 = \left\lfloor \frac{1}{4}(a+b)^2 \right\rfloor - \left\lfloor \frac{1}{4}(a-b)^2 \right\rfloor .$$

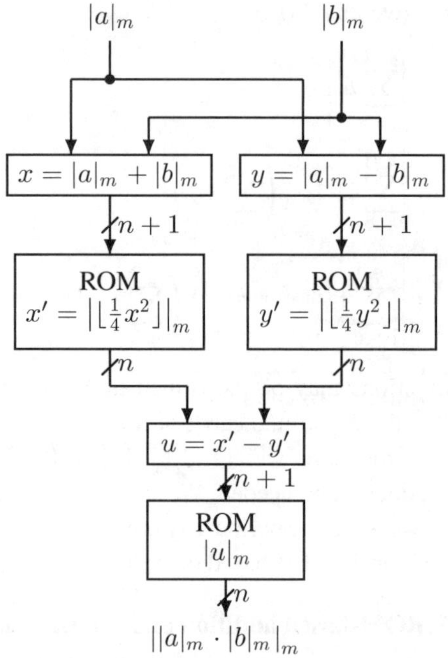

Figure 8.2.4. A ROM-based quarter-square modular multiplier.

where the last equality holds for a and b integral, since in this case the fractional parts must be identical and cancel out.

A ROM-based multiplier can then be implemented based on three carry-completing n-bit binary adders, two ROM tables with entries $\left|\lfloor \frac{1}{4}x^2 \rfloor\right|_m$ and a final ROM containing $|u|_m$, as shown in Figure 8.2.4.

An alternative approach for modular multiplication is possible when m is a prime p, in which case $\mathbb{Z}_p$ is a field isomorphic to $GF(p)$. Then it is known from number theory that there exists a *generator* $g \in GF(p)$ which multiplicatively generates all non-zero elements, thus the first $p-1$ powers of g together with 0 constitute $GF(p)$. For example, for $p = 7$ it is found that $g = 3$ is a generator and a *log–antilog table* (also called an *index table*) can be constructed:

a	Index(a)
1	6
2	2
3	1
4	4
5	5
6	3

since, for example, $3^6 = 729 \equiv 1 \pmod 7$. Hence multiplication of non-zero operands can be performed by addition of their indices ("logarithms"), followed by a table look-up of the inverse function ("antilogs"). Since by Fermat's Theorem 8.2.11 $|g^{p-1}|_p = 1$, the addition of logarithms can be performed modulo $p-1$ so that

$$\left||a|_p \cdot |b|_p\right|_p = \left|g^{i_a} \cdot g^{i_b}\right|_p = \left|g^{i_a+i_b}\right|_p = \left|g^{(i_a+i_b) \bmod (p-1)}\right|_p,$$

where i_a and i_b are the indices ("logarithms") of a, respectively b. The last table can easily incorporate the modulo $p-1$ reduction, but note that zero factors have to be handled separately from the tables. For another logarithmic representation with p of the form $p = 2^k$ see Observation 8.2.42.

8.2.6 Modular multiplication for very large moduli

If high-speed implementations of modular multiplication are needed for very large values of the modulus m (e.g., as needed for modular exponentiation in cryptosystems), it is essential to use redundant representations in the accumulation of partial products, and for space economy to keep the accumulated contents reduced modulo m. On the other hand, it is not necessary that the intermediate values are perfectly reduced, a final addition or subtraction of an appropriate multiple of m can be performed to deliver the correct residue.

The basic idea of *interleaved modular multiplication* is to interleave the accumulation steps of the multiplication with reduction steps, which based on an estimate of a quotient digit subtracts a suitable multiple of the modulus from the value in the accumulator. To avoid the gradually increasing number of digits needed when adding a shifted version of the multiplicand if (as usual) starting with the least-significant digit of the multiplier, it is possible to start with the most-significant digit. It is then straightforward to use the techniques of multioperand modular addition as described in Section 8.2.4, but here we shall show other approaches used in the implementation of cryptosystems.

To reduce the number of cycles we may use techniques known from high-radix multiplication and division, e.g., we will assume the multiplier is given in a radix $\beta = 2^k$ and quotient digits are determined in that radix also. The system modulus m then has to satisfy $2^{k(n-1)} < m < 2^{kn}$. An algorithm for modular multiplication $S := AB \bmod m$ performed most-significant digit first is shown next. Note that m is always odd for cryptosystems.

Algorithm 8.2.27 (Interleaved modular multiplication)

Stimulus: *An odd modulus $m > 2$ and integers $k \geq 1$ and $n \geq \ell \geq 1$, where m satisfies $2^{k(n-1)} < m < 2^{kn}$.*
Integers A and B such that $-m < A, B < m$, where $B = \sum_{i=0}^{\ell-1} b_i r^i$, $b_i \in D = \{-\sigma, \cdots, \sigma\}$ for $0 \leq i \leq \ell - 1$, with $2^{k-1} \leq \sigma \leq 2^k - 1$.
Normally $\ell = n$ but may be defined to be smaller. Let $r = 2^k$.

Response: *An integer S such that $-m < S < m$ and $S \equiv AB \pmod m$.*

Method: $S := 0$;
for $i := \ell - 1$ **downto** 0 **do**
$S := r \cdot S + b_i \cdot A;$
$q_i := \left\lfloor \frac{S}{m} \right\rfloor$; $\{$*or some good approximation*$\}$
$S := S - q_i \cdot m;$
end

The quotient q_i may for small values of k be determined by a table look-up like in SRT division, or by multiplying a short prefix of S by a short reciprocal of m. Note that the arithmetic in the loop may be performed on redundant representations of S and A (say the same as used for B), and that B may be significantly smaller than m ($\ell \ll n$). The correctness of the algorithm will, of course, depend on the details of the implementation, as this is only a sketch.

However, by a clever change of residue system it is possible to simplify the determination of q_i. The trick is to trade reduction modulo m with multiplication by a constant and reduction modulo r, where r is chosen such that the latter operation is simple, e.g., where r is a power of the radix of the arithmetic.

Given m choose $r > m$ such that $\gcd(m, r) = 1$ and let r^{-1} be the multiplicative inverse of r modulo m. Also determine m', $0 < m' < r$, such that

$$rr^{-1} - mm' = 1, \tag{8.2.9}$$

which is equivalent to requiring that $|-mm'|_r = 1$, so m' is the multiplicative inverse of $-m$ modulo r, which can easily be found by Algorithm 8.2.13 when r is a power of 2.

We then define a new residue operator, mapping integers into $\mathbb{Z}_m = \{i \in \mathbb{Z} \mid 0 \leq i < m\}$ as follows.

Definition 8.2.28 *The mapping* $[\cdot]_m^r : \mathbb{Z} \to \mathbb{Z}_m$ *for fixed* r *and* m, *with* $\gcd(r, m) = 1$, *is defined by*

$$[a]_m^r = i \in \mathbb{Z}_m \text{ for } a \in \mathbb{Z}$$

if and only if $a \equiv ir^{-1} \pmod{m}$, *or equivalently* $i = |ar|_m = ar \bmod m$.

We will name $[a]_m^r$ the *M-residue* after Peter Montgomery who suggested this residue system for modular multiplication.

Since $ar + br \equiv cr \pmod{m}$ iff $a + b \equiv c \pmod{m}$ it is possible to add (and subtract) M-residues just like ordinary residues:

$$[a + b]_m^r = \left| [a]_m^r + [b]_m^r \right|_m,$$

but note that the "outer" reduction is the "ordinary" reduction modulo m. Also equality/inequality tests and multiplication by an integer can be performed on M-residues as usual, but other operations are affected by the extra factor r.

At a first glance it looks like multiplication now has become more complicated since both multiplicand and multiplier contains the factor r, whereas the product is only supposed to contain a single factor r. But the product of the two residues also has to be reduced, and it turns out that this process is simplified in the sense that now only reductions modulo r are needed (and not reductions modulo m).

Algorithm 8.2.29 (M-reduce)
Stimulus: *An integer* t *such that* $0 \le t < rm$.
Integer constants m, r, m', r^{-1} *such that* $\gcd(m, r) = 1, r > m > 2$ *and* $rr^{-1} - mm' = 1$, *i.e.,* $r^{-1}r \equiv 1 \pmod{m}$ *and* $m'(-m) \equiv 1 \pmod{r}$.
Response: *An integer* $u = [t]_m^r = (tr^{-1}) \bmod m$.
Method: $q := ((t \bmod r)m') \bmod r$;
$u := (t + qm) \operatorname{div} r$;
if $u \ge m$ *then* $u := u - m$;

The simplification occurs when r is chosen as a power of the radix of the arithmetic used, so that the operations $\bmod r$ and $\operatorname{div} r$ become trivial.

Theorem 8.2.30 *Algorithm 8.2.29 correctly computes* $u = (tr^{-1}) \bmod m$ *given* t, *where* $0 \le t \le rm$.

Proof Observing that $q \equiv tm' \pmod{r}$ we have $qm \equiv tm'm \equiv -t \pmod{r}$ and hence that r divides $t + qm$. Also $ur \equiv t \pmod{m}$, so $u \equiv tr^{-1} \pmod{m}$, and finally $0 \le t + qm < rm + rm$, so $0 \le u < 2m$ before the final adjustment. □

Note that the bound on t of the algorithm and the theorem is only used in the proof to insure that at most a single subtraction of m will be needed in the final adjustment.

Now let $x = [a]_m^r$ and $y = [b]_m^r$ and compute $z = \text{M-reduce}(xy)$, then $z = xyr^{-1} \bmod m$ and

$$ab \equiv (xr^{-1})(yr^{-1}) \equiv zr^{-1} \pmod{m},$$

hence $z = [ab]_m^r$ since $0 \le z < m$.

The function M-reduce$(\cdot)$ can also be used to compute M-residues $[\cdot]_m^r$ since

$$[a]_m^r = \text{M-reduce}((a \bmod m)(r^2 \bmod m)),$$

where the constant $r^2 \bmod m$ can be precomputed for the system. By Definition 8.2.28 the function M-reduce$(\cdot)$ directly maps M-residues back into the integer domain.

If many modular multiplications are to be performed, then the cost of mapping back and forth can be amortized, and the multiplications then become simpler, since only reductions modulo r will be needed. We will now, as before, interleave the accumulation of partial products with reductions by M-reduce.

Algorithm 8.2.31 (Montgomery modular multiplication)

Stimulus: *An odd modulus* $m > 2$ *and integers* $k \geq 1, n \geq \ell \geq 1$ *such that* $2m < 2^{kn}$.

Also integers A and B are given, where $0 \leq A < 2m$ *and* $0 \leq B < 2m$ *with* $B = \sum_{i=0}^{\ell}(2^k)^i b_i$, *where* $b_i \in D = \{0, \cdots, 2^k - 1\}$ *for* $0 \leq i \leq \ell - 1$ *and* $b_\ell = 0$.

Normally $\ell = n - 1$ *but may be defined to be smaller.*

With $r = 2^{k(\ell+1)}$, *integers* r^{-1} *and* m' *are given such that* $rr^{-1} - mm' = 1$.

Response: *An integer* $S = S_{\ell+1}$ *such that* $0 \leq S < 2m$ *and* $S \equiv ABr^{-1} \pmod m$

Method: $S_0 := 0$;

for $i := 0$ **to** ℓ **do**

$L : q_i := (((S_i + b_i A) \bmod 2^k)m') \bmod 2^k$;

$S_{i+1} := (S_i + q_i m + b_i A) \text{ div } 2^k$

end

Where: *Invariant* I : $(2^{ki} S_i = A \cdot \sum_{j=0}^{i-1} b_j 2^{kj} + m \cdot \sum_{j=0}^{i-1} q_j 2^{kj})$ **and** $(0 \leq S_i < A + m)$ *holds at label L for* $i = 0, \cdots, \ell$, *and thus also on exit from the loop.*

Observe that q_i is determined directly from the least-significant k digits of $S_i + b_i A$. For an implementation also note that only the least-significant k bits of m' are needed (and r^{-1} is never used), and since m' is a constant for the system, a table look-up is feasible for moderate values of k. Note that ℓ, and thus also r, may be determined by the size of B, but usually $\ell = n - 1$ so $r = 2^{kn}$. However, we shall later utilize a value of ℓ which is of the order $n/2$. Finally, observe that the arithmetic of the loop may be performed on redundant representations of A, B, and S, where B then must be converted digitwise into the non-redundant digit set D, while being used from the least- to the most-significant digit.

Theorem 8.2.32 *Given two M-residues* $A = [a]_m^r$ *and* $B = [b]_m^r$ *with* $r = 2^{k(\ell+1)}$, *Algorithm 8.2.31 computes S, such that* $[ab]_m^r \equiv S \equiv ABr^{-1} \pmod m$ *with* $0 \leq S < 2m$.

Proof As in the proof of Theorem 8.2.30 we note that $q_i \equiv (S_i + b_i A)m' \pmod{2^k}$ so $q_i m \equiv -(S_i + b_i A) \pmod{2^k}$ and hence that 2^k divides $S_i + q_i m + b_i A$ in the updating of S. Next we want to establish that the invariant I holds at label L for all $i = 0, 1, \ldots, \ell$. The invariant holds trivially for $i = 0$. Assuming it holds for $i = h$, from the updating of S_h, $2^k S_{h+1} = S_h + q_h m + b_h A$, we obtain

$$\begin{aligned}
2^{k(h+1)}S_{h+1} &= 2^{kh}\left(S_h + q_h m + b_h A\right)\\
&= A\cdot\sum_{j=0}^{h-1} b_j 2^{kj} + m\cdot\sum_{j=0}^{h-1} q_j 2^{kj} + 2^{kh}q_h m + 2^{kh}b_h A\\
&= A\cdot\sum_{j=0}^{h} b_j 2^{kj} + m\cdot\sum_{j=0}^{h} q_j 2^{kj},
\end{aligned} \tag{8.2.10}$$

hence the first part of the invariant holds for $i = h+1$. The second part follows trivially from (8.2.10) since $0 \le b_i \le 2^k - 1$ and $0 \le q_i \le 2^k - 1$, hence I is indeed invariant during the loop. On exit from the loop

$$2^{k(\ell+1)}S_{\ell+1} = A\cdot B + m\cdot\sum_{j=0}^{\ell} q_j 2^{kj}, \tag{8.2.11}$$

so $S = S_\ell \equiv ABr^{-1} \pmod{m}$ since $r = 2^{k(\ell+1)}$.

Rewriting (8.2.11) using that $0 \le A < 2m$ and $b_\ell = 0$

$$\begin{aligned}
0 \le 2^{k(\ell+1)}S_{\ell+1} &< 2m\cdot\sum_{j=0}^{\ell-1} b_j 2^{kj} + m\cdot\sum_{j=0}^{\ell} q_j 2^{kj}\\
&\le \left(2\frac{2^{k\ell}-1}{2^k-1} + \frac{2^{k(\ell+1)}-1}{2^k-1}\right) m(2^k-1)\\
&< 2^{k(\ell+1)}(1+2^{1-k})m \le 2^{k(\ell+1)}(2m)
\end{aligned}$$

and hence $0 \le S = S_{\ell+1} < 2m$ for $k \ge 1$. □

If the implementation is based on non-redundant arithmetic, it is not very suitable for a hardware implementation, except possibly for a low-radix, systolic implementation. But it is usable for very high-radix software implementations, using multiple-precision integer representations for the operands, and digits determined by the word size of the computer, e.g., k =8 (for "smart-cards"), 16, or 32. By processing digit multiplications in order from the least- to the most-significant positions, carries can be brought forward trivially.

For hardware realizations redundancy is essential to allow constant-time additions inside the loop. This is feasible for moderate values of the multiplier radix, say 2, 4, or possibly 8 ($k = 1, 2$, or 3), where the implementation of the arithmetic in the loop may exploit any kind of redundant digit set. But if the output of a previous multiplication is to be used subsequently as a multiplier B, then the digits b_i of the multiplier must be converted into the non-redundant digit set $D = \{0, 1, \ldots, 2^k - 1\}$ when used.

Some simplifications of the loop statements. In the case that $k = 1$, i.e., the radix is 2, the determination of q_i reduces to $q_i = (S_i + b_1 A) \bmod 2$. This similarly applies for all values where $m' \bmod 2^k = 1$, where the multiplication is then avoided also for higher values of the radix 2^k. This leads to the transformation of the modulus m into a new value $\widetilde{m}$ having this property: simply set $\widetilde{m} = (m' \bmod 2^k)m$ so that $\widetilde{m} \equiv -1 \pmod{2^k}$, a transformation that only has to be performed once. The bounds on the operands are now $0 \le A, B < 2\widetilde{m}$, and on the result they are $0 \le S < 2\widetilde{m}$. The method of Algorithm 8.2.31 now becomes:

Method: $S_0 := 0;$
for $i := 0$ **to** ℓ **do**
$L:\quad q_i := (S_i + b_i A) \bmod 2^k$
$\qquad S_{i+1} := (S_i + q_i\widetilde{m} + b_i A) \operatorname{div} 2^k$
end

The correctness follows trivially from the previous proof and the definition of $\widetilde{m}$.

Next we can move the addition of the term $b_i A$ from the quotient determination to the updating of S, when substituting A by $2^k A$ to obtain the following:

Method: $S_0 := 0;\ q_i := 0;$
for $i := 0$ **to** ℓ **do**
$L:\quad S_{i+1} := (S_i + q_i\widetilde{m}) \operatorname{div} 2^k + b_i A$
$\qquad q_{i+1} := S_{i+1} \bmod 2^k$
end

where the invariant holding at label L now is

$$I : 2^{ki} S_i = 2^k A \cdot \sum_{j=0}^{i-1} b_i 2^{kj} + \widetilde{m} \cdot \sum_{j=0}^{i-1} q_i 2^{kj} \quad \textbf{and} \quad 0 \le S_i < 2^k A + \widetilde{m},$$

which is easy to prove. To verify the response note that $b_\ell = 0$ and $q_0 = 0$, so on exit

$$\begin{aligned} 0 \le 2^{k(\ell+1)} S_{\ell+1} &= 2^k A \cdot \sum_{j=0}^{\ell} b_j 2^{kj} + \widetilde{m} \cdot \sum_{j=0}^{\ell} q_j 2^{kj} \\ &= 2^k A \cdot \sum_{j=0}^{\ell-1} b_j 2^{kj} + 2^k\widetilde{m} \cdot \sum_{j=0}^{\ell-1} q_{j+1} 2^{kj} \\ &< 2^k 2(2^{k\ell} - 1)\widetilde{m} < 2^{k(\ell+1)}(2\widetilde{m}), \end{aligned}$$

proving that $S_{\ell+1} \equiv ABr^{-1} \pmod{m}$ with $r = 2^{k(\ell+1)}$ as well as $0 \le S = S_{\ell+1} < 2m$.

Finally, noting that $S_i - q_i$ and $\widetilde{m} + 1$ are both divisible by 2^k, we can rewrite the updating of S_i in the loop:

$$\begin{aligned}(S_i + q_i\widetilde{m}) \text{ div } 2^k + b_i A &= (S_i - q_i) \text{ div } 2^k + (q_i\widetilde{m} + q_i) \text{ div } 2^k + b_i A \\ &= S_i \text{ div } 2^k + (q_i(\widetilde{m} + 1)) \text{ div } 2^k + b_i A \\ &= S_i \text{ div } 2^k + q_i((\widetilde{m} + 1) \text{ div } 2^k) + b_i A,\end{aligned}$$

where $(\widetilde{m} + 1)$ div 2^k is a constant which can be precomputed once for each value of m. This rewriting eliminates the need to calculate the carry-out of the expression $S_i + q_i\widetilde{m}$ in the updating of S_i.

Combining these optimizations we arrive at the following modified algorithm:

Algorithm 8.2.33 (Optimized Montgomery modular multiplication)

Stimulus: *An odd modulus* $m > 2$ *and integers* $k \geq 1, n > \ell \geq 1$ *such that* $2m < 2^{kn}$.
Also integers A *and* B *are given where* $0 \leq A < 2m$ *and* $0 \leq B < 2m$ *with* $B = \sum_{i=0}^{\ell}(2^k)^i b_i$, *where* $b_i \in D = \{0, \ldots, 2^k - 1\}$ *for* $0 \leq i \leq \ell - 1$ *and* $b_\ell = 0$. *Normally* $\ell = n - 1$ *but may be defined to be smaller.*
With $r = 2^{k(\ell+1)}$, *integers* r^{-1} *and* m' *are given such that* $rr^{-1} - mm' = 1$, *defining constants* $\widetilde{m} = (m' \bmod 2^k)m$ *and* $M = (\widetilde{m} + 1) \text{ div } 2^k$.

Response: *An integer* $S = S_{\ell+1}$ *such that* $0 \leq S < 2m$ *and* $S \equiv ABr^{-1} \pmod{m}$

Method: $S_0 := 0$;
for $i := 0$ **to** ℓ **do**
$L: S_{i+1} := S_i \text{ div } 2^k + q_i M + b_i A$;
$q_{i+1} := S_{i+1} \bmod 2^k$
end

Where: *Invariant* I : $(2^{ki} S_i = A \cdot \sum_{j=0}^{i-1} b_j 2^{kj} + m \cdot \sum_{j=0}^{i-1} q_j 2^{kj})$ **and** $(0 \leq S_i < A + m)$ *holds at label* L *for* $i = 0, \ldots, \ell$, *and thus also on exit from the loop.*

Initially operands have to be converted into the special residue system. This can be achieved by the modular multiplication algorithm, by multiplication with the precomputed constant $r^2 \bmod m$. The algorithm can also be used for the final conversion back, since multiplication by the constant 1 maps S back into an ordinary residue.

It turns out that for this particular modular multiplication it is possible to show a stricter bound on the result, such that essentially it is fully reduced. Let us first consider the resulting bound of the non-redundant Algorithm 8.2.31, where in general the result satisfies $0 \leq S < 2m$ with $m < r = 2^{kn}$. The inputs to the conversion multiplication are $A = Y$, $0 \leq Y < 2m$, and $B = 1$. The result S satisfies $S \cdot r = Y \cdot 1 + Q \cdot m$ for some value of $Q = \sum_{j=0}^{n-1} q_j 2^{kj}$ with

$0 \leq q_i \leq 2^k - 1$, thus Q is bounded, $0 \leq Q \leq 2^{kn} - 1$. Hence we find

$$0 \leq Sr = S2^{kn} = Y + Qm < 2m + (2^{kn} - 1)m = (1 + 2^{kn})m$$

and thus $S < (1 + 2^{-kn})m$, from which it follows that $S \leq m$ since $m < 2^{kn}$ and S is integral. The only situation where the result S is not fully reduced is when $S = m$ and then a subtraction of m is needed, i.e., when $S \equiv x^e \equiv 0 \pmod{m}$.

Conversion into this residue system and back again can only be justified if several dependent multiplications are needed, so that the cost can be amortized over a more composite calculation. Modular exponentiation is an example of such a compute intensive operation, which is heavily used in cryptosystems (e.g., the RSA encryption scheme) with very large operands, typically of the order 500–2000 bits.

However, for these applications the input x to the exponentiation satisfies $0 \leq x < m$. Hence $x^e \equiv 0 \pmod{m}$ iff $x \equiv 0 \pmod{m}$, and thus no final subtraction of m will be needed in RSA applications.

Observation 8.2.34 *If Montgomery modular multiplication is being used to implement the modular exponentiation for RSA encryption or decryption, then the final conversion multiplying by the constant 1 automatically insures that the resulting ordinary residue S is in the interval* $0 \leq S < m$. *Thus only conversion into a non-redundant representation may be needed.*

Before looking at the modular exponentiation we shall first look at a trick that potentially halves the time for modular multiplication by combining in parallel the algorithms for interleaved multiplication and Montgomery modular multiplication, Algorithms 8.2.27 and 8.2.31, both operating on multipliers B of about half the number of digits that the modulus m has.

Exploiting more parallelism in modular multiplication. For this *bipartite modular multiplication* we shall now in Montgomery multiplications use $r = 2^{k(\ell+1)}$, $\ell = \alpha n$, where α is a rational number, $0 < \alpha < 1$, such that αn is an integer. With $\alpha \approx \frac{1}{2}$ and $\cdot$ denoting the ordinary integer multiply operator, define two different modular multiplication operators:

$$X \times Y = X \cdot Y \bmod m,$$
$$X \otimes Y = X \cdot Y \cdot r^{-1} \bmod m,$$

where the former can be performed by the interleaved modular multiplication (Algorithm 8.2.27) and the latter by the Montgomery modular multiplication (Algorithm 8.2.31).

To perform modular multiplication of two operands U and V first map them into the residue system $\mathbb{Z}_m$ defined by the M-residue operator, so that $A = [U]_m^r$ and $B = [V]_m^r$. Now to calculate the product $A \otimes B \in \mathbb{Z}_m$ we split the multiplier in a high and a low order part, $B = B_h \cdot r + B_l$ such that $|B_l| < r$ and evaluate

$$\begin{aligned}(A \times B_h + A \otimes B_l) \bmod m &= (A \cdot B_h + A \cdot B_l \cdot r^{-1}) \bmod m \\ &= A(B_h + B_l \cdot r^{-1}) \bmod m \\ &= A(B_h \cdot r + B_l) \cdot r^{-1} \bmod m \\ &= A \cdot B \cdot r^{-1} \bmod m \\ &= A \otimes B.\end{aligned}$$

Note that $(A \times B_h) \bmod m$ and $(A \otimes B_l) \bmod m$ can be evaluated in parallel, and the multipliers B_h and B_l are significantly smaller than B itself, and thus the two parallel multiplications can be expected to finish in about half the time of a single multiplication using the full multiplier B. The choice of α can be used to balance the time needed for the two parallel multiplications. Finally, the two products just have to be added modulo m, to form the result as the M-residue $A \otimes B$ of the product UV.

Using interleaved modular multiplication (Algorithm 8.2.27), it is possible to map an operand x into $y = [x]_m^r = x \times r = x \cdot r \bmod m$ using the definition, thus avoiding precalculation of the constant $r^2 \bmod m$.

8.2.7 Modular exponentiation

A standard way of performing the modular exponentiation $x^e \bmod m$ is by repeated modular multiplications and squarings, scanning the exponent e from the least-significant end (denoted RL for right-to-left), as described below.

Algorithm 8.2.35 (RL modular exponentiation)

Stimulus: *A modulus* $m \geq 2$, *and integers* x *and* e, *where* $0 \leq x < m$ *and* $e \geq 0$, $e = \sum_{i=0}^{n-1} e_i 2^i$.

Response: *An integer* y, *such that* $y = x^e \bmod m$.

Method:
$y := 1;\ z := x;$
for $i := 0$ **to** $n - 1$ **do**
  if $e_i = 1$ **then** $y := (y * z) \bmod m;$
  $z := (z * z) \bmod m$
end

Where: *The multiplications performed may be any of the modular multiplications on a standard or M-residue representation of* x.

Each cycle of the loop potentially requires two multiplications (noting that one factor is in common), which may be executed in parallel or pipelined through a single multiplier. In this case the product $y \times z$ may just be inhibited if it is not needed, thus the execution time will be constant, which is advantageous if used in cryptosystems for communication.

Alternatively, the exponent e may be scanned from left to right, as in the following variant of the algorithm, where we note that the factor x remains a constant and it only needs two registers (for x and y, since e may be read bit by bit), whereas the RL version requires one more (for z).

Algorithm 8.2.36 (LR modular exponentiation)

Stimulus: *A modulus $m \geq 2$, and integers x and e, where $0 \leq x < m$ and $e \geq 0$, $e = \sum_{i=0}^{n-1} e_i 2^i$.*

Response: *An integer y, such that $y = x^e \bmod m$.*

Method: $y := 1$;
    for $i := n - 1$ **downto** 0 **do**
        $y := (y * y) \bmod m$;
        if $e_i = 1$ **then** $y := (y * x) \bmod m$
    end

Where: *The multiplications performed may be any of the modular multiplications on a standard or M-residue representation of x.*

8.2.8 Inheritance and periodicity modulo 2^k

It is obvious in Algorithm 8.2.13 that the low-order k bits of the developed binary representation of $|a^{-1}|_{2^n}$ only depend on the low-order k bits of a, $k \leq n$, i.e., of $|a|_{2^k}$. In general a function with an *inheritance* property may be defined as follows.

Definition 8.2.37 *An integer function $f : \mathbb{Z} \to \mathbb{Z}$ is said to be n-digit* hereditary *if and only if for any integers a, b, $0 \leq a \leq \beta^k - 1$, there exists integers c, $0 \leq c \leq \beta^k - 1$, and d such that*

$$f(a) = c \Rightarrow f(b\beta^k + a) = d\beta^k + c \quad \text{for } k = 1, 2, \ldots, n.$$

Informally this says that the radix representation of $f(a)$ can be developed from the radix representation of a, digit-by-digit, least-significant digit first. Alternatively, one could also more compactly define f to be n-digit hereditary if and only if $|f(a)|_{\beta^k} = f(|a|_{\beta^k})$ for $1 \leq k \leq n$.

Many integer functions have this property, e.g., obviously $f(a) = a^2$ and $f(a) = |a^{-1}|_{2^n}$ as noted above. Generalizing, a similar property can be defined for functions in more than one variable, e.g., the function $f(a, b) = a + b$ satisfies $|f(a, b)|_{\beta^k} = \left||a|_{\beta^k} + |b|_{\beta^k}\right|_{\beta^k} = |f(|a|_{\beta^k}, |b|_{\beta^k})|_{\beta^k}$. But we will below only consider functions of one variable, and only the important case of $\beta = 2$, corresponding to much integer arithmetic being performed modulo 2^n, for values like $n = 8, 16, 32, 64, \ldots$.

We now want to prove that the function $f_k(i) = |3^i|_{2^k}$ is n-bit hereditary for any $n > 0$. For integers $a \geq 0$, and $0 \leq b < 2^k$ with $k = 1, 2, \ldots, n$,

$$f_k(a2^k + b) = \left|3^{(a2^k+b)}\right|_{2^k} = \left||3^{(a2^k)}|_{2^k} \cdot |3^b|_{2^k}\right|_{2^k},$$

hence $f_k(a2^k + b) = f_k(b)$ provided that $|3^{(a2^k)}|_{2^k} = 1$ for $a \geq 0$, which is easily seen to be satisfied if $|3^{(2^k)}|_{2^k} = 1$. To prove the latter, we will prove the stronger result that $|3^{(2^k)}|_{2^{(k+2)}} = 1$. Note that this holds for $k = 1$ ($|9|_8 = 1$), thus assume that it holds for $j = 1, 2, \ldots, k$ and consider the following induction assumption:

$$\begin{aligned}
&\left|3^{(2^k)}\right|_{2^{(k+2)}} = 1 \\
&\Rightarrow\ 3^{(2^k)} = u\,2^{k+2} + 1 \quad \text{for some integer } u > 0 \\
&\Rightarrow\ 3^{(2^{k+1})} = \left(3^{(2^k)}\right)^2 = \left(u\,2^{k+2} + 1\right)^2 = (u^2 2^{k+1} + u)2^{k+3} + 1,
\end{aligned}$$

hence $|3^{(2^{k+1})}|_{2^{k+3}} = 1$ and $|3^{(2^j)}|_{2^{j+2}} = 1$ holds for all $j \geq 1$. Observe that this implies $|3^{(2^i)}|_{2^k}$ is periodic as a function of i with period 2^{k-2}. We have thus proved the following lemma.

Lemma 8.2.38 *The function $f_k(i) = |3^i|_{2^k}$ is n-bit hereditary for any $n > 0$ and fixed $k = 1, 2, \ldots, n$, i.e., for any pair of non-negative integers a, b, where $b < 2^k$,*

$$\left|3^{(a2^k+b)}\right|_{2^k} = |3^b|_{2^k}.$$

Furthermore $|3^{(2^{k-2})}|_{2^k} = 1$, thus $f_k(i)$ is periodic with period 2^{k-2}.

Note that since $f_k(i) = |3^i|_{2^k}$ is a modular exponential function, its inverse function is a modular, discrete logarithm function. Obviously, it can only be defined on odd non-negative integers. As an example consider for $k = 4$:

i	0	1	2	3
$\lvert 3^i \rvert_{16}$	1	3	9	11
$\lvert -3^i \rvert_{16}$	15	13	7	5

where 3^i runs through half of the odd integers between 0 and $2^k - 1 = 15$, and -3^i runs through the other half. We leave it as an exercise to prove that this holds in general.

Although it is possible to define a discrete modular logarithm for more general values of the radix and modulus, let us restrict ourselves to the case $\beta = 3$ and modulus 2^k.

Definition 8.2.39 *The* discrete logarithm, $\mathrm{dlg}_k(r)$, *modulo 2^k for $k \geq 1$ and radix 3, of an odd residue r, $1 \leq r \leq 2^k - 1$, is the minimal exponent i, $0 \leq i < 2^{k-2}$, when it exists, such that $|3^i|_{2^k} = r$.*

For a given (moderate) value of k the function $\mathrm{dlg}_k(r)$ can be tabulated very compactly using the periodicity and the fact that it is hereditary. However, its size will still be in $O(2^k)$.

Note from the above that $\mathrm{dlg}_k(r)$ only exists for half of the odd residues modulo 2^k, the other half (the complementary set) are solutions to $|-3^i|_{2^k} = |2^{(2^k)} - 3^i|_{2^k} = r$. Thus for any odd residue r, $1 \leq r \leq 2^k - 1$, there exists a unique pair (sign, exponent), (s, i), such that

$$|(-1)^s 3^i|_{2^k} = r. \tag{8.2.12}$$

In the following we shall call (8.2.12), or the pair (s, i), the *dlg factorization* of r.

From Lemma 8.2.38 note that $|3^i|_8$ has period 2, it cycles through the set $\{1, 3\}$ and $|-3^i|_8$ through the set $\{5, 7\}$. It thus follows that $\mathrm{dlg}_k(r)$ exists if and only if $|r|_8 \in \{1, 3\}$. The sign s in the dlg factorization (8.2.12) can easily be determined from $|r|_8$, so what remains is to determine either $\mathrm{dlg}_k(r)$ or $\mathrm{dlg}_k(2^k - r)$. Note that we can determine s as the bit of weight 4 in r, as this is what distinguishes whether $|r|_8$ is in $\{1, 3\}$ or in $\{5, 7\}$, i.e., $s = |r|_8$ div 4. To simplify the discussion below, let $\mathrm{bit}(i, a) = |a|_{2^{i+1}}$ div 2^i denote the ith bit in the binary representation of a.

As an aid to obtaining $\mathrm{dlg}_k(r)$ there is the following useful lemma on an alternative factorization.

Lemma 8.2.40 *For any $k \geq 2$, every odd integer r, $1 \leq r \leq 2^k - 1$, has a unique factorization,*

$$r = \left| (-1)^s \prod_{i \in I_r} (2^i + 1) \right|_{2^k},$$

for some value of $s \in \{0, 1\}$ and index set $I_r \subseteq \{1, 2, \ldots, k-1\}$ for $k \geq 3$.

Proof Without loss of generality let us assume that $|r|_8 \in \{1, 3\}$ so $s = 0$. Let $p_0 = 1$, such that $\mathrm{bit}(0, p_0) = \mathrm{bit}(0, r) = 1$. Then for $i = 1, 2, \ldots, k-1$, if $\mathrm{bit}(i, p_{i-1}) \neq \mathrm{bit}(i, r)$ let $p_i = |(2^i + 1)p_{i-1}|_{2^k}$ and add i to the set I_r, otherwise let $p_i = p_{i-1}$. Since p_i is odd for all i, adding $2^i p_{i-1}$ to p_{i-1} will insure that $\mathrm{bit}(i, p_i) = \mathrm{bit}(i, r)$, hence $p_{k-1} = r$ when the iteration terminates. □

Note that the index value 2 will never be a member of I_r, since $\mathrm{bit}(2, r) = 0$ because $|r|_8 \in \{1, 3\}$, possibly after complementation. Also note that $|2^i + 1|_8 = 1$ for $i \geq 3$ and $i = 1 \Rightarrow |2^i + 1|_8 = 3$, hence the dlg factorization of $2^i + 1$ will have $s = 0$ for $i = 1, 3, 4, \ldots, k-1$.

It is now easy to develop the dlg factorization of r in the form of the pair $(s, \mathrm{dlg}_k(r))$ such that $r = |(-1)^s 3^{\mathrm{dlg}_k(r)}|_{2^k}$, by calculating the sum $\sum_{i \in I_r} \mathrm{dlg}_k(2^i + 1)$, provided that a table of $\mathrm{dlg}_k(2^i + 1)$ is given. Thus we have the following shift-and-add, $O(k)$-time algorithm.

Algorithm 8.2.41 (Discrete modular logarithm, DLG)
Stimulus: *An odd integer r and $k \geq 3$ where $1 \leq r \leq 2^k - 1$, together with a table T where $T[i] = \mathrm{dlg}_k(2^i + 1)$ for $i = 1, 3, 4, \ldots, k-1$.*
Response: *A pair (s, d) such that $r = |(-1)^s 3^d|_{2^k}$, $s \in \{0, 1\}$ and $0 \leq d < 2^{k-2}$, i.e., $d = \mathrm{dlg}_k(r)$ or $d = \mathrm{dlg}_k(2^k - r)$.*
Method: **if** $|r|_8 \in \{1, 3\}$ **then** $s := 0$ **else** $s := 1; r := 2^k - r;$ **end**;
$p := 1; d := 0;$

```
for i := 1, 3 to k − 1 do
    if bit(i, r) ≠ bit(i, p) then
        p := |p + p ≪ i|_{2^k};
        d := d + T[i];
    end;
end;
d := d mod 2^{k−2};
```

Where: *$p \ll i$ performs i binary left-shifts of p, thus realizing $p \times 2^i$.*

Example 8.2.5 Let $k = 6$ so $2^k = 64$, then the following table of constants can be found:

i	1	2	3	4	5
$\mathrm{dlg}_k(2^i + 1)$	1		2	4	8

For the dlg-factorization of $r = 59$, the algorithm first determines $s = 0$ since $|59|_8 = 3$, and then the following values are found:

	p		d	
i	decimal	binary	decimal	binary
0	1	1	0	0
1	3	11	1	1
3	27	11011	3	11
4	27	11011	3	11
5	59	111011	11	1011

hence $59 = \left|(-1)^0 3^{11}\right|_{64}$ which is easily checked. □

The dlg factorization may be used as a logarithmic representation of odd residue numbers modulo 2^k, thus simplifying modular multiplication in this residue system. Note that the set of odd residues is closed under multiplication, but the system cannot represent even numbers.

However, to obtain a representation for any (odd or even) integer r, $1 \le r \le 2^k - 1$, just add as a third component a power of 2, such that the triple (s, i, j) represents the factorization $|(-1)^s 3^i 2^j|_{2^k} = r$.

Observation 8.2.42 *For any $k \ge 3$ and any residue r, $1 \le r \le 2^k - 1$, there exists a factorization*

$$r = |(-1)^s 3^i 2^j|_{2^k}, \quad \text{where } s \in \{0, 1\}, \quad 0 \le i \le 2^{k-2} - 1, \quad 0 \le j \le k - 1,$$

which is unique for $0 \le i \le 2^{k-j-2}$ when $j \le k - 2$. This logarithmic representation allows modular multiplication by addition of exponents. Subsequent reduction modulo 2^k can be performed trivially by exploiting the periodicity of 3^i and 2^j modulo 2^k.

For the modulus $m = 2^k$, implementation of the modular exponentiation operation $y = x^e \bmod 2^k$ can be reduced to a sequence of redundant additions, avoiding the successive dependent squarings and multiplications employed for the general modulus m in the RL and LR Exponentiation Algorithms, Algorithms 8.2.35 and 8.2.36.

Observation 8.2.43 *For any $k \geq 3$ and residues x, e satisfying $1 \leq x \leq 2^k - 1$ and $0 \leq e \leq 2^{k-2} - 1$, the modular exponentiation operation $y = |x^e|_{2^k}$ can be implemented in $O(k)$ additions, where all additions can be performed in redundant adders.*

Proof (Sketch) From Observation 8.2.42 note that $y = |x^e|_{2^k} = |(-1)^{se}3^{ie}2^{je}|_{2^k}$. The even factors in x and y simply correspond to shifts, where having j low order zeroes in x implies y will have $(e \times j)$ low order zeroes. Hence to simplify the discussion, just consider $y = |x^e|_{2^k}$ for odd x. From Algorithm 8.2.41 we obtain the discrete log pair (s, i) corresponding to $x = |(-1)^s 3^i|_{2^k}$. In parallel, as the bits of i become available serially from right to left, they can be used for an accumulation of $e' = i \times e$. The settled low-order bits of e' may be used sequentially from right to left to determine members of the index set $I_{e'}$, such that the identity $\sum_{j \in I_{e'}} \mathrm{dlg}_k(2^j + 1) = i \times e = e'$ gradually becomes satisfied. The product $r = |\prod_{j \in I_{e'}} (2^j + 1)|_{2^k}$ may then concurrently be formed by shift-and-add operations. Thus on termination $\mathrm{dlg}(r) = e'$, so that $r = |3^{e'}|_{2^k}$, and consequently $y = |(-1)^{se} r|_{2^k} = |(-1)^{se}3^{ie}|_{2^k} = |x^e|_{2^k}$. All these accumulations can be executed concurrently in separate redundant adders. □

For the k-bit residue dlg factorization $r = |(-1)^s 3^i 2^j|_{2^k}$, the range for the ternary exponent i decreases as the value of the binary exponent j increases according to $0 \leq i < 2^{(k-2)-j}$, since by Lemma 8.2.38, $|3^i|_{2^k}$ is periodic with period 2^{k-2}. This allows the discrete logarithm triple (s, i, j) for every k-bit integer to be encoded into a unique k-bit word partitioned into three variable sized, self delimiting fields.

Definition 8.2.44 *For $k \geq 3$ and $1 \leq r \leq 2^k - 1$, let $r = |(-1)^s 3^i 2^j|_{2^k}$, with $0 \leq i < 2^{(k-2)-j}$ and $0 \leq j \leq k - 1$. Then the k-bit discrete log encoding (dl encoding) of (s, i, j) comprises a low-order, self delimiting $(j + 1)$-bit field for the encoding of 2^j, a one-bit field for the term $|s + i|_2$, and a high order $((k - 2) - j)$-bit field for i. By extension, the value $r = 0$ is encoded as all zeroes.*

Note that the even factor term 2^j effectively uses a $(j + 1)$-bit unary encoding for the value j that is self delimiting read right to left, determining the three-part partition of any k-bit dl encoding. The one-bit sign term is defined as $|s + i|_2$ rather than s, as it can be shown that the resulting dl encoding then satisfies the hereditary property. Also note that the all zeroes encoding of zero corresponds to a special encoding for $j = k$.

Example 8.2.6 Consider determining the eight-bit dl encoding of the eight-bit integer $(152_{10}) = 10011000_2 = (10011_2) \times (1000_2) = 19 \times 2^3$. Applying Algorithm 8.2.41 to the five-bit odd factor, we obtain $19 = |(-1)^0 3^5|_{32}$, so $i = 101_2$. Thus our eight-bit dl-encoding has a low order four-bit field $2^3 = 1000$, an intermediate sign field $|i + s|_2 = 1$, and a three-bit high order field for $i = 101_2$, giving 101:1:1000. Successive leading bit deletions illustrate for this case that the encoding satisfies the hereditary property.

Decimal	Binary	Dl encoding
152	1001 1000	101:1:1000
24	001 1000	01:1:1000
24	01 1000	1:1:1000
24	1 1000	:1:1000
8	1000	::1000
0	000	000

□

Lemma 8.2.45 *For any $k \geq 1$, the dl encoding of k-bit integers is a one-to-one hereditary function.*

Problems and exercises

8.2.1 Assume you were asked to compute the value of $|x^e|_p$ for some very large integer values of x and e for a prime p that is small compared to x, e. How would you proceed to minimize your computations?

8.2.2 Show that $\left|a^i\right|_m$, as a function of $i = 0, 1, \ldots$ is periodic for any a, m.

8.2.3 Compute $|21 \cdot 50|_{63}$ by application of Lemma 8.2.5.

8.2.4 Find the multiplicative inverse of 4 modulo 7 by means of Fermat's Little Theorem, as well as the Euclidean Algorithm.

8.2.5 Find the multiplicative inverse of 13 modulo 16 by means of Algorithm 8.2.13, as well as the Euclidean Algorithm. Then determine $|13^{-1}|_{2^{256}}$ using Theorem 8.2.15.

8.2.6 Show that for given integers a and b, the EEA Algorithm (Algorithm 8.2.18) can be used to solve a *Diophantine equation* of the form: $ax + by = 1$, assuming $\gcd(a, b) = 1$.

8.2.7 Design an adder network to perform a multioperand modular addition using (8.2.8) for $m = 7$ and four operands.

8.2.8 Show that the carry $c_{out} = c' \text{ div } 2^n \leq 1$ in Figure 8.2.1.

8.2.9 Design in terms of adders, etc. an internal node of a tree structure for multioperand redundant addition using Algorithm 8.2.26. Also design the root node for the final conversion and reduction of the result.

8.2.10 Compute the product $|4 \cdot 6|_7$ using the method of index tables.

8.2.11 In Montgomery multiplication by Algorithm 8.2.31 it is necessary to precompute the value of m', or rather $m' \bmod 2^k$, since that is all that is needed. Show how this can be done using Algorithm 8.2.13. (Hint: There is no need to compute r^{-1}.)

8.2.12 Prove the correctness of Algorithms 8.2.35 and 8.2.36.

8.2.13 If only one multiplier is available for Algorithm 8.2.35 or 8.2.36, the multiplications have to be executed sequentially. The timing then depends on the number of non-zero bits in the n-bit exponent e, hence the number of multiplications will vary between n and $2n$, with an average of $1.5n$. Can you devise some way of decreasing the average, as well as the maximal, number of multiplications necessary? Does it then make a difference which algorithm is used?

8.2.14 Show that the powers of 3 from 0 to 2^k cycle through half the odd residues modulo $2^{(k+2)}$, and their negative powers through the other half. Thus it is possible to efficiently enumerate all odd numbers and their inverses modulo 2^p for any p.

8.2.15 In the table of constants, $T[i] = \text{dlg}_k(2^i + 1)$ in Example 8.2.5, it appears that only powers of 2 are present. Is this always true?

8.2.16 Prove Lemma 8.2.45.

8.2.17 Prove that there are exactly 2^{2^k-1} k-bit hereditary one-to-one functions. (Hint: Consider a table look-up structure for a generic k-bit hereditary one-to-one function.)

8.3 Multiple modulus (residue) number systems

Although there are many number theoretical applications in which the simple modulus system $(\mathbb{Z}_m, \oplus_m, \otimes_m)$ is the natural system, arithmetic in such a system becomes slow and costly if m is large. If m is a prespecified, very large integer, and is either a prime or its factorization is unknown as in crypto systems, there is only the "hard way" of implementing the arithmetic operations as described in the previous section. But if the modulus can be factored into a product of a number of smaller moduli, or can be suitably chosen as such a product, then the problem of performing the basic arithmetic operations, addition and multiplication, reduces to performing the same operations (in parallel) in a number of simpler systems defined by the smaller moduli.

We thus define the *base* β of a *residue number system* (RNS) as a vector

$$\beta = [m_1, m_2, \ldots, m_n] \tag{8.3.1}$$

with

$$\gcd(m_i, m_j) = 1 \quad \text{for } i \neq j.$$

Given any integer a we can then define its *RNS representation or residue vector* $|a|_\beta$ as:

$$|a|_\beta = [|a|_{m_1}, |a|_{m_2}, \ldots, |a|_{m_n}]. \tag{8.3.2}$$

If we denote the set of such residue vectors:

$$\mathcal{R}_\beta = \{|a|_\beta \mid a \in \mathbb{Z}\}, \tag{8.3.3}$$

then the cardinality of this set is

$$M = |\mathcal{R}_\beta| = \prod_{i=1}^{n} m_i.$$

We shall postpone the discussion of mapping to and from this system until the next section.

The fundamental properties of the set $\mathcal{R}_\beta$ as a number system then follow from the next two theorems, the first of which tells us that $|a|_\beta$ is a non-redundant representation of a modulo M.

Theorem 8.3.1 *With* $\beta = [m_1, m_2, \ldots, m_n]$ *and* $M = \prod_1^n m_i$, *where* $\gcd(m_i, m_j) = 1$ *for all* $i \neq j$, *then*

$$|a|_\beta = |b|_\beta \quad \textit{iff} \quad a \equiv b \pmod{M}.$$

Proof Let $|a|_\beta = [a_1, a_2, \ldots, a_n]$, then for all i, $a_i = a - p_i m_i$ for some value of p_ι, and similarly for $|b|_\beta = [b_1, b_2, \ldots, b_n]$ for all i, there is a q_i such that $b_i = b - q_i m_i$. Hence if $|a|_\beta = |b|_\beta$, then for all i

$$a_i = b_i \Rightarrow a - b = (p_i - q_i)m_i,$$

thus $a - b$ is divisible by m_i. But since this holds for all i, $a - b$ is divisible by M as $\gcd(m_i, m_j) = 1$ for $i \neq j$.

Conversely, if $a - b$ is a multiple of $M : a - b = k \cdot M$, then

$$a - b = (a_i + p_i m_i) - (b_i + q_i m_i) = k \cdot M,$$

hence $a_i - b_i$ must be divisible by m_i. But by definition $0 \leq a_i = |a|_{m_i} < m_i$ and similarly for b_i, so

$$-m_i < a_i - b_i < m_i,$$

hence $a_i = b_i$ for all i. □

Now, with $|a|_\beta$ and $|b|_\beta$ defined by (8.3.2), define operators $\oplus_\beta$ and $\otimes_\beta$ by:

$$|a|_\beta \oplus_\beta |b|_\beta = [|a_1 + b_1|_{m_1}, |a_2 + b_2|_{m_2}, \ldots, |a_n + b_n|_{m_n}]$$

and

$$|a|_\beta \otimes_\beta |b|_\beta = [|a_1 * b_1|_{m_1}, |a_2 * b_2|_{m_2}, \ldots, |a_n * b_n|_{m_n}].$$

In the following we shall just use the ordinary notation for addition and multiplication, when the operators can unambiguously be determined.

Theorem 8.3.2 *The system* $(\mathbb{Z}_M, \oplus_M, \otimes_M)$ *is isomorphic to the system* $(\mathcal{R}_\beta, \oplus_\beta, \otimes_\beta)$*, and both are finite commutative rings.*

Proof This follows from Theorem 8.3.1 and the definition of the operators. □

As an illustration, consider the following example.

Example 8.3.1 With $\beta = [3, 5, 7]$ we have

$$|4|_\beta = [1, 4, 4], \quad |6|_\beta = [0, 1, 6], \quad |24|_\beta = [0, 4, 3] \quad \text{and } |20|_\beta = [2, 0, 6].$$

Hence we find

$$\begin{aligned}
|4 \cdot 6|_\beta &= [0, 4, 3] \ (= |24|_\beta),\\
|24 - 20|_\beta &= [1, 4, 4] \ (= |4|_\beta),\\
|24 + 20|_\beta &= [2, 4, 2] \ (= |44|_\beta = [2, 4, 2]),\\
|24 \cdot 20|_\beta &= [0, 0, 4] \ (= |480|_\beta = [0, 0, 4]),
\end{aligned}$$

but notice in the last example that $480 > M = 3 \cdot 5 \cdot 7 = 105$, so an "overflow" has occurred. Checking, we find that also $|60|_\beta = [0, 0, 4]$ but $60 \equiv 480 \pmod{105}$. □

Thus it is possible to perform addition, subtraction, and multiplication componentwise on the residue vectors, as long as we can avoid the overflow situation. Also, the operations on the residues can be performed truly in parallel, since there is no interaction between the components. By choosing a sufficiently large set of mutually prime moduli, it is possible to choose any appropriate range of representable values $0 \le a < M = \prod_1^n m_i$.

Since the speed of a parallel implementation in such separate *channels* will depend only on the magnitude of the largest channel modulus, m_i, it is then obviously advantageous to choose the moduli such that they are all of about the same order of magnitude (the set is *balanced*) . If a computational range of $[0; M)$ is wanted with n moduli, then the moduli should be chosen with values around $M^{1/n}$.

One way of choosing moduli is to select a set of consecutive primes; another is to choose numbers of the form $2^k \pm 1$, possibly together with a single power of 2. However, notice that $2^{2k} - 1 = (2^k + 1)(2^k - 1)$ so certain combinations have to be avoided.

Example 8.3.2 It has been proposed to use the set

$$\left\{2^m - 1, 2^{2^0 m} + 1, 2^{2^1 m} + 1, 2^{2^2 m} + 1, \ldots, 2^{2^k m} + 1\right\}$$

for some $k \ge 0$ and $m \ge 1$, where it can be shown that these moduli are mutually prime. It is easy here to see that the system modulus is $M = 2^{2^{k+1} m} - 1$, and note that except for the first two, the size of each modulus is about the square of the previous, and the largest modulus is very close to $\sqrt{M}$, so this system is far from balanced.

Now consider multiplication of two numbers represented in this residue system. The residues of the factors corresponding to the largest modulus could both be of a size close to that modulus (i.e., close to $\sqrt{M}$), hence their product before reduction would be close to M. Thus computations in this particular channel would have to be able to handle values in the range $[0; M)$, the same range as the whole system. Thus there would not really be any need for the remaining moduli, one could as well compute in a standard radix integer system. Similarly, addition of double-length numbers only requires one extra level in a carry look-ahead tree, compared with single-length addition. □

With ROM-based implementations of addition and multiplication, a circuitry for a computation based on these operations can be built out of n components which are identical, except for the contents of the ROMs, assuming that the system is well balanced. Since many DSP applications (e.g., filters and fast Fourier transforms) are dominated by the computation of sums of products, RNS systems are well suited for such applications.

We have so far assumed that the set of representable numbers is $\mathbb{Z}_M = \{a \in \mathbb{Z} \mid 0 \leq a < M\}$, i.e., it includes only non-negative integers. But since $M - a$ will act as an *additive inverse* in the system, subtraction is easily defined using the componentwise additive inverses:

$$\begin{aligned} |-a|_\beta &= [|M-a|_{m_1}, |M-a|_{m_2}, \ldots, |M-a|_{m_n}] \\ &= [|-a|_{m_1}, |-a|_{m_2}, \ldots, |-a|_{m_n}]. \end{aligned}$$

If we want to include negative numbers, we can just interpret numbers in the range $[\lfloor (M+1)/2 \rfloor; M)$ as negative numbers, whereas numbers in the range $[0; \lfloor (M+1)/2 \rfloor)$ are interpreted as non-negative. For odd M, any number $a \in [1; \lfloor (M+1)/2 \rfloor)$ maps into the interval $[\lfloor (M+1)/2 \rfloor, M)$ by the mapping $M - a$, and vice versa. However, for even M the two halves have different cardinalities and $a = M/2$ maps into itself, thus negation results in an overflow situation.

There are a number of other problems to consider, and they all relate to the advantage of the "carry–free" arithmetic of the componentwise computations. Since the individual components of the residue vectors are unweighted, the vectors totally lack information on the order of magnitude of the numbers represented. It is not possible to obtain any ordering information directly from the components of the residue vectors, e.g., it is difficult to determine the sign of an operand or to detect an overflow situation of an arithmetic operation.

Since overflow is fairly hard to detect, any computation has to be planned such that no undetected overflow can occur if the integer computation is to approximate a computation on reals. And since the size of the numbers in an integer computation tends to grow, it might be necessary to perform some kind of scaling during the computation. But scaling requires some kind of division which is also a difficult operation in the general case.

It turns out that most of these "difficult" operations require the conversion of operands from the residue representation into a weighted representation, so we will postpone the discussion of these to Section 8.4. However, one specific case of division can be performed componentwise on the residue vector. If it is known that the division is exact, then a divide operation can be performed using the multiplicative inverse.

If an integer a satisfies $\gcd(a, M) = 1$ or equivalently $\gcd(a, m_i) = 1$ for all i, $1 \le i \le n$, then the multiplicative inverse of a exists with respect to all the individual moduli m_i, hence we may define a residue vector

$$b = [|a^{-1}|_{m_1}, |a^{-1}|_{m_2}, \ldots, |a^{-1}|_{m_n}]$$

which by the componentwise multiplication then acts as an *multiplicative inverse* of a. But also notice that

$$|a^{-1}|_{m_i} = \left|\left(|a|_{m_i}\right)^{-1}\right|_{m_i},$$

so if

$$|a|_\beta = [a_1, a_2, \ldots, a_n],$$

then

$$|a^{-1}|_\beta = [|a_1^{-1}|_{m_1}, |a_2^{-1}|_{m_2}, \ldots, |a_n^{-1}|_{m_n}], \tag{8.3.4}$$

so reciprocation can also be performed componentwise.

It is essential here to notice that although it is possible to compute the product $|a^{-1}|_\beta \cdot |b|_\beta$ for any b, this only makes sense in this (integer) residue system if it is known by some other means that a divides b. However, as we shall see in Section 8.7, it is possible to define a rational residue system where this makes sense even if a does not divide b.

However, if $b/a = c$ is known to be an integer, then

$$|c|_\beta = |a^{-1}|_\beta |b|_\beta$$

and c is the solution to the equation $|ax|_\beta = |b|_\beta$. This is the *division-with-remainder-zero* problem, to be distinguished from the problem of finding a (quotient, remainder) pair. We shall discuss the general problem of division in Section 8.6.3.

We shall conclude this section with another observation. Since the computations described so far can all be carried through componentwise, without any interaction between the residues, any error due to a hardware failure in a single component can only affect one of the individual residue computations. Thus, if one or more additional residue computations (i.e., additional redundant moduli have been added to the residue vector) are carried through in parallel with the "ordinary" residues, then the computation of the failed component can be detected and possibly corrected. Since the ordinary RNS representation $|a|_\beta$ is unique for a in the interval $[0; M)$, adding an extra residue corresponding to a redundant modulus m^*, the value represented by $|a|_{\beta^*}$ is still a number in $[0; M)$ when β^* is the extended base. But an error in any single component of the vector $|a|_{\beta^*}$ implies that the number represented lies in the interval $[M, M \cdot m^*)$, and thus can be detected if we can map residue vectors into integers.

Problems and exercises

8.3.1 Perform the computation $13 \times 5 - 21 \times 3$ in the system $\beta = [3, 5, 7]$.

8.3.2 With $\beta = [3, 5, 7]$ scale the number $|a|_\beta = [2, 1, 3]$ by 5 such that the result is an integer "close to" $a/5$. Discuss the problem.

8.4 Mappings between residue and radix systems

Given any integer a, its residue representation or residue vector can be obtained by n divisions following the definition:

$$|a|_\beta = [|a|_{m_1}, |a|_{m_2}, \ldots, |a|_{m_n}];$$

however, there is a simpler way of obtaining the residue vector when a is given in some radix representation,

$$a = \sum_{i=0}^{\ell-1} d_i \beta^i, \quad d_i \in D,$$

since then

$$|a|_{m_j} = \left| \sum_{i=0}^{\ell-1} |d_i|_{m_j} |\beta^i|_{m_j} \right|_{m_j} \quad \text{for } j = 0, 1, \ldots, n. \tag{8.4.1}$$

If we assume that $\beta > 0$ such that $\beta \leq m_j$ for all j, and D is the standard digit set for β, $D = \{0, 1, \ldots, \beta - 1\}$, then $|d_i|_{m_j} = d_i$. By choosing β as large as possible, the number of terms to compute in (8.4.1) becomes minimal. For a given system, the set of moduli $\{m_j\}$ is known and thus the various constants $|\beta^i|_{m_j}$ can be precomputed. The n different residues $|a|_{m_j}$ can then be computed in parallel, if supplied with the digits $d_0, d_1, \ldots, d_n$, by modular multiplications and multioperand addition using the methods of Section 8.2.

For a fast, ROM-based implementation, it is possible to combine the table look-up for $|\beta^i|_{m_j}$ with the multiplication by $|d_i|_{m_j}$, which also allows us to increase β further to reduce the number of terms to be added. Assume β is of the form 2^k, so ℓ has to be chosen such that $M = \prod_1^n m_j < 2^{k\ell}$. Then the value of a digit d

encoded as k bits can be used as an address in a table T_{ij}, where the dth entry is $||d|_{m_j}|2^{ik}|_{m_j}|_{m_j}$. A total of $n \times \ell$ such tables, each of 2^k entries, are then needed for the complete conversion using (8.4.1).

Figure 8.4.1 describes a parallel implementation of the total conversion, where each chain of adders performs a multioperand modular addition following Algorithm 8.2.23 or 8.2.25, not shown in detail here. For a faster conversion but at about the same hardware cost, the linear addition chain can be rearranged into a tree structure as also discussed in Section 8.2. Another possibility is to use pipelining along the horizontal structure, by insertion of appropriate latches and skewing the data (the digits).

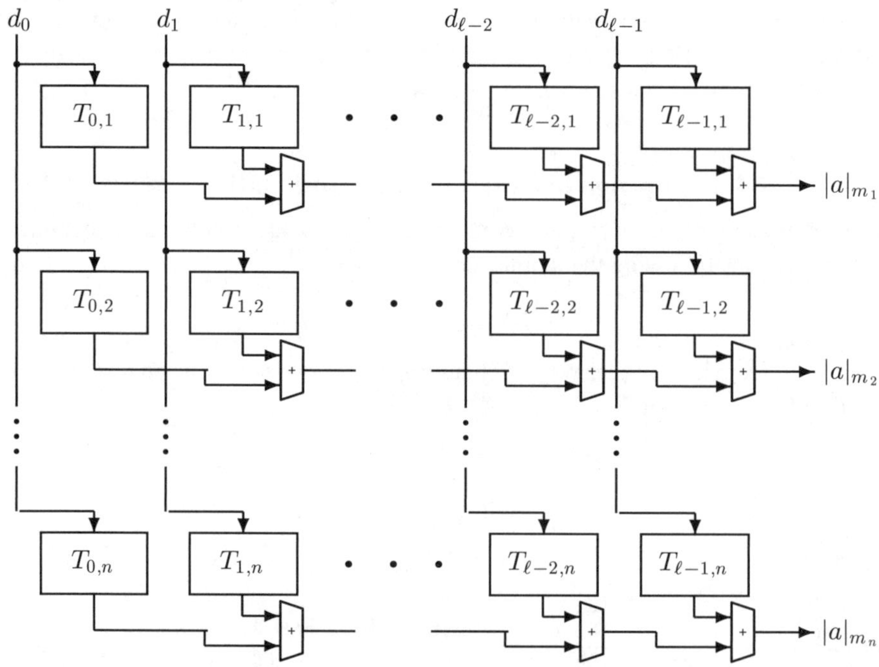

Figure 8.4.1. ROM-based, radix-to-residue converter array.

An alternative approach is to utilize the periodicity of $|2^i|_m$ for $i = 0, 1, \ldots,$ as described in Section 8.2. Consider (8.4.1) in binary, for simplicity for a single odd modulus m only:

$$|a|_m = \left| \sum_{i=0}^{\ell-1} b_i |2^i|_m \right|_m . \tag{8.4.2}$$

Due to the periodicity, the summation can now be performed in groups, simply adding up in small adder trees those bits having the same factor $|2^i|_m$. Assuming that the period is given by $P(m)$, there will be $P(m)$ such groups,

$G_j = \{b_i \mid i \bmod P(m) = j\}$ for $j = 0, 1, \ldots, P(m)-1$. Thus we can rewrite $|a|_m$ as

$$|a|_m = \left| \sum_{j=0}^{P(m)-1} \Big(\sum_{b_i \in G_j} b_i \Big) |2^j|_m \right|_m = \left| \sum_{j=0}^{P(m)-1} B_j \, |2^j|_m \right|_m = \left| \sum_{j=0}^{P(m)-1} s_j \, |2^j|_m \right|_m$$

with $B_j = \sum_{k=0}^{n_j} x_{jk} 2^k$ with $x_{jk} \in \{0, 1\}$ and $s_j \in \{0, 1\}$, where the B_j and y_j are calculated as follows (actually B_j never exists explicitly). For each value of j a small tree of (full- or half-)adders accumulates the values of $b_i \in G_j$, where carries (of weight 2^{j+1}) are moved over to be added into the tree at position $(j+1) \bmod P(m)$. At the root of the tree in position j there are only two bits to be added (the root sum and a carry), corresponding to all additions being performed in parallel with end-around carries. A carry-completion adder (say carry look-ahead), also with end-around carry, then converts the carry-save polynomial into a non-redundant polynomial $\sum_{j=0}^{P(m)-1} s_j 2^j$. This is equivalent to a tree-structured, multioperand addition with end-around carries of the operand bits arranged as $a = \lceil \ell / P(m) \rceil$ binary numbers (bit vectors), the topmost possibly 0-extended:

$$\begin{array}{cccc}
b_{ak-1} & b_{ak-2} & \cdots & b_{(a-1)k} \\
\vdots & \vdots & \vdots & \vdots \\
b_{2k-1} & b_{2k-2} & \cdots & b_k \\
b_{k-1} & b_{k-2} & \cdots & b_0 \\
\hline
s_{k-1} & s_{k-2} & \cdots & s_0
\end{array}$$

where the jth column consists of the $k = P(m)$ members of G_j.

Finally, $|a|_m = |\sum_{j=0}^{P(m)-1} y_j |2^j|_m|_m$ must be evaluated, noting that for small values of $P(m)$ (say of the order 10–12) this may be performed by table look-up, using $[s_{P(m)-1}, \ldots, s_1, s_0]$ as index. For somewhat larger values of $P(m)$ we may note that for $j \le k = \lfloor \log_2 m \rfloor$ then $|2^j|_m = 2^j$. Hence $a_1 = \sum_{j=0}^{k} s_j 2^j$ is trivially computed, and the remaining sum $a_2 = \sum_{j=k+1}^{P(m)-1} s_j |2^j|_m$ may be found by table look-up, using $[s_{P(m)-1}, \ldots, s_{k+1}]$ as index, so that $|a|_m = |a_1 + a_2|_m$ can be found by a modular carry-select adder.

Conversion from RNS representation to radix. For the conversion of a residue vector into a radix representation, the *Chinese Remainder Theorem* (CRT) formalizes an ancient method whose origin is unknown, but was described in a verse by the Chinese scholar Sun-Tsŭ sometime between AD 280 and 473.

Theorem 8.4.1 (CRT) *With* $\beta = [m_1, m_2, \ldots, m_n]$ *where* $\gcd(m_i, m_j) = 1$ *for* $i \neq j$, $M = \prod_1^n m_i$, $\widetilde{m}_i = M/m_i$, $i = 1, 2, \ldots, n$ *and* $|a|_\beta = [a_1, a_2, \ldots, a_n] = [|a|_{m_1}, |a|_{m_2}, \ldots, |a|_{m_n}]$ *then*

$$|a|_M = \left| \sum_{i=1}^{n} \widetilde{m}_i \left| \frac{a_i}{\widetilde{m}_i} \right|_{m_i} \right|_M . \tag{8.4.3}$$

Proof First note that $|\widetilde{m}_i^{-1}|_{m_i}$ exists. Then observe that the set of residue vectors $e_i = |\widetilde{m}_i|\widetilde{m}_i^{-1}|_{m_i}|_\beta = [0, \ldots, 0, 1, 0, \ldots, 0]$, with the unit in the ith position for $i = 1, 2, \ldots, n$, can act as an *orthogonal base* for RNS representations. Thus $|a|_\beta$ can be written

$$|a|_\beta = \sum_{i=1}^{n} a_i e_i = \sum_{i=1}^{n} \left|\widetilde{m}_i \left|\frac{a_i}{\widetilde{m}_i}\right|_{m_i}\right|_\beta = \left|\sum_{i=1}^{n} \widetilde{m}_i \left|\frac{a_i}{\widetilde{m}_i}\right|_{m_i}\right|_\beta$$

but then (8.4.3) follows by Theorem 8.3.1. □

The vectors $|\widetilde{m}_i|(a_i)/\widetilde{m}_i|_{m_i}|_\beta = [0, \ldots, 0, a_i, 0, \ldots, 0]$ are called *projections* of the residue vector $|a|_\beta$, based on the set of residue vectors $\{e_i\}_{i=1,\cdots,n}$ acting as the orthogonal base for the residue system.

Example 8.4.1 Let $\beta = [3, 5, 7]$ and consider $|a|_\beta = [2, 1, 3]$, then we need some constant calculations:

$$M = 3 \cdot 5 \cdot 7 = 105, \quad \widetilde{m}_1 = 5 \cdot 7 = 35, \quad \widetilde{m}_2 = 3 \cdot 7 = 21, \quad \widetilde{m}_3 = 3 \cdot 5 = 15,$$

so

$$|\widetilde{m}_1|_{m_1} = 2, \quad |\widetilde{m}_2|_{m_2} = 1, \quad |\widetilde{m}_3|_{m_3} = 1$$

and

$$|\widetilde{m}_1^{-1}|_{m_1} = 2, \quad |\widetilde{m}_2^{-1}|_{m_2} = 1, \quad |\widetilde{m}_3^{-1}|_{m_3} = 1.$$

For the conversion of $|a|_\beta = [2, 1, 3]$ we then have

$$\begin{aligned} |a|_{105} &= \left|\widetilde{m}_1|2 \cdot \widetilde{m}_1^{-1}|_{m_1} + \widetilde{m}_2|1 \cdot \widetilde{m}_2^{-1}|_{m_2} + \widetilde{m}_3|3 \cdot \widetilde{m}_3^{-1}|_{m_3}\right|_{105} \\ &= |35 \cdot |2 \cdot 2|_3 + 21 \cdot |1 \cdot 1|_5 + 15 \cdot |3 \cdot 1|_7|_{105} \\ &= |35 + 21 + 45|_{105} = 101. \end{aligned}$$

□

For moderate values of the moduli, the terms of the summation in (8.4.3) can be found by table look-up, thus avoiding the multiplications and modulo reduction in the terms $\widetilde{m}_i|a_i \cdot \widetilde{m}_i^{-1}|_{m_i}$. The terms can then be added by the methods of multi-operand modular addition, noting that the terms are bounded by M. A fast hardware implementation may then be as described in Figure 8.4.2, where the nodes of the tree are redundant modular additions along the lines of Algorithm 8.2.25.

Since the nodes of the tree can operate in constant time, the total conversion can be performed in time $O(\log n + \log\log M)$ where the $\log\log M$ term is from the final conversion into non-redundant binary.

If one of the system moduli is a power of 2, say $m_1 = 2^k$, then it is possible to perform the multioperand addition by means of ordinary binary adders followed by a single modulo M addition. Assuming again that the terms of the CRT to be added in (8.4.3) are obtained by look-ups from ROMs, then the values read from the tables could be in any other suitable representation. One such alternative representation

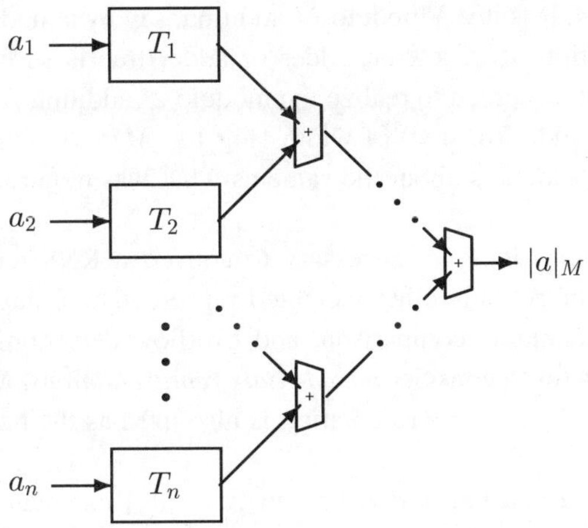

Figure 8.4.2. ROM-based CRT conversion.

could be a *quotient, remainder representation*, where a term $s_i = \widetilde{m}_i |a_i/\widetilde{m}_i|_{m_i}$ to be added in (8.4.3) is expressed as a pair (q_i, r_i), with $0 \le q_i < 2^k$ and $0 \le r_i < \widetilde{m}_1$, satisfying:

$$s_i = q_i \cdot \widetilde{m}_1 + r_i < M$$

where $\widetilde{m}_1 = M/m_1 = M \cdot 2^{-k}$.

With this representation we can rewrite (8.4.3) as

$$\begin{aligned} |a|_M &= \left| \sum_{i=1}^{n} \left(\widetilde{m}_i \left| \frac{a_i}{\widetilde{m}_i} \right|_{m_i} \right) \right|_M \\ &= \left| \widetilde{m}_1 \cdot \sum_{i=1}^{n} q_i + \sum_{i=1}^{n} r_i \right|_M \\ &= \left| \widetilde{m}_1 \cdot \left| \sum_{i=1}^{n} q_i \right|_{m_1} + \sum_{i=1}^{n} r_i \right|_M , \end{aligned} \tag{8.4.4}$$

since $m_1 = 2^k$ is a divisor of M. It then follows that

$$q = \left(\sum_{i=1}^{n} q_i \right) \bmod 2^k \Rightarrow 0 \le \widetilde{m}_1 q < M$$

and

$$r = \sum_{i=1}^{n} r_i \Rightarrow 0 \le r < n \cdot \frac{M}{m_1}.$$

Assuming that $n < m_1 = 2^k$, which is not unrealistic for practical systems, then $0 \le r < M$. Using a table look-up to determine $\widetilde{m}_1 q$ for given q, what remains to compute in (8.4.4) is just a modulo M addition, say by a modular select adder. For the computation of q, a k-bit adder or adder tree is sufficient, where the carry-outs are just discarded to realize the modulo 2^k addition. The accumulation of r_i requires an adder (or tree) of width $\lfloor \log_2(n \cdot M/m_1) + 1 \rfloor$, hence the total width of the two adders is about the same as what was required in the previous method.

In many situations it is not necessary to convert a RNS representation into ordinary radix (binary), any other weighted representation may be used in sign determination, magnitude comparison, and overflow detection. In particular, it turns out to be useful to consider *mixed-radix representations*, MRRs, where the base vector β for the RNS representation is also used as the base vector for the MRR.

Recall that with a base vector $\beta = [m_1, m_2, \ldots, m_n]$, an integer a in the interval $0 \leq a < M = \prod_1^n m_i$ can be represented uniquely in MRR as

$$a = d_1 + d_2 m_1 + d_3 m_1 m_2 + \cdots + d_n m_1 m_2 \cdots m_{n-1}, \qquad (8.4.5)$$

where $0 \leq d_i < m_i, i = 0, 1, \ldots, n$. If a number is given by its RNS representation $|a|_M = [a_1, a_2, \ldots, a_n]$ with the same base vector β satisfying $\gcd(m_i, m_j) = 1$ for $i \neq j$, then $d_1 = a_1$ since $|a|_{m_1} = a_1 = d_1$. From (8.4.5) we then note that $a - d_1$ is divisible by m_1 so that $d_2 = |(a - d_1)/m_1|_{m_2}$, etc. By a trivial rewriting of Algorithm 1.10.3 (the MRDGT Algorithm) we then obtain the following.

Algorithm 8.4.2 (RMR, residue to mixed radix)
Stimulus: *A residue number* $|a|_\beta = [a_1, a_2, \ldots, a_n]$ *with base vector* $\beta = [m_1, m_2, \ldots, m_n]$, $\gcd(m_i, m_j) = 1$ *for* $i \neq j$.
Response: *A mixed radix representation of* $|a|_M$ *as a vector*

$$a_\beta^{MR} = [d_1, d_2, \ldots, d_n], \quad 0 \leq d_i < m_i, \quad i = 1, 2, \ldots, n$$

such that $|a|_M = d_1 + d_2 m_1 + \cdots + d_n m_1 m_2 \cdots m_n$.
Method: $t_1 := |a|_\beta$; {*note that* t_i *is in RNS representation*}
$d_1 := |a|_{m_1}$; $\{d_1 := a_1\}$
for $i := 2$ **to** n **do**
$L : t_i := (t_{i-1} - d_{i-1})|m_{i-1}^{-1}|_{m_i \cdots m_n}$ {*RNS computation*}
$d_i := |t_i|_{m_i}$
end;

Note that the arithmetic operations in the updating of t_i at label L have to take place in the RNS such that at step i only the residues $m_i, m_{i+1}, \ldots, m_n$ are used, otherwise the multiplicative inverse $|m_{i-1}^{-1}|_{m_i \cdots m_n}$ would not exist.

Example 8.4.2 For the conversion of $|a|_\beta = [2, 1, 3]$ with $\beta = [3, 5, 7]$ as in the previous example, we obtain:

β	$m_1 = 3$	$m_2 = 5$	$m_3 = 7$	
$\lvert t_1\rvert_\beta = \lvert a\rvert_\beta$	$\boxed{2}$	1	3	
$\lvert d_1\rvert_\beta$	2	2	2	subtract
$\lvert t_1 - d_1\rvert_\beta$	0	4	1	
$\lvert m_1^{-1}\rvert_{\beta_1}$		2	5	multiply
$\lvert t_2\rvert_{\beta_1}$		$\boxed{3}$	5	
$\lvert d_2\rvert_{\beta_1}$		3	3	subtract
$\lvert t_2 - d_2\rvert_{\beta_1}$		0	2	
$\lvert m_2^{-1}\rvert_{\beta_2}$			3	multiply
$\lvert t_3\rvert_{\beta_2}$			$\boxed{6}$	

Hence $|a|_{105} = \boxed{2} + \boxed{3} \cdot (3) + \boxed{6} \cdot (3 \cdot 5)(= 101_{10})$. □

The residue to mixed-radix conversion can be realized by a triangular array of cells, each cell performing a subtraction followed by a multiplication with a constant, as illustrated in Figure 8.4.3.

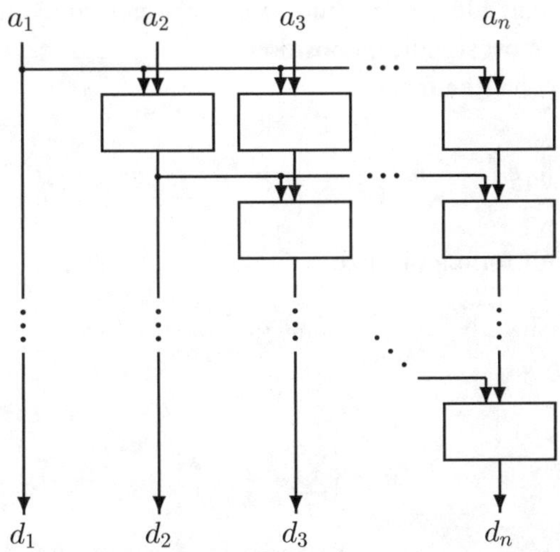

Figure 8.4.3. Classical residue to mixed-radix conversion.

For very small moduli, each cell can be realized as a single ROM, for somewhat larger moduli it may be necessary with a subtracter (which may or may not be modular) followed by a ROM to realize the multiplication. Note that the structure can easily be pipelined for increased throughput, but as the algorithm is inherently sequential, the delay is $\Omega(n)$.

The residue to mixed-radix conversion can be reformulated by looking at the calculation taking place in each "channel," which may be useful in understanding the process. Here we have to identify the components $m_{i,j}^{-1} = |m_i^{-1}|_{m_j}$ of the multiplicative inverse $|m_i^{-1}|_{m_{i+1}m_{i+2}\cdots m_n}$. Thus given a residue representation $|a|_\beta = [a_1, a_2, \ldots, a_n]$ the components of its MRR $a_\beta^{MR} = [d_1, d_2, \ldots, d_n]$ can be expressed by

$$
\begin{aligned}
d_1 &= a_1, \\
d_2 &= \left|(a_2 - d_1)m_{1,2}^{-1}\right|_{m_2}, \\
d_3 &= \left|\left(\left|(a_3 - d_1)m_{1,3}^{-1}\right|_{m_3} - d_2\right) m_{2,3}^{-1}\right|_{m_3}, \\
d_4 &= \left|\left(\left|\left(\left|(a_4 - d_1)m_{1,4}^{-1}\right|_{m_4} - d_2\right) m_{2,4}^{-1}\right|_{m_4} - d_3\right) m_{3,4}^{-1}\right|_{m_4}, \qquad (8.4.6)\\
&\vdots \\
d_n &= \left|\left(\cdots\left|\left(\left|(a_n - d_1)m_{1,n}^{-1}\right|_{m_n} - d_2\right) m_{2,n}^{-1}\right|_{m_n} - \cdots - d_{n-1}\right) m_{n-1,n}^{-1}\right|_{m_n}.
\end{aligned}
$$

We can also exploit the CRT for conversion into MRR by converting the projections of $|a|_\beta$ into MRR, and then adding the terms in this representation. This has the advantage that addition modulo M can be realized by just discarding the carry-out from the most significant position.

A projection $a_i e_i$ has the form:

$$|a_i e_i|_\beta = \left|\widetilde{m}_i \left|\frac{a_i}{\widetilde{m}_i}\right|_{m_i}\right|_\beta = [0, \ldots, a_i, \ldots, 0],$$

so its MRR in vector form looks like

$$[a_i e_i]_\beta^{MR} = [0, \ldots, 0, \overline{a}_{ii}, \overline{a}_{ii+1}, \ldots, \overline{a}_{in}], \qquad (8.4.7)$$

hence by the CRT

$$|a|_M = \left|\sum_1^n [a_i e_i]_\beta^{MR}\right|_M,$$

where the summation takes place in MRR, and potentially can be performed column by column in tree structures as shown in Figure 8.4.4.

Since the number of terms to be added in column i is one greater than the number of terms in column $i-1$, the carry-out of column $i-1$ will be available at about the time when it has to be added into column i, if sequential column summation is used. If tree structures are used, the timing is more tight, but here a carry look-ahead can be used. The conversion can thus be performed in the time of a single (parallel) table look-up, plus an adder tree of height $O(\log n)$.

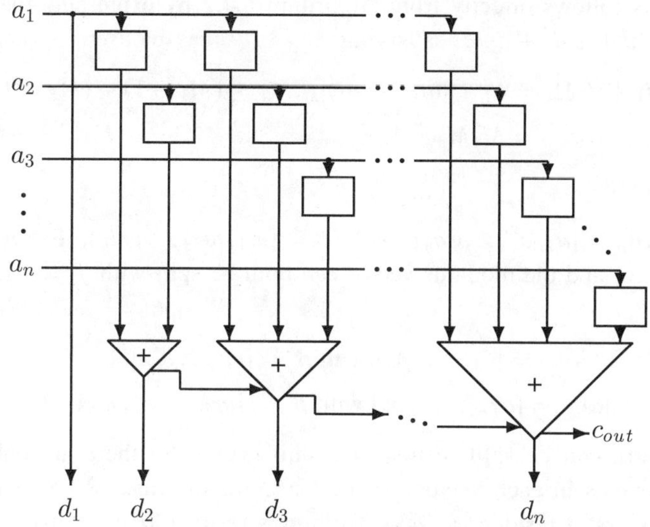

Figure 8.4.4. CRT-based residue to mixed-radix conversion.

Example 8.4.3 $|a|_\beta = [2, 1, 3]$ with $\beta = [3, 5, 7]$

$$\begin{array}{rcllllllll}
 & & & m_1 = 3 & & m_2 = 5 & & m_3 = 7 & & \\
a_1 = 2,\ [a_1 e_1]_\beta^{MR} & = & [& 2 & , & 1 & , & 2 &] & \text{(table look-up)} \\
a_2 = 1,\ [a_2 e_2]_\beta^{MR} & = & [& 0 & , & 2 & , & 1 &] & \text{(table look-up)} \\
a_3 = 3,\ [a_3 e_3]_\beta^{MR} & = & [& 0 & , & 0 & , & 3 &] & \text{(table look-up)} \\
\hline
[a]_\beta^{MR} & = & [& 2 & , & 3 & , & 6 &] &
\end{array}$$

which confirms the result from the previous example. Note that there happen to be no carries in this particular addition, something which must be taken care of in general. □

It is also possible to use the principle of the mixed-radix conversion for another logarithmic-time algorithm by a "divide and conquer" approach using Algorithm 8.4.2 for $n = 2$. The idea is to combine two residues into a residue of the combined system, by computing the integer value of that residue in radix representation from the MRR.

Observation 8.4.3 *Let* $|a|_\beta = [a_1, a_2]$ *with base vector* $\beta = [M_1, M_2]$, $\gcd(M_1, M_2) = 1$, *then*

$$|a|_{M_1 M_2} = a_1 + |q(a_2 - a_1)|_{M_2} M_1, \tag{8.4.8}$$

where $q = |M_1^{-1}|_{M_2}$.

Proof This follows directly from Algorithm 8.4.2 by expressing the value of $a_\beta^{MR} = [d_1, d_2]$ as $d_1 + d_2 M_1$ satisfying

$$0 \le d_1 + d_2 M_1 = a_1 + |q(a_2 - a_1)|_{M_2} M_1 \le (M_1 - 1) + (M_2 - 1)M_1$$
$$= M_1 M_2 - 1.$$

thus proving (8.4.8). □

Given $|a|_\beta = [a_1, a_2, \ldots, a_n]$ with $\beta = [m_1, m_2, \ldots, m_n]$, for $n \ge 3$, the residue vector and the modulus vector can both be split with $j = \lfloor n/2 \rfloor$ in two parts:

$$|a|_{\beta'} = [a_1, \ldots, a_j] \text{ with } \beta' = [m_1, \ldots, m_j],$$
$$|a|_{\beta''} = [a_{j+1}, \ldots, a_n] \text{ with } \beta'' = [m_{j+1}, \ldots, m_n],$$

which in turn can be split recursively, until eventually there are only one or two components in each vector. At the leaves of the tree, (8.4.8) can then be applied to pairs of residues a_i, a_{i+1}, forming a residue $|a|_{m_i m_{i+1}}$ using $M_1 = m_i$ and $M_2 = m_{i+1}$, etc. Continuing to combine results from below in the tree structure by Observation 8.4.3, finally at the root the results of two subtrees, $|a|_{M_1}$ and $|a|_{M_2}$, are combined into $|a|_{M_1 M_2}$ using $M_1 = \prod_1^j m_i$ and $M_2 = \prod_{j+1}^n m_i$.

This conversion then requires $O(n)$ constants of the form $q = |(M')^{-1}|_{M''}$, where M' and M'' are certain products of the original moduli, and employ modular computations of expressions of the form $|q(a'' - a')|_{M'}$ for $M' < \sqrt{\prod_1^n m_i}$. Hence all modular operations can now be performed in systems with moduli smaller than the square root of the system modulus. But note that each node in the tree requires two multiplications, one of which is modular.

Problems and exercises

8.4.1 Show that the computation of a residue $|a|_m$ in (8.4.1) becomes quite simple if $m = 2^k - 1$ and a is given in binary, choosing $\beta = 2^k$.

8.4.2 Similarly describe the computation of $|a|_m$ when $m = 2^k + 1$.

8.4.3 Prove Theorem 8.4.1 (CRT) by determining the residue vector of the right-hand side of (8.4.3), utilizing the uniqueness of the residue representation.

8.4.4 In the system with $\beta = [3, 5, 7]$ perform the conversion of $|a|_\beta = [2, 1, 3]$ into mixed radix representation, employing the divide-and-conquer approach of Observation 8.4.3.

8.5 Base extensions and scaling

As we shall see subsequently, there are many situations where it is useful to extend the base vector $\beta = [m_1, m_2, \ldots, m_n]$ with one or more additional moduli. Actually what we want is, given the residue vector of a number $|a|_\beta = [a_1, a_2, \ldots, a_n]$, to find its residue representation $|a|_{\beta'} = [a_1, a_2, \ldots, a_n, a_{n+1}]$, where $\beta' = \beta \cup \{m_{n+1}\}$. Unless more is known we can only assume that $0 \le a < M = \prod_1^n m_i$, but often it is known that $0 \le a < M' = \prod_1^{n+1} m_i$, i.e., we want to recover a "lost residue." Hence what is to be computed is just $a_{n+1} = |a|_{m_{n+1}}$ when extending with a single modulus m_{n+1}. This can be achieved by computing the value of a, and then mapping this value into $\mathbb{Z}_{m_{n+1}}$. One way of doing this is to compute the MRR of a, and from this compute $|a|_{m_{n+1}}$, which can be realized as an extension to the mixed-radix conversion.

8.5.1 Mixed-radix base extension

Let $a_\beta^{MR} = [d_1, d_2, \ldots, d_n]$ be the mixed-radix representation of a, then

$$|a|_{m_{n+1}} = |\cdots||d_1 + d_2 m_1|_{m_{n+1}} + d_3 m_1 m_2|_{m_{n+1}} + \cdots + d_n m_1 m_2 \cdots m_n|_{m_{n+1}},$$

which is easily computable in a sequential process, starting with $s_1 = d_1$ and iteratively computing

$$s_i = \left| s_{i-1} + d_i \left| \prod_1^{i-1} m_i \right|_{m_{n+1}} \right|_{m_{n+1}} \qquad i = 2, \ldots, n,$$

so $|a|_{m_{n+1}} = s_n$.

Since the d_i are computed in the same order as they are computed in the residue to mixed-radix conversion, it is easy to extend that algorithm with the computation of the s_i. For a hardware realization one extra cell for the computation of s_i then has to be added in each column of Figure 8.4.3, giving Figure 8.5.1, where the computed mixed-radix digits are also shown although they may not be needed. Each of these extra cells performs a multiplication by a constant and an addition.

Note that it is possible to add several "layers" of extra cells to compute additional residues modulo new moduli, each layer incrementing the total delay (latency) by one unit. Also, this structure is easily pipelined.

8.5.2 CRT base extension

As we have seen, it is also possible to use an approach based on the CRT (Theorem 8.4.1), however, in a slightly different form:

$$a = \sum_{i=1}^{n} \widetilde{m}_i \left| \frac{a_i}{\widetilde{m}_i} \right|_{m_i} - r_a M \tag{8.5.1}$$

(called the *alternative CRT form*), where the *rank* of a, r_a, is chosen to reconstruct a from its residue vector $|a|_\beta$ such that $0 \le a < M$. Now observe that the terms of the sum in (8.5.1) satisfy

$$0 \le \widetilde{m}_i \left| \frac{a_i}{\widetilde{m}_i} \right|_{m_i} < M,$$

so the value of the sum in (8.5.1) is bounded by nM, i.e., $r_a < n$.

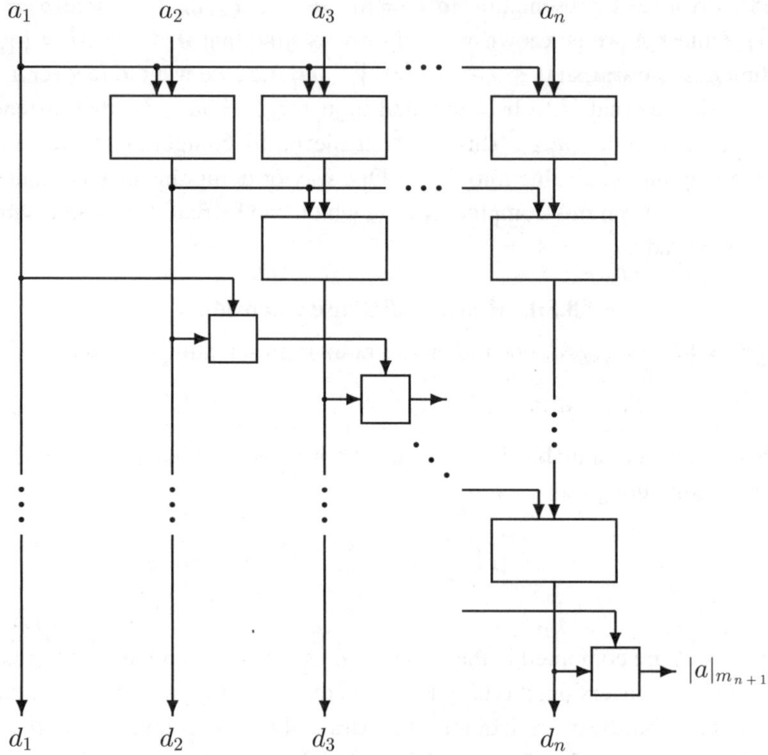

Figure 8.5.1. Mixed-radix base extension.

If we can determine r_a, it is easy to find a, from which we can then find $|a|_{m_{n+1}}$, where m_{n+1} is the extra modulus for which we want the residue. Now assume there is one extra modulus m satisfying $m \geq n$ and $\gcd(m, M) = 1$, then we can write

$$|a|_m = \left| \left| \sum_{i=1}^{n} \widetilde{m}_i \left| \frac{a_i}{\widetilde{m}_i} \right|_{m_i} \right|_m - |r_a M|_m \right|_m .$$

Rearranging terms we find

$$|r_a M|_m = \left| \left| \sum_{i=1}^{n} \widetilde{m}_i \left| \frac{a_i}{\widetilde{m}_i} \right|_{m_i} \right|_m - |a|_m \right|_m$$

and multiplying by the inverse $|M^{-1}|_m$ we obtain

$$r_a = \left| |M^{-1}|_m \left(\left| \sum_{i=1}^{n} \widetilde{m}_i \left| \frac{a_i}{\widetilde{m}_i} \right|_{m_i} \right|_m - |a|_m \right) \right|_m , \tag{8.5.2}$$

since $r_a < n \leq m$ and $|M^{-1}|_m$ exists as per the assumption on m.

To be able to compute the rank r_a from (8.5.2) it is thus necessary to have one extra (redundant) "channel" operating with a modulus m of very moderate size, since n is usually small. Then $|a|_m$ is known and with r_a computed from (8.5.2) it is possible to compute $|a|_{m_{n+1}}$ from (8.5.1) as

$$|a|_{m_{n+1}} = \left| \left| \sum_{i=1}^{n} \widetilde{m}_i \left| \frac{a_i}{\widetilde{m}_i} \right|_{m_i} \right|_{m_{n+1}} - |r_a M|_{m_{n+1}} \right|_{m_{n+1}} . \tag{8.5.3}$$

Example 8.5.1 Consider the residue system with $\beta = [3, 5]$ so

$$M = 15,\ \widetilde{m}_1 = 5,\ \widetilde{m}_2 = 3,\ |\widetilde{m}_1^{-1}|_{m_1} = 2,\ |\widetilde{m}_2^{-1}|_{m_2} = 2.$$

Assume that $|a|_\beta = [0, 4]$, but that also an extra residue is known for $m = 8$ with $|a|_8 = 1$ (the redundant residue). Let us then perform a base extension to include the modulus $m_3 = 7$.

We need $|M^{-1}|_m = |15^{-1}|_8 = 7$ and from (8.5.2) we obtain

$$r_a = |\,7 \cdot (|\,5 \cdot 0 \cdot 2 + 3 \cdot 4 \cdot 2|_8 - |a|_8)|_8 = |7 \cdot (0 - 1)|_8 = 1,$$

so from (8.5.3)

$$|a|_7 = |\,|\,5 \cdot 0 \cdot 2 + 3 \cdot 4 \cdot 2|_7 - |1 \cdot 15|_7|_7 = 2.$$

It is easy to see from $|a|_\beta = [0, 4]$ that $|a|_{15} = 9$, confirming that $|a|_7 = 2$, and assuming that $0 \le a < 15$. Actually, since an extra residue modulo 8 has been carried along, it may be possible to assume that $0 \le a < 15 \cdot 8 = 120$. □

The computations in (8.5.2) and (8.5.3) are very similar, both consisting of multioperand modular additions, which can be performed in a tree structure, followed by a computation where the result r_a from (8.5.2) is combined with the sum from (8.5.3).

In Figure 8.5.2 the initial processing nodes compute terms of the form $\widetilde{m}_i |a_i/\widetilde{m}_i|_{m_{n+1}}$ or $|M^{-1}|_m \cdot \widetilde{m}_i |a_i/\widetilde{m}_i|_m$, except for the rightmost node which computes the term $-|M^{-1}|_m \cdot |a|_m$. The computation time is $O(\log n + \log m_{n+1})$ and the structure can be pipelined. When implemented with ROMs for all nodes fewer ROMs are needed, and the latency is shorter than for a similar implementation of Figure 8.5.1. However, recall that this method requires that an extra redundant modulus is carried through in all other computations. Initially it may look strange to have an extra modulus and compute extra residues for the purpose of computing other extra residues. However, it turns out that, e.g., when base extension is needed in connection with scaling, it is torecover a residue modulo one of the original

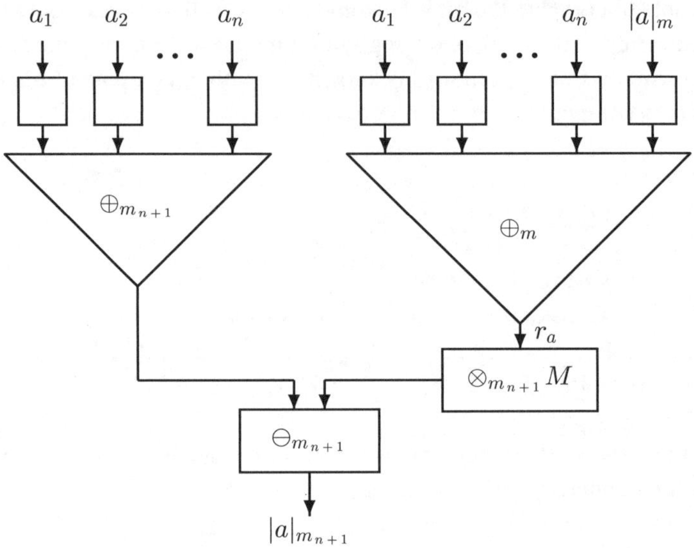

Figure 8.5.2. CRT-based base extension.

residues. Also an extra residue is needed for overflow detection and sign detection as we shall see in the next section.

8.5.3 Scaling

Since the magnitude of numbers in any composite integer computation will grow, and only a restricted range of uniquely representable numbers is available, it is necessary to plan any such computation in order that no undetected overflow occurs. When such a computation is to simulate a computation on reals, e.g., as in DSP, operands will be scaled versions of the actual variables of the problem. Occasionally during the computation a scaling will have to take place, and the total scaling then has to be taken into account when interpreting the final results.

Since scaling in an integer system implies division by some integer delivering an integer result, it necessarily implies some kind of approximation corresponding to a rounding in a floating-point system. Since it is not a radix system, there is no particular advantage in scaling being by a power of 2 or 10, and it turns out to be easiest to scale by factors which are products of some of the moduli of the system. But if the set of moduli contains moduli which are suitable powers of 2 and 5, then these could be used for scaling if this is found convenient in connection with the user interface.

The reason for the choice of a product of moduli for scaling is that it is then easy to change the scaling into the problem of exact division, as opposed to the general division problem. Suppose we want to scale the number $|a|_\beta = [a_1, \dots, a_n]$ by the modulus m_i from $\beta = [m_1, \dots, m_i, \dots, m_n]$, then obviously $a - a_i$ is divisible by m_i. If $a_i = |a|_{m_i}$, then

$$\left|\frac{a - a_i}{m_i}\right|_\beta = \left[\left|(a_1 - a_i)m_i^{-1}\right|_{m_1}, \ldots, \perp, \ldots, \left|(a_n - a_i)m_i^{-1}\right|_{m_n}\right], \tag{8.5.4}$$

where the symbol $\perp$ in position i is used to indicate that this component is undetermined, due to the fact that the inverse of m_i with respect to m_i does not exist.

Now observe that if there was no overflow in producing a, then $0 \le a < M = \prod_1^n m_i$ and hence $0 \le (a - a_i)/m_i < \frac{M}{m_i}$, and thus the value of $(a - a_i)/m_i$ is perfectly representable in the residue system with the restricted base $\beta' = \beta \backslash \{m_i\}$, i.e., the system without the modulus m_i. The value of the undetermined ith component in (8.5.4) can then be obtained by base extension of the number $|(a - a_i)/m_i|_{\beta'}$, from the β'-system into the β-system by considering m_i a "new" modulus.

When in (8.5.4) we subtract a_i before scaling by m_i we introduce a *quantization error* of absolute magnitude a_i/m_i, i.e., we effectively do a truncation. Alternatively it is possible to add $m_i - a_i$ before scaling, the choice of a_i or $m_i - a_i$ depending on which has the smaller value, thus effectively performing a round to nearest.

Example 8.5.2 Assume we want to scale the number $|a|_\beta = [2, 1, 3]$ by 3, with $\beta = [3, 5, 7]$. With $m_1 = 3$ we can either subtract 2 and then scale [0, 4, 1] by 3, or we can add 1 and scale [0,2,4] by 3. Since the latter will introduce the smaller error we will add 1:

	m_1	m_2	m_3		
$\lvert a\rvert_\beta$:	2	1	3		
$\lvert 1\rvert_\beta$:	1	1	1	add 1 to quantize (round)	
$\lvert a+1\rvert_\beta$:	0	2	4		
$\lvert m_1^{-1}\rvert_\beta$:	$\perp$	2	5	multiply by 3^{-1} to scale	
$\lvert (a+1)m_1^{-1}\rvert_\beta$:	$\perp$	$\boxed{4}$	6		
d_2:		4	4	subtract	mixed-radix conversion
		0	2		
$\lvert m_2^{-1}\rvert_{m_3}$:			3	multiply	
d_3:			$\boxed{6}$		
m_2:			5	multiply	mixed-radix base extension
$\lvert d_3 m_2\rvert_{m_1}$:			0		
$s_2 = d_2$:			4	add	
$\lvert d_2 + d_3 m_2\rvert_{m_1}$:			$\boxed{1}$		

so $|(a+1)/3|_\beta = [1, 4, 6]$. From previous examples we know that $a = 101$ so $(a+1)/3 = 34$ and it is easy to see that the result is correct, $|34|_\beta = [1, 4, 6]$. □

8.6 Sign and overflow detection, division

These operations all require information on the magnitude of the number, which in a residue representation is distributed on the residues, and they are thus collectively considered "difficult operations," as opposed to addition, subtraction, and multiplication which can be performed individually and in parallel for all component residues. As discussed in the previous section, base extension and scaling also uses information on the magnitude of the number, and forms a "bridge" to the discussions in this section, as we will use the results there as tools here. In a separate sub-section we will also present the *core function*, as a tool to handle some of these and other problems in an alternative way.

8.6.1 Overflow

First observe that any computation involving only addition, subtraction, and multiplication (no intermediate scaling or other division) may overflow in intermediate results, without damaging the interpretation of the final results, if these can be guaranteed by other means to be within *the dynamic range* of the system.

If this is not possible, a strategy may be to perform the computation in an extended system $\beta^* = \beta \cup \{m^*\}$, where m^* is an extra (redundant) modulus, provided such that particular results are sure to be in the extended range $0 \le a < M^*$, $M^* = \left(\prod_1^n m_i\right) m^*$, allowing a suitable scaling of the further computation to take place.

Computation with such an additional modulus can be used to check whether a value a is in the overflow range $M \le a < M^*$, by performing a conversion to the MRR, as is trivially seen from the uniqueness of that representation in the following observation.

Observation 8.6.1 *If* $|a|_{\beta^*} = [a_1, a_2, \ldots, a_n, a_{n+1}]$, *where* $\beta^* = [m_1, m_2, \ldots, m_n, m_{n+1}]$ *with* $m_{n+1} = m^*$ *being an extra (redundant) modulus and* $0 \le a < M^* = M \cdot m^*$, *then*

$$0 \le a < M \quad \textit{iff}\ d_{n+1} = 0,$$

where $a = d_1 + d_2 m_1 + \cdots + d_{n+1} m_1 m_2 \cdots m_n$ *is the MRR of* a.

8.6.2 Sign determination and comparison

If appropriate it is possible to associate sign information with a number in the range $0 \le a < M$, by interpreting numbers in $[1\,;\lfloor (M-1)/2 \rfloor]$ as positive and those in $[\lfloor (M+1)/2 \rfloor; M-1]$ as negative. Through conversion to the integer domain, e.g., to a radix or mixed-radix representation, the sign of a number may then be determined through a comparison in that domain.

In particular, note that by Observation 8.6.1, if the last modulus is 2, the last (most-significant) mixed-radix digit directly provides the sign, being either 0 or 1. However, modulus 2 may be unreasonably small, and any even value of the last modulus may be as convenient for the test, in particular a power of 2 can be chosen to be of the order of magnitude of the other moduli. The computation time (or delay) is then $O(n)$ or $O(\log n)$ depending on which method is used for the conversion to MRR.

Alternatively, observe that the rational $M/2$ acts as a splitting point, so by computing $s(a) = (2/M)a$ in the real domain we have

$$\lfloor s(a) \rfloor = \begin{cases} 0 & \text{if} \quad 0 \le a < M/2, \\ 1 & \text{if} \quad M/2, \le a < M, \end{cases} \tag{8.6.1}$$

where $\lfloor s(a) \rfloor$ is taken to be the integer part of $s(a)$. Exploiting the alternative formulation of the CRT given in (8.5.1) we then obtain

$$s(a) = \frac{2}{M} a = \sum_{i=1}^{n} \frac{2}{m_i} \left| \frac{a_i}{\widetilde{m}_i} \right|_{m_i} - 2r_a, \tag{8.6.2}$$

where now the computation is to be performed in the real domain. But observe that we need not concern ourselves here with the rank r_a. Its purpose in (8.5.1) was to reduce the value by subtracting a multiple of M, now it is to subtract any multiple of 2. However, that is easily accomplished if (8.6.2) is evaluated in binary, just by discarding any digits of weight 2^1 or higher.

The only problem left is to simulate the computation of (8.6.2) in a finite precision system with sufficient accuracy to be able to resolve (8.6.1), i.e., basically whether $\tilde{s}(a) < 1$ or $\tilde{s}(a) \ge 1$, where $\tilde{s}(a)$ is the "modulo 2" reduced version of $s(a)$. But note that since $s(a) = (2/M) \cdot a$ and a is integral, the values of $s(a)$ are at least $2/M$ apart, hence if the terms of (8.6.2) are computed and accumulated with a total error not exceeding $1/M$, then the sign of a can be determined. (Actually $2/M$ is sufficient for M even.)

Assume that the computation is performed in a binary arithmetic, e.g., in a redundant adder tree with a final carry completion, based on properly rounded values of the terms

$$v_i = \frac{2}{m_i} \left| \frac{a_i}{\widetilde{m}_i} \right|_{m_i}$$

given with t fractional bits. These values could be obtained by table look-up in n ROMs, one for each m_i. Adding n properly rounded values can accumulate an error of at most $n \cdot 2^{-t-1}$, so $t > \log_2(Mn) - 1$ will provide sufficient accuracy to determine the sign of a in time $O(\log n + \log M)$ by a simple adder tree in radix 2 operating in a "modulo 2" fashion. And again, t can be one less if M is even.

For an alternative algorithm see Algorithm 8.6.15, based on the core function to be discussed below.

A comparison of two residue represented numbers can be performed by a subtraction followed by a sign detection. However, care must be taken if the two operands to be compared may be of different sign, since if that is the case then the subtraction is an effective addition which may overflow. If that situation cannot be guaranteed not to occur, then a sign determination of both operands is necessary. If the operands are of different sign then the signs will determine their ordering, otherwise proceed as above.

8.6.3 The core function

Some of the difficult operations on residue represented numbers may be simplified by employing a function applied to the RNS representation, mapping the interval $[0\,;M)$ "almost linearly" to some interval $[0\,;A)$ where $A \ll M$. Here A should be chosen to be of the same order of magnitude as the individual moduli of the system, allowing the evaluation of the function to be performed by hardware similar to other residue computations. Such a mapping has been suggested to be defined as follows.

Definition 8.6.2 *For the residue system with base vector* $\beta = [m_1, \ldots, m_n]$*, let* $[w_1, \ldots, w_n]$ *be a set of fixed but arbitrary integers, not all zero, then the* core function *is defined as*

$$C(a) = \sum_{i=1}^{n} w_i \left\lfloor \frac{a}{m_i} \right\rfloor \in \mathbb{Z} \tag{8.6.3}$$

for any integer a*.*

It is not obvious from the definition that $C(a)$ will be easy to compute, since (8.6.3) assumes the value of a is known, but we shall see later that $|C(a)|_{C(M)}$ can be computed directly from the residues $|a|_\beta = [a_1, \ldots, a_n]$, as the residue of an inner product. Under suitable restrictions on the system (the weights) it is possible to construct $C(a)$ from $|C(a)|_{C(M)}$ over restricted domains, or when it is known that no overflow has occurred.

We have chosen first to present the ideas and some results, to raise awareness about their existence, only later returning to the computation of the core function.

The integers $\{w_1, \ldots, w_n\}$ are called the *core weights*. Observe that

$$C(M) = \sum_{i=1}^{n} w_i \left\lfloor \frac{M}{m_i} \right\rfloor = \sum_{i=1}^{n} w_i \widetilde{m}_i \tag{8.6.4}$$

by the definition of $\widetilde{m}_i$, from which it is noted that

$$\frac{C(M)}{M} = \sum_{i=1}^{n} \frac{w_i}{m_i}, \tag{8.6.5}$$

but also that if a particular value of $C(M)$ is wanted, then the weights must satisfy

$$w_i \equiv C(M)\widetilde{m}_i^{-1} \pmod{m_i}, \tag{8.6.6}$$

such that $w_i = 0$ if $C(M)$ is chosen as a multiple of m_i.

Note that the function is a step function. As a increases, $C(a)$ remains constant until a equals a multiple of one of the moduli, at which point $C(a)$ jumps due to the floor function. It is an "almost linear" function, as seen by rewriting for $|a|_\beta = [a_1, \ldots, a_n]$

$$\begin{aligned} C(a) &= \left(\sum_{i=1}^{n} \frac{w_i}{m_i}\right) a - \sum_{i=1}^{n} w_i \frac{a_i}{m_i} \\ &= \frac{C(M)}{M} \cdot a - \sum_{i=1}^{n} w_i \frac{a_i}{m_i}, \end{aligned} \tag{8.6.7}$$

utilizing that $a - a_i$ is divisible by m_i, together with (8.6.5). The last term here is a non-linear contribution, adding a raggedness to the function.

If the weights w_i were all positive, then obviously from (8.6.4), $C(M)$ would be quite large, but we want it to be of the same order of magnitude as the moduli m_i. It will thus be necessary to choose both positive and negative weights, and the function will not be monotonic. Based on (8.6.6) and given some wanted value of $C(M) > 0$, the weights w_i may then be chosen as either $C(M)\widetilde{m}_i^{-1} \bmod m_i$ (non-negative) or as $(C(M)\widetilde{m}_i^{-1} \bmod m_i) - m_i$ (negative), to be selected such that condition (8.6.4) is satisfied. The particular choices of w_i should be ones of small absolute value, to minimize the contribution (oscillations) from the last term in (8.6.7).

Note that this equation also allows the value of $|a|_M$ to be determined from the value of $C(a)$ and the residues $|a|_\beta = [a_1, \ldots, a_n]$, we shall return to this possibility later.

Example 8.6.1 shows the behavior of the core function $C(a)$, illustrating its global linearity, but also its local fuzziness.

Example 8.6.1 Let $\beta = [7, 9, 11]$ with $M = 693$, then the system constants are found to be

	i		
	1	2	3
m_i	7	9	11
$\widetilde{m}_i$	99	77	63
$\lvert\widetilde{m}_i^{-1}\rvert_{m_i}$	1	2	7

and by (8.6.6), the weights must satisfy $w_i \equiv C(M)\widetilde{m}_i^{-1} \pmod{m_i}$. Choosing $C(M) = 13$ we then get

$$\begin{aligned} w_1 &\equiv 13 \cdot 1 \pmod{7} \;\Rightarrow w_1 = 6 \text{ or } -1,\\ w_2 &\equiv 13 \cdot 2 \pmod{9} \;\Rightarrow w_2 = 8 \text{ or } -1,\\ w_1 &\equiv 13 \cdot 7 \pmod{11} \Rightarrow w_3 = 3 \text{ or } -8, \end{aligned}$$

where the weights $[-1, -1, 3]$ satisfy (8.6.4), since $-1 \cdot 99 - 1 \cdot 77 + 3 \cdot 63 = 13$. The graph below shows $C(a)$ for a in the interval $[0\,;M) = [0\,;692]$.

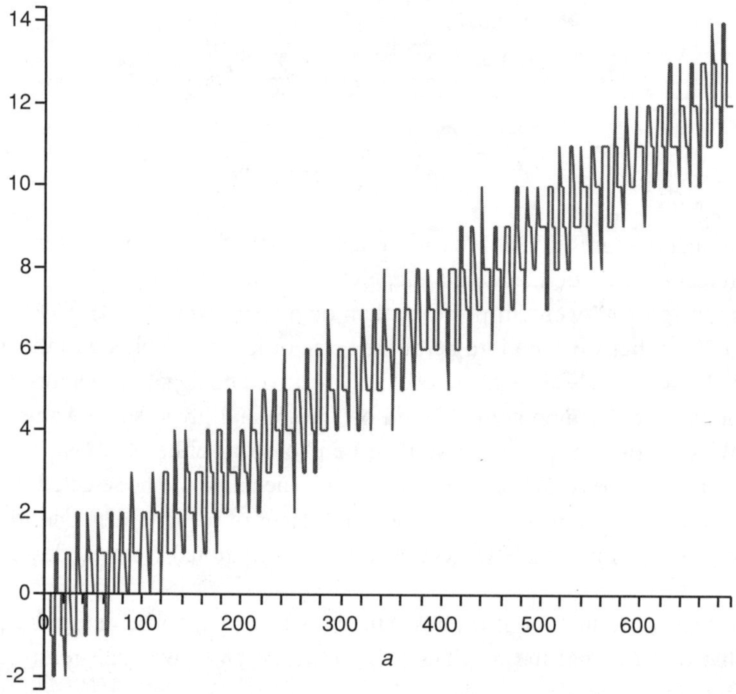

Note that $\sum_{i=1}^{n} |w_i|$ provides a bound on the oscillations, which in more realistic systems (larger values of M and $C(M)$) will not be as pronounced as in this simple example. □

The "almost linearity" can also be illustrated by the following result.

Theorem 8.6.3 *If* $|a|_\beta = [a_i, \ldots, a_n]$ *and* $|b|_\beta = [b_i, \ldots, b_n]$ *with base vector* $\beta = [m_i, \ldots, m_n]$ *and core weights* $[w_i, \ldots, w_n]$*, then*

$$C(a+b) = C(a) + C(b) + \textstyle\sum_{i=1}^{n} w_i \varepsilon_i,$$

where $\varepsilon_i = \lfloor (a_i + b_i)/m_i \rfloor \in \{0, 1\}$ *for* $i = 1, \ldots, n$.

Proof

$$
\begin{aligned}
C(a+b) &= \sum_{i=1}^{n} w_i \left\lfloor \frac{a+b}{m_i} \right\rfloor \\
&= \sum_{i=1}^{n} w_i \left(\frac{a+b-a_i-b_i}{m_i} + \varepsilon_i \right) \\
&= \sum_{i=1}^{n} w_i \left(\frac{a-a_i}{m_i} \right) + \sum_{i=1}^{n} w_i \left(\frac{b-b_i}{m_i} \right) + \sum_{i=1}^{n} w_i \varepsilon_i \\
&= C(a) + C(b) + \sum_{i=1}^{n} w_i \varepsilon_i,
\end{aligned}
$$

by the definitions of the core function and ε_i. □

The symmetry in the core function is expressed in the following lemma.

Lemma 8.6.4 *For all a,*

$$C(M-1-a) = \left(C(M) - \sum_{i=1}^{n} w_i \right) - C(a), \tag{8.6.8}$$

and

$$C\left(2\left\lfloor \frac{M}{2} \right\rfloor - a\right) = 2C\left(\left\lfloor \frac{M}{2} \right\rfloor\right) - C(a). \tag{8.6.9}$$

Proof Writing $a = k_i m_i + a_i$ with $0 \le a_i < m_i$ for $i = 1, \ldots, n$, then

$$M - a - 1 = (\widetilde{m}_i - k_i - 1)m_i + (m_i - a_i - 1)$$

with $0 \le m_i - a_i - 1 < m_i$, hence

$$
\begin{aligned}
C(M-1-a) &= \sum_{i=1}^{n} w_i \left\lfloor \frac{M-a-1}{m_i} \right\rfloor \\
&= \sum_{i=1}^{n} w_i (\widetilde{m}_i - k_i - 1) \\
&= C(M) - C(a) - \sum_{i=1}^{m} w_i .
\end{aligned}
$$

For the second result first note two equalities (exercise), $C(-a) = -C(a) - \sum_{i=1}^{n} w_i$ and $2C(M/2) = C(M) - \sum_{i=1}^{n} w_i$ for M even. So first assume that M

is even, then from Theorem 8.6.3 it follows that

$$C\left(2\frac{M}{2}-a\right) = C(M) + C(-a)$$
$$= C(M) - C(a) - \sum_{i=1}^{n} w_i$$
$$= 2C\left(\frac{M}{2}\right) - C(a).$$

For M odd, $\lfloor M/2 \rfloor = (M-1)/2$, hence from (8.6.8)

$$C\left(2\left\lfloor\frac{M}{2}\right\rfloor - a\right) = C(M-1-a) = C(M) - \sum_{i=1}^{n} w_i - C(a).$$

Choosing $a = \lfloor M/2 \rfloor = (M-1)/2$ in (8.6.8) it follows that $2C((M-1)/2) = C(M) - \sum_{i=1}^{n} w_i$, hence (8.6.9) also holds for odd M. □

Now let us look at some applications of the core function, assuming that we have an efficient way of evaluating it.

Theorem 8.6.5 (Overflow check) *Assume $C(M) \neq 0$, where M is the product of the moduli, let $|a|_\beta = [a_i, \ldots, a_n]$ with $a < M$, $|b|_\beta = [b_i, \ldots, b_n]$ with $b < M$, and let c be defined as their RNS sum $|c|_\beta = [|a_i + b_i|_{m_i}, \ldots, |a_n + b_n|_{m_i}]$. Then the sum $a + b < M$ if and only if*

$$C(c) = C(a) + C(b) + \textstyle\sum_{i=1}^{n} w_i \varepsilon_i,$$

where $\varepsilon_i = \lfloor (a_i + b_i)/m_i \rfloor \in \{0, 1\}$ for $i = 1, \ldots, n$.

Proof Obviously, $a + b < M$ if and only if $a + b = c$, so the necessary part follows from the previous lemma. For the sufficiency, assume that $c = a + b - M$. Then by the lemma, utilizing that $|M|_{m_i} = 0$,

$$C(c) = C(a+b-M)$$
$$= C(a+b) - C(M)$$
$$= C(a) + C(b) + \sum_{i=1}^{n} w_i \varepsilon_i - C(M),$$

hence since $C(M) \neq 0$, $C(c) \neq C(a) + C(b) + \sum_{i=1}^{n} w_i \varepsilon_i$, a contradiction. □

Theorem 8.6.6 (Parity check) *Assume all the moduli m_i are odd, and the core $C(M)$ is odd, $M = \prod_{i=1}^{n} m_i$. Let $|a|_\beta = [a_1, \ldots, a_n]$ and δ_i be the parity function $\delta(x) = |x|_2$ applied to a_i. Then a is even if and only if $C(a)$ and $\sum_{i=1}^{n} w_i \delta_i$ have the same parity.*

Proof Since $C(M)$ is assumed odd, a and $a \cdot C(M)$ have the same parity. Also $C(a)$ and $M \cdot C(a)$ have the same parity since $M = \prod m_i$ is odd. The sums

$\sum_{i=1}^{n} w_i \delta_i$, $\sum_{i=1}^{n} w_i a_i$, and $\sum w_i a_i \widetilde{m}_i$ must all have the same parity, since a_i, δ_i, and $w_i a_i$ are of the same parity. From $\lfloor a/m_i \rfloor = (a - a_i)/m_i$

$$\begin{aligned} M \cdot C(a) &= \sum_{i=1}^{n} w_i M \frac{a - a_i}{m_i} \\ &= a \sum_{i=1}^{n} w_i \widetilde{m}_i - \sum_{i=1}^{n} w_i a_i \widetilde{m}_i \\ &= a\, C(M) - \sum_{i=1}^{n} w_i a_i \widetilde{m}_i, \end{aligned}$$

hence $a\, C(M) = M \cdot C(a) + \sum_{i=1}^{n} w_i a_i \widetilde{m}_i$. But the sum of two terms is even if and only if the terms have the same parity, hence the result. □

This theorem may be used to scale a number $|a|_\beta$ by 2 in the case that all moduli are odd, by subtracting the parity from each a_i, and multiplying by $|2^{-1}|_\beta$.

When one of the moduli is even, the parity is immediately known, but scaling the number by 2 would normally require extension of the base vector, to obtain the residue modulo the even modulus. Theorem 8.6.8 below shows for the special case of scaling by 2 how to compute the residue of $a/2$ relative to the even modulus. But first we need a lemma.

Lemma 8.6.7 *If a with $|a|_\beta = [a_1, \ldots, a_n]$ is divisible by k, and $|a/k|_\beta = [b_1, \ldots, b_n]$, then*

$$C\left(\frac{a}{k}\right) = \frac{C(a) - \sum_{i=1}^{n} w_i \dfrac{k b_i - a_i}{m_i}}{k}.$$

Proof By iterating Theorem 8.6.3 for k summands, a/k, we have

$$C(a) = k \cdot C\left(\frac{a}{k}\right) + \sum_{i=1}^{n} w_i \frac{k b_i - a_i}{m_i}$$

from which the result follows. □

In general, scaling by one of the moduli can be realized by the core function using the following result on how to recover the corresponding residue without employing base extension as discussed in Section 8.5.3.

Theorem 8.6.8 (Residue recovery) *Let* $\gcd(w_j, m_j) = 1$ *and* $|a|_\beta = [a_1, \ldots, a_n]$ *be divisible by m_j (so $a_j = 0$), then the residue modulo m_j after scaling by m_j can be found as*

$$\left|\frac{a}{m_j}\right|_{m_j} = \left|\left|\frac{1}{w_j}\right|_{m_j} \left(C(a) + \sum_{i \neq j} \left|\frac{w_i}{m_i}\right|_{m_j} a_i\right)\right|_{m_j}.$$

Proof By the previous lemma with $k = m_j$ we obtain

$$C\left(\frac{a}{m_j}\right) = \frac{C(a) - w_j b_j - \sum_{i \neq j} w_i \dfrac{m_j b_i - a_i}{m_i}}{m_j},$$

where $|a/m_j|_\beta = [b_1, \ldots, b_n]$. Hence

$$b_j = \frac{C(a) - m_j C\left(\dfrac{a}{m_j}\right) - \sum_{i \neq j} w_i \dfrac{m_j b_i - a_i}{m_i}}{w_j}.$$

Reducing modulo m_j the result is obtained since $0 \leq b_j < m_j$. □

Calculating the core function. The foundation for the core function computation is the orthogonal base for the RNS system, as defined in the proof of Theorem 8.4.1 (CRT), employing the set of residue vectors defined by

$$e_i = \left|\widetilde{m}_i |\widetilde{m}_i^{-1}|_{m_i}\right|_\beta$$

of the form $[0, \ldots, 0, 1, 0, \ldots, 0]$, with the unit in the ith position. The core function utilizes the *core base* of the RNS system, the set of integer constants $\{E_i\}_{i=1,\cdots,n}$ defined by $|E_i|_\beta = e_i$:

$$E_i = \widetilde{m}_i |\widetilde{m}_i^{-1}|_{m_i}, \tag{8.6.10}$$

easily seen to satisfy $0 \leq E_i < M$. The core base is then used to define a set of precomputable constants

$$\begin{aligned} C(E_i) &= \sum_{j=1}^{n} w_j \left\lfloor \frac{E_i}{m_j} \right\rfloor \\ &= E_i \sum_{j=1}^{n} \frac{w_j}{m_j} - \frac{w_i}{m_i} \\ &= E_i \frac{C(M)}{M} - \frac{w_i}{m_i}, \end{aligned} \tag{8.6.11}$$

utilizing the definition of the core function, (8.6.3), and (8.6.5) together with

$$\left\lfloor \frac{E_i}{m_j} \right\rfloor = \begin{cases} \dfrac{E_i}{m_j} & \text{for} \quad j \neq i, \\ \dfrac{E_i - 1}{m_i} & \text{for} \quad j = i, \end{cases}$$

since $E_i - 1 = (\widetilde{m}_i - |\widetilde{m}_i|_{m_i})|\widetilde{m}_i^{-1}|_{m_i}$ is divisible by m_i.

The computation of the *core function* can now be expressed based on the alternative form of the CRT, using the rank r_a of a, as given by (8.5.1).

Theorem 8.6.9 (Alternative CRT for core functions) *If a is given by its RNS representation $|a|_\beta = [a_1, \ldots, a_n]$, then its core can be expressed as*

$$C(a) = \sum_{i=1}^{n} a_i C(E_i) - C(M) \cdot r_a.$$

Proof Using (8.6.5) the core is

$$\begin{aligned} C(a) &= \sum_{j=1}^{n} w_j \left\lfloor \frac{a}{m_j} \right\rfloor = \sum_{j=1}^{n} w_j \frac{a - a_j}{m_j} \\ &= \sum_{j=1}^{n} \frac{w_j}{m_j} \left(\sum_{i=1}^{n} a_i E_i - M \cdot r_a - a_j \right) \\ &= \sum_{i=1}^{n} a_i \left(E_i \sum_{j=1}^{n} \frac{w_j}{m_j} - \frac{w_i}{m_i} \right) - r_a \sum_{j=1}^{n} w_j \widetilde{m}_j \\ &= \sum_{i=1}^{n} a_i C(E_i) - r_a \cdot C(M). \end{aligned}$$

□

Reducing modulo $C(M)$, the term containing the rank r_a disappears, yielding a much simpler expression, but unfortunately an ambiguous result.

Corollary 8.6.10 *Under the conditions of Theorem 8.6.9, the residue modulo $C(M)$ of a core can be computed as*

$$|C(a)|_{C(M)} = \left| \sum_{i=1}^{n} a_i C(E_i) \right|_{C(M)}. \tag{8.6.12}$$

Furthermore, with $C_{min} = \min\{C(a) | a \in [0; M)\}$ and $C_{max} = \max\{C(a) | a \in [0; M)\}$, if

$$L = C_{max} - C(M) < |C(a)|_{C(M)} < C_{min} + C(M) = U, \tag{8.6.13}$$

then

$$C(a) = |C(a)|_{C(M)}.$$

Proof Equation (8.6.12) is a trivial consequence of the theorem. With $C(a) = |C(a)|_{C(M)} + kC(M)$ it is easily seen that (8.6.13) implies

$$C(a) - C(M) < C(a) - kC(M) < C(a) + C(M),$$

hence $k = 0$ and the last result follows. □

If a is in one of two "critical regions" close to 0 or $C(M)$, where $|\sum_{i=1}^{n} a_i C(E_i)|_{C(M)}$ does not satisfy condition (8.6.13), the core is called *critical*. Otherwise, when the condition is satisfied, a has a *non-critical core*.

Note that the calculation of the terms $a_i C(E_i)$ can be implemented by a table look-up with index a_i, so what remains is just an accumulation and a modular reduction.

Example 8.6.2 Let the residue system be defined as in Example 8.6.1, with $\beta = [7, 9, 11]$ so $M = 693$, and with weights $[-1, -1, 3]$ chosen for $C(M) = 13$. The following table of system constants can then be calculated:

	i		
	1	2	3
m_i	7	9	11
w_i	−1	−1	3
$\widetilde{m}_i$	99	77	63
$\lvert\widetilde{m}_i^{-1}\rvert_{m_i}$	1	2	7
E_i	99	154	441
$C(E_i)$	2	3	8

The figure below shows $|C(a)|_{C(M)}$, illustrating the critical cores where the modulo operation has possibly added or subtracted $C(M) = 13$. Here $C_{min} = -2$ and $C_{max} = 14$, so cores for which $2 \le |C(a)|_{C(M)} \le 10$ are non-critical, where $C(a) = |C(a)|_{C(M)}$. The critical cores here are those for which $|C(a)|_{C(M)} \in \{0, 1, 11, 12\}$.

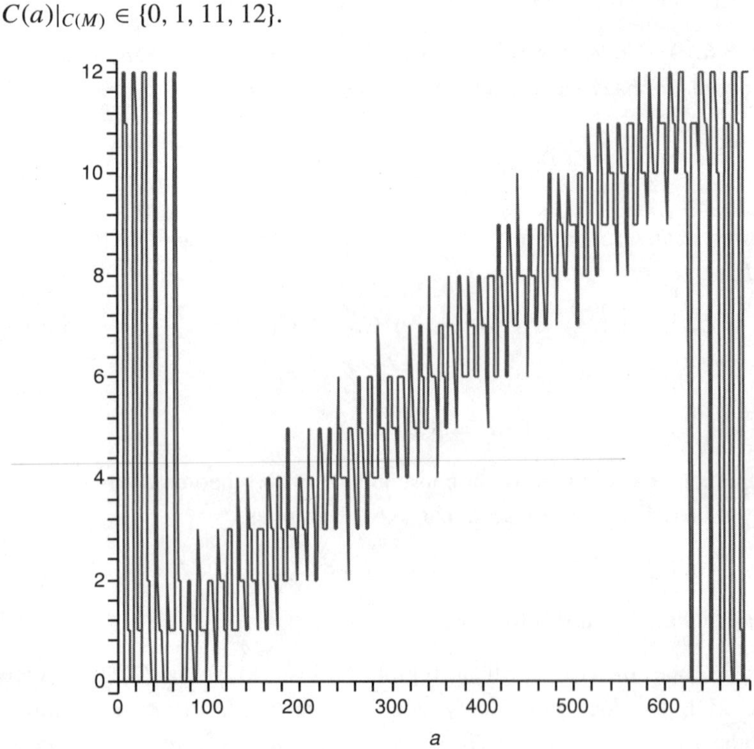

□

However, under fairly relaxed conditions on the choice of parameters, it is possible to insure that arguments in the middle third of the range $[0; M-1]$ have non-critical cores. Then it is possible by Theorem 8.6.3 to translate arguments in the lower part of the range into the middle range for evaluation, and then by the theorem to map the result back to the original range. Thus for a restricted range it is quite simple to obtain $C(a)$.

Lemma 8.6.11 *Let the parameters of the system be chosen such that for some constant m, $0 < m \leq M/3$, all a, $m \leq a \leq M-m$, have non-critical cores. For bounded values of a, $0 \leq a \leq M-m$, $C(a)$ can be determined from its residue representation $|a|_\beta = [a_1, \ldots, a_n]$ as*

$$C(a) = \begin{cases} \gamma(a) & \text{if } \gamma(a) \text{ is non-critical,} \\ \gamma(a+m) - \sigma(a,m) - \gamma(m) & \text{otherwise,} \end{cases}$$

where $\gamma(a) = |C(a)|_{C(M)} = \left|\sum_{i=1}^n a_i C(E_i)\right|_{C(M)}$ *and* $\sigma(a,b) = \sum_{i=1}^n w_i \lfloor (a_i + b_i)/m_i \rfloor$.

If it is known for the system that $C(a) \geq 0$ for all $a \in [0; M)$, then $|C(a)|_{C(M)}$ can only differ from $C(a)$ for a close to M, in which case the lemma simplifies to $C(a) = |C(a)|_{C(M)}$.

Obviously, if it is known that a is bounded away from zero, for $m \leq a \leq M-1$, there is an equivalent lemma, the proof of which we leave as an exercise.

Example 8.6.3 Continuing the previous example it is found that the value $m = 176 \leq 213 = \lfloor M/3 \rfloor$ may be used, so that all a in the interval $[176, 516]$ have non-critical cores (175 and 517 have critical cores), and $C(a)$ can then be evaluated by the lemma for $0 \leq a \leq 516$. □

For the general case it is possible to scale the argument a down by a factor 2 applying Theorem 8.6.8, to find the core by Lemma 8.6.11, and then to scale it back up by adding it to itself using Theorem 8.6.3.

However, after scaling there is no longer any ambiguity, since the scaled argument $\lfloor a/2 \rfloor$ is then known to be less than $M/2$, hence it may be possible to know when to subtract $C(M)$ from $|C(a)|_{C(M)} = \left|\sum_{i=1}^n a_i C(E_i)\right|_{C(M)}$ to obtain $C(a)$. A sufficient condition for this to be possible is given by $\sum_{i=1}^n |w_i| < C(M)/2$, thus it is possible to substitute the use of Lemma 8.6.11 by a simple test on $|C(a)|_{C(M)}$.

Scaling by 2 requires the parity of the argument to be known, but this easily obtained while no overflows occur by using a redundant modulus $m_{n+1} = 2$, which is simple to maintain during additions and multiplications by parallel XOR, respectively AND, operations on the parities of the arguments.

Theorem 8.6.12 *Let an RNS system be given, $\beta = \{m_1, \ldots, m_n, 2\}$ with odd moduli m_i, $M = \prod_{i=1}^n m_i$, and a redundant modulus $m_{i+1} = 2$. Core computation weights are given as $\{w_1, \ldots, w_n\}$, with $C(M)$ odd and weights satisfying*

$\sum_{i=1}^{n} |w_i| < C(M)/2$. *Let predefined constants* E_i *and* $C(E_i)$ *for* $i = 1, \ldots, n$ *be given by* $E_i = \widetilde{m}_i |\widetilde{m}_i^{-1}|_{m_i}$ *and* $C(E_i) = e_i C(M)/M - w_i/m_i$. *With the reduced core defined as*

$$|C(x)|_{C(M)} = \left| \sum_{i=1}^{n} a_i C(E_i) \right|_{C(M)},$$

let $L = \max_{x \in [0;M)} C(x) - C(M)$ *and* $U = \min_{x \in [0;M)} C(x) + C(M)$ *be constants given such that whenever* x *satisfies* $L < |C(x)|_{C(M)} < U$ *the core of* x *is non-critical, i.e.,* $C(x) = |C(x)|_{C(M)}$.

Let a *be given by its extended residue representation* $|a|_\beta = [a_1, \ldots, a_n, p]$, *where* p *is the parity of* a. *Provided that there has not been an overflow in producing* a, $C(a)$ *can be determined by the following algorithm:*

> **if** $L < |C(a)|_{C(M)} < U$ **then** $C(a) := |C(a)|_{C(M)}$
> **else** $|b|_\beta := \left[|(a_1 - p)2^{-1}|_{m_1}, \ldots, |(a_n - p)2^{-1}|_{m_n} \right]$;
> $h := \left| \sum_{i=1}^{n} b_i C(E_i) \right|_{C(M)}$;
>   **if** $h \geq U$ **then** $h := h - C(M)$;
> $C(a) := 2h + \sum_{i=1}^{n} w_i |a_i + p|_2$;
> **end if;**

Notes The extended RNS representation $[a_1, \ldots, a_n, p]$ corresponds to a range $[0; 2M)$, but the algorithm only evaluates $C(a)$ correctly over the range $[0; M)$. If adding or multiplying two operands from $[0; M)$ creates an overflow situation in the basic system specified by the modulus vector $\beta = [w_1, \ldots, w_n]$, then the parity of the result may be incorrect since some multiple of the odd M has been subtracted. Also note that the condition $\sum_{i=1}^{n} |w_i| < C(M)/2$ is simple to evaluate, but overly restrictive.

Proof The correctness follows from the preceding comments, noting that $b \sim \lfloor a/2 \rfloor$, so when mapping back by Theorem 8.6.3, $\varepsilon_i = \lfloor 2b_i/m_i \rfloor \in \{0, 1\}$ in the corrective sum $\sum_{i=1}^{n} w_i \varepsilon_i$. Now the core of b can only be critical if b is close to 0, in which case the reduced core must be corrected. By the definition of b_i we have $2b_i = a_i - p$ if $a_i - p$ is even and $2b_i = a_i - p + m_i$ when $a_i - p$ is odd, thus the final expression for $C(a)$. □

Example 8.6.4 To illustrate the problem with overflow, consider the system as above with $\beta = [7, 9, 11]$ and $M = 693$, let $a = 349$, then $|349|_\beta = [6, 7, 8]$. Now forming $a + a$ in RNS arithmetic yields $|698|_\beta = [5, 5, 5]$ after an overflow corresponding to $[5, 5, 5]$ representing $|698|_{693} = 5$. But including the parity corresponding to the redundant modulus 2, the result will be represented as $[5, 5, 5, 0]$, and the algorithm of Theorem 8.6.12 will then return 13 instead of the correct core value $C(5) = 0$. □

Mapping a core back to the integer domain. It was previously observed that (8.6.7) allows the value of $|a|_M$ to be determined from $C(a)$ and $\{a_1, \ldots, a_n\}$, i.e., it performs a mapping from RNS representation to the integer domain, hence we have the following theorem.

Theorem 8.6.13 *Given $C(a)$ and the RNS representation of a, $|a|_\beta = [m_1, \ldots, m_n]$, then the value of $|a|_M$ can be determined as*

$$|a|_M = \frac{M \cdot C(a) + \sum_{i=1}^{n} w_i \widetilde{m}_i a_i}{C(M)}. \tag{8.6.14}$$

Alternatively, the following equivalent of the CRT employs $|C(a)|_{C(M)}$ instead of $C(a)$.

Theorem 8.6.14 (CRT for core functions) *Given $|C(a)|_{C(M)}$ and the RNS representation of a, $|a|_\beta = [m_1, \ldots, m_n]$, then the value of $|a|_M$ can be determined as*

$$|a|_M = \left| \frac{M \cdot |C(a)|_{C(M)} + \sum_{i=1}^{n} w_i \widetilde{m}_i a_i}{C(M)} \right|_M.$$

Proof Let $C(a) = |C(a)|_{C(M)} + kC(M)$, then the result follows from (8.6.14). □

With C_{max} and C_{min} defined as in Corollary 8.6.10 it is almost always possible to choose weights and $C(M)$ such that $C_{max} - C_{min} < 2C(M)$. Then it is easy to see that if k is defined as in the proof, then $k \in \{-1, 0, 1\}$. Assuming that 2's complement arithmetic is used, the modular reduction can be performed based on simple bit tests as follows. Let n be defined such that $2^{n-1} < M < 2^n$ and $p = 2^k - M$, then the need for subtraction of M can be decided by adding p and testing bit n, following the idea used in the modular select adder of Section 8.2 and the need for addition of M can be decided simply by testing the sign bit.

In concluding this discussion on the core function, note that the applicability of the results in this section is limited due to the complications in calculating the value of the core. For example, testing for overflow by Theorem 8.6.5 is impossible, since neither Lemma 8.6.11 nor Theorem 8.6.12 may be used to obtain the value of the core. Similarly, the parity of an RNS represented value can only be determined by Theorem 8.6.6 if the argument is known to be in a restricted range so that Lemma 8.6.11 (or equivalent) may be applied.

A core-based approach to sign-determination. Due to the "almost linearity" of the core function, for some ranges of core values $C(a)$ close to 0 or close to $C(M)$, it is possible to state the sign of a, interpreting numbers in $[1\,;\lfloor (M-1)/2 \rfloor]$ as positive and in $[\lfloor (M+1)/2 \rfloor\,; M-1]$ as negative. To be able to calculate core values by Theorem 8.6.12 we assume that M is odd, in which case the interval $[1; \lfloor M/2 \rfloor]$ represents positive and $[\lceil M/2 \rceil; M-1]$ negative values (the set $\{-\lceil M/2 \rceil, \ldots, -1\}$), and thus the set of representable values is symmetric.

The sign of a may be determined from $C(a)$ (evaluated by Theorem 8.6.12) whenever $C(a) < C_{low} = \min\{C(x) \mid x > \lfloor M/2 \rfloor\}$ or $C(a) > C_{high} = \max\{C(x) \mid x < \lfloor M/2 \rfloor\}$. The problem lies in the middle of the range where the argument a is close to $\lfloor M/2 \rfloor$ so that $\lfloor M/2 \rfloor - a$ has a small absolute value, but has the same sign as a when interpreted as a signed value. Thus for some interval $[a_{low}; a_{high}]$ (symmetric around $\lfloor M/2 \rfloor$) another method must be applied, except for $a = \lfloor M/2 \rfloor$, where the sign is known to be positive by definition.

Hence form $\delta = \lfloor M/2 \rfloor - a$, satisfying $|\delta| \le \lfloor M/2 \rfloor - a_{low}$ and having the same sign as a. We may now evaluate the reduced core $|C(\delta)|_{C(M)}$, which allows the sign to be determined if $|\delta|$ is sufficiently large so that $|C(\delta)|_{C(M)}$ is non-critical. But to improve the possibility of success, we can multiply a non-zero δ by a maximal integer constant $k_1 \ge 1$ chosen such that $|k_1\delta| < \lfloor M/2 \rfloor$. Thus let $\delta_1 = k_1\delta$, choosing

$$k_1 = \left\lfloor \frac{\lfloor M/2 \rfloor}{\lfloor M/2 \rfloor - \max\{a \mid C(a) < C_{low}\}} \right\rfloor,$$

and check if the sign can be determined from $|C(\delta_1)|_{C(M)}$.

If $|C(\delta_1)|_{C(M)}$ is critical, then we may again multiply by a constant $k_2 \ge 2$ chosen to improve the likelihood of $|C(k_2\delta_1)|_{C(M)}$ being non-critical. This time the maximal possible factor is determined by

$$k_2 = \left\lfloor \frac{\lfloor M/2 \rfloor}{\min\{a \mid C(a) > L\}} \right\rfloor,$$

insuring that $|k_2\delta_1| < \lfloor M/2 \rfloor$, since $|C(\delta_1)|_{C(M)}$ was critical. If necessary, this may be repeated recursively on $\delta_3, \delta_4, \ldots, \delta_i = k_2^{i-1}\delta_1$, until eventually terminated at some δ_ℓ, in the worst case (for $|\delta| = 1$) for $\ell = O(\log M)$. To insure $k_2 \ge 2$ we must require $\min\{a \mid C(a) > L\} \le \lfloor M/2 \rfloor / 2$.

Note that it is sufficient to use the reduced core, $|C(\delta_i)|_{C(M)}$, since all that is needed is to know whether its value is "high" or "low." Actually, it would not even be possible to apply Theorem 8.6.12 since an overflow will occur when forming δ.

Algorithm 8.6.15 (Core-based sign determination)

Stimulus: *An extended residue number* $|a|_\beta = [a_1, a_2, \ldots, a_n, p]$, *where* p *is the parity of* a, *base vector* $\beta = [m_1, m_2, \ldots, m_n, 2]$, *with* $M = \prod_1^n m_i$ *odd. To allow evaluation of the core* $C(a)$ *by Theorem 8.6.12, the value of the constant* $C(M)$ *and suitable weights* $[w_1, w_2, \ldots, w_n]$ *must be given, such that the system allows the sign of* a *to be determined from* $C(a)$ *whenever* $C(a) < C_{low} = \min\{C(x) \mid x > \lfloor M/2 \rfloor\}$ *and* $C(a) > C_{high} = \max\{C(x) \mid x < \lfloor M/2 \rfloor\}$.

Constants L, U such that $|C(a)|_{C(M)}$ is non-critical for $L < a < U$. Also constants $C(\lfloor M/2 \rfloor)$, $C(M)$ and $C(E_i)$ ($E_i = \widetilde{m}_i|\widetilde{m}_i^{-1}|_{m_i}$, $i = 1, \ldots, n$), and factors $k_1 \geq 1$ and $k_2 \geq 2$ (determined as above).

Response: *The value of the sign function,* sgn(a).

Method:

if $a = \lfloor M/2 \rfloor$ **then return** 1; {*Evaluated in RNS arithmetic*}
if $a = 0$ **then return** 0; {*Evaluated in RNS arithmetic*}
$c := C(a)$; {*Using Theorem 8.6.12 employing parity*}
if $c < C_{low}$ **then return** 1;
if $c > C_{high}$ **then return** -1;
$\delta := k_1 \cdot (\lfloor M/2 \rfloor - a)$; {*Evaluated in RNS arithmetic*}
$c := |C(\delta)|_{C(M)}$; {*Evaluated by* $\left|\sum_{i=1}^{n} w_i C(E_i)\right|_{C(M)}$}
while $(c \leq L) \vee (c \geq U) \vee (c = C(\lfloor M/2 \rfloor))$ **do**
  $\delta := k_2 \cdot \delta$; {*Evaluated in RNS arithmetic*}
  $c := |C(\delta)|_{C(M)}$; {*Evaluated by* $\left|\sum_{i=1}^{n} w_i C(E_i)\right|_{C(M)}$}
end;
if $c < C(\lfloor M/2 \rfloor)$ **then return** 1 **else return** -1;
end;

Example 8.6.5 For the previously discussed system with $\beta = [7, 9, 11]$ and weights $[-1, -1, 3]$, the constants $C_{low} = 4$, $C_{high} = 8$, $C(\lfloor M/2 \rfloor) = 6$, $k_1 = 1$ and $k_2 = 2$ can be determined beyond those already known. Algorithm 8.6.15 then determines the sign function as shown:

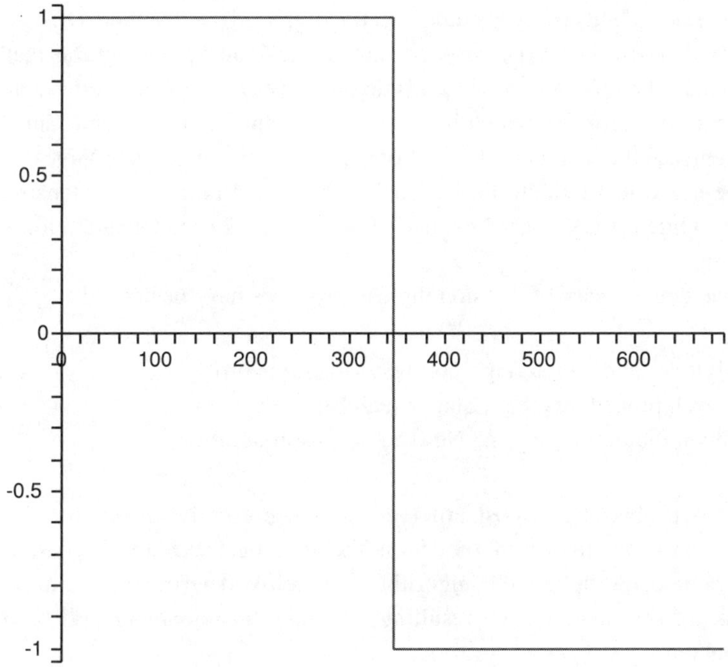

Except for the two special cases $a = 0$ and $a = \lfloor M/2 \rfloor$, $C(a)$ has to be evaluated for each call of the algorithm, based on which signs may be resolved. If this fails, δ_i and $|C(\delta_i)|_{C(M)}$ must be evaluated a number of times. For this system it turns out that in the 693 cases used to plot this diagram δ_i and $|C(\delta_i)|_{C(M)}$ had to be evaluated 638 times, corresponding to an average frequency of .922. For other systems the frequency may be much smaller, e.g., for the slightly extended system with $\beta = \{7, 9, 11, 13\}$, weights $\{1, 0, 1, -1\}$, $M = 9009$ and $C(M) = 27$, the frequency was found to be 0.259. □

8.6.4 General division

The problem of determining a quotient, remainder pair for an arbitrary given dividend and divisor has turned out to be the hardest problem in residue arithmetic, and many algorithms for it have been proposed. The problem is that division algorithms, in general, are of an iterative nature, employing sign detection or comparison in each step. And as seen above such operations are generally "expensive," hence the proposed algorithms for division attempt to simplify them or minimize their number.

However, before discussing the most general case, let us assume that the divisor D is a constant which is not a product of some of the moduli. We can then turn the division problem into a scaling problem by performing a base extension to include D as an extra modulus, and thus determine the integer $|N|_D = N \bmod D$, where N is the dividend, given by its residue vector $|N|_\beta$. If D is smaller than the smallest modulus in the modulus vector β, $||N|_D|_\beta = |N \bmod D|_\beta$ is directly known, otherwise it can be mapped into the residue system by the method in Section 8.4. Then $|N - N \bmod D|_\beta$ is divisible by D and what is left is a division-remainder-zero problem which is solved by multiplying by the constant $|D^{-1}|_\beta$, equivalently to the scaling problem. Note, however, that when D is large compared with the individual moduli, then probably the mixed-radix base extension is not feasible, while the CRT-based method of Section 8.5.2 is better suited for the base extension.

For the general case of division three approaches have been used:

- digit-serial division (e.g., non-restoring division),
- search procedures (e.g., binary search),
- functional iteration (e.g., Newton–Raphson iteration).

As it is probably the most efficient method presently known we shall here describe only an algorithm based on the first of these approaches, a digit-serial method. Although the algorithm implicitly determines a binary representation of the quotient, the resulting quotient and remainder will be in RNS representation.

The problem is, given N and D, $0 \le N < M$ and $0 < D < M$, M being the system modulus, to determine Q and R such that

$$N = Q \cdot D + R,$$

where $0 \le R < D$. Since the algorithm implicitly determines the quotient in binary, the ith step consists of finding a $k_i \ge 0$ such that with $R_0 = N$,

$$R_i = 2^{k_i} D + R_{i+1}, \qquad \text{where} \quad 0 \le R_{i+1},$$

but (8.6.15)

$$R_i = 2^{k_i+1} D + R'_{i+1}, \qquad \text{where} \quad R'_{i+1} < 0,$$

terminating when $R_{i+1} < D$. Since the operands are given in the RNS representation we will perform the comparison between R_i and D, by comparing instead the ratios R_i/M and D/M, both of which are proper fractions. Estimates of such fractions can be obtained from the alternative form of the CRT, as given by (8.5.1) for any X in the RNS representation with $X < M$:

$$\begin{aligned} X &= \sum_{i=1}^{n} \widetilde{m}_i \left| \frac{x_i}{\widetilde{m}_i} \right|_{m_i} - r_x M \\ &= \sum_{i=1}^{n} \frac{M}{m_i} \left| \frac{x_i}{\widetilde{m}_i} \right|_{m_i} - r_x M, \end{aligned}$$

where r_x is some integer. Dividing through by M we have

$$\frac{X}{M} = \sum_{i=1}^{n} \frac{1}{m_i} \left| \frac{x_i}{\widetilde{m}_i} \right|_{m_i} - r_x, \tag{8.6.16}$$

which can be computed by multioperand addition of binary values found by table look-up given $X = [x_1, x_2, \ldots, x_n]$. Since it is known that X/M is a proper fraction, any carry-out in the summation can be discarded, corresponding to the value of

$$r_x = \left\lfloor \sum_{i=1}^{n} \frac{1}{m_i} \left| \frac{x_i}{\widetilde{m}_i} \right|_{m_i} \right\rfloor$$

being of no interest. The table values to be accumulated (all less than 1) have to be properly rounded values of say t fractional bits, hence the total error in the sum is at most $n2^{-t-1}$. Since we want to be able to compare numbers X/M and Y/M which (if different) are at least $1/M$ apart, we must require that $t > \log_2(Mn)$.

Returning to the problem of determining k_i of (8.6.15), it is sufficient to compare the positions of the leading non-zero bits of R_i/M and D/M, so, we define

$$h(x) = \begin{cases} 0 & \text{if } x = 0, \\ \lfloor \log_2 x \rfloor & \text{if } 0 < x < 1. \end{cases}$$

This function is fairly easy to compute in hardware (a priority encoder), so we are now able to formulate the division algorithm.

Algorithm 8.6.16 (Digit-serial RNS division)

Stimulus: *Integers N and D in RNS representation ,* $0 \le N < M$ *and* $0 < D < M$*, where M is the system modulus.*

Response: *Integers Q and R in RNS representation such that* $N = Q \cdot D + R$ *where* $0 \le R < D$.

Method:
$R := N;\ Q := 0;$
$\ell := h(D/M);$
$k = h(R/M);$
while $k > \ell$ **do**
  $Q := Q + 2^{k-\ell-1};$
  $R := R - 2^{k-\ell-1}D;$
  $k = h(R/M)$
end;
if $(k = \ell)$ **and** $(h((R-D)/M) \neq -1)$ **then**
  $R := R - D;$
  $Q := Q + 1$
end

Where: *All arithmetic is performed in RNS (employing RNS constants of the form* 2^i *found by table look-up), except for the ratios* X/M *which are determined in binary by (8.6.4).*

It is easy here to see that the while loop determines pairs of Q, R satisfying the invariant $N = QD + R$, terminating with $k \le \ell$. Also $R \ge 0$ since the new R is constructed in the loop from a right-hand side satisfying $R - 2^{k-\ell-1}D \ge M(2^k - 2^{k-\ell-1}2^{\ell+1}) = 0$ by the definition of $h(x)$. If $k < \ell$, then certainly $0 \le R < D$ and the result is correct; the only tricky case left is when $k = \ell$ after termination of the while loop.

If $k = \ell$, then $2^\ell < R/M < 2^{\ell+1}$ and $2^\ell < D/M < 2^{\ell+1}$, so $0 < R < 2D$ and there are three cases to consider:

$R > D$: Here $(R-D)/M < 2^\ell$ implies $h((R-D)/M) < \ell \le -1$, and hence $h((R-D)/M) \le -2$.

$R = D$: Here $h((R-D)/M) = 0$.

$R < D$: The subtraction $R - D$ in RNS arithmetic results in a negative value, however represented as $R' = M + R - D$ with $R' < M$. Here $h(R'/M) = h(1 - (D-R)/M)$ and $0 < (D-R)/M < 2^\ell$, but $\ell \le -1$ implies that $h(R'/M) = h((R-D)/M) = -1$.

In the first two cases $D \le R < 2D$, thus D has to be subtracted from R, and Q correspondingly increased by 1, to obtain the correct remainder and quotient. In the third case there is nothing to correct.

Example 8.6.6 Consider the division of $|N|_\beta = [1, 2, 6]$ (~ 97) by $|D|_\beta = [1, 0, 3]$ (~ 10) in the system with $\beta = [3, 5, 7]$. First we need some constant calculations:

$$M = 3 \cdot 5 \cdot 7 = 105, \quad \widetilde{m}_1 = 5 \cdot 7 = 35, \quad \widetilde{m}_2 = 3 \cdot 7 = 21, \quad \widetilde{m}_3 = 3 \cdot 5 = 15,$$

so

$$|\widetilde{m}_1|_{m_1} = 2, \quad |\widetilde{m}_2|_{m_2} = 1, \quad |\widetilde{m}_3|_{m_3} = 1$$

and

$$|\widetilde{m}_1^{-1}|_{m_1} = 2, \quad |\widetilde{m}_2^{-1}|_{m_2} = 1, \quad |\widetilde{m}_3^{-1}|_{m_3} = 1.$$

To perform the algorithm we need an estimate of D/M computed to t bits, where $t > \log_2(Mn) = \log_2 315$, so $t = 9$ will suffice. From (8.6.4) we then find for D/M:

$$\begin{array}{rcr} \frac{1}{3} \cdot 1 \cdot 2 & \approx & 0.101010101 \\ \frac{1}{5} \cdot 0 \cdot 1 & \approx & 0.000000000 \\ \frac{1}{7} \cdot 3 \cdot 1 & \approx & \underline{0.011011011} \\ & & 1.000110000 \end{array}$$

so $D/M \approx 0.000110000$ and $\ell = h(D/M) = -4$.

For $R_0 = N = [1, 2, 6]$ we similarly approximate $R_0/M \approx 0.111011001$, so $k_0 = -1$, and since $2^{k_0 - \ell - 1} = 4 \sim [1, 4, 4]$ we have:

$$\begin{aligned} Q_1 &= [0, 0, 0] + [1, 4, 4] = [1, 4, 4], \\ R_1 &= [1, 2, 6] - [1, 4, 4] \cdot [1, 0, 3] = [0, 2, 1], \\ R_1/M &\approx 0.100010110 \;\Rightarrow\; k_1 = -1, \end{aligned}$$

and for the next cycle with $2^{k_1 - \ell - 1} = 4 \sim [1, 4, 4]$:

$$\begin{aligned} Q_2 &= [1, 4, 4] + [1, 4, 4] = [2, 3, 1], \\ R_2 &= [0, 2, 1] - [1, 4, 4] \cdot [1, 0, 3] = [2, 2, 3], \\ R_2/M &\approx 0.001010011 \;\Rightarrow\; k_2 = -3, \end{aligned}$$

and finally with $2^{k_2 - \ell - 1} = 1 \sim [1, 1, 1]$:

$$\begin{aligned} Q_3 &= [2, 3, 1] + [1, 1, 1] = [0, 4, 2], \\ R_3 &= [2, 2, 3] - [1, 1, 1] \cdot [1, 0, 3] = [1, 2, 0], \\ R_3/M &\approx 0.000100010 \;\Rightarrow\; k_3 = -4. \end{aligned}$$

Since the loop terminates with $k_3 = \ell$, then $h((R - D)/M)$ must be computed. But $R - D = [1, 2, 0] - [1, 0, 3] = [0, 2, 4]$, so $(R - D)/M \approx 0.111110001$ and

$h((R-D)/M) = -1$, thus there is no correction to make. The result is then

$$[1,2,6] = [0,4,2] \cdot [1,0,3] + [1,2,0] \quad \sim \quad 97 = 9 \cdot 10 + 7. \qquad \square$$

Finally, note that since the core function is an "almost" linear function of its argument, it may be used to provide an approximation to the quotient of two operands. However, due to the "fuzziness" of the core, the operands should be fairly large so that the deviations only play minor roles in providing an estimate of their ratio, which will have to be checked by other means, for example, checking that the remainder is smaller than the divisor is non-trivial.

Problems and exercises

8.6.1 Prove that $2C(M/2) = C(M) - \sum_{i=1}^{n} w_i$ for M even, and $C(-a) = -C(a) - \sum_{i=1}^{n} w_i$ for all a.

8.6.2 In using Theorem 8.6.8 for the general case of an even modulus, it is implicitly assumed that any system with an even modulus is equivalent to a system with a modulus of value 2. Show why this is the case.

8.6.3 Show an analogous result to Lemma 8.6.11, valid for arguments a known to satisfy $m \le a \le M-1$.

8.6.4 Find a simple expression for the difference $C_{max} - C_{min}$ as defined in Corollary 8.6.10.

8.6.5 Show an alternative to the CRT for core functions, Theorem 8.6.14, yielding the following expression to be evaluated:

$$a = \left| \sum_{i=1}^{n} E_i a_i \right|_M ,$$

from which $C(a)$ may be evaluated. This appears to be much simpler than the expression in the theorem. But what are the complications in using this expression?

8.7 Single-modulus rational systems

Recall from Observation 8.2.20 that when a multiplicative inverse $|b^{-1}|_m$ exists, division is possible in the sense that $|a \cdot b^{-1}|_m$ acts algebraically as the rational a/b. This leads naturally to the definition of *generalized residue classes of rationals*:

$$Q_a^m = \left\{ \frac{p}{q} \in \hat{\mathbb{Q}}_m \;\middle|\; |pq^{-1}|_m = a \right\}, \quad a \in \mathbb{Z}_m, \tag{8.7.1}$$

where

$$\hat{\mathbb{Q}}_m = \left\{ \frac{p}{q} \in \mathbb{Q} \;\middle|\; gcd(q,m) = 1 \right\}. \tag{8.7.2}$$

With these we can generalize the residue mapping $|\cdot|_m$ to rationals.

Definition 8.7.1 *The mapping* $|\cdot|_m : \hat{\mathbb{Q}}_m \to \mathbb{Z}_m$ *is defined by*

$$\left|\frac{p}{q}\right|_m = |p \cdot q^{-1}|_m = k \in \mathbb{Z}_m.$$

For $k \neq 0$, Q_k^m is the set of rationals (including integers) mapped onto k, and Q_0^m consists of those rationals p/q for which $\gcd(q, m) = 1$ and p is a multiple of m. If $x = p/q$ and $y = r/s$ are rational numbers such that q^{-1} and s^{-1} exists, then $|x|_m = |y|_m$ if and only if

$$ps \equiv qr \pmod{m}. \tag{8.7.3}$$

Thus two distinct rational numbers belong to the same residue class Q_k^m if and only if (8.7.3) is satisfied. But when $\gcd(q, m) \neq 1$ the multiplicative inverse does not exist and $|p/q|_m$ is not defined. Obviously, not every element of $\mathbb{Q}$ lies in one of the generalized residue classes, but since

$$\hat{\mathbb{Q}}_m = \bigcup_{i=0}^{m-1} Q_i^m,$$

then $(\hat{\mathbb{Q}}_m, \oplus_m, \otimes_m)$ is a commutative ring with identity, and the mapping $|\cdot|_m = \hat{\mathbb{Q}}_m \to \mathbb{Z}_m$ is a homomorphism with respect to addition and multiplication. Since $\mathbb{Z}_m$ is a homomorphic image of $\hat{\mathbb{Q}}_m$, arithmetic operations in $(\hat{\mathbb{Q}}_m, \oplus_m, \otimes_m)$ correspond to arithmetic operations in $(\mathbb{Z}_m, \oplus_m, \otimes_m)$. If m is a prime, then $(\hat{\mathbb{Q}}_m, \oplus_m, \otimes_m)$ is a field.

The mapping $|\cdot|_m$ is onto, but not one-to-one, since each integer $a \in \mathbb{Z}_m$ is the image of the infinite set Q_a^m. But with N specified as the largest integer satisfying

$$N \leq \sqrt{\frac{m-1}{2}}, \tag{8.7.4}$$

it turns out that the set of *order-N Farey fractions*:

$$\mathcal{F}_N = \left\{ \frac{p}{q} \;\middle|\; \gcd(p, q) = 1,\ 0 \leq |p| \leq N \text{ and } 0 < q \leq N \right\}$$

is a set of rationals such that

$$\hat{\mathbb{Z}}_m = \left\{ \left|\frac{p}{q}\right|_m \;\middle|\; \frac{p}{q} \in \mathcal{F}_N,\ \gcd(q, m) = 1 \right\} \tag{8.7.5}$$

satisfies $\hat{\mathbb{Z}}_m \subset \mathbb{Z}_m$.

Example 8.7.1 Let $m = 19$, then $N = 3$ satisfies (8.7.4) and the set $\mathcal{F}_3$ together with its images in $\hat{\mathbb{Z}}_{19}$ can be tabulated:

$\frac{p}{q}$	$\left\|\frac{p}{q}\right\|_{19}$	$\frac{p}{q}$	$\left\|\frac{p}{q}\right\|_{19}$
0	0		
$\frac{1}{3}$	13	$-\frac{1}{3}$	6
$\frac{1}{2}$	10	$-\frac{1}{2}$	9
$\frac{2}{3}$	7	$-\frac{2}{3}$	12
1	1	-1	18
$\frac{3}{2}$	11	$-\frac{3}{2}$	8
2	2	-2	17
3	3	-3	16

Note from the table that the additive inverse is obtained as $|-x|_m = |m-x|_m$ and e.g., $|\frac{1}{3}+\frac{2}{3}|_{19} = \left||\frac{1}{3}|_{19} + |\frac{2}{3}|_{19}\right|_{19} = |13+7|_{19} = 1 = |1|_{19}$. Also $|3 \cdot \frac{-2}{3}|_{19} = \left||3|_{19} \cdot |\frac{-2}{3}|_{19}\right|_{19} = |3 \cdot 12|_{19} = 17 = |-2|_{19}$. □

Observe from the example that $\hat{\mathbb{Z}}_m$ is a proper subset of $\mathbb{Z}_m$, i.e., 4, 5, 14, and 15 are not in $\hat{\mathbb{Z}}_{19}$. On the other hand, $\mathcal{F}_4$ is too large a set as can be seen since both 4 and $\frac{-3}{4}$ are in $\mathcal{F}_4$ but have the same image $|4|_{19} = \left|\frac{-3}{4}\right|_{19} = 4$. Also $\left|\frac{1}{4}\right|_{19} = \left|\frac{-4}{3}\right|_{19} = 5$, $\left|\frac{4}{3}\right|_{19} = \left|\frac{-1}{4}\right|_{19} = 14$ and $|-4|_{19} = \left|\frac{3}{4}\right|_{19} = 15$, so there seems to be no simple way of describing a set of rationals in a one-to-one correspondence with $\mathbb{Z}_m$. However, $\mathcal{F}_N$ seems to be a convenient set of rationals to handle, due to the simplicity of its description in terms of numerator, denominator limitations. The set $\mathcal{F}_N$ has been extensively studied in number theory, and we shall also return to it in Chapter 9 in the context of another rational system.

Theorem 8.7.2 *If $N \leq \sqrt{(m-1)/2}$ and $p/q \in \mathcal{F}_N$ with $|p/q|_m = k$, then p/q is the only member in $\mathcal{F}_N \cap Q_k^m$, i.e., p/q is the unique order-N Farey fraction for which $|p/q|_m = k$.*

Proof Assume p/q and r/s both are in $\mathcal{F}_N \cap Q_k^m$, then

$$\left|\frac{p}{q}\right|_m = \left|\frac{r}{s}\right|_m = k$$

and hence $ps \equiv qr \pmod{m}$ or equivalently $|ps - qr|_m = 0$. But

$$0 \leq |ps - qr| \leq |p| \cdot |s| + |q| \cdot |r| \leq 2N^2 < m,$$

which implies $|ps - qr| = 0$, i.e., $p/q = r/s$. □

With $\hat{\mathbb{Z}}_m$ defined by (8.7.5) as the set of images of rationals in $\mathcal{F}_N$, we thus have the following result, assuming that N is determined by m satisfying (8.7.4).

Corollary 8.7.3 *The mapping $|\cdot|_m : \mathcal{F}_N \to \hat{\mathbb{Z}}_m$ is a bijection and thus has an inverse.*

Having established the mapping $|\cdot|_m$ as a homomorphism we are able to map operands in $\mathcal{F}_N$ into $\hat{\mathbb{Z}}_m$ and perform operations in the ring $(\mathbb{Z}_m, +, \times)$ and then map the results back into $\mathcal{F}_N$. If it is known (by some other means) that the result is in $\mathcal{F}_N$, we are able to perform the arithmetic operations in the integer domain instead of in the domain of rationals. Observe also that the result will be correct, even if some intermediate results during the computation have "overflown" (their rational equivalents not being in $\mathcal{F}_N$), as long as it can be guaranteed that the final result is in $\mathcal{F}_N$. We will term such an "intermediate overflow" a *pseudo-overflow* . Note that the value of N, as determined by m through (8.7.4), determines not only the "range" of values, but also the "resolution" of the system, as pictured when marking the members of $\mathcal{F}_N$ on the real line, e.g., for the very small system $\mathcal{F}_3$:

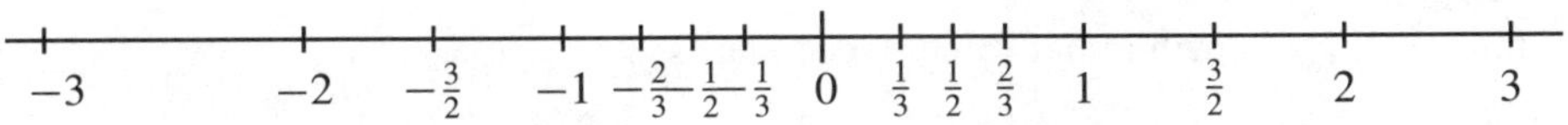

For any practical computation in such a system we will need a fairly large value of N, and hence a large value of m which also ought to be a prime in order that $|q^{-1}|_m$ exists for all $|q| \leq N$.

In a single modulus system, as described so far, it is obviously not then feasible to rely on precomputed tables (ROMs) for realizing the mapping $|\cdot|_m$ and its inverse. What is needed are efficient algorithms for these mappings, and in fact we have already the foundation for these in the EEA Algorithm, as used to find the multiplicative inverse. By a trivial extension of Theorem 8.2.19 we have the following for the mapping $|p/q|_m$ (the *forward mapping*).

Theorem 8.7.4 *When the EEA Algorithm is applied to the seed matrix*

$$\begin{bmatrix} m & 0 \\ q & p \end{bmatrix},$$

generating the sequence $\{(a_i, b_i)\}_{i=0,1,\ldots,n}$*, terminating with* $a_n = 0$*, then whenever* $a_{n-1} = 1$ $(= \gcd(q, m))$ *the multiplicative inverse of* q *exists and*

$$\left|\frac{p}{q}\right|_m = |b_{n-1}|_m.$$

Example 8.7.2 In the system with $m = 19$ $(N = 3)$ for $p = 2$ and $q = 3$, EEA Algorithm yields

f_i	a_i	b_i
	19	0
	3	2
6	1	−12
3	0	38

Since $a_1 = 0$ and $a_0 = \gcd(19, 3) = 1$, we have

$$\left|\frac{2}{3}\right|_{19} = |b_0|_{19} = |-12|_{19} = 7. \qquad \square$$

Observe that it is not necessary that $p/q \in \mathcal{F}_N$ for the algorithm to work: in fact applying it to $m = 19$, $p = -5$ and $q = 2$ yields $\left|\frac{-5}{2}\right|_{19} = 7$ so $\frac{-5}{2}$ and $\frac{3}{2}$ both belong to the same generalized residue class Q_7^{19}. By Lemma 8.2.17, it is possible to generate an arbitrary sequence of rationals from the same generalized residue class by a recursion of the form

$$\begin{aligned} a_i &= a_{i-2} - f_i a_{i-1}, \\ b_i &= b_{i-2} - f_i b_{i-1} \end{aligned} \tag{8.7.6}$$

for $i = 0, 1, \ldots$ with initial values (seed matrix)

$$\begin{bmatrix} a_{-2} & b_{-2} \\ a_{-1} & b_{-1} \end{bmatrix} = \begin{bmatrix} a & b \\ c & d \end{bmatrix},$$

using any arbitrary sequence $\{f_0, f_1, \ldots\}$. Lemma 8.2.17 then states that the sequence $\{(a_i, v_i)\}$ generated by (8.7.6) satisfies $a_i b_j = a_j b_i \pmod{m}$ when the seed matrix satisfies $ad = bc \pmod{m}$.

In particular, if we choose the seed matrix

$$\begin{bmatrix} m & 0 \\ k & 1 \end{bmatrix},$$

it is possible to generate an arbitrary sequence of members of the generalized residue class Q_k^m.

Example 8.7.3 Continuing the previous example we may choose the seed matrix

$$\begin{bmatrix} 19 & 0 \\ 7 & 1 \end{bmatrix}$$

and use the particular sequence $f_i = \lfloor a_{i-2}/a_{i-1} \rfloor$, i.e., the quotients obtained by applying the Euclidean algorithm to the pair 19, 7:

f_i	a_i	b_i
	19	0
	7	1
2	5	−2
1	2	3
2	1	−8
2	0	19

so $\frac{-5}{2}$, $\frac{2}{3}$, and -8 all belong to the residue class Q_7^{19}. Note that we happened to find $\frac{2}{3}$ among the fractions a_i/b_i generated by the algorithm. $\square$

To prove that the EEA Algorithm always will recover the unique order-N Farey fraction $p/q \in Q_k^m$ with $|p/q|_m = k$ when seeded as above, we will need some results on *continued fractions* from number theory, so we will have to make a small digression into this topic.

Assume p/q is any rational with $p \geq 0$ and $q > 0$. Define an extended seed matrix as follows:

$$\begin{bmatrix} a_{-2} & b_{-2} & c_{-2} \\ a_{-1} & b_{-1} & c_{-1} \end{bmatrix} = \begin{bmatrix} p & 0 & 1 \\ q & 1 & 0 \end{bmatrix} \tag{8.7.7}$$

and perform the EEA Algorithm on this by the recursion

$$f_i = \left\lfloor \frac{a_{i-2}}{a_{i-1}} \right\rfloor \quad \text{and} \quad \begin{cases} a_i = a_{i-2} - f_i a_{i-1}, \\ b_i = b_{i-2} + f_i b_{i-1}, \\ c_i = c_{i-2} + f_i c_{i-1}, \end{cases} \tag{8.7.8}$$

then as usual the algorithm will terminate for some value $i = n$ with $a_n = 0$ and $a_{n-1} = gcd(p, q)$. Since $f_0 = \lfloor p/q \rfloor$ and thus

$$\frac{p}{q} = \frac{a_{-2}}{a_{-1}} = f_0 + \frac{1}{\frac{a_{-1}}{a_0}}$$

we obtain recursively the *continued fraction expansion* of p/q:

$$\frac{p}{q} = f_0 + \cfrac{1}{f_1 + \cfrac{1}{f_2 + \cfrac{}{\ddots + \cfrac{1}{f_n}}}} \tag{8.7.9}$$

for which we will introduce the following shorthand notation:

$$\frac{p}{q} = [f_0/f_1/\ldots/f_n],$$

which is unique (exercise) if we require $f_i \geq 1$ for $0 < i < n$ and $f_n \geq 2$ if $n \geq 1$. Note that f_0 may be negative or zero. The integers f_i, $i = 0, 1, \ldots, n$ are called *partial quotients* of the continued fraction expansion.

Truncated versions of (8.7.9) obtained as

$$\frac{p_i}{q_i} = [f_0/f_1/\ldots/f_i], \quad i = 0, 1, \ldots, n$$

are termed *convergents* of the expansion. From (8.7.8) and (8.7.9) it can be seen that $p_i = b_i$ and $q_i = c_i$ with $\gcd(b_i, c_i) = 1$, satisfying $b_i c_{i+1} - b_{i+1} c_i = (-1)^{i+1}$ (exercise).

Example 8.7.4 Performing the EEA Algorithm on $p/q = \frac{17}{39}$ using (8.7.8) on the seed matrix (8.7.7) we obtain:

f_i	a_i	b_i	c_i
	17	0	1
	39	1	0
0	17	0	1
2	5	1	2
3	2	3	7
2	1	7	16
2	0	17	39

so the sequence of convergents is $\frac{0}{1}, \frac{1}{2}, \frac{3}{7}, \frac{7}{16}, \frac{17}{39}$. □

Observe that for p/q irreducible the triples (a_i, b_i, c_i) satisfy the relation

$$pc_i - qb_i = (-1)^i a_i \tag{8.7.10}$$

as seen by induction from (8.7.7) and (8.7.8). Finally, we need the following result from the theory of continued fractions.

Theorem 8.7.5 *If p/q is an order-N Farey fraction satisfying*

$$\left| x - \frac{p}{q} \right| < \frac{1}{2q^2},$$

then p/q is a convergent of x.

Proof The proof is somewhat involved, so we refer the interested reader to the literature on continued fractions, e.g., [HW79, page 153]. □

We are now ready to prove that the EEA Algorithm can be used for the *"inverse mapping."*

Lemma 8.7.6 *If p/q is an order-N Farey fraction satisfying*

$$\left| \frac{p}{q} \right|_m = k,$$

where $k \in \hat{\mathbb{Z}}_m$ and $p > 0$, then there exists an i such that $(p, q) = (a_i, b_i)$, where $\{(a_j, b_j)\}_{j=0,1,\ldots,n}$ is the sequence of pairs generated by the EEA Algorithm seeded with the matrix

$$\begin{bmatrix} m & 0 \\ k & 1 \end{bmatrix}.$$

Proof For the purpose of the proof we will assume that the seed matrix is extended as in (8.7.7) defining also a sequence $\{c_j\}_{j=0,1,\ldots,n}$ such that the sequence

$$\frac{|b_0|}{|c_0|}, \frac{|b_1|}{|c_1|}, \ldots, \frac{|b_n|}{|c_n|} \tag{8.7.11}$$

is the complete sequence of convergents of m/k, where $k \neq 0$ since $p > 0$. From $|p/q|_m = k$ it follows that $p \equiv kq \pmod{m}$ so there is a unique integer t such that

$p = kq - tm$, and hence that

$$\left|\frac{k}{m} - \frac{t}{q}\right| = \left|\frac{kq - tm}{mq}\right| = \left|\frac{p}{mq}\right| \le \frac{1}{q^2}\frac{|q| \cdot N}{2N^2+1} \le \frac{1}{q^2}\frac{N^2}{2N^2+1} < \frac{1}{2q^2},$$

since $2N^2 + 1 \le m$. By the previous theorem either t/q or $-t/-q$ must be a convergent of k/m. Since (8.7.11) is the complete sequence of convergents of m/k, by reciprocation there is an i such that

$$\frac{t}{q} = \frac{c_i}{b_i} \quad \text{with } q = |b_i|.$$

From (8.7.10) we finally obtain

$$\frac{a_i}{b_i} = k - m\frac{c_i}{b_i} = k - m\frac{t}{q} = \frac{p}{q},$$

noting that a_i is positive, so $(a_i, b_i) = (p, q)$. □

This lemma is almost what is needed for mapping a result of a computation in $\mathbb{Z}_m$ back into $\mathcal{F}_N$. But it is required that it is known that $k \in \hat{\mathbb{Z}}_m$ which is a (proper) subset of $\mathbb{Z}_m$, hence we need a somewhat stronger result.

Theorem 8.7.7 *Let k be any integer, $0 \le k \le m - 1$ and let $\{(a_j, b_j)\}_{j=0,1,\dots,n}$ be the sequence generated by the EEA Algorithm seeded with*

$$\begin{bmatrix} m & 0 \\ k & 1 \end{bmatrix}.$$

Then $k \in \hat{\mathbb{Z}}_m$ if and only if there exists an i, $-1 \le i \le n$ such that $a_i/b_i \in \mathcal{F}_N$, in which case:

$$\gcd(m, b_i) = 1 \quad \textit{and} \quad \left|\frac{a_i}{b_i}\right|_m = k.$$

Proof Notice that the theorem is trivially true if $k = 0$ with $n = -1$. If $k \ne 0$ with $k \in \hat{\mathbb{Z}}_m$ by definition of $\hat{\mathbb{Z}}_m$, there exists an order-N Farey fraction p/q such that

$$\gcd(m, q) = 1 \quad \text{and} \quad \left|\frac{p}{q}\right|_m = k,$$

hence the previous lemma applies by choosing the appropriate sign for p.

To prove the other part, note that $\gcd(b_i, m)$ must be a divisor of a_i since (8.7.10) with this seed matrix states that $a_i = kb_i - mc_i$. By definition $a_i/b_i \in \mathcal{F}_N$ implies $\gcd(a_i, b_i) = 1$ so $\gcd(m, b_i) = 1$, but $a_i = kb_i - mc_i$ also implies that $a_i \equiv kb_i \pmod m$ so

$$\left|\frac{a_i}{b_i}\right|_m = k$$

thus $k \in \hat{\mathbb{Z}}_m$. □

We have then finally established the results that the EEA Algorithm when seeded with the matrix

$$\begin{bmatrix} m & 0 \\ q & p \end{bmatrix}$$

realizes the forward mapping $|\cdot|_m : \mathcal{F}_N \to \hat{\mathbb{Z}}_m$, and when seeded with

$$\begin{bmatrix} m & 0 \\ k & 1 \end{bmatrix}$$

realizes the inverse mapping $|\cdot|_m^{-1} : \hat{\mathbb{Z}}_m \to \mathcal{F}_N$.

Example 8.7.5 As an example where the algorithm determines that there is no value of i for which $a_i/b_i \in \mathcal{F}_N$, let us consider the case of $m = 19$ so $N = 3$, and $k = 4$ where we find:

f_i	a_i	b_i
	19	0
	4	1
4	3	−4
1	1	5
3	0	−19

confirming that 4 is not in $\hat{\mathbb{Z}}_{19}$ since there is no $a_i/b_i \in \mathcal{F}_3$. □

It is thus possible to map operands from $\mathcal{F}_N$ into $\hat{\mathbb{Z}}_m$ and perform arithmetic operations in $\mathbb{Z}_m$ by modular arithmetic, finally mapping results back into $\mathcal{F}_N$. If m is a prime, then even division is possible since it can be accomplished by multiplication with the multiplicative inverse in $\mathbb{Z}_m$.

However, to realize a specific (exact) rational computation in this system it is necessary to insure that operands as well as results are rationals in $\mathcal{F}_N$, i.e., N and hence m has to be chosen sufficiently large, m preferably being a prime. One such possibility is to choose m as the Mersenne prime $2^{61} - 1$ for which $N = 2^{30} - 1$ satisfies (8.7.4), so that $\mathcal{F}_N$ consists of rationals with numerators and denominators that are 30-bit integers, and the modular arithmetic can be performed on 61-bit integers using Lemma 8.2.5. Observe that with m prime, then

$$\frac{p}{q} \in \mathcal{F}_N \Rightarrow \gcd(q, m) = 1$$

since $q \leq N < m$ for $m \geq 3$ by (8.7.4), and thus $|p/q|_m$ is always defined for $p/q \in \mathcal{F}_N$.

Note that although we can perform (exact) arithmetic on the residues in $\mathbb{Z}_m$ as representations of rationals in $\mathcal{F}_N$, the results must be members of $\mathcal{F}_N$, i.e., numerators and denominators must be bounded by N in order that their rational equivalents can be determined. Also there is no simple way of "scaling," except possibly before some intermediate value of large numerator/denominator value is to be calculated, to substitute one or both operands with rational approximations of smaller numerator/denominator values.

We deliberately allowed m to be a composite number $m = m_1 \cdot m_2 \cdot \cdots \cdot m_n$ or even $m = p^r$ for p prime and $r \geq 1$, but with the implication that the system cannot directly handle rationals whose denominator is not relatively prime to m. The systems where m is a product of mutually prime moduli naturally leads to systems analogous to the multiple modulus systems treated in Section 8.3, whereas a modulus of the form $m = p^r$ corresponds to "truncated p-adic systems," also termed *Hensel codes*. In the following two sections we shall briefly introduce these systems.

Problems and exercises

8.7.1 With $m = 617$ (a prime) map the fraction $\frac{10}{13}$ into $\hat{\mathbb{Z}}_m$, i.e., find $\left|\frac{10}{13}\right|_{617}$.

8.7.2 With $m = 617$, find the unique Farey fraction $p/q \in \mathcal{F}_N$ such that $|p/q|_{617} = 124$.

8.7.3 Perform the computation $\frac{10}{13} \times \frac{3}{5}$ in the system with $m = 617$.

8.7.4 Prove that if the partial quotients $f_i \geq 1$ for $1 \leq i \leq n$, then they are unique if it is required that the last one satisfies $f_n \geq 2$ for $n \geq 1$.

8.7.5 Given the sequence of triples $\{a_i, b_i, c_i\}$ defined by (8.7.7) and (8.7.8), prove that $b_i c_{i+1} - b_{i+1} c_i = (-1)^i$ and $pc_i - qb_i = (-1)^{i+1} a_i$ for $i = 0, 1, 2, \ldots, n-1$ when p/q is irreducible.

8.8 Multiple-modulus rational systems

In complete analogy with the definition of the integer residue systems of Section 8.3 we can define residue systems over rationals. Given a base vector $\beta = [m_1, m_2, \ldots, m_n]$ where $\gcd(m_i, m_j) = 1$ for $i \neq j$, let $M = \prod_{i=1}^{n} m_i$.

Assuming $\gcd(q, m_i) = 1, i = 1, \ldots, n$, then $|p/q|_M = |pq^{-1}|_M$ is defined, and we define

$$\left|\frac{p}{q}\right|_\beta = \left[\left|\frac{p}{q}\right|_{m_1}, \left|\frac{p}{q}\right|_{m_2}, \ldots, \left|\frac{p}{q}\right|_{m_n}\right], \tag{8.8.1}$$

and thus given operation in this representation, arithmetic operations in $\hat{\mathbb{Q}}_M$ can be performed as parallel arithmetic operations in the systems $\mathbb{Z}_{m_i}, i = 1, 2, \ldots, n$. When the result is to be mapped back to $\mathcal{F}_N$, $N \leq \sqrt{(M-1)/2}$, the residue vector representing the result must first be mapped from $\prod_{i=1}^{n} \mathbb{Z}_{m_i}$ to $\mathbb{Z}_M$, using one of the methods of Section 8.4, e.g., through a mixed-radix representation or using the CRT. Then finally, the result can be mapped back into $\mathcal{F}_N$ using the inverse mapping of Theorem 8.7.7.

The assumption that $\gcd(q, m_i) = 1, i = 1, \ldots, n$ turns out only to be satisfiable for all denominators q of fractions $p/q \in \mathcal{F}_N$ for certain two-modulus systems by the following two lemmas.

Lemma 8.8.1 *Let $M = m_1 m_2$, where m_1 and m_2 are successive primes, and let $N > 0$ be any integer satisfying the inequality*

$$N \le \sqrt{\frac{M-1}{2}},$$

then m_1 and m_2 are both greater than N. Thus $\gcd(q, m_1) = \gcd(q, m_2) = 1$ *for all denominators q of fractions $p/q \in \mathcal{F}_N$.*

Proof Let m_1 and m_2 be successive primes, then from Bertran's Postulate (see, e.g., [HW79, page 343]) it follows that

$$m_1 < m_2 < 2m_1,$$

hence

$$2m_1^2 > m_1 m_2 = M \ge 2N^2 + 1,$$

which implies $m_1 > N$. □

For systems with more than two moduli, as found below, one of these must be smaller than N, and hence there are fractions in $\mathcal{F}_N$ with denominators for which the inverses do not exist.

Lemma 8.8.2 *Let $M = m_1 m_2 \cdots m_n$, where $m_1 < m_2 < \cdots < m_n$ for $n > 2$ are successive primes, and let N be the largest integer satisfying the inequality*

$$N \le \sqrt{\frac{M-1}{2}},$$

then m_1 is less than N.

Proof By assumption

$$N \le \sqrt{\frac{m_1 m_2 \cdots m_n - 1}{2}} < N + 1,$$

hence

$$m_1 m_2 \cdots m_n < 2N^2 + 4N + 3.$$

If $m_2 \le N$, the lemma is proved, so assume that $m_2 > N$, hence

$$m_1 < \frac{2N^2 + 4N + 3}{N^{n-1}} = \frac{2}{N^{n-3}} + \frac{4}{N^{n-2}} + \frac{3}{N^{n-1}}.$$

Since the first three primes are 2, 3, and 5, $N \ge 3$, hence if $n > 2$

$$m_1 < 2 + \tfrac{4}{3} + \tfrac{1}{3} = \tfrac{11}{3}$$

and thus $m_1 \le 3 \le N$. □

As we shall see below there are ways of handling systems with more than two moduli, by explicitly dealing with moduli occurring as factors in the numerators and denominators. But let us first show an example of a computation in a two-modulus system.

Example 8.8.1 Consider the system with $\beta = [5, 7]$, so $M = 35$ and $N = 4$, and perform the computation

$$x = \tfrac{1}{2} + (-\tfrac{3}{4}).$$

In $\mathbb{Z}_5$ we find

$$|x|_5 = \left||\tfrac{1}{2}|_5 + |-\tfrac{3}{4}|_5\right|_5 = |3 + 3|_5 = 1,$$

and in $\mathbb{Z}_7$ we compute

$$|x|_7 = \left||\tfrac{1}{2}|_7 + |-\tfrac{3}{4}|_7\right|_7 = |4 + 1|_7 = 5,$$

thus $|x|_\beta = [1, 5]$. This result now has to be mapped to $\mathbb{Z}_M = \mathbb{Z}_{35}$ which can be done using the CRT (Theorem 8.4.1), with

$$\widetilde{m}_1 = 7 \Rightarrow |\widetilde{m}_1|_{m_1} = 2 \Rightarrow |\widetilde{m}_1^{-1}|_{m_1} = 3,$$
$$\widetilde{m}_2 = 5 \Rightarrow |\widetilde{m}_2|_{m_2} = 5 \Rightarrow |\widetilde{m}_2^{-1}|_{m_2} = 3,$$

so from $|x|_\beta = [1, 5]$ we get

$$|x|_{35} = |7 \cdot |1 \cdot 3|_5 + 5 \cdot |5 \cdot 3|_7|_{35} = 26.$$

Finally through the EEA Algorithm and Theorem 8.7.7

f_i	a_i	b_i
	35	0
	26	1
1	9	−1
2	8	3
1	1	−4
8	0	35

the result is found to be $x = -\frac{1}{4}$, the only member in $\mathcal{F}_N = \mathcal{F}_4$. □

In cases where $n > 2$ by Lemma 8.8.2 it is possible that a fraction $a/b \in \mathcal{F}_N$ contains one or more of the moduli m_i as factors in the denominator so that $|a/b|_\beta$ is not defined. Thus it is necessary to handle such factors explicitly, which we can do by "normalizing," i.e., extracting these factors from the fraction, from the numerator and the denominator.

For a system with $\beta = [m_1, m_2, \ldots, m_n]$ where $\gcd(m_i, m_j) = 1$ for $i \neq j$ and $M = \prod_{i=1}^{n} m_i$, any non-zero a/b can be written as

$$\frac{a}{b} = \frac{a_i}{b_i}\, m_i^{e_i},$$

where e_i is chosen such that $\gcd(a_i, b_i) = \gcd(a_i, m_i) = \gcd(b_i, m_i) = 1$, and we may choose $e_i = 0$ for $a/b = 0$. By normalizing each residue and writing it as a pair:

$$\left|\frac{a}{b}\right|^*_{m_i} = \left(\left|\frac{a_i}{b_i}\right|_{m_i}, e_i\right) = (r_i, e_i),$$

we can define the *normalized residue vector* of a/b

$$\left|\frac{a}{b}\right|^*_{\beta} = \left[\left|\frac{a}{b}\right|^*_{m_1}, \left|\frac{a}{b}\right|^*_{m_2}, \cdots, \left|\frac{a}{b}\right|^*_{m_n}\right]$$
$$= [(r_1, e_1), (r_2, e_2), \cdots, (r_n, e_n)].$$

It is then possible to perform arithmetic operations componentwise on numbers in this representation. Addition or subtraction is performed componentwise by proper "alignment" of first components in analogy with floating-point operations, where the additive inverse is found by negating the first component of each pair (r_i, e_i). Multiplication is simply performed componentwise by multiplying first components and adding second components, whereas division requires multiplication with the multiplicative inverse, found by inverting the first component and negating the second.

The only remaining problem is to map results back into the rationals, assuming these belong to $\mathcal{F}_N$. For any given normalized residue vector $|a/b|^*_\beta$ we define the following

$$M_+ = \prod_{\{i|e_i>0\}} m_i, \qquad M_0 = \prod_{\{i|e_i=0\}} m_i, \qquad M_- = \prod_{\{i|e_i<0\}} m_i,$$

so $M = M_+ M_0 M_-$, where empty products have the value 1. Also we determine the unique integer $q \in \mathbb{Z}_{M_0} = \{0, 1, \ldots, M_0 - 1\}$ such that

$$|q|_{m_i} = \left|\frac{a}{b}\right|_{m_i} \quad \text{for all } m_i \text{ such that } e_i = 0, \tag{8.8.2}$$

and from q determine

$$q' = \left|qM_-|M_+^{-1}|_{M_0}\right|_{M_0} = \left|q\,\frac{M_-}{M_+}\right|_{M_0}. \tag{8.8.3}$$

With M_+, M_0, M_-, and q' thus determined from a given normalized residue vector, it is then possible to recover the equivalent fraction by the following theorem.

Theorem 8.8.3 *Given* $|a/b|^*_\beta \in \left\{|p/q|^*_\beta | p/q \in \mathcal{F}_N\right\}$, *the EEA Algorithm finds the unique order-N Farey fraction* a/b, *when seeded with*

$$\begin{bmatrix} M_+ M_0 & 0 \\ M_+ q' & M_- \end{bmatrix}. \tag{8.8.4}$$

Proof (Sketch only) Observe that for the fraction a/b, a is divisible by M_+ and b is divisible by M_-. By the constructions (8.8.2) of $q \in \mathbb{Z}_{M_0}$ and (8.8.3) of $q' \in \mathbb{Z}_{M_0}$, the fraction

$$\frac{u}{v} = \frac{a/M_+}{b/M_-},$$

belongs to the generalized residue class (8.7.1)

$$Q_{q'}^{M_0} = \left\{ \frac{u}{v} \in \hat{\mathbb{Q}}_{M_0} \;\middle|\; |uv^{-1}|_{M_0} = q' \right\},$$

members u_i/v_i of which can be generated by the EEA Algorithm with seed

$$\begin{bmatrix} M_0 & 0 \\ q' & 1 \end{bmatrix}.$$

Applying the EEA Algorithm to the seed matrix (8.8.4), obtained from the above seed matrix by multiplying the first column by M_- and the second by M_+, a sequence of tuples

$$(a_i, b_i) = (u_i M_-, v_i M_+), \; i = 0, 1, \ldots, n$$

is obtained. Assuming that $a/b \in \mathcal{F}_M$, this sequence contains a member $a_i/b_i \in \mathcal{F}_M$, and by Theorem 8.7.7 it must be the fraction a/b wanted. □

Example 8.8.2 In the system specified by $\beta = [2, 3, 5, 7]$, $M = 210$, and $N = 10$, let the following normalized residue vector be given:

$$\left| \frac{a}{b} \right|_\beta^* = [(1, 2), (2, 0), (3, 0), (3, -1)]$$

from which we find

$$M_+ = 2 \qquad M_0 = 3 \cdot 5 = 15 \qquad M_- = 7.$$

We then determine $q \in \mathbb{Z}_{M_0} = \{0, 1, \ldots, 14\}$ from

$$|q|_3 = 2 \text{ and } |q|_5 = 3 \Rightarrow q = 8$$

by the CRT or through mixed-radix conversion. Then q' is found:

$$q' = \left| q \frac{M_-}{M_+} \right|_{M_0} = |8 \cdot \tfrac{7}{2}|_{15} = 13,$$

and finally using the seed

$$\begin{bmatrix} M_+ M_0 & 0 \\ M_+ q' & M_- \end{bmatrix} = \begin{bmatrix} 30 & 0 \\ 26 & 7 \end{bmatrix}$$

by the EEA Algorithm and Theorem 8.7.7:

f_i	a_i	b_i
	30	0
	26	7
1	4	−7
6	2	49
2	0	−105

the result is found to be $a/b = -\frac{4}{7}$, the only member in $\mathcal{F}_N = \mathcal{F}_{10}$. □

8.9 *p*-adic expansions and Hensel codes

A residue system based on a (single) modulus of the form $m = p^r$ corresponds to a "truncated p-adic system," also called a *Hensel code*. It was introduced by Krishnamurthy, Rao, and Subramanian in 1975 for the purpose of being able to perform exact computations on rational matrices. It is also applicable to polynomial computations, but we will not discuss these here.

Since number representations in the system are derived from truncating (in general infinite) p-adic representations, we shall first very briefly introduce these without proofs.

8.9.1 *p*-adic numbers

Any rational $\alpha = a/b$ can be uniquely represented as

$$\alpha = \frac{c}{d}\, p^n,$$

where p is a given prime and $\gcd(c, d) = \gcd(c, p) = \gcd(d, p) = 1$. The *p-adic norm* of α is then defined as

$$||\alpha||_p = \begin{cases} p^{-n} & \text{when } \alpha = \frac{c}{d}\, p^n \neq 0, \\ 0 & \text{when } \alpha = 0, \end{cases} \tag{8.9.1}$$

and then the *p-adic metric* is defined by

$$d(\alpha, \beta) = ||\alpha - \beta||_p. \tag{8.9.2}$$

Note that this norm and its metric are counterintuitive, e.g., the sequence $p, p^2, p^3, \ldots$ is convergent (with limit zero) since $\|p^i\|_p = p^{-i}$. In analogy with the construction of the reals as the set of equivalence classes of Cauchy sequences over $\mathbb{Q}$ using the absolute value metric, it is possible to construct $\mathbb{Q}_p$, the *set of p-adic numbers*, from $\mathbb{Q}$ and d defined above. With addition, multiplication, and multiplicative inverses properly defined, the system $(\mathbb{Q}_p, +, \times)$ constitutes a field.

Theorem 8.9.1 *Any p-adic number $\alpha \in \mathbb{Q}_p$ has a p-adic expansion*

$$\alpha = \sum_{j=n}^{\infty} a_j \, p^j,$$

where $a_j \in \mathbb{Z}$ and n is such that $\|\alpha\|_p = p^{-n}$. If $a_j \in \{0, 1, \ldots, p-1\}$, then the expansion is unique (canonical).

As we shall be particularly interested in representations of the rationals we note the following corollary.

Corollary 8.9.2 *Any rational $\alpha = (c/d)\, p^n$ has a unique p-adic expansion*

$$\alpha = \sum_{j=n}^{\infty} a_j \, p^j,$$

where $0 \le a_j \le p-1$, and convergence of the infinite series is in the p-adic metric.

In analogy with radix notation, it is customary to use a string notation for writing p-adic numbers, e.g., for $\alpha = \sum_{j=n}^{\infty} a_j \, p^j$ to write

$$\alpha \sim \begin{cases} a_n a_{n+1} \cdots a_{-1}.a_0 a_1 \cdots & \text{for } n < 0, \\ .a_0 a_1 \cdots & \text{for } n = 0, \\ .0 \cdots 0 a_n a_{n+1} \cdots & \text{for } n > 0. \end{cases}$$

It can be shown that:

- The p-adic expansion is periodic if and only if α is rational.
- The p-adic expansion is finite if and only if α is a positive radix fraction, i.e., $\alpha = a/b$, where $b = p^i$, $i \ge 0$, and $\gcd(a, p) = 1$.
- If α is a positive integer, then α has a finite p-adic expansion.
- If $\alpha = \sum_{j=n}^{\infty} a_j \, p^j$, then the additive inverse is $-\alpha = \sum_{j=n}^{\infty} b_j \, p^j$, where $b_n = p - a_n$ and $b_j = (p-1) - a_j$ for $j > n$.

Example 8.9.1 With $p = 5$ we have the following p-adic expansions

$$\begin{aligned} 199 &\sim .442100\cdots, \\ -199 &\sim .102344\cdots, \\ \tfrac{199}{125} &\sim 442.100\cdots, \\ -\tfrac{199}{125} &\sim 102.344\cdots, \end{aligned}$$

where we note that the p-adic string representation is essentially the radix-p representation with the digits written in reverse order, e.g., $199_{10} = 1244_5$. □

The p-adic expansion of an integer can be obtained through an algorithm similar to the classical one for obtaining a radix representation. For a rational number the process is quite similar, except that when determining a digit it has to be found as a residue of a rational number modulo p, i.e., employing the forward mapping of Theorem 8.7.4. As the expansion, in general, is infinite, the process can be terminated when the period has been determined.

Theorem 8.9.3 *Given $\alpha = a/b$, determine n such that $\alpha = a/b = (c/d)p^n$ where $\gcd(c, d) = \gcd(c, p) = \gcd(d, p) = 1$. Let $\alpha_0 = c/d$ and recursively compute the digits a_i of α_0 by*

$$a_i = |\alpha_i|_p \quad \textit{and} \quad \alpha_{i+1} = (\alpha_i - a_i)/p \quad \textit{for } i = 0, 1, 2, \ldots.$$

The canonical p-adic expansion of α is then obtained by scaling with p^n.

Addition of p-adic numbers can be performed digitwise, using their expansions as for radix representations, just noting that the carry now has to ripple to the right. Hence with n chosen for alignment such that $a_n + b_n \neq 0$ for

$$\alpha = \sum_{j=n}^{\infty} a_j p^j \quad \text{and} \quad \beta = \sum_{j=n}^{\infty} b_j p^j,$$

with $c_{n-1} = 0$, we recursively rewrite $a_i + b_i + c_{i-1} = s_i + c_i$ for $i = n, n+1, \ldots$ such that $0 \leq s_i \leq p - 1$, then

$$\alpha + \beta = \sum_{j=n}^{\infty} s_j p^j.$$

Example 8.9.2 For $p = 5$

$$\begin{array}{r} \frac{2}{3} \sim .41313131\cdots \\ \frac{5}{6} \sim .01404040\cdots \\ \hline \frac{2}{3} + \frac{5}{6} \sim .42222222\cdots \end{array}$$

□

Subtraction is obtained by addition of the additive inverse, and multiplication by accumulating partial products in p-adic representation, in complete analogy with radix representations. Note that we can always assume that operands are normalized such that $n = 0$, by treating the scale factors separately.

Example 8.9.3 Again for $p = 5$, the product of $\frac{2}{3}$ by $\frac{5}{6} = 5^1 \cdot \frac{1}{6}$, where $\frac{1}{6} \sim .140404$, may be found as

$$\begin{array}{l} .413131\cdots \times .140404\cdots \\ \hline 413131\cdots \\ 12313\cdots \\ 0000\cdots \\ 123\cdots \\ 00\cdots \\ 1\cdots \\ \hline .420124\cdots \end{array}$$

corresponding to $\frac{2}{3} \times \frac{5}{6} = \frac{5}{9} \sim .0420124320\cdots$. □

Division can similarly be performed in analogy with radix division, but with the interesting difference that digit selection here is deterministic, i.e., a non-restoring algorithm results in a non-redundant representation. To see this consider normalized operands

$$\alpha = \sum_{j=0}^{\infty} a_j\, p^j \quad \text{and} \quad \beta = \sum_{j=0}^{\infty} b_j\, p^j, \quad a_0 \neq 0,\ b_0 \neq 0$$

and let

$$\frac{\alpha}{\beta} = \sum_{j=0}^{\infty} d_j\, p^j, \quad \text{where} \quad 0 \leq d_j \leq p-1.$$

Then $d_0 b_0 = a_0$ so $d_0 = |a_0 b_0^{-1}|_p$ and introducing the partial remainder $\gamma_i = \sum_{j=i}^{\infty} g_j^{(i)} p^j$,

$$\gamma_{i+1} = \gamma_i - d_i p^i \beta, \quad \text{where} \quad d_i = |g_i^{(i)} b_0^{-1}|_p,$$

hence the quotient digit d_i is obtained by a simple modular multiplication by the inverse of the first digit of the divisor β.

Example 8.9.4 Let $\alpha = \frac{2}{3} \sim .413131\cdots$ be the dividend and $\beta = \frac{1}{12} \sim .342424\cdots$ be the divisor, so $|b_0^{-1}|_5 = |3^{-1}|_5 = 2$ and

$$\begin{aligned}
\gamma_0 &\sim .413131\ \cdots \quad \Rightarrow d_0 = |4 \cdot 2|_5 = 3\\
3 \cdot 5^0 \cdot \beta &\sim .433333\ \cdots\\
&\Downarrow\\
\gamma_0 &\sim .413131\ \cdots\\
-3 \cdot 5^0 \cdot \beta &\sim \underline{.111111\ \cdots}\\
\gamma_1 &\sim .034242\ \cdots \quad \Rightarrow d_1 = |3 \cdot 2|_5 = 1\\
3 \cdot 5^1 \cdot \beta &\sim .034242\ \cdots\\
&\Downarrow\\
\gamma_2 &\sim .000000\ \cdots \quad \Rightarrow 0 = d_2 = d_3 = \cdots
\end{aligned}$$

hence $\frac{2}{3}/\frac{1}{12} \sim .31$, where $.31 \sim 8$. □

8.9.2 Hensel codes

The idea introduced by Krishnamurthy is to truncate p-adic representations and thus operate in a finite number system, using the notation $H(p, r, \alpha)$ for the r-digit truncated p-adic representation of $\alpha \in \mathbb{Q}$. For example,

$$\tfrac{1}{3} \sim .231313\cdots \ (p = 5) \ \text{yields} \ H(5, 4, \tfrac{1}{3}) = .2313.$$

The first and obvious question to ask is: which set of rationals is then uniquely represented by this finite set of Hensel codes? This is partially answered by the following lemma.

Lemma 8.9.4 *Let* $\alpha = a/b$ *where the denominator* b *is relatively prime to* p, $\gcd(d, p) = 1$, *and let*

$$H\left(p, r, \frac{a}{b}\right) = .a_0 a_1 \cdots a_{r-1},$$

then $a_{r-1} \cdots a_1 a_0$ *is the radix-*p *representation of* $|a/b|_{p^r}$, *i.e.,*

$$\left|\frac{a}{b}\right|_{p^r} = a_{r-1} p^{r-1} + \cdots + a_1 p + a_0.$$

Proof Let a/b have the p-adic expansion

$$\frac{a}{b} = \sum_{j=0}^{\infty} a_j p^j = \sum_{j=0}^{r-1} a_j p^j + p^r \gamma$$

with $\gamma \in \mathbb{Q}_p$, then

$$\left|\frac{a}{b}\right|_{p^r} = \sum_{j=0}^{r-1} a_j p^j.$$

□

Hence the set of Hensel codes $\{H(p, r, a/b) \mid \gcd(b, p) = 1\}$ is isomorphic to the subset of Farey fractions, $\hat{\mathbb{Z}}_{p^r} \subset F_N$, introduced in (8.7.5), where N is the largest integer such that $N < \sqrt{(p^r - 1)/2}$. Thus it corresponds to the single modulus rational system with $m = p^r$, and it is possible to map a member of the set to and from the equivalent fraction by the EEA Algorithm.

To avoid the problems with factors of p in the denominator (and also to handle such factors in the numerator), it is then natural to introduce the so-called *floating-point Hensel codes*, $\hat{H}(p, r, \alpha)$, by "normalization" and scale factors. With $\alpha = a/b = (c/d)p^n$, where $\gcd(c, d) = \gcd(c, p) = \gcd(d, p) = 1$ we define

$$\hat{H}(p, r, \alpha) = (m_\alpha, e_\alpha) \quad \text{with} \quad m_\alpha = H(p, r, \tfrac{c}{d}) \quad \text{and} \quad e_\alpha = n,$$

i.e., we represent the fraction by a (mantissa, exponent) pair.

Example 8.9.5

$$\hat{H}(5, 4, \tfrac{2}{3}) = (.4131, 0),$$
$$\hat{H}(5, 4, \tfrac{2}{15}) = (.4131, -1),$$
$$\hat{H}(5, 4, \tfrac{10}{3}) = (.4131, 1),$$
$$\hat{H}(5, 4, -\tfrac{2}{3}) = (.1313, 0).$$

□

Let us then look at some examples of arithmetic operations.

Example 8.9.6 For addition or subtraction of operands with different scale factors (exponents) it is necessary to align operands. Let

$$\hat{H}(5, 4, \tfrac{2}{3}) = (.4131, 0),$$
$$\hat{H}(5, 4, \tfrac{1}{5}) = (.1000, -1),$$

hence for addition

$$\begin{array}{l} .04131 \\ \underline{.1000} \\ .1413 \end{array} \quad \text{i.e., } \hat{H}(5, 4, \tfrac{2}{3} + \tfrac{1}{5}) = (.1413, -1).$$

Another example is

$$\hat{H}(5, 4, \tfrac{1}{2}) = (.3222, 0),$$
$$\hat{H}(5, 4, \tfrac{1}{8}) = (.2414, 0),$$

and adding

$$\begin{array}{l} .3222 \\ \underline{.2414} \\ .0241 \end{array} \quad \text{i.e., } \hat{H}(5, 4, \tfrac{1}{8} + \tfrac{1}{2}) = (.0241, 0),$$

where we notice that the result is not normalized. The problem obviously is that the result ($\frac{5}{8}$) contains a factor 5, but we do not know what to shift in for the missing digit! □

As the last example points out, when adding (or subtracting) it is possible to obtain results which are not normalized. Continuing with such operands may yield results that are not even Hensel codes, and it is not possible to divide with a non-normalized operand.

One way to be able to continue operations would be to carry along additional digits, where it can be shown that a bounded number will be sufficient. Another possibility is to recover the fraction by the EEA Algorithm, and then map it back into its Hensel code in normalized form. But since the system of Hensel codes is equivalent to a single modulus rational system with $m = p^r$, p prime, there does not seem to be much reason not to use the latter system, with m chosen as a suitable large prime.

8.10 Notes on the literature

Much of the notation used in this chapter, and some basic results on modular operations can be found in ordinary textbooks on abstract algebra. Also Knuth [Knu98] provides many results useful in this chapter, in particular the EEA Algorithm (8.2.18), which he traces back to fifth-century India. Algorithm 8.2.13 for the multiplicative inverse modulo 2^k can be traced back to Hensel in 1913 [Hen13]. Hardware solutions for these algorithms can be found in [Tak93, Kor94]. Lemma 8.2.14 and Theorem 8.2.15 are from [Knu98, third edition, Problem 17, Section 4.5.2]. These results were "rediscovered" in [AQ08]. Algorithm 8.2.21 is from [KT03] and is based on the binary GCD algorithm also found in [Knu98]; a somewhat similar algorithm is described in [Sav05]. [FFDT05c] describes a very compact table structure for tabulating inverses (and other function values) modulo 2^k.

The idea of using a bias as in the modular addition select adder of Figure 8.2.1 is due to Jenkins [Jen73], and is used in various proposals for multioperand addition in [Zha87, EB92, ZSY93]. A different approach to multioperand addition, based on (8.2.8), was presented by Piestrak in [Pie94]. A fast adder was proposed by Hiasat [Hia02]. The quarter-square method is from [Che71]. Another multiplier is found in [AM91], and Jullien discussed various ROM-based methods in [Jul78]. Soderstrand [Sod83] described end-around-carry, modulo adders in the context of the eight-bit RNS system with base vector [253, 255, 256].

Due to the importance of the RSA public key cryptosystem [RSA78] and other cryptographic algorithms that employ modular multiplication on very large operands to implement modular exponentiation, a vast number of software and hardware implementations have been proposed. Algorithms based on remainder determination through division have been around for a long time, e.g., [Bri82, Bla83, Mor90] and many more recent papers, and form the foundation for our interleaved modular multiplication as described in Algorithm 8.2.27. The *M-residue* and the *M-reduce* algorithm (Algorithm 8.2.29), based on [Mon85] by Montgomery, is probably the most used technique for modular multiplication and is exemplified in our Algorithm 8.2.31. It appears in a vast number of publications, e.g., [DJ91, Eve91, Tak92, EW93, Wal93, SV93, Kor93, Kor94, Oru95a, Oru95b] to cite just a few, and other algorithms based on the RNS implementation of the arithmetic operations [PP95, BDK98, SKS00, BDK01]. The optimizations leading to Algorithm 8.2.33 are from Orup [Oru95a], and Observation 8.2.34 is Walter [Wal99]. The idea of bipartite modular multiplication is from [KT05]. [CPO95] proposes the use of the *M-residue* representation for DSP applications. Actually, Montgomery's Algorithm 8.2.29 is identical to Hensel's 1913 algorithm for odd integer division in p-adic representation [Hen13], and for the case of $p = 2$ it is identical to Algorithm 8.2.13.

The material on inheritance and periodicity in Subsection 8.2.8 was to a large extent developed by Fit-Florea and Matula in [FFM04, FFDT05a, FFDT05b, FFDT05c]. The residue representation in Observation 8.2.42 is from a patent by Benschop [Ben99].

RNSs originated with the work of Svoboda and Valach in Czechoslovakia [SV55] and independently the work of Garner in the USA [Gar59]. A standard reference is [ST67] by Szabo and Tanaka, which contains the basic foundations of RNS, including the classical conversions based on the CRT, as well as these through mixed-radix representation (due to Garner). The idea of using RNS systems for fault tolerance was proposed by Garner in his Ph.D. dissertation and in [Gar58]. A collection of papers on RNS representation and arithmetic, with applications in DSP has been reproduced in [SJJF86]. Only recently have other textbooks [Moh02, OP07] on RNS systems become available.

The argument for using RNS systems is generally stated as the "carry-free" nature of the arithmetic, due to the parallel operations in the different channels being completely independent.

The mapping from radix to RNS as described in Figure 8.4.1 is based on [AM93]. The alternative method based on the periodicity of $|2^i|_m$ is from [Pie94]; [Moh99, Pre02] and many others discuss further developments. The quotient, remainder representation was introduced in [SVC83], and the conversion algorithm on page 573 based upon it is from [Vu85]. The CRT-based conversion algorithm of Figure 8.4.4 is from [Hua83, BP94a] and the "divide and conquer" algorithm based on Observation 8.4.3 is from [WA96]. Conversion from a RNS to a weighted representation has been and still is the subject of much research; some further references are [EB92, BP94b, Pie94, PY97, SBC98].

Specialized RNS systems based on the three moduli $2^k - 1$, 2^k, and $2^k + 1$ (or more generally $2n - 1, 2n$, and $2n + 1$) are subject to much research, due to certain simplifications of algorithms for arithmetic operations in such systems, see e.g., [Ma98, Zim99, VEN01], and conversions, see e.g., [GPS97, PY97, Ska98, SA99, BSC99]. However, it seems questionable to what extent there is much to be gained in speed by using as few as three moduli, as a modular channel addition requires quite a number of additional logic levels beyond those of a carry-completion adder. It is not obvious then that a three-channel RNS system addition is faster than a triple-width integer system. However, since the multiplier area grows with the square of the operand width, there are likely to be area savings for a three-channel system, but again not necessarily a speed advantage. However, there seem to be advantages for fault tolerance, despite the overhead costs of mapping in and out of an RNS system. We shall not, however, pursue the subject of fault tolerance here, but refer the reader to, e.g., [Moh02, OP07] and research papers on the subject.

The classical mixed-radix base extension of Section 8.5.1 appears in [ST67]. Our CRT-based algorithm from [SK89], employing an extra, redundant modulus, is an efficient alternative Gregory and Matula in [GM77] discussed the problem of base conversion, i.e., given a residue vector for one base vector β, converting it into the residue vector for another base vector β'. They showed that when a modulus m'_i from β' is not relatively prime to the system modulus M_β, the computations are simplified in the sense that reductions modulo m'_i can be substituted by reductions modulo a smaller modulus m''_i. Our discussion of scaling is based on [ST67]; some other references are [Jul78, SL90, HP94]. The method for sign detection and comparison presented in Section 8.6.2 is based on [Vu85].

The core function was originally presented by Akushskii Burcev, and Pak in [ABP77], but later translated into English and expanded in [MAKP86]. The presentation here is largely based on the latter reference, but also to some extent, in particular Theorem 8.6.12, on [Bur97, Bur03] by Burgess, dealing with conversion and scaling. He also presented an algorithm for scaling by a product of a subset of the moduli. Algorithm 8.6.15 was inspired by [AS05].

Finding an efficient algorithm for general division has been the subject of much research, see, e.g., [KCT62, KKK73, BCG81, LLM84, Chr90, CL91, Gam91, HK95, BDM96]. Our presentation is based on [HZ95, HZ97]. It is interesting to notice that although the division problem is considered difficult in RNS, it has been shown [DL91] that for ordinary division there is an asymptotically time optimal algorithm ($\mathcal{O}\log(n)$ depth) based on RNS arithmetic.

Rational number systems based on modular/residue representations and Hensel codes were promoted by Gregory in [Gre78, Gre81, Gre83], and by Krishnamurthy and Gregory in the monograph [GK84]. The algorithm for mapping between rationals and Farey fractions is from [KG03] by Gregory and Kornerup, but was also independently found around the same time by others [WGD82, Mio82]. Farey fractions are discussed in many standard texts on number theory, e.g., [HW79]. Hensel codes based on Hensel's p-adic representations [Hen08, Hen13, Kob77] were introduced by Krishnamurthy, Rao, and Subramanian in [KRS75, Kri77]. [GRK86] points out that there are limitations in their arithmetic algorithms, and [ZC90] further investigates these, pointing out that there are no advantages in using Hensel codes, the problems can be avoided by working on the equivalent Farey fractions.

Another representation based on p-adic expansions was proposed by Hehner and Horspool in [HH79], in which finiteness for the representations of rationals is obtained by exploiting that their expansions are cyclic. But unfortunately, the cycle length for a fraction p/q can be as large as $q-1$. The authors claim from simulations to obtain an average bit-length of only about six bits using a variable length encoding, thus necessarily having to address data indirectly, adding significant overhead in storage space and access time.

References

[ABP77] I. J. Akushskii, V. M. Burcev, and I. T. Pak. A new positional characteristic of non-positional codes and its applications. In V. M. Amerbaev, editor, *Coding Theory and the Optimization of Complex Systems*. 'Nauka' Kazah, 1977.

[AM91] G. Alia and E. Martinelli. A VLSI modulo multiplier. *IEEE Trans. Computers*, C-40(7):873–978, 1991.

[AM93] G. Alia and E. Martinelli. On the lower bound to the VLSI complexity of number conversion from weighted to residue representation. *IEEE Trans. Computers*, C-42(8):962–967, 1993.

[AQ08] O. Arazi and H. Qi. On calculating multiplicative inverses modulo 2^m. *IEEE Trans. Computers*, 57(10):1435–1438, October 2008.

[AS05] M. Abtahi and P. Siy. The non-linear characteristic of core function of RNS numbers and its effect on RNS to binary conversion and sign detection algorithms. In *North American Fuzzy Information Processing Society, 2005*, pages 731–736. IEEE, 2005.

[BCG81] D. Banerji, T. Cheung, and V. Ganesan. A high-speed division method in residue arithmetic. In *Proc. 5th IEEE Symposium on Computer Arithmetic*, pages 158–164. IEEE Computer Society, 1981.

[BDK98] J.-C. Bajard, L.-S. Didier, and P. Kornerup. An RNS Montgomery modular multiplication algorithm. *IEEE Trans. Computers*, 46(7):766–776, July 1998.

[BDK01] J.-C. Bajard, L.-S. Didier, and P. Kornerup. Modular multiplication and base extensions in residue number systems. In *Proc. 15th IEEE Symposium on Computer Arithmetic*, pages 59–65. IEEE Computer Society, 2001.

[BDM96] J.-C. Bajard, L.-S. Didier, and J.-M. Muller. A new Euclidean division algorithm for residue number systems. In *Proceedings of Application-Specific Systems, Architectures, and Processors*. IEEE Computer Society, 1996.

[Ben99] N. F. Benschop. Multiplier for the Multiplication of at Least Two Figures in an Original Format. US Patent N. 5,923,888, July 13 1999.

[Bla83] G. R. Blakley. A computer algorithm for calculating the product *AB* modulo *M*. *IEEE Trans. Computers*, C-32:497–500, 1983.

[BP94a] F. Barsi and M. C. Pinotti. A fully parallel algorithm for residue to binary conversion. *Information Processing Lett.*, 50:1–8, 1994.

[BP94b] F. Barsi and M. C. Pinotti. Time optimal mixed radix conversion for residue number applications. *The Computer Journal*, 37(10):907–916, 1994.

[Bri82] E. F. Brickell. A fast modular multiplication algorithm with applications to two key cryptography. In D. Schaum, R. L. Rivest, and A. T. Sherman, editors, *Advances in Cryptology, Proceedings of Crypto '82*, pages 51–60, Plenum Press, 1982.

[BSC99] M. Bhardwaj, T. Shrikantan, and C. T. Clarke. A reverse converter for the 4-moduli superset $\{2^n - 1, 2^n, 2^n + 1, 2^{n+1} + 1\}$. In *Proc. 14th IEEE Symposium on Computer Arithmetic*, pages 168–175. IEEE Computer Society 1999.

[Bur97] N. Burgess. Scaled and unscaled residue number to binary conversion techniques using the core function. In *Proc. 13th IEEE Symposium on Computer Arithmetic*, page 250–257. IEEE Computer Society, 1997.

[Bur03] N. Burgess. Scaling an RNS number using the core function. In *Proc. 16th IEEE Symposium on Computer Arithmetic*, page 262–271. IEEE Computer Society, 2003.

[Che71] T. C. Chen. A binary multiplication scheme based on squaring. *IEEE Trans. Computers*, C-20:678–680, 1971.

[Chr90] W. Chren Jr. A new residue number division algorithm. *Computer Math. Appl.*, 19(7):13–29, 1990.

[CL91] J.-S. Chiang and Mi Lu. A general division algorithm for residue number systems. In P. Kornerup and D. W. Matula, editors, *Proc. 10th IEEE Symposium on Computer Arithmetic*, pages 76–83. IEEE Computer Society, 1991.

[CPO95] E. D. Di Claudio, F. Piazza, and G. Orlando. Fast combinatorial RNS processors of DSP Applications. *IEEE Trans. Computers*, 44(5):624–633, May 1995.

[DJ91] S. R Dussé and B. S. Kaliski Jr. A cryptographic library for the Motorola DSP56000. In I. B. Damgård, editor, *Advances in Cryptology – EUROCRYPT '90*, pages 230–244. Springer-Verlag, 1991.

[DL91] G. I. Davida and B. Litow. Fast parallel arithmetic via modular representation. *SIAM J. Comput.*, 20(4):756–765, August 1991.

[EB92] K. M. Elleithy and M. A. Bayoumi. Fast and flexible architectures for RNS arithmetic decoding. *IEEE Trans. Circuits and Systems-II*, 39(4):226–235, April 1992.

[Eve91] S. Even. Systolic modular multiplication. In I. B. Damgård, editor, *Advances in Cryptology – EUROCRYPT '90*, pages 619–624. Springer-Verlag, 1991.

[EW93] S. E. Eldridge and C. D. Walker. Hardware implementation of Montgomery's modular multiplication algorithm. *IEEE Trans. Computers*, C-42(6):693–699, June 1993.

[FFDT05a] A. Fit-Florea, D. W. Matula, and M. Thornton. Addition based expoentiation modulo 2^k. *IEE Electronics Lett.*, 41(2):56–57, Jan. 2005.

[FFDT05b] A. Fit-Florea, D. W. Matula, and M. Thornton. Additive bit-serial algorithm for the discrete logarithm modulo 2^k. *IEE Electronics Lett.*, 41(2):57–59, Jan. 2005.

[FFDT05c] A. Fit-Florea, D. W. Matula, and M. Thornton. Look-up table structures for multiplicative inverses modulo 2^k. In *Proc. 17th IEEE Symposium on Computer Arithmetic*, pages 130–135. IEEE Computer Society, 2005.

[FFM04] A. Fit-Florea and D. W. Matula. A digit-serial algorithm for the discrete logarithm modulo 2^k. In *Proc. Application-Specific Systems, Architectures and Processors (ASAP2004)*. IEEE, 2004.

[Gam91] D. Gamberger. New approach to integer division in residue number systems. In *Proc. 10th IEEE Symposium on Computer Arithmetic*, pages 84–91. IEEE Computer Society, 1991.

[Gar58] H. L. Garner. Generalized parity checking. *IRE Trans. Electronic Computers*, EC-7(3):207–213, September 1958.

[Gar59] H. L. Garner. The residue number system. *IRE Trans. Electronic Computers*, EC-8:140–147, June 1959.

[GK84] R. T. Gregory and E. V. Krishnamurthy. *Methods and Applications of Error-Free Computations*. Springer-Verlag, 1984.

[GM77] R. T. Gregory and D. W. Matula. Base conversion in residue number systems. *BIT*, 17:286–302, 1977.

[GPS97] D. Galaher, F. E. Petry, and P. Shrinivasan. The digit parallel method for fast RNS to weighted number system conversion for specific moduli ($2^k - 1, 2^k, 2^k + 1$). *IEEE Trans. Computers*, 44(1):53–57, 1997.

[Gre78] R. T. Gregory. The use of finite-segment p-adic arithmetic for exact computation. *BIT*, 18:282–300, 1978.

[Gre81] R. T. Gregory. Error-free computation with rational numbers. *BIT*, 21(2):194–202, 1981.

[Gre83] R. T. Gregory. *A Two-Modulus Residue Number System for Exact Computation Using Rational Numbers*. Technical report, CS-83-54 University of Tennessee, Knoxville, June 1983.

[GRK86] R. N. Gorgui-Naguiband and R. A. King. Comments on "Matrix processors using p-adic arithmetic for exact linear computations." *IEEE Trans. Computers*, C-35:928–930, October 1986.

[Hen08] K. Hensel. *Theorie der Algebraischen Zahlen*. B. G. Teubner, 1908.

[Hen13] K. Hensel. *Zahlentheorie*. G. J. Göschen, 1913.

[HH79] E. C. R. Hehner and R. N. S. Horspool. A new representation of the rational numbers for fast easy arithmetic. *SIAM J. Computing*, 8(2):124–134, 1979.

[Hia00] A. A. Hiasat. New efficient structure for a modular multiplier for RNS. *IEEE Trans. Computers*, 49(2):170–174, February 2000.

[Hia02] A. A. Hiasat. High-speed and reduced area modular adder structures for RNS. *IEEE Trans. Computers*, pages 84–89, 2002.

[HK95] M. Hitz and E. Kaltofen. Integer division in residue number system. *IEEE Trans. Computers*, 44(8):983–989, 1995.

[HP94] C. Y. Hung and B. Parhami. An approximate sign detection method for residue numbers and its application to RNS division. *Computers Math. Applic.*, 27(4):23–35, 1994.

[Hua83] C. H. Huang. A fully mixed-radix conversion algorithm for residue number applications. *IEEE Trans. Computers*, C-32(4):398–402, April 1983.

[HW79] C. H. Hardy and E. M. Wright. *An Introduction to the Theory of Numbers*. Oxford University Press, fifth edition, 1979.

[HZ95] A. A. Hiasat and H. S. Zohdy. A high speed division algorithm for residue number system. In *Proc. IEEE Symposium on Circuits and Systems*, pages 1996–1999. IEEE, 1995.

[HZ97] A. A. Hiasat and H. S. Zohdy. Design and implementation of an RNS division algorithm. In *Proc. 13th IEEE Symposium on Computer Arithmetic*, pages 240–249. IEEE Computer Society, 1997.

[Jen73] W. K. Jenkins. A technique for efficient generation of projections for error correcting residue codes. *IEEE Trans. Computers*, C-22:762–767, 1973.

[Jul78] G. A. Jullien. Residue number scaling and other operations using ROM arrays. *IEEE Transactions on Computers*, C-27(4):325–336, April 1978.

[KCT62] Y. Kier, P. Cheney, and M. Tannenbaum. Division and overflow detection in residue number systems. *IRE Trans. Electronic Computers*, EC-11:501–507, 1962.

[KG03] P. Kornerup and R. T. Gregory. Mapping integers and Hensel codes onto Farey fractions. *BIT*, 23(1):9–20, 1983. DAIMI-PB 149, Aarhus University, July 1982.

[KKK73] E. Kinoshita, H. Kosako, and Y. Kojima. General division in the symmetric residue number system. *IEEE Trans. Computers*, C-22:134–142, 1973.

[Knu98] D. E. Knuth. *Seminumerical Algorithms*, volume 2 of *The Art of Computer Programming*. Addison Wesley, first edition 1969 and second edition 1981, third edition, 1998.

[Kob77] N. Koblitz. *p-adic Numbers, p-adic Analysis an Zeta Functions*. Springer Verlag, 1977.

[Kor93] P. Kornerup. High-radix modular multiplication for cryptosystems. In *Proc. 11th IEEE Symposium on Computer Arithmetic*, pages 277–283. IEEE Computer Society, 1993.

[Kor94] P. Kornerup. A systolic, linear-array multiplier for a class of right-shift algorithms. *IEEE Trans. Computers*, C-43(8):892–898, August 1994.

[Kri77] E. V. Krishnamurthy. Matrix processors using p-adic arithmetic for exact linear computations. *IEEE Trans. Computers*, C-26(7):633–639, 1977.

[KRS75] E. V. Krishnamurthy, T. M. Rao, and K. Subramanian. Finite segment p-adic number systems with applications to exact computation. *Proc. Indian Acad. Sci.*, 81A:58–79, 1975.

[KT03] M. E. Kaihara and N. Takagi. A VLSI algorithm for modular multiplication and division. In *Proc. 16th IEEE Symposium on Computer Arithmetic*, pages 220–227. IEEE Computer Society, 2003.

[KT05] M. E. Kaihara and N. Takagi. Bipartite modular multiplication. In *Proc. of Cryptographic Hardware and Embedded Systems (CHES 2005)*, pages 201–210, LNCS 3659. Springer, 2005.

[LLM84] L. Lin, E. Leiss, and B. McInnis. Division and sign detection algorithm for residue number systems. *Computer Math. Appl.*, 10:331–342, 1984.

[Ma98] Yutai Ma. A simplified architecture for modulo $(2^n + 1)$ multiplication. *IEEE Trans. Computers*, 47(3):333–337, 1998.

[MAKP86] D. E. Miller, R. E. Altschul, J. R. King, and J. N. Polky. Analysis of the residue class core function of Akushskii, Burchef, and Pak. In M. A. Soderstrand, W. K. Jenkins, G. A. Jullien, and F. J. Taylor, editors, *Residue Number System Arithmetic: Modern Applications in Digital Signal Processing*, pages 390–401. IEEE Press, 1986.

[Mio82] A. M. Miola. The conversion of Hensel codes to their rational equivalents (or how to solve Gregory's open problem). *Sigsam Bulletin*, 16(4):24–26, November 1982.

[Moh99] P. V. Ananda Mohan. Efficient design of binary to RNS converters. *J. Circuits, Systems and Computers*, 9(3&4):145–154, 1999.

[Moh02] P. V. A. Mohan. *Residue Number Systems – Algorithms And Architectures*. Kluwer Academic Publishers, 2002.

[Mon85] P. L. Montgomery. Modular multiplication without trial division. *Math. Computation*, 44(170):519–521, April 1985.

[Mor90] H. Morita. A fast modular-multiplication algorithm based on a higher radix. In G. Brassard, editor, *Advances in Cryptology – CRYPTO '89*, pages 387–399. Springer-Verlag, 1990.

[OP07] A. Omondi and B. Premkumar. *Residue Number Systems: Theory and Implementation. Advances in Computer Science and Engineering*. Imperial College Press, 2007.

[Oru95a] H. Orup. Simplifying quotient determination in high-radix modular multiplication. In S. Knowles and W. H. McAllister, editors, *Proc. 12th IEEE Symposium on Computer Arithmetic*, pages 193–199. IEEE Computer Society, 1995.

[Oru95b] H. Orup. Exponentiation, modular multiplication and VLSI implementation of high-speed RSA cryptography. Ph.D. thesis, Aarhus University, 1995. DAIMI PB-499.

[Pie94] S. J. Piestrak. Design of residue generators and multioperand modular adders using carry-save adders. *IEEE Trans. Computers*, C-43:68–77, Jan. 1994.

[PP95] K. C. Posch and R. Posch. RNS-modulo reduction in residue number systems. *IEEE Trans. Parallel and Distributed Systems*, 6(5):449–454, May 1995.

[Pre02] A. B. Premkumar. A formal framework for conversion from binary to residue numbers. *IEEE Trans. Circuits and Systems, II*, 49(2):135–144, 2002.

[PY97] F. Pourbigharaz and H. M. Yassine. A signed-digit architecture for residue to binary transformation. *IEEE Trans. Computers*, 46(10):1146–1150, October 1997.

[RSA78] R. L. Rivest, A. Shamir, and L. Adleman. A method for obtaining digital signatures and public-key cryptosystems. *Commun. the ACM*, 21(2):120–126, February 1978.

[SA99] A. Skavantzos and M. Abdallah. Implementation issues of the two-level residue number system with pairs of conjugate moduli. *IEEE Trans. Signal Processing*, 47(3):826–838, 1999.

[Sav05] E. Savas. A carry-free architecture for Montgomery inversion. *IEEE Trans. Computers*, 54(12):1508–1519, Dec. 2005.

[SBC98] T. Srikanthan, M. Bhardwaj, and C. T. Clarke. Area-time-efficient VLSI residue-to-binary converters. *IEE Proc. – Computers and Digital Techniques*, 145(3):229–235, May 1998.

[SJJF86] M. A. Soderstrand, W. K. Jenkins, G. A. Jullien, and F. D. Taylor. *Residue Number System Arithmetic: Modern Applications in Digital Signal Processing*. IEEE Press, 1986.

[SK89] A. P. Shenoy and R. Kumaresan. Fast base extension using a redundant modulus in RNS. *IEEE Trans. Computers*, C-38(2):292–297, February 1989.

[Ska98] A. Skavantzos. An efficient residue to weighted converter for a new residue number system. In *Proc. 8th Great Lakes VLSI Symposium*, pages 185–191. IEEE, 1998.

[SKS00] F. Sano S. Kawamura, M. Koike and A. Shimbo. Cox–Rower architecture for fast montgomery multiplication. In *Proc. EUROCRYPT 2000*, pages 523–538, LNCS 1807. Springer Verlag, 2000.

[SL90] C.-C. Su and H.-Y. Lo. An algorithm for scaling and single residue error-correction in residue number systems. *IEEE Trans. Computers*, C-39(9):1053–1064, August 1990.

[Sod83] M. A. Soderstrand. A new hardware implementation of modulo adders for residue number systems. In *Proc. 26th Midwest Symposium on Circuits and Systems*, pages 412–415. Western Periodicals, 1983. Reprinted in [SJJF86].

[ST67] N. Szabo and R. I. Tanaka. *Residue Arithmetic and Its Application to Computer Technology*. McGraw-Hill, 1967.

[SV55] A. Svoboda and M. Valach. Operátorové Obvody (Operational Circuits). *Stoje na Zpracování Informací*, Sborník III, Nakl. ČSAV, Praha, 1955.

[SV93] M. Shand and J. Vuillemin. Fast implementations of RSA cryptography. In *Proc. 11th IEEE Symposium on Computer Arithmetic*, pages 252–259. IEEE Computer Society, 1993.

[SVC83] M. A. Soderstrand, C. Vernia, and J. H. Chang. An improved residue number system digital-to-analog converter. *IEEE Tran. Circuits and Systems*, CAS-30:903–907, Dec. 1983.

[Tak92] N. Takagi. A radix-4 modular multiplication hardware algorithm for modular exponentiation. *IEEE Trans. Computers*, C-41(8):949–956, August 1992.

[Tak93] N. Takagi. A modular inversion hardware algorithm with a redundant binary representation. *IEICE Trans. Information and Systems*, E76-D(8):863–869, August 1993.

[VEN01] H. T. Vergos, C. Efstathiou, and D. Nikolos. High-speed parallel-prefix modulo $(2^n + 1)$ adder for diminished-one operands. In *Proc. 15th IEEE Symposium on Computer Arithmetic*, pages 211– 217. IEEE Computer Society, 2001.

[Vu85] Thu Van Vu. Efficient implementations of the Chinese remainder theorem for sign detection and residue decoding. *IEEE Trans. Computers*, C-34(7):646–651, July 1985.

[WA96] Y. Wang and M. Abd-El-Barr. A new algorithm for RNS decoding. *IEEE Trans. Circuits and Systems – I*, 43(12):998–1001, December 1996.

[Wal93] C. D. Walter. Systolic modular multiplication. *IEEE Trans. Computers*, C-42(3):376–378, March 1993.

[Wal99] C. D. Walter. Montgomery exponentiation needs no final subtraction. *Electron. Lett.*, 35(21):1831–1832, October 1999.

[WGD82] P. S. Wang, M. J. T. Guy, and J. H. Davenport. *P*-adic reconstruction of rational numbers. *Sigsam Bulletin*, 16(2):2–3, 1982.

[ZC90] C. J. Zarowski and H. C. Card. On addition and multiplication with Hensel codes. *IEEE Trans. Computers*, C-39(12):1417–1423, 1990.

[Zha87] C. N. Zhang. VLSI-based algorithms and designs for cryptography and residue arithmetic. Ph.D. thesis, Southern Methodist University, 1987.

[Zim99] R. Zimmermann. Efficient VLSI implementation of modulo $(2^n \pm 1)$ addition and multiplication. In *Proc. 14th IEEE Symposium on Computer Arithmetic*, pages 158–168. IEEE Computer Society, 1999.

[ZSY93] C. N. Zhang, B. Shirazi, and D. Y. Y. Yun. An efficient algorithm and parallel implementation for binary and residue number systems. *J. Symbolic Computation*, 15(4):451–462, 1993.

9

Rational arithmetic

9.1 Introduction

This chapter presents a detailed introduction to various finite precision representations of rational numbers and associated arithmetic. It is the result of more than a decade of collaborative research by the authors, and as such has not so far been given a cohesive presentation, but has only been available in a sequence of conference presentations and journal papers. These representations are well founded in classical mathematics (number theory), and many problems are naturally stated in and computed in the domain of rationals. Such problems can be solved exactly in systems based on unbounded integer representations, and possibly also in residue representations as discussed in the previous chapter. However, in many cases an approximate solution based on a finite representation may be sufficient.

The first class of representations is based on the traditional way of using pairs of integers of bounded size. Two natural bounds are proposed: the first simply applying a common bound on the values of numerator and denominator, the other a bound on the product of these. The latter representation will allow a larger range of representable rational values, but for practical reasons it will be approximated by bounding the total number of digits (bits) employed for representing the numerator and denominator using radix polynomials.

It is essential in all such finite representation systems to have an associated rounding rule which allows the mapping of any value from the underlying infinite mathematical domain into the chosen finite set of representable values. Fortunately, there is a natural rounding rule based on the concept of a "best rational approximation," known from number theory. It can be applied to fractions where the numerator and/or the denominator cannot be represented in the finite system, e.g., as obtained by an arithmetic operation on operands in the system, but also to any irrational real value.

This rounding rule is based on the use of the extended Euclidean Algorithm, providing a sequence of fractions of increasing numerator and denominator values, and of gradually better approximations, among which simply the last representable fraction is chosen. These fractions are consecutive members of the continued fraction expansion of the number to be approximated (rational or irrational). It turns out that arithmetic operations like $+$, $-$, $\times$, and / can be combined with this rounding rule in such a way that while rounding a previous result, another arithmetic operation can be performed for "free." As a continued fraction expansion can be described by a sequence of integer values (partial quotients), it is then natural to consider possible representations of such sequences.

Until the middle of the twentieth century, no algorithms were known to perform general arithmetic operations on values represented by continued fractions. In 1945 Hall published a first approach to performing addition and multiplication using a bi-linear (homographic) form, $y = (ax + b)/(cx + d)$, and around 1970 Gosper discovered that bi-homographic forms are also invariant under variable substitutions of the form $x' = a + 1/x$. This allows operands to be entered into the computation of such forms. Also the value of such a form can be rewritten in a similar way, allowing the result to be emitted in continued fraction form. In the bi-homographic form

$$z(x, y) = \frac{axy + bx + cy + d}{exy + fx + gy + h}$$

the eight coefficients a, b, c, d, e, f, g, and h can be initialized such that $z(x, y)$ expresses the sum, difference, product, and quotient of the variables x and y, plus other useful expressions in the two variables, provided in the form of continued fractions.

We will thus describe ways of representing continued fraction expansions, based on some special radix-2 "self-delimiting" representations of the integer-valued partial quotients. These can be used concatenated to represent the sequence of partial quotients, first as a non-redundant, binary representation with the useful property that the usual lexicographic ordering of bit strings corresponds to the numerical ordering of the rational values represented. Then a very useful redundant representation will be presented, allowing implementation of on-line algorithms for the standard arithmetic operations on finite, and also on infinite, expansions, which can be used to represent all computable reals.

9.2 Numerator–denominator representation systems

The traditional way of representing rational numbers is in terms of pairs of integers. Formally a *fraction* denoted p/q is an ordered pair composed of a non-negative integer *numerator* p and a non-negative *denominator* q. Whenever q is non-zero, the *quotient* of p/q is the rational number determined by the ratio of p to q.

Allowing q to be zero in a natural way includes infinity, provided p is non-zero, and the pair $0/0$ can similarly denote *Not-a-Number (NaN)*. A fraction p/q is called *irreducible* if $\gcd(p, q) = 1$. A sign can be associated with the pair, or it can be associated with either one or both of the numerator and the denominator, whichever is convenient. For the subsequent analysis of the properties of rational number systems we will usually not be concerned with the sign, i.e., we will assume as above that p and q are non-negative.

Basic dyadic operators can be realized in finite precision by the well known rules for rational arithmetic:

$$\frac{p}{q} \oplus \frac{r}{s} = \Phi\left(\frac{ps + qr}{qs}\right),$$
$$\frac{p}{q} \ominus \frac{r}{s} = \Phi\left(\frac{ps - qr}{qs}\right),$$
$$\frac{p}{q} \otimes \frac{r}{s} = \Phi\left(\frac{pr}{qs}\right),$$
$$\frac{p}{q} \oslash \frac{r}{s} = \Phi\left(\frac{ps}{qr}\right),$$

where Φ is some *rounding rule* mapping the exact result into the particular finite precision system considered. The choice of the set of representable fractions and of an associated rounding rule becomes crucial in specifying a finite precision rational number system. Fortunately there exists a very natural rule for determining a canonical mapping Φ, which is applicable to a very wide class of sets of fractions. In the next section we shall see that for this rounding rule it is possible to combine the arithmetic operation with the subsequent rounding into one unified algorithm.

In order to define such sets of fractions we say that the fraction r/s is *simpler than* the fraction p/q (denoted $r/s \lessdot p/q$) if $r \le p$, $s \le q$, and at least one of the inequalities is strict.

Definition 9.2.1 *A finite ordered set F of irreducible fractions, including $\frac{0}{1}$ and $\frac{1}{0}$:*

$$F = \left\{\frac{0}{1} = \frac{p^{(1)}}{q^{(1)}}, \frac{p^{(2)}}{q^{(2)}}, \dots, \frac{p^{(k-1)}}{q^{(k-1)}}, \frac{p^{(k)}}{q^{(k)}} = \frac{1}{0}\right\}$$

is a simple chain *if and only if*

$$\frac{p^{(i-1)}}{q^{(i-1)}} \le \frac{p^{(i)}}{q^{(i)}}$$

and

$$\frac{r}{s} \lessdot \frac{p^{(i)}}{q^{(i)}} \in F \Longrightarrow \frac{r}{s} \in F.$$

Any reasonable way of representing fractions as pairs of integers would probably employ an integer representation of numerators and denominators such that if p/q is representable, then r/s will also be representable if $r/s \preccurlyeq p/q$. Thus the set of irreducible, representable fractions will form a simple chain.

The simplest set of fractions is obtained if numerators and denominators are both chosen as integers between 0 and n:

$$FXS(n) = \left\{ \pm \frac{p}{q} \middle| 0 \le p, q \le n \right\},$$

where n is determined by some computer representation of integers, e.g., $n = 2^k - 1$ for k-bit binary. This *fixed-slash* representation of fractions can be envisioned as a two-word, binary encoding of a rational:

(9.2.1)

where s contains the sign and the a-bit optionally could be used to signal exact/inexact. By enforcing symmetry in the ranges of p and q the rational p/q is representable if and only if q/p is representable. Thus the two monadic operators negation and reciprocation are both trivially implemented.

The set

$$FXS^*(n) = \left\{ \frac{p}{q} \middle| 0 \le p, q \le n,\ \gcd(p, q) = 1 \right\}$$

then forms a simple chain. The set $\mathcal{F}_n$ of irreducible fractions in $[0, 1]$, with numerators and denominators bounded by n, has been extensively studied in classical number theory under the name of *Farey fractions*.

However, the fixed format of (9.2.1) is not the most economical for representing very large (or very small) values. A more flexible representation is obtained if the boundary between the space allocated for p and q is allowed to "float:"

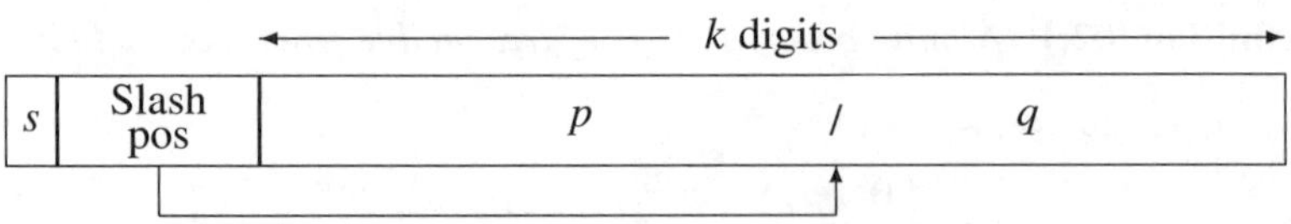

(9.2.2)

for which the set of irreducible fractions in radix β, k digits is

$$FLS^*(\beta, k) = \left\{ \frac{0}{1} \right\} \cup \left\{ \frac{1}{0} \right\} \cup \left\{ \left\{ \frac{p}{q} \right\} \middle| p, q \ge 1, \gcd(p, q) = 1, \right.$$
$$\left. \lfloor \log_\beta p \rfloor + \lfloor \log_\beta q \rfloor \le k - 2 \right\},$$

which is termed the *floating-slash* representation. The set of representable rationals is here radix-dependent, a related but radix-independent set is the *hyperbolic chain*:

$$H^*(n) = \left\{ \frac{p}{q} \middle| pq \le n, \gcd(p, q) = 1 \right\}.$$

This set has a "smoother" boundary, and can be used to enclose floating-slash systems in the following way:

$$H^*(\beta^{k-1} - 1) \subset FLS^*(\beta, k) \subset H^*(\beta^k - 1). \tag{9.2.3}$$

The efficiency of these representations is determined by the ratio of irreducible fractions to the total number of fractions representable, which is known from the literature to be asymptotically $6/\pi^2 = 0.6079\cdots$ for the Farey sets and the hyperbolic chains. Note that this corresponds to less than one bit loss of representation efficiency, irrespective of word-length used.

Depending on the radix and the number of available digits in a computer word, there is a further loss of representation efficiency in the floating-slash system due to a possible unutilized range of the slash position field. For a binary representation the total loss of efficiency can be kept below 2 bits, i.e., a k-bit representation yields at least 2^{k-2} distinct representable rational values.

With such a modest loss of representation efficiency the next question is: how are the representable numbers distributed on the real line, or equivalently which spacings (gaps) are there between these numbers?

Example 9.2.1 The members of the simple chain $F_5 = FXS^*(5)$ falling in the unit interval are shown below to scale:

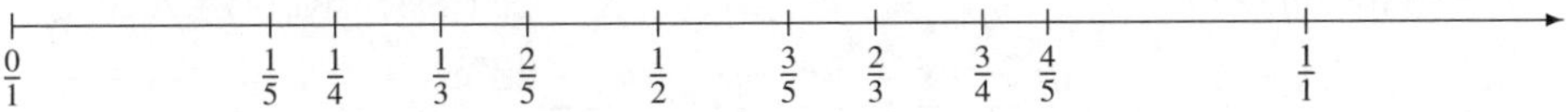

and similarly for members of $H^*(18)$:

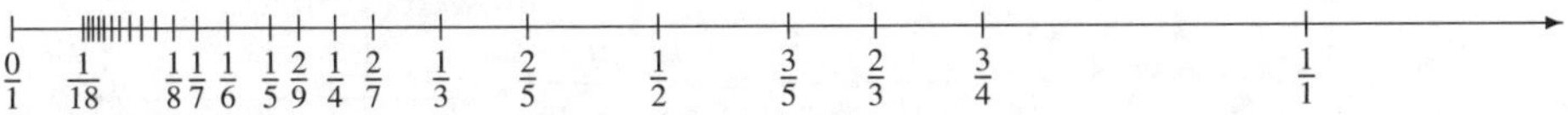

Both diagrams show a somewhat irregular spacing between representable values. Note that the intervals are widest when one of the end points of the interval is a simple fraction, in the sense that its denominator is small. □

The size of the interval $[p/q; r/s]$ between finite fractions is $|ps - rq|/qs$, which for distinct fractions will be minimum when $|ps - rq| = 1$. We will term p/q and r/s *adjacent* if $|ps - rq| = 1$, in which case the size of the interval is $1/(qs)$. The following lemma shows that two members of a simple chain which are consecutive in the ordering of the chain are also adjacent in the above sense.

Lemma 9.2.2 *For the fractions p/q and r/s the following two properties are equivalent and serve as definitions of adjacency*

(i) $|ps - rq| = 1$.
(ii) p/q *and* r/s *are both irreducible, and are both simpler than any other fraction in the interval bounded by* p/q *and* r/s.

To prove the lemma we will need some properties of the *mediant* of two adjacent fractions p/q and r/s, which is defined to be $(p+r)/(q+s)$. The mediant satisfies a number of properties which will also be needed in the following analysis, in particular it is of importance in defining the rounding rule for the number systems.

Theorem 9.2.3 *The mediant* $(p+r)/(q+s)$ *of the adjacent fractions* p/q, r/s, $p/q < r/s$ *satisfies:*

(i) *Irreducibility:* $(p+r)/(q+s)$ *is irreducible,*
(ii) *Quotient ordering:* $p/q < (p+r)/(q+s) < r/s$,
(iii) *Precision ordering:* p/q *and* r/s *are both simpler than* $(p+r)/(q+s)$,
(iv) *Adjacency:* $(p+r)/(q+s)$ *is adjacent to* p/q *and* r/s,
(v) *Simplicity:* $p/q < t/u < r/s$ *implies* $(p+r)/(q+s)$ *is simpler than or identical to* t/u.

Proof Property (iii) is trivial. From

$$\frac{p+r}{q+s} - \frac{p}{q} = \frac{1}{(q+s)q} > 0$$

it follows that $p/q < (p+r)/(q+s)$ and that p/q and $(p+r)/(q+s)$ are adjacent, and by symmetry (ii) and (iv) have been established (the case $r/s = \frac{1}{0}$ is treated separately). Now (i) then follows from (iv). To prove (v) let $p/q < t/u < r/s$ so for finite r/s

$$\frac{qr - ps}{qs} = \frac{ur - ts}{us} + \frac{qt - up}{qu}$$

with $ur - ts \geq 1$ and $qt - up \geq 1$. Thus we have

$$\frac{1}{qs} \geq \frac{1}{us} + \frac{1}{qu} = \frac{1}{qs} \cdot \frac{q+s}{u}$$

so $u \geq q + s$. If $r/s = \frac{1}{0}$ then $q = 1$ so also $u \geq q + s = 1$. And from $s/r < u/t < q/p$ it similarly follows that $t \geq p + r$, proving (v). □

Lemma 9.2.4 *Every finite non-zero irreducible fraction is the mediant of precisely one pair of adjacent fractions.*

Proof Let $\frac{0}{1} < p/q < \frac{1}{0}$ be irreducible and form the sequence of sets $\{F_i(p/q)\}_{i=0,1,2,\ldots,N(p,q)}$, with $F_0(p/q) = \{\frac{0}{1}, \frac{1}{0}\}$, $F_1(p/q) = \{\frac{0}{1}, \frac{1}{1}, \frac{1}{0}\}$, and recursively form $F_i(p/q)$ from $F_{i-1}(p/q)$ by inserting the mediant of the two consecutive fractions that enclose the fraction p/q. If the mediant inserted is p/q, the process terminates with $N(p, q) = i$, otherwise it continues. Each F_i contains a monotonically increasing sequence of irreducible adjacent fractions, and before termination p/q must fall in a unique interval between these. By Theorem 9.2.3(iii) the procedure must terminate, determining a unique adjacent pair of fractions $(r/s, t/u)$, of which p/q is the mediant. □

The unique pair of adjacent fractions $(r/s, t/u)$ of which p/q is the mediant are called the *parents* of p/q, satisfying $r/s < p/q < t/u$, both parents being simpler than $p/q = (r+t)/(s+u)$. We can now return to the proof of Lemma 9.2.2:

Proof of Lemma 9.2.2 From Theorem 9.2.3(v) it follows that (i) implies (ii), hence assume (ii) so that $p/q < r/s$ are both irreducible with no fractions simpler than p/q and r/s in the open interval between these fractions. Let p^*/q^* be the larger parent of p/q and r^*/s^* be the smaller parent of r/s, thus

$$\frac{r^*}{s^*} \leq \frac{p}{q} < \frac{r}{s} \leq \frac{p^*}{q^*}, \tag{9.2.4}$$

since the parents are simpler and hence cannot fall in the open interval $(p/q; r/s)$ by assumption. Now assume $p/q \lessdot r/s$, then it follows that $r^*/s^* = p/q$, since by Theorem 9.2.3(v) $r^*/s^* < p/q < r/s$ would imply $(r^*+r)/(s^*+s) \lessdot p/q$, a contradiction. Similarly $r/s \lessdot p/q$ implies $r/s = p^*/q^*$, so both inequalities in (9.2.4) cannot be strict, and thus one of the fractions p/q and r/s must be a parent of the other and hence they must be adjacent. □

We can now return to the sets $FXS^*(n)$ and $FLS^*(\beta, k)$ of fixed- and floating-slash numbers, noting that they are simple chains, for which we have the following result.

Theorem 9.2.5 *Any two consecutive fractions $p/q < r/s$ of a simple chain are adjacent, and hence the gap between them is $1/qs$.*

Proof The chain contains all irreducible fractions simpler than p/q and r/s, and none of these fall between p/q and r/s. By Lemma 9.2.2, p/q and r/s are then adjacent. □

Corollary 9.2.6 *For the interval* $[0; 1]$ *and any* $n \geq 2$ *the gap between consecutive representable values of $FXS^*(n)$ varies from a minimum of $1/n(n-1)$ to a maximum of* $1/n$.

Proof First note that $s = q$ for two adjacent fractions p/q and r/s is only possible if $s = q = 1$. Hence the minimum gap size $1/qs$ is $1/n(n-1)$, found between $1/n$ and $1/(n-1)$, and the maximum $1/qs = 1/q \cdot 1$ is found between $\frac{0}{1}$ and $1/n$, noting that $q + s > n$ in the interval $[0; 1]$. □

The absolute values of the gaps are mostly of interest for the system $FXS^*(n)$ over the interval $[0; 1]$, where the previous result may be loosely interpreted as a variation in gap sizes between "half" and "full" precision, considering that $FXS^*(n)$ contains $\Omega(n^2)$ representable numbers.

For floating-slash numbers we shall show the same variation in the relative spacing of numbers, where we define the relative size of the gap as

$$\operatorname{relgap}\left(\frac{p}{q}, \frac{r}{s}\right) = \int_{p/q}^{r/s} \frac{1}{x} dx$$

from which we obtain

$$\operatorname{relgap}\left(\frac{p}{q}, \frac{r}{s}\right) = \operatorname{relgap}\left(\frac{s}{r}, \frac{q}{p}\right) = \log\left(1 + \frac{1}{ps}\right), \tag{9.2.5}$$

and the following bounds:

$$\frac{1}{1+ps} = \frac{1}{qr} < \operatorname{relgap}\left(\frac{p}{q}, \frac{r}{s}\right) < \frac{1}{ps}. \tag{9.2.6}$$

We leave it as an exercise to show that our definition of relgap conforms to the usual definition.

Theorem 9.2.7 *For any $n \geq 2$ the relative gap size between a consecutive pair $p/q < r/s$ of finite, non-zero fractions of $H^*(n)$ satisfies*

$$\frac{1}{n} < \operatorname{relgap}\left(\frac{p}{q}, \frac{r}{s}\right) < \frac{2}{\sqrt{n}}.$$

Proof From $p/q, r/s \in H^*(n)$ it follows that $pqrs \leq n^2$, hence $ps(ps+1) \leq n^2$ so $ps \leq n-1$ since ps is integral, and the left-hand inequality then follows from the lower bound of (9.2.6). For the right-hand inequality we may by (9.2.5) assume that $\frac{1}{1} < p/q < r/s$ (the case $\frac{1}{1} = p/q$ is trivial). Thus $p \geq q+1$ and $r \geq s+1$ implies $p+q-2 \geq r+s$ and

$$(p+r-1)^2 > (p+r)(p+r-2) \geq (p+r)(q+s) > n,$$

hence $p+r-1 > \sqrt{n}$. Using the right-hand side of (9.2.6) and the definition of relgap:

$$\begin{aligned}
\operatorname{relgap}\left(\frac{p}{q}, \frac{r}{s}\right) &= \operatorname{relgap}\left(\frac{p}{q}, \frac{p+r}{q+s}\right) + \operatorname{relgap}\left(\frac{p+r}{q+s}, \frac{r}{s}\right) \\
&< \frac{1}{p(q+s)} + \frac{1}{(p+r)s} \\
&= \frac{1}{(p+r)q-1} + \frac{1}{(p+r)s} \\
&< \frac{2}{\sqrt{n}}
\end{aligned}$$

since $q \geq 1$ and $s \geq 1$. □

Due to the bounds (9.2.3) on $FLS^*(\beta, n)$ in terms of hyperbolic chains, the floating-slash systems exhibit a similar variation in the relative gap sizes.

Problems and exercises

9.2.1 Starting with the fractions $\frac{0}{1}$ and $\frac{1}{0}$, construct the set F_7 as the tree of mediants (a subset of the infinite *Stern–Brocot tree*,) derived from $\frac{1}{1}$ (with $\frac{1}{1}$ as root) by successive insertion of mediants.

9.2.2 Prove (9.2.5) and (9.2.6), and show that our definition conforms to the usual definition of relative errors.

9.3 The mediant rounding

As was noted in the previous section, the monadic operators negation and reciprocation are always exact for fixed- and floating-slash systems. Furthermore, the standard dyadic operators, through integer arithmetic on numerators and denominators, yield exact results in higher-precision number systems, which then normally must be mapped back into the same system as the operands. For this mapping we need a rounding procedure which will yield good approximations. We shall now show that there exists a canonical choice for this rounding, derived from the theory of continued fractions and best rational approximation.

For completeness we shall here repeat from Chapter 8 the definition of a *continued fraction* $[a_0/a_1/a_2/\cdots]$ as a notation for the finite or infinite expression

$$a_0 + \cfrac{1}{a_1 + \cfrac{1}{a_2 + \ddots}},$$

where the *partial quotients* $a_i \geq 0$ are assumed to be integral. Any non-negative rational number p/q then has a finite expansion

$$\frac{p}{q} = [a_0/a_1/\cdots/a_n],$$

which is unique (*canonical*) with the added requirements $a_0 \geq 0$, $a_i \geq 1$ for $i = 1, 2, \ldots, n-1$ and $a_n \geq 2$ when $n \geq 1$. Note that for $a_n \geq 2$, $[a_0/a_1/\cdots/a_n] = [a_0/a_1/\cdots/a_n - 1/1]$ so there is an alternative expansion, which is then also unique.

The truncated continued fractions

$$\frac{p_i}{q_i} = [a_0/a_1/\cdots/a_i] \quad i = 0, 1, \ldots, n$$

yield rational numbers (called *convergents*) forming a sequence of approximations to p/q, whose properties can be summarized from the classical material on continued fractions as follows.

Theorem 9.3.1 *The convergents* $p_i/q_i = [a_0/a_1/\cdots/a_i]$ *of any continued fraction* $p/q = [a_1/a_2/\cdots/a_n]$ *for* $i = 0, 1, \ldots, n$ *satisfy the following properties:*

(i) recursive ancestry: with $p_{-2} = 0$, $p_{-1} = 1$, $q_{-2} = 1$, *and* $q_{-1} = 0$,

$$p_i = a_i p_{i-1} + p_{i-2},$$
$$q_i = a_i q_{i-1} + q_{i-2},$$

or in matrix form:

$$\begin{Bmatrix} p_{i-1} & p_i \\ q_{i-1} & q_i \end{Bmatrix} = \begin{Bmatrix} p_{i-2} & p_{i-1} \\ q_{i-2} & q_{i-1} \end{Bmatrix} \begin{Bmatrix} 0 & 1 \\ 1 & a_i \end{Bmatrix}$$

and

$$\begin{Bmatrix} p_{i-1} & p_i \\ q_{i-1} & q_i \end{Bmatrix} = \begin{Bmatrix} 0 & 1 \\ 1 & 0 \end{Bmatrix} \prod_{j=0}^{i} \begin{Bmatrix} 0 & 1 \\ 1 & a_i \end{Bmatrix};$$

(ii) irreducibility:

$$\gcd(p_i, q_i) = 1;$$

(iii) adjacency:

$$q_i p_{i-1} - p_i q_{i-1} = (-1)^i;$$

(iv) simplicity:

$$\frac{p_i}{q_i} \lessgtr \frac{p_{i+1}}{q_{i+1}} \text{ for } i \leq n-1;$$

(v) alternating convergence:

$$\frac{p_0}{q_0} < \frac{p_2}{q_2} < \cdots \frac{p_{2i}}{q_{2i}} < \cdots \leq \frac{p}{q} \leq \cdots < \frac{p_{2i-1}}{q_{2i-1}} < \cdots < \frac{p_1}{q_1};$$

(vi) best rational approximation:

$$\frac{r}{s} \lessgtr \frac{p_i}{q_i} \Longrightarrow \left|\frac{r}{s} - \frac{p}{q}\right| > \left|\frac{p_i}{q_i} - \frac{p}{q}\right|;$$

(vii) quadratic convergence:

$$\frac{1}{q_i(q_{i+1}+q_i)} < \left|\frac{p_i}{q_i} - \frac{p}{q}\right| \leq \frac{1}{q_i q_{i+1}} \text{ for } i \leq n-1;$$

(viii) real approximation: $|x - p/q| < 1/2q^2$ *for irreducible* p/q *implies that* p/q *is a convergent of a (possibly infinite) continued fraction expansion of* x.

Since the partial quotients a_i of the expansion $p/q = [a_0/a_1/\cdots/a_n]$ are the quotients obtained from the standard Euclidean GCD algorithm applied to p and q, we can extend the Euclidean algorithm by incorporating the calculation of p_i and q_i, in parallel with the computation of the a_i, and thus obtain an algorithm for determination of the convergents p_i/q_i, $i = 0, 1, \ldots, n$:

Algorithm 9.3.2 (Euclidean convergent algorithm (ECA))
Stimulus: *Integers* $p \geq 0$ *and* $q \geq 0$.
Response: *An integer* $n \geq 0$ *and sequences* $\{a_i\}$, $\{p_i/q_i\}$, $i = 0, 1, 2, \ldots, n$.
Method: $i := 0$;
$b_{-2} := p;\ p_{-2} := 0;\ q_{-2} := 1;$
$b_{-1} := q;\ p_{-1} := 1;\ q_{-1} := 0;$
repeat
$a_i := \left\lfloor \frac{b_{i-2}}{b_{i-1}} \right\rfloor;$ {*quotient*}
$b_i := b_{i-2} - a_i b_{i-1};$ {*remainder*}
$p_i := a_i p_{i-1} + p_{i-2};$
$q_i := a_i q_{i-1} + q_{i-2};$
until $b_i = 0$;

Example 9.3.1 To illustrate the ECA Algorithm let us apply it to $p/q = \frac{277}{642}$. The calculations are conveniently displayed in a table:

i	b_i	a_i	p_i	q_i
−2	277		0	1
−1	642		1	0
0	277	0	0	1
1	88	2	1	2
2	13	3	3	7
3	10	6	19	44
4	3	1	22	51
5	1	3	85	197
6	0	3	277	642

So $\frac{277}{642} = [0/2/3/6/1/3/3]$ and the convergents of $\frac{277}{642}$ are

$$\frac{0}{1}, \frac{1}{2}, \frac{3}{7}, \frac{19}{44}, \frac{22}{51}, \frac{85}{197}, \frac{277}{642}.$$

Furthermore, notice that $\gcd(277, 642) = 1 (= b_5)$. □

For any rational $p/q = [a_0/a_1/\cdots/a_n]$ or irrational $x = [a_0/a_1/\cdots]$, the convergents $p_0/q_0, p_1/q_1, \ldots$ represent successively more accurate approximations, where every convergent is simpler than its successor in the sequence. Thus for any simple chain F there is a last convergent $p_i/q_i \in F$ such that no subsequent convergent is in F. This provides the foundation for a mapping of arbitrary rationals (or reals) into the rationals of a simple chain F:

Definition 9.3.3 (The mediant rounding) *Let $\pm F$ be the set of all signed fractions corresponding to the fractions of a simple chain F. The mapping $\Phi_F : \mathbb{R} \to \pm F$ is defined for any real x with convergents $p_0/q_0, p_1/q_1, \ldots$ by*

$$\Phi_F(x) = \begin{cases} \dfrac{p_n}{q_n} & \textit{if } x = \dfrac{p_n}{q_n} \in F, \\ \dfrac{p_i}{q_i} & \textit{if } x > 0, \ \dfrac{p_i}{q_i} \in F \textit{ and } \dfrac{p_{i+1}}{q_{i+1}} \notin F, \\ -\Phi_F(-x) & \textit{if } x < 0. \end{cases}$$

Referring to the previous example, if we wish to round $\frac{277}{642}$ to a fraction where numerators and denominators are limited to two decimal digits, then with $F = FXS^*(99)$ we find $\Phi_F\left(\frac{277}{642}\right) = \frac{22}{51} = 0.431372\cdots$, compared with $\frac{266}{642} = 0.4314641\cdots$, i.e., the error is less than 10^{-4}. Consider as naive alternatives that the numerator and denominator were just truncated to two decimal digits, i.e., to $\frac{27}{64} = 0.4218\cdots$, or rounded to $\frac{28}{64} = 0.4375$; both approximations are much more distant from the correct value.

There are certain desirable properties a rounding should satisfy to ease the analysis of the error behavior of numerical algorithms. The *mediant rounding* defined above can be shown to satisfy the following properties usually required of a rounding:

(i) monotonic: $x < y \to \Phi_F(x) \le \Phi_F(y)$,
(ii) antisymmetric: $\Phi_F(-x) = -\Phi_F(x)$,
(iii) fixed points: $x \in F \Rightarrow \Phi_F(x) = x$,

but, furthermore, the mediant rounding also satisfies:

(iv) inverse symmetry: $\Phi_F\big(1/x\big) = 1/\Phi_F(x)$.

The reason for the name mediant rounding is apparent from the following theorem which we state without proof:

Theorem 9.3.4 *Let $p/q < r/s$ be consecutive members of a simple chain F. Then for $x \in [p/q; r/s]$,*

$$\Phi_F(x) = \begin{cases} \dfrac{p}{q} & \text{if} \quad x \in \left[\dfrac{p}{q}; \dfrac{p+r}{q+s}\right) \quad \text{or if} \quad x = \dfrac{p+r}{q+s} \text{ and } \dfrac{p}{q} \preccurlyeq \dfrac{r}{s}, \\[2ex] \dfrac{r}{s} & \text{if} \quad x \in \left(\dfrac{p+q}{q+s}; \dfrac{r}{s}\right] \quad \text{or if} \quad x = \dfrac{p+r}{q+s} \text{ and } \dfrac{r}{s} \preccurlyeq \dfrac{p}{q}. \end{cases}$$

Thus the mediant acts as a "splitting point" between what is rounded down to p/q or up to r/s, with the mediant itself being rounded to the simpler of the two end points.

We have already observed that the gaps next to a simple fraction are larger than those between fractions with large denominators. Now also note that the mediant $(p+r)/(q+s)$ of the adjacent pair $p/q \preccurlyeq r/s$ is positioned closest to the "least simple" fraction r/s. For example, consider the three consecutive fractions $\frac{5}{8} < \frac{2}{3} < \frac{5}{7}$ of $FXS^*(8)$ drawn to scale:

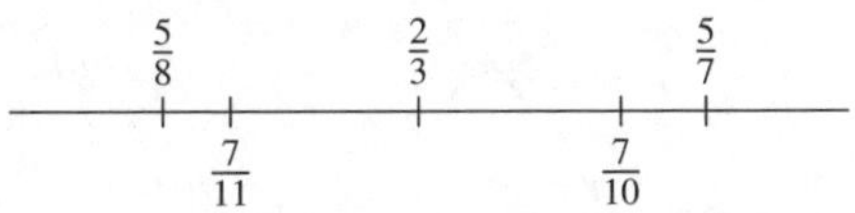

together with the two mediants. It is now evident that the simpler fractions act as "magnets," in the sense that much larger intervals of numbers are rounded into the simpler fractions than are into the less simple fractions.

This implies that in a composite computation where roundings take place, a result which happens to be close to a simple rational is likely to incur a large rounding error. On the other hand, if the exact result is a simple rational, then it is now more likely that the exact result will be recovered, even if some rounding errors have been introduced. We can illustrate this "bias towards simple fractions" with an example.

Example 9.3.2 The figure below illustrates (part of) the computation of the determinant

$$D = \begin{vmatrix} \frac{10}{13} & \frac{20}{17} & \frac{1}{13} \\[1ex] \frac{11}{19} & \frac{7}{11} & \frac{77}{95} \\[1ex] \frac{69}{91} & \frac{4}{17} & \frac{56}{65} \end{vmatrix} = \frac{5}{13}$$

in an arithmetic where at most three decimal digits are allowed in the numerators and denominators (i.e., in $FXS^*(999)$):

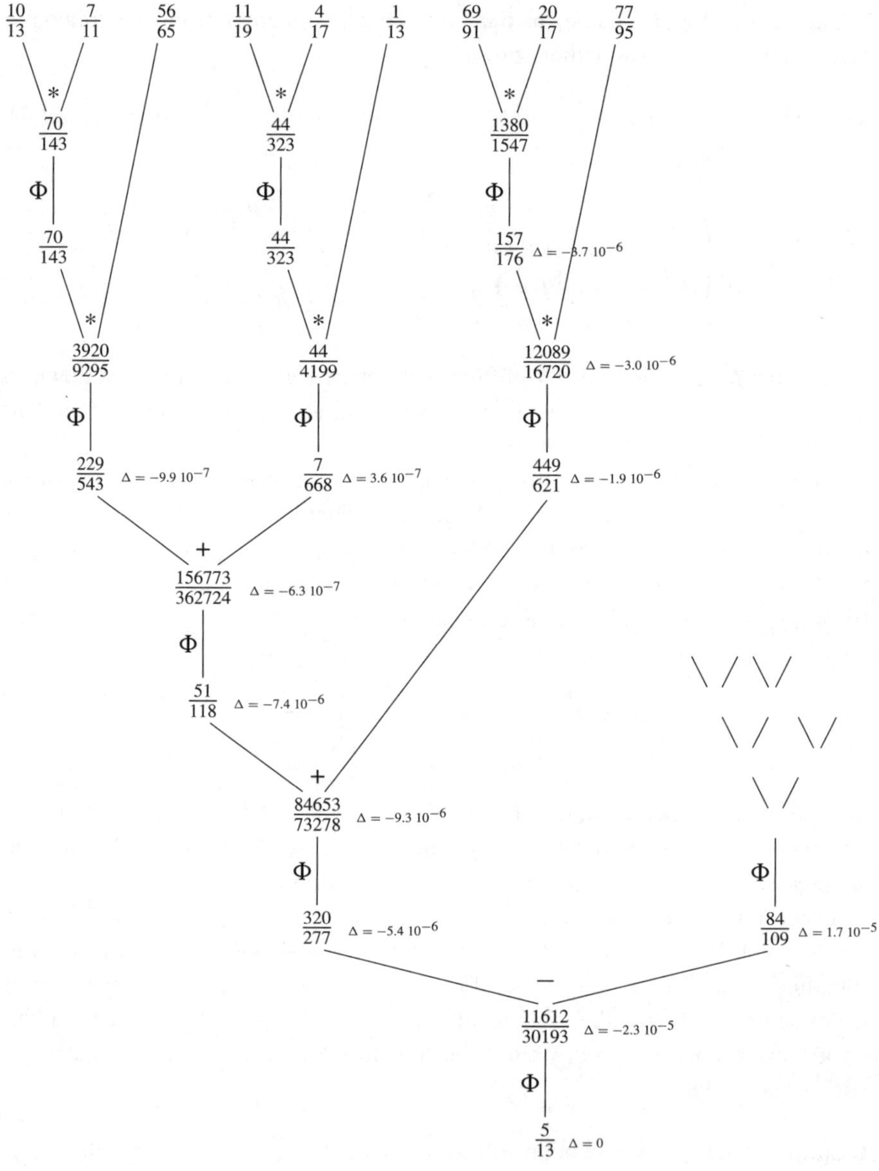

In the final rounding of $\frac{320}{277} - \frac{84}{109} = \frac{11612}{30193}$ the convergents are found to be

$$\frac{0}{1}, \frac{1}{2}, \frac{1}{3}, \frac{2}{5}, \frac{3}{8}, \frac{5}{13}, \frac{1288}{6711}, \frac{11612}{30193},$$

thus $\Phi_F\left(\frac{11612}{30193}\right) = \frac{5}{13}$ and the exact result is recovered, despite the intermediate roundings. □

Problems and exercises

9.3.1 Show the correctness of Theorem 9.3.1(i). (Hint: Use the identity: $p_i/q_i = [a_0/a_1/\cdots/a_{i-1}/a_i] = \lfloor a_0/a_1/\cdots/a_{i-1} + 1/a_i \rfloor$.)

9.3.2 Let $p/q = [a_0/a_1/\cdots/a_n] \geq 1$. Show that $[a_n/a_{n-1}/\cdots/a_0] = p_n/p_{n-1}$.

9.3.3 Rewrite the continued fraction, eliminating the partial quotient $a_i = 1$ in

$$[a_0/a_1/\cdots/a_{i-1}/1/a_{i+1}/\cdots/a_n]$$

by permitting non-positive partial quotients in the expansion.

9.3.4 Similarly rewrite $[a_0/a_1/\cdots/a_{i-1}/0/a_{i+1}/\cdots/a_n]$, eliminating $a_i = 0$.

9.3.5 Show that the sequence of remainders $\{b_i\}_{i=0,1,\cdots n}$ in Algorithm 9.3.2 (ECA) satisfies $b_i = (-1)^i(pq_i - qp_i)$ together with

$$q_{i+1} \leq \frac{q}{b_i} \leq q_i + q_{i+1} \quad \text{and} \quad p_{i+1} \leq \frac{p}{b_i} \leq p_i + p_{i+1}.$$

9.3.6 Show that $p/q \lessdot r/s$ are adjacent if and only if

$$\frac{r}{s} = [a_0/a_1/\cdots/a_n] \quad \text{on canonical form}$$

and

$$\frac{p}{q} = \begin{cases} [a_0/a_1/\cdots/a_{n-1}] \\ \text{or} \\ [a_0/a_1/\cdots/a_{n-1}/a_n - 1]. \end{cases}$$

Note that this condition then forms a third characterization of adjacency, extending the result of Lemma 9.2.2.

9.4 Arithmetic on fixed- and floating-slash operands

A finite precision, rational arithmetic can now be realized by forming integer expressions for the numerator and denominator of the result of a dyadic operator applied to two rational operands, and then subsequently applying the mediant rounding to map the result back into the system. However, it turns out that the standard algebraic rules involving multiplications need not be applied: the computation can be embedded in the Euclidean Algorithm applied in the final rounding.

Notice that in the ECA Algorithm, the initialization matrix

$$\begin{Bmatrix} p_{-2} & q_{-2} \\ p_{-1} & q_{-1} \end{Bmatrix} = \begin{Bmatrix} 0 & 1 \\ 1 & 0 \end{Bmatrix}$$

can be substituted by an arbitrary matrix (*seed matrix*)

$$\begin{Bmatrix} u_{-2} & v_{-2} \\ u_{-1} & v_{-1} \end{Bmatrix} = \begin{Bmatrix} a & c \\ b & d \end{Bmatrix}$$

to yield a sequence of pairs (u_i, v_i) instead of the (p_i, q_i) pairs, where the u_i and v_i are linear combinations of p_i and q_i:

$$u_i = aq_i + bp_i,$$
$$v_i = cq_i + dp_i.$$

In particular, with the bilinear form

$$f(x) = \frac{a + bx}{c + dx}$$

we find that

$$f\left(\frac{p_i}{q_i}\right) = \frac{aq_i + bp_i}{cq_i + dp_i} = \frac{u_i}{v_i},$$

so the (u_i, v_i) pairs are the numerators and denominators of the bi-linear (or Möbius) function $f(x)$, taken at the values of the convergents of p/q.

However, notice that the u_i/v_i obtained need not be in reduced form, nor do they, in general, form the sequence of convergents of $f(p/q)$. But

$$\begin{aligned}\frac{u_i}{v_i} - \frac{u_{i-1}}{v_{i-1}} &= \frac{u_i v_i - u_{i-1} v_i}{v_i v_{i-1}}\\ &= \frac{(ad - bc)(q_i p_{i-1} - q_{i-1} p_i)}{v_i v_{i-1}}\\ &= (-1)^i \frac{ad - bc}{v_i v_{i-1}},\end{aligned}$$

where $ad - bc$ is the determinant of the seed matrix. Thus if $f(x)$ is monotone on the interval $[p_i/q_i; p_{i+1}/q_{i+1}]$, or equivalently it has no pole in this interval, then

$$\left|\frac{u_i}{v_i} - f\left(\frac{p}{q}\right)\right| \le \left|\frac{u_i}{v_i} - \frac{u_{i+1}}{v_{i+1}}\right| \le \left|\frac{ad - bc}{v_i v_{i+1}}\right|$$

for $i < n$, due to the alternating convergence of the sequence $\{p_i/q_i\}$.

Furthermore note that the ECA Algorithm works as well if p and/or q are negative, $q \neq 0$, by choosing the remainder b_i of the same sign as the dividend. If $q = 0$, then $p_{-1}/q_{-1} = \frac{1}{0}$ may be considered the only convergent.

It now follows that we can realize the standard arithmetic operations $+, -, \times$, and /, by choosing appropriate seed matrices

$$\begin{Bmatrix} r & s \\ s & 0 \end{Bmatrix}, \begin{Bmatrix} -r & s \\ s & 0 \end{Bmatrix}, \begin{Bmatrix} 0 & s \\ r & 0 \end{Bmatrix}, \text{ and } \begin{Bmatrix} 0 & r \\ s & 0 \end{Bmatrix}$$

respectively in the following algorithm.

Algorithm 9.4.1 (Euclidean arithmetic algorithm, EAA)
Stimulus: *A rational number p/q and a seed matrix*

$$\begin{Bmatrix} u_{-2} & v_{-2} \\ u_{-1} & v_{-1} \end{Bmatrix} = \begin{Bmatrix} a & c \\ b & d \end{Bmatrix}$$

Response: *An integer* $n \geq 0$ *and a sequence of pairs* (u_i, v_i), $i = 1, 2, \ldots, n$ *such that*

$$\frac{u_i}{v_i} = f\left(\frac{p_i}{q_i}\right), \text{ where } f(x) = \frac{a+bx}{c+dx}$$

and $\{p_i/q_i\}$ *is the sequence of convergents of* p/q.

Method: $i := -1$;
$b_{-2} = p$; $b_{-1} = q$;
while $b_i \neq 0$ **do**
  $i := i + 1$;
  $\{$*choose* (a_i, b_i) *such that* $b_{i-2} = b_{i-1} \cdot a_i + b_i$
  with $|b_i| < |b_{i-1}|$ *and* $\mathrm{sgn}(b_i) = \mathrm{sgn}(b_{i-2})\}$
  $u_i := u_{i-1} \cdot a_i + u_{i-2}$
  $v_i := v_{i-1} \cdot a_i + v_{i-2}$
end

Example 9.4.1 To illustrate an arithmetic operation realized by the EAA Algorithm, consider the subtraction

$$\frac{371}{243} - \frac{26}{17}\left(= -\frac{11}{4131}\right)$$

as obtained through initialization by the seed matrix

$$\begin{Bmatrix} -26 & 17 \\ 17 & 0 \end{Bmatrix},$$

describing the function $f(x) = (-26 + 17x)/(17 + 0x) = x - \frac{26}{17}$, and the following table of intermediate results:

i	b_i	a_i	u_i	v_i
−2	371		−26	17
−1	243		17	0
0	128	1	−9	17
1	115	1	8	17
2	13	1	−1	34
3	11	8	0	289
4	2	1	−1	323
5	1	5	−5	1904
6	0	2	−11	4131

Note that

$$\frac{u_3}{v_3} = \frac{0}{289} = 0,$$

which is caused by the fact that $\frac{26}{17}$ happens to be a convergent of $\frac{371}{243}$. □

We can now make some observations on the behavior of the EAA Algorithm:

- Applying an identity operator (e.g., add $\frac{0}{1}$ or multiply by $\frac{1}{1}$) yields the convergents of p/q, i.e., it performs the ECA Algorithm.
- The sequence of pairs (u_i, v_i) are precisely the very same numerators and denominators that would have been obtained by applying the standard algebraic rules on the convergents p_i/q_i and r/s.
- The rational p/q need not be in reduced form: it could be an unreduced result from a previous arithmetic operation, as obtained by the EAA Algorithm.

For an implementation we may think of an arithmetic unit containing six registers organized and initialized the following way:

$$\begin{array}{c|cc} p & a & c \\ q & b & d \end{array}$$

where p, q represents one operand p/q, and a, b, c, d is the seed representing the operator and the other operand r/s. The EAA Algorithm then subtracts a suitable multiple of q from p, while at the same time adding the same multiples of b and d to a and c, respectively. The bottom row is then shifted up into the top three registers discarding the previous contents, and the new values are loaded into the bottom three registers. This process is then repeated so after the ith step of the algorithm the contents of the six registers are

$$\begin{array}{c|cc} b_{i-1} & u_{i-1} & v_{i-1} \\ b_i & u_i & v_i \end{array}$$

in the notation of the EAA Algorithm. The process is terminated when b_i becomes zero, or when an attempt is made to compute a pair (u_i, v_i) whose size exceeds the capacity of the registers. The resulting pair (u_i, v_i) may then be loaded into the registers originally containing p, q to be used as an operand for a new arithmetic operation, or for a final rounding before being stored in the memory. In the latter case of rounding, the algorithm has to be terminated when a pair (u_i, v_i) has been found which does *not* fit in the storage representation, and the pair (u_{i-1}, v_{i-1}) is then packed into the storage format.

Assuming that the unit is supplied with "double-length" registers, then intermediate results are being built in an extended precision which can be exploited in a subsequent arithmetic operation. The algorithm is effectively breaking the original operand p/q down into a sequence of partial quotients $\{a_i\}$, which is then used in the transformation of the original seed matrix to build the sequence of approximations to $f(p/q)$, where $f(x)$ is the bilinear form defined by the seed matrix. Thus if the entries of the seed matrix are "small" integers, more partial quotients of p/q can be utilized before the capacity of the registers is being exceeded.

As noted it is sufficient for a hardware realization to use six registers for storing two consecutive triples (b_i, u_i, v_i); it is not necessary to record the values of the quotients a_i, as they only need to exist temporarily. The EAA Algorithm actually performs repeated divisions of b_{i-2} by b_{i-1} which can be performed in binary, determining individual bits of the (partial) quotients a_i, which can then be used immediately for constructing and accumulating multiples of u_{i-1} and v_{i-1}, thus building up u_i and v_i. Utilizing hardware parallelism, the subtractions of the divisions and the additions in the multiplications can be executed truly in parallel.

Standard techniques for digit-serial division may thus be used for a binary implementation of the EAA Algorithm, employing the hardware structure sketched in Figure 9.4.1.

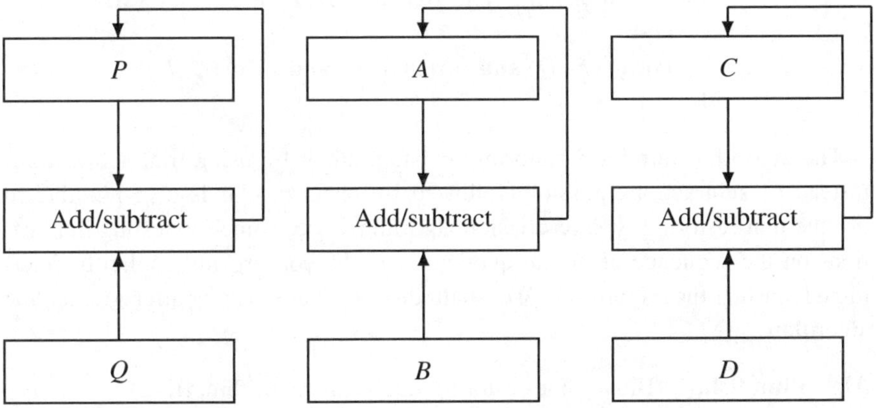

Figure 9.4.1. Basic hardware for the EAA Algorithm.

The leftmost adder structure performs a division (like say a non-restoring division) by subtracting shifted versions of the contents of the Q-register from the P-register, while the two adders perform equivalent additions of shifted versions of the B- and D-registers from the A- and C-registers respectively. When a complete partial quotient a_i has been found (though it never explicitly exists) the contents of the register triple (P, A, C) are swapped with those of the triple (Q, B, D), unless the algorithm is terminated because the remainder (in P) is zero, or the contents in the B or C register is overflowing.

In the following algorithm this is formalized, although for simplicity we assume that the operands are non-negative and that the registers have sufficient capacity. The register K is used to keep track of any misalignment of the contents of the Q-register. The keyword **and** is used to denote the parallel execution of statements.

Algorithm 9.4.2 (Binary shift/subtract Euclidean algorithm, BSSE)

Stimulus: *A rational operand p/q in registers P and Q. A seed matrix (a, b, c, d) in registers A, B, C, D.*

Response: *A rational u/v with u, v in registers B, D such that $u/v = f(p/q) = (aq' + bp')/(cq' + dp')$, where $p'/q' = p/q$, with* $\gcd(p', q') = 1$.

Method: $K := 1;$
while {*P and Q not normalized*} **do** {*leftshift Q,B,D and K*}
while $Q \neq 0$ **do**
  while {*Q not normalized*} **do** {*leftshift Q, B, D and K*}
  loop
    $P' := P - Q$ **and** $A' := A + B$ **and** $C' := C + D;$
    if $P' \geq 0$ **then**
    $\{P := P'$ **and** $A := A'$ **and** $B := B'\};$
    exitloop if $K = 1;$
    {*leftshift P* **and** *rightshift B, D, K*}
  end;
  {*swap* (P, Q) **and** *swap* (A, B) **and** *swap* (C, D)}
end.

The algorithm mimics a restoring division, albeit by using trial subtractions instead of subtractions possibly followed by restorings. It is also possible to use the non-restoring type of division algorithms with some interesting implications on the sequence of partial quotients and the convergents implicitly determined during the algorithm. We shall discuss these below after stating the algorithm.

Algorithm 9.4.3 (Binary non-restoring Euclidean algorithm, BNE)
Stimulus: *As for the BSSE Algorithm.*
Response: *As for the BSSE Algorithm.*
Method: $K := 1$
while {*P and Q not normalized*} **do** {*leftshift P and Q*}
while $Q \neq 0$ **do**
  while {*Q not normalized*} **do** {*leftshift Q, B, D, K*}
  loop
   while {*P not normalized*} **do**
    exitloop if $K = 1;$
    {*leftshift P* **and** *rightshift B, D, K*}
   end;
   if $\{\mathrm{sgn}(P) = \mathrm{sgn}(Q)\}$
    then $\{P := P - Q$ **and** $A := A + B$ **and** $C := C + D\}$
    else $\{P := P + Q$ **and** $A := A - B$ **and** $C := C - D\}$
  end
  {*swap* (P, Q) **and** *swap* (A, B) **and** *swap* (C, D)
end.

This algorithm also accepts p and/or q negative, and negative as well as positive remainders may occur during the computation, thus negative as well as positive partial quotients may (implicitly) be determined.

The BNE Algorithm computes pairs (u_i, v_i) based on "convergents" p_i/q_i (defined in the usual way), but based on a *signed continued fraction* expansion:

$$\frac{p}{q} = a_0 + \cfrac{1}{a_1 + \cfrac{1}{a_2 + \ddots + \cfrac{1}{a_n}}}$$

where the partial quotients a_i are arbitrary integers, except that $|[a_i/a_{i+1}/\cdots/a_n]| > 1$ for $1 \le i \le n$.

Signed continued fractions are redundant as is easily seen from examples:

$$\frac{11}{4} = \begin{cases} [2/1/3], \\ [2/2/-1/-2], \\ [2/2/-2/2], \\ [3/-4]. \end{cases}$$

In the following we shall call such redundant expansions *signed continued fractions* if they satisfy the properties stated in Lemma 9.4.4. Other classes could be defined, e.g., allowing zero-valued partial quotients would allow more redundancy and restricting $|a_i| \ge 2$ for $i \ge 1$ would allow less.

Lemma 9.4.4 *The sequence of partial quotients a_i, implicitly computed by the BNE Algorithm, satisfies the following:*

(i) $|a_i| \ge 1$ *for* $1 \le i \le n$,
(ii) $|a_i| = 1 \Rightarrow \operatorname{sgn}(a_i) = \operatorname{sgn}(a_{i+1})$ *for* $1 \le i \le n-1$,
(iii) $|a_i - [a_i/a_{i+1}/\cdots/a_n]| < 1$ *for* $1 \le i \le n-1$, *and for* $i = 0$ *when* $n \ge 2$.

The proof is based on an analysis of the sequence of partial quotients a_i and remainders b_i computed during the algorithm, but we will omit it. Similarly we state without proof the following theorem on the properties of the sequence of convergents that the algorithm computes when seeded with the unit matrix.

Theorem 9.4.5 *Let $[\hat{a}_0/\hat{a}_1/\cdots/\hat{a}_n]$ be the canonical continued fraction expansion of p/q, and*

$$\frac{\hat{p}_0}{\hat{q}_0}, \frac{\hat{p}_1}{\hat{q}_1}, \ldots, \frac{\hat{p}_n}{\hat{q}_n}$$

be the canonical convergents of p/q. The BNE Algorithm, seeded with the unit matrix, computes a signed continued fraction expansion $[a_0/a_1/\cdots/a_m]$ and corresponding convergents

$$\frac{p_0}{q_0}, \frac{p_1}{q_1}, \ldots, \frac{p_m}{q_m},$$

where $p_i/q_i = u_i/v_i$, $i = 0, 1, 2, \ldots, m$ *satisfy the following:*

(i) $gcd(p_i, q_i) = 1$,

(ii) $p_m/q_m = \hat{p}_n/\hat{q}_n$.

(iii) *If for some* i, $1 \leq i \leq m-1$, $p_i/q_i \neq \hat{p}_j/\hat{q}_j$ *for any* $j = 0, 1, \ldots, n$, *then there exists a* k, $0 \leq k \leq n-1$ *such that*

$$\frac{p_{i-1}}{q_{i-1}} = \frac{\hat{p}_k}{\hat{q}_k} \text{ and } \frac{p_{i+1}}{q_{i+1}} = \frac{\hat{p}_{k+1}}{\hat{q}_{k+1}}.$$

(iv) *If for some* i, $1 \leq i \leq n-1$, $\hat{p}_i/\hat{q}_i \neq p_j/q_j$ *for any* $j = 0, 1, \ldots, m$, *then there exists a* k, $0 \leq k \leq m-1$ *such that*

$$\frac{\hat{p}_{i-1}}{\hat{q}_{i-1}} = \frac{p_k}{q_k} \text{ and } \frac{\hat{p}_{i+1}}{\hat{q}_{i+1}} = \frac{p_{k+1}}{q_{k+1}}.$$

(v) *If* $\operatorname{sgn}(a_i) = \operatorname{sgn}(a_{i+1})$, *then there exists a* k *such that*

$$\frac{p_i}{q_i} = \frac{\hat{p}_k}{\hat{q}_k}$$

(vi) *If* $\operatorname{sgn}(a_i) \neq \operatorname{sgn}(a_{i+1})$, *then there exists a* k *such that*

$$\frac{p_i - p_{i-1} \cdot \operatorname{sgn}(p_i p_{i-1})}{q_i - q_{i-1} \cdot \operatorname{sgn}(q_i q_{i-1})} = \frac{\hat{p}_k}{\hat{q}_k}.$$

The theorem states that the BNE Algorithm, when used to perform mediant rounding, computes a sequence of convergents which may differ from the sequence of canonical convergents in the sense that some extra convergents may be present and some may be missing. However, it also states how a missing canonical convergent may be recovered, but note that by (ii) this is only necessary if the algorithm is terminated prematurely.

At first sight, it may seem difficult to adapt the algorithm to use a redundant representation of the operands, and thus to speed it up by avoiding carry-completion adders. The apparent problem is that the algorithm needs sign- and zero-detection, as well as normalization. However, if the normalization problem is solved, then testing for the sign of P becomes trivial, and zero-detection (of Q) can be tied into the normalization since the unit position is known.

In the presence of redundancy it is, in general, not possible to left normalize a number such that the value is in the interval $\left(-1; -\frac{1}{2}\right]$ or $\left[\frac{1}{2}; 1\right)$ (the unit position is considered to be at the left end of the register), just by inspection of the leading bit positions. However, by inspection of the two leading positions it is possible to scale the number to be in the interval $\left(-1; -\frac{1}{4}\right)$ or $\left(\frac{1}{4}; 1\right)$. Thus *quasi-normalization* turns out to be almost sufficient, but there is a possibility that the BNE Algorithm will enter an infinite loop where it will alternate between adding and subtracting without any intermediate shifting.

The algorithm computes a "distance"

$$d(p, q) = p - q \cdot \mathrm{sgn}(pq),$$

which is then substituted for p in the P-register. If $p \in \left(\frac{1}{4}; \frac{3}{8}\right)$ and $q \in \left(\frac{7}{8}; 1\right)$ then $d(p, q) \in \left(-\frac{3}{4}; -\frac{1}{2}\right)$ and will be considered quasi-normalized when substituted for p. The following step will then compute

$$d(q, d(p, q)) = q + (p - q) = p$$

and thus the algorithm will loop forever. The missing shift of p can easily be detected and this can be used to break the loop.

We shall conclude this section with some remarks on the complexity of these algorithms. It is known from the theory of continued fractions that when (p, q) is chosen uniformly $0 \le p \le N$ and $0 \le q \le N$, then the average number of partial quotients a_i in the continued fraction expansion of $p/q = [a_0, a_1, \ldots, a_n]$ grows asymptotically as

$$\frac{12(\ln 2)^2}{\pi^2} \log_2 N + O(1),$$

where the constant of the leading term is approximately 0.5842.

This expression then also provides the average number of cycles of the outer loop of the BSSE Algorithm, when applied to the mediant rounding mapping a uniformly chosen number in [0; 1] into a fixed-slash system. The average total number of executions of the inner loop can then be shown to be

$$1 + \frac{12(\ln 2)^2}{\pi^2} \log_2 N + O(1).$$

When the BSSE Algorithm is applied to n-bit floating-slash numbers under log-uniform distribution it can be shown that the average, asymptotic number of operations is:

$$\text{outer loop cycles:} \quad \frac{3(\ln 2)^2}{\pi^2} \cdot n \approx 0.1460n,$$

$$\text{inner loop cycles:} \quad \left(\frac{3}{4} + \frac{3(\ln 2)^2}{\pi^2}\right) n \approx 0.8960n,$$

$$\text{shift operations:} \quad \frac{3}{2} n = 1.5n.$$

It is to be expected that the BNE Algorithm will perform slightly better, since by not being forced to choose the remainder of the same sign as the dividend it may occasionally choose the remainder of opposite sign but of smaller absolute value, and thus reduce the subsequent work to be done. This has been confirmed by simulations as shown in Table 9.4.1 for mediant roundings of log-uniformly-distributed floating-slash operands.

Table 9.4.1. *Simulation of the BNE Algorithm on log-uniform distribution*

	2's complement	Carry-save	Borrow-save
Outer loop	$0.11n$	$0.11n$	$0.11n$
Inner loop	$0.41n$	$0.44n$	$0.44n$
Shifts	$1.50n$	$1.50n$	$1.50n$

The slightly higher number of inner loop cycles for redundant representations is due to the previously mentioned situations, where shifts are enforced due to alternating add/subtract cycles, without intermediate shifts.

The inner loop count is the number of add/subtract cycles, which may be compared with the average number of add/subtract operations in an average divide operation. It is known that the optimal selection of signed digits here leads to an average of $\frac{1}{3}n = 0.33n$, hence the complexity of the BNE Algorithm is comparable to that of a divide operation.

Problems and exercises

9.4.1 What is the value of the continued fraction $[2/-2/\cdots/2/-2]$ with n pairs of the form $2/-2$?

9.4.2 When applying the EAA Algorithm to certain seed matrices, there may be situations where simplifications of the matrices can lead to a reduction in the work to be done. Consider the case where the two operands for a division happen to be integers, e.g., respectively $p/q = i/1$ and $r/s = j/1$, and show that the problem can be transformed into an equivalent, but simpler, problem. Identify other general classes of seed matrices allowing similar simplifications?

9.5 A binary representation of the rationals based on continued fractions

The continued fraction expansion of a rational $p/q = [a_0/a_1/\cdots/a_n]$ provides a representation of the rationals, or if infinite expansions are permitted also a representation of the computable reals. But the partial quotients a_i can be of arbitrary size, hence there is no obvious convenient or economical way of using such a list of integers (possibly truncated) directly as a representation of a rational in a fixed-length word. It is known from the theory of continued fractions that large values of partial quotients are rare, and it can be shown that for uniformly distributed rationals on the unit interval a partial quotient takes the value k with probability given by

$$p_k = \log_2\left(1 + \frac{1}{k(k+2)}\right), \tag{9.5.1}$$

hence

$$p_1 = 0.415,\ p_2 = 0.170,\ p_3 = 0.093,\ p_4 = 0.059, \ldots,$$

confirming that the majority of partial quotients are small.

We have seen in the BSSE Algorithm that the information is "pulled out" of the operand p/q and used to transform the other operand (given by the seed matrix) in the form of very primitive "instructions." The operand q (or b_{n-1}) is first left shifted a number of places, and then it is conditionally subtracted from p (or b_{n-1}) while being gradually shifted back into its original position. Then the transformed remainder p (or b_{n-1}) is swapped with q, and the process is repeated. We shall see that this sequence of primitive operations can conveniently be encoded in binary, thus forming a binary representation of a rational number.

An alternative approach is to construct such a representation as a concatenation of binary strings, each representing a partial quotient encoded in binary in such a way that these encoded bit strings are *self delimiting*. Thus these strings should in some way contain an encoding of their own length, or contain some kind of "end marker."

We shall use such an approach and thus define a representation of an integer $p \geq 1$ having binary radix representation $1b_{n-1}b_{n-2}\cdots b_1b_0$ of length $n+1$, by the following string of length $2n+1$:

$$\ell(p) \equiv 1^n \circ 0 \circ b_{n-1}b_{n-2}\cdots b_1b_0. \tag{9.5.2}$$

When read from left to right it starts with a (possibly vacuous) *unary part* 1^n, delimited by a *switch bit* 0, followed by the (possibly vacuous) *binary part*. The unary part thus encodes the length, and the rest encodes the value of the integer represented. Notice that the unary part may conveniently be interpreted as an encoding of the initial left-shifting of q in the BSSE Algorithm, the switch bit corresponding to the first subtraction and the ones of the binary part to further subtractions taking place while right-shifting. The reason that the unary part is encoded with ones and the switch bit as the complement of the leading bit is that the encoding thus chosen is order preserving, i.e., the lexicographic ordering of strings then corresponds to the numeric ordering of the values encoded. This is also apparent from Table 9.5.1, which lists the encodings of selected integers.

For this reason we shall term this integer representation the *lexibinary encoding*. We shall carry this ordering property over to the representation of rationals by defining the *lexicographic continued fraction* (LCF) expansion of $p/q \geq 0$ as the (infinite) bit-string obtained by concatenation of the lexibinary encoding of the

Table 9.5.1. *Binary and lexibinary encodings*

Integer	Binary	Lexibinary
1	1.	0
2	10.	100
3	11.	101
4	100.	11000
5	101.	11001
6	110.	11010
7	111.	11011
8	1000.	1110000
16	10000.	111100000
32	100000.	11111000000
100	1100100.	1111110100100
200	11001000.	111111101001000
1000	1111101000.	1111111110111101000

partial quotients:

$$LCF\left(\frac{p}{q}\right) = \begin{cases} 1 \circ \ell(a_0) \circ \overline{\ell(a_1)} \circ \cdots \circ \ell(a_{2m}) \circ \overline{\ell(\infty)} \text{ for } 1 \leq \dfrac{p}{q} \\ 0 \circ \overline{\ell(a_1)} \circ \ell(a_2) \circ \cdots \circ \ell(a_{2m}) \circ \overline{\ell(\infty)} \text{ for } 0 \leq \dfrac{p}{q} < 1. \end{cases}$$

Here we have used the fact that the continued fraction can always be chosen such that it terminates with an even index and an appended infinity symbol:

$$\frac{p}{q} = [a_0/a_1/\cdots/a_{2m}/\infty].$$

The bars over the odd indexed encoded partial quotients indicate that the bit pattern is to be inverted. Using an infinite string of ones for the encoding of ∞, the terminating $\overline{\ell(\infty)}$ then becomes an infinite string of zeroes.

It is easily seen from the definition of a continued fraction that $[a_0/a_1/\cdots/a_n]$ is an increasing function of any even indexed partial quotient, and a decreasing function of any odd indexed partial quotient.

With this encoding of rationals it now follows that the lexicographic ordering of strings corresponds to the numeric ordering of the rationals, which is the reason for the name chosen. Notice that this ordering property does not depend on whether or not the trailing string of zeroes is explicitly represented. The termination of the string can just be interpreted as the infinite string of zeroes. Also any trailing zeroes of the encoding of $\ell(a_{2m})$ can then be disregarded, and we say that the LCF expansion $b_0 b_1 \cdots b_{k-1} 10^\infty$ has *order* k, with $\frac{0}{1} \sim 0^\infty$ having order zero.

The leading bit is called the *reciprocal bit* or *switch bit*, and all subsequent bits may then uniquely be identified as members of the unary, switch, or binary portion of some partial quotient, when the string is read from left to right (i.e., it is being parsed).

Example 9.5.1

$$\begin{aligned}
\tfrac{22}{7} &= \tfrac{22}{7} && \text{irreducible fraction form}\\
&= [3/6/1] && \text{even order continued fraction form}\\
&= [3/6/1/\infty] && \text{infinite extension}\\
&= 1 \circ \ell(3) \circ \overline{\ell(6)} \circ \ell(1) \circ \overline{\ell(\infty)} && \\
&= 1 \circ 101 \circ \overline{11010} \circ 0 \circ \overline{1^\infty} && \Bigg\} \text{ equivalent parsed LCF expansions}\\
&= 1 \circ 101 \circ 00101 \circ 0 \circ 0^\infty && \\
&= 110100101 && \text{minimal LCF expansion (order 8)} \quad \square
\end{aligned}$$

The LCF representation being non-redundant implies that there are situations where the next partial quotient of a value cannot be determined, i.e., more information on the value is continously needed to determine whether the next partial quotient is say r or $r+1$. As an example consider that $\sqrt{2} = [1/2/2/\cdots]$, then when trying to determine the square of it, it is not possible to determine whether the result is $[2/k]$ or $[1/1/m]$, where m and k are some large integers. The corresponding LCF representations are

$$\begin{aligned}
LCF([2/m]) &= 1 \circ 100 \circ \overline{11\cdots 10\cdots} &&= 11000\cdots 01\cdots,\\
LCF([1/1/k]) &= 1 \circ 0 \circ \overline{0} \circ 11\cdots 10\cdots &&= 10111\cdots 10\cdots,
\end{aligned}$$

which are approximations of LCF(2) $= 1100\cdots$ from above, respectively below. Note that these LCF representations are very close in the lexicographic ordering.

It is now also apparent that any finite bit-string can be parsed and interpreted as an LCF representation of some rational number, if sufficient trailing zeroes are appended.

The LCF representation can easily be extended to cover negative rationals by the following definition:

$$SLCF\left(\frac{p}{q}\right) = \begin{cases} 1 \circ LCF\left(\dfrac{p}{q}\right) & \text{for } \dfrac{p}{q} \geq 0,\\[2ex] 0 \circ \widetilde{LCF}\left(-\dfrac{p}{q}\right) & \text{for } \dfrac{p}{q} < 0. \end{cases}$$

where $\widetilde{LCF}$ denotes the 2's complement of a finite LCF bitstring and the 1's complement of an infinite bit-string. Since this extension of the domain follows trivially we shall limit our discussion to the LCF representation of the non-negative numbers.

From the definition of the LCF representation and noting that $p/q = [a_0/a_1/\cdots/a_n] \geq 1 \Rightarrow q/p = [0/a_0/a_1/\cdots/a_n]$ it follows that given the representation of q/p, $LCF(p/q) = \overline{b_0 b_1 \cdots b_{k-1}}$.

Let us now look at the irreducible fraction p_k/q_k, where $LCF(p_k/q_k) = b_0b_1\cdots b_{k-1}1$ for $0 \le k \le n$, obtained by truncating the LCF representation of p/q, $LCF(p/q) = b_0b_1\cdots b_{n-1}1$ at position k, and appending a unit bit. We will term p_k/q_k the kth order *biconvergent* (*binary convergent*) of p/q, in analogy with convergents being truncated versions of a continued fraction expansion. Each biconvergent in the sequence $p_0/q_0, p_1/q_1, \ldots, p_n/q_n$ then provides an improved upper or lower bound on p/q, determined by

$$\frac{p_k}{q_k} = b_0, b_1\cdots b_{k-1}1 \begin{cases} \ge \dfrac{p}{q} & \text{if } b_k = 0, \\ \le \dfrac{p}{q} & \text{if } b_k = 1. \end{cases}$$

Example 9.5.2 Continuing the last example we find the biconvergents of $LCF(\frac{22}{7}) = 110100101$ to be:

Order	Parsed biconvergent	Continued fraction	Ordinary fraction	Decimal
0	$1 \circ 0$	[1]	$\frac{1}{1}$	1.0000
1	$1 \circ 100$	[2]	$\frac{2}{1}$	2.0000
2	$1 \circ 1100$	[4]	$\frac{4}{1}$	4.0000
3	$1 \circ 101$	[3]	$\frac{3}{1}$	3.0000
4	$1 \circ 101 \circ \bar{0} \circ 0$	[3/1/1]	$\frac{7}{2}$	3.5000
5	$1 \circ 101 \circ \overline{101} \circ 0$	[3/3/1]	$\frac{13}{4}$	3.2500
6	$1 \circ 101 \circ \overline{11011} \circ 0$	[3/7/1]	$\frac{25}{8}$	3.1250
7	$1 \circ 101 \circ \overline{11001} \circ 0$	[3/5/1]	$\frac{19}{6}$	3.1667
8	$1 \circ 101 \circ \overline{11010} \circ 0$	[3/6/1]	$\frac{22}{7}$	3.1428

□

Since there is a one-to-one correspondence between finite bit-strings and rationals we can enumerate this mapping in a binary tree, associating the rational p/q with the node reached from the root along a path denoted by the LCF expansion. Thus the label p/q with $LCF(p/q) = b_0, b_1\cdots b_{k-1}1$ is found at depth k by proceeding to the left child if $b_i = 0$ and to the right child if $b_i = 1$ for $i = 0, 1, 2, \ldots, k-1$. Note that the path does not include the terminating unit bit.

The left half of the *LCF tree* truncated at depth 5 is illustrated in Figure 9.5.1, where it can be noted that the corresponding right half of the LCF tree is just the reciprocals reached by the complemented bitstring.

Note that the rationals found along the path to a rational p/q are precisely the biconvergents of p/q, and that the set Q_k of rationals found in the LCF tree of depth k are the rationals representable in $k+1$ bits when $\frac{0}{1}$ is added. An in-order transversal of the tree will list the members of Q_k in numerical order, thus

$$Q_2 = \left\{\frac{0}{1}, \frac{1}{4}, \frac{1}{2}, \frac{2}{3}, \frac{1}{1}, \frac{3}{2}, \frac{2}{1}, \frac{4}{1}\right\},$$

and in the following we shall interpret Q_k as an ordered set.

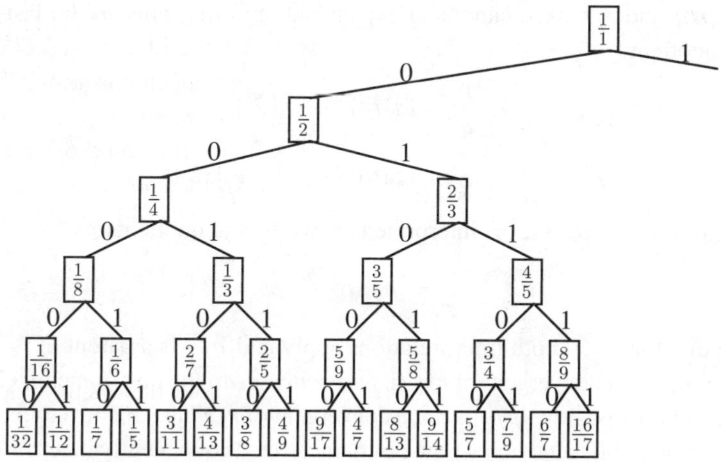

Figure 9.5.1. The left half of the LCF tree to depth 5.

From the LCF tree it follows that neighbors in the set Q_k are in some way related since they share a path down the LCF tree, but also that this relation is radix-dependent. In the following we want to explore this relation by defining it as a "binary adjacency" which we will show to be sufficient to characterize all neighbor pairs in Q_k.

Obviously, neighbors in Q_k have LCF representations that differ by a unit in the last place (ulp) when considering these as bit-strings. We define two finite bit-strings α and β *lexicographically adjacent* (with α *preceding* β) if and only if

$$\alpha = \sigma \circ 0 \circ 1^j,$$
$$\beta = \sigma \circ 1 \circ 0^j$$

for some prefix string σ and some $j \geq 0$. However, it turns out to be most convenient to define the wanted relation in terms of continued fractions expansions.

The fractions $p/q \lessgtr r/s$ are *bijacent* (of positive, zero or negative degree i) if and only if p/q and r/s are both irreducible and

$$\frac{p}{q} = [a_0/a_1/\cdots/a_n],$$
$$\frac{r}{s} = [a_0/a_1/\cdots/a_n + 2^i], \tag{9.5.3}$$

where $a_n = k2^i \geq 2$ when $i \geq 1$. The definition includes different situations, depending on whether i is positive, zero or negative, which can be characterized as follows:

$i \geq 0$: p/q and r/s have canonical expansions differing only in the last partial quotient:

$$\frac{p}{q} = [a_0/a_1/\cdots/k2^i],$$
$$\frac{r}{s} = [a_0/a_1/\cdots/(k+1)2^i].$$

$i \leq 0$: In this case $a_n + 2^i$ is interpreted as two partial quotients:

$$\frac{r}{s} = [a_0/a_1/\cdots/a_n/2^{-i}].$$

Note that for $i = 0$ both interpretations apply and p/q is a parent of r/s. For i negative, p/q is a convergent of r/s (its *preconvergent*, the one immediately prior to r/s), and for i positive p/q is not a convergent of r/s but a biconvergent to r/s.

The notion of bijacency may now be characterized equivalently in terms of LCF bit-strings, fraction form, or continued fraction form.

Theorem 9.5.1 *Each of the following three properties implies the other two and serves as an equivalent definition of bijacency.*

(i) Continued fraction form: $p/q \lessdot r/s$ are bijacent of degree i as specified in (9.5.3).
(ii) LCF form: The irreducible fractions p/q and r/s have lexicographically adjacent LCF expansions (and thus are neighbors in Q_k for some k).
(iii) Fraction form: The irreducible fractions $p/q \lessdot r/s$ are bijacent of degree i if and only if for

$i \geq 0$: $|ps - qr| = 2^i = gcd(r-p, s-q)$ *and* $r/s \lessdot 2p/2q$,
$i \leq 0$: $|ps - qr| = 1$ *and* $2^{-i}p/2^{-i}q \lessdot r/s \lessdot (2^{-i}+1)p/(2^{-i}+1)q$ *when* $p/q \neq 0/1$, *and* $r/s = 1/2^{-i}$ *when* $p/q = 0/1$.

Proof (Sketch only) It is straightforward to prove (ii) and (iii) from (i). The proof of (iii) $\Rightarrow$ (i) is split in three cases. For negative i it follows from Theorem 9.3.1(viii) that p/q is a convergent of r/s, but it cannot be a parent. The case for $i = 0$ follows similarly, and for i positive it follows from the existence of a/b such that $r/s = (2^i a + p)/(2^i b + q)$ with a/b adjacent to p/q and r/s so a/b must be a convergent of both, where a/b cannot be the parent of p/q. The proof of (ii) implying (i) can be based on a lengthy analysis of two automatons decoding the bit strings $\sigma 01^j$ and $\sigma 10^j$. □

Having described the relation between neighbors in Q_k in analogy with neighbors in the Farey set F_n, it is natural to ask whether it is possible to determine the members of Q_{k+1} given Q_k, i.e., whether there is an analog of the mediant construction. Specifically, given two bijacent neighbors from Q_k with LCF representations $\sigma \circ 0 \circ 1^j$ and $\sigma \circ 1 \circ 0^j$, we want to characterize the unique rational with LCF representation $\sigma \circ 0 \circ 1^{j+1}$ falling between these.

Recall that the mediant $(p+r)/(q+s)$ of two adjacent fractions $p/q < r/s$ from F_n is adjacent to both and satisfies $p/q < (p+r)/(q+s) < r/s$. However, when say p/q is a very "simple fraction" whereas r/s is not (e.g., $p \ll r$ and $q \ll s$), their mediant will be of numeric value very close to r/s and rather distant from p/q. Now if p and q were multiplied by some common factor c, chosen such that cp and cq were of the same order of magnitude as r and s, respectively, then the value of the expression $(cp+r)/(cq+r)$ would split the interval between p/q and r/s in two intervals of about the same width.

This leads us to the introduction of an alternative form of mediant, more nearly bisecting the interval between consecutive members of Q_k. The *binary mediant* of two bijacent fractions $p/q \lessdot r/s$ is the irreducible fraction u/v with value

$$u/v = (2^j p + r)/(2^j q + s),$$

where j is the largest integer such that $2^j p/2^j q \lessdot r/s$. The binary mediant is defined to be in reduced form, but $\gcd(2^j p + r, 2^j q + s)$ can only have the value 1 or 2.

The following theorem can be obtained by extending the proof of the previous theorem, and provides an alternative characterizations of the binary mediant.

Theorem 9.5.2 *The binary mediant of two bijacent fractions $p/q, r/s$ is the irreducible fraction u/v equivalently determined by any of the following three conditions:*

(i) Continued fraction form: If

$$\frac{p}{q} = [a_0/a_1/\cdots/a_{n-1}/a_n],$$

$$\frac{r}{s} = [a_0/a_1/\cdots/a_{n-1}/a_n + 2^i],$$

with $a_n = k2^i \geq 2$ when $i \geq 1$, then

$$\frac{u}{v} = [a_0/a_1/\cdots/a_{n-1}/a_n + 2^{i-1}].$$

(ii) LCF expansion form: If $p/q < r/s$ and for some bit string σ and integer $j \geq 0$,

$$\frac{p}{q} = \sigma \circ 0 \circ 1^j,$$

$$\frac{r}{s} = \sigma \circ 1 \circ 0^j,$$

then

$$\frac{u}{v} = \sigma \circ 0 \circ 1^{j+1}.$$

(iii) Fraction form: If $p/q \lessdot r/s$ are bijacent of degree i, then

$$\frac{u}{v} = \begin{cases} \dfrac{2^{-i}p + r}{2^{-i}q + s} & \textit{for } i \leq 0, \\ \dfrac{(p+r)/2}{(q+s)/2} & \textit{for } i > 0. \end{cases}$$

Lemma 9.5.3 *The binary mediant u/v of the two bijacent fractions $p/q, r/s$ is bijacent to both p/q and r/s.*

The LCF tree can thus be built to any depth from the fraction $\frac{1}{1}$ assigned to the root, and the fraction assigned to any other node is the binary mediant of the nearest left ancestor (or $\frac{0}{1}$ if none exists). With edges labelled 0 for the left branch and 1 for the right branch, the bit-string composed along the path from the root to any particular node provides the LCF representation of the fraction at that node when terminated with a unit bit.

We will conclude this description of the LCF representation with some considerations on the *gaps* between neighbors in Q_k. Since there are $2^k + 1$ members of Q_k falling in the interval [0; 1], the average gap size is 2^{-k} over the unit interval. As discussed above, the binary mediant more closely bisects the gap between two bijacent rationals than found for the classical mediant. However, forming the binary mediant between two bijacent fractions $p/q \lessdot r/s$ of degree $i \geq 0$ we find

$$\Delta_1 = \frac{p}{q} - \frac{(p+r)/2}{(q+s)/2} = \frac{ps - qr}{q(q+s)},$$
$$\Delta_2 = \frac{(p+r)/2}{(q+s)/2} - \frac{r}{s} = \frac{ps - qr}{s(q+s)},$$

and since by Theorem 9.5.1(iii), $1 \leq s/q \leq 2$ for $i \geq 0$, we have

$$1 \leq \frac{\Delta_1}{\Delta_2} \leq 2,$$

and for $i \leq 0$ we find similarly

$$1 \leq \frac{\Delta_1}{\Delta_2} \leq 1 + 2^i.$$

The extreme value 2 is easily seen to occur for the binary mediant $\frac{2}{3}$ of $\frac{1}{2}$ and $\frac{1}{1}$, and in general for $(1 + 6a)/3 \cdot 2^{i-1}$ as the binary mediant of $(1 + 4a)/2^i$ and $(1 + 8a)/2^{i+1}$ for $i > 0$ and $a = 0$, or $a = 2^{i-2}$ and $i \geq 2$.

In k-bit fixed-point radix representations of the form $0.b_1b_2 \cdots b_k$ the gaps between representable numbers are of uniform size 2^{-k} over the unit interval [0; 1]. Since the gaps between rationals of Q_k over the same interval [0; 1] vary to accommodate the $2^k + 1$ rationals representable, but have the same average gap size, 2^{-k}, it is natural to measure gap sizes relative to the number of bits used, i.e.,

$$\text{gap} = 2^{-\alpha k} \quad \text{where } \alpha = -\frac{1}{k}\log_2(\text{gap}).$$

Exhaustive searches for $k \le 24$ have shown that minimum gaps in Q_k occur between biconvergents of $z = \frac{\sqrt{5}-1}{2} = [0/1/1/1\cdots]$ having LCF expansion $010101\cdots$. The sequence of biconvergents $\frac{1}{1}, \frac{1}{2}, \frac{2}{3}, \frac{3}{5}, \frac{5}{8}$ (with the well known Fibonacci numbers in the numerators and denominators) coincides with the convergents of z, and the gaps are thus

$$\frac{1}{1\cdot 2}, \frac{1}{2\cdot 3}, \frac{1}{3\cdot 5}, \frac{1}{5\cdot 8}, \cdots.$$

Since the Fibonacci numbers grow asymptotically at the rate $(1+\sqrt{5})/2$, the gaps decrease at a rate approaching

$$\frac{4}{(1+\sqrt{5})^2} = \frac{2}{3+\sqrt{5}},$$

hence for any $\varepsilon > 0$ and k sufficiently large, the minimum gap size in Q_k will be no bigger than $2^{-(a-\varepsilon)k}$ for

$$a = \log_2\left(\frac{3+\sqrt{5}}{2}\right) = 1.38848\cdots.$$

For the maximum gap it has similarly been observed that the largest gaps in Q_k occur next to a rational p/q whose LCF expansion contains replications of the bit pattern 11110000. This pattern can be identified from replications of partial quotients $\cdots 2/4/2/4/\cdots$ since $\overline{\ell(2)} \circ \ell(4) = \overline{1001100} = 01111000$, as in $y = \sqrt{6}-2 = [0/2/4/2/4\cdots]$ with $LCF(y) = 0011110000111100000\cdots$. Here y has a sequence of convergents in which the rate of increase in numerators and denominators in two steps p_{i+2}/p_i approaches $5+2\sqrt{6} = 9.898979\cdots$. Since having eight bits in the LCF expansion corresponds to the encoding of another pair 2/4, the bounding interval will decrease at a rate of $1/(5+2\sqrt{6})$ over a sequence of eight biconvergents. Thus for any $\varepsilon > 0$ and k sufficiently large the maximum gap size will be at least $2^{-(b+\varepsilon)k}$ for

$$b = \tfrac{1}{4}\log_2(5+2\sqrt{6}) = 0.82682\cdots.$$

It is conjectured that with $\alpha_k = -(1/k)\log_2(\text{max gap in } Q_k \text{ over } [0,1])$ then $\lim_{k\to\infty}\alpha_x = \frac{1}{4}\log_2(5+2\sqrt{6}) = 0.82682\cdots$, but it has only been proven that for any $\varepsilon > 0$ and k sufficiently large the maximum gap size in Q_k is not greater than $2^{-(c-\varepsilon)k}$ for $c = \frac{1}{4}\log_2(16/153) = 0.814347\cdots$.

Simulations have shown that the distribution of $-(1/k)\log_2(\text{gap in } Q_k$ over [0,1]) is bell-shaped between approximately 0.8 and 1.3, with the bell becoming narrower and more highly peaked for increasing value of k. Thus gaps between LCF representable values vary in size corresponding to a loss of approximately 19% and a gain of 38% in precision, compared to the equivalent binary radix fixed-point system.

9.6 Gosper's Algorithm

Given the LCF representation it is natural to ask whether operands given as encoded continued fractions can be used directly as operands for arithmetic operations, delivering results in the same representation. The binary LCF representation was derived from the observation that the sequence of primitive steps of Algorithm 9.4.2 (the BSSE Algorithm) provides an encoding of one of the operands p/q. Thus implicitly the operand p/q is entered into the computation in such a representation, whereas the other operand is represented in numerator/denominator form in the seed matrix. The algorithm performs a computation of the expression

$$f(x) = \frac{a + bx}{c + dx},$$

where implicitly x is delivered piecewise in the form of partial quotients in the original Algorithm 9.4.1 (the EAA Algorithm), and in terms of more primitive (binary) steps in Algorithm 9.4.2.

Generalization to two variables then leads naturally to considering an expression of the form (a *bi-homographic* form)

$$z(x, y) = \frac{axy + bx + cy + d}{exy + fx + gy + h}, \tag{9.6.1}$$

which allows us to compute the sum, difference, product, and quotient of x and y as special cases, by choosing appropriate values of the coefficients $a, b, \ldots, h$. For example, with $b = c = h = 1$ and zero for the remaining coefficients, we obtain $z(x, y) = x + y$, but many other functions can be computed by defining the coefficients suitably. For a rational arithmetic it is sufficient to consider integer-valued coefficients, and in this section we develop an algorithm for the evaluation of (9.6.1) for x and y given by their continued fraction expressions, which delivers the result $z(x, y)$ as an equivalent expansion.

Gosper observed that the form (9.6.1) is preserved when operands x and y are provided in the form of continued fraction expansions. Let $x = [a_0/a_1/\cdots/a_n]$, so

$$x = a_0 + \frac{1}{x'}$$

where $x' = [a_1/a_2/\cdots/a_n]$, so in general substituting $x = p + 1/x'$ into (9.6.1) we find

$$\begin{aligned} z(x, y) = z\left(p + \frac{1}{x'}, y\right) &= \frac{(pa + c)x'y + (pb + d)x' + ay + b}{(pe + g)x'y + (pf + h)x' + ey + f} \\ &= z_1(x', y), \end{aligned}$$

which is an expression of the same form as (9.6.1) with new coefficients obtained by simple, linear transformations of the original coefficients. Similarly entering a leading partial quotient q of y corresponds to substituting $y = q + 1/y'$ into

(9.6.1) obtaining

$$z_2(x, y') = \frac{(qa + b)xy' + ax + (qc + d)y' + c}{(qe + f)xy' + ex + (qg + h)y' + g}.$$

Any sequence of leading partial quotients from two operands can thus be "input" serially in (9.6.1), changing it into a new function specified in terms of a transformed 8-tuple of coefficients $(a', b', \dots, h')$, and some transformed variables x' and y' representing the so-far-not-seen parts of x and y (the "tails" of x and y):

$$z'(x', y') = \frac{a'x'y' + bx' + cy' + d'}{e'x'y' + fx' + gy' + h'}. \tag{9.6.2}$$

The values of the tails are unknown, but their effects as corrections of the previous partial quotient have the form of terms $1/x'$ and $1/y'$, and are thus bounded. In the case of a canonical (non-redundant) continued fraction we have

$$1 < x' \leq \infty \quad \text{and } 1 < y' \leq \infty,$$

where the value ∞ corresponds to the case of an empty tail ($1/x' = 0$ and/or $1/y' = 0$). For the redundant (signed) continued fractions the equivalent bounds are

$$1 < |x'| \leq \infty \quad \text{and } 1 < |y'| \leq \infty.$$

Given such bounds on x' and y', it is then possible to find bounds on $z'(x', y')$; in particular it is easily seen that if all partial quotients have been read, then the transformed function is a constant:

$$z(x, y) = z'(\infty, \infty) = \frac{a'}{e'}.$$

Consequently, if only some prefixes of the two lists of partial quotients have been read, then a/e is the value of the original function $z(\cdot, \cdot)$ taken at the point determined by the corresponding convergents of x and y.

After the input of a few partial quotients from x and y it may be possible to bound $z'(x', y')$ as

$$r \leq z'(x', y') < r + 1$$

for some integer r, in which case r is the first partial quotient of the canonical continued fraction expansion of $z(x, y)$. Similarly, if z' can be bounded for some integer r by

$$r - 1 < z'(x', y') < r + 1,$$

then r is the first partial quotient of a redundant (signed) continued fraction expansion.

Now it is possible to "output r" and rewrite z' in the form

$$z'(x', y') = r + \frac{1}{z''(x', y')}. \qquad (9.6.3)$$

Solving this equation for z'', we obtain

$$z''(x', y') = \frac{e'x'y' + f'x' + g'y' + h'}{(a' - re')x'y' + (b' - rf')x' + (c' - rg')y' + (d' - rh')}.$$

Hence again we find that the transformed function $z''(\cdot, \cdot)$ has the same form as (9.6.1), with coefficients obtained by simple linear transformations.

We can depict the process of performing the input of partial quotients from x or y, and the output of a partial quotient of the result $z(x, y)$, as a *cube* with the eight coefficients in the corners of a cube:

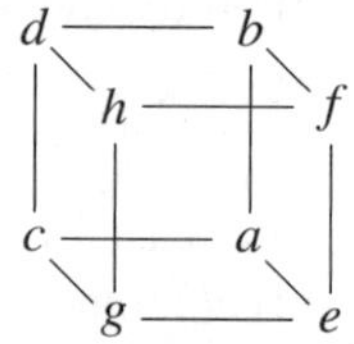

A transformation corresponding to the input or output of a partial quotient then substitutes one face of the cube by a linear combination of that face and the opposite face, while "shifting out" the latter face, as illustrated in Figure 9.6.1.

Note that the transformations performed on the 4-tuples at the faces of the cube upon input of partial quotients mirror the transformations in Algorithm 9.4.1 on pairs of tuples (u_i, v_i). Now input is taking place in two different directions (x and y), and output is performed in the third direction (z). Based on the range of $z(x, y)$, quotients can be determined by a Euclidean algorithm performed on the faces of the cube in the z-direction, i.e., numerator/denominator 4-tuples.

To be able to determine the quotients we need a selection procedure, which, given the 8-tuple of coefficients of an updated cube, can specify a partial quotient to be output. It is thus necessary to determine the range of $z(x, y)$ over appropriate domains $x \in D_x$ and $y \in D_y$. Initially $D_x = D_y = \mathbb{R}$, the set of real numbers, but when the first partial quotients of x and y have been read the domains D_x and D_y become restricted. For canonical continued fractions (non-redundant) $D_x = D_y = (1, \infty]$. However, recall from Section 9.4, the conditions of Lemma 9.4.4 state that for redundant expansions normally $D_x = D_y = (1, -1)$, the affine interval $|x| > 1$ with infinity included. Since $|a_i| = 1$ implies that a_i and a_{i+1} must have the same sign, then following, say, the input of a partial quotient $a_i = 1$ from x, $D_x = (1, \infty)$, infinity not included.

To determine the range of

$$z(x, y) = \frac{axy + bx + cy + d}{exy + fx + gy + h}$$

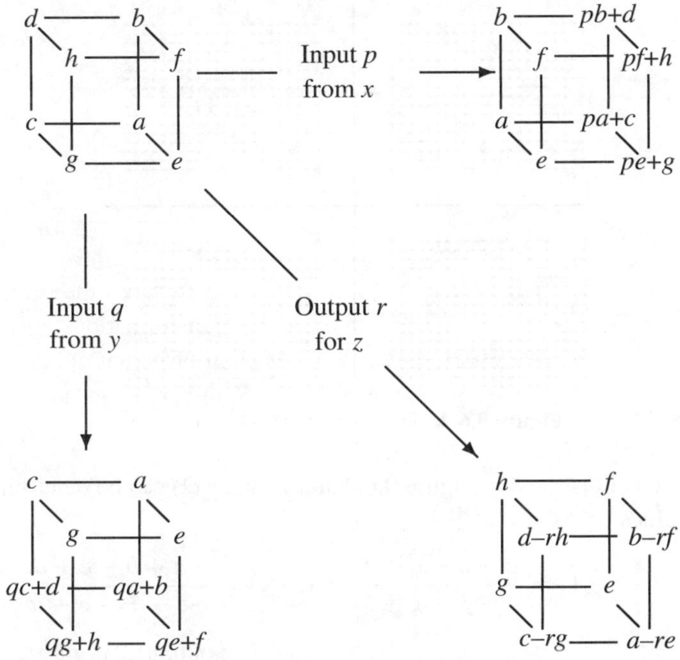

Figure 9.6.1. The coefficient cube and its transformations by partial quotient input and output.

over the domain $x \in D_x$ and $y \in D_y$, note that $z(x, y)$ is a monotone function of x and y, provided that the denominator is non-zero over the domain $D_x \times D_y$. The root curves of the denominator are hyperbolas, and it turns out that the well definedness of $z(x, y)$ can be checked by an analysis of the signs of the denominator evaluated at the four corners of the domain $D_x \times D_y$. Since the intervals D_x and D_y are open at finite end points, we must assume that $z(x, y)$ is well defined on the intervals closed at such end points. If this is the case, then

$$\text{Zrange}(Q) = \{z(x, y) \mid x \in D_x \text{ and } y \in D_y\}$$

can be determined as a finite interval, given the 8-tuple Q of coefficients of the cube. Provided that there exists an r such that

$$\text{Zrange}(Q) \subseteq [r; r+1) \quad \text{for non-redundant output}$$

or

$$\text{Zrange}(Q) \subseteq (r-1; r+1) \quad \text{for redundant output,}$$

then r is the next partial quotient to be output, otherwise more input must be taken from the x and/or y direction to reduce the interval Zrange(Q).

For redundant (signed) continued fractions the domains are state-dependent, but normally $D_x = D_y = (1; -1)$ with infinity included. The domain $D_x \times D_y$

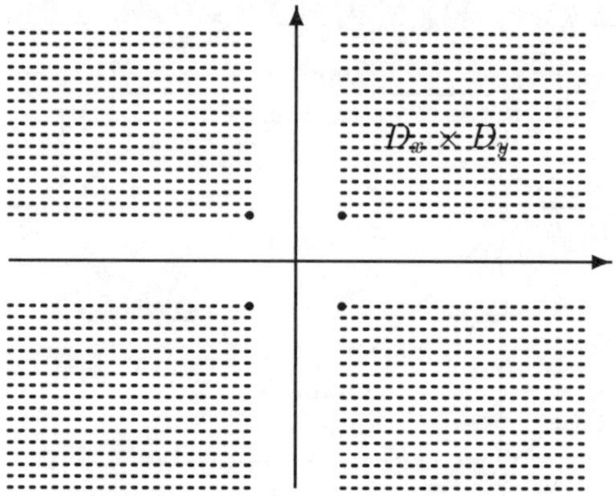

Figure 9.6.2. The domain $(1; -1) \times (1; -1)$.

for this case is illustrated in Figure 9.6.2 and Zrange(Q) can be determined from the four values

$$z(-1,-1) = \frac{a-b-c+d}{e-f-g+h}, \; z(1,-1) = \frac{-a+b-c+d}{-e+f-g+h},$$

$$z(-1,1) = \frac{-a-b+c+d}{-e-f+g+h}, \; z(1,1) = \frac{a+b+c+d}{e+f+g+h}.$$

provided that the root curves of the denominator are outside $D_x \times D_y$.

If either x or y, but not both have become exhausted, only two ratios determine the range of $z(x, y)$. Say if x has terminated so $D_x = \infty$ and $D_y = (1; -1)$, then $z(\infty, 1)$ and $z(\infty, -1)$ determine Zrange(Q). If both x and y have terminated then Zrange(Q) reduces to the single point $z(\infty, \infty) = \frac{a}{e}$.

Example 9.6.1 The mechanics of the algorithm for a signed (redundant) continued fraction is illustrated in Figure 9.6.3 for the computation $\frac{18}{11} + \frac{14}{11}$ with $\frac{18}{11} = [2/-3/4]$ and $\frac{14}{11} = [1/4/-3]$, displayed in terms of coefficient cube transformations in three dimensions. Each transformation creates a new face in the x, y, or z direction. The new face together with the previous face in that direction form the new cube.

In particular, observe that after the input of two partial quotients from the x as well as from the y direction the cube represents the function

$$z'(x', y') = \frac{-35x'y' - 8x' + 13y' + 3}{-12x'y' - 3x' + 4y' + 1}$$

with

$$z'(-1,-1) = \tfrac{37}{12}, \; z'(1,-1) = \tfrac{17}{6},$$
$$z'(-1,1) = \tfrac{59}{20}, \quad z(1,1) = \tfrac{27}{10},$$

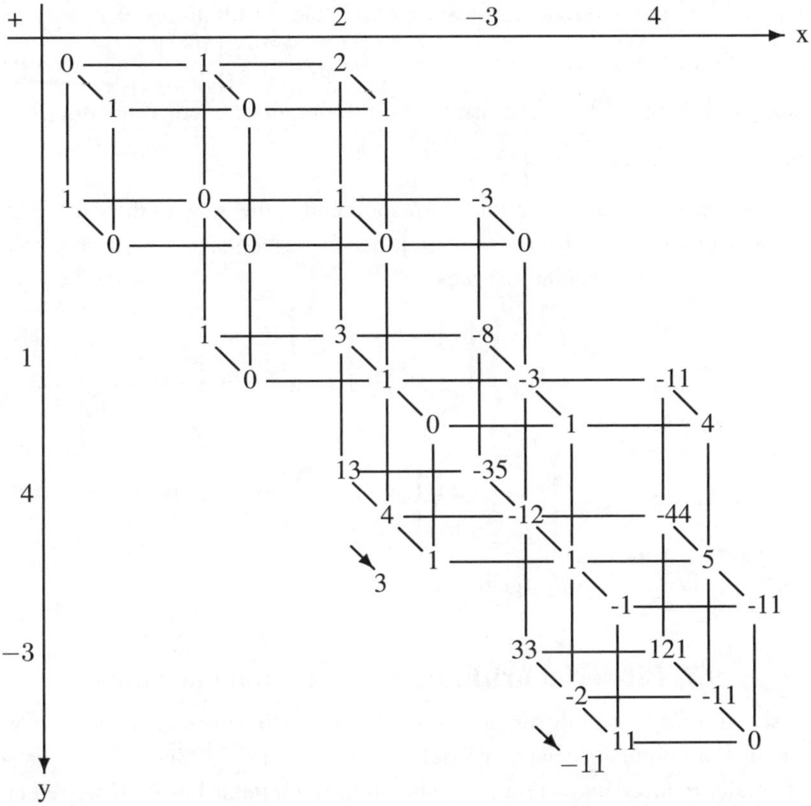

Figure 9.6.3. Coefficient cube transformations for the computation $\frac{32}{11} = \frac{18}{11} + \frac{14}{11}$.

hence $\text{Zrange}(Q) = (\frac{27}{10}; \frac{37}{12}) \subseteq (2; 4)$, so $r = 3$ can be output and the cube appropriately transformed. After the remaining partial quotients have been input, one further partial quotient of value -11 can be output, and the algorithm terminates with the result $[3/-11] = \frac{32}{11}$ since now $\text{Zrange}(Q) = \{a/e\} = \{\infty\}$. □

We shall conclude this section by introducing a matrix notation for the transformations. Observe that the cube is a $2 \times 2 \times 2$ array (tensor) of integers, and by a simple generalized notion of matrix multiplication (tensor product) we may describe the input of a partial quotient p in the x direction as the product

$$\left\{\left\{\begin{matrix} d & & b & \\ & h & & f \\ c & & a & \\ & g & & e \end{matrix}\right\}\right\} \left\{\begin{matrix} 0 & 1 \\ 1 & p \end{matrix}\right\} = \left\{\left\{\begin{matrix} b & & pb+d & \\ & f & & pf+h \\ a & & pa+c & \\ & e & & pe+g \end{matrix}\right\}\right\}.$$

The input of partial quotient q corresponding to the substitution $y = q + 1/y$ can similarly be described as a multiplication by a matrix $\begin{Bmatrix} 0 & 1 \\ 1 & q \end{Bmatrix}$ on a properly transposed version of the cube array, and the output of r correspondingly as a multiplication by the matrix $\begin{Bmatrix} 0 & 1 \\ 1 & -r \end{Bmatrix}$.

Any sequence of transformations corresponding to the input of the k first partial quotients from, say, $x = [a_0/a_1/\cdots/a_n]$ then corresponds to a multiplication by the product of the equivalent matrices:

$$T_k = \begin{Bmatrix} 0 & 1 \\ 1 & a_0 \end{Bmatrix} \begin{Bmatrix} 0 & 1 \\ 1 & a_1 \end{Bmatrix} \cdots \begin{Bmatrix} 0 & 1 \\ 1 & a_k \end{Bmatrix}.$$

It is here easily seen that

$$T_k = \begin{Bmatrix} p_{k-1} & p_k \\ q_{k-1} & q_k \end{Bmatrix}, \qquad k = 0, 1, \ldots, n,$$

where p_k/q_k is the kth convergent of x.

9.7 Bit-serial arithmetic on rational operands

Heuristic arguments on information contents tells us that we must anticipate large delays in the output of a large partial quotient of $z(x, y)$. Such delays can be substantially reduced and a relatively smooth throughput achieved if we read and emit partial quotients bitwise, e.g., using the LCF representation. For this purpose consider first a more general variable transformation

$$x = \frac{\gamma + \alpha x'}{\delta + \beta x'}, \tag{9.7.1}$$

which when substituted into (9.6.1) again yields an expression of the same form as (9.6.1). Also here the transformation of the $2 \times 2 \times 2$ cube can be described as a generalized matrix multiplication, or *tensor product*:

$$\left\{\left\{\begin{matrix} d & & b & \\ & h & & f \\ c & & a & \\ & g & & e \end{matrix}\right\}\right\} \begin{Bmatrix} \delta & \beta \\ \gamma & \alpha \end{Bmatrix} = \left\{\left\{\begin{matrix} \gamma b + \delta d & & \alpha b + \beta d & \\ & \gamma f + \delta h & & \alpha f + \beta h \\ \gamma a + \delta c & & \alpha a + \beta c & \\ & \gamma e + \delta g & & \alpha e + \beta g \end{matrix}\right\}\right\},$$

Now consider the input of a partial quotient p from x, i.e., the substitution $x = p + 1/x$, where $p = 2^k + b_{k-1}2^{k-1} + \cdots + b_0$ with lexibinary representation $\ell(p) = 1^k 0 b_{k-1} \cdots b$, and note the following fundamental factorization:

$$\begin{Bmatrix} 0 & 1 \\ 1 & p \end{Bmatrix} = \begin{Bmatrix} 1 & 0 \\ 0 & 2 \end{Bmatrix}^{k-1} \begin{Bmatrix} 0 & 1 \\ 2 & 2b_k \end{Bmatrix} \begin{Bmatrix} \frac{1}{2} & \frac{b_{k-1}}{2} \\ 0 & 1 \end{Bmatrix} \cdots \begin{Bmatrix} \frac{1}{2} & \frac{b_0}{2} \\ 0 & 1 \end{Bmatrix}. \tag{9.7.2}$$

Thus instead of multiplying the coefficient cube by the matrix $\begin{Bmatrix} 0 & 1 \\ 1 & p \end{Bmatrix}$ it is possible to multiply by a sequence of matrices, each of which is in a one-to-one correspondence with one of the bits of $\ell(p)$. Also note that multiplication by any of these matrices is easily realized by shifts and adds of coefficients of the cube. Hence input operands x and y into the computation of $z(x, y)$ can be provided in the form of the LCF representations of x and y, reading individual bits of these. A simple state machine can keep track of the interpretation of each bit from each operand, and determine the appropriate matrix transformation to be performed on the coefficient cube. The variable substitution corresponding to each matrix is easily obtained from (9.7.1).

For output there is an equivalent factorization

$$\begin{Bmatrix} 0 & 1 \\ 1 & -r \end{Bmatrix} = \begin{Bmatrix} 1 & 0 \\ 0 & 2 \end{Bmatrix}^{k-1} \begin{Bmatrix} 0 & 1 \\ 2 & -2b_k \end{Bmatrix} \begin{Bmatrix} \frac{1}{2} & -\frac{b_{k-1}}{2} \\ 0 & 1 \end{Bmatrix} \cdots \begin{Bmatrix} \frac{1}{2} & -\frac{b_0}{2} \\ 0 & 1 \end{Bmatrix}, \tag{9.7.3}$$

corresponding to the output of a partial quotient r with $\ell(r) = 1^k 0 b_{k-1} \cdots b_0$, enabling bits of the LCF representation of the result $z(x, y)$ to be output bitwise. Note that input of bits from x and y can be interleaved arbitrarily, and output of the result performed whenever possible. This is due to the fact that variable transformations in x and y can be made in any order, and that these can be performed independently of the rewriting of $z(x, y)$ on output, where multiplication by

$$\begin{Bmatrix} \delta & \beta \\ \gamma & \alpha \end{Bmatrix}$$

corresponds to a rewriting:

$$z(x, y) = \frac{-\gamma + \alpha z'(x, y)}{\delta - \beta z'(x, y)} \quad \text{or } z'(x, y) = \frac{\gamma + \delta z(x, y)}{\alpha + \beta z(x, y)}.$$

To be able to determine output bits it is necessary to evaluate Zrange(Q) and compare it against certain intervals, which depend on the state of the output finite state machine, i.e., where in the LCF representation of $z(x, y)$ the bit to be omitted is located.

For this purpose we need an analysis of the intervals representing the “tails” of an LCF representation, i.e., the part not yet seen during reading. For convenience we shall assume that the LCF representation contains an end marker corresponding to the value infinity, otherwise we would have to deal with the infinite string $\overline{\ell(\infty)}$. Also when dealing with the encoding of a partial quotient $\ell(a_i)$ we assume that the bit-string appears in its proper (non-complemented) form.

We are here only concerned with non-negative numbers, so initially the domain is $[0; \infty]$. If the leading bit is zero, a reciprocation takes place corresponding to the substitution $x = 1/x'$, as realized by the matrix $\begin{Bmatrix} 0 & 1 \\ 1 & 0 \end{Bmatrix}$, and the domain of the tail is $[1;\infty]$, which is the general situation whenever the reading of a new partial quotient is to begin. However, the value 1 can only occur if the tail is $[1/\infty]$ with LCF expansion 0#, where # is the end marker, and the value ∞ can only occur at the end marker.

So assume the domain is $(1;\infty)$ and let us analyze the reading of $\ell(a_i) = 1^k 0 b_{k-1} \cdots b_0$. The reading of 1 from the unary part implies that the domain is $[2;\infty)$, and reading 0 implies the unary part is empty and the domain is $(1;2)$. In each case an appropriate variable transformation is performed and a new domain is obtained. The complete parsing of $1^k 0 b_{k-1} \cdots b_0$ is most conveniently displayed in a diagram as shown in Figure 9.7.1, depicting the state transitions with corresponding variable substitutions and resulting domain transformations.

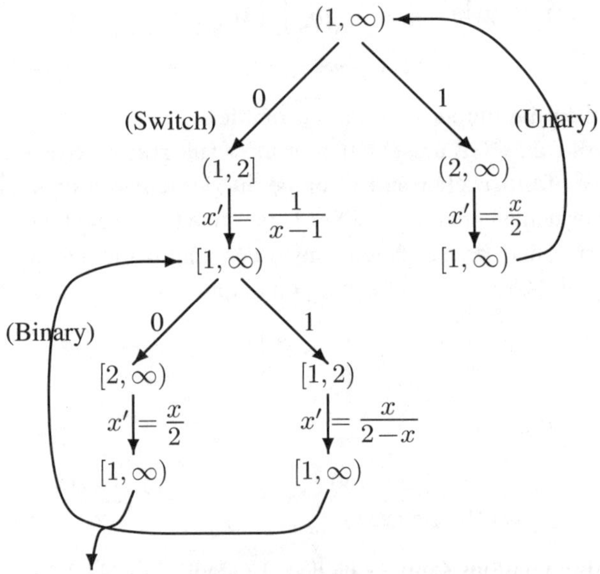

Figure 9.7.1. Domain transformations during reading of $\ell(a_i)$.

We can now return to the problem of determining bits of the output, where we note from the above analysis that we can assume that $D_x = D_y = [1;\infty]$, since we will assume that some input has been taken from both x and y before any attempt to perform output.

Thus it is possible to determine Zrange(Q) by evaluating $z(x, y)$ at the four corners of $D_x \times D_y$, i.e., at $(1, 1)$, $(1, -1)$, $(-1, 1)$, and $(-1, -1)$, provided that $z(x, y)$ is well defined over $D_x \times D_y$. Initially, if it can be determined that Zrange(Q) $\subseteq [0; 1)$, then a zero bit can be output, and if Zrange(Q) $\subseteq [1;\infty)$ a unit bit can be omitted. If neither of these decisions can be made, more input must be taken until Zrange(Q) falls in one of the decision intervals. Subsequently, starting to emit further bits of the encoding of a partial quotient, unary bits can be emitted while it can be determined that Zrange(Q) $\subseteq (2, \infty)$ as seen from Figure 9.7.1. When it is determined that Zrange(Q) $\subseteq (1;2]$ the switch bit is emitted, followed by the appropriate number of bits from the binary part as these can be selected. From Figure 9.7.1 it is found that Zrange(Q) $\leq [2;\infty)$ implies that a zero, and Zrange$\subseteq$ $[1;2)$ implies that a unit bit can be selected. Every time a bit is selected the appropriate matrix is multiplied onto the cube in the z direction, and whenever no decision is possible more input is taken from x and/or y.

Having sketched an algorithm for bit-serial arithmetic on rational operands in LCF representation, let us take a look at the performance of this algorithm. By representing each partial quotient a_i by its lexibinary encoding $\ell(a_i)$, the algorithm can digest information from its operands and produce the result bitwise. Each bit of the unary part doubles a lower bound on the partial quotient, the switch bit establishes an upper bound, and subsequent bits of the binary part in effect perform a binary search, narrowing the interval established. Thus the information flow has been granularized, which is a condition for a smooth behavior of the algorithm.

However, there is still a fundamental problem when the algorithm has to emit the LCF representation of a result, since the selection of output bits is based on the decision on whether Zrange(Q) is included in one or the other of two disjoint intervals. An arbitrary amount of input from both operands perhaps even until their termination, may be required before such a decision can be made.

Example 9.7.1 Let x and y be some rational approximations of $\sqrt{2} = [1/2/2/2\cdots]$. Until x or y terminates it is not possible to determine whether their product is $[2/k]$ or $[1/1/m]$, where k and m are (large) integers. Their LCF representations are

$$LCF([2/k]) = 1 \circ 100 \circ \overline{11\cdots 10\cdots} = 110\cdots 01\cdots,$$
$$LCF([1/1/m]) = 1 \circ 0 \circ \bar{0} \circ 11\cdots 10\cdots = 101\cdots 10\cdots,$$

which are approximations from above and below of

$$LCF([2]) = 1 \circ 100 = 110\cdots.$$

The algorithm will become stuck after the output of the first (reciprocation) bit, not knowing whether to output a unary bit or a switch bit. And the situation cannot be resolved before one of the operands has been terminated and it is known whether the result xy is below or above 2. □

The problem is obviously that the LCF representation is non-redundant; this is the case for the continued fraction expansion used, as well as the lexibinary representation used for the individual partial quotients. The example above illustrates the former case, but similar situations may occur within a partial quotient. In the latter case the ambiguity will be resolved when the complete partial quotient can be determined, but again since this can be of any value the delay can be of arbitrary size. The solution is then to introduce redundancy at the quotient level, as well as in the encoding of the individual quotients.

Recall from Section 9.4 that the redundant signed continued fractions defined by the restriction $|[a_i/a_{i+1}/\cdots/a_n]| > 1$ for $1 \le i \le n$ by Lemma 9.4.4 satisfy the constraint:

$$|a_i| = 1 \Rightarrow \operatorname{sgn}(a_i) = \operatorname{sgn}(a_{i+1}) \quad \text{for } 1 \le i \le n-1. \tag{9.7.4}$$

The domain of the tail $t_i = [a_{i+1}/a_{i+2}/\cdots/a_n]$ is $|t_i| \in (1;-1)$, the affine interval with infinity included, but restricted when $|a_i| = 1$ for $1 \le i \le n-1$ such that

$$\begin{aligned} a_i &= 1 \Longrightarrow t_i \in (1;\infty] \\ a_i &= -1 \Longrightarrow t_i \in [-\infty;-1). \end{aligned}$$

Note, however, that restriction (9.7.4) does not apply for $i = 0$ where $|a_0| = 1 \Rightarrow t_0 \in (1,-1)$.

Since we are dealing with affine intervals where $-\infty = \infty$ it is convenient when discussing such domains to picture these using the *stereographic projection* of the reals upon the unit circle (in the complex plane):

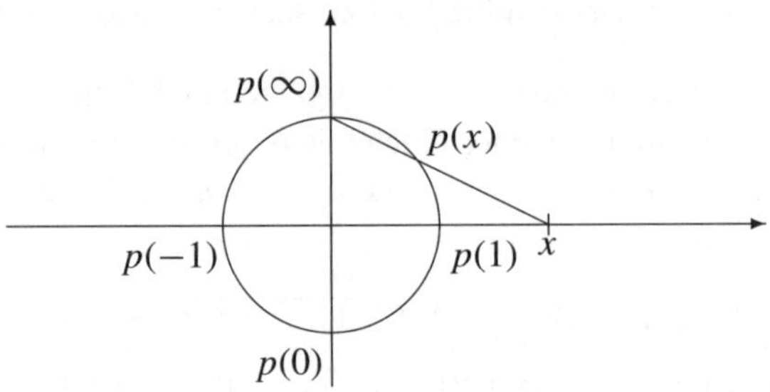

whereby $-\infty$ and ∞ are mapped into the same point. As an illustration the following figure shows the mapping of some reals:

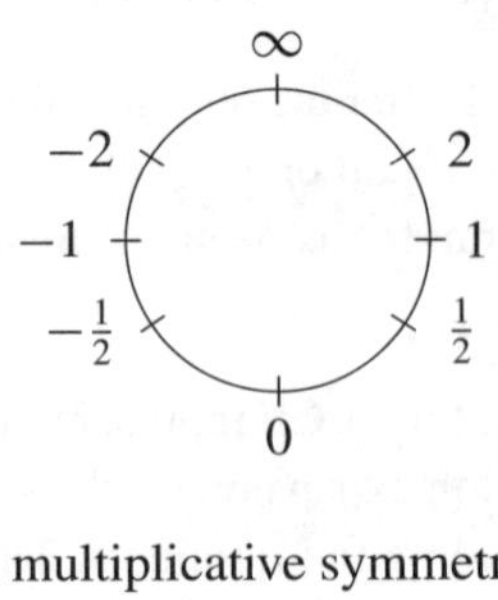

illustrating the additive and multiplicative symmetries of the rationals.

For the representation of partial quotients of a signed continued fraction we need an encoding of signed partial quotients, which we want also to be redundant. Using the digit set $\{-1, 0, 1\}$ we can use the principles of the lexibinary representation if we employ an additional symbol u for the unary part encoding the length of the representation. Thus if the (signed) integer p can be represented as $b_k b_{k-1} \cdots b_0$ with $|b_k| = 1$ and $b_i \in \{-1, 0, 1\}$ for $i = 0, \cdots, k-1$, we can introduce the following *admissible* self delimiting representation of the value p:

$$R(p) = u^{k-1} b_k b_{k-1} \cdots b_0 \quad \text{for } p \neq 0,$$
$$R(0) = 0$$

with $|b_k| = 1$ and the further range restriction

$$2^{k-1} + 1 \leq |p| \leq 2^{k+1} - 1 \quad \text{for } p \neq 0 \text{ and } k \geq 2.$$

This restriction prevents excessive redundancy in the representation in the prefix of the binary part, e.g., as in $1\bar{1}\bar{1} \cdots \bar{1} = 00 \cdots 01$, but does allow $R(3) = 11$ as well as $R(3) = u1\bar{1}1$, and in general strings with $b_k b_{k-1} = 1\bar{1}$ or $\bar{1}1$, but only if the sign of $b_{k-2} \cdots b_0$ agrees with that of b_k. Note that 0 and $\pm 2^i$, $i \geq 0$ have unique representations, e.g., $0 \sim 0,\ 1 \sim 1\bar{1},\ 2 \sim 10,\ 4 \sim u100, \cdots$.

For any (redundant) signed continued fraction $x = [a_0/a_1/\cdots/a_n]$ we then obtain an *admissible string* for x by concatenation

$$R(x) = R(a_0) \circ R(a_1) \circ \cdots \circ R(a_n),$$

whenever $R(a_i)$ is an admissible representation of a_i, $0 \leq i \leq n$. We now have redundancy at the quotient level, as well as in the representation of the individual quotients in what we will call the *RPQ representation*.

The value represented by an admissible string is uniquely determined by reading it from left to right; however, any rational except 0 and ± 1 has a multitude of representations as admissible strings. Some examples of admissible strings representing $\frac{18}{11}$ are

$$\begin{aligned} R\left(\tfrac{18}{11}\right) &= R([2/-3/4]) &&= 10u\bar{1}1\bar{1}u100 \\ & &&= 10\bar{1}\bar{1}u100, \\ &= R([1/1/2/-4]) &&= 1\bar{1}1\bar{1}10u\bar{1}00. \end{aligned}$$

For the purpose of digit-serial processing an end marker can be added to represent a final partial quotient of infinite value, or any string of us followed by the end marker will have the same effect.

It is now possible to construct a digit-serial algorithm for arithmetic operations in this representation. In analogy with the previous algorithm we need a factorization of the matrix $\begin{Bmatrix} 0 & 1 \\ 1 & p \end{Bmatrix}$ for a (signed) partial quotient corresponding to an admissible representation $R(p) = u^{k-1}b_k b_{k-1} \cdots b_0$ for $p \neq 0$

$$\begin{Bmatrix} 0 & 1 \\ 1 & p \end{Bmatrix} = \begin{Bmatrix} 1 & 0 \\ 0 & 2 \end{Bmatrix}^{k-1} \begin{Bmatrix} 0 & 1 \\ 2 & 2b_n \end{Bmatrix} \begin{Bmatrix} \frac{1}{2} & \frac{b_{n-1}}{2} \\ 0 & 1 \end{Bmatrix} \cdots \begin{Bmatrix} \frac{1}{2} & \frac{b_0}{2} \\ 0 & 1 \end{Bmatrix},$$

where again there is a one-to-one correspondence between the digits of $R(p)$ and matrices in the factorization. For $0 < |x| < 1$, the initial partial quotient $a_0 = 0$ is encoded as 0 with corresponding matrix $\begin{Bmatrix} 0 & 1 \\ 1 & 0 \end{Bmatrix}$, but note that a value in this interval can also be represented by an initial partial quotient $|a_0| = 1$ followed by a tail of the opposite sign.

We are now able to state an algorithm which, given a function

$$z(x, y) = \frac{axy + bx + cy + d}{exy + fx + gy + h}$$

in terms of initialization of the cube Q by the 8-tuple of integer coefficients, and an appropriate function Zrange(Q) over the domains of the tails of x and y, by repeated applications determines the signed continued fraction expansion of $z(x, y)$ in the form of an admissible string $R(z(x, y))$. Our algorithm uses a loop construct where the guards must be evaluated in the order listed.

Algorithm 9.7.1 Redundant partial quotient (RPQ) algorithm
{*This algorithm determines a string* $u^{k-1}b_k b_{k-1} \cdots b_0$ *which is an admissible representation* $R(r)$ *of the next partial quotient* r *of* $z(x, y)$, *as specified by the given coefficient cube* Q, *and admissible input strings* $R(x)$ *and* $R(y)$.}
k:=1;
loop

Zrange(Q) $\subset (-1; 1)$: {*output 0; k:=0;*
perform transf. $\begin{Bmatrix} 0 & 1 \\ 1 & 0 \end{Bmatrix}$;
exit loop};

Zrange(Q) $\subset (-4; 0)$: {*output* $\bar{1}$;
perform transf. $\begin{Bmatrix} 0 & 1 \\ 2 & 2 \end{Bmatrix}$;
exit loop};

Zrange(Q) $\subset (0; 4)$: {*output 1;*
perform transf. $\begin{Bmatrix} 0 & 1 \\ 2 & -2 \end{Bmatrix}$;
exit loop};

Zrange$(Q) \subset (2;-2)$: {*output u; k:=k+1;*

perform transf. $\begin{Bmatrix} 1 & 0 \\ 0 & 2 \end{Bmatrix}$};

true : {*take more input from x or y*};

end loop;

Assert{Zrange$(Q) \subset (1;-1)$ }

loop

k=0 : {*exit loop*};

Zrange$(Q) \subset (-\infty;-1)$: {*output* $\bar{1}$*; k:=k-1;*

perform transf. $\begin{Bmatrix} \frac{1}{2} & \frac{1}{2} \\ 0 & 1 \end{Bmatrix}$};

Zrange$(Q) \subset (1;\infty)$: {*output 1; k:=k-1;*

perform transf. $\begin{Bmatrix} \frac{1}{2} & -\frac{1}{2} \\ 0 & 1 \end{Bmatrix}$};

Zrange$(Q) \subset (2;-2)$: {*output 0; k:=k-1;*

perform transf. $\begin{Bmatrix} \frac{1}{2} & 0 \\ 0 & 1 \end{Bmatrix}$};

true : {*take more input from x or y*};

end loop;

Assert{*Zrange*$(Q) \subset (1,-1)$ }

After completion of the algorithm Q has been transformed into a new cube Q' representing $z'(x,y) = 1/|z(x,y) - r)$ obtained by multiplication in the z-direction by the matrix

$$\begin{Bmatrix} 0 & 1 \\ 1 & -r \end{Bmatrix}.$$

Theorem 9.7.2 *The output* $R(r) = u^{k-1} b_k b_{k-1} \cdots b_0$ *of the RPQ Algorithm is an admissible string for the integer* r*. By repeated applications of the algorithm until termination of input* $R(x)$ *and* $R(y)$ *an admissible string* $R(z(x,y)) = R(a_0) \circ R(a_1) \circ \cdots \circ R(a_n)$ *is obtained.*

Proof Since we shall later see that Zrange$(Q) \subset (1;-1)$ upon exit of the algorithm, it is only possible at the very first call to find Zrange$(Q) \subset (-1;1)$, in which case 0 is emitted and the roles of the numerators and the denominators are switched. Now Zrange$(Q) \subset (1;-1)$ with $k = 0$, so the second loop terminates immediately.

If either 1 or $\bar{1}$ is emitted, then the first loop exits with $k = 1$ and the second loop will output $\bar{1}, 0$ or 1. Any of the resulting strings $\bar{1}\bar{1}$, $\bar{1}0$, $\bar{1}1$, $1\bar{1}$, 10, 11 is then admissible. If in the first loop the guard Zrange$(Q) \subset (-4;0)$ was satisfied, then by the transformation performed, $z' = 2/(z+2)$, the interval $(-4;0)$ is mapped into $(1;-1)$. Similarly all intervals are mapped into $(1;-1)$ as illustrated in the following diagram of the transformation in the first loop:

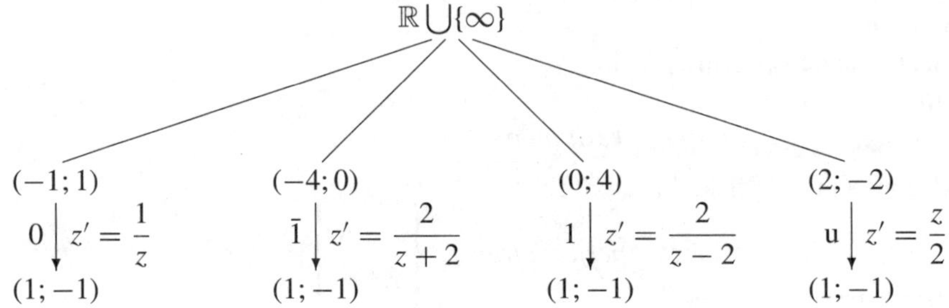

During the second loop intervals are mapped as follows:

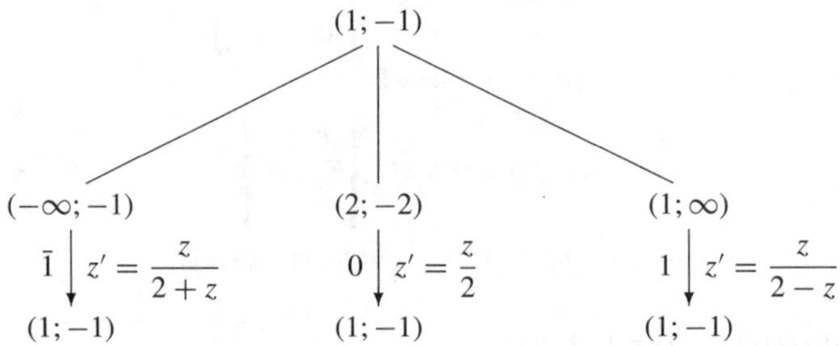

Hence in all cases the transformed cube Q' also satisfies Zrange$(Q') \subset (1;-1)$. If during the first loop one or more us are emitted, the loop exits emitting 1 or $\bar{1}$ with $k \geq 2$, and the second loop completes by emitting a string of symbols from $\{\bar{1}, 0, 1\}$. In these cases it is found that an admissible string has been emitted for the first partial quotient.

Observe that for later applications of the algorithm, and in the case when the first loop cycles back after emitting u, then the interval in the start of the first loop is $(1, -1)$. Thus we have the following diagram with restricted intervals:

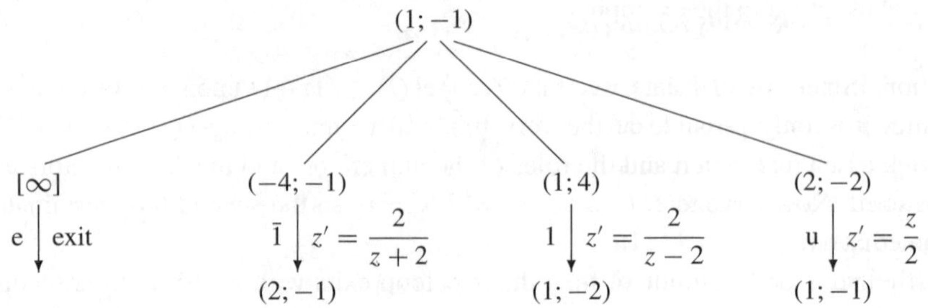

Now we can observe that if 1 is emitted in the first loop and if in the second loop Zrange(Q) $\subset (1; -2) \cap (-\infty; -1) = (-\infty; -2)$, then $\bar{1}$ will be emitted and Zrange(Q') $\subset (1; \infty)$ for the transformed Q'. Hence, following $1\bar{1}$, the tail of $z(x, y)$ is in $(1; \infty)$. If $1\bar{1}$ is the complete partial quotient (of value 1), then the following partial quotient will have the same sign, and if $1\bar{1}$ is part of a string $u^{k-1}1\bar{1}b_{k-2}\cdots b_0$, then this string is admissible since the string $b_{k-2}\cdots b_0$ must represent a positive value. If $\bar{1}$ is emitted in the first loop followed by 1 in the second, the same argument applies, and we have thus proven that the algorithm generates admissible representations of the individual partial quotients, and consecutive calls emits admissible strings. □

Note that decisions now are based on overlapping intervals, thus it will never be necessary to restrict Zrange(Q) to a single point before a decision can be made. The decision intervals for the RPQ Algorithm can be summarized in the following diagrams using the stereographic projection:

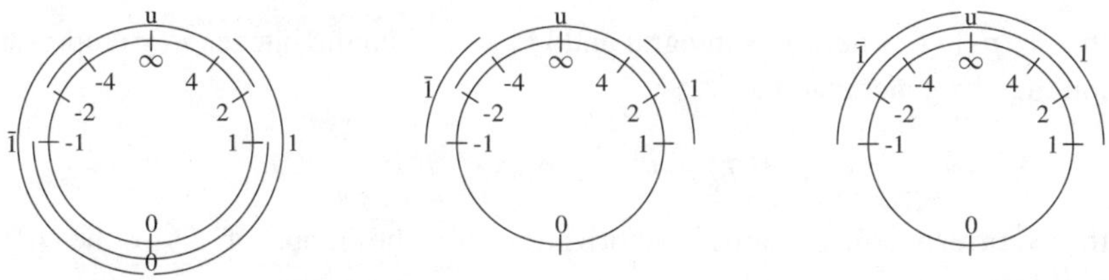

From the analysis of the previous proof we can also make some observations on the domains D_x and D_y to be used in the computation of Zrange(Q). Since we assume that some input from x and y has been taken before any attempt to perform output is made, it is obvious that $D_x \subset (1; -1)$ and $D_y \subset (1; -1)$. A simple Zrange(Q) evaluation may thus be based on the assumption that $(x, y) \in (1; -1) \times (1; -1)$. However, as noted in the proof, after the string $u^{k-1}1\bar{1}$, $k \geq 1$, the tail belongs to the domain $D = (1; \infty)$, and similarly the tail is in $(-\infty; -1)$ following the string $u^{k-1}\bar{1}1$, $k \geq 1$. Finally, when one of the operands x or y has terminated, the corresponding domain is reduced to the single point ∞.

Let us return to the example of forming the product of two rational approximations of $\sqrt{2} = [1/2/2/2/\cdots] = 1\bar{1}10101010\cdots$, but now in the RPQ representation. In the case of the LCF representation we saw that, due to the lack of redundancy, it is not possible to determine whether to emit a first partial quotient of value 1 or 2, i.e., yielding a result of either the form $[2/k/\cdots]$ or the form $[1/1/m\cdots]$ for some (large) values of k and m. Note that this is possible in the RPQ representation, emitting either $1\bar{1}1\bar{1}uuu\cdots$ or $10uuu\cdots$, where the number of us depends on the order of magnitude of k and m. If computing is in the real domain, then the product of the two infinite expansions of $\sqrt{2} = [1/2/2/2/\cdots] = 1\bar{1}10101010\cdots$

will yield a result in one of the two above forms, with an infinite number of us, corresponding to either $[1/1/\infty]$ or $[2/\infty]$, both being valid representations of 2. In general, any rational having a finite continued fraction expansion, can in the real domain be represented in RPQ with an infinite string of trailing us.

Observation 9.7.3 *The RPQ representation allows for infinite expansions to represent all computable reals, with rational numbers terminating in infinite strings of us. The RPQ cell (allowing arbitrary integer values as coefficients) is able to perform all standard arithmetic operations $+, -, \times, /$ on operands from the real domain.*

We conclude this section with some observations generalizing the integer representation $R(p)$ and the factorization (9.7.2) to non-integral, binary radix representations. First consider a (possibly redundant) binary representation

$$x = b_n b_{n-1} \cdots b_0 b_{-1} \cdots b_{-k},$$

with $x = p + f$, where p is integral and $|f| < 1$, then the integer $R(p)$ representation may be generalized to

$$R'(p+f) = u^{n-1} b_n b_{n-1} \cdots b_0 b_1 \cdots b_{-k}$$

with a corresponding matrix factorization for the input of x to the RPQ Algorithm:

$$\left\{\begin{matrix} 0 & 2^k \\ 1 & 2^k x \end{matrix}\right\} = \left\{\begin{matrix} 1 & 0 \\ 0 & 2 \end{matrix}\right\}^{n-1} \left\{\begin{matrix} 0 & 1 \\ 2 & 2b_n \end{matrix}\right\} \left\{\begin{matrix} \frac{1}{2} & \frac{b_{n-1}}{2} \\ 0 & 1 \end{matrix}\right\}$$
$$\cdots \left\{\begin{matrix} \frac{1}{2} & \frac{b_0}{2} \\ 0 & 1 \end{matrix}\right\} \left\{\begin{matrix} 1 & b_{-1} \\ 0 & 2 \end{matrix}\right\} \cdots \left\{\begin{matrix} 1 & b_{-k} \\ 0 & 2 \end{matrix}\right\}. \qquad (9.7.5)$$

Note that the matrix product has integral values, due to a compensatory scaling of the matrix by 2^k, which does not affect the value of $z(x, y)$ as given by (9.6.1). The RPQ Algorithm is to be modified such that the second loop emitting the binary part is allowed to continue beyond $k = 0$, when each factor matrix has been multiplied by 2. Also observe that we have implicitly assumed that the integral part p is non-zero, but that does not prevent us from representing a number x with $0 < |x| < 1$. A number in the interval $0 < x < 1$ with the binary representation $x = 0 \cdot 0 \cdots 01 b_{-j} \cdot b_{-k}$ can be represented as $R'(X) = 1\bar{1}\bar{1} \cdots \bar{1} b_{-j} \cdots b_{-k}$, since the RPQ Algorithm in the initial (and only) call can emit $b_1 b_0 = 1\bar{1}$ with a transformed Zrange$(Q) \subset (-\infty; -1)$, just as $a_0 = 1$ can be followed by a negative a_1 in a redundant signed continued fraction. Thus the RPQ Algorithm can be modified to take either or both of its operands in the radix representation $R'(\cdot)$ as well as in the continued fraction representation $R(\cdot)$, and that output can be produced in either of these representations.

9.8 The RPQ cell and its operation

A computation cell performing the RPQ Algorithm, given its operands x, y and producing its result $z(x, y)$ in digit serial form, can be depicted as

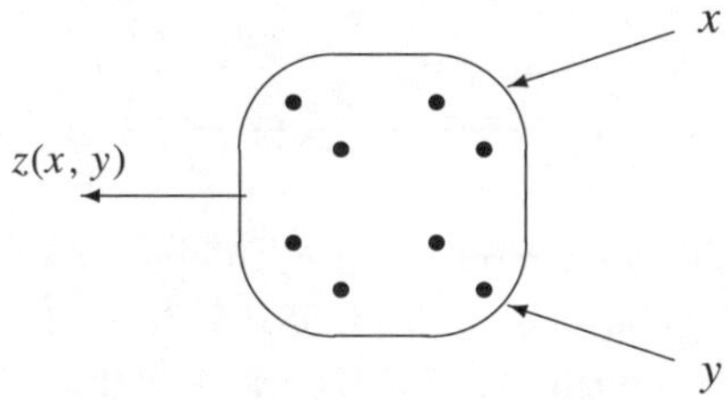

The function $z(x, y)$ computed is specified by the initialization of the 8-tuple of registers of the cell (the cube). The operands and the result can be in any combination of the redundant $R'(\cdot)$ radix or the $R(\cdot)$ rational representation, as specified in the state machines controlling the input interpretation and the output selection (including here the Zrange computation). Operands or result could even be in the non-redundant LCF representation, if there were appropriate changes in the state machines.

An on-line arithmetic unit is intended to operate by inputting digits of the arguments and outputting digits of the result with little delay and considerable regularity. One characterization of a standard on-line unit is that in order to generate the kth digit of the result, it is necessary and sufficient to input up to $k + \delta$ digits of the operands, where δ is some small positive integer. Typical values for δ are of the order 1–4, depending on the arithmetic operation $(+, -, \times, /)$ and on the radix of the redundant representation employed. Units based on this property will impose an initial delay of δ digits and then produce output on a regular basis, with one output digit per input digit tuple (i.e., one from each operand).

A cell modeled on the RPQ Algorithm will always favor output. A digit pipelined sequence of operations will then have units downstream initiated earlier in the computation process. Some loss of regularity occurs in this model. The availability of output given additional input can vary, even though the average number of output digits per input digit is close to unity. Regularity can be measured quite simply by investigating the distribution of local delay, i.e., the frequency of the size $0, 1, 2, \cdots$ of the number of additional digits of input between successive output digits.

Delays due specifically to the nature of the rational representations can be analyzed by considering the conversion of a binary radix string $R'(x)$ to an admissible string $R(x)$. The unit is implicitly a binary radix-to-rational converter when

Table 9.8.1. *Delay distribution (in %) for radix-to-rational conversion*

Delay	RPQ	LCF
0	34.1	41.8
1	49.8	36.8
2	13.2	11.6
3	2.5	5.3
4	0.4	2.5
5	0	1.1
≥6	0	0.9

adding $R'(0) = 0e$ to $R'(x)$ with rational string output $R(x)$. A simulation of the RPQ Algorithm for the conversion of $55005/65536 = .1101011011011101$, where digits are from the alphabet {u,0,1,m,e} with m denoting $\bar{1}$ and e denoting termination, yields:

```
 a>0e
 b>..0.1..1.0.1.0.1.10...1..101..1..10.1...e
 c>   0 1m u 1 1 m u  u10 mm   m0 uu  u 1m1 1m1mu11m1m
    123456789 123456789 123456789 123456789 123456789 12
```

Note that the gaps between the 21 output digits of $R(x)$ before input termination showed 1 three-digit delay, 2 two-digit delays, 9 one-digit delays, and 8 zero-digit delays (no further input before output). The percentage distribution of such binary-to-rational on-line delays for 1000 conversions of 16-bit numbers (selected randomly over $[0, 1 - 2^{-16}]$) is shown in Table 9.8.1.

For comparison we have also included delay distributions for the non-redundant LCF rational representation. Note that the LCF representation shows k-digit delays occurring with a frequency of approximately $1/2^{k+1}$. This distribution is characteristic of non-redundant binary delays as illustrated by the equivalent of the wait to determine (parametrically in k) which of the two k-digit strings, $011\cdots 1$ or $100\cdots 0$ is next to be output. The RPQ admissible strings show quite reasonable regularity, with no delays over four digits, and delays over two digits in less than 3% of all output. This performance approaches that of redundant binary radix representation, where delays over two digits can be shown never to occur.

The delay will become somewhat more irregular when rational representation $R(\cdot)$ is used for both input and output with varying arithmetic operations, particularly for units downline in a tree pipelined computation. This is best shown by some examples, the first being a simple multiplication $\frac{328}{145} \times \frac{27}{101} = \frac{8856}{14645}$. Here the operands employ the following representations:

$$R(\tfrac{328}{145}) = 11\bar{1}1\bar{1}\bar{1}u11\bar{1}11\bar{1}1\bar{1}0 \text{ and } R(\tfrac{27}{101}) = 0111\bar{1}11u\bar{1}\bar{1}\bar{1}.$$

A simulation of the RPQ Algorithm then gives the following output:

```
a>.1..1..m...1...m.m..u...1.1....m.1...1.m.1m.0.e
b>0.1..1...1...m..1.1...u..m.m....m...e
c>   1  m u m 0 1    1 0 u    u10   00  u u  m 0 m1uummm111e
  123456789 123456789 123456789 123456789 123456789 123456789
```

Each line here represents the flow along the arcs into or out of the cell, where the horizontal position represents the time step at which the signed digit is on the arc (i.e., consumed). For example, at step 4 the first digit of the result is produced, based on the availability of the 1 digit consumed from operand b at step 3. The periods indicate that a signed digit is available, but has not been consumed. As may be noticed, if both operands are available the individual cells are servicing their input arcs in a round-robin manner.

As the second example we consider a tree-pipelined computation of the expression $\frac{8}{5} \times \frac{3}{2} + \frac{7}{9} \times \frac{1}{5} = \frac{23}{9}$, as depicted in the tree below:

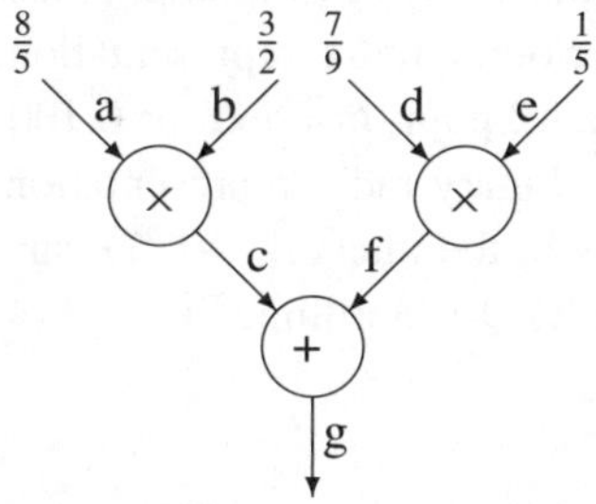

The redundant representations of the operands used are:

$$R(\tfrac{8}{5}) = 1\bar{1}1\bar{1}1\bar{1}10,\ R(\tfrac{3}{2}) = 1\bar{1}10,$$
$$R(\tfrac{7}{9}) = 01\bar{1}1110,\quad R(\tfrac{1}{5}) = 0u11\bar{1}.$$

The computation in the cells performing multiplications may proceed in parallel, producing the input operands for the cell doing the addition. In each cell the eight registers of the coefficient cube are initialized to realize the appropriate operation for that cell. The result of a simulation illustrates the flow of information into and out of the cells.

```
a>.1.m..1.m..1...m........10e
b>1.m..1.0..e
c>    1    1  ...m .1..m..0   1.0.e
d>.0....1.m..1.......1....1...0e
e>0..u...1.1...m.....e
f>  0 .u    u1  ...0  m.1 .m0  u.1.......00..e
g>:: : :::   u10 m  m   0   :     um00uu  uu e
  123456789 123456789 123456789 123456789 123456789
```

Note that at step 37 the last digit of the result has actually been produced, but since not all information produced on arc f has been consumed the add node continues to emit u's until all information on arc f has been input. This is due to the algorithm preferring to produce output when possible, rather than reading more input. The colons indicate that the cell is idle, waiting for input and not being able to output. Note that the final cell was idle for only 7 of the 45 cycles, and only once after initiating its own output.

As a final example we will repeat the first example, $\frac{328}{145} \times \frac{27}{101} = \frac{8856}{14645}$, however, this time producing the result in radix representation:

```
a>.1..1..m...1..m.m..u..1...1..m..1...1..m1.m.0e
b>0.1..1...1..m..1..1..u..m..m...m..e
c>   1  m 0 m  m   0  m  1 0  m 0  m 0 10  m 1  1mm1m111111mm
  123456789 123456789 123456789 123456789 123456789 123456789
```

where the simulation has been forced to a stop, since the result obviously has a non-terminating (but periodic) radix representation. Note that the result is $1\bar{1}.0\bar{1}\bar{1}0\bar{1}10\cdots$ with the radix point inserted, or $0.1001101\cdots$ when converted to non-redundant, standard binary radix representation. This example then also demonstrates how numbers in the interval $(-1, 1)$ can be generated in the $R'(\cdot)$ radix representation by the RPQ Algorithm.

9.9 Notes on the literature

The presentation in this chapter is to a large extent based directly on a series of publications by the authors.

Continued fractions have been extensively studied in classical mathematics, e.g., [Khi35, Wal48], and so has the Euclidean Algorithm [Hei68, Dix70, Ran73, Bre76], and Knuth [Knu98] covers continued fractions and the Euclidean Algorithm, including its complexity quite extensively. Trivedi investigated a special class of continued fractions in [Tri77].

The original ideas for fixed- and floating-slash representations and mediant rounding were published by Matula in [Mat75b, Mat75a]. These were further elaborated with detailed studies of the number systems and arithmetic in [MK78, KM78, MK80]. The idea of an arithmetic unit combining a rounding of a previous result with a following arithmetic operation is described by Kornerup and Matula in [KM81, KM83], and the number systems are described in [MK85]. [Sco89] describes an algorithm for the mediant rounding in multiprecision arithmetic, and [PM91] investigates the CGD algorithm applied to redundant representations corresponding to Algorithm 9.4.3. [FM85] analyzes the special properties of the mediant rounding, based on simulations of slash arithmetic used for matrix computations.

In [Hor78] Horn proposed the simple rounding rule just to truncate numerators and denominators, as an alternative to the mediant rounding, for use in minicomputers without floating-point arithmetic. As discussed in connection with the mediant rounding, this is a very poor choice. The mediant rounding has been used in the muMath [RS79] computer algebra system for an approximate rational arithmetic. [Sco89] describes an implementation of the mediant rounding in multiprecision arithmetic. Thill in [Thi08] investigated the use of "intermediate fractions" as alternative "best approximations."

In 1935 Khinchin said in his famous short monograph [Khi35]

"for continued fractions there are no practically applicable rules for arithmetical operations; even the problem of finding the continued fraction for the sum from the continued fraction representation representing the addends is exceedingly complicated, and unworkable in computational practice."

However, in 1947 Hall[Hal47] discovered that this is, in fact, possible, using the bi-linear form (or Möbius transformation) $y = (ax + b)/(cx + d)$, as described in Section 9.4.

Gosper continued from there, and came up with the idea of using the bi-homographic form, as published in [Gos72] among other memos in a collection of "Hackers Memos" from MIT. These were further elaborated in [Gos78], a typescript which unfortunately was never finished and properly published, but photocopies have been circulated informally. Here he also suggested a binary representation called "Continued Logarithms," based on a sequence of integer terms k_i (encoded in unary), developed from $a = a_0$ defined by $a_i = 2^{k_i}(1 + 1/a_{i+1})$, $k_i = \lfloor \log_2 a_i \rfloor$. This representation has been investigated in detail by Brabec [Bra10], looking in particular at how redundancy may be exploited. Seidensticker [Sei83] also investigated continued fraction arithmetic based on Gosper's work.

The LCF representation was presented in [MK83, MK85], whereas the detailed analysis of the LCF tree, bijacency, etc., was published much later in [KM95]. As the LCF representation is non-redundant, despite many useful properties, it is not very suitable for an on-line arithmetic unit. Such possibilities were described in [KM88] and in particular the coefficient cube and the RPQ Algorithm in [KM89, KM90]. [NK95] develops a formal account of digit serial MSB-first number representations, by describing them as strings from a language. A prefix of a string then represents an interval approximating a number by enclosure, and prefixes of increasing length must represent contracting intervals. [Hec02] also discusses the contractivity of such representations based on tensor products.

The tree of all Farey fractions obtained by taking mediants as an infinite tree is called a Stern–Brocot tree, discussed by Knuth and coworkers in [GKP94, Knu98], and as a foundation for verifiable numerical calculations and exact arithmetic by Niqui in [Niq04].

Continued fractions are also being used as representations of the computable reals in computer algebra systems. In [Vui90] J. Vuillemin discussed various continued fraction representations and arithmetic for this purpose, using redundancy by allowing zero-valued partial quotients in the expansions. He applied a "normalization" process to expansions containing such partial quotients, but this redundancy allows any previously emitted sequence to be annihilated. [MM94] studies the same problems, and in [Les01, LG03] it is proved that under suitable conditions on the use of zero-valued partial quotients, it is possible to insure progress in the sense that the intervals represented are strictly contracting. Potts and Edalat at Imperial College, London, also have a series of publications on exact real arithmetic, e.g., [PE97, Pot98]. These efforts are all based on integer-valued partial quotients, where the only way of emitting "partial information" on a quotient value is to emit a preliminary value, and then modify it by subsequently emitting a zero followed by a correction value.

References

[Bra10] T. Brabec. Speculatively redundant continued logarithm representation. *IEEE Trans. Computers*, to appear, 2010.

[Bre76] R. P. Brent. Analysis of the binary Euclidean algorithm. In J. Traub, editor, *Algorithms and Complexity*, pages 321–355. Academic Press, 1976.

[Dix70] J. D. Dixon. The number of steps in the Euclidean Algorithm. *J. Number Theory*, 2:414–422, 1970.

[FM85] W. E. Ferguson and D. W. Matula. Rationally biased arithmetic. In *Proc. 7th IEEE Symposium on Computer Arithmetic*, pages 194–202. IEEE Computer Society, 1985.

[GKP94] R. E. Graham, D. E. Knuth, and O. Patashnik. *Concrete Mathematics*. Addison-Wesley, second edition, 1994.

[Gos72] R. W. Gosper. Item 101 in Hakmem. *AIM239, MIT*, pages 37–44, April 1972. Available at: www.inwap.com/pdp10/baker/hakmem/hakmem.html.

[Gos78] R. W. Gosper. Continued fraction arithmetic. Unpublished draft paper, available at: http://www.tweedledum.com/rwg/cfup.htm, 1978.

[Hal47] M. Hall. On the sum and product of continued fractions. *Ann. Math.*, 48(4):966–993, October 1947.

[Hec02] R. Heckmann. Contractivity of linear fractional transformations. *Theor. Computer Sci.*, 279:65–81, 2002.

[Hei68] H. Heilbronn. On the average length of a class of finite continued fractions. In P. Turan, editor, *Abhandlungen aus Zahlentheorie und Analysis*, pages 89–96. VEB Deutcher Verlag der Wissenschaften, 1968. Also published by Plenum Press, 1968.

[Hor78] B. K. P. Horn. Rational arithmetic for minicomputers. *Software-Practice and Experience*, 8:171–176, 1978.

[Khi35] A. Y. Khinchin. *Continued Fractions*. State Publishing House of Physical-Mathematical Literature, Moscow., 1935. Translation from Russian by P. Wynn,

P. Noordhoff Ltd., 1963. Also by H. Eagle, The University of Chicago Press, 1964.

[KM78] P. Kornerup and D. W. Matula. A feasibility analysis of fixed-slash and floating-slash arithmetic. In *Proc. 4th IEEE Symposium on Computer Arithmetic*, pages 39–47. IEEE Computer Society, 1978.

[KM81] P. Kornerup and D. W. Matula. An integrated rational arithmetic unit. In *Proc. 5th IEEE Symposium on Computer Arithmetic*, pages 233–240. IEEE Computer Society, 1981.

[KM83] P. Kornerup and D. W. Matula. Finite precision rational arithmetc: an arithmetic unit. *IEEE Trans. Computers*, C-32(4):378–387, April 1983.

[KM88] P. Kornerup and D. W. Matula. An on-line arithmetic unit for bit-pipelined rational arithmetic. *J. Parallel and Distributed Computing*, 5(3):310–330, May 1988.

[KM89] P. Kornerup and D. W. Matula. Exploiting redundancy in bit-pipelined rational arithmetic. In *Proc. 9th IEEE Symposium on Computer Arithmetic*, pages 119–126. IEEE Computer Society, 1989.

[KM90] P. Kornerup and D. W. Matula. An algorithm for redundant binary bit-pipelined rational arithmetic. *IEEE Trans. Computers*, C-39(8):1106–1115, August 1990.

[KM95] P. Kornerup and D. W. Matula. LCF: a lexicographic continued fraction representation of the rationals. *J. Universal Computer Sci.*, 1(7), July 1995.

[Knu98] D. E. Knuth. *Seminumerical Algorithms*, volume 2 of *The Art of Computer Programming*. Addison Wesley, first edition 1969, second edition 1981, third edition, 1998.

[Les01] D. Lester. Effective continued fractions. In *Proc. 15th IEEE Symposium on Computer Arithmetic*, pages 163–170. IEEE Computer Society, 2001.

[LG03] D. Lester and P. Gowland. Using VPS to validate the algorithm of an exact arithmetic. *Theoret. Computer Sci.*, 291, 2003.

[Mat75a] D. W. Matula. Fixed-slash and floating slash arithmetic. In *Proc. 3rd IEEE Symposium on Computer Arithmetic*, pages 90–91. IEEE Computer Society, 1975.

[Mat75b] D. W. Matula. *Radix/Residue/Rational: The Three Rs of Computer Arithmetic and Associated Computer Architecture*. Technical report, Department of Applied Mathematics and Computer Science. Washington University, St. Louis, Mo. 63130, 1975.

[MK78] D. W. Matula and P. Kornerup. A feasibility analysis of fixed-slash and floating-slash number systems. In *Proc. 4th IEEE Symposium on Computer Arithmetic*, pages 29–38. IEEE Computer Society, 1978.

[MK80] D. W. Matula and P. Kornerup. Foundations of finite precision rational arithmetic. *Computing, Suppl. 2*, pages 88–111, February 1980.

[MK83] D. W. Matula and P. Kornerup. An order preserving finite binary encoding of the rationals. *Proc. 6th IEEE Symposium on Computer Arithmetic*, pages 201–209. IEEE Computer Society, 1983.

[MK85] D. W. Matula and P. Kornerup. Finite precision rational arithmetic: slash number systems. *IEEE Trans. Computers*, C-34(8):3–18, January 1985. Reprinted in [Swa90].

[MM94] V. Ménissier-Morain. Arithmétique Exacte. Ph.D. thesis, L'Univerrsité Paris VII, 1994.

[Niq04] M. Niqui. Formalizing exact arithmetic: representations, algorithms and proofs. Ph.D. thesis, Radboud Universiteit 2004.

[NK95] A. Munk Nielsen and P. Kornerup. MSB-first digit serial arithmetic. *J. Universal Computer Sci.*, 1(7), July 1995.

[PE97] P. J. Potts and A. Edalat. *Exact Real Computer Arithmetic*. Technical Report 97/9, Imperial College, London, http://www.doc.ic.ac.uk/deptechrep/tc9710.html#97-9, 1997.

[PM91] S. N. Parikh and D. W. Matula. A redundant binary Euclidean GCD Algorithm. In *Proc. 10th IEEE Symposium on Computer Arithmetic*, pages 220–225. IEEE Computer Society, 1991.

[Pot98] P. J. Potts. Exact Real Arithmetic using Möbius transformations. Ph.D. thesis, Imperial College, University of London, July 1998.

[Ran73] G. N. Raney. On continued fractions and finite automata. *Math. Ann.*, 206:265–283, 1973.

[RS79] A. Rich and D. R. Stoutemeyer. Capabilities of the MUMATH-78 computer algebra system. In E. W. Ng, editor, *Symbolic & Algebraic Computation, Proceedings of EUROSAM 79*, LNCS 72, pages 241–248. Springer Verlag, 1979.

[Sco89] M. Scott. Fast rounding in multiprecision floating-slash arithmetic. *IEEE Trans. Computers*, C-38(7):1049–1052, July 1989.

[Sei83] R. B. Seidensticker. Continued fractions for high-speed and high-accuracy computer arithmetic. In *Proc. 6th IEEE Symposium on Computer Arithmetic*, pages 184–193. IEEE Computer Society, 1983.

[Swa90] E. E. Swartzlander, editor. *Computer Arithmetic,* volume II. IEEE Computer Society Press, 1990.

[Thi08] M. Thill. A more precise rounding algorithm for rational numbers. *Computing*, 82:189–198, 2008. *Erratum*, pp. 261–262.

[Tri77] K. S. Trivedi. On the use of continued fractions for digital computer arithmetic. *IEEE Trans. Computers*, C-26(7):700–704, 1977.

[Vui90] J. Vuillemin. Exact real computer arithmetic with continued fractions. *IEEE Trans. Computers*, C-39(8):1087–1105, August 1990.

[Wal48] H. S. Wall. *Analytic Theory of Continued Fractions*. D. Van Nostrand Co., Inc, 1948. Reprinted in 1973 by Chelsea Publishing Co.

AUTHOR INDEX

INDEX